IONIC LIQUIDS - NEW ASPECTS FOR THE FUTURE

Edited by **Jun-ichi Kadokawa**

Ionic Liquids - New Aspects for the Future
http://dx.doi.org/10.5772/45605
Edited by Jun-ichi Kadokawa

Contributors

Anna Martinelli, Jun-ichi Kadokawa, Tateki Ishida, Fatemeh Moosavi, Elaheh Kowsari, Clarissa Piccinin Frizzo, Dayse N. Moreira, Marcos Antonio Pinto Martins, Izabelle Gindri, Aniele Tier, Lilian Buriol, Madhulata Shukla, Satyen Saha, Hisashi Miyafuji, Yinghuai Zhu, Narayan Hosmane, Hidetaka Noritomi, Chuan-Pei Lee, Te-Chun Chu, Ling-Yu Chang, Jiang-Jen Lin, Kuo-Chuan Ho, Ana P.M. Tavares, Oscar Rodriguez, Eugénia A. Macedo, Nicholas Gathergood, Rohitkumar G. Gore, Jefferson Trindade Filho, Guilherme Caleffi, Toshiro Kaneko, Rikizo Hatakeyama, Shinya Sasaki, Yuriko Kondo, Takahiro Koyama, Takaya Sato, Takashi Morinaga, Takeo Ishizuka, Changwei Hu, Liangfang Zhu, Mihai Putz, Ana-Maria Putz (n. Lacrama), Pedro Mancini, María N. Kneeteman, Claudia D. Della Rosa, Carla Ormachea, Adam McCluskey, Ahmed Al Otaibi, Susumu Kuwabata, Iuliana Cota, Rafael Gonzalez-Olmos, Miguel Iglesias, Francesc Medina, Vicky Lange, Pete Licence, Barry Azzopardi, Ana P. C. Ribeiro, Carlos Nieto De Castro, Salomé I.C. Vieira, Maria J.V. Lourenço, Fernando J. V. Santos, Sohel S.M. Murshed, Peter Goodrich, Christopher Hardacre, João M. França

Published by InTech

Janeza Trdine 9, 51000 Rijeka, Croatia

Publishing Process Manager Iva Lipovic
Technical Editor InTech DTP team
Cover InTech Design team

Additional hard copies can be obtained from orders@intechopen.com

Ionic Liquids - New Aspects for the Future, Edited by Jun-ichi Kadokawa
p. cm.
ISBN 978-953-51-0937-2

For EU product safety concerns:
IN TECH d.o.o., Prolaz Marije Krucifikse Kozulić 3, 51000 Rijeka, Croatia,
info@intechopen.com or visit our website at intechopen.com.

Meet the editor

Jun-ichi Kadokawa was born in Matsuyama, Japan, in 1964. He studied applied and materials chemistriy at Tohoku University, where he received his Ph.D. in 1992. He then joined Yamagata University as Research Associate. From 1996 to 1997 he worked as a visiting scientist at Max-Planck-Institute for Polymer Research in Germany. In 1999 he became an Associate Professor at Yamagata University and moved to Tohoku University in 2002. He was appointed Professor of Kagoshima University in 2004. His research interests focus on polysaccharide-based composite materials with ionic liquids and enzymatic synthesis of nanostructured polysaccharides. He received the Award for Encouragement of Research in Polymer Science (1997) and the Cellulose Society of Japan Award (2009).

Contents

Preface

Concerns with ionic liquids are one of the most interesting and rapidly developing areas in modern physical chemistry, materials science, technologies, and engineering. Increasing attention has also been paid to the use of ionic liquids in the research fields of biological aspects and natural resources. This book provides the forum for dissemination and exchange of up-to-date scientific information on theoretical, generic, and applied areas of ionic liquids. It, therefore, tends to review recent progresses in ionic liquid research on fundamental properties, solvents and catalysts in organic reactions, biological applications, providing energies and fuels, biomass conversions, functional materials, and other applications. I trust that this book will provide an active source of information for research in ionic liquid science and engineering.

Prof. Dr. Jun-ichi Kadokawa
Kagoshima University
Kagoshima, Japan

Section 1

Fundamental Properties

Chapter 1

The Dynamical Properties on Ionic Liquids: Insights from Molecular Dynamics Study

Tateki Ishida

Additional information is available at the end of the chapter

http://dx.doi.org/10.5772/51652

1. Introduction

Ionic liquids (ILs) have attracted many researchers in the areas including physics and chemistry because of their characteristics that are different from conventional molecular liquids and, today, ILs have been one of interesting subjects of scientific study. ILs are consisted of ions and liquids at or near room temperature, and show negligible vapor pressure, thermal and chemical properties, and so on [1-8]. Also, ILs have been widely used as solvents for organic reactions with the expectation of high yields [1]. Most interesting features of ILs can be attributed to remarkable interionic interactions, and these can be an important key factor to study the characteristics of ILs at molecular level. From the results of both experimental and theoretical investigations, it has been recognized that the interionic interaction of ILs could determine physical and chemical properties.

From an experimental side, both femtosecond optically heterodyne-detected Raman-induced Kerr effect spectroscopy (OHD-RIKES)[3, 9, 10] and THz time-domain spectroscopy (THz-TDS)[11, 12] have been applied to investigate the intermolecular vibrational dynamics in ILs. In particular, with OHD-RIKES studies, the possibility to control a property such as shear viscosity by substituting an atomic element in an ionic unit has been reported [13, 14]. On the other hand, from theoretical and computational viewpoints in recent years, ILs have been chosen to study static properties such as structural and thermophysical properties[15-18] and novel interionic dynamics under solvation dynamics [19-21], dynamical properties [22-26], and Kerr spectra analyses [14, 27]. It has been suggested from the simulation studies by Ishida et al. [14] that interionic properties in ILs could be effectively adjustable by substituting an atomic unit in an ion unit in addition to a combination of cations and anions. Also, it has been pointed out that the interplay of motions between cation and anion species could play an important role in specific interionic interactions of ILs.[28]

Now, we can consider two factors important to understand specific interionic interactions in ILs. One is the interionic interaction depending on specific correlations such as cross-correlation terms between cations and anions, and the other is polarization effects due to many-body interactions caused by cations and anions in ILs. While a large number of experimental approaches have been applied to investigate these subjects, molecular-level understanding of many specific properties of ILs has been left unresolved. Obviously, theoretical researches are suitable to tackle these problems to which experimental procedures are not accessible. Thus, it is expected that such computational method as molecular dynamics (MD) simulations enables us to obtain significant information of ILs, utilizing the force field with well-parameterized potential functions and partial charges [29].

With the MD simulation procedures, it is possible to study the effects of the cross-correlations on dynamic collective motions of ions in ILs which are considered to govern the strength and behavior of couplings between ionic motions through interionic interactions [28]. The computation of the time correlations of velocity and momentum between a tagged ion and other unlike ions at different distances provides cross-correlation and momentum correlation functions [28, 30, 31]. Utilizing these calculated functions, we can investigate not only interionic interactions at molecular level but also how collective motions in ILs can proceed accompanied by the momentum transfer between ions in the target IL system.

On the other hand, electrostatic interactions between ions in ILs could be modulated due to many-body interactions and, then, it could emerge as polarization effects caused by the distortion of electron densities under anisotropic environment in ILs. It has been pointed out that the inclusion of polarization effects is significant to investigate the characteristics of ILs in structural and dynamical properties [23, 24]. Therefore, it is required for us to carry out the MD simulation, introducing such procedures as a point dipole model and a polarization energy term into the total potential energy representation of the system [32]. In addition to electrostatic interactions, describing the variation and relaxation of the polarizability anisotropy of ILs is important to investigate dynamics in ILs. For achieving this, we need to compute time-dependent polarization effects on a target system due to environmental effects in ILs. Theoretically, when we would take polarizability anisotropy into account, to track the change of molecular polarizability tensor partly dependent on molecular orientations such as rotational motions would be required. With the calculation of the time correlation function (TCF) of off-diagonal elements of the total polarizability of the system, we can investigate collective properties with the result of the polarizability anisotropy relaxation of the system.

In this chapter, we choose a 1-butyl-3-methylimidazolium cation based ILs with hexafluorophosphate anion, [BMIm][PF_6], and bis(trifluoromethylsulfonyl)amide anion, [BMIm][NTf_2], as target systems. (See Figure 1 for all the molecular structures of $[BMIm]^+$, $[PF_6]^-$ and $[NTf_2]^-$.) Firstly, we focus on the collective properties of [BMIm][PF_6] with the cross-correlation functions of ionic species in the IL and polarization effects on ionic motions. As a second, we investigate dynamical properties of [BMIm][NTf_2]. Below, we start from the explanation for the velocity cross-correlations of a central atom with neighboring atoms. Then, we show how to evaluate both cross-correlation and momentum correlation functions of the target IL. In addition, we describe the introduction of a polarization energy term and the procedure to calculate induced dipole moments and the polarization energy. Following those, the theoretical background of polarization TCFs is given. Computational details are also summarized. In later sections, we discuss polarization effects on interionic interactions and specific properties related to those based on the MD simulation results. We show computation results obtained from performing MD simulation and computing the polarization TCF with the dipole-dipole (DID) approximation [4,5]. Also, we examine the relation between the anisotropic polarizability relaxation and collective motions of ionic species. Lastly, we discuss relaxation processes of ILs including an explanation of important points in studying dynamical properties on ILs.

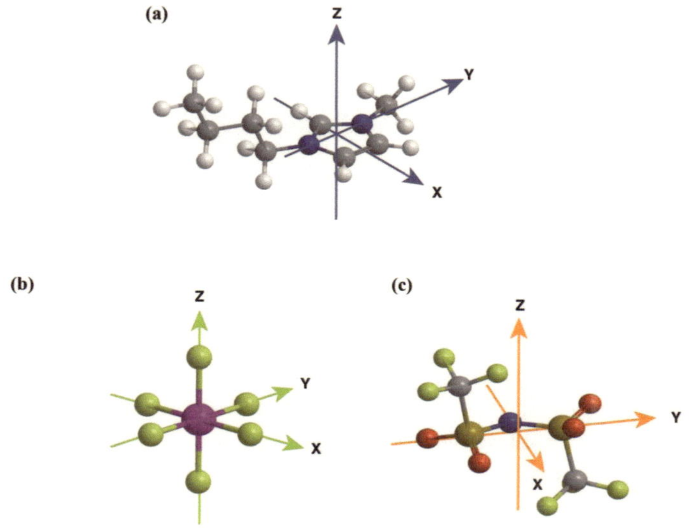

Figure 1. Molecular structures and definitions of body-fixed coordinate axes: (a) $[BMIm]^+$, (b) $[PF_6]^-$ and (c) $[NTf_2]^-$. (See text.) In $[BMIm]^+$, the *Y* direction is along the line connecting two nitrogen (blue colored) atoms in the ring, and the *Z* direction is set perpendicular to the ring plane and the *Y* direction axis. The *X* directionis set in the ring plane orthogonal to both the *Y* and the *Z* axes. In $[PF6]^-$, the *X*, *Y*, and *Z* direction axes are set equivalently. In $[NTf_2]^-$, the Y axis is along the line connecting two sulfurs (dark yellow colored), and the Z axis is set perpendicular the S-N-S plane and the X axis. The X axis is set in the S-N-S plane orthogonal to both the Y and Z axes.

2. Tracking Ionic Motions Through Interionic Interactions: Cross-Correlation, Polarization Effects, and Dynamical Properties

As mentioned in the previous section, for many problems of ILs to which experimental procedures are not accessible, theoretical investigations with computer simulation procedures are promising and suitable to obtain detailed information at molecular level. In particular, specific correlations such as cross-correlations between cations and anions seem not to be feasible to detect experimentally, but the MD simulations enable us to evaluate those. Also, for tracing dynamics in ILs such as librational and reorientational dynamics of ionic species, cross-correlated ionic motions, and the influence of polarization coming from many-body effects caused by cations and anions in ILs from a microscopic point of view, the MD simulations are considered to be one of useful and powerful tools. In this chapter, we will show and explain the dynamical properties on ILs based on MD simulation results taking into consideration mainly following points:

1. how can cross-correlated ionic motions in ILs be modified by interionic dynamics and electronic polarizability effects ?
2. how can the collective dynamics through interionic interactions in ILs be tracked by computer simulation procedures ?
3. what kinds of properties with simulation data do we have to check and care in the investigation of dynamical properties of ILs ?
4. what kinds of subjects of ILs can be or should be investigated theoretically, considering important properties of ILs experimentally observed ?

The items 1 and 2 are main parts in this chapter, including the key results mentioned above. The item 3 is considered to be important for suggesting (or notifying) attention to researchers who are working on analyzing the dynamical properties on ILs. The item 4 includes future perspectives. In particular, we will show and discuss new aspects of ILs in each section related to the items 1, 2, and 3. Also, some of theoretical backgrounds and computational procedures for studying the dynamical properties of ILs are given in each related section, below.

At first, let us consider following points:

1. how can the collective dynamics through interionic interactions cause the unique physical and chemical properties of ILs ?
2. how can interionic dynamics be modified by electronic polarizability effects ?

The former includes the investigation of the contribution of ionic motions due to Coulombic interactions to velocity cross-correlation functions. In particular, through the analysis of the longitudinal and nonlongitudinal contributions to the velocity cross-correlation function in ILs, we will be able to investigate interionic interactions in detail. Also, important properties for physical and chemical interests such as case effects seems to be within the scope of unique collective dynamics in ILs. The later covers the relation between polarizability correlation functions and interionic interactions for ILs. To investigate these points, we can utilize useful information for static properties obtained from computer simulations, but those are not often enough to extract the details of specific interactions.

Here, we give an example that it is difficult for us to find the importance of the interactions between cation and anion species only from static properties. Figure 2 displays the computed radial distribution functions, g(r), comparing non-polarizable model with polarizable model [28]. As shown in the figure, we can observe only small difference in RDFs between two models, except that the g(r) of cation-anion is different from those of cation-cation and anion-anion as easily deduced. But, obviously, it is not feasible for us to observe information other than strong spatial correlations and sequential ordering of cation and anion pairs. In addition, the

http://dx.doi.org/10.5772/51652

analysis of averaged static structures such as RDFs is not enough to investigate remarkable polarization effects such as screening effects influenced by polarization. Below, we give a schematic explanation and discuss that to consider interionic interactions with cross-correlation analyses is important to investigate the interplay between cation and anion species in ILs, and it is shown that the cross-correlation analyses provide contrastive features.

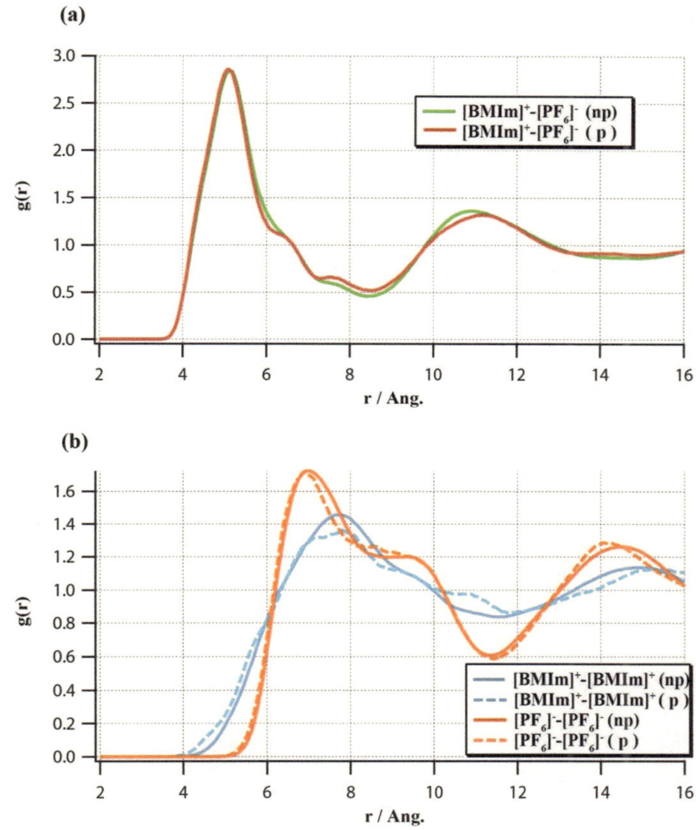

Figure 2. Radial distribution functions (rdfs) for the center of mass of $[BMIm][PF_6]$ for both nonpolarizable (np) and polarizable (p) models: (a) $[BMIm]^+$ - $[PF_6]^-$ and (b) $[BMIm]^+$ - $[BMIm]^+$ and $[PF_6]^-$ - $[PF_6]^-$.

Different from a usual velocity autocorrelation functions (VACF), cross-correlation functions describe interactions between unlike (ionic) species (that is, between a cation and an anion in ILs) and show opposite features to VACFs. In Figure 3, these features are explained schematically. Here, it should be emphasized that cross-correlation functions provide more information on interionic interactions than that static properties such as RDFs include. As seen in Figure 3 ((a) and (b)), the variation of cross-correlation function correlates with the alteration of the VACF. In particular, the cross-correlation function shows the increasing toward the maximum peak where the VACF approaches a minimum point.

These features indicate that it is possible for us to track the time evolution of cross-correlations (between cation and anion species in ILs) as collective dynamics. In addition, it is observed in Figure 3 ((c)) that a particle (ionic species) is bouncing back and forth between like and unlike particles (ions). This implies collective (ionic) motions between coordination shells, thus, it is indicated that not only interionic interactions but also momentum transferring among ionic species could be extracted by analyzing cross-correlation functions. Also, considering that these cross-correlation functions could be modulated by the strength of interionic interactions and the coupling between cation and anion motions, cross-correlation functions would be largely influenced by polarization effects.

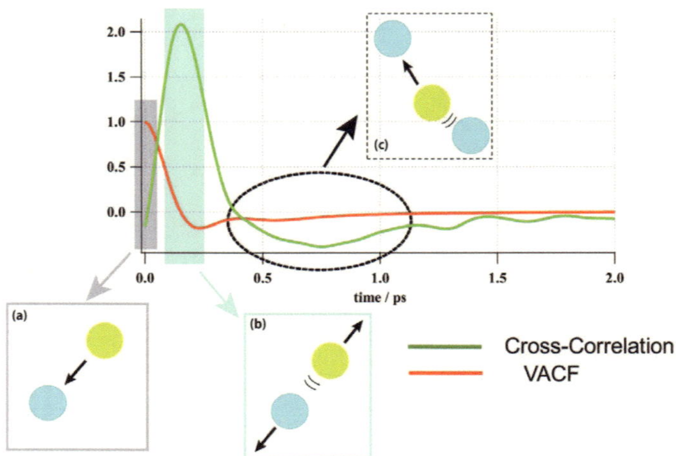

Figure 3. Schematic explanation for a cross-correlation and a VACF: (a) short-time region: an ion approaches a counter ion, and, at the same time, the VACF decreases, (b) the cross-correlation functions start decreasing after the maximum (see text) and (c) bounced ions again approaches counter ions.

On the other hand, as another type of the appearance of interplay between cross-correlation and polarization effects, we consider the polarizability anisotropy and its relaxation of an IL system. These correspond to the variation of the sum of molecular polarizability depending on time. Therefore, it is required to compute the change of molecular polarizablity on each molecule due to interionic interactions and interaction-induced multipole effects. Obviously, it is expected that molecular polarizabilites are influenced by interionic cross-correlations. In later sections, we introduce the theoretical background of polarizability TCF and its application to the study of ILs, and show how degree these are effective and discuss the importance of considering cross-correlations.

3. Theoretical Background

Here, we introduce cross-correlation functions, and then, give an explanation of a polarizable model and polarizability time correlation function. Computational details are also given.

3.1. Cross-Correlation Function

A velocity autocorrelation function (VACF) is defined for calculating the velocity correlation of a same particle as follows,

$$C(t) = \langle v_i(t) \cdot v_i(0) \rangle \langle v_i(0) \cdot v_i(0) \rangle^{-1} \tag{1}$$

where $v_i(t)$ is the velocity of i-th species. Different from the VACF, the time cross-correlation functions between the velocity of a central species i and the velocity of a neighboring species j different from the particle i is defined as follows [30, 31],

$$\mathrm{CR}_n^{ij}(t) = \langle v_i(0) \cdot v_j(t) \rangle_n \left(\langle v_i^2 \rangle \langle v_j^2 \rangle \right)^{-1/2} \tag{2}$$

where $v_i(t)$ is the velocity of the species i and $v_j(t)$ is the velocity of the species j. $\langle v_i^2 \rangle$and $\langle v_j^2 \rangle$ represent the mean square velocities. Here, it should be noted that $\langle\rangle_n$ represents a restricted statistical average [30, 31],

$$\langle v_i(0) \cdot v_j(t) \rangle_n = \left\langle v_i(0) \sum_j v_j(0) \cdot \theta\left(r_{ij}(0) - a_n\right) \cdot \theta\left(b_n - r_{ij}(0)\right) \right\rangle / N_n \tag{3}$$

where θ represents the step function. a_n and b_n are set as the positions of the nth minima of the radial distribution function of the system, $g(r)$. The N_n in Eqs. (2) and (3) is the coordination number in the region between $r = a_n$ and b_n. as follows,

$$N_n = 4\pi\rho \int_{a_n}^{b_n} r^2 g(r)\, dr \tag{4}$$

where ρ is the number density of the system.

Here, we can define the cross-correlation in ILs as that between a centered cation and the total contribution of other anions [28, 30],

$$\mathrm{CR}_n^{\mathrm{CA}}(t) = N_n \langle v_{\mathrm{C}}(0) \cdot v_{\mathrm{A}}(t) \rangle_n \left(\langle v_{\mathrm{C}}^2 \rangle \langle v_{\mathrm{A}}^2 \rangle \right)^{-1/2} \tag{5}$$

where "C" and "A" in super- and subscript represent cation and anion, respectively. By interchanging "C" and "A" in super- and subscript in the above equation, we can easily derive the formulation for the cross-correlation between a centered anion and other cations.

Also, with the computation of the cross-correlation functions between the velocity of a central ion and velocities of neighboring distinct ions, we can analyze the momentum transfer between distinct ions in ILs. Introducing the momentum correlation function [28, 31], the transfer of the momentum of a cation to distinct cations and anions is defined as follows,

$$P_n^{C(total)}(t) = \left(N_n^{CC}\langle p_+(0)\cdot p_+(t)\rangle_n + N_n^{CA}\langle p_+(0)\cdot p_-(t)\rangle_n\right)\left(p_+^2\right)^{-1}$$
$$= P_n^{cation\to cation(CC)}(t) + P_n^{cation\to anion(CA)}(t) \tag{6}$$

where p_+ and p_- mean the momentums of cation and anion, respectively. The transference of momentum of an anion to distinct anions and cations, $P_n^{A(total)}(t)$, is also given as follows,

$$P_n^{A(total)}(t) = \left(N_n^{AA}\langle p_+(0)\cdot p_+(t)\rangle_n + N_n^{AC}\langle p_+(0)\cdot p_-(t)\rangle_n\right)\left(p_+^2\right)^{-1}$$
$$= P_n^{anion\to anion(AA)}(t) + P_n^{anion\to cation(AC)}(t) \tag{7}$$

3.2. Polarization Effects

3.2.1. A Polarizable Model

Here, we introduce a polarizable model considering induced dipole moments and explain the procedure for computing induced dipole moments and the many-body polarization energy [33-35].

The total potential energy of the system under the resulting polarizable force field is defined as follows,

$$V_{tot} = V_{bond} + V_{nonbond} + V_{pol} \tag{8}$$

where the terms V_{bond} and $V_{nonbond}$ are intra- and intermolecular interaction energies. The polarization energy, V_{pol}, is decomposed into the three terms as follows,

$$V_{pol} = V_{charge-dipole} + V_{dipole-dipole} + V_{self} \tag{9}$$

where the charge-dipole contribution, $V_{charge-dipole}$, the dipole-dipole contribution, $V_{dipole-dipole}$, and the self-polarizability term, V_{self}, are defined, respectively, as

$$V_{charge-dipole} = -\sum_i \mu_i \cdot E_i \tag{10}$$

$$V_{dipole-dipole} = \sum_{i>j} \mu_i \cdot T_{ij} \cdot \mu_j \tag{11}$$

$$V_{self} = \sum_i \frac{1}{2\alpha_i} \mu_i \cdot \mu_i \tag{12}$$

In the above equations, E_i is the electric field on atom *i*, produced by the partial charges of all other surrounding atoms, and μ_i represents the induced dipole moment on atom *i*. Also, $\alpha_i$means the isotropic atomic polarizability of atom *i*. In Eq. (11), the T_{ij}, is the dipole field tensor element defined as follows,

$$T_{ij} = \frac{1}{r_{ij}^3}\left[1 - \frac{3\mathbf{r}_{ij}\mathbf{r}_{ij}}{r_{ij}^2}\right] \tag{13}$$

where $r_{ij} = r_i - r_j$. The induced dipole moment on atom *i* is given by

$$\mu_i = \alpha_i \left(E_i - \sum_{j \neq i} T_{ij} \mu_j \right) \tag{14}$$

With Eqs. (10), (11), and (12), the polarization energy, E_{pol}, is summarized as follows,

$$V_{pol} = -\frac{1}{2}\sum_i \mu_i \cdot E_i \tag{15}$$

3.2.2. Polarizability Time-Correlation Function (TCF)

The polarizability anisotropy of the system can be tractable by calculating the TCF of off-diagonal elements of the total polarizability. Here, the theoretical background of the polarizability TCF is summarized briefly.

At first, we define the total polarizability of the system, $\Pi(t)$, that is the sum of the molecular polarizability, $\Pi^M(t)$, and the interaction-induced polarizability, $\Pi^{II}(t)$, as follows,

$$\Pi(t) = \Pi^{\mathrm{M}}(t) + \Pi^{\mathrm{II}}(t) \tag{16}$$

where *t* represents the time dependence, and subscripts M and II mean the molecular part and the interaction-induced part, respectively. The molecular part is given by the sum of the polarizability tensors of isolated gas phase molecular polarizability in the laboratory frame,

$$\Pi^{\mathrm{M}}(t) = \sum_{i=1}^{N} \alpha_i(t) \tag{17}$$

where N is the number of molecules, and α_i is the polarizability tensor of molecule i. For the formulation of the interaction-induced part, we employ the dipole-induced-dipole (DID) model approximation [36, 37], which assumes that the molecular polarizabilities are modified due to a dipolar coupling with the influence of higher order unconsidered. The interaction-induced polarizability in the DID approximation is given as follows,

$$\Pi^{\mathrm{II}}(t) = \sum_{i=1}^{N} \sum_{j \neq i}^{N} \alpha_i(t) \cdot T_{ij}(t) \cdot \tilde{\alpha}_j(t) \tag{18}$$

where T_{ij} means the dipole interaction tensor between molecules i and j. $\tilde{\alpha}_i(t)$is the effective polarizability for molecule i defined by the following equation including the interaction-induced effects,

$$\tilde{\alpha}_i(t) = \alpha_i(t) + \sum_{j \neq i}^{N} \alpha_i(t) \cdot T_{ij}(t) \cdot \tilde{\alpha}_j(t) \tag{19}$$

Equation (19) can be solved by the calculation procedure that is called the all-orders DID approximation [37]. It should be noted that the DID model employed here assumes a center-center DID model, where it recognizes the polarizability as concentrated in the center of mass of the molecule.

Here, we give the representation of the total system polarizability including the cationic and anionic components [14, 27],

$$\Pi(t) = \Pi^{\mathrm{C}}(t) + \Pi^{\mathrm{A}}(t) \tag{20}$$

where the superscripts C and A represent cation and anion species, respectively. Referring to Equations. (16), (17), and (18), the $\Pi^{C}(t)$ and $\Pi^{A}(t)$ defined as follows,

$$\Pi^{\mathrm{C}}(t) = \sum_{i=1} (\alpha_i^{\mathrm{C(M)}}(t) + \alpha_i^{\mathrm{C(II)}}(t)) \tag{21}$$

$$\Pi^{\mathrm{A}}(t) = \sum_{j=1} (\alpha_j^{\mathrm{A(M)}}(t) + \alpha_j^{\mathrm{A(II)}}(t)) \tag{22}$$

http://dx.doi.org/10.5772/51652

where the indices k and l go over all cations and anions, respectively. Then, the total polarizability of the system, $\Pi(t)$, can be rewritten, as follows,

$$\begin{aligned}\Pi(t) &= \frac{1}{3}\mathrm{Tr}\left(\Pi^{\mathrm{C}}(t)+\Pi^{\mathrm{A}}(t)\right)\mathbf{I} \\ &+\left\{\Pi^{\mathrm{C}}(t)-\frac{1}{3}\mathrm{Tr}\left(\Pi^{\mathrm{C}}(t)\right)\mathbf{I}\right\}+\left\{\Pi^{\mathrm{A}}(t)-\frac{1}{3}\mathrm{Tr}\left(\Pi^{\mathrm{A}}(t)\right)\mathbf{I}\right\} \\ &=\frac{1}{3}\mathrm{Tr}\left(\Pi^{\mathrm{C}}(t)+\Pi^{\mathrm{A}}(t)\right)\mathbf{I}+\beta^{\mathrm{C}}(t)+\beta^{\mathrm{A}}(t)\end{aligned} \tag{23}$$

Where $\beta^{\mathrm{C}}(t)=\{\Pi^{\mathrm{C}}(t)-\frac{1}{3}\mathrm{Tr}(\Pi^{\mathrm{C}}(t))\boldsymbol{I}\}$ and $\beta^{\mathrm{A}}(t)=\{\Pi^{\mathrm{A}}(t)-\frac{1}{3}\mathrm{Tr}(\Pi^{\mathrm{A}}(t))\boldsymbol{I}\}$ and $\boldsymbol{I}$ is the unit tensor. Finally, the polarizability TCF, $\phi(t)$, can be rewritten as follows,

$$\begin{aligned}\phi(t) &= \left\langle \mathrm{Tr}\left[\left(\beta^{\mathrm{C}}(0)+\beta^{\mathrm{A}}(0)\right)\cdot\left(\beta^{\mathrm{C}}(t)+\beta^{\mathrm{A}}(t)\right)\right]\right\rangle \\ &= \left\langle \mathrm{Tr}\left(\beta^{\mathrm{C}}(0)\cdot\beta^{\mathrm{C}}(t)\right)\right\rangle + \left\langle \mathrm{Tr}\left(\beta^{\mathrm{A}}(0)\cdot\beta^{\mathrm{A}}(t)\right)\right\rangle \\ &+ \left\langle \mathrm{Tr}\left(\beta^{\mathrm{C}}(0)\cdot\beta^{\mathrm{A}}(t)\right)\right\rangle + \left\langle \mathrm{Tr}\left(\beta^{\mathrm{A}}(0)\cdot\beta^{\mathrm{C}}(t)\right)\right\rangle \\ &= \phi^{\mathrm{C}}(t)+\phi^{\mathrm{A}}(t)+\phi^{\mathrm{C\text{-}A}}(t)\end{aligned} \tag{24}$$

where

$$\begin{aligned}\phi^{C}(t)&=\langle \mathrm{Tr}(\beta^{C}(0)\cdot\beta^{C}(t))\rangle \\ \phi^{A}(t)&=\langle \mathrm{Tr}(\beta^{A}(0)\cdot\beta^{A}(t))\rangle \\ \phi^{C-A}(t)&=\langle \mathrm{Tr}(\beta^{C}(0)\cdot\beta^{A}(t))\rangle+\langle \mathrm{Tr}(\beta^{A}(0)\cdot\beta^{C}(t))\rangle\end{aligned}$$

The nuclear response function, $R(t)$, is represented as the time derivative of the polarizability TCF,

$$R(t) = -\frac{1}{k_B T}\frac{\partial}{\partial t}\phi(t) \tag{25}$$

3.3. Computational Details

Utilizing the sets of force field parameters for $[BMIm]^{+}$, $[PF_6]^{-}$ and $[NTf_2]^{-}$ [38-40], MD simulations have been performed with the DL_POLY molecular dynamics suite [41]. In all the simulations, all the stretching bonds were constrained with the SHAKE algorithm [42]. 125 ion pairs (4000 atoms for $[BMIm][PF_6]$ and 5000 for $[BMIm][NTf_2]$) were set in a cubic box under the periodic boundary condition. The lengths of cubic box size were set to be 35.47

and 38.46 Å to reproduce the experimental data of the densities of [BMIm][PF_6] and [BMIm][NTf_2] at 298 K [13, 43]. 12 Å was set as cutoff length. The time step was 2 fs. The long-range Coulomb and polarization terms were computed with the Ewald's summation technique [44]. Firstly, each system was equilibrated at 600 K for 15 ns in the NVE run and then successively cooled down to 298 K in several stages using velocity scaling. Then, 20 ns NVT run for equilibration at 298 K was carried out. After these equilibration runs, trajectories were recorded every 20 fs (50 fs for [BMIm][NTf_2]) and collected during 10 ns (20 ns for [BMIm][NTf_2]) production runs. For atomic polarizabilities, from the literature [45, 46], we adopted 1.152, 0.705 and 0.0885 Å^3 for the C, N, and H atoms of the cation, respectively, and 0.121 and 3.630 Å^3 for the F and P atoms of the anion, respectively. We considered the distance and vector between the atomic sites of distinct molecules in the dipole interaction tensor, $T=(\boldsymbol{I}-3\hat{\boldsymbol{r}}\hat{\boldsymbol{r}})/r^3$, where$\hat{\boldsymbol{r}}=\boldsymbol{r}/r$Then, the range of the attenuation of dipolar interactions at short distances, s, was evaluated with the Thole's definition [47], $s = 1.662(\alpha_i\alpha_j)^{1/6}$ with atomic polarizabilities, $\alpha_i$andα_j. Also, the dipole interaction tensor we used is given by [47]

$$T = (\mathbf{I} - 3\hat{\mathbf{r}}\hat{\mathbf{r}}) / r^3 \ (r > s) \tag{28}$$

$$T = \left[(4a^3 - 3a^4)\mathbf{I} - 3a^4\hat{\mathbf{r}}\hat{\mathbf{r}} \right] / r^3 \ (r \le s) \tag{29}$$

where a = r/s. Equation (14) is solved with an iterative procedure at each time step, and then, the criterion value of iterative solution for induced dipole moment was set to 0.001 D. For the computation of the polarizability TCF of [BMIm][NTf_2], we used molecular polarizabilities of 14.372 Å^3 for [BMIm]$^+$ and 11.259 Å^3 for [NTf_2]$^-$.[48] The body-fixed coordinate axes set in the cation and the anion are shown in Figure 2. For more computational procedures in detail, interested readers should refer to references [14] and [28].

4. Cross-Correlation, Momentum Correlation, and Polarization Effects

The computed velocity cross-correlation functions [28] are shown in Figures 4 and 5. These figures show the comparison of velocity cross-correlation functions observed around the cation, [BMIm]$^+$, and the anion, [PF_6]$^-$, placed at the center, for both the non-polarizable and polarizable models. From the result of the cation-anion RDF in Figure 2, we selected 3.5, 8.4, 8.4, and 14.75 Å for the value of a_1, b_1, a_2, and b_2 in Equation (4) for the nonpolarizable model, respectively, and 3.5, 8.5, 8.5, and 15.0 Å for the polarizable model, respectively. With these values, the first ($C_1(t)$) and second ($C_2(t)$) coordination shells were specified. It should be noted that the $C_1(t)$ between the center anion and distinct anions is not shown in Figure 5 because the $C_1(t)$ is almost zero corresponding to the result that the anion-anion RDF result is almost zero at the region of the specified first coordination shell (see RDFs in Figure 2(b)). Also, it should be noted that the initial values, $C_n(0)$, are negative since the system size is finite as have been pointed out [30].

http://dx.doi.org/10.5772/51652

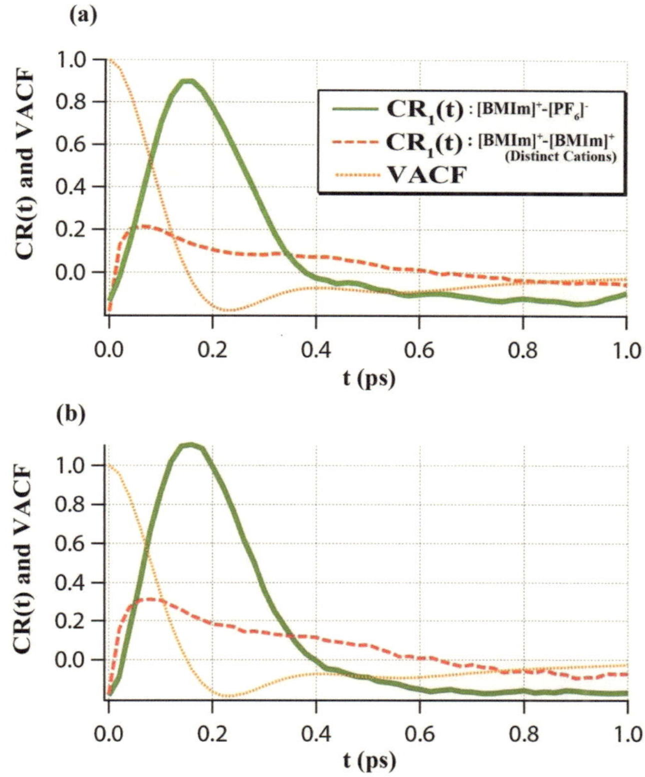

Figure 4. The Calculated velocity cross-correlation functions between the $[BMIm]^+$ set at the center in the first coordination shell and $[PF_6]^-$, and between the center cation and other cations, including VACF: (a) in the nonpolarizable model and (b) in the polarizable model.

As shown in Figures 4 and 5, the VACFs are also displayed for comparison. As briefly explained in Section 2 with Figure 3, the initial rise of the cross-correlation function, $C_1(t)$, appears toward the maximum for both models, corresponding to the decay of the VACF to the minimum (see the (a) and (b) in Figure 3). These results indicate that the initial momentum of the central ion is gained by neighboring ions immediately after $t = 0$. In addition, following decay profiles (see the (c) in Figure 3) are seen. These are ascribed to the spread of transferred momentum to the outer coordination shells. The peak height of the $C_1(t)$ in the polarizable model is larger than that in the nonpolarizable model. These significant results indicate that the momentum transfer can be intensified from the more distant coordination shells to the first ones due to both the charge-dipole and dipole-dipole interactions by polarization effects in addition to charge-charge Coulomb interactions. As the characteristics of the VACF profile, it is noted that the minima of the VACF are located at the positions later than those of the maxima of the velocity cross-correlation functions for the first coordination shell. These features imply that the momentum transmitted to the neighbors at the first shell is regained partly by the central cation or anion, bouncing back and forth for some time, as have been pointed out in the literature on the computer simulations of simple liquid binary mixtures [28, 31].

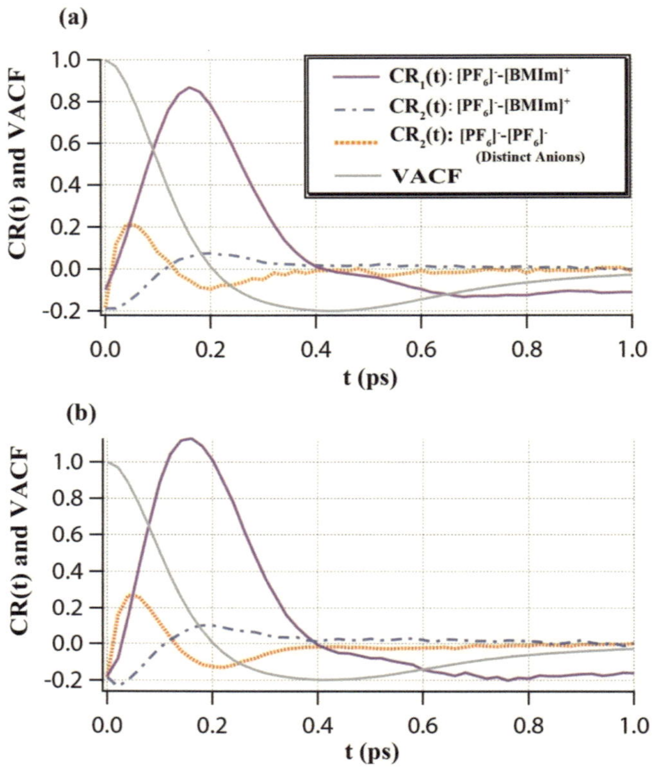

Figure 5. The Calculated velocity cross-correlation functions between the $[PF_6]^-$ set at the center in the first and second coordination shells and $[BMIm]^+$, and between the center anion and anions, including VACF: (a) in the nonpolarizable model and (b) in the polarizable model.

On the other hand, an interesting feature is seen in the $C_1(t)$ between distinct cations (or the $C_2(t)$ for distinct anions). As indicated in Figures 4 and 5, the cross-correlation function for distinct ions reaches the maximum point earlier than the $C_1(t)$ for the cation-anion cross-correlation, even though the peak height is smaller than that in the cation-anion cross-correlation function. On these results, it is considered that the alteration of cage effects increases the probability that a cation (or an anion) meets a distinct cation (or anion) at early time region.

Also, in Figures 4 and 5, it is clearly observed that the cross-correlation function between distinct like ions is influenced by the modulation of cage effects due to polarization effects and that its peak height is enhanced in the polarizable model [28]. As shown in Figures 4 and 5, when VACFs pass the value of zero to negative, the cross-correlation functions reach the maximum point. Thus, this implies that the central cation (or anion) is likely to lose its initial momentum. In addition, as seen in Figure 5, the peak position of $C_2(t)$ is shifted to the

http://dx.doi.org/10.5772/51652

time region later than that of $C_1(t)$. This is consistent with the consideration that the momentum of the central ion is transferred from the first coordination shell to the second.

Here, we examine the characteristics of the cross-correlation functions. As shown in Figure 6, we consider the decomposition of a cross-correlation function into the longitudinal (denoted as L in the figure), $C_n{}^L(t)$, and nonlogitudinal (denoted as NL in the figure), $C_n{}^{NL}(t)$, contributions as follows [28, 31, 49],

$$CR_n(t) = CR_n^L(t) + CR_n^{NL}(t) \tag{30}$$

where $CR_n{}^L(t)$ is represented as the velocity cross-correlation along the direction designated by the center of masses of distinct ions at t = 0. We can compute $CR_n{}^L(t)$ with the following equation [28, 31, 49],

$$CR_n^L(t) = N_n \left\langle v_i{}^L(0) \cdot v_j{}^L(t) \right\rangle_n \left(\left\langle v_i{}^2 \right\rangle \left\langle v_j{}^2 \right\rangle \right)^{-1/2} \tag{31}$$

where $v_i{}^L(t) = v_i(t)[r_{ij}(0)/r_{ij}(0)]$ and $v_i(t)$ is the velocity of the ionic species *i*. Also, $r_{ij}(0)$ means the direction vector between the center of masses of the distinct ions i and j. $\langle v_i{}^2\rangle$and $\langle v_j{}^2\rangle$ represent the mean square velocities. $CR_n{}^{NL}(t)$ can be computed with $CR_n(t)$ and $CR_n{}^L(t)$. Figure 7 shows the computed $C_n{}^L(t)$ and $C_n{}^{NL}(t)$ functions for a centered cation at the first coordination shell (n = 1) both for the nonpolarizable and polarizable models [28]. All the results indicate that the velocity cross-correlations at the short time region up to 0.4 ps are predominantly governed by the longitudinal function, $C_n{}^L(t)$.

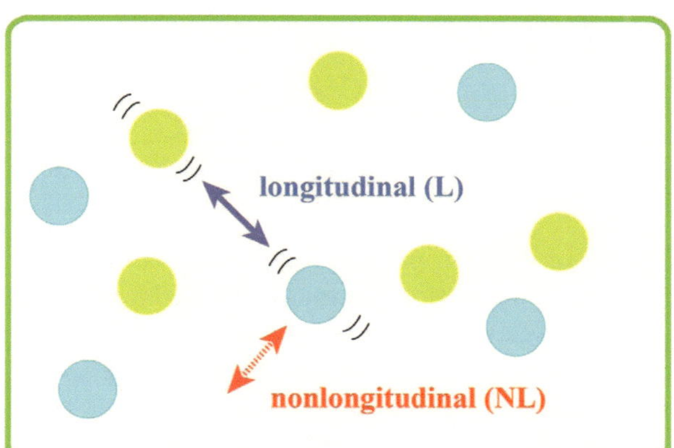

Figure 6. Schematic explanation of the longitudinal (L) and nonlongitudinal (NL) of a cross-correlation function.

These results clearly indicate that the Coulomb interactions between neighboring ions are mainly effective on the ionic motions as "driving forces" at the short time. In addition, the magnitude of the longitudinal correlation of ions in the polarizable model is larger than that in the nonpolarizable model. Therefore, it is considered that the longitudinal motions are strongly affected by polarization effects. On the decay behavior, similar features are obtained in both models. These indicate that polarization effects on the longitudinal motion complete mostly at the short time region.

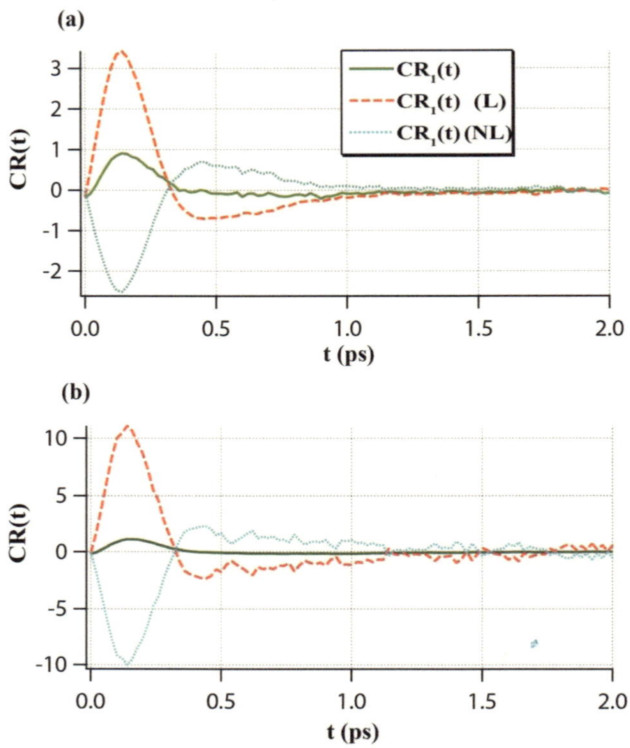

Figure 7. The simulated longitudinal and nonlongitudinal contributions to the velocity cross-correlation function: (a) in the nonpolarizable model and (b) in the polarizable model.

Next, the computed results of momentum correlation functions [28] are displayed in Figures 8 and 9. As shown in Figures 8 and 9 in common, the initial momentum of the cation (or anion) is mainly transferred to the close anions (or cations) while the momentum correlation with distinct cations (or anions) is smaller. These results are consistent with the consideration that strong Coulomb interactions between the cations and the anions enhance the possibility of approaching or attracting each other. Also, while the contribution of the momentum correlation between distinct cations, $P_1^{CC}(t)$, is smaller than that between the cation and anion molecules, $P_1^{CA}(t)$, the peak height of the $P_1^{CA}(t)$ in the polarizable model is larger than that in the nonpolarizable model. These results indicate that the transference of the momentum between distinct cations could be intensified by both charge-dipole and dipole-dipole interactions coming from polarization effects. As we could observe in the cross-correlation functions (see Figures 4 and 5), the maxima of the momentum correlation functions between distinct cations emerge at earlier time region in comparison with those of the momentum correlations between cation and anion species.

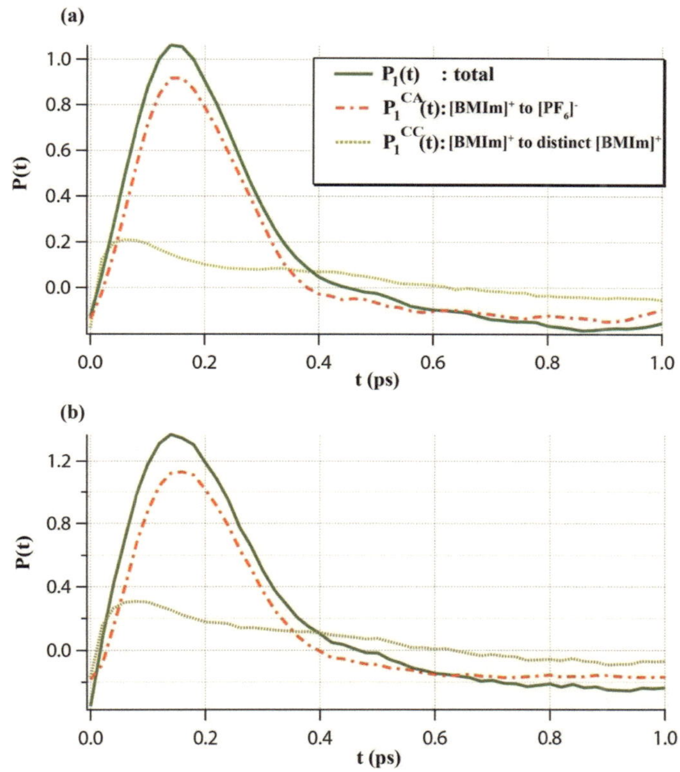

Figure 8. The calculated momentum correlation functions for the $[BMIm]^+$ at the center in the first coordination shell with the VACF: (a) in the nonpolarizable model and (b) in the polarizable model.

Based on these results, obviously, it is indicated that the variation of cage effects promotes the transfer of the initial momentum of the central cation to distinct cations rather than anions. Then, the cage effects is considered to be weakened by polarization effects, though the degree of this effect is likely to be relatively small as deduced from the figures.

On the other hand, the momentum correlations for the center anion in Figure 9 show distinct features from those for the center cation. The momentum correlation function between distinct anions, $P_1^{AA}(t)$, indicates much smaller contributions to the total momentum correlations, $P_1(t)$, in both the nonpolarizable and polarizable models, compared with the $P_1^{AC}(t)$. Therefore, these indicate that the initial momentum of the center anion is mainly transferred to neighboring cations, and the transference between distinct anions is not enhanced. These are consistent with the consideration that the cation-anion interactions promote the propagation of the momentum one after another, as mentioned in the cation case. In addition, Figure 9 shows more interesting features. Compared with the results for the central cation in Figure 8, while the $P_1{}^{AA}(t)$ does not contribute negligibly to the total momentum correlation in the nonpolarizable model, it indicates characteristic oscillating behavior in the short time region up to 0.4 ps in the polarizable model, and the latter has a relatively larger contribution to the $P_1(t)$ up to 0.1 ps than in the nonpolarizable model. This clearly indicates the decrease of cage effects through polarization effects and implies that the interionic interactions between distinct anions could become effective by the charge-charge and dipole-dipole interactions [28].

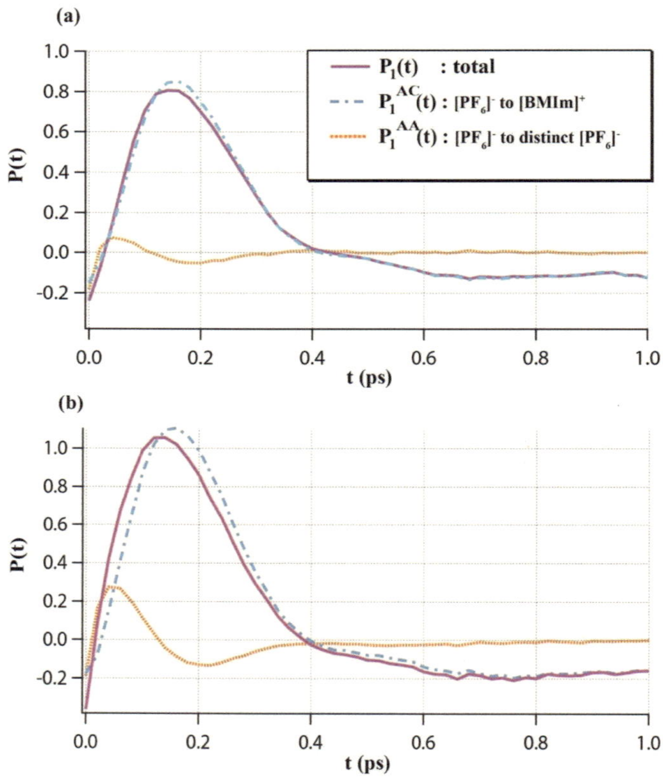

Figure 9. The calculated momentum correlation functions for the $[PF_6]^-$ at the center in the first coordination shell with the VACF: (a) in the nonpolarizable model and (b) in the polarizable model.

5. Relaxation Processes and Dynamical Properties

Firstly, the relaxation feature of IL system is examined with the computed polarizability TCF and its time derivative, and then, we discuss dynamical behavior of cations and anions in ILs, and consider what kinds of properties with simulation data we have to check and care in the study of dynamical properties of ILs.

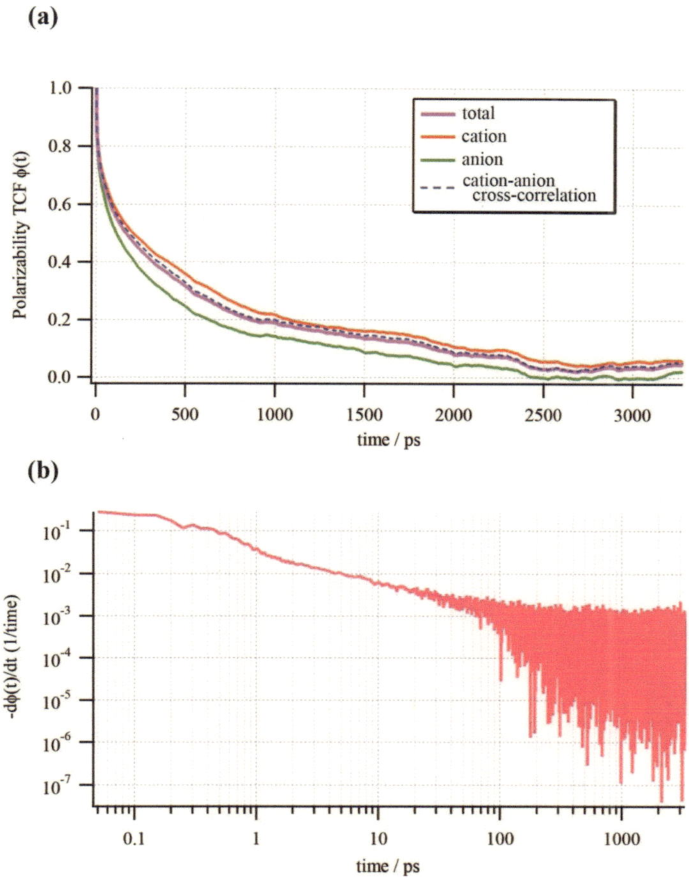

Figure 10. The computed polarizability TCF for each component (a) and the time derivative of the total polarizability TCF (b) for [BMIm][NTf_2] up to about 3 ns.

Figure 10 shows the computed polarizability TCF for each component and the time derivative of the total polarizability TCF for [BMIm][NTf_2]. As seen in Figure 10(a), the relaxation of the polarizability TCFs seems to be slower than in usual liquids. In particular, the relaxation of the anion shows faster decay than that of the cation. Also, for each component, it takes about 2~3 ns to approach toward zero. Thus, this indicates that it is required for us to perform a production MD simulation run for, at least, a few ns to study the relaxation behavior of the polarizability TCF. Also, as seen in Figure 10(a), the cation-anion cross-correlation indicates similar variation to the total polarizability TCF. Therefore, the relaxation behavior of the system strongly correlates with the cation-anion cross-correlation, and it is emphasized that tracking cross-correlation terms is very important. On the other hand, corresponding to this result, the time derivative of the polarizability TCF shows long-decay feature extending up to the nanosecond times, as seen in Figure 10(b). Considering that the time derivative of the polarizability TCF is directly related to optical Kerr effect (OKE) response,[14] these results are indicative of how long a MD simulation has to be carried out to examine the relaxation of the total system polarizability and the OKE response.

The relaxation process of ILs implies long-time dynamics of ILs. In particular, it has been known that diffusive motion of each ion in ILs at room temperature is usually much slower than in usual liquids.[1, 16] Therefore, a MD simulation would need to be run for a few ns or more. Then, we are able to use a procedure to check whether a system is in the diffusive regime. The procedure is to compute β(t) = dlog(MSD)/dlog(t),[16] where MSD means the mean-squared displacement (MSD) of the center of mass of ions, $\left\langle \sum_{i=1}^{N} [\boldsymbol{r}_i(t)-\boldsymbol{r}(0)]^2 \right\rangle$. In the case that β(t) = 1, the system is in the diffusive regime, while when b(t) < 1, the system is in the sub-diffusive regime. In Figure 11, computed β(t)s for [BMIm][NTf_2] are shown. These results indicate that both the cation and the anion reach the β(t) region between 0.8 and 1 (the shadowed area in Figure 11) after about 16 ns. This β(t) region is considered to be almost in the diffusive regime. Therefore, Figure 11 obviously indicates that, for confirming reliable self-diffusivities carefully, we need to perform a longer MD simulation than about 15 ~ 20 ns.

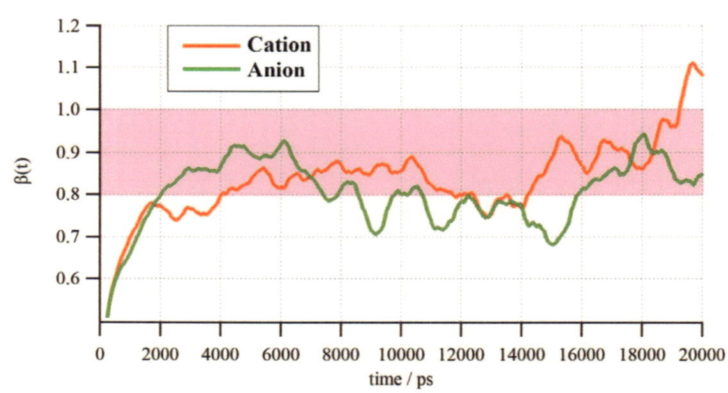

Figure 11. β(t) (see text) for $[BMIm]^+$ and $[NTf_2]^-$. Also, see text for the shadowed area.

Figure 12 displays the results calculated for the MSD and, in comparison, the non-Gaussian parameter, $\alpha(t)$, [50, 51]

$$\alpha(t) = 3\left\langle r^4(t)\right\rangle \Big/ 5\left\langle r^2(t)\right\rangle^2 - 1 \tag{32}$$

where $r(t)$ is the displacement of an ion at time t with respect to its position at t = 0. In Figure 12, the MSDs of $[PMIm]^+$ and $[NTf_2]^-$ indicate three typical dynamic ranges (regions), respec-

http://dx.doi.org/10.5772/51652

tively. Also, the behavior of the short- and long-time regimes of MSD for both the cation and the anion is similar to each other.

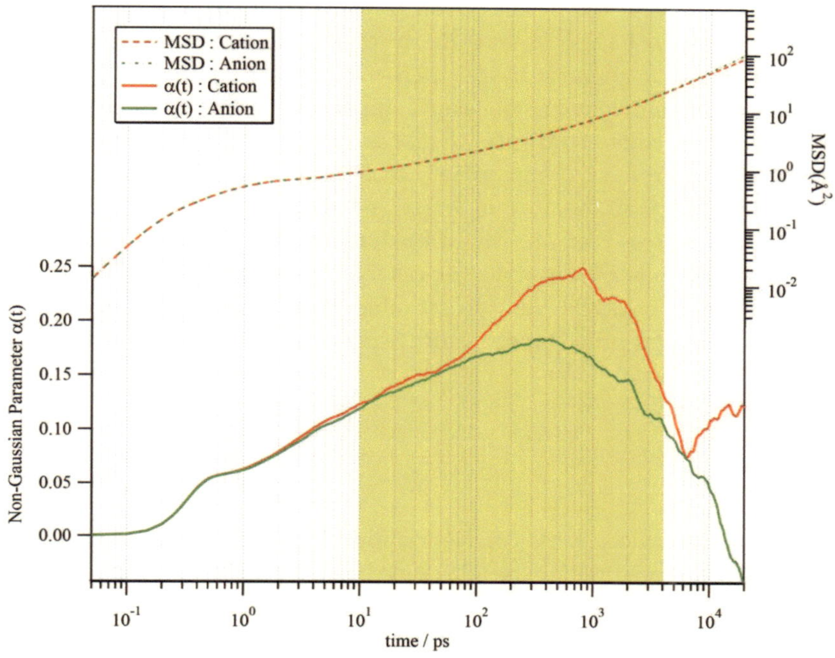

Figure 12. MSDs and non-Gaussian parameters for $[BMIm]^+$ and $[NTf_2]^-$. See text for the shadowed area.

For $[BMIm]^+$, first region is a microscopic regime until about 10 ps. Second is a crossover regime (the shadowed area in the Figure 12) until about 3-5 ns, and third regime is a sub-linear time dependence (sub-diffusive region) toward 20 ns. Also, $\alpha(t)$ has a double peak structure. One is at about 0.5 ps and the other at about 800 ps. In addition, a small shoulder structure emerges at between two peaks. As clearly seen in Figure 12, the short-time maximum peak corresponds to the microscopic region of the MSD, while the long-time maximum is located at around the center of the crossover regime of the MSD. On the other hand, for the anion, $[NTf_2]^-$, $\alpha(t)$ has a double peak structure similar to that of the cation. In particular, the long-time maximum peak is slightly shifted to about 300 ps, while the short-time maximum is locacted at the almost same time region as the cation. The time regime where the long-time maximum appear, is in good accord with those of $[BMIm]^+$, as shown in Figure 12. Therefore, considering that the magnitude of the deviation from the Gaussian behavior is remarkably larger for $[BMIm]^+$ than for $[NTf_2]^-$ as shown in Figure 12, it is expected that the relaxation behavior of the cation in perturbed $[BMIm][NTf_2]$ IL might be largely different from that of the anion. This consideration implies the possibility for us to find a distinct relaxation process in the OKE response (compare Figure 10 with Figure 12). Also, it should be noticed that the diffusive regime (linear time-dependence region) after about 10 ns in Figure 12 coincides with the diffusive region in Figure 11 (after about 16 ns) for both the cation and the anion.

In particular, the shadowed area in Figure 12 covers the time range of the full width at half maximum (FWHM) of the second peak of $\alpha(t)$. As pointed out previously,[52] this FWHM range is likely to include a slow decay region usually known as the α relaxation which is characteristic of glass-forming supercooled liquids in general. Also, recently, it has been reported that similar behavior could appear for ILs at room temperature.[51, 53, 54] Therefore, we had better study the α relaxation in ILs with the observations of the non-Gaussian parameter in addition to information on spatial relaxation such as intermediate scattering function. Also, it is suggested that, when we investigate dynamical properties of ILs, the crossover, the sub-diffusive, and the diffusive regimes have to be carefully examined, as one of criterions for reliable research. Thus, only with a long-time MD simulation, it is considered that we are able to investigate dynamical properties of ILs to which experimental procedures are not accessible.

6. Conclusion

In this chapter, we introduced cross-correlation function analyses with a polarizable model, polarizability TCFs for investigating the relaxation behavior of ILs, and some of procedures for studying dynamical properties. We showed the importance of considering cross-correlation functions and related properties, showing some of examples. Firstly, we employed the polarizable model based on point dipole treatment to the investigation of the polarization effect on the target IL, [BMIm][PF_6]. With the MD simulation data for both the nonpolarizable and polarizable models, velocity cross-correlation analyses were shown, and we presented the momentum correlation functions between the cation and anion species in the IL. Next, we computed polarizability TCFs of [BMIm][NTf_2] and discussed their relaxation behavior. In addition, we investigated dynamical properties of ILs such MSDs and non-Gaussian parameters and considered what kinds of properties with simulation data we have to check and care in the study of dynamical properties of ILs. This chapter is summarized as follows:

1. In the study and discussion of velocity cross-correlation functions, it was shown that polarization effects could enhance the cross-correlations between $[BMIm]^+$ and $[PF_6]^-$ in the polarizable model in comparison with that in the nonpolarizable model. These features are ascribed to interionic interactions through attractive forces coming from the charge-dipole and dipole-dipole interactions caused by polarization effects in addition to charge-charge Coulomb interactions. Based on the results of computed cross-correlation between distinct cations (or anions), it was shown that, at early time region, the modulation of cage effects through polarization effects could improve the probability of approach between like ions. Also, by decomposing the cross-correlation function into the longitudinal and nonlongitudinal components, it was indicated that, at the short time region, the velocity cross-correlation is predominantly controlled by the longitudinal contribution. In addition, it was indicated that, compared with the

longitudinal correlation in the nonpolarizable model, the longitudinal component is further modified in the polarizable model.

2. On the momentum correlation functions between $[BMIm]^+$ and $[PF_6]^-$, it was exhibited that the correlation between $[BMIm]^+$ and $[PF_6]^-$ plays a important role. Also, it was shown that the contribution of the cross-correlation between distinct anions could be enhanced in the polarizable model. This result indicates that cage effects could be diminished with polarization effects, implying that the interionic interactions between distinct anions could be intensified by the charge-charge and dipole-dipole interactions related to polarization effects. Therefore, as has been pointed out in the literature [23, 24, 29], it is considered that the cage effect in ILs could be reduced by many-body polarization effects.

3. Both the computed polarizability TCF for each component and the time derivative of the total polarizability TCF for $[BMIm][NTf_2]$ showed long-time decay behavior. It was indicated that those took about 2~3 ns to approach toward zero. Therefore, this indicates that it is required for us to perform a production MD simulation run. Also, the relaxation behavior of ILs was investigated with the calculation of $\beta(t) = dlog(MSD)/dlog(t)$ as an indicator of the diffusive region. Our results suggested that long-time dynamics of ILs has to be studied with a longer MD simulation than about 15 ~ 20 ns for confirming reliable self-diffusivities in ILs, carefully. Furthermore, we examined MSDs and non-Gaussian parameters for $[BMIm][NTf_2]$. From our studies, it was exhibited that the magnitude of the deviation from the Gaussian behavior is remarkably larger for $[BMIm]^+$ than for $[NTf_2]^-$, and that, comparing with corresponding MSDs, it is possible to study diffusive motion of the cation and the anion. In addition, it was suggested that these studies imply the possibility for us to find a distinct relaxation process in the OKE response. Therefore, it is suggested that we had better study the α relaxation in ILs with the observations of the non-Gaussian parameter in addition to information on spatial relaxation such as intermediate scattering function. Lastly, it is concluded, only with a long-time MD simulation (> 15 ~ 20 ns), that we are able to investigate the dynamical properties of ILs to which experimental procedures are not accessible.

Acknowledgements

This work was supported in part by the Ministry of Education, Culture, Sports, Science and Technology (MEXT) of Japan (Grant-in Aid Scientific Research (C): 23550029).

Author details

Tateki Ishida*

Address all correspondence to: ishida@ims.ac.jp

Department of Theoretical and Computational Molecular Science, Institute for Molecular Science, Japan

References

[1] Wasserscheid, P., & Welton, T. (2008). *Ionic Liquids in Synthesis*, Weinheim, Wiley-VCH.

[2] Wasserscheid, P., & Keim, W. (2000). Ionic liquids- New "solutions" for transition metal catalysis. *Angew. Chem., Int. Ed.*, 39, 3773-3789.

[3] Castner, E. W. Jr, Wishart, J. F., & Shirota, H. (2007). Intermolecular Dynamics, Interactions, and Solvation in Ionic Liquids. *Acc. Chem. Res.*, 40, 1217-1227.

[4] Weingaertner, H. (2007). Understanding ionic liquids at the molecular level: facts, problems, and controversies. *Angew. Chem., Int. Ed.*, 47, 654-670.

[5] Ohno, H. (2005). *Electrochemical Aspects of Ionic Liquids*, Hoboken, Wiley-Interscience.

[6] Rogers, R. D., & Voth, G. A. (2007). Special Issue on Ionic Liquids. *Special Issue on Ionic Liquids. Acc. Chem. Res.*, 40(11).

[7] Wishart, J. F., & Castner, E. W. Jr. (2007). Special Issue on Physical Chemistry of Ionic Liquids.Special Issue on Physical Chemistry of Ionic Liquids. *Special Issue on Physical Chemistry of Ionic Liquids. J. Phys. Chem. B*, 111(18).

[8] Ohno, H., & Fukumoto, K. (2008). Progress in ionic liquids for electrochemical reaction matrices. *Electrochemistry*, 76, 16-23.

[9] Shirota, H., Wishart, J. F., Castner, E. W., & Jr , . (2007). Intermolecular Interactions and Dynamics of Room Temperature Ionic Liquids That Have Silyl- and Siloxy-Substituted Imidazolium Cations. *J. Phys. Chem. B*, 111, 4819-4829.

[10] Xiao, D., Rajian, J. R., Hines, J., , L. G., Li, S., Bartsch, R. A., & Quitevis, E. L. (2008). Nanostructural Organization and anion effects in the optical Kerr effect spectra of binary ionic liquids mixtures. *J. Phys. Chem. B*, 112, 13316-13325.

[11] Koeberga, M., Wu, C.-C., Kimc, D., & Bonn, M. (2007). THz dielectric relaxation of ionic liquid:water mixtures. *Chem. Phys. Lett.*, 439, 60-64.

[12] Yamamoto, K., Tani, M., & Hangyo, M. (2007). Terahertz Time-Domain Spectroscopy of Imidazolium Ionic Liquids. *J. Phys. Chem. B*, 111, 4854-4859.

[13] Shirota, H., Nishikawa, K., & Ishida, T. (2009). Atom substitution effects of $[XF6]^-$ in ionic liquids. 1. Experimental study. *J. Phys. Chem. B*, 113, 9831-9839.

[14] Ishida, T., Nishikawa, K., & Shirota, H. (2009). Atom substitution effects of $[XF6]^-$ in ionic liquids. 2. Theoretical study. *J. Phys. Chem. B*, 113, 9840-9851.

[15] Refer to the sec.4 in ref.1 and references therein.

[16] Maginn, E. J. (2007). Atomistic Simulation of the Thermodynamic and Transport Properties of Ionic Liquids. *Acc. Chem. Res.*, 40, 1200-1207.

[17] Lopes, J. N. A. C., & Padua, A. A. H. (2006). Nanostructural Organization in Ionic Liquids. *J. Phys. Chem. B*, 110, 3330-3335.

[18] Bhargava, B. L., & Balasubramanian, S. (2007). Refined potential model for atomistic simulations of ionic liquid [bmim][PF_6]. *J. Chem. Phys.*, 127(114510), 1-6.

[19] Znamenskiy, V., & Kobrak, M. N. (2004). Molecular Dynamics Study of Polarity in Room-Temperature Ionic Liquids. *J. Phys. Chem. B*, 108, 1072-1079.

[20] Shim, Y., Duan, J., Choi, M. Y., & Kim, H. J. (2003). Solvation in molecular ionic liquids. *J. Chem. Phys.*, 119, 6411-6414.

[21] Kobrak, M. N. (2006). Characterization of the solvation dynamics of an ionic liquid via molecular dynamics simulation. *J. Chem. Phys.*, 125(064502), 1-11.

[22] Hu, Z. H., & Margulis, C. J. (2006). Heterogeneity in a room-temperature ionic liquid: Persistent local environments and the red-edge effect. *Proc. Natl. Acad. Sci.*, U.S.A., 103, 831-836.

[23] Yan, T., Burnham, C. J., Del Popolo, M. G., & Voth, G. A. (2004). Molecular Dynamics Simulation of Ionic Liquids: The Effect of Electronic Polarizability. *J. Phys. Chem. B*, 108, 11877-11881.

[24] Jiang, W., Yan, T., Wang, Y., & Voth, G. A. (2008). Molecular Dynamics Simulation of the Energetic Room-Temperature Ionic Liquid, 1-Hydroxyethyl-4-amino-1,2,4-triazolium Nitrate (HEATN). *J. Phys. Chem. B*, 112, 3121-3131.

[25] Urahata, S. M., & Ribeiro, M. C. C. (2005). Single particle dynamics in ionic liquids of 1-alkyl-3-methylimidazolium cations. *J. Chem. Phys.*, 122(024511), 1-9.

[26] Urahata, S. M., & Ribeiro, M. C. C. (2006). Collective excitations in an ionic liquid. *J. Chem. Phys.*, 124(074513), 1-8.

[27] Hu, Z., Huang, X., Annapureddy, H. V. R., & Margulis, C. J. (2008). Molecular Dynamics Study of the Temperature-Dependent Optical Kerr Effect Spectra and Intermolecular Dynamics of Room Temperature Ionic Liquid 1-Methoxyethylpyridinium Dicyanoamide. *J. Phys. Chem. B*, 112, 7837-7849.

[28] Ishida, T. (2011). Molecular dynamics study of the dynamical behavior in ionic liquids through interionic interactions. *J. Non-Cryst. Solids*, 357, 454-462.

[29] Leach, A. R. (2001). *Molecular Modeling, Principles and Applications*, Harlow, Pearson Education.

[30] Verdaguer, A., Padró, J. A., & Trullàs, J. (1998). Molecular dynamics study of the velocity cross-correlations in liquids. *J. Chem. Phys.*, 109, 228-234.

[31] Verdaguer, A., & Padró, J. A. (2001). Computer simulation study of the velocity cross correlations between neighboring atoms in simple liquid binary mixtures. *J. Chem. Phys.*, 114, 2738-2744.

[32] Frenkel, D., & Smit, B. (2002). *Understanding Molecular Simulation, From Algorithms to Applications*, London, Academic Press.

[33] Bernardo, D. N., Ding, Y., Krogh-Jespersen, K., & Levy, R. M. (1994). An Anisotropic Polarizable Water Model: Incorporation of All-Atom Polarizabilities into Molecular Mechanics Force Fields. *J. Phys. Chem.*, 98, 4180-4187.

[34] Předota, M., Cummings, P. T., & Chialvo, A. A. (2002). Pair approximation for polarization interaction and adiabatic nuclear and electronic sampling method for fluids with dipole polarizability. *Mol. Phys.*, 100, 2703-2717.

[35] Ahlström, P., Wallqvist, A., Engström, S., & Jönsson, B. (1989). A molecular dynamics study of polarizable water. *Mol. Phys.*, 68, 563-581.

[36] Frenkel, D., & Mc Tague, J. P. (1980). Molecular dynamics studies of orientational and collision-induced light scattering in molecular fluids. *J. Chem. Phys.*, 72, 2801-2818.

[37] Geiger, L. C., & Ladanyi, B. M. (1987). Higher order interaction-induced effects on Rayleigh light scattering by molecular liquids. *J. Chem. Phys.*, 87, 191-202.

[38] Lopes, J. N. C., Deschamps, J, & Padua, A. A. H. (2004). Modeling Ionic Liquids Using a Systematic All-Atom Force Field. *J. Phys. Chem. B*, 108, 2038-2047.

[39] Lopes, J. N. C., Deschamps, J., & Padua, A. A. H. (2004). Additions and Corrections: Modeling Ionic Liquids Using a Systematic All-Atom Force Field. *J. Phys. Chem. B*, 108, 11250.

[40] Köddermann, T, Paschek, D, & Ludwig, R. (2007). Molecular Dynamic Simulations of Ionic Liquids: A Reliable Description of Structure, Thermodynamics and Dynamics. *ChemPhysChem*, 8, 2464-2470.

[41] Smith, W., & Forster, T. R. (2001). *The DL_POLY_2 User Manual*, Daresbury, United Kingdom, Daresbury Laboratory.

[42] Allen, M. P., & Tildesley, D. J. (1987). *Computer Simulation of Liquids*, Clarendon, Oxford.

[43] Shirota, H, Mandai, T, Fukazawa, H, & Kato, T. (2011). Comparison between Dicationic and Monocationic Ionic Liquids: Liquid Density, Thermal Properties, Surface Tension, and Shear Viscosity. *J. Chem. Eng. Data*, 56, 2453-2459.

[44] Nymand, T. M., & Linse, P. (2000). Ewald summation and reaction field methods for potentials with atomic charges, dipoles, and polarizabilities. *J. Chem. Phys.*, 112, 6152-6160.

[45] Van Duijnen, P. T., & Swart, M. (1998). Molecular and Atomic Polarizabilities: Thole's Model Revisited. *J. Phys. Chem. A*, 102, 2399-2407.

[46] Lide, D. R. (2006). *CRC Handbook of Chemistry and Physics*, Boca Raton, CRC Press.

[47] Thole, B. T. (1981). Molecular Polarizabilities Calculated with a Modified Dipole Interaction. *Chem. Phys.*, 59, 341-350.

[48] Molecular polarizabilities of [BMIm]+ and $[NTf2]^-$ were calculated with the same method in [14]. For molecular polarizability tensor elements of [BMIm]+, the reported values were used [14]. With the same procedures as in [14], molecular polarizability tensor elements of $[NTf2]^-$ were calculated, and the values of 9.8855, 1.4440 x 10^{-3}, 13.577-5x 10^{-4},-1.0539, and 10.315 Å^3 were used for α_{XX}, α_{XY}, α_{YY}, α_{XZ}, α_{YZ}, and α_{ZZ}, respectively.

[49] Verdaguer, A., & Padró, J. A. (2000). Velocity cross-correlations and atomic momentum transfer in simple liquids with different potential cores. *Phys. Rev. E*, 62, 532-537.

[50] Rahman, A. (1964). Correlations in the Motion of Atoms in Liquid Argon. *Phys. Rev.*, 136, A405-A411.

[51] Del Pópolo, M. G., & Voth, G. A. (2004). On the Structure and Dynamics of Ionic Liquids. *J. Phys. Chem. B*, 108, 1744-1752.

[52] Colmenero, J., Alvarez, F., & Arbe, A. (2002). Self-motion and the α relaxation in a simulated glass-forming polymer: Crossover from Gaussian to non-Gaussian dynamic behavior. *Phys. Rev. E*, 65(041804), 1-12.

[53] Ngai, K. L. (2011). *Relaxation and Diffusion in Complex Systems*, New York, Springer.

[54] Bhargava, B. L., Klein, M. L., & Balasubramanian, S. (2008). Structural Correlations and Charge Ordering in a Room-Temperature Ionic Liquid. *ChemPhysChem*, 9, 67-70.

Chapter 2

Modeling of Ionic Liquid Systems: Phase Equilibria and Physical Properties

Filipa M. Maia, Noelia Calvar, Emilio J. González,
Aristides P. Carneiro, Oscar Rodriguez and
Eugénia A. Macedo

Additional information is available at the end of the chapter

http://dx.doi.org/10.5772/51812

1. Introduction

Ionic liquids (ILs) are a class of salts with a melting temperature below 100 °C, and the study of these compounds is considered priority by the U.S. Environmental Protection Agency. Due to their specific properties, which can be adjusted by changing either the cation or the anion, ILs have received great attention by the scientific community as potential replacements for volatile organic solvents (VOCs), and nowadays, ILs are starting to leave academic labs and find their way into a wide variety of industrial applications [1]. For example, ILs are used for the dispersion of nano-materials at IOLITEC, Air Products uses ILs instead of pressurized cylinders as a transport medium for reactive gases, ION Engineering is commercializing technology using ILs and amines for CO_2 capture and natural gas sweetening, and many others.

In order to apply these new compounds in different processes, the study of their physical properties, pure or mixed with other solvents, and phase equilibria (vapor-liquid, liquid-liquid, and solid-liquid equilibria) is crucial from a technological point of view. For example, density is fundamental to develop equations of state and it is also required for the design of different equipments, while viscosity is necessary for the design of processing units and pumping systems, and to study heat and mass transfer processes [2]. On the other hand, the refractive index can be used as a measure of the electronic polarizability of a molecule and can provide useful information when studying the forces between molecules or their behavior in solution [3].

As known, ILs also show an interesting potential to be used in separation processes and extraction media. Therefore, the knowledge of the mutual solubilities of molecular solvents and ILs prior to their industrial applications is also of primary importance. Moreover, many factors that control the phase behavior of these ionic salts with molecular solvents may be described from the phase equilibrium data.

However, as the number of possible ILs is enormous, this cannot be accomplished via experimental determination. Thus, it is very important to obtain models or empirical equations able to describe satisfactorily the experimental data.

In this chapter, a revision of the different equations applied for the modeling of physical properties of pure ILs and their mixtures, and phase equilibria of binary and ternary mixtures containing ILs, is presented and discussed. Future trends regarding the use of new models, namely equations of state accounting for association effects, are also focused.

2. Physical properties

2.1. Pure ionic liquids

Since ILs are relatively new compounds, experimental data on physical properties, such as density, viscosity, or refractive index of pure ILs and its mixtures with other solvents are required for the design of different equipment and processing units and very useful for developing accurate theoretical models.

Due to innumerable number of ILs that can be synthesized, experimental measurements are impractical for selection of a suitable IL for a specific application. Therefore, development of correlations and theoretical approaches allowing accurate modeling of IL-based systems is essential. This section shows the most common empirical equations used to correlate the temperature dependence of some of the physical properties of ILs.

For pure ILs, temperature dependence of physical properties such as density, speed of sound, or refractive index is very important for the successful and large-scale use of these compounds. Usually, this dependence is described using simple polynomial expressions, mainly equations of first, second and third order [4-6].

Several papers were also published concerning the experimental densities of pure ILs as a function of temperature and pressure [6-9]. The Tait equation [10] with four adjustable parameters is commonly used to fit these experimental data [6,8]. This equation is an integrated form of an empirical equation representative of the isothermal compressibility behavior versus pressure, and it can be expressed as:

$$\rho\left(T,\ p,\ C,\ B\right) = \frac{\rho(T,\ 0.1\ \text{MPa})}{1 - C \cdot Ln\left(\frac{B(T) + p}{B(T) + 0.1\ \text{Mpa}}\right)} \tag{1}$$

where ρ (T, 0.1 MPa) represents the temperature dependence of density at 0.1 MPa. C is an adjustable parameter, and $B(T)$ is commonly expressed as a second-order polynomial:

$$B(T)=\sum_{i=0}^{2} B_i \cdot T^{\,i} \tag{2}$$

Regarding the variation of viscosity with temperature for pure ILs, a large number of empirical equations for correlating this property of pure fluids and mixtures can be found in literature [4,5,11-14]. The most commonly used equation is an Arrhenius-like law:

$$\eta=\eta \cdot exp\left(-\frac{E_{\mathrm{a}}}{RT}\right) \tag{3}$$

where the viscosity at infinite temperature (η_∞) and the activation energy (E_a) are characteristic parameters generally adjusted from experimental data.

According to Seddon *et al.* [15], the Arrhenius law can generally be applied when the cation presents only a limited symmetry. If it is not the case, and especially in the presence of small and symmetrical cations with low molar mass, other equations such as the Vogel–Fulcher–Tamman (VFT) equation [16-18], or the modified VFT (mVFT) equation are recommended. This kind of expressions includes an additional adjustable temperature parameter (T_0) to the exponential term:

$$\eta=A \cdot exp\left(\frac{B}{T - T_0}\right) \tag{4}$$

$$\eta=A \cdot T^{\,0.5} \cdot exp\left(\frac{B}{T - T_0}\right) \tag{5}$$

where A and B are also adjustable parameters.

Another empirical equation to correlate viscosity data with temperature was proposed by Litovitz [19]:

$$\eta=A \cdot exp\left(\frac{B}{RT^{\,3}}\right) \tag{6}$$

This equation is used at ambient pressures and has the advantage of containing only two fitting parameters. Comparing all these equations, in general, the best fits for the variation of viscosity with temperature for pure ILs are obtained with the VFT equation [5].

As reflected by Harris *et al.* [11], the general form of the pressure dependence of the shear viscosity of liquids is greater than exponential at moderate pressures and less than exponential at very high pressures. Taking this into account, these authors modified the VFT and Litovitz equations in order to include the temperature and pressure dependence of the viscosity for several pure imidazolium-based ILs. The new equations were defined as:

$$\eta=exp\left(A + Bp + \frac{(C + Dp + E\,p^{2})}{T^{\,3}}\right) \tag{7}$$

$$\eta = exp\left(A' + B' + \frac{(C' + D'p + E'p^2)}{T - T_0}\right) \tag{8}$$

As for the density, a Tait-form equation is also used to correlate the pressure dependence of viscosity demonstrating good correlations [12,13].

$$\frac{\eta(T,\ p,\ C,\ B)}{\eta(T,\ 0.1\ MPa)} = \exp\left(C \cdot Ln\left(\frac{B(T) + p}{B(T) + 0.1\ \text{Mpa}}\right)\right) \tag{9}$$

where $\eta(T, 0.1\ \text{MPa})$ represents the temperature dependence of viscosity at 0.1 MPa. C is an adjustable parameter, and $B(T)$ is also commonly expressed as a second-order polynomial.

A hybrid Tait–Litovitz equation at elevated pressures (up to 126 MPa) was also presented in the literature to correlate the viscosity data for a series of room-temperature ILs [14]. The Litovitz equation is firstly used to correlate the data at ambient pressure, and then the Tait parameters are fitted for the higher pressures. This equation has the advantages of containing fewer fitting parameters than other models and simplicity of data analysis. The results show a good fit between the experimental data and those predicted by this equation.

2.2. Binary and ternary mixtures

In order to better understand the nature of ILs and design any future technological processes, detailed knowledge on the physical properties of ILs mixed with other solvents is required. During the last few years, the number of studies on thermophysical and thermodynamic properties of pure ILs and their mixtures with molecular solvents has increased significantly [4,20-24].

As for pure ILs, the dependence of the physical properties with temperature and composition is also correlated using empirical equations. In general, the change of density, speed of sound, refractive index and viscosity with composition is typically fitted to a polynomial expression although other more specific equations can be also found in literature. As example, the Connors and Wright equation [25] is employed to describe the variation of density with composition:

$$\rho = \rho_1 - \left[1 + \frac{bx_1}{(1 - ax_1)}\right](1 - x_1)(\rho_1 - \rho_2) \tag{10}$$

where ρ_1 and ρ_2 are densities of pure compounds; x_i is the mole fraction of component i of the mixture; and a and b are fitting parameters. Although this equation was initially adopted only for the concentration dependencies of surface tension, Geppert-Rybczynska *et al.* [20] employed this equation for density getting a fit better than using any other simple polynomial.

As it is known, the above mentioned physical properties can be used to obtain the corresponding excess properties, which are generally fitted to a Redlich-Kister type equation [26]:

$$\delta Q = x_1(1 - x_1)\sum_{i=0}^{M} A_i(2x_1 - 1)^i \tag{11}$$

where δQ is the excess property, x is the mole fraction, A_i are the adjustable parameters and M is the degree of the polynomial expansion.

An extended version of the Redlich-Kister equation, which takes into account the dependence on composition and temperature simultaneously, is also used to fit the excess properties [22]:

$$\delta Q = x_1(1 - x_1)\frac{\sum_{i=0}^{M} A_i(2x_1 - 1)^i}{1 + \sum_{j=1}^{N} B_j(2x_1 - 1)^j} \tag{12}$$

In order to take into account the influence of temperature on the excess properties, all the coefficients A_i and B_i are usually expressed as a second-order polynomial.

Density, refractive index, and viscosity data for ternary mixtures containing ILs are also common in the literature [27-29] and they are usually fitted to polynomial expressions. In these cases, weight fractions are often used instead of mole fractions, due to the large difference of molar mass between ILs and most organic solvents.

For the modeling of the excess properties, such as excess molar volumes, viscosity deviations, refractive index deviations or excess free energies of activation of viscous flow, the use of empirical equations is commonly adopted. Although the Redlich-Kister equation is also applied to correlate the excess properties for ternary systems containing ILs [29], the equations more widely employed are those proposed by Cibulka [30], Singh *et al.* [31] and Nagata and Sakura [32]. All the correlative models are capable of representing the behavior of the ternary mixtures with a higher or lesser degree of accuracy, although that developed by Cibulka usually leads to a better agreement with experimental data [28].

Cibulka equation:

$$Q_{123}^{E} = Q_{12}^{E} + Q_{13}^{E} + Q_{23}^{E} + x_1x_2x_3(A + Bx_1 + Cx_2) \tag{13}$$

Singh et al. equation:

$$Q_{123}^{E} = Q_{12}^{E} + Q_{13}^{E} + Q_{23}^{E} + Ax_1x_2x_3 + Bx_1(x_2 - x_3) + Cx_1^2(x_2 - x_3)^2 \tag{14}$$

Nagata and Sakura equation:

$$Q_{123}^{E} = Q_{12}^{E} + Q_{13}^{E} + Q_{23}^{E} + x_1x_2x_3A \tag{15}$$

where A, B and C are fit parameters and Q_{ij}^{E} is the contribution to the excess property of the constituent binary mixtures evaluated by Redlich-Kister equation.

3. Phase equilibria

Despite the large number of published articles and the broad fields of applications, there has not been a model explicitly derived for phase equilibria of ILs. This lack of models intended for systems containing ILs has forced researchers to use the equations available. But these equations were intended for ionic solutions, or not intended for ions at all. Thus, a model to account for a medium which is composed of ions, without a molecular solvent, is still needed. Nevertheless, the phase equilibria of ILs and their mixtures are being modeled in the literature. The models used will be described ahead. In general, those based on the excess Gibbs energy such as Wilson, NRTL or UNIQUAC were the first to appear. Lately, models with modifications for association (UNIQUAC ASM, NRTL1) and for electrolytes (PDH, e-NRTL) were also applied for this kind of systems in order to improve the results obtained with the initial models. Besides, Equations of State (EoS) have also been applied, especially for mixtures with gases and for broad ranges of pressure.

3.1. g^E-based models

Most of the g^E models available in the literature are for non-electrolytes. Thus, many authors have been using these models, or models for electrolyte solutions, for the phase equilibria of systems containing ILs. For example, for binary systems containing ILs it is common to use the model developed by Debye-Hückel (which was derived for small salt concentrations), although it is not recommendable for solutions at high ionic concentration.

The models for the correlation of these experimental data can be split into two main groups: *i*) ion-interaction Pitzer models and *ii*) local composition models.

i. The model developed by Pitzer has created a new generation of theories which use multi-parameter regression. In the ion-interaction Pitzer model [33] the ion-interaction parameters are dependent on temperature and pressure, and it takes into account the Debye-Hückel constant.

The three-parameter Pitzer-ion interaction model has been successfully used for modeling vapor-liquid data of mixtures of ILs with water [34-37] or with alcohol [38] and has the following form for a binary 1:1 electrolyte solution:

$$\phi - 1 = f^{\phi} + \mathrm{m}B^{\phi} + \mathrm{m}^2 C^{\phi}, \tag{16}$$

where

$$f^{\phi} = -A_{\phi}\, I^{1/2} / (1 + bI^{1/2}), \tag{17}$$

$$A_{\phi} = (1/3)(2\pi N_A d_s)^{1/2} (e^2 / 4\pi\varepsilon_0\varepsilon kT)^{3/2}, \tag{18}$$

$$B^{\phi} = \beta^{(0)} + \beta^{(1)} \exp(-\alpha_1 I^{1/2}) \tag{19}$$

In these equations, $\beta^{(0)}$, $\beta^{(1)}$ and C^{ϕ} are ion-interaction parameters of the Pitzer model that are dependent on temperature and pressure, and A_{ϕ} is the Debye-Hückel constant for the osmotic coefficient on the molal scale. Also, N_A is the Avogadro number, e is the proton charge, ε_0 is the permittivity of vacuum, and k is the Boltzman constant. The term I is the ionic strength in molality, $I = 1/2\ \Sigma m_i z_i^2$, where m_i is the molality of ith ion and z_i is the absolute value for i^{th} ionic charge. The remaining symbols have their usual meanings. The values for constants b and α were $b = 1.2$ kg$^{1/2}$ mol$^{-1/2}$ and $\alpha = 2$ kg$^{1/2}$ mol$^{-1/2}$, respectively.

In the last years, the Extended Pitzer model of Archer [39,40], in which the third adjustable parameter in the Pitzer model is replaced by a two-parameter function depending on the ionic strength, has demonstrated its accuracy in modeling binary Vapor-Liquid Equilibria (VLE) of systems containing ILs. In this model, the equation for B^{ϕ} is extended to:

$$B^{\phi} = \beta^{(0)} + \beta^{(1)} \exp(-\alpha_1 I^{1/2}) + \beta^{(2)} \exp(-\alpha_2 I^{1/2}), \tag{20}$$

and a new equation is introduced:

$$C^{\phi} = C^{(0)} + C^{(1)} \exp(-\alpha_3 I^{1/2}) \tag{21}$$

In the previous equations, the ion interaction parameters of the extended Pitzer model of Archer are $\beta^{(0)}$, $\beta^{(1)}$, $\beta^{(2)}$, $C^{(0)}$ and $C^{(1)}$, dependent on temperature and pressure, and α_1, α_2, α_3, and b can be adjustable parameters or kept fixed at constant values; they are usually fixed to the values: $\alpha_1 = 2$ kg$^{1/2}$ mol$^{-1/2}$, $\alpha_2 = 7$ kg$^{1/2}$ mol$^{-1/2}$, $\alpha_3 = 1$ kg$^{1/2}$ mol$^{-1/2}$ and $b = 3.2$ [41-44], although other values can be used [45]. This model is widely used in literature for binary mixtures containing ILs and water or alcohol or acetonitrile [46-55] with very satisfactory results.

ii. The local composition (LC) models give a better empirical description and have physical meaning for the correlation of osmotic and activity coefficients. There are several LC models reported in literature for the modeling of phase equilibria experimental data; such as UNIversal QUAsiChemical (UNIQUAC) [56], Non-Random Two Liquids (NRTL) [57], electrolyte NRTL (e-NRTL) [58], Non-Random Factor (NRF) [59], modified NRTL (MNRTL) [60], Mean Spherical Approximation NRTL (MSA-NRTL) [61] or Extended Wilson (EW) [62]. Furthermore, models with modifications for association such as UNIQUAC ASM or NRTL 1 are also applied for this kind of systems, although their use is less common [63,64].

In local composition models it is assumed that the activity coefficient is composed by two terms: a long-range contribution (LR) and a short-range contribution (SR):

$$\ln\gamma_i = \ln\gamma_i^{LR} + \ln\gamma_i^{SR} \tag{22}$$

The well-known models of UNIQUAC and NRTL calculate the activity coefficients as follows:

ii.i In the UNIQUAC model [56], the long-range contribution term is expressed as:

$$\ln \gamma_i^{\mathrm{LR,UNIQUAC}} = \ln\left(\frac{\phi}{x_i}\right) + \frac{z}{2} q_i \ln\left(\frac{\theta_i}{\phi_i}\right) + l_i - \frac{\phi_i}{x_i}\sum_{j=1}^{N} x_j l_j \tag{23}$$

where the term l_i is defined as a function of the external surface of the molecule and the bond segments:

$$l_i = \frac{z}{2}(r_i - q_i) - (r_i - 1) \tag{24}$$

and the coordination number, z, is set to 10.

The short-range contribution in this model is expressed as:

$$\ln \gamma_i^{\mathrm{SR,UNIQUAC}} = q_i\left[1 - \ln\left(\sum_j \theta_j \tau_{ji}\right) - \sum_{j=1}^{N} \frac{\theta_j \tau_{ij}}{\sum_k \theta_k \tau_{kj}}\right] \tag{25}$$

The expression for the energy parameter, τ_{ij}, the volume fraction, ϕ_i, and the surface fraction, θ_i, are given for the next equations:

$$\tau_{ji} = \exp\left(-\frac{u_{ji} - u_{ii}}{RT}\right) = \exp\left(\frac{\Delta u_{ij}}{RT}\right) \tag{26}$$

$$\theta_i = \frac{q_i x_i}{\sum_j q_j x_j} \quad \phi_i = \frac{r_i x_i}{\sum_j r_j x_j} \tag{27}$$

where u_{ij} and u_{ji} are the energetic parameters, and r_i and q_i are the structural parameters corresponding to relative volume and surface area of the component *i*, respectively.

This model has been used for VLE of binary systems containing ILs [65]. The equation has also been used successfully for Liquid-Liquid Equilibria (LLE) of ternary systems including an IL [66,67], and even for Solid-Liquid Equilibria (SLE) of binary systems [68-70]. The main problem using the UNIQUAC model is the need for structural parameters r_i and q_i, the volume and surface area of the component. Despite for molecular components these parameters can be obtained easily (even from group-contribution data), in the case of ILs values for both anion and cation are calculated and summed. Different procedures have been proposed [66,68] and used successfully.

ii.ii In the NRTL model [57], the activity coefficients are calculated as follows:

$$\ln \gamma_i^{NRTL} = \left(\frac{\sum_{j=1}^{n} \tau_{ji} x_j G_{ji}}{\sum_{k=1}^{n} x_k G_{ki}} \right) + \sum_{j=1}^{n} \left(\frac{x_j G_{ij}}{\sum_{k=1}^{n} x_k G_{kj}} \right) \left(\tau_{ij} - \frac{\sum_{m=1}^{n} \tau_{mi} x_m G_{mi}}{\sum_{k=1}^{n} x_k G_{kj}} \right) \tag{28}$$

where

$$G_{ij} = \exp\left(-\alpha_{ij} \tau_{ij}\right) \tag{29}$$

$$\tau_{ij} = \frac{g_{ij} - g_{jj}}{RT} = \frac{\Delta g_{ij}}{RT} \quad \tau_{ji} = \frac{g_{ji} - g_{ii}}{RT} = \frac{\Delta g_{ji}}{RT} \tag{30}$$

where x represents the mole fraction, g_{ij} is an energy parameter that characterizes the interaction of species i and j, R is the gas constant, T is the absolute temperature, and the parameter $\alpha_{ij} = \alpha_{ji}$ is related to the nonrandomness in the mixture. Although α_{ij} can be adjusted, it can also be considered fixed in a value, usually between 0.2 and 0.47.

This is the local composition model most widely used in literature for binary and ternary systems containing ILs, regarding VLE [71-79] with alcohols or water, LLE of binary [80,81] and ternary systems [67,82-85] with ethanol or hydrocarbons or water and also SLE [68-70].

For the explanation of the following models, the Pitzer-Debye-Hückel (PDH) equation [33] has been used as the long-range term on a mole fraction scale as proposed by Chen *et al.* [86], assuming that the studied mixtures involve completely dissociated electrolytes. For the solvent, the formulation of this equation is:

$$\ln \gamma_1^{PDH} = \frac{2 A_\phi I_x^{3/2}}{\sqrt{M_S} \left(1 + \rho I_x^{1/2}\right)} \tag{31}$$

in which ρ is the number density of ionic species and the term I is the ionic strength on a mole fraction basis, $I = 1/2 \Sigma x_i z_i^2$.

The short-range contribution calculated using different models, such as electrolyte NRTL (e-NRTL) [58], non-random factor (NRF) [59], modified NRTL (MNRTL) [60], mean spherical approximation NRTL (MSA-NRTL) [61] or Extended Wilson (EW) [62] are explained below.

ii.iii In the e-NRTL model [58], the short-range contribution for the activity coefficient is calculated as:

$$\ln \gamma_1^{e-NRTL} = 2X_c^2 \left[\frac{2\tau_{ca,m} \exp\left(-2\alpha\tau_{ca,m}\right)}{\left(X_1 + 2X_c \exp\left(-\alpha\tau_{ca,m}\right)\right)^2} + \frac{\tau_{m,ca} \exp\left(-\alpha\tau_{m,ca}\right)}{\left(X_1 \exp\left(-\alpha\tau_{m,ca}\right) + X_c\right)^2} \right] \quad (32)$$

where α is the nonrandomness factor (usually set to 0.2) and X_i is the effective mole fraction of the component i, calculated as $X_i = j_i x_i$ with $j_i = z_i$ for ions and $j_i = 1$ for solvent. In this model, $\tau_{ca,m}$ and $\tau_{m,ca}$ are the adjustable parameters.

Examples of the correlation of VLE for binary mixtures [38,73-75] and ternary mixtures [73-75] containing ILs and ethanol or water can be found in literature. Nevertheless, the literature is scarce in examples for SLE [70] and LLE [80,81].

ii.iv The NRF model [59] calculates the short-range contribution for the activity coefficient of the solvent as:

$$\ln \gamma_1^{NRF} = X_c \left\{ \begin{array}{l} \left(\frac{\nu X_c \lambda_E}{z_c \nu_c}\right)\left(1 + \frac{\exp(-\lambda_E / Z)}{\left(X_c \exp(-\lambda_E / Z) + X_1\right)^2}\right) - \\ \frac{X_1 \lambda_S \left((\nu / z_c \nu_c) - 2\exp(-\lambda_S / Z)\right)}{2X_c \exp(-\lambda_S / Z) + X_1} \cdot \left(2 - X_1 \left(1 + \frac{1}{2X_c \exp(-\lambda_S / Z) + X_1}\right)\right) \end{array} \right\} \quad (33)$$

where ν and ν_c are the total number of ions into which the salt dissociates and the number of cations in one mole of the salt, respectively, and Z is the coordination number (usually set to 8). In this model λ_E and λ_S are the adjustable parameters; its use is not common for describing the vapor-liquid equilibria of systems containing ILs [38,77].

ii.v The short-range contribution in the MNRTL model has been developed by Jaretum and Aly [60], and Sardroodi *et al.* [87] used it in the presented form for the first time:

$$ln\gamma_1^{MNRTL} = 4X_c^2[\tau_{ca,m} W_{ca,m} / \left(X_1 + 2W_{ca,m} X_c\right)^2 + \tau_{m,ca}\left(W_{m,ca} - 1\right) / \left(X_c + X_1 W_{ca,m}\right)^2] \quad (34)$$

where $\tau_{ca,m}$ and $\tau_{m,ca}$ are the parameters of the model, X_i is the effective mole fraction, and the next expression is assumed:

$$W_i = exp\ (-\alpha\tau_i + w_i) \quad (35)$$

where α is the nonrandomness factor (usually fixed to 0.2) and $\omega_{ca,m}$ and $\omega_{m,ca}$ are the adjustable parameters.

Among the local composition models, it is one of those giving lower deviations for the modeling of VLE of systems with ILs [34-37,47-54,77].

ii.vi The equation for calculating the short-range contribution for the activity coefficient of the solvent given by the MSA-NRTL model [61] is as follows:

$$\ln \gamma_1^{MSA-NRTL} = \frac{A}{B^2} + \nu_C \nu_a \left(C_c + C_a \right) + \nu \tau_{mc,ac}^2 x_S^2 x_S^i \left(D_c + D_a \right) \tag{36}$$

taking into account that:

$$A = (\nu_c p_{cm} \tau_{cm} + \nu_a p_{am} \tau_{am})(\nu_c p_{cm} + \nu_a p_{am}) \tag{37}$$

$$B = \nu_c p_{cm} + \nu_a p_{am} + x_S^1 \tag{38}$$

$$C_c = p_{cm,ac} \tau_{mc,ac} / \left(\nu_a + x_S^i p_{mc,ac} \right)^2 \tag{39}$$

$$C_a = p_{ma,ca} \tau_{ma,ca} / \left(\nu_c + x_S^i p_{ma,ca} \right)^2 \tag{40}$$

$$D_c = \frac{\nu_c p_{mc,ac}}{\left(\nu_a + x_S^i p_{mc,ac} \right)^2 \left(1 - \frac{\nu_a \alpha \tau_{mc,ac}}{\nu_a + x_S^i p_{mc,ac}} \right)} \tag{41}$$

$$D_a = \frac{\nu_a p_{ma,ca}}{\left(\nu_c + x_S^i p_{ma,ca} \right)^2 \left(1 - \frac{\nu_c \alpha \tau_{ma,ca}}{\nu_c + x_S^i p_{ma,ca}} \right)} \tag{42}$$

being

$$\mathrm{p_i} = \exp(-\alpha \tau_\mathrm{i}) \tag{43}$$

$$\tau_{\mathrm{ma,ca}} = \tau_{\mathrm{mc,ac}} + \tau_{\mathrm{am}} - \tau_{\mathrm{cm}} \tag{44}$$

$$\tau_{\mathrm{mc,ac}} = \tau_{\mathrm{mc,ac}}^1 + \tau_{\mathrm{mc,ac}}^2 X_1 \tag{45}$$

$$x^i = x_1 / x_S = 1 / x_S - \nu \tag{46}$$

in which x_1 and x_S are the mole fraction of solvent and salt, respectively.

The correlation using this model it is not common for the treatment of the VLE data of systems containing ILs [77].

ii.vii The EW model was presented by Zhao *et al.* [62], and it describes the short-range contribution for the activity coefficient as:

$$\ln \gamma_1^{EW} = -C\left(\ln\left(\frac{X_1+2X_c\exp\left(-E_{e1}/CRT\right)}{X_1+2X_c}\right)\right)+ \\ 2X_c\left(\frac{\left(1-\exp\left(-E_{e1}/CRT\right)\right)X_1}{\left(X_1+2X_c\exp\left(-E_{e1}/CRT\right)\right)\left(X_1+2X_c\right)}+\frac{X_c\left(\exp\left(-E_{1e}/CRT\right)-1\right)}{\left(X_1\exp\left(-E_{1e}/CRT\right)+X_c\right)\left(X_1+X_c\right)}\right) \quad (47)$$

where C is the coordination number (set to 10 [77]). In this model, E_{e1} and E_{1e} are the adjustable parameters. This model has been checked for its use in VLE of aqueous systems with ILs [77].

Model	References		
	VLE	LLE	SLE
Pitzer	[34-38]		
Extended Pitzer of Archer	[46-55]		
UNIQUAC	[65]	[66,67]	[68-70]
NRTL	[71-79]	[67,80-85]	[70-72]
e-NRTL	[38,73-75]	[80,81]	[70]
NRF	[38,77]		
MNRTL	[34-37,47-54,77]		
MSA-NRTL	[77]		
EW	[77]		

Table 1. Literature examples for the modeling of phase equilibria in systems containing ILs

A list of representative examples found in literature for the modeling of phase equilibria in systems containing ILs with the above mentioned models is presented in Table 1.

Among these correlation models, those which have demonstrated to give the best results in VLE are the Extended Pitzer model of Archer, the NRTL and the MNRTL models. Regarding LLE and SLE, there are less examples available, specially comparing different models. Nevertheless, it is clear that NRTL is, by far, the most used equation. It is also important to highlight that differences in performance among the models are small.

3.2. Equations of state

Equations of state (EoS) are powerful tools, which can be used to describe the properties of pure fluids or their mixtures. In the last 10 years, this kind of models has been widely applied to describe the properties of pure ILs, as well as to model the phase equilibrium (VLE and LLE) of mixtures containing them.

3.2.1. Peng-Robinson

The Peng-Robinson EoS was developed in 1976 by Peng and Robinson [88] and can be expressed as:

$$P = \frac{RT}{V - b} - \frac{a}{V(V + b) + b(V - b)} \tag{48}$$

with:

$$b = 0.077796 \frac{RT_c}{P_c} \tag{49}$$

$$a = 0.457235 \frac{R^2 T_c^2}{P_c} \alpha \tag{50}$$

$$\alpha = \left[1 + m\left(1 - \sqrt{\frac{T}{T_c}}\right)\right]^2 \tag{51}$$

$$m = 0.37464 + 1.54226\omega - 0.26992\omega^2 \tag{52}$$

where T_c is the critical temperature, P_c is the critical pressure and ω is the acentric factor. For most ILs, the critical properties and the acentric factor are impossible to determine experimentally, because they start to decompose even before the temperature reaches the boiling point. For this reason, Valderrama and co-workers have applied an extended group contribution method, the modified Lydersen-Joback-Reid method, to determine the critical properties, boiling temperatures and acentric factors of several ILs [89-92]. This method only requires knowledge of the structure of the ILs and its molecular weight. Since there are no experimental data available for these properties, the accuracy of the method is verified by comparing calculated liquid densities of the ILs to experimental data available. The results show that the method is sufficiently accurate for several applications, with average absolute deviations (AAD) between calculated and experimental liquid densities in the range of 5 to 6%. The properties determined by this research group have been widely used by the scientific community when modeling the properties of ILs and its mixtures, using this or other equations.

For mixtures of fluids, mixing rules have to be applied to parameters *a* and *b*, which imply the use of one or more binary interaction parameters. For systems containing ILs, the Wong-

Sandler mixing rules have been widely used [93-96], as well as the quadratic [97,98], van der Waals [94], [99,100] and Mathias-Klots-Prausnitz mixing rules [101,102].

The application of the Peng-Robinson EoS to systems with ILs has been mainly focusing on the VLE with CO_2 and other gases. For example, Shin *et al.* [103] modeled the high-pressure solubilities of CO_2 in ILs of the family $[C_n mim][Tf_2N]$ (1-alkyl-3-methylimidazolium bis(trifluoromethylsulfonyl) imide) using this EoS and quadratic mixing rules with two temperature-dependent binary interaction parameters. They obtained AAD between calculated and experimental equilibrium pressures between 10 and 14%, for which they conclude that the model can satisfactorily predict the solubility of high-pressure CO_2 in this family of ILs over a wide range of pressures up to the supercritical region of CO_2. Also Álvarez and Aznar [93] used the Peng-Robinson EoS to model the VLE of binary systems composed of IL + supercritical CO_2 or CHF_3 and IL + hydrocarbons. In this work, the van der Waals and Wong-Sandler mixing rules were used, with UNIQUAC and NRTL models used as excess Gibbs energy models in the Wong-Sandler mixing rules. Their results present AAD in pressure between 2 and 24% for low pressures and between 7 and 52% for high pressures. The authors conclude that the EoS is not able to represent the data at high pressures; however, it performs well at low pressures. Later, Álvarez *et al.* [104] used the Peng-Robinson EoS with the Wong-Sandler mixing rules, once again with UNIQUAC and NRTL models for the excess Gibbs energy, but also using the COSMO-SAC model. They modeled the isobaric VLE of 1-ethyl-3-methylimidazolium ethylsulfate ($[C_2mim][EtSO_4]$) with propionaldehyde or valeraldehyde. When using UNIQUAC or NRTL models, they obtain AAD below 0.15% for both systems. Since the COSMO-SAC model is a predictive model, *i.e.*, it does not require any adjustable binary interaction parameters, the authors regard the application of this model with the Peng-Robinson EoS and the Wong-Sandler mixing rules as pure predictive results. The AAD obtained in this way was 0.3 and 2% for each system, respectively. As a final example, Ren and Scurto [99] used the Peng-Robinson EoS with the van der Waals one fluid mixing rule to model the VLE and VLLE of imidazolium-based ILs and the refrigerant gas 1,1,1,2-tetrafluoroethane. The authors obtained AAD values between 0.4 and 7.4% for VLE and concluded that the model is able to represent the bubble-point data with excellent agreement. However, when using the interaction parameters regressed from the VLE data alone to predict compositions of the VLLE transition to LLE, only satisfactory results were obtained for all systems.

3.2.2. *Soave-Redlich-Kwong*

A Soave modification of the Redlich-Kwong EoS [105] has been frequently applied to systems with ILs. The Soave-Redlich-Kwogn (SRK) EoS was introduced in 1972 by Giorgio Soave, and can be expressed as:

$$P = \frac{RT}{V - b} - \frac{a(T)}{V(V + b)} \tag{53}$$

with:

$$a(T)=0.427480\frac{R^2T_c^2}{P_c}\alpha(T) \tag{54}$$

$$b=0.08664\frac{RT_c}{P_c} \tag{55}$$

In the works involving systems with ILs, the temperature dependent part of the a parameter has been modeled by the empirical form:

$$\alpha(T)=\sum_{k=0}^{\leq 3}\beta_k\left(\frac{T_c}{T}-\frac{T}{T_c}\right)^k \tag{56}$$

Coefficients β_K are usually determined to reproduce the vapor pressure of the pure compound. However, since there are no available experimental vapor pressure data for most ILs, the coefficients are treated as adjustable fitting parameters in the calculations [106]. The authors have found that for ILs, only one adjustable parameter, β_1, is sufficient (with β_0=1 and β_2= β_3=0). In the case of this EoS, modified van der Waals-Berthelot mixing rules have been used, with four binary interaction parameters.

Shiflett and Yokozeki [106] first used this EoS to correlate the solubility of CO_2 in ILs l-butyl-3-methylimidazolium hexafluorophosphate ([C_4mim][PF_6]) and l-butyl-3-methylimidazolium tetrafluoroborate ([C_4mim][BF_4]). The authors obtained an excellent fit between experimental and calculated solubility data, with standard deviations of 17 and 10.5 kPa for each system, respectively. They used the same model to correlate the VLE of CO_2 and other gases (like ammonia [107,108] or SO_2 [109]) in different ILs. Later, the same authors modeled the solubility of water in several different ILs using the same EoS [110]. The four binary interaction parameters were determined using binary VLE data from the literature, having obtained standard deviations lower than 0.6 kPa for all systems. The graphical results presented by the authors show a very good agreement between experimental and calculated solubilities of water in all ILs considered. The same research group has also used the SRK EoS to study the phase behavior of ternary mixtures CO_2/H_2/[C_4mim][PF_6] [111], CO_2/SO_2/1-butyl-3-methylimidazolium methyl sulfate ([C_4mim][MeSO4]) [112], CO_2/H_2S/[C_4mim][PF_6] [113] and CO_2/H_2S/[C_4mim][$MeSO_4$] [114]. For all the previous studies the authors obtained good agreement between calculated and experimental data. Very recently, Shiflett *et al.* [115] have also modeled the ternary system N_2O/CO_2/1-butyl-3-methylimidazolium acetate ([C_4mim][Ac]) with the purpose of understanding the separation of N_2O and CO_2 using ILs. They determined the binary interaction parameters using VLE data (either their own or from the literature) for the pairs N_2O/[C_4mim][Ac], CO_2/[C_4mim][Ac] and N_2O/CO_2. Unlike what happened in the previously mentioned ternary systems, the phase behavior prediction of the ternary system of N_2O/CO_2/[C_4mim][Ac] may not be guaranteed based on the binary interaction parameters alone. This happens because for systems containing mixtures with the chemical complex formation (hydrogen-bonding, charge-transfer complex, etc.) as in the case of the binary CO_2/[C_4mim][Ac] here present, the third component (in this case, N_2O) may interfere with the binary interactions of CO_2/[C_4mim][Ac], originating a decrease or an increase of the complex formation and changing the binary interaction parameters of this pair in the ternary system. Consequently, one of the binary interaction parameters of the pair CO_2/[C_4mim][Ac] was modified so as to fit the experimental ternary VLE data. This way, the model provides an excellent agreement between experimental and calculated data for the ternary system.

3.2.3. Statistical associating fluid theory

The original Statistical Associating Fluid Theory (SAFT) was developed in 1989 [116] based on Wertheim's first-order thermodynamic perturbation theory [117-120]. Its main advantage over the traditional cubic EoS, is that it takes into account the structure of the molecule, similarly to group contribution models. It regards the molecules as chains of hard-spheres, which contain multiple association sites. Several variations of the SAFT EoS have been used to model the phase behavior of systems containing ILs: tPC-PSAFT [121-125], soft-SAFT [126-128], hetero-SAFT [129], PCP-SAFT [130] and PC-SAFT [131].

Kroon *et al.* [121] applied the tPC-PSAFT model to the phase behavior of IL + CO_2 systems and later Karakatsani *et al.* [122] applied it to the correlation of the solubility of CO_2, CO, O_2 and CHF_3 in the same ILs. Pure component parameters for the ILs were estimated from ILs experimental thermodynamic data (density, enthalpy, and entropy of dissolution of CO_2) and physicochemical data for the constituent ions (size, polarizability, number of electrons). The cross-association parameters were estimated from enthalpy and entropy data for the dissolution of CO_2 in the ILs. A temperature-dependent binary interaction parameter k_{ij} was adjusted in order to fit the model to experimental VLE data. The results of the model were in good agreement with experimental data. Later, Karakatsani and Economou [123] used the same parameters to model the VLE of the ternary system CO_2/acetone/[C_4mim][PF_6], also obtaining satisfactory agreement between calculated and experimental data. Additionally, Economou *et al.* [124] applied the same model to describe the VLE of the binary [C_8mim][BF_4]/benzene and the ternary CO_2/H_2O/[C_4mim][NO_3] (1-butyl-3-methylimidazolium nitrate) systems. For binary mixture calculations, a binary interaction parameter was fitted to the experimental data and the model was capable of accurately correlating the phase equilibria of the binary and of the ternary IL mixtures with polar solvents. In a more recent work, Karakatsani *et al.* [125] proved that the model can accurately predict the phase equilibrium of non-polar solvent/IL mixtures, without the use of any adjustable binary interaction parameter, by applying it to binary and ternary mixtures of ILs with organic solvents and water. They also concluded that in the case of aqueous solvents, the dissociation of IL has to be incorporated explicitly into the model in order to obtain a good correlation with the experimental data.

Andreu and Vega [126] used the soft-SAFT EoS to describe the solubility of CO_2 in ILs. They modeled the families of ILs [$C_n$mim][BF_4] and [$C_n$mim][PF_6] as Lennard-Jones chains with one associating site in each molecule. The chain length, size and energy parameters of the ILs were obtained by fitting the model predictions to available density data, obtaining AAD values lower than 0.2%. For the association parameters of ILs, values previously used for alkanols were adopted. The authors found that the model correlations and experimental data for VLE are in good agreement. They later used the same model to describe the solubility of hydrogen, CO_2 and xenon in ILs of the family [$C_n$mim][Tf_2N] [127]. In this case, the ILs were modeled as Lennard-Jones chains with three associating sites in each molecule. The pure component parameters for ILs were obtained as in the previous work, and a good description of experimental solubilities was obtained. Recently, Llovell *et al.* [128] re-parameterized the model for the [$C_n$mim][Tf_2N] family of ILs and modeled the solubilities of methanol and ethanol. Good agreement was found between the predictions of the model and the experimental data, with AAD values below 5% for methanol and around 10% for ethanol. They also modeled the VLE of IL/water systems, with the model providing good agreement with the experimental data, with AAD values between 6 and 12%. As for the LLE of mixtures of ILs with water, the authors were not able to obtain reasonable predictions, having obtained significant deviations in the water-rich phase. Quantitative agreement was achieved by using two adjustable parameters, which were adjusted only to the water-rich phase of the aqueous [C_4mim][Tf_2N] mixture and used in a predictive manner for the IL-rich phase and for the aqueous mixtures with [C_2mim][Tf_2N] and [C_6mim][Tf_2N].

Ji and Adidharma [129] used the heterosegmented SAFT (hetero-SAFT) to describe the solubility of CO_2 in the families of ILs [$C_n$mim][BF_4], [$C_n$mim][PF_6] and [$C_n$mim][Tf_2N]. The molecules of ILs were divided into groups representing the alkyl chain, the cation head and the anion. To account for the electrostatic/polar interactions between cation and anion, the spherical segments representing the cation head and the anion were assumed to have one association site each, which can only associate to each other. The parameters for the alkyl chains were obtained from those of the corresponding n-alkanes and the parameters for groups representing the cation head and the anion, including the two association parameters, were fitted to experimental IL density data. The model was capable of satisfactorily describing the solubility of CO_2 in the ILs studied.

Finally, Paduszynski *et al.* [130] used the Perturbed-Chain Polar Statistical Associating Fluid Theory (PCP-SAFT) to model the LLE of the IL 1-methyl-1-propylpiperidinium bis(trifluoromethylsulfonyl)imide [C_3mpip][NTf_2] with several alkan-1-ols. They modeled the IL as strongly associating molecules, with symbol A_1 representing a positive site which corresponds to the nitrogen atom on the cation and its proximity, and symbol B_1 representing a negative site which corresponds to the delocalized charge due to the oxygen molecules on the anion. They defined each type of associating site in an identical way; however, they only allowed A_1B_1 interactions to take place. Additionally, they assumed that each molecule has 5 positive sites of type A_1 and 5 negative sites of type B_1. The pure component parameters of the IL were determined by fitting to liquid density and total solubility parameter data. Self-association of IL and alkan-1-ols, as well as cross-association, were accounted for and one linearly temperature-dependent binary interaction parameter was needed in order to obtain qualitative agreement between calculated and experimental data. AAD values obtained were between 0.6 and 5%. On a different work, Paduszynski and Domanska [131] modeled the LLE of systems composed by piperidinium-based ILs ([C_3mpip][NTf_2] and [C_4mpip][NTf_2]) and several aliphatic hydrocarbons using the Perturbed-Chain Statistical Associating Fluid Theory (PC-SAFT) model. Pure component parameters for ILs were obtained as in the previous work and they used activity coefficients at infinite dilution of hydrocarbons in ILs reported in the literature to optimize the binary interaction parameter, which was again considered to be linearly temperature-dependent, with two adjustable parameters. In the calculation of the LLE of the systems, they obtained AAD values between 1.5 and 8%. Additionally, the authors decided to test their approach for the cross-associating systems [C_3mpip][NTf_2] + 1-pentanol and [C_3mpip][NTf_2] + water. Regarding the LLE of [C_3mpip][NTf_2] + 1-pentanol, the authors claim that the experimental data is well described by the model, although deviations between calculated and experimental compositions increase as the temperature increases and the model ends up over predicting the upper critical solution temperature by about 15 K. The AAD values obtained for this system were 52% for the IL-rich phase and 1.7% for the alcohol-rich phase. As for the system [C_3mpip][NTf_2] + water, the authors conclude that the PC-SAFT model is surprisingly good at describing the experimental data, but only when the IL-rich phase is considered alone. For the water-rich phase, the model predicts solubilities of IL much lower than those observed experimentally by several orders of magnitude. The authors justify this fact because the molecular model chosen for the IL is not appropriate for dilute solutions of IL in water, in which case the cation and anion of the IL are probably dissociated, which results on increased solubility.

3.2.4. Other EoS

Several other EoS have been used to model systems with ILs. For example, Tsioptsias et al. [132] used the Non-Random Hydrogen-Bonding (NRHB) model [133] to describe the phase behavior of binary systems containing ILs of the family [$C_n$mim][Tf_2N], obtaining good agreement between model correlations and experimental data. Wang et al. [134] used the square well chain fluid (SWCF) EoS [135] to model the solubilities of gases such as CO_2, C_3H_6, C_3H_8 and C_4H_{10} in several ILs. Breure et al. [136] used a group contribution EoS to study the phase behavior of binary systems of ILs of the families [$C_n$mim][PF_6] and [$C_n$mim][BF_4] with CO_2, also obtaining good agreement between model predictions and experimental data. Very recently, Maia et al. [137] applied the Cubic Plus Association (CPA) EoS [138], which combines the SRK EoS with an advanced association term similar to that of the SAFT type models, to describe the VLE with CO_2 and the LLE with water of ILs [C_2mim][Tf_2N] and [C_4mim][Tf_2N]. Good agreement was obtained between calculated and experimental data, even though smaller AAD percentage values were obtained for the VLE than for the LLE.

4. Conclusions

Regarding the treatment of physical properties of pure ILs, temperature dependence of physical properties such as density, speed of sound, or refractive index is described using simple polynomial expressions, mainly equations of first, second and third order. For the viscosity, usually the VFT or mVFT equations are strongly recommended. For binary mixtures containing ILs, the dependence of the physical properties with temperature and composition is also correlated using empirical equations, and their excess properties are generally fitted to a Redlich-Kister type equation. For the fitting of the excess properties of ternary systems containing ILs, the most widely used equation is that proposed by Cibulka.

For the correlation of experimental data concerning phase equilibria of mixtures containing ILs, several g^E-based models have been applied in literature (Pitzer, Extended Pitzer model of Archer, UNIQUAC, NRTL, e-NRTL, NRF, MNRTL, MSA-NRTL, EW), being the NRTL model the one that unifies simplicity and satisfactory results for the treatment of vapor-liquid, liquid-liquid and solid-liquid equilibria.

The use of EoS for the modeling of phase equilibria involving ILs is frequent. Unlike what happens with g^E models, most of the literature with EoS involve VLE data, rather than LLE or SLE. Nevertheless, many authors have proved that excellent results can be obtained in data correlation or, for some cases, even prediction. The main difficulty with the application of EoS is the calculation of pure component parameters for ILs. Up to date, a general procedure has not yet been defined. Very recently, a complete review on the use of EoS with ILs, with special emphasis on the obtention of model parameters, has been published by Maia and co-workers [137]. The interested reader is directed to that work for further details.

Author details

Filipa M. Maia, Noelia Calvar, Emilio J. González, Aristides P. Carneiro,
Oscar Rodriguez and Eugénia A. Macedo

LSRE-Laboratory of Separation and Reaction Engineering, Associate Laboratory LSRE/LCM, Department of Chemical Engineering, Faculty of Engineering, University of Porto, Rua Dr. Roberto Frias, Porto, Portugal

References

[1] Short PL. Out of the ivory tower. Chemical and Engineering News 2006;84(17) 15-21.

[2] Aparicio S, Atilhan M, Karadas F. thermophysical properties of pure ionic liquids: review of present situation. Industial and Engineering Chemistry Research. 2010;49(20) 9580-9595.

[3] Tariq M, Forte PAS, Gomes MFC, Lopes JNC, Rebelo LPN. Densities and refractive indices of imidazolium and phosphonium based ionic liquids: effect of temperature, alkyl chain length, and anion. Journal of Chemical Thermodynamics. 2009;41(6) 790-798.

[4] Widegren JA, Magee JW. Density, viscosity, speed of sound, and electrolytic conductivity for the ionic liquid 1-hexyl-3-methylimidazolium bis(trifluoromethylsulfonyl)imide and its mixtures with water. Journal Chemical and Engineering Data 2007;52(6) 2331-2338.

[5] Seoane RG, Corderí S, Gómez E, Calvar N, González EJ, Macedo EA, Domínguez A. Temperature dependence and structural influence on the thermophysical properties of eleven commercial ionic liquids. Industrial and Engineering Chemical Research 2012;51(5) 2492–2504.

[6] Gu Z, Brennecke JF. Volume expansivities and isothermal compressibilities of imidazolium and pyridinium-based ionic liquids. Journal Chemical and Engineering Data 2002;47(2) 339-345.

[7] Paduszyński K, Domańska U, Chang H. Thermodynamic modeling of ionic liquid systems: development and detailed overview of novel methodology based on the PC-SAFT. Journal of Physical Chemistry B - Condensed Phase 2012;116(16) 5002–5018.

[8] Tome LIN, Gardas RL, Carvalho PJ, Pastoriza-Gallego MJ, Piñeiro MM, Coutinho JAP. Measurements and correlation of high-pressure densities of phosphonium based ionic liquids. Journal Chemical and Engineering Data 2011;56(5) 2205–2217.

[9] Machida H, Sato Y, Smith RL. Pressure–volume–temperature (PVT) measurements of ionic liquids ([bmim+][PF6−], [bmim+][BF4−], [bmim+][OcSO4−]) and analysis with the Sanchez–Lacombe equation of state. Fluid Phase Equilibria 2008;264(1-2) 147–155.

[10] Dymond JH, Malhotra R. The Tait equation: 100 years on. International Journal of Thermophysics 1988;9(6) 941-951.

[11] Harris KR, Kanakubo M, Woolf LA. Temperature and pressure dependence of the viscosity of the ionic liquids 1-methyl-3-octylimidazolium hexafluorophosphate and 1-methyl-3-octylimidazolium tetrafluoroborate. Journal Chemical and Engineering Data 2006;51(3) 1161-1167.

[12] Tomida D, Kumagai A, Kenmochi S, Qiao K, Yokoyama C. Viscosity of 1-hexyl-3-methylimidazolium hexafluorophosphate and 1-octyl-3-methylimidazolium hexafluorophosphate at high pressure. Journal Chemical and Engineering Data 2007;52(2) 577-579.

[13] Bandres I, Alcalde R, Lafuente C, Atilhan M, Aparicio S. On the viscosity of pyridinium based ionic liquids: an experimental and computational study. Journal of Physical Chemistry B 2011;115(43) 12499–12513.

[14] Ahosseini A, Scurto AM. Viscosity of imidazolium-based ionic liquids at elevated pressures: cation and anion effects. International Journal of Thermophysics 2008;29(4) 1222-1243.

[15] Seddon KR, Starck AS, Torres, MJ. ACS Symposium Series 901: Washington DC, 2004.

[16] Vogel H. The law of the relation between the viscosity of liquids and the temperature. Phys. Z. 1921;22, 645-646.

[17] Fulcher GS. Analysis of recent measurements of the viscosity of glasses. Journal of the American Ceramic Society 1925;8(6) 339-355.

[18] Tammann G, Hesse W. The dependence of viscosity upon the temperature of super-cooled liquids. Zeitschrift für anorganische und allgemeine Chemie 1926;156(1) 245-257.

[19] Litovitz, TA. Temperature dependence of the viscosity of associated liquids. Journal of Chemical Physics 1952;20(7) 1088-1089.

[20] Geppert-Rybczyńska M, Heintz A, Lehmann JK, Golus A. Volumetric properties of binary mixtures containing ionic liquids and some aprotic solvents. Journal Chemical and Engineering Data 2010;55(9) 4114–4120.

[21] Heintz A, Klasen D, Lehmann JK, Wertz C. Excess molar volumes and liquid-liquid equilibria of the ionic liquid 1-methyl- 3-octyl-imidazolium tetrafluoroborate mixed with butan-1-ol and pentan-1-ol. Journal of Solution Chemistry 2005;34(10) 1135–1144.

[22] Vercher E, Llopis FJ, González-Alfaro MV, Martínez-Andreu A. Density, speed of sound, and refractive index of 1-ethyl-3-methylimidazolium triiifluoromethanesulfonate with acetone, methyl acetate, and ethyl acetate at temperatures from (278.15 to 328.15) K. Journal Chemical and Engineering Data 2010;55 (3) 1377–1388.

[23] González EJ, Domínguez A, Macedo EA. Excess properties of binary mixtures containing 1-hexyl-3-methylimidazoliumbis(trifluoromethylsulfonyl)imide ionic liquid and polar organic compounds. Journal of Chemical Thermodynamics 2012;47(1) 300-311.

[24] García-Miaja G, Troncoso J, Romaní L. Excess enthalpy, density, and heat capacity for binary systems of alkylimidazolium-based ionic liquids + water. Journal of Chemical Thermodynamics 2009;41(2) 161–166.

[25] Connors KA, Wright JL. Dependence of surface tension on composition of binary aqueous-organic solutions. Analytical Chemistry 1989;61(3) 194-198.

[26] Redlich O, Kister AT. Algebraic representation of thermodynamic properties and the classification of solutions. Industrial and Engineering Chemistry 1948;40(2) 345–348.

[27] González B, Calvar N, González E, Domínguez A. Density and viscosity experimental data of the ternary mixtures 1-propanol or 2-propanol + water + 1-ethyl-3-methylimidazolium ethylsulfate. Correlation and prediction of physical properties of the ternary systems. Journal of Chemical and Engineering Data 2008;53(3) 881–887.

[28] Gómez E, González B, Calvar N, Domínguez A. Excess molar properties of ternary system (ethanol + water + 1,3-dimethylimidazolium methylsulphate) and its binary mixtures at several temperatures. Journal of Chemical Thermodynamics 2008;40(8) 1208–1216.

[29] Andreatta AE, Arce A, Rodil E, Soto A. Physico-chemical properties of binary and ternary mixtures of ethyl acetate + ethanol +1-butyl-3-methyl-imidazolium bis(tri-

fluoromethylsulfonyl)imide at 298.15 K an atmospheric pressure. Journal of Solution Chemistry 2010;39(3) 371–383.

[30] Cibulka I. Estimation of excess volume and density of ternary liquid mixtures of non-electrolytes from binary data. Collection of Czechoslovak Chemical Communication 1982;47(5) 1414–1419.

[31] Singh P, Nigam R, Sharma S, Aggarwal S. Molar excess volumes of ternary mixtures of nonelectrolytes. Fluid Phase Equilibria 1984;18(3) 333–344.

[32] Nagata M, Sakura J. Refractive index and excess volume for binary liquid mixtures. Part 1. Analyses of new and old data for binary mixtures. Journal of the Chemical Society, Faraday Transactions 1 1987;83(8) 2449–2457.

[33] Pitzer KS. Activity Coefficients in Electrolyte Solutions. Ed. CRC Press: Boca Raton, FL; 1991.

[34] Shekaari H, Mousavi SS. Influence of alkyl chain on the thermodynamic properties of aqueous solutions of ionic liquids 1-alkyl-3-methylimidazolium bromide at different temperatures. The Journal of Chemical Thermodynamics 2009;41 90–96.

[35] Shekaari H, Mousavi SS, Mansoori Y. Thermophysical properties of ionic liquid, 1-pentyl-3-methylimidazolium chloride in water at different temperatures. International Journal of Thermophysics 2009;30 499–514.

[36] Shekaari H, Mousavi SS. Osmotic coefficients and refractive indices of aqueous solutions of ionic liquids containing 1-butyl-3-methylimidazolium halide at T = (298.15 to 328.15) K. Journal of Chemical and Engineering Data 2009;54 823–829.

[37] Shekaari H, Armanfar E. Physical properties of aqueous solutions of ionic liquid, 1-propyl-3-methylimidazolium methyl sulfate, at T = (298.15 to 328.15) K. Journal of Chemical and Engineering Data 2010;55 765–772.

[38] Sardroodi JJ, Atabay M, Azamat J. Isopiestic determination of the osmotic coefficient and vapour pressure of N-R-4-(N,N-dimethylamino)pyridinium tetrafluoroborate (R = C4H9, C5H11, C6H13) in the ethanol solution at T = 298.15 K. The Journal of Chemical Thermodynamics 2012;49 70–74.

[39] Archer DG. Thermodynamic properties of the NaBr+H2O System. Journal of Physical and Chemical Reference Data 1991;20 509–555.

[40] Archer DG. Thermodynamic properties of the NaCl+H2O system. II. Thermodynamic properties of NaCl(aq), NaCl·2H2(cr), and phase equilibria. Journal of Physical and Chemical Reference Data 1992;21 793–829.

[41] Shekaari H, Zafarani-Moattar MT. Osmotic coefficients of some imidazolium based ionic liquids in water and acetonitrile at temperature 318.15K. Fluid Phase Equilibria 2007;254 198–203.

[42] Nasirzadeh K, Neueder R, Kunz W. Vapor pressures, osmotic and activity coefficients for (LiBr + acetonitrile) between the temperatures (298.15 and 343.15) K. The Journal of Chemical Thermodynamics 2004;36 511–517.

[43] Nasirzadeh K, Neueder R, Kunz W. Vapor pressures, osmotic and activity coefficients of electrolytes in protic solvents at different temperatures. 2. Lithium bromide in ethanol. Journal of Solution Chemistry 2004;33 1429-1446.

[44] Nasirzadeh K, Neueder R, Kunz W. Vapor pressures, osmotic and activity coefficients of electrolytes in protic solvents at different temperatures. 3. Lithium bromide in 2-propanol. Journal of Solution Chemistry. 2005;34 9-24.

[45] Nasirzadeh K, Neueder R, Kunz W. Vapor pressures and osmotic coefficients of aqueous LiOH solutions at temperatures ranging from 298.15 to 363.15 K. Industrial and Engineering Chemistry Research 2005;44 3807-3814.

[46] Sadeghi R, Ebrahimi N. Ionic association and solvation of the ionic liquid 1-hexyl-3-methylimidazolium chloride in molecular solvents revealed by vapor pressure osmometry, conductometry, volumetry, and acoustic measurements. The Journal of Physical Chemistry B 2011;115 13227–13240.

[47] González B, Calvar N, Domínguez A, Macedo EA. Osmotic coefficients of aqueous solutions of four ionic liquids at T = (313.15 and 333.15) K. The Journal of Chemical Thermodynamics 2008;40 1346-1351.

[48] Calvar N, González B, Domínguez A, Macedo EA. Osmotic coefficients of binary mixtures of 1-butyl-3-methylimidazolium methylsulfate and 1,3-dimethylimidazolium methylsulfate with alcohols at T = 323.15 K. The Journal of Chemical Thermodynamics 2009;41 617-622.

[49] Calvar N, González B, Domínguez A, Macedo EA. Vapour pressures and osmotic coefficients of binary mixtures of 1-ethyl-3-methylimidazolium ethylsulfate and 1-ethyl-3-methylpyridinium ethylsulfate with alcohols at T = 323.15 K. The Journal of Chemical Thermodynamics 2009;41 1439-1445.

[50] Calvar N, González B, Domínguez A, Macedo EA. Osmotic coefficients of binary mixtures of four ionic liquids with ethanol or water at T = (313.15 and 333.15) K. The Journal of Chemical Thermodynamics 2009;41 11-16.

[51] Calvar N, Gómez E, Domínguez A, Macedo EA. Vapour pressures, osmotic and activity coefficients for binary mixtures containing (1-ethylpyridinium ethylsulfate + several alcohols) at T = 323.15 K. The Journal of Chemical Thermodynamics 2010;42 625-630.

[52] Gómez E, Calvar N, Domínguez A, Macedo EA. Measurement and modeling of osmotic coefficients of binary mixtures (alcohol + 1,3-dimethylpyridinium methylsulfate) at T = 323.15 K. The Journal of Chemical Thermodynamics 2011;43 908-913.

[53] Calvar N, Gómez E, Domínguez Á, Macedo EA. Determination and modelling of osmotic coefficients and vapour pressures of binary systems 1- and 2-propanol with

CnmimNTf2 ionic liquids (n = 2, 3, and 4) at T = 323.15 K. The Journal of Chemical Thermodynamics 2011;43 1256-1262.

[54] Calvar N, Gómez E, Domínguez A, Macedo EA. Study of the influence of the structure of the alcohol on vapor pressures and osmotic coefficients of binary mixtures alcohol + 1-hexyl-3-methylimidazolium bis(trifluoromethylsulfonyl)imide at T = 323.15 K. Fluid Phase Equilibria 2012;313 38-45.

[55] Calvar N, González EJ, Domínguez A, Macedo EA. Acoustic, volumetric and osmotic properties of binary mixtures containing the ionic liquid 1-butyl-3-methylimidazolium dicyanamide mixed with primary and secondary alcohols. The Journal of Chemical Thermodynamics 2012;50 19-29.

[56] Abrams DS, Prausnitz, JM. Statistical thermodynamics of liquid mixtures: A new expression for the excess Gibbs energy of partly or completely miscible systems. AIChE Journal 1975;21 116-128.

[57] Renon, H.; Prausnitz, JM. Local compositions in thermodynamic excess functions for liquid mixtures. AIChE Journal 1968;14 135-144.

[58] Chen CC, Evans LB. A local composition model for the excess Gibbs energy of aqueous electrolyte systems. AIChE Journal 1986;32 444-454.

[59] Haghtalab A, Vera J. A nonrandom factor model for the excess Gibbs energy of electrolyte solutions. AIChE Journal 1988;34 803-812.

[60] Jaretum A, Aly G. New local composition model for electrolyte solutions: single solvent, single electrolyte systems. Fluid Phase Equilibria 1999;163 175-193.

[61] Papiconomou N, Simonin JP, Bernard O, Kunz W. MSA-NRTL model for the description of the thermodynamic properties of electrolyte solutions. Physical Chemistry Chemical Physics 2002;4 4435-4443.

[62] Zhao E, Yu M, Sauve RE, Khoshkbarchi MK. Extension of the Wilson model to electrolyte solutions. Fluid Phase Equilibria 2000;173 161-175.

[63] Domanska U, Bogel-Łukasik E. Measurements and correlation of the (solid + liquid) equilibria of [1-decyl-3-methylimidazolium chloride + alcohols (C2-C12)]. Industrial and Engineering Chemistry Research 2003;42, 6986-6992.

[64] Domanska U, Bogel-Łukasik E, Bogel-Łukasik R. Solubility of 1-dodecyl-3-methylimidazolium chloride in alcohols (C2-C12). Journal of Physical Chemistry B 2003;107 1858-1863.

[65] Kato R, Krummen M, Gmehling J. Measurement and correlation of vapor–liquid equilibria and excess enthalpies of binary systems containing ionic liquids and hydrocarbons. Fluid Phase Equilibria 2004;224 47-54.

[66] Banerjee T, Singh MK, Sahoo RK, Khanna A. Volume, surface and UNIQUAC interaction parameters for imidazolium based ionic liquids via Polarizable Continuum Model. Fluid Phase Equilibria 2005;234 64–76.

[67] Simoni LD, Lin Y, Brennecke JF, Stadtherr MA. Modeling liquid-liquid equilibrium of ionic liquid systems with NRTL, electrolyte-NRTL, and UNIQUAC. Industrial and Engineering Chemistry Research 2008;47 256-272.

[68] Domanska D, Mazurowska L. Solubility of 1,3-dialkylimidazolium chloride or hexafluorophosphate or methylsulfonate in organic solvents: effect of the anions on solubility. Fluid Phase Equilibria 2004;221 73–82.

[69] Carneiro AP, Rodríguez O, Macedo EA. Solubility of monosaccharides in ionic liquids – Experimental data and modeling. Fluid Phase Equilibria 2012;314 22– 28.

[70] Carneiro AP, Rodríguez O, Macedo EA. Solubility of xylitol and sorbitol in ionic liquids – Experimental data and modeling. The Journal of Chemical Thermodynamics 2012; http://dx.doi.org/10.1016/j.jct.2012.05.020

[71] Calvar N, González B, Gómez E, Domínguez A. Vapor-liquid equilibria for the ternary system ethanol + water + 1-butyl-3-methylimidazolium chloride and the corresponding binary systems at 101.3 kPa. Journal of Chemical and Engineering Data 2006;51 2178-2181.

[72] Calvar N, González B, Gómez E, Domínguez A. Study of the behavior of the azeotropic mixture ethanol–water with imidazolium-based ionic liquids. Fluid Phase Equilibria 2007;259 51-56

[73] Calvar N, González B, Gómez E, Domínguez A. Vapor–liquid equilibria for the ternary system ethanol + water + 1-ethyl-3-methylimidazolium ethylsulfate and the corresponding binary systems containing the ionic liquid at 101.3 kPa. Journal of Chemical and Engineering Data 2008;53 820-825.

[74] Calvar N, González B, Gómez E, Domínguez A. Vapor-liquid equilibria for the ternary system ethanol + water + 1-butyl-3-methylimidazolium methylsulfate and the corresponding binary systems at 101.3 kPa. Journal of Chemical and Engineering Data 2009;54 1004-1008.

[75] Calvar N, Gómez E, González B, Domínguez A. Experimental vapor-liquid equilibria for the ternary system ethanol + water + 1-ethyl-3-methylpyridinium ethylsulfate and the corresponding binary systems at 101.3 kPa: Study of the Effect of the Cation. Journal of Chemical and Engineering Data 2010;55 2786-2791.

[76] Safarov J, Verevkin SP, Bich E, Heintz A. Vapor pressures and activity coefficients of n-alcohols and benzene in binary mixtures with 1-methyl-3-butylimidazolium octyl sulfate and 1-methyl-3-octylimidazolium tetrafluoroborate. Journal of Chemical and Engineering Data 2006;51 518-525.

[77] Shekaari H, Mousavi SS. Measurement and modeling of osmotic coefficients of aqueous solution of ionic liquids using vapor pressure osmometry method. Fluid Phase Equilibria 2009;279 73-79.

[78] Orchillés AV, Miguel PJ, González-Alfaro V, Vercher E, Martínez-Andreu A. 1-Ethyl-3-methylimidazolium dicyanamide as a very efficient entrainer for the extractive

distillation of the acetone + methanol system. Journal of Chemical and Engineering Data 2012;57 394-399.

[79] Jongmans MTG, Schuur B, de Haan AB. Binary and ternary LLE data of the system (ethylbenzene + styrene + 1-ethyl-3-methylimidazolium thiocyanate) and binary VLE data of the system (styrene + 1-ethyl-3-methylimidazolium thiocyanate). The Journal of Chemical Thermodynamics 2012;47 234-240.

[80] Domanska U, Marciniak A. Liquid phase behaviour of 1-butyl-3-methylimidazolium 2-(2-methoxyethoxy)-ethylsulfate with organic solvents and water. Green Chemistry 2007;9 262–266.

[81] Maia FM, Rodríguez O, Macedo EA. LLE for (water + ionic liquid) binary systems using [Cxmim][BF4] (x = 6, 8) ionic liquids. Fluid Phase Equilibria 2010;296 184–191.

[82] Arce A, Rodríguez O, Soto A. Tert-Amyl Ethyl Ether separation from its mixtures with ethanol using the 1-butyl-3-methylimidazolium trifluoromethanesulfonate ionic liquid: Liquid-Liquid Equilibrium. Industrial and Engineering Chemistry Research 2004;43 8323-8327.

[83] Arce A, Rodríguez O, Soto A. A comparative study on solvents for separation of tert-amyl ethyl ether and ethanol mixtures. New experimental data for 1-ethyl-3-methyl imidazolium ethyl sulfate ionic liquid", Chemical Engineering Science 2006;61 6929–6935.

[84] Wytze Meindersma G, Podt AJG, de Haan AB. Ternary liquid–liquid equilibria for mixtures of toluene + n-heptane + an ionic liquid. Fluid Phase Equilibria 2006;247 158–168.

[85] Hu X, Yu J, Liu H. Liquid-Liquid Equilibria of the system 1-(2-Hydroxyethyl)-3-methylimidozolium tetrafluoroborate or 1-(2-hydroxyethyl)-2,3-dimethylimidozolium tetrafluoroborate + water + 1-butanol at 293.15 K. Journal of Chemical and Engineering Data 2006;51 691-695.

[86] Chen CC, Boston JF, Evans LB, Britt HI. Local composition model for excess Gibbs energy of electrolyte systems. Part I: Single solvent, single completely dissociated electrolyte systems. AICHE Journal 1982;28 588-596.

[87] Sardroodi JJ, Zafarani-Moattar MT. Vapor pressures and apparent molal volumes of the solutions of ZnCl2 in ethanol at 298.15 K. Fluid Phase Equilibria 2005;230 64-71.

[88] Peng DY, Robinson DB. A New Two-Constant Equation of State. Industrial & Engineering Chemistry Fundamentals 1976;15 59-64.

[89] Valderrama JO, Robles PA. Critical properties, normal boiling temperatures, and acentric factors of fifty ionic liquids. Industrial & Engineering Chemistry Research 2007;46 1338-1344.

[90] Valderrama JO, Sanga WW, Lazzu JA. Critical properties, normal boiling temperature, and acentric factor of another 200 ionic liquids. Industrial & Engineering Chemistry Research 2008;47 1318-1330.

[91] Valderrama JO, Rojas RE. Critical properties of ionic liquids. Revisited. Industrial & Engineering Chemistry Research 2009;48 6890-6900.

[92] Valderrama JO, Forero LA, Rojas RE, Serena L. Critical properties and normal boiling temperature of ionic liquids. Update and a new consistency test. Industrial & Engineering Chemistry Research 2012;51 7838-7844.

[93] Alvarez VH, Aznar M. Thermodynamic modeling of vapor–liquid equilibrium of binary systems ionic liquid+supercritical {CO2 or CHF3} and ionic liquid+hydrocarbons using Peng–Robinson equation of state. Journal of the Chinese Institute of Chemical Engineers 2008;39 353-360.

[94] Arce PF, Robles PA, Graber TA, Aznar M. Modeling of high-pressure vapor–liquid equilibrium in ionic liquids+gas systems using the PRSV equation of state. Fluid Phase Equilibria 2010;295 9-16.

[95] Yazdizadeh M, Rahmani F, Forghani AA. Thermodynamic modeling of CO2 solubility in ionic liquid ([Cnmim] [Tf2N]; n=2, 4, 6, 8) with using Wong-Sandler mixing rule, Peng-Rabinson equation of state (EOS) and differential evolution (DE) method. Korean Journal of Chemical Engineering 2011;28 246-251.

[96] Mattedi S, Carvalho PJ, Coutinho JAP, Alvarez VH, Iglesias M. High pressure CO2 solubility in N-methyl-2-hydroxyethylammonium protic ionic liquids. Journal of Supercritical Fluids 2011;56 224-230.

[97] Shariati A, Peters CJ. High-pressure phase behavior of systems with ionic liquids: measurements and modeling of the binary system fluoroform + 1-ethyl-3-methylimidazolium hexafluorophosphate. Journal of Supercritical Fluids 2003;25 109-117.

[98] Song HN, Lee BC, Lim JS. Measurement of CO2 solubility in ionic liquids: [BMP] [TfO] and [P14,6,6,6][Tf2N] by measuring bubble-point pressure. Journal of Chemical Engineering Data 2010;55 891-896.

[99] Ren W, Scurto AM. Phase equilibria of imidazolium ionic liquids and the refrigerant gas, 1,1,1,2-tetrafluoroethane (R-134a). Fluid Phase Equilibria 2009;286 1-7.

[100] Hwang S, Park Y, Park K. Measurement and prediction of phase behaviour for 1-alkyl-3-methylimidazolium tetrafluoroborate and carbon dioxide: Effect of alkyl chain length in imidazolium cation. Journal of Chemical Thermodynamics 2011;43 339-343.

[101] Bogel-Łukasik R, Matkowska D, Bogel-Łukasik E, Hofman T. Isothermal vapour–liquid equilibria in the binary and ternary systems consisting of an ionic liquid, 1-propanol and CO2. Fluid Phase Equilibria 2010;293 168-174.

[102] Bogel-Łukasik R, Matkowska D, Zakrzewska ME, Bogel-Łukasik E, Hofman T. The phase envelopes of alternative solvents (ionic liquid, CO2) and building blocks of biomass origin (lactic acid, propionic acid). Fluid Phase Equilibria 2010;295 177-185.

[103] Shin EK, Lee BC, Lim JS. High-pressure solubilities of carbon dioxide in ionic liquids: 1-Alkyl-3-methylimidazolium bis(trifluoromethylsulfonyl)imide. Journal of Supercritical Fluids 2008;45 282-292.

[104] Alvarez VH, Mattedi S, Aznar M. Isobaric (vapor+liquid) equilibria of 1-ethyl-3-methylimidazolium ethylsulfate plus (propionaldehyde or valeraldehyde): Experimental data and prediction. Journal of Chemical Thermodynamics 2011;43 895-900.

[105] Soave G. Equilibrium constants from a modified Redlich-Kwong equation of state. Chemical Engineering Science 1972;27 1197-1203.

[106] Shiflett MB, Yokozeki A. Solubilities and diffusivities of carbon dioxide in ionic liquids: [bmim][PF6] and [bmim][BF4]. Industrial & Engineering Chemistry Research 2005;44 4453-4464.

[107] Yokozeki A, Shiflett MB. Ammonia solubilities in room-temperature ionic liquids. Industrial & Engineering Chemistry Research 2007;46 1605-1610.

[108] Yokozeki A, Shiflett MB. Vapor–liquid equilibria of ammonia+ionic liquid mixtures. Applied Energy 2007;84 1258-1273.

[109] Shiflett MB, Yokozeki A. Chemical absorption of sulfur dioxide in room-temperature ionic liquids. Industrial & Engineering Chemistry Research 2010;49 1370-1377.

[110] Yokozeki A, Shiflett MB. Water solubility in ionic liquids and application to absorption cycles. Industrial & Engineering Chemistry Research 2010;49 9496-9503.

[111] Yokozeki A, Shiflett MB. Hydrogen purification using room-temperature ionic liquids. Applied Energy 2007;84 351-361.

[112] Shiflett MB, Yokozeki A. Separation of carbon dioxide and sulfur dioxide using room-temperature ionic liquid [bmim][MeSO4]. Energy Fuels 2010;24 1001-1008.

[113] Shiflett MB, Yokozeki A. Separation of CO2 and H2S using room-temperature ionic liquid [bmim][PF6]. Fluid Phase Equilibria 2010;294 105-113.

[114] Shiflett MB, Niehaus AMS, Yokozeki A. Separation of CO2 and H2S using room-temperature ionic liquid [bmim][MeSO4]. Journal of Chemical Engineering Data 2010;5 4785-4793.

[115] Shiflett MB, Elliott BA, Niehaus AMS, Yokozeki A. Separation of N2O and CO2 using room-temperature ionic liquid [bmim][Ac]. Separation Science and Technology 2012;47 411-421.

[116] Chapman WG, Gubbins KE, Jackson G, Radosz M. SAFT: Equation-of-state solution model for associating fluids. Fluid Phase Equilibria 1989;52 31-38.

[117] Wertheim MS. Fluids with highly directional attractive forces. I. Statistical thermodynamics. Journal of Statistical Physics 1984;35 19-34.

[118] Wertheim MS. Fluids with highly directional attractive forces. II. Thermodynamic perturbation theory and integral equations. Journal of Statistical Physics 1984;35 35-47.

[119] Wertheim MS. Fluids with highly directional attractive forces. III. Multiple attraction sites. Journal of Statistical Physics 1986;42 459-476.

[120] Wertheim MS. Fluids with highly directional attractive forces. IV. Equilibrium polymerization. Journal of Statistical Physics 1986;42 477-492.

[121] Kroon MC, Karakatsani EK, Economou IG, Witkamp GJ, Peters CJ. Modeling of the carbon dioxide solubility in imidazolium-based ionic liquids with the tPC-PSAFT equation of state. Journal of Physical Chemistry B 2006;110 9262-9269.

[122] Karakatsani EK, Economou IG, Kroon MC, Peters CJ, Witkamp GJ. tPC-PSAFT Modeling of gas solubility in imidazolium-based ionic liquids. Journal of Physical Chemistry C 2007;111 15487-15492.

[123] Karakatsani EK, Economou IG. Phase equilibrium calculations for multi-component polar fluid mixtures with tPC-PSAFT. Fluid Phase Equilibria 2007;261 265-271.

[124] Economou IG, Karakatsani EK, Logotheti G, Ramos J, Vanin, A.A. Multi-scale modeling of structure, dynamic and thermodynamic properties of imidazolium-based ionic liquids: Ab initio DFT calculations, molecular simulation and equation of state predictions. Oil & Gas Science and Technology 2008;63 283-293.

[125] Karakatsani EK, Economou IG, Kroon MC, Bermejo MD, Peters CJ, Witkamp GJ. Equation of state modeling of the phase equilibria of ionic liquid mixtures at low and high pressure. Physical Chemistry Chemical Physics 2008;10 6160-6168.

[126] Andreu JS, Vega LF. Capturing the solubility behavior of CO2 in ionic liquids by a simple model. Journal of Physical Chemistry C 2007;111 16028-16034.

[127] Andreu JS, Vega LF. Modeling the solubility behavior of CO(2), H(2), and Xe in [C(n)-mim][Tf(2)N] ionic liquids. Journal of Physical Chemistry B 2008;112 15398-15406.

[128] Llovell F, Valente E, Vilaseca O, Vega LF. Modeling complex associating mixtures with [Cn-mim][Tf2N] ionic liquids: predictions from the soft-SAFT equation. Journal of Physical Chemistry B 2011;115 4387-4398.

[129] Ji X, Adidharma H. Thermodynamic modeling of CO2 solubility in ionic liquid with heterosegmented statistical associating fluid theory. Fluid Phase Equilibria 2010;293 141-150.

[130] Paduszyński K, Chiyen J, Ramjugernath D, Letcher TM, Domańska U. Liquid–liquid phase equilibrium of (piperidinium-based ionic liquid+an alcohol) binary systems and modelling with NRHB and PCP-SAFT. Fluid Phase Equilibria 2011;305 43-52.

[131] Paduszyński K, Domańska U. Solubility of aliphatic hydrocarbons in piperidinium ionic liquids: measurements and modeling in terms of perturbed-chain statistical associating fluid theory and nonrandom hydrogen-bonding theory. Journal of Physical Chemistry B 2011;115 12537-12548.

[132] Tsioptsias C, Tsivintzelis I, Panayiotou C. Equation-of-state modeling of mixtures with ionic liquids. Physical Chemistry Chemical Physics 2010;12 4843-4851.

[133] Panayiotou C, Tsivintzelis I, Economou IG. Nonrandom hydrogen-bonding model of fluids and their mixtures. 2. Multicomponent mixtures. Industrial & Engineering Chemistry Research 2007;46 2628-2636.

[134] Wang T, Peng C, Liu H, Hu Y. Description of the pVT behavior of ionic liquids and the solubility of gases in ionic liquids using an equation of state. Fluid Phase Equilibria 2006;250 150-157.

[135] Hu Y, Liu H, Prausnitz, JM. Equation of state for fluids containing chainlike molecules. Journal of Chemical Physics 1996;104 396-404.

[136] Breure B, Bottini SB, Witkamp GJ, Peters CJ. Thermodynamic modeling of the phase behavior of binary systems of ionic liquids and carbon dioxide with the group contribution equation of state. Journal of Physical Chemistry B 2007;111 14265-14270.

[137] Maia FM, Tsivintzelis I, Rodrígez O, Macedo EA, Kontogeorgis GM. Equation of state modelling of systems with ionic liquids: Literature review and application with the Cubic Plus Association (CPA) model. Fluid Phase Equilibria; 332, 128–143.

[138] Kontogeorgis GM, Voutsas EC, Yakoumis IV, Tassios DP. An equation of state for associating fluids. Industrial & Engineering Chemistry Research 1996;35 4310-4318.

Chapter 3

A Comparative Study of Piperidinium and Imidazolium Based Ionic Liquids: Thermal, Spectroscopic and Theoretical Studies

Madhulata Shukla and Satyen Saha

Additional information is available at the end of the chapter

http://dx.doi.org/10.5772/51797

1. Introduction

Ionic liquids (ILs) comprise an extremely broad class of molten salts that are attractive for many practical applications because of their useful combinations of properties [1-3]. The ability to mix and match the cationic and anionic constituents of ILs and functionalize their side chains. These allow amazing tenability of IL properties, including conductivity, viscosity, solubility of diverse solutes and miscibility/ immiscibility with a wide range of solvents. [4] Over the past several years, room temperature ILs (RTILs) has generated considerable excitement, as they consist entirely of ions, yet in liquid state and possess minimal vapour pressure. Consequently, ILs can be recycled, thus making synthetic processes less expensive and potentially more efficient and environmentally friendly. Considerable progress has been made using ILs as solvents in the areas of monophasic and biphasic catalysis (homogeneus and heterogeneous).[5-6] The ILs investigated herein provides real practical advantages over earlier molten salt (high temperature) systems because of their relative insensitivity to air and water. [6-7] A great deal of progress has been made during last five years towards identifying the factors that cause these salts to have low melting points and other useful properties.[8] ILs are subject of intense current interest within the physical chemistry community as well. There have been quite a lot of photophysical studies in ionic liquids. [8] The most important properties of ionic liquids are: thermal stability, low vapour pressure, electric conductivity, liquid crystal structures, high electro-elasticity, high heat capacity and inflammability properties enable the use of ionic liquids in a wide range of applications, as shown in Figure 1. It is also a suitable solvent for synthesis, [5, 8, 9-12] catalysis [6, 8, 13] and purification. [14-18] It is also used in electrochemical devices and processes, such as rechargeable lithium batteries and electrochemical capacitors, etc.[19] Rechargeable Lithium batteries are a ubiquitous energy device that is being worldwide in many types of portable electronic equipment, such as cellular phones, laptop computers, and digital cameras and many more devices.[20] Recently, it has been realized that variation of the type of cationic core is a very valuable approach to get more number of ILs.[8]

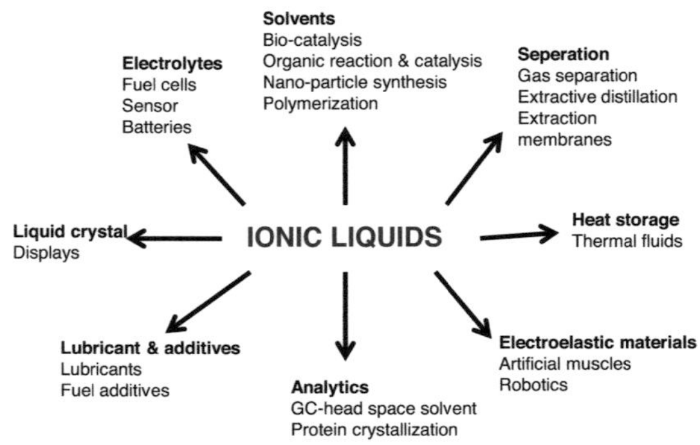

Figure 1. Applications of Ionic Liquids

A major difference between imidazolium salt on one hand and piperidinium salts on the other hand is that the positive charge is delocalized over the aromatic ring in imidazolium salt, whereas the positive charge is localized on the nitrogen atom of piperidinium salts. Next to imidazolium, piperidinium based ILs are the most popular and versatile. Several literatures are available on study of imidazolium based ILs, where x-ray crystallography studies [21-22], theory [23-28], viscosity [8], spectroscopic studies including IR and Raman spectra [27] have been discussed. In comparison to this, very few literatures exist related to piperidinium based ILs. During the last five years interest towards piperidinium based ILs has increased considerably. Cyclic alkyl quaternary ammonium (QA) cations, N-alkyl-N-methylpiperidinium (PIP_{1n}; where 1 indicates CH_3 and n= number of carbon in another alkyl substitution) are class of cations whose room temperature ILs (RTILs) are very promising in the field of electrochemical applications due to their high thermal and electrochemical stabilities.[29-43] Recently, PIP_{1n} ILs found to be potentially useful for electrochemical applications due to their water immiscibility, high conductivities, thermal stabilities and wide electrochemical windows. For example, 1-butyl-1-mehylpiperidinium bis(trifluoromethylsulfonyl)imide ($PIP_{14}NTf_2$; where, NTf_2^-: bis(trifluoromethylsulfonyl)imide anion) improves the stabilization of the chemical composition and structure of the sulphur cathode in Li/S cells during charge-discharge cycles. In the state of the art technologies of 4V-class rechargeable Li batteries, a mixture of organic aprotic solvents and LiPF6 is generally used.[44] Recent studies have shown that the highly fluid and conductive 1, 3-dialkylimidazolium salts cannot be used as electrolyte for 4V-class Li batteries, because of very positive cathodic potential of the 1,3-dialkylimidazolium cations. [45-47] On the other hand, it was revealed that the ILs based on quaternary ammonium cations with the electrochemically stable and weakly coordinating anion, NTf_2^- offer some promising properties.[48-50] These promising properties mainly include i) wide electrochemical windows on account of the low cathodic potential of the saturated QA cations and the high anodic potential of the NTf_2^-, ii) low viscosities on account of the high flexibility and good charge distribution of the NTf_2^-, and iii) wide stable liquid ranges on account of the low melting point and high thermal stability of the NTf_2^- salts.[51-54] The free NTf_2 anion itself has several interesting features.[55] The negative charge can be expected to delocalized over five atoms (four oxygen and the nitrogen), implicating a weak coordinating power. This is of vital importance in the applications where the formation of ion pairs would reduce the number of charge carriers and hence the ionic conductivity. The possibility of rotations around the two central S-N bonds would provide a mechanical flexibility resulting in a plasticizing effect of the polymer electrolyte, making the system more conductive due to the large internal mobility of the dissolved species. Both these effects have been observed for polymer electrolytes based on PEO/$LiNTf_2$ [56-57] and PEO/alkaline salts.[58] As for example, $PIP_{1n}NTf_2$, have been recently proposed for high-voltage super capacitors and lithium batteries.[31-33, 36] Particularly $PIP_{13}NTf_2$ has been found to be useful for the use in Li-batteries [38] with a superior reversible discharge capacity of 340-350 mA-h/g with only a small irreversible capacity loss per cycle [59]. Despite this tremendous interest, the properties, molecular structure and theoretical calculation of PIP_{1n} cation based ILs have not been investigated in detail so far. As the properties of any material depends on the structure of molecules in different phases, it is important to understand the structural features of ILs in depth. In general, liquids are much less understood than gases and crystals. While structure in the gas phase can be accurately determined by electron diffraction or high-resolution rotationally-resolved spectroscopy, solid/crystal structure can be determined by X-ray or neutron diffraction. On the other hand, diffraction and spectroscopic techniques has limited applicability to elucidate liquids structure. Structural information available for liquids is thus much less. Thus theoretical calculations are of very much important in predicting the structure of different room temperature ionic liquids (RTILs). In particular, Density Functional Theory (DFT) calculation found to be very useful in predicting structure of various RTILs. [23-24, 27] DFT calculation also helps us to understand the interaction present among cation and anion in the molecule as well as the type of bonding present in the molecule. Magnetic moment, dipole moment and many other physical properties as well as wavelength of various electronic transition of ILs can also be calculated by DFT calculation. In this chapter we addressed the following very specific issues related to important class of piperidinium based ILs. Synthetic procedure for different piperidinium based ILs has been described and it was found that with variation of anion, cation being the same, physical state of ILs changes drastically.[23] Thermophysical properties of imidazolium and piperidinium based cation with similar anion has been compared briefly. Melting point, viscosity and cyclic voltammetry properties with variation of alkyl chain as well as variation of several anions has been reported for both types of cation. Further we have compared the optimized molecular geometry of bmimBr, bmimI and bmimNTf2 ion pairs with $PIP_{14}Br$, $PIP_{14}I$ and $PIP_{14}NTf_2$ respectively in gaseous phase using theoretical calculations. bmimBr and bmimI ILs, exist as solid of very low melting point or in liquid state [22], whereas its analogous piperidinium based ILs ($PIP_{14}Br$ and $PIP_{14}I$ respectively) exist in solid having high melting point. In addition, the calculated vibrational frequency of the molecule gives us a strong base to analyze the experimental spectra and also the effect of interaction causing shifting in vibrational bands. Further, experimental IR frequencies of $PIP_{14}NTf_2$ and its correlation with theoretical (DFT and HF methods) vibrational frequencies have been reported.

2. Reagents and instrumentation

N-methylpiperidine (Sigma Aldrich, >99%), bromobutane, Iodobutane (Merck, Germany), bis(trifluoromethanesulfonyl)imide (Sigma Aldrich, >99%) were used as received for the synthesis. Acetonitrile (HPLC grade) was procured from Merck, Germany and were used after purification following standard procedures. UV-Visible spectra were measured by CARY 100 BIO UV- Visible Spectrophotometer, which has photometric linearity till absorbance 3.5. Infrared spectra were was measured with Varian FTIR 3100 in the region 400 cm^{-1} – 3500 cm^{-1} using neat sample. 300MHz NMR (JEOL) was used to measure the ^{1}H NMR and ^{13}C NMR. Melting point of the synthesized samples were recorded using automatic digital melting point apparatus (Optimelt).

3. Experimental

3.1. Conventional preparation for ammonium cation based ILs

The general synthetic path for preparing ammonium based ILs is shown in Figure 2. The first step usually is a quaternization reaction, where an amine (NR_3) is alkylated with an appropriate alkylation reagent (R’X) e.g. alkyl halide, resulting in the corresponding IL. When an IL with a desired anion cannot be formed via this reaction, an anion exchange reaction is needed. A previously formed IL is used as precursor and the anion is changed by a metathesis reaction. This metathesis reaction can be performed by using a metal salt (MA).

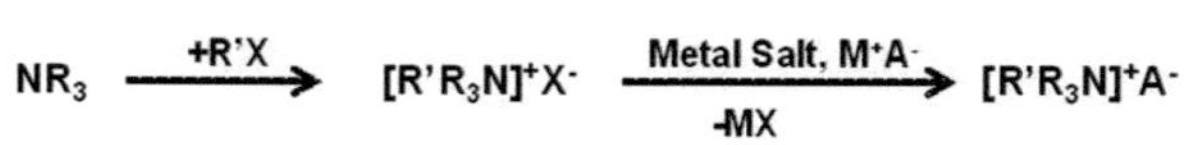

Figure 2. Synthetic path for preparing ammonium based ILs

3.2. Synthesis of 1-butyl-3-methylimidazolium halide (bmimX, where X^- = Br, I) and $bmimNTf_2$

Synthetic Procedure for bmimX and $bmimNTf_2$ have been reported by us earlier [24] Since we have used considerable less temperature than generally reported [60], the time required for the reactions are also much higher. Nevertheless, this low temperature reaction has been found to provide much purer ILs.

3.3. Synthesis of N-butyl-N-methylpiperidinium halide ($PIP_{1n}X$, where X^- = Br, I) and $PIP_{14}NTf_2$

3.3.1. *Synthesis of N-butyl-N-methylpiperidinium bromide ($PIP_{14}Br$)*

A general synthesis procedure for synthesizing $PIP_{14}Br$ was reported in literature.[40] A modified form of that reported procedure is followed; instead of using high temperature, lower temperature (50 °C) is used with longer time of stirring. This excludes major possibilities of inclusion of impurities in ionic liquids. Scheme for the synthesis of $PIP_{14}Br$ is shown in Figure 3. 20 mL of ethyl acetate was taken in 100 mL RB flask. To it 10 mL (82.3 mmol) N-methyl piperidine was added with stirring and then 9.7 mL (90.5 mmol) of bromobutane was added slowly with continuous stirring at 25 °C. Mixture was stirred for 16 h at RT in nitrogen atmosphere and then stirring was made at 50 °C for 3 h. The solution was washed with 150 mL dry distilled ethylacetate and remaining solvent was evaporated on a rotavapour. White solid product was kept under high vaccum at 50 °C for 3 h. (yield= 90%). Melting point (mp) found to be 241°C. The product was confirmed by ^1H NMR (δ, ppm, 1.01 (t, 3H), 1.47 (q, 2H), 1.72 (8H), 3.63 (s, 3H), 3.66 (4H) and 3.81 (2H); IR: 569, 673, 904, 940, 1030, 1227, 1369, 1464, 2874 and 2959 cm^{-1}.

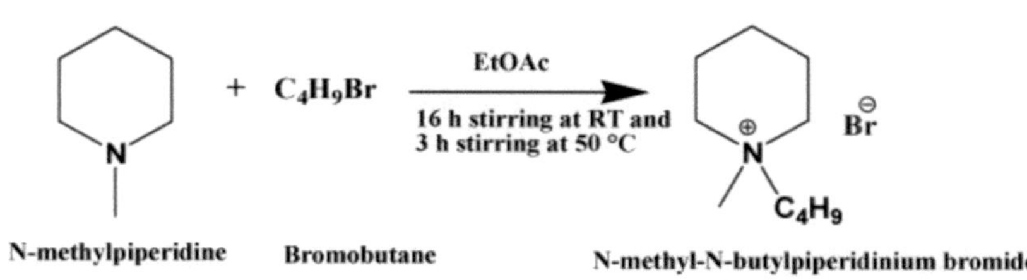

Figure 3. Scheme for the synthesis of $PIP_{14}Br$

3.3.2. *Synthesis of N-alkyl-N-methylpiperidinium Iodide ($PIP_{1n}I$)*

A general synthesis procedure for synthesising $PIP_{1n}I$ (where n=1, 3, 4, 6, 8) is shown in Figure 4. As above mentioned process, here also instead of using high temperature, RT is preferred with longer time of stirring to exclude the major possibilities of inclusion of impurities in ionic liquids. All the reactions were carried out with 1:1.1 molar ratios. 10 mL of Ethyl acetate was taken in 100 mL RB flask. To it 2 mL (16 mmol) N-methyl piperidine was added with stirring and then X mL (X mmol) of iodoalkane was added slowly with continuous stirring at 25 °C. Mixture was stirred for 24-48 h at RT in nitrogen atmosphere (except for $PIP_{11}I$, where stirring was done for 4 h only). The solution was washed with 150 mL dry distilled ethylacetate and remaining solvent was evaporated on a rotavapour. White solid product was kept under high vacuum for 3 h. Yield for $PIP_{11}I$ = 100% and for rest of the salt, it is found to 70%. Melting point (mp) found to be 181°C, 198°C, 124°C and 131°C for $PIP_{13}I$, $PIP_{14}I$, $PIP_{16}I$ and $PIP_{18}I$ respectively. The products were confirmed by ^{1}H NMR, ^{13}C NMR and IR.

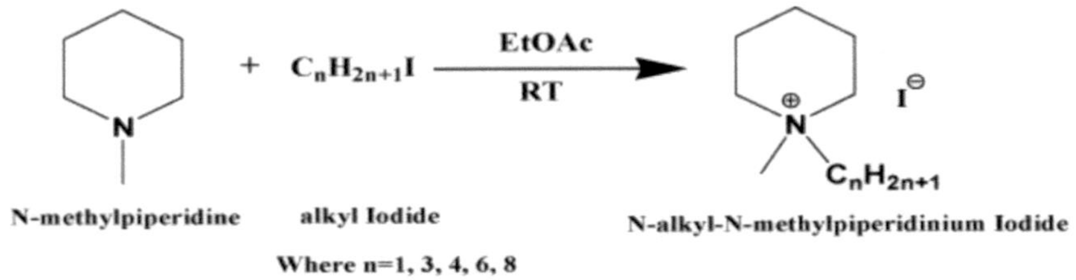

Figure 4. Scheme for the synthesis of $PIP_{1n}I$

3.3.3. Synthesis of N-butyl-N-methylpiperidiniumbis(trifluoromethanesulfonyl)imide ($PIP_{14}NTf_2$)

$PIP_{14}NTf_2$ was done following the similar procedure as described in literarture.[40] Scheme for preparation of $PIP_{14}NTf_2$ is shown in Figure 5. 4.6 g (19.5 mmole) of $PIP_{14}Br$ was taken in a RB and to it 10 mL of triple distilled (TD) water was added. 6.1 g (21.4 mmole) $LiNTf_2$ dissolved in 10 mL TD water added to it. Stirring was done for 4 h. 150 mL dichloromethane (DCM) solvent was used to wash followed by the cold distilled water. DCM was evaporated on rotavapour and after that it was kept under high vaccum for 2h at 60 °C. Light yellow colour liquid was obtained with a yield of 88%. This light yellow colored liquid was further dissolved in 10 mL of pure predistilled acetonitrile (ACN) and treated with activated celite for decolorization. Room temperature stirring was done for 4 h followed by filtration through a column packed with fresh charcoal and activated alumina. The resultant solution was evaporated on rotavapour at reduced pressure. Completely colorless liquid was obtained. The product was confirmed by ^{1}H NMR (δ, ppm, 1.01 (t, 3H), 1.44 (q, 2H), 1.74 (8H), 3.42 (s, 3H), 3.56 (4H) and 3.85 (2H) and IR: 570, 619, 1054, 1139, 1197, 1348, 1474, 2881 and 2966 cm^{-1}.

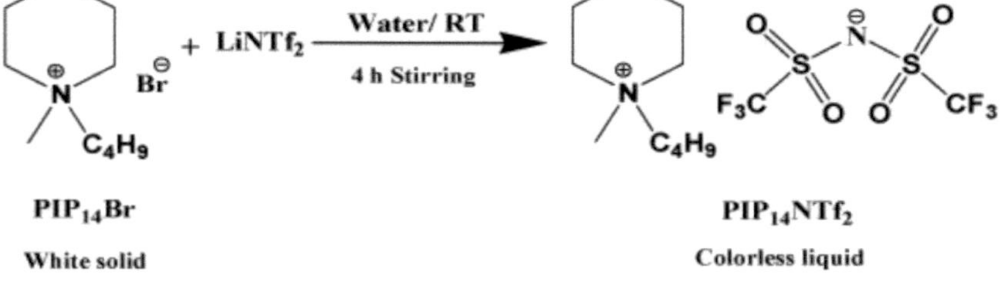

Figure 5. Scheme for the synthesis of $PIP_{14}NTf_2$

4. Thermo-physical properties study

The thermal stability of an ionic liquid is determined by the strength of the formed heteroatom-carbon or heteroatom-hydrogen bonds and the stability of the formed ion species. Although there have been extensive studies of RTILs, relations between their structure and physicochemical properties has not yet been fully understood. The primary research on the properties of pure ILs has focused on understanding and developing the relationship between the structures of cation and anion and the physical properties. To optimize the use of ILs and design the desirable ILs, knowledge of physical and chemical properties of ILs is essentially important. Physical properties such as melting point, viscosity, glass transition temperature, density, surface tension etc. must be known before utilizing them as either green solvent for chemical reactions or for the usage as new materials for various applications.[61] In addition, basic thermodynamic properties are also vital for design and evaluation of its application.

The melting point (T_m) of an organic molecular compound is determined by the strength of its crystal lattice, which is in turn controlled by three main factors: molecular symmetry, intermolecular forces, and conformational degrees of freedom of the molecule. This principle is also applicable to the ILs, as intensively described in a very large number of well-characterized imidazolium and QA salts [62-63]. For ILs, melting points is one of the most important physical properties and have been studied with interest. The melting point primarily indicates whether a salt should be considered as IL or not. The melting point of an ionic liquid depends on its cation/anion composition [61]. Generally, symmetric ions with a localized charge and strong interactions between ions result in good packing efficiency and hence a high melting point (e.g. mp of NaCl: 801ºC) [64]. Ionic liquids based on large, asymmetric cations with a delocalized charge often have low melting points. Packing efficiency depends on interactions between ions. Hydrogen bonding (or similar non-bonded interactions) increases the order of the system and thus raises the melting point [65].The relations between density with melting temperature, along with glass transition decomposition temperature and heat capacities for a series of imidazolium based ILs have been reported [66-68]. A series of hydrophilic and hydrophobic 1-alkyl-3-methylimidazolium and 1-alkyl-1-methylpiperidinium salts of NTf_2^-, PF_6^-, BF_4^-, Br^- and I^- has been presented in Table 1. Data presented in Table 1 shows that most of the imidazolium based ILs found to exist in liquid at room temperature or having low melting point. In comparision to this, piperidinium based ILs have very high melting behaviour. Melting points reported here clearly shows that for a different cation, anion being same piperidinium based ILs found to have quite higher than those of imidazolium based ILs. As for example, melting point of bmimBr (79ºC), hmimBr (-54.9ºC) are quite low as compared to its corresponding $PIP_{14}Br$ (241ºC) and $PIP_{16}Br$ (201ºC) salts. Similarly for bmimI (-72ºC) and hmimI (-72ºC), compared with $PIP_{14}I$ (198ºC) and $PIP_{16}I$ (124ºC) have quite high difference in melting point value. Compared to $bmimBF_4$ (-81ºC), corresponding $PIP_{14}BF_4$ (146ºC) salt have huge difference in melting point. This significant difference in melting point behaviour is expected due to molecular interaction present between cation and anion, which has been discussed further in following sections. Melting point of $PIP_{14}NTf_2$ as well as $bmimNTf_2$ reported to be -25°C [30-32]. It is interesting to note that when anion is larger with having multiple interacting sites (such as NTf_2), the difference of mp between piperidinium and imidazlium cation based ILs fades away. Therefore, one can have two drastically different types of ILs having same mp.

Imidazolium ILs	Melting point (t_m°C) [ref]	Viscosity (η)/ cP	Piperidinium ILs	Melting point (t_m°C) [ref]	Viscosity (η)/ cP
bmimBr	79 [22]	solid	$PIP_{14}Br$	241[40]	solid
hmimBr	-54.9 [69]	3986	$PIP_{16}Br$	201 *	solid
pmimI	-56 [70]	35	$PIP_{13}I$	181 *	solid
bmimI	-72 [66]	1110[66]	$PIP_{14}I$	198 *	solid
hmimI	-72 [66]	771	$PIP_{16}I$	124 *	solid
$bmimBF_4$	-81 [66]	219 [66]	$PIP_{14}BF4$	146 [40]	solid
$bmimPF_6$	4 [66]	450 [66]	$PIP_{14}PF_6$	-	solid
$hmimPF_6$	-61 [66]	585 [66]	$PIP_{16}PF_6$	188.7 *	solid
$emimNTf_2$	4 [66]	28 [66]	$PIP_{12}NTf_2$	84.3 [71]	solid
$bmimNTf_2$	-25 [66]	69 [66]	$PIP_{14}NTf_2$	-25 [21, 40]	182 [40]

* our results

Table 1. Melting point of some popular imidazolium and piperidinium cation based ILs.

4.1. Viscosity

The viscosity of an ionic liquid is a very important parameter in electrochemical studies due to its strong effect on the rate of mass transport within solution. The type of the anion and the cation which compose the ionic liquid have a huge effect on the viscosity of the ionic liquid. With respect to the anionic species, higher capacity and relative basicity to form hydrogen bonds result in more viscous RTILs. Viscosity of several imidazolium ILs has been presented in Table 1 but piperidinium based ILs (almost all of them except $PIP_{14}NTf_2$) are solid at room temperature and hence their viscosity data are not available. Viscosity of bmimPF6 (450cP) found to nearly double when compared with viscosity of $bmimBF_4$ (219cP).[66] This may be due to presence of more number of hydrogen bonding present in $bmimPF_6$, than $bmimBF_4$. Whereas, ionic liquids containing BF_4^- anions are much more viscous than those formed with NTf_2^- anions, where the negative charge is delocalized. An increase in the viscosity of the various anion/cation combinations was attributed to an increase in van der Waals forces over hydrogen bonding.[72] However, hydrogen bonding between cationic protons and anionic halides has been noted in the crystalline state from X-ray diffraction studies [73-74] and may represent an additional factor. In addition, the increased symmetry of the inorganic anions (e.g., PF_6^- or BF_4^-) compared to the organic anion (NTf_2^-) may play an important role. The data of Table 1 seems to indicate that the geometry and molar mass of the anions have a strong influence on the viscosity of this class of IL, since

http://dx.doi.org/10.5772/51797

$[bmim]^+$ combined with either PF_6^- or NTf_2^-, produces ILs with significantly different viscosities. These results suggest a complex relationship of cation–anion interactions. ILs containing NTf_2 anions are the most widely used for electrochemical applications due to their low viscosity and, consequently, improved mass transport. Viscosity of $bmimNTf_2$ (62cP) [66] found to be much lower than that of $PIP_{14}NTf_2$ (182cP) [40]. Hence the viscosity of ILs is controlled by the number of hydrogen bonding between cation and anion, as well the van der Waals interaction. Hence $bmimNTf_2$ seems to have wide application in electrochemistry as compared to $PIP_{14}NTf_2$.

4.2. Cyclic voltammetry

The ionic conductivity of an IL is another most important property, aiming its application as electrolyte for electrochemical devices. It is expected that ILs possesses a large conductivity, since they are composed exclusively by ions. However, in addition to the number of charge carriers, their mobility should also be taken into account. Cyclic Voltammogram of both ILs ($bmimNTf_2$ and $PIP_{14}NTf_2$) using carbon glass electrode as working electrode and current density limit at 150μA cm^{-2} is shown in Figure 6. The utility of a liquid for electrochemical applications is frequently reflected in the width of the electrochemical windows (EW). The electrochemical window is defined as the potential range where the limiting current density is reached. EW for $bmimNTf_2$ found to be only 3V, whereas EW for $PIP_{14}NTf_2$ found to be 4.5V i.e $PIP_{14}NTf_2$ has 1.5V wider EW as compared to $bmimNTf_2$.This 1.5 V advantage over the $bmimNTf_2$ is due to the fact that the aromatic imidazolium core is much more readily reduced than the piperidnium system, which contain no vacant orbital as explained by Belhocine et al. for azepanium and 3-methylpiperidinium based ILs. [75] They have also shown that azepanium and 3-methylpiperidinium based ILs have higher EW as compared to imidazolium based ILs. Hence EW is one of the fundamental properties required for evaluating the ILs as electrolyte in many electrochemical devices. Hence this property of $PIP_{14}NTf_2$ leads to its extensive use in electrochemical processes. This is the most important advantages of piperidinium based ILs over imidazolium based ILs.

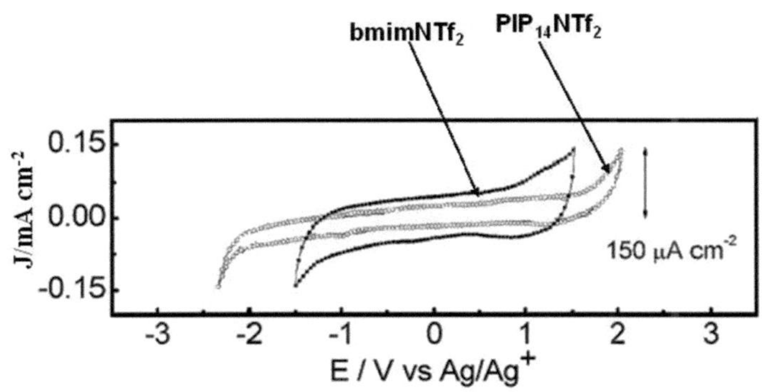

Figure 6. Cyclic Voltammogram of $bmimNTf_2$ (•) and $PIP_{14}NTf_2$ (o)

5. Computational details

5.1. Studies of interactions in molecular structure of different ILs using DFT calculations

Optimization of ion pairs was done using density functional theory (DFT), MP2 and Hartee-Fock (HF) methods to get the most stable structure. DFT computations were performed at Becke's three-parameter hybrid model using Lee-Yang-Parr correlation functional (B3LYP) level of theory. [76-77] 6-31++ G (d,p) basis set used to get the optimized geometry and IR bands using Gaussian 03 programme.[78] IR frequencies calculations was performed using DFT and HF method [27], to compare which method produces better experimental results. All optimized structures were confirmed to be minimum energy conformation, as no imaginary frequencies are obtained.

Molecular geometry optimization of $PIP_{14}X$ and bmimX ion pairs carried out at B3LYP/6-31++ G (d, p) level in gaseous phase and has been shown in Figure 7. DGDZVP basis set was used for iodine atom, as explained in our previous paper. [23-24] The optimized structure reveals that piperidinium ring is stable in chair conformation and butyl group in trans conformation, as reported by Reichert et al. for $PIP_{1n}I$ crystal.[42]. Hydrogen bonding present in these moieties is shown with the dotted line. H---Br distances in PIP14Br found to be 2.57Å, 2.45 Å and 2.54 Å (shown in Figure 7a) which are smaller than van der Waal radius and also C17-H18----Br, C2-H6----Br and C22-H27---Br angle found to be 155°, 153° and 156° respectively. C-H---Br bond length and angle presented in Table 2 satisfied the conditions of H-bonding definition. Hence H-bonding present in piperidinium ILs or salts have strong control over the physical property of the moiety. When compared with 1-butyl-3-methylimidazolium bromide (bmimBr, mp = 79 °C) IL [22], its melting point found to be quite high (241°C). This can be explained as, in bmimBr IL, only one H-bonding is present between cation and anion (shown in Figure 7b) with C2-H---Br distance of 2.19 Å and C2-H---Br bond angle of 154°. While in $PIP_{14}Br$ moiety, three H-bonding are observed with single Br- ion, leading to its higher melting point. In bmimBr, no H- bonding was observed between anion and alkyl chain, whereas in $PIP_{14}Br$ moiety, two H-bonding with alkyl chain along with one H-bonding with hydrogen of piperidine ring was observed. Hence this higher number of H-bonding present in $PIP_{14}Br$ led to its higher melting point. Similar is the reason for the higher melting point of $PIP_{1n}I$ salts, when compared to those of bmimI. In $PIP_{14}I$, H---I H-bonding found to be 2.78, 2.86 and 2.98 Å, (shown in Figure 7c) whereas, in bmimI, single H-bonding of bond length 2.50 Å exist between cation (C2-H) and iodide anion (shown in Figure 7d). Here also three H-bonding between cation and anion lead to its higher melting point when compared with bmimI. Theoretical calculation on NTf2 anion based ILs have attracted attention in recent times. Several literatures are available for $bmimNTf_2$ IL, where DFT calculation performed to explain the conformation of cation as well as anion. [79] According to Fujii et al., DFT calculation followed by frequency analyses confirms two conformations (cis and trans) of NTf_2^- anion with an energy difference of 2.2- 3.3kJ mol^{-1} and trans conformation reported to be more stable than the cis conformation. Keeping these stable conformations in mind, we further carried out theoretical calculation for $PIP_{14}NTf_2$ ion pair in gaseous phase. The selected structural parameters for $PIP_{14}Br$ and $PIP_{14}NTf_2$ are shown in

Table 2. Optimized structure of $PIP_{14}NTf_2$ (shown in Figure 7e) explains three H-bonding of bond length 2.34, 2.55, 2.64 Å between cation and anion, parallel as $bmimNTf_2$ (shown in Figure 7f). Hence in case of NTf_2 anionic ILs, melting point of both imidazolium and piperidinium based ILs found to be nearly comparable to each other. It is due to the fact that number of H-bonding is same in both cases. Further, discussion on $PIP_{14}NTf_2$ IL has been done in detail, as it (NTf_2^- based ILs) exist in liquid state at room temperature. From Table 2, it is clear that, N-S, S=O, S-C and C-F bond length calculated using DFT found to be 1.62, 1.46, 1.89 and 1.33 Å, which deviate by 0.05, 0.04, 0.07 and 0.01 Å from the crystal data reported in literature.[79] Whereas bond lengths calculated using HF method are in good agreement with those in the crystals, its bond angle and dihedral angles are largely different. On the other hand, bond length calculated at B3LYP level of theory are appreciably larger than those in crystal data, whereas the S-N-S bond angle, C-S-S-C and S-N-S-C dihedral angles are reproduced fairly well. Indeed MP2 calculation also produces similar result as DFT but its time consumption is nearly thrice of that of DFT method. Hence DFT proves to be the best among these three methods to reproduce the crystal data.

Parameter	PIP14Br	Parameter/ $PIP_{14}NTf_2$	crystal data (reported) [79]	DFT results	MP2 results	HF results
H18-Br34	2.57 Å	N34-S35	1.57 Å	1.62 Å	1.62 Å	1.57 Å
H27-Br34	2.54 Å	S35-O38	1.42 Å	1.46 Å	1.46 Å	1.42 Å
H6-Br34	2.45 Å	S35-C39	1.83 Å	1.89 Å	1.88 Å	1.83 Å
C17-H18---Br	155°	C39-F44	1.32 Å	1.33Å	1.34 Å	1.32 Å
C2-H6---Br	153°	S35-N34-S36	125°	126°	124°	129°
C22-H27---Br	156°	S36-N34-S35-C39	92.6°	88.9°	101°	103°
		S35-N34-S36-C40	92.3°	88.4°	93°	89°
		C39-S35-S36-C40	172°	165°	171°	161°

Table 2. Selected bond lengths (Å), bond angles (°) and dihedral angles (°) for optimized structure of $PIP_{14}Br$ using DFT method and $PIP_{14}NTf_2$ using DFT, MP2 and HF methods as well as its reported crystal data:

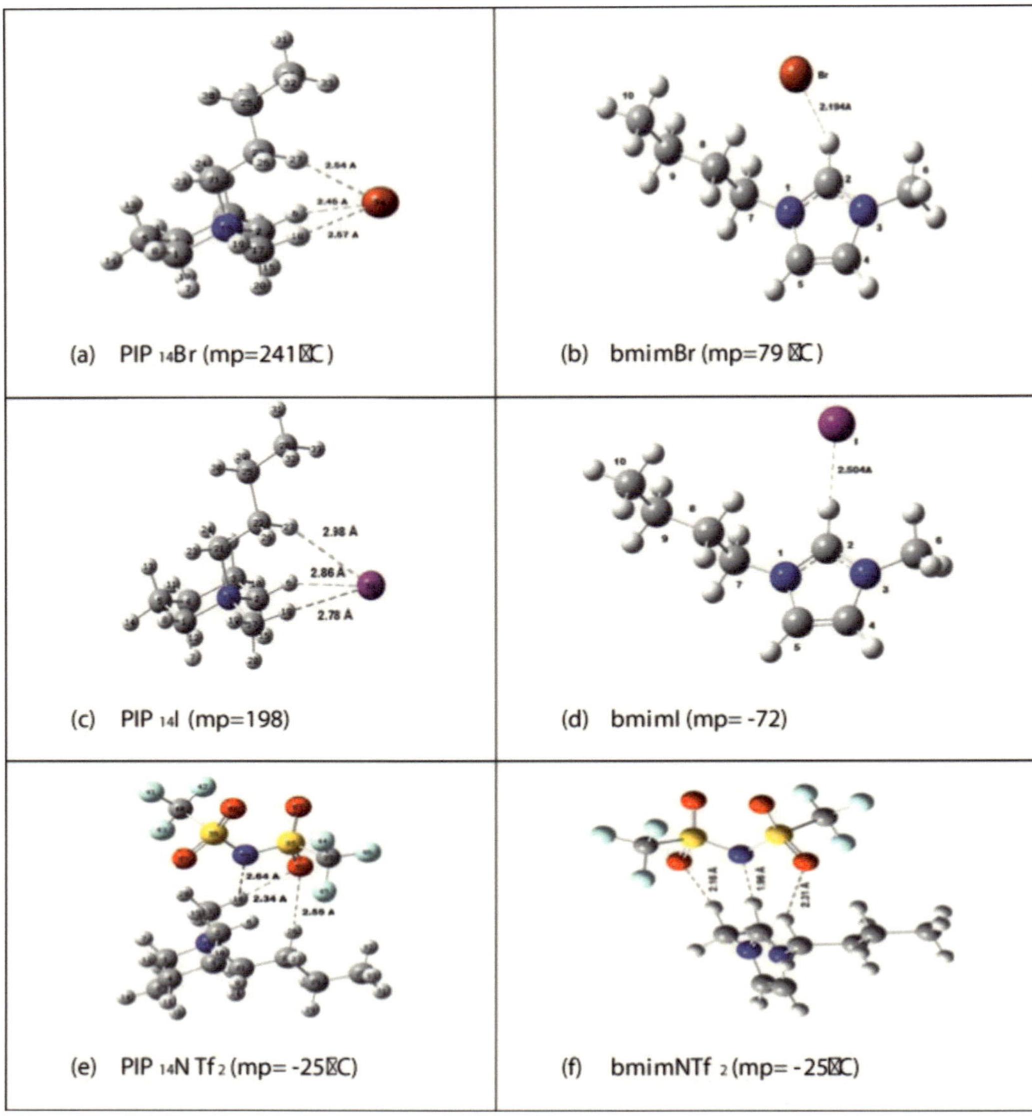

Figure 7. Optimized structure of (a) N-methyl-N-butylpiperidinium bromide ($PIP_{14}Br$) and (b)1-butyl-3-methylimidazolium bromide (bmimBr), (c) N-methyl-N-butylpiperidinium iodide ($PIP_{14}I$), (d) 1-butyl-3-methylimidazolium iodide (bmimI), (e) N-methyl-N-butylpiperidinium bis(trifluoromethanesulfonyl)imide ($PIP_{14}NTf_2$), (f) 1-butyl-3-methylimidazolium bis(trifluoromethanesulfonyl)imide ($bmimNTf_2$)

5.2. Experimental and theoretical infrared spectrum of $PIP_{14}NTf_2$

The analysis of IR spectra also expected to bring its contribution to the general debate of the ILs intermolecular structure. Experimental Infrared (IR) spectrum for synthesized $PIP_{14}NTf_2$ has been shown in Figure 8. It has been observed that the major peaks appeared at 570, 619, 1054,

1139, 1197, 1348, 1474, 2881 and 2966 cm^{-1} which are quite intense. These peaks are analysed with the help of frequency calculation using DFT and HF methods. Theoretical investigations were done to find out, which method reproduces better correlation with the experimental result. We performed vibrational frequency calculation of $PIP_{14}NTf_2$ ion pair in gaseous phase using DFT/B3LYP and HF methods. Figure 9 contains a plot of combined experimental versus theoretical vibrational frequencies using B3LYP and HF methods. It is very clear from Figure 9 that DFT/B3LYP method correlate better with the experimental results, when compared with HF method. $PIP_{14}NTf_2$ ion pair requires a scale factor of 0.966 (DFT) in higher wavenumber region (above 1500 cm^{-1}) to produce the experimental vibrational frequencies. The HF method overestimates the vibrational frequencies to a greater degree than B3LYP method and requires a scale factor of 0.915 to reproduce the experimental frequencies. Similar theoretical results have been obtained for a wide range of imidazolium based ionic liquids.[26-27] The overall correlation of band positions in calculated vibrational frequencies at B3LYP level agree reasonably well with the obtained experimental frequencies. On the basis of our DFT calculation performed, experimental peaks have been assigned and are presented in Table 3. Band at 570 and 1054 cm^{-1} assign to be out of plane bending of N and S=O symmetric stretching weakly coupled with S-N asymmetric stretching respectively. Band at 1139 and 1197 cm^{-1} correspond to C-F stretching and C-F symmetric bending. Band at 2881 and 2966 cm^{-1} correspond to symmetric and asymmetric C-H stretching in cation. The overall correlation of band positions in calculated vibrational frequencies at B3LYP level agree reasonably well with the obtained experimental frequencies. In addition, theoretically determined relative intensities are also found to be in good agreement with experimental intensity of IR absorption band.

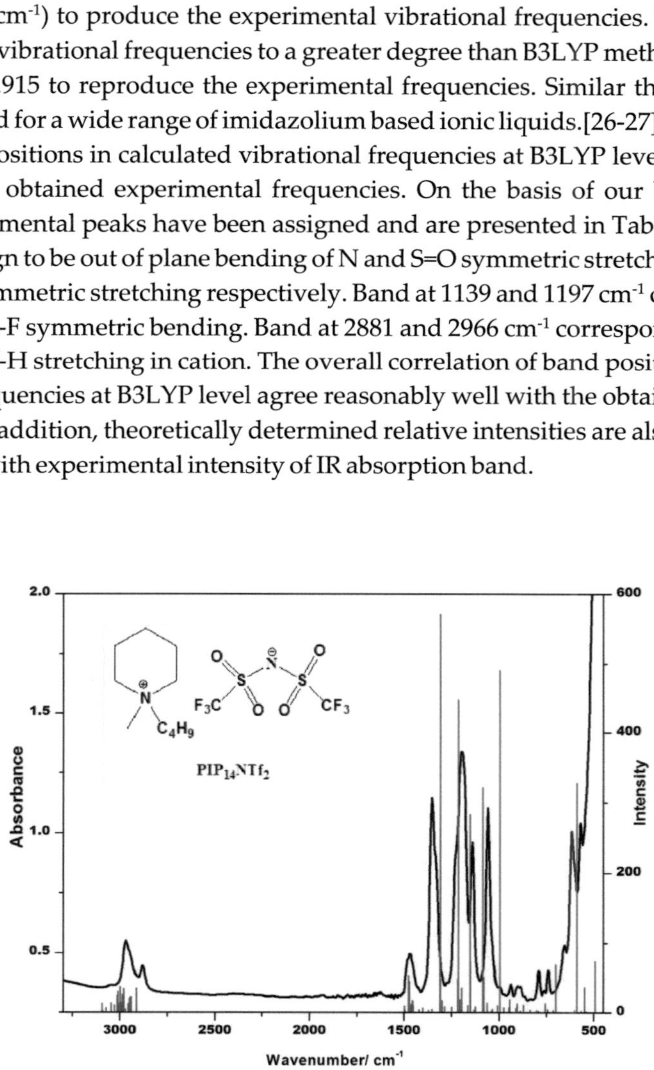

Figure 8. Infrared spectrum of neat $PIP_{14}NTf_2$ correlated with calculated vibrational bands (vertical lines). A scaling factor of 0.9664 was required to reproduce the experimental observations at higher wavenumber region.

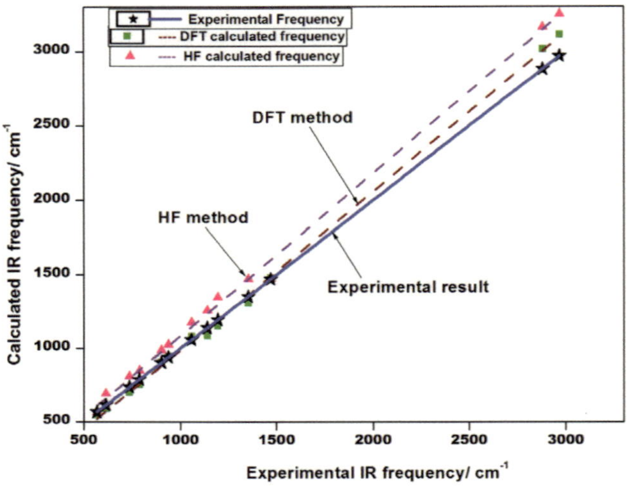

Figure 9. Correlation diagram for IR spectrum of $PIP_{14}NTf_2$; experimental versus calculated IR frequencies. (DFT and HF method). Scaling factors = 0.966 (B3LYP) and 0.915 (HF).

Vibrational frequency calculation of cation and anion was also performed independently using their optimized structure at same level of calculation to analyse the shifting of band due to presence of cation/anion interaction. Selected vibrational bands for cation, anion and its ion pair are depicted in Table 4. It has been observed from DFT calculation that due to presence of anion in ion pair, frequencies of cation are shifted to a greater extent, mainly the C-H stretching bands. Symmetric and asymmetric stretching of H-C2-H observed at 3113 and 3196 cm^{-1} in $PIP_{14}NTf_2$, whereas it found to be at 3082 and 3154 cm^{-1} respectively for PIP_{14}^+ cation, showing a shifting of 31 and 41 cm^{-1} respectively due to presence of interaction with anion. Symmetric C-H stretching of methyl group shifted from 3087 in PIP_{14}^+ cation to 3083 in $PIP_{14}NTf_2$ ion pair, shows an insignificant deviation of 4 cm^{-1}. Vibrational bands arising from anion shows very less shifting due to presence of cation. This may be due to delocalization of charge on NTf_2^- anion, whereas in cation charge is totally localized on nitrogen atom. Hence due to cation-anion interaction, cationic species shows significant shifting in their peak position.

http://dx.doi.org/10.5772/51797

Wavenumber/ cm^{-1}/ DFT (Expt.)	Intensity	Wavenumber/ cm^{-1}/HF	Intensity	Band assignment in PIPNTf2
549(568)	36	629	69	Scissoring in O=S=O and CF3
589 (570)	327	694	435	Out of plane bending of N in NTf_2 anion
701 (739)	69	810	84	N-S sym stretching
755 (790)	13	847	21	C-F and N-S sym stretching
905 (905)	12	987	22	N-C(CH3) stretching and twisting in H-C-H
945 (938)	18	1025	20	N-C stretching and rocking of H-C-H
997	490	1176	416	S-N asym stretching and S=O sym stretching
1086 (1054)	321	1257	565	S=O sym stretching and S-N asym stretching
1149	74	1341	272	C-F asym stretching
1152 (1139)	283	1346	103	C-F stretching
1196 (1197)	35	1379	338	C-F sym bending in NTf2 (umberalla bending)
1213	447	1405	374	C-F asym stretching and N-S stretching and twisting of H-C-H
1309	571	1470	715	S=O asym stretching, Twisting of H-C-H in PIP ring and wagging of H-C-H in Bu group
1480 (1469)	10	1569	5	H-C-H wagging and umbrella bending in CH3 group
1523	44	1653	77	Scissoring of H-C-H
3015 (2881)	35	3162	43	sym C25-H stretching in Bu group
3041	22	3182	27	sym C-H stretching in terminal CH3 group
3083	34	3238	6	sym C-H stretching in N-Me group and C1-H
3089 (2966)	26	3241	15	sym C1-H and asym C5-H stretching in PIP ring
3100	36	3247	16	asym C3-H and C4-H stretching in PIP ring
3113	29	3250	52	sym C2-H stretching in PIP ring
3118	23	3255	46	asym C28-H stretching in Bu group
3131	10	3292	14	asym C1-H stretching in Pip ring
3151	13	3326	20	asym C21-H stretching in Bu group
3196	11	3356	6	asym C2-H stretching
3202	12	3388	7	asym C-H stretching in N-Me group

Sym-symmetric; asym- asymmetric

Table 3. Selected IR frequencies of $PIP_{14}NTf_2$ calculated using DFT and HF methods. Experimental results are written in parentheses:

Wavenumber/ cm-1/ DFT	Intensity	Wavenumber (ion pair)/ cm-1/ DFT	Deviation due to cation anion interaction (Δυ)	Band assignment
PIP_{14}^+ Cation				
3082	4	3113	31	sym C2-H stretching in PIP ring
3087	9	3083	4	sym C-H stretching in N-Me group and C1-H
3127	17	3118	9	asym C28-H stretching in Bu group
3154	5	3196	42	Asym H-C2-H stretching
3167	8	3151	16	asym C21-H stretching in Bu group
3194	2	3202	8	Asym C-H stretching in N-CH3
NTf_2^- anion				
700	8	701	1	N-S sym stretching
992	600	997	5	S-N asym stretching
1303	500	1309	6	S=O asym stretching

Table 4. Selected vibrational frequencies of PIP_{14}^+ cation, NTf_2^- anion and its ion pair

5.3. UV-visible spectra of $PIP_{14}NTf_2$ and $bmimNTf_2$

Since application of the ILs as media of photophysical studies depend on how transparent these substances are in the optical region, we have characterised the UV-visible absorption behaviour of $PIP_{14}NTf_2$ and $bmimNTf_2$ ILs. The UV-Visible spectrum of neat $PIP_{14}NTf_2$ and $bmimNTf_2$ ILs is presented in Figure 10. Contrary to imidazolium based ILs, piperidinium based ILs shows wide range of transparency in UV region. $bmimNTf_2$ IL shows non-negligible absorption in the UV region, with an absorption tail extending well into the visible region. Initially, the band of the spectrum for charcoal treated $PIP_{14}NTf_2$ appears at 258 nm, with its absorption tail extended upto 325 nm. But on further charcoal treatment, absorbance as well as its extended tail diminish sharply. As it is well known fact that during the process of synthesising these ILs, often colour impurity makes the resultant ILs coloured. In most of the cases improperly purified or unpurified ILs shows pale colour whereas on persistent and proper purifications, the final IL comes out to be colourless. So, to understand whether the $PIP_{14}NTf_2$ is inherently light yellow coloured or the colour is due to presence of impurity, charcoal treatment of synthesized IL was carried out repeatedly. Successive recording of

http://dx.doi.org/10.5772/51797

UV-visible spectra was done after each charcoal treatment and it was observed that absorbance diminishes sharply. Hence we obtained a wide transparent window in UV-visible region for $PIP_{14}NTf_2$ IL. This gives us an idea that phtophysical studies or electron transfer reaction of several donor-acceptor complexes can be carried out in $PIP_{14}NTf_2$ IL whose absorption is in UV region as well. This very unique and interesting property of $PIP_{14}NTf_2$ IL revealed the extensive study of piperidinium based ILs in fluorescence study also.

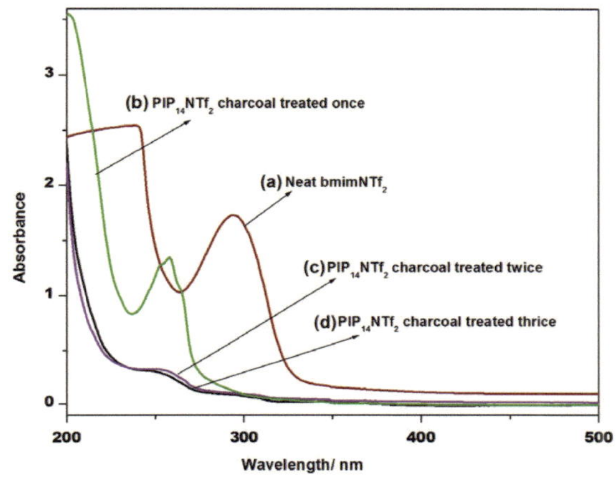

Figure 10. UV-visible spectrum of $PIP_{14}NTf_2$ compared with that of $bmimNTf_2$

6. Conclusion

Synthesis of several imidazolium and piperidinium based ILs and their physical properties are reported. Melting point of piperidinium based ILs found to be higher than its imdazolium based ILs, anion being identical. This was explained by DFT calculation which clearly shows the higher number of hydrogen bonding present in piperidinium based ILs than imidazolium based ILs. Whereas, in bmimBr or bmimI, no H- bonding was observed between anion and alkyl chain, while two H-bonding with alkyl chain along with one H-bonding with hydrogen of piperidine ring was observed. Hence this higher number of H-bonding present in $PIP_{14}Br$ and $PIP_{14}I$ led to its higher melting point than its corresponding analogues ILs, bmimBr or bmimI respectively. Viscosity of ILs is also controlled by number of hydrogen bonding between cation and anion, as well as vander Waal interactions present in it. Though the viscosity of $PIP_{14}NTf_2$ is higher than that of $bmimNTf_2$, the wider (by 1.5V) electrochemical window for the former makes it more useful in electrochemical applications. DFT calculation reproduces the experimental IR spectrum very well as compared to HF method. Finally, piperidinium based ILs shows wide range of optical transparency making it markedly superior for photophysical studies (e.g. electron transfer reaction) in it.

Acknowledgement

Financial assistance from CSIR, India (01(2210)/08/EMR-II) and DST, India (SR/FTP/CS-70/2006) are gratefully acknowledged. Authors also thank the department of Chemistry, Banaras Hindu University for infrastructural instrumentation facilities. Hyderabad Central University, Hyderabad is gratefully acknowledged for computational facility (CMSD) for providing computational facility. MLS thanks University Grant Commission, New Delhi for providing fellowship.

Author details

Madhulata Shukla and Satyen Saha

Department of Chemistry, Faculty of Science, Banaras Hindu University, Varanasi, India

References

[1] Wishart J F. (2009). Energy applications of ionic liquids. Energy Environ. Sci., , 2, 956-961.

[2] Plechkova N V, Seddon K R. (2008). Applications of ionic liquids in the chemical industry. Chem. Soc. Rev., , 37, 123-150.

[3] Metlen M S A, Rogers R D. (2007). The second evolution of ionic liquids: from solvents and separations to advanced materials: energetic examples from the ionic liquid cookbook. Acc. Chem. Res., , 40, 1182-1192.

[4] Fei, Z. F., Geldbach, T. J., Zhao, D. B., & Dyson, P. J. (2006). From dysfunction to bisfunction: on the design and applications of functionalised ionic liquids. Chem. Eur. J., , 12, 2123-2130.

[5] Wasserscheid, P., & Keim, W. (2000). Ionic Liquids- New solutions for transition metal catalysis. Angew. Chem., Int. Ed., , 39, 3772-3789.

[6] Welton, T. (1999). Room-temperature ionic liquids. Solvents for synthesis and catalysis. Chem. Rev., , 99, 2071-2083.

[7] Wilkes J S.(2002). A short history of ionic liquids-from molten salts to neoteric solvents Green Chem., , 4, 73-80.

[8] Castner Jr, E. W., Margulis, C. J., Maroncelli, M., & Wishart, J. F. (2011). Ionic Liquids: Structure and Photochemical Reactions, Annu. Rev. Phys. Chem., , 62, 85-105.

[9] Dupont, J., Suarez, P. A. Z., Umpierre, A. P., & de Souza, R. F. (2000). Pd(II)-dissolved in ionic liquids: a recyclable catalytic system for the selective biphasic hydrogenation of dienes to monoenes. J. Braz. Chem. Soc., , 11, 293-297.

[10] Xu L J, Chen W P, Xiao J L,(2000). Heck Reaction in Ionic Liquids and the in Situ Identification of N-Heterocyclic Carbene Complexes of Palladium. Organometallics, , 19, 1123-1127.

[11] Carmichael A J, Earle M J, Holbrey J D, McCormac P B, Seddon K R. (1999). The Heck Reaction in Ionic Liquids: A Multiphasic Catalyst System. Org. Lett., , 1-997.

[12] Brown, R. A., Pollet, P., Mc Koon, E., Eckert, C. A., Liotta, C. L., & Jessop, P. G. (2001). Asymmetric Hydrogenation and Catalyst Recycling Using Ionic Liquid and Supercritical Carbon Dioxide. J. Am. Chem. Soc., , 123-1254.

[13] Olivier, H. (1999). Recent developments in the use of non-aqueous ionic liquids for two-phase catalysis. J. Mol. Catal. A: Chem., , 146-285.

[14] Huddleston J G, Willauer H D, Swatloski R P, Visser A E, Rogers R D.(1998). Room temperature ionic liquids as novel media for 'clean' liquid-liquid extraction. Chem. Commun., , 1765-1766.

[15] Song C E, Roh E J. (2000). Practical method to recycle a chiral (salen)Mn epoxidation catalyst by using an ionic liquid. Chem. Commun., , 837-838.

[16] Anderson J L, Armstrong D W. (2003). High-Stability Ionic Liquids. A New Class of Stationary Phases for Gas Chromatography. Anal. Chem., , 75-4851.

[17] Armstrong D W, He L F, Liu Y S, (1999). Examination of Ionic Liquids and Their Interaction with Molecules, When Used as Stationary Phases in Gas Chromatography. Anal. Chem., , 71, 3873-3876.

[18] Bhatt, A. I., May, I., Volkovich, V. A., Hetherington, M. E., Lewin, B., Thied, R. C., & Ertok, N. (2002). Group 15 quaternary alkyl bistriflimides: ionic liquids with potential application in electropositive metal deposition and as supporting electrolytes. J. Chem. Soc. Dalton Trans., , 24, 4532-4534.

[19] Suarez, P. A. Z., Einloft, S., Dukkius, J. L., de Souza, R. F., & Dupont, J. (1998). Synthesis and physical-chemical properties of ionic liquids based on 1- n-butyl-3-methylimidazolium cation J. Chim. Phys., , 95, 1626-1639.

[20] Holbrey J D, Seddon K R. (1999). The phase behaviour of alkyl-3-methylimidazolium tetrafluoroborates; ionic liquids and ionic liquid crystals, J. Chem. Soc., Dalton Trans., 2133-2139., 1.

[21] Triolo, A., Russina, O., Bleif, H. J., & Cola, E. D. (2007). Nanoscale Segregation in Room Temperature Ionic Liquids. J. Phys. Chem. B, , 111, 4641-4644.

[22] Ozawa, R., Hayashi, S., Saha, S., Kobayashi, A., & Hamaguchi, H. (2003). Rotational Isomerism and Structure of the 1-Butyl-3-methylimidazolium Cation in the Ionic Liquid State. Chem. Lett., , 32, 948-949.

[23] Shukla, M., Srivastava, N., & Saha, S. (2010). Theoretical and spectroscopic studies of 1-butyl-3-methylimidazolium iodide room temperature ionic liquid: Its differences with chloride and bromide derivatives, J. Mol. Struct., , , 975, 349-356.

[24] Shukla, M., Srivastava, N., & Saha, S. (2011). Interactions and Transitions in Imidazolium Cation Based Ionic Liquids in Ionic Liquids-classes and properties, , 150-170.

[25] Tsuzuki, S., Katoh, R., & Mikami, M. (2008). Analysis of interactions between 1-butyl-3-methylimidazolium cation and halide anions (Cl-, Br- and I-) by ab initio calculations: anion size effects on preferential locations of anions Mol. Phys., , 106, 1621-1629.

[26] Krossing, I., Slattery, J. M., Daguenet, C., Dyson, P. J., Oleinikova, A., & Weinga, H. (2006). Why Are Ionic Liquids Liquid? A Simple Explanation Based on Lattice and Solvation Energies, J. Am. Chem. Soc., , 128, 13427-13434.

[27] Talaty, E. R., Raja, S., Storhaug, V. J., Dolle, A., & Carper, W. R. (2004). Raman and Infrared Spectra and ab Initio Calculations of C2-4MIM Imidazolium Hexafluorophosphate Ionic Liquids. J. Phys. Chem. B, 108, 13177-13184.

[28] Fumino, K., Wulf, A., & Ludwig, R. (2009). The potential role of hydrogen bonding in aprotic and protic ionic liquids. Phys. Chem. Chem. Phys., 11, 8790-8794.

[29] Tsuda, T., Kondo, K., Tomioka, T., Takahashi, Y., Matsumoto, H., Kuwabata, S., & Hussey, C. L. (2011). Design, Synthesis, and Electrochemistry of Room-Temperature Ionic Liquids Functionalized with Propylene Carbonate, Angew. Chem. Int. Ed., 50, 1310-1313.

[30] Lava, K., Binnemans, K., & Cardinaels, T. (2009). Piperidinium, Piperazinium and Morpholinium Ionic Liquid Crystals, J. Phys. Chem. B , 113, 9506-9511.

[31] Egashira, M., Okada, S., Yamaki, J., Yoshimoto, N., & Morita, M. (2005). Effect of small cation addition on the conductivity of quaternary ammonium ionic liquids Electrochimica Acta, 50, 3708-3712.

[32] Triolo, A., Russina, O., Fazio, B., Battista, G., Carewska, A. M., & Passerini, S. (2009). Nanoscale organization in piperidinium-based room temperature ionic liquids J. Chem. Phys. 130, 164521 (1-6).

[33] Salminen, J., Papaiconomou, N., Kumara, R. A., Lee, J. M., Kerr, J., Newmana, J., & Prausnitz, J. M. (2007). Physicochemical properties and toxicities of hydrophobic piperidinium and pyrrolidinium ionic liquids" Fluid Phase Equilibria , 261, 421-426.

[34] Yuan, L. X., Feng, J. K., Ai, X. P., Cao, Y. L., Chen, S. L., & Yang, H. X. (2006). Improved dischargeability and reversibility of sulfur cathode. Electrochem. Comm., 8, 610-614.

[35] Matsumoto, K., & Hagiwara, R. (2009). A New Series of Ionic Liquids Based on the Difluorophosphate Anion. Inorg. Chem., 48, 7350-7358.

[36] Rao, C. J., Krishnan, R. V., Venkatesan, K. A., Nagarajan, K., & Srinivasan, T. G. (2009). Thermochemical properties of some bis(trifluoromethyl sulfonyl)imide based room temperature ionic liquids. J. Therm. Anal. Calorim., 97, 937-943.

[37] Lewandowski, A., & Widerska, A. S. (2004). New composite solid electrolytes based on a polymer and ionic liquids. Solid State Ionics, 169, 21-24.

[38] Sakaebe, H., & Matsumoto, H. (2003). N-Methyl-N-propylpiperidinium bis(trifluoromethanesulfonyl)imide (novel electrolyte base for Li battery. Electrochem. Comm. 5, 594-598., 13.

[39] Fernanda, F., Bazito, C., Kawano, Y., & Torresi, R. M. (2007). Synthesis and characterization of two ionic liquids with emphasis on their chemical stability towards metallic lithium. Electrochimica Acta , 52, 6427-6437.

[40] Zhou, Z. B., Matsumoto, H., & Tatsumi, K. (2006). Cyclic Quaternary Ammonium Ionic Liquids with Perfluoroalkyltrifluoroborates: Synthesis, Characterization, and Properties. Chem. Eur. J., 12, 2196-2212.

[41] Cho T Y, Yoon S G, Sekhon S S, Han C H(2011). Effect of Ionic Liquids with Different Cations in I-/I3- Redox Electrolyte on the Performance of Dye-sensitized Solar Cells Bull. Korean Chem. Soc., , 32, 2058-2062.

[42] Reichert, W. M., Henderson, W. A., Trulove, P. C., Urban, J. J., & De Long, H. C. (2010). Effects of Crystal Packing on the Thermal Behavior of N,N′-alkylpiperidinium and N,N′-alkylmorpholinium Iodide Salts. ECS Transactions , 33, 667-677.

[43] Haddad, B., Villemin, D., & Belarbi, E. (2012). Synthesis, Differential Scanning Calorimetry (DSC) and Dielectric Relaxation Spectroscopy (DRS) Studies of N-methyl-N NpropylpiperidiniumBis(trifluoromethylsulfonyl)imide J. Mater. Environ. Sci., , 3, 312-319.

[44] Xu, K. (2004). Nonaqueous Liquid Electrolytes for Lithium-Based Rechargeable Batteries. Chem. Rev., , 104, 4303-4417.

[45] Nakagawa, N., Izuchi, S., Kuwana, K., Nukuda, T., & Aihara, Y. (2003). Liquid and Polymer Gel Electrolytes for Lithium Batteries Composed of Room-Temperature Molten Salt Doped by Lithium Salt J. Electrochem. Soc., 150, AA700;, 695.

[46] Garcia, B., Lavallee, S., Perron, G., Michot, C., & Armand, M. (2004). Room temperature molten salts as lithium battery electrolyte, Electrochim. Acta, , 49, 4583-4588.

[47] Matsumoto, H., Sakaebe, H., & Tatsumi, K. (2005). Preparation of room temperature ionic liquids based on aliphatic onium cations and asymmetric amide anions and their electrochemical properties as a lithium battery electrolyte J. Power Sources, , 146, 45-50.

[48] Sakaebe, H., & Matsumoto, H. (2003). N-Methyl-N-propylpiperidinium bis(trifluoromethanesulfonyl)imide (novel electrolyte base for Li battery Electrochem. Commun., 5, 594-598., 13.

[49] Shin, J. H., Henderson, W. A., & Passerini, S. (2003). Ionic liquids to the rescue: Overcoming the ionic conductivity limitations of polymer electrolytes Electrochem. Commun., , 5, 1016-1020.

[50] Sato, T., Maruo, T., Marukane, S., & Takagi, K. (2004). Ionic liquids containing carbonate solvent as electrolytes for lithium ion cells J. Power Sources , 138, 253-261.

[51] Howlett P C, MacFarlane D R, Hollenkamp A F,(2004). High Lithium Metal Cycling Efficiency in a Room-Temperature Ionic Liquid. Electrochem. Solid-State Lett., 7, AA101., 97.

[52] Matsumoto, H., Yanagida, M., Tanimoto, K., Nomura, M., Kitagawa, Y., & Miyazaki, Y. (2000). Highly Conductive Room Temperature Molten Salts Based on Small Trimethylalkylammonium Cations and Bis(trifluoromethylsulfonyl)imide Chem. Lett., , 29, 922-923.

[53] Baranchugov, V., Markevich, E., Pollak, E., Salitra, G., & Aurbach, D. (2007). Amorphous silicon thin films as a high capacity anodes for Li-ion batteries in ionic liquid electrolytes, Amorphous silicon thin films as a high capacity anodes for Li-ion batteries in ionic liquid electrolytes Electrochem. Commun., , 9, 796-800.

[54] Yuan, L. X., Feng, J. K., Ai, X. P., Cao, Y. L., Chen, S. L., & Yang, H. X. (2006). Improved dischargeability and reversibility of sulfur cathode in a novel ionic liquid electrolyte, Electrochem. Commun., , 8, 610-614.

[55] Fuller, J., Breda, A. C., & Carlin, R. T. (1997). Ionic liquid-polymer gel electrolytes, J. Electrochem. Soc., 144, L667.

[56] Lee, J. M., Ruckes, S., & Prausnitz, J. M. (2008). Solvent polarities and Kamletâ˘Taft parameters for ionic liquids containing a pyridinium cation, J. Phys. Chem. B, , 112-1473.

[57] Hinckley, G., Mozhaev, V. V., Budde, C., & Khmelnitsky, Y. L. (2002). Oxidative enzymes possess catalytic activity in systems with ionic liquids, Biotechnol. Lett., , 24, 2083-2087.

[58] Peter, W., & Wilhelm, K. (2000). Ionic liquids-new ldquoSolutionsrdquo for transition metal catalysis, Angew. Chem., , 39-3772.

[59] Reiter, J., & Nadherna, M. (2012). N-Allyl-N-methylpiperidinium bis(trifluoromethanesulfonyl)imide-A film forming ionic liquid for graphite anode of Li-ion batteries, Electrochimica Acta, , 71, 22-26.

[60] Holbrey, J. D., Reichert, W. M., Nieuwenhyuzen, M., Johnston, S., Seddon, K. R., & Rogers, R. D. (2003). Crystal polymorphism in butyl-3-methylimidazolium halides: supporting ionic liquid formation by inhibition of crystallization, Chem. Commun., 1636-1637., 1.

[61] Anthony, J. L., Brennecke, J. F., Holbrey, J. D., Maginn, E. J., Mantz, R. A., Rogers, R. D., Trulove, P. C., Visser, A. E., & Welton, T. (2003). Physicochemiocal properties of

ionic liquids. In Wasserscheid P & Welton T (Eds) Ionic Liquids in Synthesis, Wiley-WCH, Weinheim: , 41-126.

[62] Mac, Farlane. D. R., Meakin, P., Sun, J., Amini, N., & Forsyth, M. (1999). Pyrrolidinium Imides:A New Family of Molten Salts and Conductive Plastic Crystal Phases, J. Phys. Chem. B, , 103, 4164-4170.

[63] Zhou, Z. B., Matsumoto, H., & Tatsumi, K. (2004). Low-Melting, Low-Viscous, Hydrophobic Ionic Liquids: 1-Alkyl(Alkyl Ether)-3-methylimidazolium Perfluoroalkyltrifluoroborate, Chem. Eur. J., , 10, 6581-6591.

[64] Lide DR. (2003). CRC Handbook of chemistry and physics, 84 edition, CRC press, Boca Raton, New York.

[65] Stenzel, O., Raubenheimer, H. G., & Esterhuysen, C. (2002). Biphasic hydroformulation in new molten salts- analogies and differences to organic solvents. J. Chem. Soc. Dalton Trans, , 1132-1138.

[66] Huddleston J G, Visser A E, Reichert W M, Willauer H D, Broker G A, Rogers R. D. (2001). Characterization and comparision of hydrophilic and hydropbobic room temperature ionic liquids incorporating the imidazolium cation, Green Chem., , 3, 156-164.

[67] Alavi, S., & Thompson, D. L. (2005). Molecular dynamics studies of melting and some liquid-state properties of ethyl-3-methylimidazolium hexafluorophosphate [emim][PF6], J. Chem. Phys., 122, 154704., 1.

[68] Sun, J., Forsyth, M., & Mac, Farlane. D. R. (1998). Room-Temperature Molten Salts Based on the Quaternary Ammonium Ion. J. Phys. Chem. B, , 102, 8858-8864.

[69] Wang, J., Wang, H., Zhang, S., Zhang, H., & Zhao, J. Y. (2007). Conductivities, Volumes, Fluorescence, and Aggregation Behavior of Ionic Liquids [C4mim][BF4] and [Cnmim]Br (n = 4, 6, 8, 10, 12) in Aqueous Solutions J. Phys. Chem. B, , 111, 6181-6188.

[70] Turner E A, Pye C C, Singer R D.(2003). Use of ab Initio calculation towards the rational design of room temperature ionic liquids, J. Phys. Chem. A, , 107, 2277-2288.

[71] Furlani, M., Albinsson, I., Mellander, B. E., Appetecchi, G. B., & Passerini, S. (2011). Annealing protocols for pyrrolidinium bis(trifluoromethylsulfonyl)imide type ionic liquids, Electrochemica Acta , 57, 220-227.

[72] Bonhote, P., Dias, A. P., Papageogiou, N., Kalyanasundaram, K., & Gratzel, M. (1996). Hydrophobic, Highly Conductive Ambient-Temperature Molten Salts, Inorg. Chem., , 35, 1168-1178.

[73] Rogers R D, Visser A E, Swatloski R P, Hartman D H.(2000). in Metal Ion Separations Beyond: Integrating Novel Chemistry with Processing, ed.

[74] Liddell K C, Chaiko D J.(1999). The Minerals, Metals and Materials Society, Warrendale, PA, , 139-147.

[75] Belhocine, T., Forsyth, S. A., Gunaratne, H. Q. N., Nieuwenhuyzen, M., Puga, A. V., Seddon, K. R., Srinivasan, G., & Whiston, K. (2011). New ionic liquids from azepane and 3-methylpiperidine exhibiting wide electrochemical windows, Green Chem, , 13, 59-63.

[76] Becke A D.(1988). Density-functional exchange-energy approximation with correct asymptotic behaviour, Phys. Rev. A, , 38, 3098-3100.

[77] Lee, C., Yang, W., & Parr, R. G. (1988). Development of the Colle-Salvetti correlation-energy formula into a functional of the electron density, Phys. Rev. B, , 37, 785-789.

[78] Frisch, M. J., et al. (2004). Gaussian 03, Revision C.02, Gaussian, Inc., Wallingford, CT.

[79] Fujii, K., Fujimori, T., Takamuku, T., Kanzaki, R., Umebayashi, Y., & Ishiguro, S. (2006). Conformational Equilibrium of Bis(trifluoromethanesulfonyl) Imide Anion of a Room-Temperature Irequency onic Liquid: Raman Spectroscopic Study and DFT Calculations J. Phys. Chem. B, , 110(16), 8179-8183.

Chapter 4

Spectral-Structure Activity Relationship (Spectral-SAR) Assessment of Ionic Liquids' in Silico Ecotoxicity

Ana-Maria Putz and Mihai V. Putz

Additional information is available at the end of the chapter

http://dx.doi.org/10.5772/51657

1. Introduction

The use of green solvents, including ionic liquids (IL), in the synthesis of new materials is currently highlighted worldwide [1-5]. The green high thermal stabilities, and can remain in the liquid state over a wide range of temperatures (from -50°C to 200°C). Furthermore, ILs exhibit low toxicity because of their exceptionally low volatility. This property is why ILs could be used as "green" alternatives to volatile organic solvents in many processes [6,7].

Accordingly, ILs have been used as a reaction medium for inorganic materials, which demonstrates the pre-organized structure of the ILs as a template for generating porous nanomaterials [8,9]. It is well-known that the properties and structures of mesoporous materials vary by the choice of synthesis conditions, such as temperature, reagents, base concentrations, and the nature of the organic groups [10]. By organic groups, one means the organic chains and structures that affect the functionality of the precursor; for example, where there is Si-C bond, depending on the silica to surfactant ratio, different pore topologies (hexagonal, cubic and lamellar) are present, and when only ionic liquid was used as the surfactant, mesoporous silica with an irregular shape may be obtained [11]. However, when a small amount of cetyl trimethylammonium bromide (CTAB) was added to the reaction, mesoporous silica with a spherical shape was obtained. When a mixture of ionic liquids and CTAB was used as the surfactant templates, larger mesopores were formed in the silica spheres [12]; however, ILs can generally dramatically decrease the polarity and dielectric constant of water and increase the solubility of surfactants when used as co-solvents [13,14]. However, the interaction between CTAB and IL has also been observed, which causes the IL to behave as a co-surfactant [14]. Starting from these new concepts, ILs with different cations (e.g., imidazolium, pyridinium and ammonium) and anions (e.g., hexafluorophosphate, tetrafluoroborate, octyl sulfate and bromide) and the ordinary CTAB and the mixture of CTAB and ILs are used in the synthesis of mesoporous silica.

Furthermore, ILs provide an interesting method for controlling the chemical contents of the collapsing bubble through the reduction of the solvent vapor pressure [15] because the rates of sonochemical reactions can be increased by decreasing the vapor pressure of the solvent or by choosing a less volatile one [16]. ILs are usually more viscous and denser than other organic solvents, and producing cavitations should be more difficult under such conditions with large cohesive forces [16,17].

Studies of the thermal decomposition of ILs confined in nanopores compared to bulky pores are also currently on the forefront of nanomaterials research. Many physicochemical properties of ILs have been observed to change when confined in nanopores, which results from the interaction of the IL cations/anions with the pore walls, which could be weak, as in silica nanopores [18,19]. These studies influence the next step, i.e., the surface functionalization of mesoporous silica materials through the covalent bonding of organic groups, which is achieved through the comparative co-condensation and post-grafting of amino and alkyl functionalized silica alkoxide precursors [20]. Importantly, the comparative study is necessary to optimize the synthesis methods for nanomaterials because, even if the post-grafting method produces well-ordered functionalized mesostructured materials, it often produces non-uniformly distributed organic groups [20]. Furthermore, the co-condensation synthetic method for mesoporous materials involves a one-step procedure and allows better control of the loading and distribution of the organic groups [20], as was detailed in the recent review on the advances in (bio)responsive nanomaterials [21]. Supplementary information on the effects of ILs on the texture of gels can be obtained from the pore size distribution. This aspect is interesting because the ionic liquid concentration changes the pore volume due to the low volatility of the IL, which reduces the surface tension associated with the pore collapse [22-25].

The next step in the ecotoxicological evaluation of the synthetized mesoporous silica is to use the logistic enzyme kinetic and Spectral-SAR methods [26-30] with acetylcholine esterase as a working enzyme. Acetylcholine is a notably good marker for monitoring the evolution in ionic liquids, due to the similar chemical structures of many ionic liquids cations and acetylcholine and especially because this enzyme does not follow the standard Michaelis-Menten kinetics mechanism. The logistic enzyme kinetics [31-35] is an analytical substitute for the Michaelis-Menten mechanism, and it is based on some probabilistic considerations that better accommodate the QSSA (quasi-steady state approximation) conditions, while providing the analytical transformation of the W-Lambert function into a logarithmic function; this so-called logistic transformation was successfully evaluated for different cases of enzyme kinetics [31-39]. The analytical information contained in the logistic enzymatic temporal curves could be further combined with a QSAR algorithm, such as the Spectral-SAR, for designing molecular mechanisms of the ecotoxicity; this procedure has already been applied for bare ILs [27,28,38,40]. Eventually, this procedure is intended to be used for some toxicity tests [41,42] or even for obtaining the efficiency of the material as a drug carrier in biological media, if the proper conditions are satisfied.

Regarding the action/toxicity mechanism of ionic liquids (M.O.A.), possible mechanisms of toxic actions are identified through membrane disruption because they have structural similarity with detergents, pesticides and antibiotics [43-46] that induce polar narcosis, due to their interfacial properties, and may cause membrane-bond protein disruption [47-49]. As a possible mode of action, disrupted membranes and hydrophobic molecules have a greater ability to accumulate at this interface [50]. The low K_{ow} values of imidazolium ionic liquids indicate the low permeability of these ionic liquids [47]. However, other mechanisms arise from acethylcolinesterase inhibition [51], relate to a common cellular structure or process [45], involve structural DNA damage [47,52], or are mediated by certain bacteria that could potentially break down imidazolium into different metabolites that could be inhibited by the ionic liquid itself [43,44]. Nevertheless, knowing if the chemicals could reach the target site in the organism and by which mechanism is of paramount importance, and at minimum, a conceptual extrapolation from *in vitro* to *in vivo* studies is required [51] because other factors, such as biodegradation and bioaccumulation, are necessary before conclusions can be drawn [49].

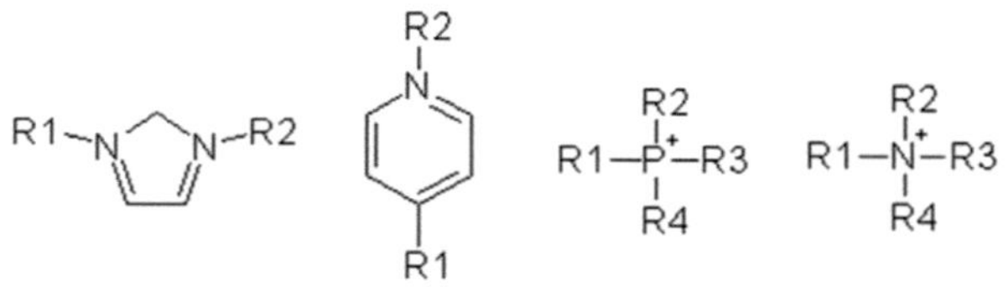

Figure 1. The four most important classes of ionic liquids based on the reference cation: (1) imidazolium, (2) pyridinium-, (3) phosphonium-, (4) ammonium- [39].

The dose-response approach for estimating the lethal effects of toxicants on organisms is being criticized because of the lack of real ecological meaning; furthermore, regulatory norms have been established around LC50 values that can be compared between different toxicants and organisms [47,53]. In this respect, it was observed that the side chain toxic mode of action of ionic liquids is not specific to species, and a similar relationship has been observed [47]. However, the systematic variation of, for instance, R1 and R2 at identical head groups and anions from the molecular to individual organism level lead to the conclusion in most published data that the shorter the chain lengths of the side chains, the lower the cytotoxicity (higher EC50 values) [50,54]. The electronic portion of the bond factor extends along a chain that has no more than 5 alkane carbons, which results in a decrease in the overall hydrophobicity; therefore, chains longer than 5-6 carbon atoms will have a considerably greater permeability through the cell membrane (for example, [DMIM][BF4]), see ref. [55]. Therefore, chemical transformations of the side chains of ionic liquids may reduce their toxicity because the metabolites are less toxic compared to their parent chemicals [54]. With respect to the cationic effects, see Figure 1, the imidazolium ring plays a major role due its structure as a delocalized aromatic system with a high electron acceptor potential; therefore, the nitrogen atoms are not capable of forming any hydrogen bonds, which results in a rigid and sterically hindered system. The elongation of the R2 residue causes a continuous increase of flexibility, which implies more conformational freedom [56]. The chemical nature of a head group also influences the biological activity [54]. The fragmental hydrophobicity of each carbon connected to a quaternary amine combines a geometric bond factor that applies to the neutral solute together with a negative electronic bond factor, which decreases in magnitude with the square of the distance from the central nitrogen atom [55,57]. Complementarily, the anion could play a central role as a technicophore because it exhibits a high potential for changing the materials technological properties (solubility, viscosity); however, there are reports that the anion presents no significant effects [44,45,55]. The first study reporting the influence of the anion was by the Jastorff group [54], where it was reported that the anion in the ion pair can partially decompose, as was previously observed [55,58,59].

Finally, at the computational level, the detailed examination of the relative energies and structural interactions (such as ion position, H-bonding and variability in the anion conformation) in gas-phase ion-pairs has emphasized how these quantities can be used to construct a picture of the local structure and interactions that occur in ionic liquids. For instance, [BMIM][Cl] (1-n-butyl-3-methylimidazolium chloride) forms a highly connected liquid with relatively strong interactions, [BMIM][TFMSi] (1-n-butyl-3-methylimidazolium bis(trifluoromethylsulfonyl)imide) forms a low connectivity network of weakly linked ions, whereas [BMIM][BF4] lies between these extremes and forms a weak but more regular network. The melting points and viscosity are somewhat dependent on the local interactions between an ion and other molecules in the first solvation shell. For imidazolium based cations, nine sites of interactions are preferred by the anion [60-63]. Hunt et al. demonstrated that the hydrogen bond is primarily ionic with a moderate covalent character. The fact that the Coulombic attraction is the dominant stabilization force was demonstrated by analyzing the charge distribution, molecular orbitals and electron density of the dimer complex [BMIM][Cl] [63]. The interactions governing the top conformers are notably different compared to those where the Cl anion remains in-plane (where the Cl anion interacts with the π-manifold of the orbitals); the LUMO of the ion-pair is the cation antibonding orbital; therefore, electron acceptance is not favorable [63]. The effect of the chloride anion on the rotation of the butyl chain was consequently investigated and observed to lower certain rotational barriers while enhancing others [61]. The Coulombic forces in an ionic liquid, and hence the point charges or charge distribution on the constituent ions are anticipated to be more important than in liquids composed of neutrally charged molecules [60]; these ions allow a type of semi-classical analysis for the toxicity of ILs, which is well-adapted for quantitative structure-activity studies (QSAR), to which the present review belongs. Nevertheless, even the well-regarded QSAR methodology has to be refined and adapted or extended to include the chemical electrostatic interactions present in the IL and of their transduction into various organism triggering (or not) toxicity. These aspects will be reviewed and exemplified in the following.

2. QSAR/Spectral-SAR Modeling of the Structure-EcoToxicity of Ionic Liquids

It is an already established fact that the costs of all approaches for sustainable product design can be reduced using the SAR and QSAR methods [56]. For instance, the anti-microbial activity of quaternary ammonium chlorides is dependent on the lipophilicity, and the 1-octanol-water partition coefficient, K_{ow}, could provide a good approximation for the lipophilicity of the compound. The relationship between the chemical structure and anti-microbial activity of several new choline-like quaternary ammonium derivatives was analyzed using the QSAR method, and all of the studied compounds were active against the microorganisms. The statistically significant correlation of the obtained QSAR equations confirms that the lipophilicity is the primary factor that governs anti-microbial activity [43]. The octanol-water partition coefficient has been correlated with bioaccumulation and toxicity in fish and to sorption in soil and sediments. The K_{ow} values for the BMIM cation range from 0.003 to 1.1, depending on the choice of anion, and the K_{ow} value increases with increasing length of the alkyl chain on the cation. Replacing the H atom between the 2 nitrogen atoms in the ring with a methyl group has almost no effect on the K_{ow} value. Because all of the measured K_{ow} values in the analyzed ionic liquids, i.e., imidazolium ionic liquids with butyl, ethyl, dimethyl, hexyl and octyl side chains with different anions, such as tetrafluoroborate, nitrate, bis(trifluoromethylsulfonyl)imide), are less than 15, these ionic liquids will not accumulate or concentrate in the environment. It was also observed that the K_{ow} values for ionic liquids with the TFMSi (bis(trifluoromethylsulfonyl)imide) anion (a notably hydrophobic anion) are dependent on concentration, even at the diluted concentrations studied [64].

However, QSPRs (Quantitative Structure Properties Relationships) methods could be used to correlate and predict the toxicity of ionic liquids. With at most 4 molecular descriptors, log EC50 and log EC5 data are reproduced with a R^2 (statistical Pearson correlation coefficient) of 0.78-0.88 using tests with *Vibrio fischeri* and *Daphnia magna.* These methods calculated the electronic, spatial, structural, thermodynamic and topological descriptors for both the cation and anion *separately,* whereas the geometries were also optimized *separately*. Concerning *Vibrio fischeri,* the QSPR equations that used cationic (first equation) and cationic + anionic (second equations) descriptors fit the training set well, with R^2=0.887 and R^2=0.782, respectively. However, the majority of the training set compounds were imidazolium and pyridinium based ionic liquids. Then, Couling and co-workers [49] used the equation to predict the toxicity of new compounds, which contained ammonium and phosphonium cations, but the toxicity of these new additional compounds was not predicted with the same accuracy, which led to the conclusion that the equation is only good for compounds of the same class. The best accuracy for a new predicted compound (using the first QSPR equation) was for [BdMAPy][Br] (1-n-butyl-4-dimethylaminopyridinium bromide) because it contains an aromatic pyridinium cation, and it also contains an amino group, unlike all of the other training set compounds. Concerning *Daphnia magna,* the QSPR equation fits the training data set well (R^2=0.862), and the descriptors are similar to those provided for *Vibrio fischeri,* which suggests that there may be similar indicators of toxicity observed in many different species. The equation was used to predict the toxicity of another compound, and contrary to the case of *Vibrio fischeri,* the correlation was good, R^2=0.775; however, the neglected compounds were more similar to those of the training set compounds [49]. Such results further motivated the use of the actual Spectral-SAR method for these two paradigmatic species response upon IL action because it is an important step in constructing a consistent ecotoxicological test battery. The reliability of the Spectral-SAR/QSAR studies primarily relies on the lack of readily accessible ecotoxicological data [65]. Note that because the principles of green chemistry state that one should consider the entire process (life cycle analysis), rather than the individual components of a reaction (single issue sustainability), the acute toxicity measurements do not fully characterize the full impact of the release of a substance into the environment, but they are only part of the environmental impact assessment [65,66]. In particular, ionic liquids with cations, such as pyridinium, imidazolium and pyrrolidinium, have been nominalized by the United States National Toxicology Program (NTP) for toxicological testing because of their potential use as new solvents and potential to enter the aquatic system [67]: if an accidental discharge of ionic liquids into water were to occur, there may be an environmental risk to aquatic plants and animals because many of the ILs are water-soluble [48]. The recorded averaged descriptions of the acute toxicity based on the LC50 (mg/mL) are very highly toxic ($<10^{-4}$), highly toxic (10^{-4}-10^{-3}), moderately toxic (10^{-3}-10^{-2}), slightly toxic (10^{-2}-10^{-1}) and not acutely toxic ($>10^{-1}$) [55,68].

3. Algebraic QSAR: Spectral-SAR [69,70]

The key concept in the SAR discussion concerns the algebraic consideration of biological activity and the structural parameters in Table 1. As a consequence, we may further employ this feature to quantify the basic SAR through an *orthogonal* space. The idea is to transform the columns of structural data in Table 1 into an abstract orthogonal space, where all of the predictor variables are independent, solve the SAR problem in orthogonal space and subsequently compare the result to the initial data using a coordinate transformation. Because QSAR models aim to develop correlations between the molecular structures of interest and the measured (or otherwise evaluated) activity, it naturally appears that the *structure* part of the problem is accommodated within quantum theory and its formalisms.

Activity	**Structural predictor variables**					
$\|Y_{OBS(ERVED)}\rangle$	$\|X_0\rangle$	$\|X_1\rangle$	...	$\|X_k\rangle$	...	$\|X_M\rangle$
$y_{1\text{-OBS}}$	1	x_{11}	...	x_{1k}	...	x_{1M}
$y_{2\text{-OBS}}$	1	x_{21}	...	x_{2k}	...	x_{2M}
⋮	⋮	⋮	⋮	⋮	⋮	⋮
$y_{N\text{-OBS}}$	1	x_{N1}	...	x_{Nk}	...	x_{NM}

Table 1. The vectorial descriptors in a Spectral-SAR analysis.

In fact, there are a few quantum characteristics that we are using within the present approach [29]:

- Any molecular structural state (dynamical because it undergoes interactions with the species and organisms) may be represented by a $|ket\rangle$ *state vector* in an abstract Hilbert space, which follows the $\langle bra|ket\rangle$ Dirac formalism [71]; such states are to be represented using any reliable molecular index, or specifically, in our study, by the hydrophobicity $|LogP\rangle$, polarizability $|POL\rangle$, and the total optimized energy $|E_{tot}\rangle$. These parameters are only the so-called Hansch parameters, which are usually employed to account for the diffusion, electrostatic and steric effects for molecules acting, for instance, within the cells of organisms, respectively;
- The (quantum) *superposition principle*, which ensures that the summation of molecular states map onto another resulting molecular state, which is interpreted here as the bio-, eco- or toxico- logical activity, e.g., $|Y\rangle = |X_0\rangle + c_1|X_1\rangle + ... + c_M|X_M\rangle$, where $|X_0\rangle$ represents the "noise" activity (present even when all other influences are absent);
- The *orthogonalization feature* of quantum states, which is a crucial condition where the superimposed molecular states generate other molecular states (here quantified as the activity of the molecular ligand - linking receptor); analytically, the orthogonalization condition is represented by the $\langle bra|ket\rangle$ scalar product of two envisaged states (molecular indices). If $\langle bra|ket\rangle=0$, then the states are said to be orthogonal, and the molecular descriptors are independent; therefore, they are suitable to be added as contributing states in the resulting activity and as molecular indices in the activity correlation.

As such, the analytical procedure is decomposed into three fundamental steps.

I. Given a set of *N* molecules that are being examined for their biological activity, they produce through the considered *M* structural indicators all of the input information (the states) that may be vectorially expressed by the columns in Table 1 and correlated through the following equation:

$$\begin{aligned}|Y_{OBS(ERVED)}\rangle &= b_0|X_0\rangle + b_1|X_1\rangle + ... + b_k|X_k\rangle + ... + b_M|X_M\rangle + |prediction\ \ error\rangle \\ &= |Y_{PRED(ICTED)}\rangle + |prediction\ \ error\rangle\end{aligned} \quad (1)$$

with

$$|X_0\rangle = |1\,1 \cdots 1_N\rangle \quad (2)$$

added to account for the noise term. For equation (1) to represent a reliable model of the given activities, the assumed molecular states (indices) should constitute an orthogonal set; having this constraint is a quantum mechanical fundamental, as described above. However, unlike other important studies that have addressed this problem [72,73], the present use of the Spectral-SAR method assumes that the prediction error vector in eq. (1) arises from the predicted activity's being orthogonal because it cannot consider the input data otherwise

$$\langle Y_{PRED} | prediction\ \ error \rangle = 0 \tag{3}$$

being not known *a priori* any correlation is made. In this manner, it follows from eqs. (1) and (3) that the prediction error vector has to be orthogonal on all other descriptor states of the predicted activity.

$$\langle X_{i=\overline{0,M}} | prediction\ \ error \rangle = 0 \tag{4}$$

In other words, conditions (3) and (4) confirm the form of (1) in the sense that the prediction vector and the prediction activity $| Y_{PRED} \rangle$ (with all of its sub-intended states $| X_{i=\overline{0,M}} \rangle$) belong to disjointed (thus orthogonal) Hilbert spaces; that is, one can state that the Hilbert space of the observed activity $| Y_{OBS} \rangle$ may be decomposed into a predicted and an error independent Hilbert sub-spaces of states. Therefore, within the *Spectral-SAR* procedure, the first step in the orthogonalization procedure orthogonalizes the predicted activity to its prediction error, whereas the remaining orthogonalization algorithm does not search for optimizing the minimization of errors, but it searches for the optimum method for producing the ideal correlation between $| Y_{PRED} \rangle$ and the given descriptors $| X_{i=\overline{0,M}} \rangle$.

II. Next, the Gram-Schmidt orthogonalization algorithm is applied by constructing the orthogonal set of descriptors using the consecrated iteration [74-76]

$$|\Omega_0\rangle = |X_0\rangle \tag{5}$$

$$|\Omega_k\rangle = |X_k\rangle - \sum_{i=0}^{k-1} r_i^k |\Omega_i\rangle \tag{6}$$

$$r_i^k = \frac{\langle X_k | \Omega_i \rangle}{\langle \Omega_i | \Omega_i \rangle}, k = \overline{1, M} \tag{7}$$

providing the orthogonal correlation

$$|Y_{PRED}\rangle = \omega_0 |\Omega_0\rangle + \omega_1 |\Omega_1\rangle + \ldots + \omega_k |\Omega_k\rangle + \ldots + \omega_M |\Omega_M\rangle \tag{8}$$

$$\omega_k = \frac{\langle \Omega_k | Y \rangle}{\langle \Omega_k | \Omega_k \rangle}, k = \overline{0, M} \tag{9}$$

http://dx.doi.org/10.5772/51657

III. Remarkably, while the studies dedicated to the orthogonal problem usually stop at this stage, the Spectral-SAR method uses this stage to provide the solution for the original searched correlation, eq. (1). This process can be adequately achieved by rearranging eqs. (6) and (8) such that the system of all descriptors in Table 1 can be written in terms of orthogonal descriptors

$$\begin{cases} |Y_{PRED}\rangle = \omega_0|\Omega_0\rangle + \omega_1|\Omega_1\rangle + ... + \omega_k|\Omega_k\rangle + ... + \omega_M|\Omega_M\rangle \\ |X_0\rangle = 1\cdot|\Omega_0\rangle + 0\cdot|\Omega_1\rangle + ... + 0\cdot|\Omega_k\rangle + ... + 0\cdot|\Omega_M\rangle \\ |X_1\rangle = r_0^1|\Omega_0\rangle + 1\cdot|\Omega_1\rangle + ... + 0\cdot|\Omega_k\rangle + ... + 0\cdot|\Omega_M\rangle \\ \cdots\cdots \\ |X_k\rangle = r_0^k|\Omega_0\rangle + r_1^k|\Omega_1\rangle + ... + 1\cdot|\Omega_k\rangle + ... + 0\cdot|\Omega_M\rangle \\ \cdots\cdots \\ |X_M\rangle = r_0^M|\Omega_0\rangle + r_1^M|\Omega_1\rangle + ... + r_k^M|\Omega_k\rangle + ... + 1\cdot|\Omega_M\rangle \end{cases} \tag{10}$$

System (10) has no trivial (orthogonal) solution if and only if the associated extended determinant vanishes; this condition introduces the Spectral-SAR determinant and its equation [26]

$$\begin{vmatrix} |Y_{PRED}\rangle & \omega_0 & \omega_1 & \cdots & \omega_k & \cdots & \omega_M \\ |X_0\rangle & 1 & 0 & \cdots & 0 & \cdots & 0 \\ |X_1\rangle & r_0^1 & 1 & \cdots & 0 & \cdots & 0 \\ \vdots & \vdots & \vdots & \vdots & & \vdots & \\ |X_k\rangle & r_0^k & r_1^k & \cdots & 1 & \cdots & 0 \\ \vdots & \vdots & \vdots & \vdots & & \vdots & \\ |X_M\rangle & r_0^M & r_1^M & \cdots & r_k^M & \cdots & 1 \end{vmatrix} = 0 \tag{11}$$

If the determinant of eq. (11) is expanded on its first column, and the result rearranged so that to have $|Y_{PRED}\rangle$ on the left side and the rest of states/indicators on the right side, the searched QSAR solution of the initial problem of eq. (1) is obtained as the Spectral-SAR vectorial expansion (from where the "spectral" name is also justified) with the error vector already absorbed in the orthogonalization procedure. In fact, the Spectral-SAR procedure uses double conversion passages: one forward from the given problem of eq. (1) to the orthogonal one of eq. (8) where the error vector is orthogonally "dissolved"; and the reverse one, back from the orthogonal to the real descriptors throughout the system (10), which provides the determinant (11) to be expanded as the QSAR solution.

The result is that the QSAR/Spectral-SAR equation is now directly delivered by the determinant (11) and not through matrices products, as in the statistical Pearson approach, while directly providing the Spectral-SAR correlation equation and not only the parameters of multi-variate correlation [77-83]. Furthermore, the Spectral-SAR algorithm is also *invariant* to the order of descriptors that are chosen in orthogonalization procedure, which provides equivalent determinants only with rearranged lines; this is a matter that was not previously achieved by other orthogonalization techniques [84-87].

However, it is worth being convinced by comparing the present Spectral-SAR method with the standard statistical one by specializing the general problem (1) to the linear case

$$|Y_{PRED}\rangle = b_0|X_0\rangle + b_1|X_1\rangle \tag{12}$$

and to determine whether it furnishes the linear regression parameters given by the consecrated least squares analysis through the Spectra-SAR equation (11) [81-83]. In this respect, we actually deal with the particular equation

$$0 = \begin{vmatrix} |Y_{PRED}\rangle & \omega_0 & \omega_1 \\ |X_0\rangle & 1 & 0 \\ |X_1\rangle & r_0^1 & 1 \end{vmatrix} = |Y_{PRED}\rangle \begin{vmatrix} 1 & 0 \\ r_0^1 & 1 \end{vmatrix} - |X_0\rangle \begin{vmatrix} \omega_0 & \omega_1 \\ r_0^1 & 1 \end{vmatrix} + |X_1\rangle \begin{vmatrix} \omega_0 & \omega_1 \\ 1 & 0 \end{vmatrix} \tag{13}$$

which is immediately rearranged to

$$|Y_{PRED}\rangle = \underbrace{\left(\omega_0 - r_0^1\omega_1\right)}_{b}|X_0\rangle + \underbrace{\omega_1}_{a}|X_1\rangle \tag{14}$$

such that identifying the actual linear coefficients

$$a = \omega_1 \tag{15}$$

$$b = \omega_0 - r_0^1\omega_1 \tag{16}$$

When evaluating expressions (15) and (16) within the Spectral-SAR algorithm, there are instructions to identify only the relevant variables from Table 1 using the convenient notation

$\|Y_{PRED}\rangle$	$\|X_0\rangle$	$\|X_1\rangle$
y_1	1	x_1
y_2	1	x_2
⋮	⋮	⋮
y_N	1	x_N

Other working tools are the zero-th and the first orthogonal vectors, which are accordingly considered and computed as (5) with (2) and respectively by.

$$\begin{aligned}|\Omega_1\rangle &= |X_1\rangle - r_0^1|\Omega_0\rangle \\ &= |x_1\ x_2\ \dots\ x_N\rangle - \frac{1}{N}\sum_{i=1}^{N} x_i |1\,1\,1\ \dots\ 1\rangle = \left| x_1 - \frac{1}{N}\sum_{i=1}^{N} x_i\ \dots\ x_N - \frac{1}{N}\sum_{i=1}^{N} x_i \right\rangle\end{aligned} \tag{17}$$

with the help of the coefficient

$$r_0^1 = \frac{\langle X_1|\Omega_0\rangle}{\langle\Omega_0|\Omega_0\rangle} = \frac{1}{N}\sum_{i=1}^{N} x_i \tag{18}$$

specialized from the general definition (7).

In the same manner, the other specific Spectral-SAR coefficients from the general orthogonal recipe (9) are now computed as the zero-th order contribution for linear regression

$$\omega_0 = \frac{\langle\Omega_0|Y\rangle}{\langle\Omega_0|\Omega_0\rangle} = \frac{1}{N}\sum_i y_i \tag{19}$$

whereas the first order contribution precisely recovers the linear slope [69,70]

$$\begin{aligned}\omega_1 &= \frac{\langle\Omega_1|Y\rangle}{\langle\Omega_1|\Omega_1\rangle} \\ &= \frac{\left\langle x_1 - N^{-1}\sum x_i\ \dots\ x_N - N^{-1}\sum x_i \middle| y_1\ \dots\ y_N\right\rangle}{\sum_i\left(x_i - N^{-1}\sum_i x_i\right)^2} \\ &= \frac{\sum_i y_i\left(x_i - N^{-1}\sum_i x_i\right)}{\sum_i\left(x_i - N^{-1}\sum_i x_i\right)^2} = \frac{\sum_i y_i x_i - N^{-1}\left(\sum_i y_i\right)\left(\sum_i x_i\right)}{\sum_i\left[x_i^2 + N^{-2}\left(\sum_i x_i\right)^2 - 2N^{-1}x_i\sum_i x_i\right]} \\ &= \frac{N\sum_i y_i x_i - \left(\sum_i y_i\right)\left(\sum_i x_i\right)}{N\sum_i x_i^2 - \left(\sum_i x_i\right)^2} = a\end{aligned} \tag{20}$$

as prescribed by the correspondence of (15). Additionally, its companion – the noise term coefficient of eq. (14) - may be now directly evaluated with (16)

$$\begin{aligned} b &= \omega_0 - r_0^1\omega_1 \\ &= \frac{1}{N}\sum_i y_i - \frac{1}{N}\left(\sum_i x_i\right)\frac{N\sum_i y_i x_i - \left(\sum_i y_i\right)\left(\sum_i x_i\right)}{N\sum_i x_i^2 - \left(\sum_i x_i\right)^2} \\ &= \frac{\left(\sum_i y_i\right)\left(\sum_i x_i\right)^2 - \left(\sum_i x_i\right)\left(\sum_i y_i x_i\right)}{N\sum_i x_i^2 - \left(\sum_i x_i\right)^2} \end{aligned} \tag{21}$$

which successfully regains the noise term that is otherwise consecrated by means of a variational statistical (optimization of errors' squares summation) procedure.

With this description, it is clear that the SPECTRAL-SAR methodology not only recovers the standard statistical QSAR least square correlation results but also *generalizes* it analytically to a great extent toward a better assessment of mechanistically ordering and influences in practical eco- and bio- logical applications.

4. From in Cerebro to in Silico Principles of Ecotoxicity

When utilizing the analytical model of QSAR/Spectral-SAR for environmental interactions, one should consider the framework of the principles both at the general and applied levels. As such, regarding the general principles or green chemistry and engineering, they are provided in Table 2 to ensure that they can be readily compared [88]. Note that the fundamental principles constitute the background or the general framework or desiderate that is eventually supported by the associate engineering principle; the green engineering principle is primarily based on "minimizing" or "maximizing" the time, space, energy and costs, and it is constituted either in an economical enterprise and an extension of the main principles of nature or in terms of optimizing mass-energy and time-space. From this point forward, the most basic physical and chemical principles are observed as those acting on each process, system, or state to be created, maintained or modified [89-91].

However, while restraining the analysis to the specific interactions between chemical structures and biological species, the Organization for Economic Co-operation and Development (OECD) advanced a set of standard principles for the validation and for regulatory purposes of the (quantitative) structure-activity relationship models [92-95]:

- QSAR-1: a defined endpoint
- QSAR-2: an unambiguous algorithm
- QSAR-3: a defined domain of applicability
- QSAR-4: appropriate measures of goodness-of–fit, robustness and predictivity

• QSAR-5: a mechanistic interpretation, if possible

No.	Principle of Green Chemistry	Principle of Green Engineering
1.	Prevention of waste that must cleaned afterwards	Prevention rather than treatment
2.	Inherently safer chemistry for accident prevention such as releasing, explosions, and fires	Inherent rather than circumstantial processes and components to prevent hazard
3.	Atom economy in maximizing the incorporation of all material used	Conserving complexity of embedded entropy for minimizing the recycling process
4.	Less hazardous chemical systems should be designed with little or no toxicity	Design for commercial "afterlife" through their nontoxic availability
5.	Designing safer chemicals to minimize their toxicity	Durability rather than immortality because whatever compound should be degradable
6.	Safer solvents and auxiliars (separation agents)	Integrate material and energy flows allowing interconnectivity in components
7.	Designing for energy efficiency while synthetic methods should be conducted at ambient temperature and pressure whenever possible.	Maximizing efficiency in producing products through minimizing mass, energy, space, and time consumption
8.	Use of renewable raw materials and feedstocks rather than depleting them	Design for separation and purification operations should maximize recycling
9.	Reducing derivatives as those modifying physical-chemical processes because they are virtually converted into waste	Minimizing material diversity in multicomponent products towards promoting easiest disassembly process
10.	The use of catalytic rather than stoichiometric reagents is desirable for maintaining control over the selectivity	Output-pulled of reaction products rather than input-pushed reactants as additional starting material
11.	Design for degradation targeting biodegradability and not persistent components in environment	Renewable rather than depleting of material and energy inputs
12.	Real time analysis for pollution prevention by means of in-process monitoring and analytical methodologies	Meet need while minimizing the excess of unnecessary capacities or capabilities for bio-physicochemical systems

Table 2. The twelve principles of Green Chemistry and Engineering [89-91].

Within this context, the present QSAR-Spectra-SAR (QSAR-SSAR) approach "responds" to these OECD-QSAR principles by the present Spectral-SAR ecotoxicological principles' realization [96], with special reference to ionic liquids (Principle 3):

• *ECOTOX-SSAR Principle 1 is assured by:* the "length" of the predicted/measured (eco)biological action follows the self-scalar product rule of the computed endpoint activity

$$\left\| \left| Y \right\rangle^{ENDPOINT(MEASURED/PREDICTED)} \right\| = \sqrt{\sum_{i=1}^{N} \left(y_i^2 \right)^{MEASURED/PREDICTED}} \tag{22}$$

- *ECOTOX-SSAR Principle 2 is assured by:* the "orthogonality" of assumed molecular factors that correlate with eco- and bio-effects is assured by the spectral decomposition of the associate activity respecting them (see eq. (8)), and the orthogonal-real space transformation given by the Spectral-determinant, eq. (11), giving in fact the searched structure-activity relationship model;
- *ECOTOX-SSAR Principle 3 is assured by:* the method of considering the structural parameters and the activities with which they should be correlated, for specific (target) class of molecules; for instance, the consecrated Hansch quantitative structure activity relationships (QSARs) generally prescribes the activity expansion under the generic minimal but meaningful form:

$$A = B_0 + B_1 \begin{pmatrix} electronic \\ parameter \end{pmatrix} + B_2 \begin{pmatrix} hydrophobic \\ parameter \end{pmatrix} + B_3 \begin{pmatrix} steric \\ parameter \end{pmatrix} \tag{23}$$

which provides sufficient information about the transport, electronic affinity and specific interaction at the molecular level, respectively; whereas the hydrophobicity index, *LogP*, describes, at best, the quality of molecular transport through cellular membranes. For the electronic and steric contributions, many structural parameters may be considered [97,98]; among them, the polarizability (*POL*) measures the inductive electronic effect that reflects the long range or van der Waals bonding, whereas for the steric component, the total energy (E_{TOT}) is assumed to be the representative index because it is calculated at the optimum molecular geometry at which the stereo-specificity is included. These parameters have been demonstrated to be quite reliable in modeling the ecotoxicological interactions [26, 29, 30], and they will also be used in the present IL applications. To this aim, when information about the eco-biological influence of ionic liquids (*IL*) is desired, the particular anionic-cationic structure has to be properly considered because almost all structural information about ionic liquids is based on the superposition of the separate anionic and cationic contributions. In this situation, two different additive models for modeling anionic-cationic interaction can be considered. The *first* model is based on the vectorial summation of the produced anionic and cationic biological effects. In other words, this so-called $|1+\rangle$ model is constructed from the superposition of the anionic (subscripted with *A*) and cationic (subscripted with *C*) activities, and can be formally represented as [27, 30]:

$$\left| Y_{AC} \right\rangle^{1+} = \left| 1+ \right\rangle = \left| Y_A \right\rangle + \left| Y_C \right\rangle = \hat{O}_{S-SAR} \left[g\left(\left\{ \left| X_A \right\rangle \right\} \right) + g\left(\left\{ \left| X_C \right\rangle \right\} \right) \right] \tag{24}$$

with Hansch combinations

$$\{|X_{A,C}\rangle\} = \{LogP_{A,C}, POL_{A,C}, E_{TOT(A,C)}\} \tag{25}$$

Practically, with the |1+> model, the SPECTRAL-SAR procedure is separately performed for the anionic and cationic subsystems, and it is subsequently summed in the resulting *IL*-activity. The *second* SPECTRAL-SAR model can be advanced here when the additive stage is considered at the incipient stage of the SPECTRAL-SAR operator (1) such that the considered Hansch factors, for instance, are first combined to produce the anionic-cationic (subscripted with *AC*) indices that are further used to produce the spectral mechanistic map of the concerned interaction, producing the so-called |0+> model [28, 30]:

$$|Y_{AC}\rangle^{0+} = \hat{O}_{S-SAR}|0+\rangle = \hat{O}_{S-SAR}\, f\left(\{|X_A\rangle\},\{|X_C\rangle\}\right) \tag{26}$$

with the Hansch specification of the spectral vectors:

$$f(LogP_A, LogP_C) \equiv LogP_{AC} = \log\left(e^{LogP_A} + e^{LogP_C}\right) \in \{|X_{1AC}\rangle\} \tag{27}$$

$$f(POL_A, POL_C) \equiv POL_{AC} = \left(POL_A^{1/3} + POL_C^{1/3}\right)^3 \in \{|X_{2AC}\rangle\}\left[Å^3\right] \tag{28}$$

$$f(E_A, E_C) \equiv E_{AC} = E_A + E_C - 627.71\frac{q_A q_C}{POL_{AC}^{1/3}} \in \{|X_{3AC}\rangle\}[\mathrm{kcal/mol}] \tag{29}$$

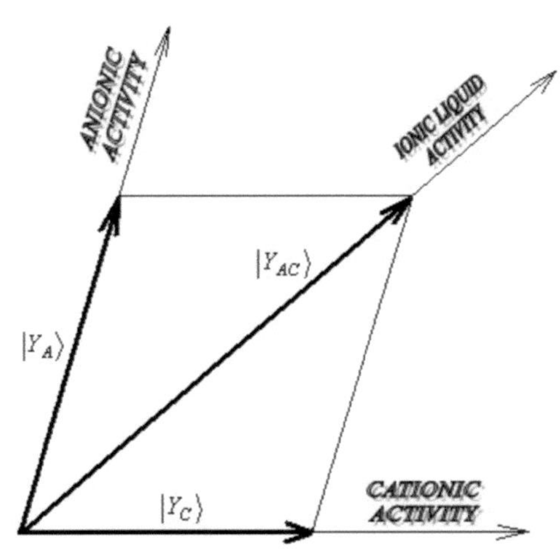

Figure 2. The vectorial composition of anionic and cationic sub-system activities in the ionic liquid global system, leading with the internal angle formation, with the trigonometric expression of eq. (30) [27,30].

The open issue addresses whether the |0+> & |1+> states leave with the same results or what aspects of the SPECTRAL-SAR operator (1) might differ in the *IL* ecotoxicity, a matter that is solved by computing the so-called *ionic liquid internal angle* between the anion-cationic activity vectors, Figure 2, with y_{iA}, y_{iC}, $i=\overline{1,\ N}$ components following the prescription [27,28,30,40]:

$$\cos\theta_{AC} = \frac{\langle Y_C | Y_A \rangle}{\left\| Y_C \rangle \right\| \left\| Y_A \rangle \right\|} = \frac{\sum_{i=1}^{N} y_{iC} y_{iA}}{\sqrt{\sum_{i=1}^{N} y_{iC}^2 \sum_{i=1}^{N} y_{iA}^2}} \begin{cases} \geq 0.707107 \ldots |0+\rangle \ MODEL \\ < 0.707107 \ldots |1+\rangle \ MODEL \end{cases} \tag{30}$$

- *ECOTOX-SSAR Principle 4 is assured by* the "intensity" of the chemical-eco-bio-interaction that is determined by the ratio of the expected to measured activity norms [26,29,69]

$$RA \equiv r_{S-SAR}^{ALGEBRAIC} = \sqrt{\frac{\sum_{i=1}^{N} y_{i-PRED}^2}{\sum_{i=1}^{N} y_{i-OBS}^2}} = \frac{\left\| Y_{PRED} \rangle \right\|}{\left\| Y_{OBS} \rangle \right\|} \leq 1 \tag{31}$$

as a counterpart of the classical statistical correlation factor [69,98,99]

$$R \equiv r_{QSAR}^{STATISTIC} = \sqrt{1 - \frac{\sum_{i=1}^{N} \left(y_{i-OBS} - y_{i-PRED} \right)^2}{\sum_{i=1}^{N} \left(y_{i-OBS} - \frac{1}{N} \sum_{i=1}^{N} y_{i-OBS} \right)^2}} \tag{32}$$

- *ECOTOX-SSAR Principle 5 is assured by* the "selection" of the manifested chemical-eco-(bio-)binding that parallels the minimum distances of paths [26,96,97]

$$\delta[A,B] = 0 \tag{33}$$

connecting all possible endpoints in the norm-correlation hyperspace

$$[A,B] = \sqrt{\left(\left\| Y_B \rangle \right\| - \left\| Y_A \rangle \right\| \right)^2 + \left(r_B^{STATISTIC/ALGEBRAIC} - r_A^{STATISTIC/ALGEBRAIC} \right)^2} \tag{34}$$

In this manner, the "validation" of the obtained mechanistic picture is achieved by requiring that the influential minimum paths are numbered by the cardinal of the input structural fac-

http://dx.doi.org/10.5772/51657

tors set such that, excepting that the final endpoint that is always considered as the final evolution target, all other endpoints are activated one time and one time only.

The present chapter reviews the presented Spectral-SAR-IL models as applied to studying the ecotoxicity of the aquatic species *Vibrio fischeri* and *Daphnia magna* by the tested ionic liquids, which were appropriately chosen such that they contained a wide variety of heads, side chains, and anions.

5. Two Cases of Ionic Liquids Ecotoxicity

Long and short term tests are used to explore the toxicity of the ionic liquids [100-102].

NAME	A_{exp} $\|Y_{EXP}\rangle$	Log P		Polarizability		TOTAL ENERGY	
		CAT. $\|X_{1C}\rangle$	AN. $\|X_{1A}\rangle$	CAT. $\|X_{2C}\rangle$	AN. $\|X_{2A}\rangle$	CAT. $\|X_{3C}\rangle$	AN. $\|X_{3A}\rangle$
1-n-butylpyridinium chloride	0.41*	2.85	0.63	17.51	2.32	-250008.14	-285190.78
1-n-butylpyridinium dicyanoamide	0.31*	2.85	0.43	17.51	5.51	-250008.14	-147935.98
1-n-butyl-3-methylpyridinium dicyanoamide	-0.34*	3.32	0.43	19.35	5.51	-274222.62	-147935.98
1-n-butyl-3,5-dimethylpyridinium dicyanoamide	-0.62*	3.78	0.43	21.18	5.51	-298437.03	-147935.98
1-n-butylpyridinium bromide	0.40*	2.85	0.94	17.51	3.01	-250008.14	-1596918.25
1-n-butyl-3-methylpyridinium bromide	-0.25*	3.32	0.94	19.35	3.01	-274222.62	-1596918.25
1-n-butyl-3,5-dimethylpyridinium bromide	-0.31*	3.78	0.94	21.18	3.01	-298437.03	-1596918.25
1-n-hexyl-3-methylpyridinium bromide	-0.94*	4.11	0.94	23.02	3.01	-322641.81	-1596918.25
1-n-octyl-3-methylpyridinium bromide	-2.21*	4.90	0.94	26.69	3.01	-371060.81	-1596918.25

NAME	A_{exp} $\vert Y_{EXP}\rangle$	Log P		Polarizability		TOTAL ENERGY	
		CAT. $\vert X_{1C}\rangle$	AN. $\vert X_{1A}\rangle$	CAT. $\vert X_{2C}\rangle$	AN. $\vert X_{2A}\rangle$	CAT. $\vert X_{3C}\rangle$	AN. $\vert X_{3A}\rangle$
1-n-butyl-4-dimethylaminopyridinium bromide	-0.68	3.11	0.94	22.53	3.01	-332525.97	-1596918.25
1-n-butyl-3-methylimidazolium dicyanoamide	0.67*	0.68	0.43	17.22	5.51	-260646.64	-147935.98
1-n-butyl-3-methylimidazolium chloride	0.71*	0.68	0.63	17.22	2.32	-260646.64	-285190.78
1-n-butyl-3-methylimidazolium bromide	1.01*	0.68	0.94	17.22	3.01	-260646.64	-1596918.25
1-n-butyl-3-methylimidazolium bis(trifluoromethanesulfonyl)imide	0.39	0.68	3.05	17.22	7.20	-260646.64	-1128283.62
1-n-hexyl-3-methylimidazolium bromide	-1.58*	1.47	0.94	20.89	3.01	-309065.84	-1596918.25
1-n-octyl-3-methylimidazolium bromide	-2.37*	2.26	0.94	24.56	3.01	-357484.59	-1596918.25
tetrabutylammonium bromide	0.27	4.51	0.94	30.91	3.01	-422421.97	-1596918.25
hexyltriethylammonium bromide	-0.54	2.71	0.94	23.57	3.01	-325587.25	-1596918.25
tetrabutylphosphonium bromide	-0.29	2.89	0.94	30.91	3.01	-600149.62	-1596918.25
tributylethylphosphonium diethylphosphate	0.07	2.02	2.63	27.24	10.53	-551729.87	-494172.37
Trihexyl(tetradecyl) phosphoniumbromide	0.41	9.23	0.94	60.27	3.01	-987499.25	-1596918.25
Cholinebis(trifluoro methanesulfonyl)imide	1.15	-0.76	3.05	11.36	7.20	-202450.36	-1128283.62

Table 3. The series of ionic liquids whose toxic activities $A = \text{Log}(EC_{50})$ on *Vibrio fischeri* were considered [49], with the marked values being taken from [45] along with the structural parameters *LogP*, *POL* (Å^3), and E_{TOT} (kcal/mol) that account for the hydrophobicity, electronic (polarizability) and steric (total energy at optimized 3D geometry) effects, computed with the help of the HyperChem program [103], for each cation and anion containing ionic liquid, respectively [27].

The results from using the Ames test for mutagenity with the *Salmonella typhyimurium* species indicated that none of the imidazolium, pyridinium and quaternary ammonium ionic liquids caused mutations. However, the Ames test is a short-term test, and it cannot be possible to fully predict the carcinogenicity in animals [46]. The antimicrobial activity (against strains of Gram-positive and –negative bacteria and fungi) increase with increasing alkyl chain lengths in pyridinium, imidazolium and quaternary ammonium salts [43]. However, because bacteria have a short generation time, they are an ideal starting point for investigating the structure-activity relationship in ionic liquids and serve as a basis for further toxicity tests to higher organisms and more complex systems; a few examples are listed below [45,67]:

- [BMIM][Br] (1-n-butyl-3-methylimidayolium bromide) was observed to be considerably less toxic than [BMPy][Br] (1-n-butyl-3-methylpyridinium bromide);
- Cation toxicity increases with the increase in the hydrophobicity of the molecules, from Bpy (1-n-butyl-3-pyridinium) to BMPy (1-n-butyl-3-methylpyridinium) to BdMPy (1-n-butyl-3-dimethylpyridinium);
- Butyl substituted ionic liquids have considerably lower logP values and are more water soluble;
- In comparison with some commonly used industrial solvents (acetone, methanol, ethyl acetate, and dichloromethane), the octyl and hexyl substituted ionic liquids are more toxic.
- [BMIM][Cl] (1-n-butyl-3-methylimidayolium bis(trifluoromethylsulfonyl)imide chloride) and [BPy][Cl] (1-n-butyl-3-pyridinium chloride) may have similar activity to the pesticide parquet.

Mode	Vectors	Cationic S-SAR			Anionic S-SAR			Ionic Liquid S-SAR		
	(C/A/IL)	$\lVert \lvert Y_C\rangle^{Mode}\rVert$	$r_{S-SAR}^{STATISTIC}$	$r_{S-SAR}^{ALGEBRAIC}$	$\lVert \lvert Y_A\rangle^{Mode}\rVert$	$r_{S-SAR}^{STATISTIC}$	$r_{S-SAR}^{ALGEBRAIC}$	$\lVert \lvert Y_{AC}\rangle^{Mode}\rVert$	$r_{S-SAR}^{STATISTIC}$	$r_{S-SAR}^{ALGEBRAIC}$
Ia	$\lvert X_0\rangle, \lvert X_1\rangle$	1.47807	0.267342	0.334755	1.38453	0.238974	0.313569	2.58965	0.185959	0.586507
Ib	$\lvert X_0\rangle, \lvert X_2\rangle$	1.08531	0.132169	0.245803	1.48745	0.270118	0.33688	2.30825	0.179482	0.522776
Ic	$\lvert X_0\rangle, \lvert X_3\rangle$	0.9452	0.0469985	0.21407	1.75453	0.345553	0.397368	2.41575	0.259502	0.547123
IIa	$\lvert X_0\rangle, \lvert X_1\rangle, \lvert X_2\rangle$	1.64279	0.314715	0.372062	1.52849	0.282139	0.346174	2.88368	0.219986	0.653101
IIb	$\lvert X_0\rangle, \lvert X_1\rangle, \lvert X_3\rangle$	1.72651	0.3379	0.391023	2.15865	0.451919	0.488893	3.48056	0.352356	0.788283
IIc	$\lvert X_0\rangle, \lvert X_2\rangle, \lvert X_3\rangle$	1.71867	0.335748	0.389246	1.82903	0.365689	0.414242	3.17318	0.299925	0.718667
III	$\lvert X_0\rangle, \lvert X_1\rangle, \lvert X_2\rangle, \lvert X_3\rangle$	1.79053	0.355322	0.405522	2.36461	0.504184	0.53554	3.7151	0.397148	0.841402

Table 4. Spectral structure activity relationships (S-SAR) of the ionic liquids in Table 3 against their toxicity to *Vibrio fischeri*, and the associated computed spectral norms with $\lVert \lvert Y_{EXP}\rangle \rVert$ =4.41537, statistical and algebraic correlation factors, computed upon the relations (22), (31), and (32), throughout the possible correlation models considered from the *anionic, cationic and composed ionic liquids data* from Table 3, respectively [27]

Mode	*Ia*	*Ib*	*Ic*	*IIa*	*IIb*	*IIc*	*III*
$cos\theta_{AC}$	0.66397	0.600124	0.562018	0.653248	0.60019	0.599635	0.591015

Table 5. The variation of the cosines of the anion-cationic correlation angle in vectorial space based on Eq. (30) for all considered modes of action of the ionic liquids in Table 3 with the cationic and anionic subsystems S-SAR predicted activities, respectively [27].

Path	**Value**					
	Cationic		**Anionic**		**Ionic Liquid**	
	statistic	***algebraic***	***statistic***	***algebraic***	***statistic***	***algebraic***
Ia-IIa-III	0.324618	**0.320376 α**	1.0154	1.0049	1.14608	**1.15396 α**
Ia-IIb-III	0.324616	0.320376	**1.01536 γ**	**1.0049 γ**	**1.1451 α**	1.15396
Ia-IIc-III	**0.324616 α**	0.320376	1.01541	1.0049	1.14513	1.15396
Ib-IIa-III	0.739827	0.723082	**0.907864 β**	**0.899373 β**	**1.42694 γ**	1.44248
Ib-IIb-III	**0.739746 β**	0.723082	0.907871	0.899373	1.42377	**1.44248 γ**
Ib-IIc-III	0.739754	**0.723082 β**	0.907893	0.899373	1.42385	1.44248
Ic-IIa-III	**0.900418 γ**	0.86674	1.09986	1.08906	1.31968	1.33226
Ic-IIb-III	0.900057	**0.86674 γ**	0.630373	0.625533	1.30763	1.33226
Ic-IIc-III	0.90009	0.86674	**0.630371 α**	**0.625533 α**	**1.30908 β**	**1.33226 β**

Table 6. Synopsis of the statistic and algebraic values of paths connecting the S-SAR models of Table 4 in the norm-correlation spectral-space of eq. (34) for the ionic liquids in Table 3 against the toxicity to *Vibrio fischeri*. The primary, secondary and tertiary - the so called alpha (α), beta (β) and gamma (γ) paths - are indicated according to the least path principle in spectral norm-correlation space with the statistical and algebraic variants of the correlation factors used, respectively [27].

The Microtox Acute Toxicity Test is often used to determine the toxicity of single compounds in sediment contamination studies and for monitoring industrial effluents in environmental water quality surveys [49,100]. Before releasing the ionic liquids into the environment, the antimicrobial properties using *Vibrio fischeri* in the Microtox method were thoroughly examined [45]. Equally, by other tests with *Vibrio fisheri*, an increase in the toxicity corresponding to an increase in the chain length and an increase in the number of alkyl groups substituted on the cation ring was also demonstrated [45,101].

Salts used for anion substitution, such as sodium bromide and sodium dicyanoamide, were less toxic to *Vibrio fisheri* than the compounds used for synthesizing the cation. For instance, 3-methyl pyridine and 1-methylimidazole may be used as starting compounds. The addition of a butyl chain to the C1 (carbon atom) of the pyridinium and imidazolium cation slightly increased the toxicity of the ionic liquid. Furthermore, the addition of a hexyl or octyl chain will also increase the toxicity. The same observation holds for quaternary ammonium salts. All of these observations suggest that the effects are related to the lipophilicity of the cation, thus explaining why many of the imidazolium and pyridinium ionic liquids were more toxic than acetonitrile, acetone and methanol. In contrast, monatomic anions (such as bromide, chloride) are predicted to be less toxic than large anions that contain regions of positive charge, e.g., TFMSi (bis(trifluoromethylsulfonyl)imide) [49].

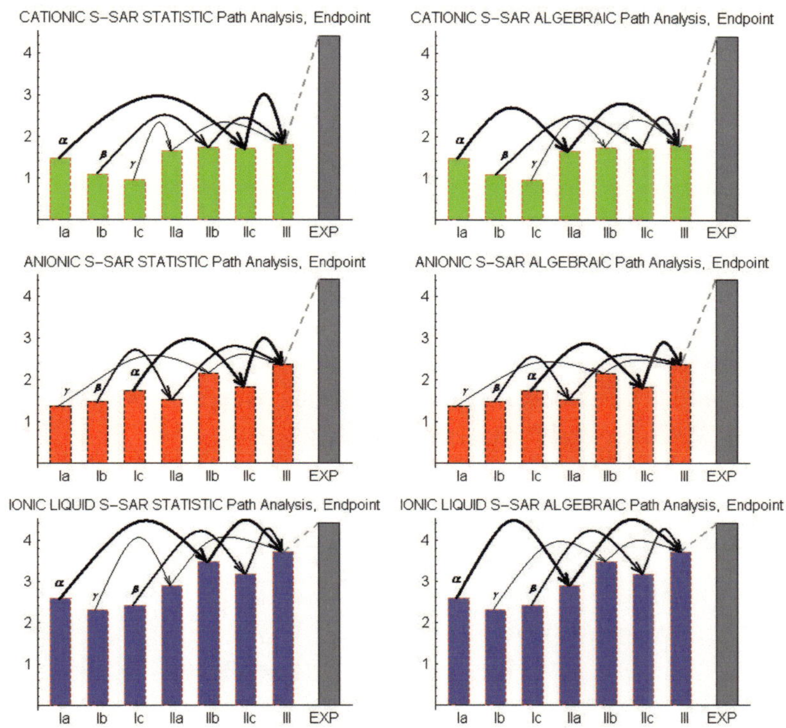

Figure 3. The spectral representation of the chemical-biological interaction paths across the S-SAR to the modeled endpoints of the *Vibrio fischeri*, according to the least (shortest) path rule within the spectral norm-correlation space applied to the data in Table 6 for the cationic, anionic and resulting ionic liquid norms of Table 4 for the statistical and algebraic versions of correlation factors from up to down and left to right, respectively. The primary-alpha, secondary-beta and tertiary-gamma path hierarchies of Table 6 are indicated by decreasing thicknesses of the connecting lines [27].

At the organism level, which are *Vibrio fisheri* and *Daphnia magna* in this review, the toxicity generally increases with the cation type as ammonium<pyridinium<imidazolium<triazolium<tetrazolium. The choline ionic liquids (with a negatively charged oxygen atom) and the quaternary ammonium ionic liquids with short chain lengths were relatively non-toxic to *Vibrio fisheri*. Furthermore, the methylation of the aromatic ring of the cation reduces the toxicity to *Vibrio fisheri* and *Daphnia magna* [49]. The toxicity increases with the number of nitrogen atoms, whereas the anions play a secondary role in the toxicity (even if the presence of positively charged atoms on the anion is predicted to slightly increase the toxicity) [49].

Note that *Daphnia* is an important link between the microbial and higher trophic levels [49,102]. By acting on *Daphnia magna*, ionic liquids with longer alkyl chain substituents have toxicities comparable to phenol, whereas those with shorter substituents (for example, [BMIM][Br]) were more toxic to *Daphnia magna* than benzene and methanol [47,49].

The lethal concentration to *D. magna* (also called the acute toxicity) was observed to be considerably lower for ILs that employed imidazolium as cation than salts with Na^+; therefore, the toxicity was related to the imidazolium cation and not to the anions. With respect to the life history of the freshwater crustacean, salts with a sodium cation ($NaPF_6$ sodium hexafluorophosphate and $NaBF_4$, sodium tetrafluoroborate) affect the reproduction of *D. magna* at high concentrations. Sub-lethal effects on the life-history traits occurred when the concentrations of the ionic liquids were an order of magnitude lower than those for acute effects. The clonally variation was a possible reason why the average brood size in controls with those for ionic liquids. This was the first study on aquatic eukaryotes [47]. The EC50 values on the toxicity to a single species are necessary for further studies at the community and ecosystems levels [47]. Overall, qualitatively, the toxic effects were not observed to be notably different for the imidazolium and pyridinium ionic liquids, although the toxicity was greater with an increase in the length of the side chain. The K_{ow} values for the IL were lower than those of chemicals that bioaccumulate in the tissue of organisms. Quantitative analysis by the presented QSAR/Spectral-SAR algorithm follows.

Ionic Liquid Compound		A_{exp}	**LogP**			**POL [Å³]**			E_{TOT} **[kcal/mol]**		
Structure	*Name*	$\|Y_{exp}>$	$\|X_{1C}>$	$\|X_{1A}>$	$\|X_{1AC}>$	$\|X_{2C}>$	$\|X_{2A}>$	$\|X_{2AC}>$	$\|X_{3C}>$	$\|X_{3A}>$	$\|X_{3AC}>$
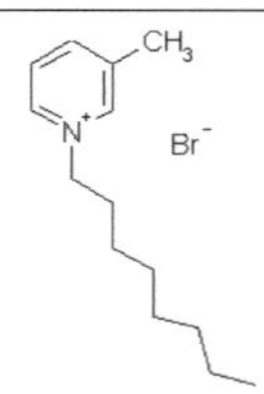	1-n-octyl-3-methylpyridinium bromide	-2.60	4.90	0.94	4.92	26.69	3.01	87.08	-371060.81	-1596918.25	-1967840
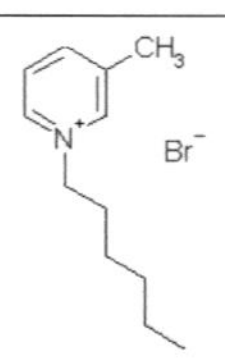	1-n-hexyl-3-methylpyridinium bromide	-2.41	4.11	0.94	4.15	23.02	3.01	78.87	-322641.81	-1596918.25	-1919410
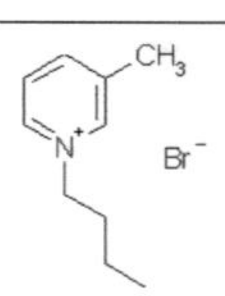	1-n-butyl-3-methylpyridinium bromide	-1.24	3.32	0.94	3.41	19.35	3.01	70.37	-274222.62	-1596918.25	-1870990
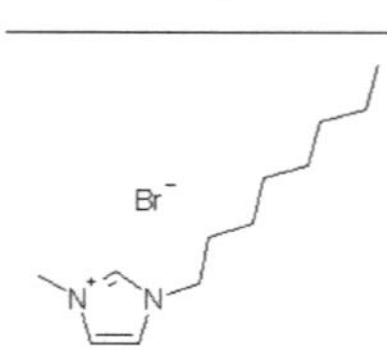	1-n-octyl-3-methylimidazolium bromide	-4.33	2.26	0.94	2.5	24.56	3.01	82.35	-357484.59	-1596918.25	-1954260

Ionic Liquid Compound		A_{exp}	LogP			POL [Å³]			E_{TOT} [kcal/mol]		
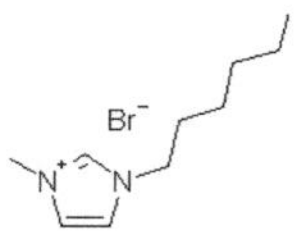	1-n-hexyl-3-methylimidazolium bromide	-2.22	1.47	0.94	1.93	20.89	3.01	73.98	-309065.84	-1596918.25	-1905830
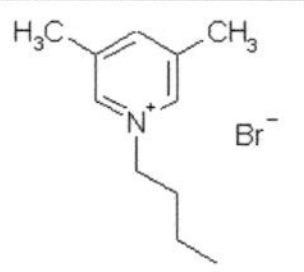	1-n-butyl-3,5-dimethylpyridinium bromide	-1.01	3.78	0.94	3.84	21.18	3.01	74.65	-298437.03	-1596918.25	-1895210
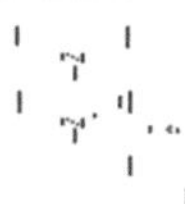	1-n-hexyl-4-piperidino pyridinium bromide	-3.66	4.63	0.94	4.65	30.93	3.01	96.25	-452857.03	-1596918.25	-2049640
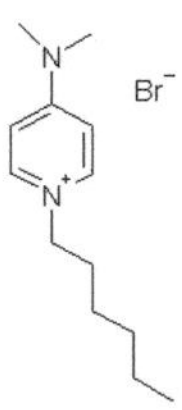	1-n-hexyl-4-dimethylamino pyridinium bromide	-3.28	3.91	0.94	3.96	26.2	3.01	86.00	-380945.12	-1596918.25	-1977720
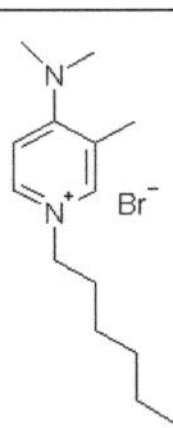	1-n-hexyl-3-methyl-4-dimethylamino pyridinium bromide	-2.79	4.37	0.94	4.40	28.04	3.01	90.03	-405145.97	-1596918.25	-2001920
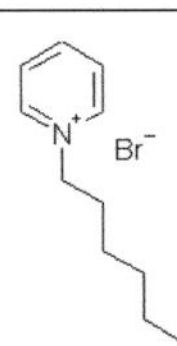	1-n-hexylpyridinium bromide	-1.93	3.64	0.94	3.71	21.18	3.01	74.65	-298427.37	-1596918.25	-1895200
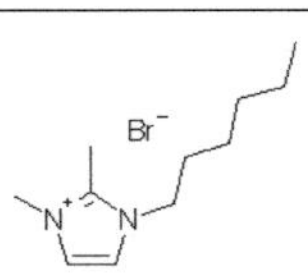	1-n-hexyl-2,3-dimethylimidazolium bromide	-2.19	1.67	0.94	2.06	22.72	3.01	78.19	-333284.94	-1596918.25	-1930060
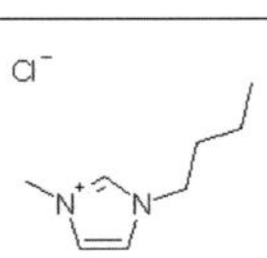	1-n-butyl-3-methylimidazolium * chloride	-1.07	0.68	0.63	1.34	17.22	2.32	59.60	-260646.64	-285190.78	-545677

Ionic Liquid Compound		A_{exp}	LogP			POL [Å³]			E_{TOT} [kcal/mol]		
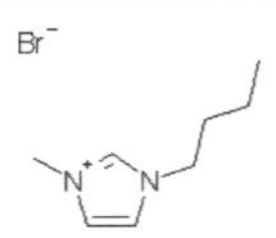	1-n-butyl-3-methylimidazolium bromide	-1.43 *	0.68	0.94	1.51	17.22	3.01	65.26	-260646.64	-1596918.25	-1857410
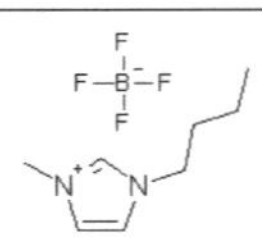	1-n-butyl-3-methylimidazolium tetrafluoroborate	-1.32 *	0.68	1.37	1.78	17.22	2.46	60.80	-260646.64	-261310.59	-521798
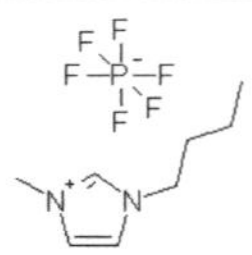	1-n-butyl-3-methylimidazolium hexafluorophosphate	-1.15 *	0.68	2.06	2.28	17.22	1.78	54.62	-260646.64	-580264.94	-840746
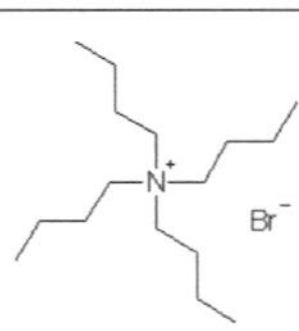	Tetrabutyl ammonium bromide	-1.53	4.51	0.94	4.54	30.91	3.01	96.21	-422421.97	-1596918.25	-2019200
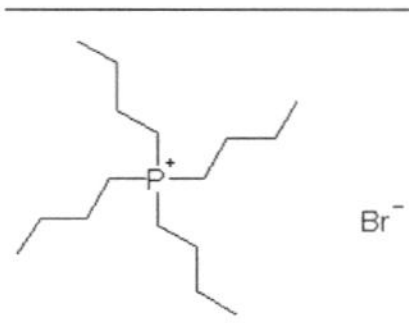	Tetrabutyl phosphonium bromide	-2.05	2.89	0.94	3.02	30.91	3.01	96.21	-600149.625	-1596918.25	-2196930

Table 7. The actions of the studied ionic liquids on the *Daphnia magna* species with the toxic activities Aexp= Log(EC50) [49], while the marked values were taken from [47] along with the structural parameters LogP, POL, and ETOT to account for the hydrophobicity, electronic (polarizability) and steric (total energy at optimized 3D geometry) effects, computed with the HyperChem program [103], for each cation and anion fragment, and for the anionic-cationic |0+> composed state by means of equations (27)-(29), respectively [28].

The application of S-SAR-IL to *Vibrio Fischeri* conforms with the |1+> model presented above for the data presented in Table 3. Accordingly, the Spectral-SAR models for anions and cations are reported in Table 4, and the internal angle for all of the possible models are listed in Table 5, which confirms that model |1+>, with the results at the global level of ionic liquids being shown on the last column of Table 4.

Picturing a mechanistically mode of action for ionic liquids containing cations and anions and of their summed effects on the considered *Vibrio fischeri* species remains as the final purpose of the QSAR method, OECD-QSAR normative, and the Spectral-SAR algorithm, as extensively presented. In this regard, the spectral paths analysis is presented in Table 6, which is based on eqs. (33) and (34) with the cationic, anionic and ionic liquids data of Table 4. The minimum path procedure assumes the identification of the minimum paths (equal in numbers with the number of structural parameters considered, thus *alpha, beta,* and *gamma* for the present case), in an "ergodic manner"; this means that the overall path is firstly identified; if two or more paths appear with equal lengths, the one to be chosen is the one that contains the "minimum first movement", which is that path that links the closest first two norms that belong to successive norms; therefore, the second overall minimum path is identified such that it does not contain the models "touched" by the *alpha* path, thereby excluding the common final endpoint (model III here). In this manner, the remaining *beta* and *gamma* paths are also selected. Such analysis is performed for both statistical and algebraic norms; see Table 6. In this manner, interesting results are obtained, allowing conceptual interpretation, such as the following [27]:

- While in cationic case, the statistical and algebraic paths do not coincide (e.g., what the alpha *Ia-IIc-III* path in statistic differs than the alpha *Ia-IIa-III* path in algebraic views), in the anionic case, they are identically predicted (except the fact that in the algebraic case, the shortened paths are registered), which together provide a mixed behavior for the resulting ionic liquid (only the beta path *Ic-IIc-III* is overlapping between statistical and algebraic views);
- While in the cationic and anionic subsystems, the path hierarchies are reversed as $\alpha\rightarrow\beta\rightarrow\gamma$ and $\gamma\rightarrow\beta\rightarrow\alpha$. In the resulting ionic liquid, the mixture effect is again observed because the succession $\alpha\rightarrow\gamma\rightarrow\beta$ against the successions of starting endpoints $Ia\rightarrow Ib\rightarrow Ic$, respectively;
- The cationic alpha path is started on the lipophylicity causes (*Ia*), which is the same as for the containing ionic liquid. A different situation arises for the alpha anionic path that begins with the steric influence (*Ic*); in this way, the previously noted dominance of the cationic influence when correlated with lipophilicity and the observed anionic influence related with steric effects are theoretically confirmed in this picture;
- Figure 3 clearly illustrates the fact that while anionic and cationic activity tendencies are somewhat complementary, they do not cancel each other in the ionic liquid that contains them but add up to attain the overall observed toxicity, in the spectral norm – correlation factor space. Furthermore, other useful data can be extracted from Figure 3 concerning the major path of structural causes in manifested toxicological action, as revealed below;
- While the algebraic paths are systematically lower that the corresponding statistical ones for cationic and anionic subsystems, in the ionic liquid case, the situation is reversed; the interpretation is that it also confirms that the chemical-biological ionic liquid dispersive (not specific) actions in environment are merely through its subsystem components than from itself as a whole;
- The ecotoxicological paths in the cationic and anionic subsystems are summed up in the paths of the corresponding ionic liquids in a nontrivial manner: the anionic gamma path effect is marginal over the cationic alpha path

$$\alpha_C + \gamma_A = \alpha_{AC} \tag{35}$$

- the cationic and anionic beta paths decay into the gamma ionic liquid path when joined such that recording a sort of reciprocal cancellation of their effects

$$\beta_C + \beta_A = \gamma_{AC} \tag{36}$$

- the anionic alpha path effect is reinforced over the cationic gamma path averaging both at the beta path level of the resulting ionic liquid

$$\gamma_C + \alpha_A = \beta_{AC} \tag{37}$$

Mode	Vectors' Predicted	$\| \|Y\rangle^{Mode} \|$	$r_{S-SAR}^{STATISTIC}$	$r_{S-SAR}^{ALGEBRAIC}$
Ia	$\|Y_{A-Ia}\rangle = f(\|X_0\rangle, \|X_{1A}\rangle)$	8.83127	0.266552	0.920421
	$\|Y_{C-Ia}\rangle = f(\|X_0\rangle, \|X_{1C}\rangle)$	8.92169	0.420761	0.929845
	$\|Y_{AC-Ia}\rangle^{0+} = f(\|X_0\rangle, \|X_{1AC}\rangle)$	8.89048	0.374616	0.926593
	$\|Y_{AC-Ia}\rangle^{1+} = \|Y_{A-Ia}\rangle + \|Y_{C-Ia}\rangle$	17.6883	2.21964 ***i***	1.84353
Ib	$\|Y_{A-Ib}\rangle = f(\|X_0\rangle, \|X_{2A}\rangle)$	8.94784	0.455964	0.932572
	$\|Y_{C-Ib}\rangle = f(\|X_0\rangle, \|X_{2C}\rangle)$	9.06691	0.59121	0.944981
	$\|Y_{AC-Ib}\rangle^{0+} = f(\|X_0\rangle, \|X_{2AC}\rangle)$	*9.08979*	*0.613973*	*0.947366*
	$\|Y_{AC-Ib}\rangle^{1+} = \|Y_{A-Ib}\rangle + \|Y_{C-Ib}\rangle$	17.9079	2.19638 ***i***	1.86641
Ic	$\|Y_{A-Ic}\rangle = f(\|X_0\rangle, \|X_{3A}\rangle)$	8.96309	0.475327	0.934161
	$\|Y_{C-Ic}\rangle = f(\|X_0\rangle, \|X_{3C}\rangle)$	8.95817	0.469161	0.933648
	$\|Y_{AC-Ic}\rangle^{0+} = f(\|X_0\rangle, \|X_{3AC}\rangle)$	*8.99267*	*0.510889*	*0.937244*
	$\|Y_{AC-Ic}\rangle^{1+} = \|Y_{A-Ic}\rangle + \|Y_{C-Ic}\rangle$	17.8233	2.20167 ***i***	1.8576
IIa	$\|Y_{A-IIa}\rangle = f(\|X_0\rangle, \|X_{1A}\rangle, \|X_{2A}\rangle)$	8.96021	0.47173	0.933861
	$\|Y_{C-IIa}\rangle = f(\|X_0\rangle, \|X_{1C}\rangle, \|X_{2C}\rangle)$	9.06885	0.59317	0.945183
	$\|Y_{AC-IIa}\rangle^{0+} = f(\|X_0\rangle, \|X_{1AC}\rangle, \|X_{2AC}\rangle)$	*9.1014*	*0.62522*	*0.948575*
	$\|Y_{AC-IIa}\rangle^{1+} = \|Y_{A-IIa}\rangle + \|Y_{C-IIa}\rangle$	17.9161	2.1931 ***i***	1.86727
IIb	$\|Y_{A-IIb}\rangle = f(\|X_0\rangle, \|X_{1A}\rangle, \|X_{3A}\rangle)$	8.96426	0.476781	0.934283
	$\|Y_{C-IIb}\rangle = f(\|X_0\rangle, \|X_{1C}\rangle, \|X_{3C}\rangle)$	8.99112	0.50908	0.937082
	$\|Y_{AC-IIb}\rangle^{0+} = f(\|X_0\rangle, \|X_{1AC}\rangle, \|X_{3AC}\rangle)$	*8.99774*	*0.51675*	*0.937772*
	$\|Y_{AC-IIb}\rangle^{1+} = \|Y_{A-IIb}\rangle + \|Y_{C-IIb}\rangle$	17.8793	2.21324 ***i***	1.86343
IIc	$\|Y_{A-IIc}\rangle = f(\|X_0\rangle, \|X_{2A}\rangle, \|X_{3A}\rangle)$	8.96808	0.481497	0.93468
	$\|Y_{C-IIc}\rangle = f(\|X_0\rangle, \|X_{2C}\rangle, \|X_{3C}\rangle)$	9.09155	0.615686	0.947549
	$\|Y_{AC-IIc}\rangle^{0+} = f(\|X_0\rangle, \|X_{2AC}\rangle, \|X_{3AC}\rangle)$	9.09116	0.615307	0.947508
	$\|Y_{AC-IIc}\rangle^{1+} = \|Y_{A-IIc}\rangle + \|Y_{C-IIc}\rangle$	17.9586	2.19937 ***i***	1.8717

Mode	Vectors' Predicted	$\Vert \lvert Y \rangle^{Mode} \Vert$	$r_{S-SAR}^{STATISTIC}$	$r_{S-SAR}^{ALGEBRAIC}$
III	$\lvert Y_{A\text{-}III}\rangle = f(\lvert X_0\rangle, \lvert X_{1A}\rangle, \lvert X_{2A}\rangle, \lvert X_{3A}\rangle)$	8.96926	0.482946	0.934803
	$\lvert Y_{C\text{-}III}\rangle = f(\lvert X_0\rangle, \lvert X_{1C}\rangle, \lvert X_{2C}\rangle, \lvert X_{3C}\rangle)$	9.12145	0.644232	0.950666
	$\lvert Y_{AC\text{-}III}\rangle^{0+} = f(\lvert X_0\rangle, \lvert X_{1AC}\rangle, \lvert X_{2AC}\rangle, \lvert X_{3AC}\rangle)$	9.10319	0.62694	0.948762
	$\lvert Y_{AC\text{-}III}\rangle^{1+} = \lvert Y_{A\text{-}III}\rangle + \lvert Y_{C\text{-}III}\rangle$	17.9531	2.17933 ***i***	1.87113

Table 8. Spectral structure activity relationships (SPECTRAL-SAR) of the ionic liquids toxicity of Table 7 against the *Daphnia magna* species, and the associated computed spectral norms, with ‖ |YEXP> ‖ =9.59481, statistic and algebraic correlation factors [26,30], throughout the possible correlation models considered from the anionic, cationic, and ionic liquid |1+> and |0+> states, respectively [28].

Mode	***Ia***	***Ib***	***Ic***	***IIa***	***IIb***	***IIc***	***III***
$\boldsymbol{\cos\theta_{AC}}$	0.985468	0.976338	0.978196	0.975018	0.983081	0.97768	0.969683

Table 9. The values of the cosines of the anion-cationic vectorial angles [27,30] for all considered modes of action of Table 8 indicating the |0+> states of the considered ionic liquids [28].

Path	**Cationic**		**Anionic**		**Ionic Liquid**			
					state |0+>		**state |1+>**	
	statistic	*algebraic*	*statistic*	*algebraic*	*statistic*	*algebraic*	*statistic*	*Algebraic*
Ia-IIa-III	0.299988	0.200851	**0.256742γ**	0.13874	0.330033	**0.213862 γ**	0.260535	0.266181
Ia-IIb-III	**0.300103γ**	0.200851	0.2567	0.13874	**0.330581 γ**	0.213862	**0.25639γ**	**0.266181γ**
Ia-IIc-III	0.299895	**0.200851 γ**	0.25666	**0.13874 γ**	0.33011	0.213862	0.269477+ R* ***i***	0.277223
Ib-IIa-III	**0.07607α**	**0.0548409 α**	0.034447	**0.0215298 β**	0.0186427	0.0134673	**0.0418672 α**	**0.0454552α**
Ib-IIb-III	0.299514	0.207241	0.0344468	0.0215298	0.286398	0.198562	0.0886683	0.102966
Ib-IIc-III	0.0760732	0.0548409	**0.0344464β**	0.0215298	**0.0186427 α**	**0.0134673 α**	0.0506137+ R* ***i***	0.0564973
Ic-IIa-III	0.23953	0.16417	0.0190146	0.0119873	**0.160257 β**	0.111113	0.126723	0.130484
Ic-IIb-III	0.23952	**0.16417 β**	**0.00980164 α**	**0.00619966α**	0.160264	**0.111113 β**	**0.120323 β**	**0.130484β**
Ic-IIc-III	**0.239484 β**	0.16417	0.00980164	0.00619966	0.16027	0.111113	0.135252+ R* ***i***	0.141526

Table 10. Synopsis of the statistical and algebraic values of the paths connecting the SPECTRAL-SAR models of Table 8, in the norm-correlation spectral-space, for *Daphnia magna* species against the ionic liquids toxicity of Table 7; the primary, secondary and tertiary - the so called alpha (α), beta (β) and gamma (γ) paths, are indicated according to the "selection" and "validation" principles in norm-correlation spectral space when the statistic and algebraic variants of the correlation factors are respectively used [28].

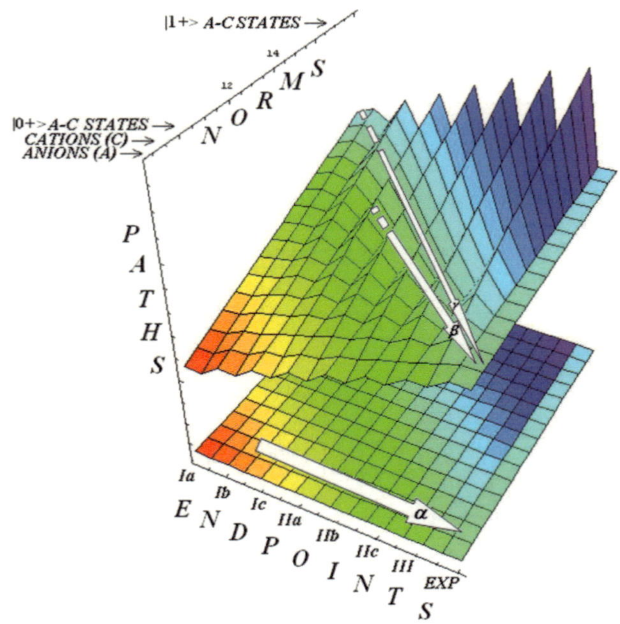

Figure 4. The spectral hypersurface of the structural hierarchical paths toward the recorded (*EXP*) ecotoxicological activity (in the extreme right hypersurface region) of the ionic liquids of Table 7 on *Daphnia magna* species: the alpha path (*α*) initiates on the polarizability (*Ib*) anionic-cationic interaction (in the left-bottom hypersurface region), being followed by the beta path (*β*), which originates on the steric (*Ic*) anionic-cationic interaction (in the left-top hypersurface region hypersurface region), and successively by the gamma path (*γ*) based on the hydrophobic (*Ia*) anionic-cationic interaction (in the extreme left-top hypersurface region) of the norm-correlation spectral space of Table 8 with the decaying order of the thickness of the connecting arrows, respectively [28].

When turning to the study of the action of ionic liquids on the *Daphnia* species, the pool of molecules in Table 7 are employed for assessing the SAR-Ionic Liquid Ecotoxicological |0+> Model for IL-*Daphnia* chemical-biological interaction. This fact is confirmed by employing eqs. (27)-(29) for the cationic and anionic data of Table 7 and later computing the S-SAR determinants (11) associated with all models and combinations presented in Table 8, which leads to the internal angle computations and the results in Table 9 and assures the application of the model |0+> is in accordance with the prescription given by eq. (30). Within the Spectral-SAR algorithm and allied eco-toxicological principles, the results allow the specific conclusions [28]:

- From the toxicological actions of Table 8, it can be observed that both anionic and cationic fragments have important contributions to the "length" and "intensity" of the ionic liquids ecotoxicity through the computed spectral norms and algebraic correlation factors, respectively, which are close to the experimental one, i.e., to 9.59481;
- In all cases, the mode of action where all three Hansch factors were considered (mode *III* with $LogP{+}POL{+}E_{TOT}$) records the best norm and correlations that are the closest description of the ionic liquids-*Daphnia magna* chemical-biological interaction;

http://dx.doi.org/10.5772/51657

- The cationic influence is observed with the dominant contribution over the anionic effects in ecotoxicity, in all considered Hansch modes of action;
- The statistical correlation factors always yield smaller values than the corresponding algebraically ones, see Table 8;
- There are recorded imaginary statistical correlations of the computed endpoints $|Y_{AC\text{-}Mode}>^{1+}$ that indicate certain limitations of its use for activity modeling in ecotoxicology; for these cases, the algebraically outputs provide almost the sum of the anionic and cationic length and intensity endpoint activity. This result can be phenomenologically explained by the so-called "resonance effect" when the angles between the anionic and cationic endpoint vectors are almost zero, as clearly evidenced by the cosine values of Table 9;
- Within the $|0+>$ model, all of the lengths and intensities of the endpoints $|Y_{AC\text{-}Mode}>^{0+}$ behave as an average of the anionic and cationic ecotoxicological effects with a smooth increase over the individual cationic effects for the modes *Ib* (*POL*), *Ic* (E_{TOT}), *IIa* (*LogP* +*POL*), and *IIb* (*LogP*+E_{TOT}); however, further selection for the binding mechanism is performed by identifying the minimum analysis of the Spectral-paths, Table 10.

The Spectral path analysis is unfolded in the same manner as previously used for the $|1+>$ models, i.e., by "ergodic" selection of the models per paths, with the ecotoxicological results in Table 10 and correspondingly interpreted as follows [28]:

- the additive parametric and endpoint models, $|0+>$ and $|1+>$, provide the same hierarchies of the paths for the chemical-biological actions;
- the statistical imaginary correlation values for the ionic liquids $|1+>$ are avoided from the mechanistic principle and do not belong to any selected path in Table 10;
- the dominant cationic effects can also be noted here at the least paths level because the nature of the cationic mechanism is preserved to the ionic liquids nature according with the spectral path equations:

$$\alpha_C + \beta_A = \alpha_{IL} \tag{38}$$

$$\beta_C + \alpha_A = \beta_{IL} \tag{39}$$

$$\gamma_C + \gamma_A = \gamma_{IL} \tag{40}$$

- the results of all SPECTRAL-SAR ecotoxicological principles applied to ionic liquids-*Daphnia magna* case of chemical-eco-biological interaction can be unitarily presented in the Figure 4, where the spectral hypersurface was generated by the 3D interpolation of all lengths (norms) for all the endpoint modes of Table 8, for all cationic, anionic, $|0+>$ and $|1+>$ states of ionic liquids of Table 7. The alpha dominant paths are easily identified according to Table 10, as originating in the *Ib*, i.e., on *POL*arizability or van der Waals molecular mode of action, while the beta and gamma ones starts with the steric (*Ic*: E_{TOT}) and hydrophobic (*Ia*: *Log*P) specific chemical-biological binding, respectively.

Overall, one may assess the sets of eqs. (35)-(37) and (38)-(40) as specific for ionic liquids action over biological marine species within the additive and parametric models |1+> and |0+>, respectively, that should be further confirmed or extended by future studies with other structural parameters (beyond Hansch descriptors, i.e., by topological and quantum molecular factors) and/or with other species on similar congeneric ILs and working algebraic chemical-biological interaction models.

6. Conclusions

Since their emergence a decade ago, ionic liquids (ILs) have had a constantly increasing influence on organic, bio- and green chemistry, due to their unique physico-chemical properties manifested by their typical salt structure: a heterocyclic nitrogen-containing organic cation (in general) and an inorganic or organic anion [43] with melting points below 100 ºC and no vapor pressure [47]. The latter property leads to the practical replacement of conventional volatile organic compounds (VOCs) from the point of view of atmospheric emissions, though they do present the serious drawback that a small amount of IL could enter the environment through groundwater [104]. This risk makes it necessary to perform further eco-toxicological studies of IL on various species to improve the "design rules" for synthesized ILs with minimal toxicity to the environmental integrated organisms. Ionic liquids display variable stability in terms of moisture and solubility in water, polar and nonpolar organic solvents [45]. Various values of ionic liquid hydrophobicity and polarity may be tailored [104] with the help of nucleoside chemistry [105] according to the main principles of green chemistry [54,66]: the new chemicals must be designed to preserve effectiveness of function while reducing toxicity and not persisting in the environment at the end of their usage but rather breaking down into inoffensive degradation products.

In this respect, the costs of all approaches for sustainable product design can be reduced using the SAR and QSAR methods [26-30]. While the 1-octanol-water partition coefficient could be observed only as the first approximation for compound lipophilicity, bioaccumulation and toxicity in fish, and sorption to soil and sediments, it assumes that lipophilicity is the main factor of anti-microbial activity [44,56]. Nevertheless, aiming at a deeper understanding of the specific mechanistic description of IL eco-toxicity, it is worth considering that the ionic liquid properties are more comprehensively quantified through lipophilicity, polarizability and total energy as a unitarily complex of factors in developing appropriate structure-activity relationship (SAR) studies. However, the main problem in assessing the viable QSAR studies for predicting ionic liquid toxicities concerns the *anionic-cationic interaction* superimposed on the anionic and cationic subsystems containing ionic liquids. There are two main complementary ways of attaining this goal [27,28,30]. One may address the search of special rules for assessing the anionic-cationic structural separately from the individual anionic and cationic ones and later generating the QSAR models – the so-called model |1+>. Otherwise, when the so-called internal angle of ionic liquids sub-systems (cation and anion), eq. (30), is high enough that a sort of "resonance effects" would appear between them, that outcome should be avoided by considering structural parameter composition (superposition, as orthogonal states in quantum mechanics) such that it constitutes the way of parameter depending the predicted endpoints, that is, the so-called causal model |0+>. Together, these properties may be unitarily expressed by the operatorial equation

$$\hat{O}_{S-SAR}\left|0+\right\rangle \begin{cases} =\left|1+\right\rangle, \text{ fixed endpoints} \\ =\left|0+\right\rangle, \text{ path endpoints} \end{cases} \quad (41)$$

with the Spectral-SAR operator, consistently defined by the successive rules:

$$\hat{O}_{S-SAR}: \left\{ \begin{array}{l} Det\left(\left|Y\right\rangle,\left|X_0\right\rangle,\left|X_1\right\rangle,\ldots,\left|X_M\right\rangle\right)=0, \left\|\left|Y\right\rangle\right\|, r_{S-SAR}^{ALGEBRAIC}, \\ \qquad\qquad\qquad\qquad\qquad\qquad A,B: ENDPOINTS \\ \delta\left[A_{(\|\bullet\rangle\|,r)}, B_{(\|\bullet\rangle\|,r)}\right]=0, \langle\alpha,\beta,\gamma,\ldots\rangle, \quad \alpha,\beta,\gamma,\ldots: SPECTRAL\ \ PATHS \end{array} \right\} \quad (42)$$

that widely fulfill the principles of green chemistry specialized to the OECD-QSAR principles and here manifested as Spectral-SAR ecotoxicological principles. These equations are able to offer the complete picture of molecular specific interaction between a series of chemicals with certain species by generating the molecular mechanisms of actions; also, they are opening room for preventing and controlling the envisaged bio- and ecological systems.

Note that the presented Spectral-SAR methodology provides the possibility of analytical characterizing the bio-and eco- activity of other species against given a set of trained or new synthesized chemicals and for the inter-species correlations; it has, beyond giving quantitatively similar results as the already traditional regression QSAR methods [106], many practical advantages, namely [97]:

- it has the strength of no dependency on the way in which the input data are considered, thus being largely independent of the outliers detection [26,30];
- it uses the algebraically instead of statistically recipe to furnish a generalized view for the "intensity" of chemical-biological interaction, through the vectors, predicted norms $\|\,|Y\rangle\|$, and of their properties in generalized multi-dimensional orthogonal spaces [69];
- it is also easily applicable to the case where the number of structural parameters exceeds those of the available biological activities, a situation more often observed in actual practice but being still an open problem in QSAR, due to the statistically forbidden condition that such situations imply [98];
- it is also able to furnish the key in treating the so-called spectral analysis of the activity itself through *action norm* and its *least activity path principle* (i.e., $\delta\|\,|Y\rangle\|=0$ over many

possible predicted end-points), thus providing the appropriate mechanistic picture of the envisaged ecotoxicity [26,30].

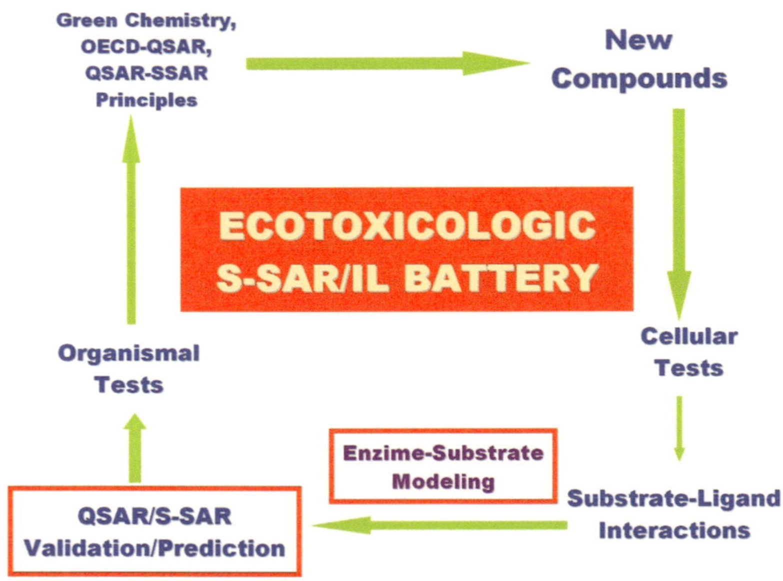

Figure 5. Conceptual description of an ecotoxicological battery constructed with the aid of QSAR/Spectral-SAR modeling of ionic liquids (IL) ligand interactions over cellular and later on organismal substrates to validate the toxicological safety of the newly synthesized compounds in accordance with the green chemistry, OECD-QSAR and, eventually, the QSAR/SSAR ecotoxicological principles.

In this way, the presented S-SAR model appears to provide a uniform picture of the anionic-cationic interaction in ionic liquids as conciliating the anionic and cationic effects observed to date. However, further studies on different species with diverse computational schemes and parameters are required, especially those that employ the quantum and algebraic features of the general operatorial equation derived in this study (41). Furthermore, by involving the enzyme-substrate modeling for receptor-effector interactions for biological activity driving chemical reactivity, as was recently reported [107], this study conceptually assessed a definitive theory of ionic liquid inter- and intra- mode of action such that it can be used for being integrated over a wide range of organisms toward designing specific eco-toxicological batteries [97], see Figure 5, for the ionic liquid's chemical-biological interactions.

Acknowledgements

Work supported by the Romanian CNCS-UEFISCDI research agency through the project PN II-RU TE16/2010-2013.

Author details

Ana-Maria Putz[1,2] and Mihai V. Putz[2*]

*Address all correspondence to: mvputz@cbg.uvt.ro

1 Institute of Chemistry Timişoara of the Romanian Academy, Timişoara, Romania

2 Laboratory of Computational and Structural Physical Chemistry, Biology-Chemistry Department, West University of Timişoara, Romania

References

[1] Dahl, J. A., Maddux, B. L. S., & Hutchison, J. E. (2007). Toward Greener Nanosynthesis. *Chem Rev*, 107, 2228-2269.

[2] Liu, Z. M., & Sun, Z. Y. (2010). Green Solvent-Based Approaches for Synthesis of Nanomaterials. *Science China Chemistry*, 53, 372-382.

[3] Wilkes, J. S. (2002). A Short History of Ionic Liquids-From Molten Salts to Neoteric Solvents. *Green Chemistry*, 4, 73-80.

[4] Valkenburg, M. V., Vaughn, R. L., Williams, M., & Wilkes, J. S. (2005). Thermochemistry of Ionic Liquid Heat-Transfer Fluids. *Thermochimica Acta*, 425, 181-188.

[5] Zhou, Y. (2005). Recent Advances in Ionic Liquids for Synthesis of Inorganic Nanomaterials. *Current Nanoscience*, 1, 35-42.

[6] Sarbu, T., & Matyjaszewski, K. (2001). Macromolecular ATRP of Methyl Methacrylate in the Presence of Ionic Liquids with Ferrous and Cuprous Anions. *Chemical Physics*, 202, 3379-3391.

[7] Shamsipur, M., Beigi, A. A. M., Teymouri, M., Pourmortazavi, S. M., & Irandoust, M. (2010). Physical and Electrochemical Properties of Ionic Liquids 1-Ethyl-3-Methylimidazolium Tetrafluoroborate, 1-Butyl-3-Methylimidazolium Trifluoromethanesulfonate and 1-Butyl-1-Methylpyrrolidinium Bis (Trifluoromethylsulfonyl) Imide. *Journal of Molecular Liquids*, 157, 43-50.

[8] Karout, A., & Pierre, A. C. (2009). Silica Gelation Catalysis by Ionic Liquids. *Catalysis Communications*, 10, 359-361.

[9] Karout, A., & Pierre, A. C. (2009). Porous Texture of Silica Aerogels Made with Ionic Liquids as Gelation Catalysts. *Journal of Sol-Gel Science and Technology*, 49, 364-372.

[10] Wahab, M. A., Kim, I. I., & Ha, C. S. (2004). Hybrid Periodic Mesoporous Organosilica Materials Prepared from 1,2-bis(triethoxysilyl)ethane and (3- cyanopropyl)triethoxysilane. *Microporous and Mesoporous Materials*, 69, 19-27.

[11] Lebold, T. (2010). Mesoporous Silica Nanostructures: A Versatile Platform in Drug-Delivery and Material Science. *Dissertation zur Erlangung des Doktorgrades der Fakultät für Chemie und Pharmazie der Ludwig-Maximilians-Universität München*, http://edoc.ub.uni-muenchen.de/11819/1/Lebold_Timo.pdf, accessed 2 May 2011.

[12] Li, S., Liu, M., Zhang, A., & Guo, X. (2010). Spherical Mesoporous Silica Templated With Ionic Liquid and Cetyltrimethylammonium Bromide and Its Conversion to Hollow Spheres. *Materials Letters*, 64, 599-601.

[13] Wu, J., Liu, C., Jiang, Y., Hu, M., Li, S., & Zhai, Q. (2010). Synthesis of Chiral Epichlorohydrin By Chloroperoxidase-Catalyzed Epoxidation of 3-Chloropropene in the Presence of an Ionic Liquid as Co-Solvent. *Catalysis Communications*, 11, 727-731.

[14] Hu, J., Gao, F., Shang, Y., Peng, C., Liu, H., & Hu, Y. (2011). One-Step Synthesis of Micro/Mesoporous Material Templated by CTAB and Imidazole Ionic Liquid in Aqueous Solution. *Microporous and Mesoporous Materials*, 142, 268-275.

[15] Flannigan, D. J., Hopkins, S. D., & Suslick, K. S. (2005). Sonochemistry and Sonoluminescence in Ionic Liquids, Molten Salts, and Concentrated Electrolyte Solutions. *Journal of Organometallic Chemistry*, 690, 3513-3517.

[16] Bravo, J. L., Lopez, I., Cintas, P., Silvero, G., & Arevalo, M. J. (2006). Sonochemical Cycloadditions in Ionic Liquids. Lessons from Model Cases Involving Common Dienes and Carbonyl Dienophiles. *Ultrasonics Sonochemistry*, 13, 408-414.

[17] Mason, T. J. (1991). *Practical Sonochemistry. Users Guide to Applications in Chemistry and Chemical Engineering*, Chichester, Ellis Horwood, 22-23.

[18] Gobel, R., Hesemann, P., Weber, J., Moller, E., Friedrich, A., Beuermann, S., & Taubert, A. (2009). Surprisingly High, Bulk Liquid-Like Mobility of Silica-Confined Ionic Liquids. *Phys. Chem. Chem. Phys.*, 11, 3653-3662.

[19] Singh, M. P., Singh, R. K., & Chandra, S. (2010). Thermal Stability of Ionic Liquid in Confined Geometry. *Journal of Physics D: Applied Physics*, 43, 1-4.

[20] Wang, G., Otuonye, A. N., Blair, E. A., Denton, K., Tao, Z., & Asefa, T. (2009). Functionalized Mesoporous Materials for Adsorption and Release of Different Drug Molecules: A Comparative Study. *Journal of Solid State Chemistry*, 182, 1649-1660.

[21] Savii, C., & Putz, A. M. (2011). Advances In (Bio)Responsive Nanomaterials. *In: Putz M.V. (ed.) Carbon Bonding and Structures: Advances in Physics and Chemistry*, Berlin-London-New York, Springer Verlag, 379-435.

[22] Putz, A. M., Ianasi, C., Dascalu, D., Savii, C., & Sfirloaga, P. (2008, 6-7 November 2008). Comparative Studies of Sonogels and Xerogels Synthesized with 1-Butyl-4-Methylpyridinium Tetrafluoroborate Ionic Liquid. Timișoara, Romania. *Proceedings of New Trends and Strategies in the Chemistry of Advanced materials with Relevance in Biological Systems, Technique and Environmental Protection*, Timișoara, Mirton Publishing House, 90-100.

[23] Putz, A. M., Ianăşi, C., Dascălu, D., & Savii, C. (2010). Acid Catalysed Silica Xerogels and Sonogels Synthesized with Buty-4-Methypyridinium Telrafluoroborate Ionic Liquid. *International Journal of Environmental Sciences*, 1, 79-88.

[24] Ianasi, C., Putz, A. M., Dascalu, D., & Savii, C. (2009, 28 September 2009). ILs Assisted Silica Mesoporous Sonogels Synthesis And Porosity Characterization. Szeged, Hungary. *In: Zoltán Galbács (ed.) Proceedings of the 16h Symposium on Analytical and Environmental Problems.*

[25] Ianasi, C., Putz, A. M., & Savii, C. (2009, 5-6 November). Approximation of Silica Xerogels And Sonogels Texture Parameters Dependence On Synthesis Variables: ILs:Si mole ratio. Timişoara, Romania. *Proceedings of the Symposium New trends and strategies in the chemistry of advanced materials with relevance in biological systems, technique and enviromental protection,* Timişoara, Mirton Publishing House, 161-164.

[26] Putz, M. V., & Lacrămă, A. M. (2007). Introducing Spectral Structure Activity Relationship (S-SAR) Analysis. Application to Ecotoxicology. *International Journal of Molecular Sciences*, 8, 363-391.

[27] Lacrămă, A.-M., Putz, M. V., & Ostafe, V. (2007). A Spectral-SAR Model for the Anionic-Cationic Interaction in Ionic Liquids: Application to Vibrio fischeri Ecotoxicity. *International Journal of Molecular Sciences*, 8, 842-863.

[28] Putz, M. V., Lacrămă, A.-M., & Ostafe, V. (2007). *Spectral-SAR Ecotoxicology of Ionic Liquids. The Daphnia magna Case, International Journal of Ecology (former Research Letters in Ecology),* Article ID12813, 5 pages, DOI:10.1155/2007/12813.

[29] Chicu, S. A., & Putz, M. V. (2009). Köln-Timişoara Molecular Activity Combined Models toward Interspecies Toxicity Assessment. *International Journal of Molecular Sciences*, 10, 4474-4497.

[30] Putz, M. V. (2012). *QSAR & SPECTRAL-SAR in Computational Ecotoxicology,* Toronto & New Jersey, Apple Academics & CRC Press- Taylor & Francis Group.

[31] Putz, M. V., Lacrămă, A. M., & Ostafe, V. (2006). Full Analytic Progress Curves of Enzymic Reaction in Vitro. *International Journal of Molecular Sciences*, 7, 469-484.

[32] Putz, M. V., & Lacrămă, A.-M. (2007). Enzymatic Control of The Bio-Inspired Nanomaterials at The Spectroscopic Level. *Journal of Optoelectronics and Advanced Materials*, 9, 2529-2534.

[33] Putz, M. V., Lacrămă, A.M., & Ostafe, V. (2007). Introducing Logistic Enzyme Kinetics. *Journal of Optoelectronics and Advanced Materials*, 9, 2910-2916.

[34] Putz, M. V. (2011). On Reducible Character of Haldane-Radić Enzyme Kinetics to Conventional and Logistic Michaelis-Menten Models. *Molecules*, 16, 3128-3145.

[35] Putz, M. V., & Putz, A. M. (2011). Logistic vs. W-Lambert Information in Quantum Modeling of Enzyme Kinetics. *International Journal of Chemoinformatics and Chemical Engineering*, 1, 42-60.

[36] Lacrămă, A. M., Putz, M. V., & Ostafe, V. (2005). New Enzymatic Kinetic Relating Michaelis-Menten Mechanisms. *Annals of West University of Timişoara-Series of Chemistry*, 14, 179-190.

[37] Lacrămă, A. M., Putz, M. V., & Ostafe, V. (2006). Studies of Lactate Dehidrogenase from Different Species and Use of The Enzyme In Ecotoxicologically Test Batteries. *Monographs Series of Annals of West University of Timisoara, Series of Biochemistry* [5], 1584-1227.

[38] Lacrămă, A. M., Popet, L., & Ostafe, V. (2007). Use of Catalase from Spinach for Testing at Molecular Level the Toxicity of Some Ionic Liquids. *Annals of West University of Timişoara- Series of Chemistry*, 16, 191-200.

[39] Lacrămă, A. M. (2007). Ecotoxicological Batteries with Organisms from Different Species. *PhD Thesis*, West University of Timisoara.

[40] Putz, M. V., Putz, A.-M., Ostafe, V., & Chiriac, A. (2010). Spectral-SAR Ecotoxicology of Ionic Liquids-Acetylcholine Interaction on E. Electricus Species. *International Journal of Chemical Modeling*, 2, 85-96.

[41] Mincea, M., Lacrămă, A. M., & Ostafe, V. (2004). Use of Bovine Liver Alkaline Phosphatase for Testing at Molecular Level the Toxicity of Chemical Compunds. *Annals of West University of Timişoara-Series Chemistry*, 13, 87-98.

[42] Mincea, M., Lacrămă, A. M., Stoian, C., Baicu, I., Nemes, N., Popet, L., & Ostafe, V. (2005, 24-25 February). Multienzimatic Test Battery-a Model for Testing at Molecular Level the Toxicity of Chemical Compounds. Timişoara, Romania. *In: Ionel I. (ed.) Humboldt Sustainability for Humanity and Environment in the Extended Connection Field Science-Economy-Policy*, Timișoara, Polytechnic Publishing House, II, 211-214.

[43] Pernak, J., & Chwala, P. (2003). Synthesis and Anti-Microbial Activities of Choline-Like Quaternary Ammonium Chlorides. *European Journal of Medicinal Chemistry*, 38, 1035-1042.

[44] Pernak, J., Sobaszkiewicz, K., & Mirska, I. (2003). Antimicrobial Activities of Ionic Liquids. *Green Chemistry*, 5, 52-56.

[45] Docherty, K. M., & Kulpa, C. F., Jr. (2005). Toxicity and Antimicrobial Activity of Imidazolium and Pyridinium Ionic Liquids. *Green Chemistry*, 7, 185-189.

[46] Docherty, K. M., Hebbeler, S. Z., & Kulpa, C. F., Jr. (2006). An Assessment of Ionic Liquid Mutagenicity Using the Ames Test. *Green Chemistry*, 8, 560-567.

[47] Bernot, R. J., Brueseke, M. A., Evans-White, M. A., & Lamberti, G. A. (2005). Acute and Chronic Toxicity of Imidazolium-Based Ionic Liquids on Daphnia Magna. *Environmental Toxicology and Chemistry*, 24, 87-92.

[48] Bernot, R. J., Kennedy, E. E., & Lamberti, G. A. (2005). Effects of Ionic Liquids on the Survival, Movement, and Feeding Behavior of the Freshwater Snail, Physa Acuta. *Environmental Toxicology and Chemistry*, 24, 1759-1765.

[49] Couling, D. J., Bernot, A. R., Docherty, K. M., Dixon, J. K., & Maginn, E. J. (2006). Assessing the Factors Responsible for Ionic Liquid Toxicity to Aquatic Organisms Via Quantitative Structure- Property Relationship Modeling. *Green Chemistry*, 8, 82-90.

[50] Garcia, M. T., Gathergood, N., & Scammells, P. J. (2005). Biodegradable Ionic Liquids. Part II. Effect of the Anion and Toxicology. *Green Chemistry*, 7, 9-14.

[51] Stock, F., Hoffmann, J., Ranke, J., Stormann, R., Ondruschka, B., & Jastorff, B. (2004). Effects of Ionic Liquids on the Acetylcholinesterase- A Structure-Activity Relationship Consideration. *Green Chemistry*, 6, 286-290.

[52] Shugart, L. (1996). Molecular Markers to Toxic Agents. *In: Newman Mc, Jagoe Ch. (eds.), Ecotoxicology: A Hierarchical Treatment*, Boca Raton, Lewis, 133-161.

[53] Newman, M., & Dixon, P. (1996). Ecologically Meaningful Estimates of Lethal Effect in Individuals. *In: Newman Mc, Jagoe Ch. (eds.), Ecotoxicology: A Hierarchical Treatment*, Boca Raton, Lewis, 225-253.

[54] Jastorff, B., Molter, K., Behrend, P., Bottin-Weber, U., Filser, J., Heimers, A., Ondurschka, B., Ranke, J., Scaefer, M., Schroder, H., Stark, A., Stepnowski, P., Stock, F., Stormann, R., Stolte, S., Welz-Biermann, U., Ziegert, S., & Thoming, J. (2005). Progress in Evaluation of Risk Potential of Ionic Liquids-Basis for an Eco-design of Sustainable Products. *Green Chemistry*, 7, 362-372.

[55] Stepnowski, P., Skladanowski, A. C., Ludwiczak, A., & Laczynska, E. (2004). Evaluating the Cytotoxicity of Ionic Liquids Using Human Cell Line Hela. *Human and Experimental Toxicology*, 23, 513-517.

[56] Jastorff, B., Stormann, R., Ranke, J., Molter, K., Stock, F., Oberheitmann, B., Hoffmann, W., Hoffmann, J., Nuchter, M., Ondruschka, B., & Filser, J. (2003). How Hazardous are Ionic Liquids? Structure- Activity Relationship and Biologic Testing as Important Elements for Sustainability Evaluation. *Green Chemistry*, 5, 136-142.

[57] Hansh, C., & Leo, A. (1995). Exploring QSAR Washington ACS Professional Reference Book.

[58] Swatloski, R. P., Holbrey, J. D., & Rogers, R. D. (2003). Ionic Liquids Are Not Always Green: Hydrolysis of 1-Butyl-3-Methylimidazolium Hexafluorophosphate. *Green Chemistry*, 5, 361-363.

[59] Swatloski, R. P., Holbrey, J. D., Memon, S. B., Caldwell, G. A., Caldwell, K. A., & Rogers, R. D. (2004). Using Caenorhabditis Elegans to Probe Toxicity Of 1-Alkyl-3-Methylimidazolium Chloride Based Ionic Liquids. . ChemInform DOI:10.1002/chin.200428226 , 35

[60] Hunt, P. A. (2006). The Simulation of Imidazolium-Based Ionic Liquids. *Molecular Simulation*, 32, 1-10.

[61] Hunt, P. A., & Gould, I. R. (2006). Structural Characterization of the 1-Butyl-3-Methylimidazolium Chloride Ion Pair Using Ab Initio Methods. *Journal of Physical Chemistry A*, 110, 2269-2282.

[62] Hunt, P. A., Gould, I. R., & Kirchner, B. (2007). The Structure of Imidazolium-Based Ionic Liquids: Insights from Ion-Pair Interactions. *Aust. J. Chem.*, 60, 9-14.

[63] Hunt, P. A., Kirchner, B., & Welton, T. (2006). Characterising the Electronic Structure of Ionic Liquids: An Examination of the 1-Butyl-3-ethylimidazolium Chloride Ion Pair. *Chemical European Journal*, 12, 6762-6775.

[64] Ropel, R., Belveze, L. S., Aki, S. N. V. K., Stadtherr, M. A., & Brennecke, J. F. (2005). Octanol-Water Partition Coefficients of Imidazolium-Based Ionic Liquids. *Green Chemistry*, 7, 83-90.

[65] Wells, A. S., & Coombe, V. T. (2006). On the Freshwater Ecotoxicity and Biodegradation Properties of Some Common Ionic Liquids. *Organic Process Research and Development*, 10, 794-798.

[66] Anastas, P. T., & Warner, J. C. (1998). *Green Chemistry Theory and Practice*, New York, Oxford University Press.

[67] National Toxicology Program (NTP) and National Institute of Environmental Health Sciences (NIEHS). (2004). *Review of Toxicological Literature for Ionic Liquids*, Prepared By Integrated Laboratory Systems Inc., Research Triangle Park, NC.

[68] Kamrin, M. A. (1997). *Pesticide Profiles: Toxicity, Environmental Impact, and Fate*, Boca Raton, Lewis Publishers.

[69] Putz, M. V., & Putz, A. M. (2011). Timişoara Spectral- Structure Activity Relationship (Spectral-SAR) Algorithm: From Statistical and Algebraic Fundamentals to Quantum Consequences. *In: Mihai V. Putz (ed.), Quantum Frontiers of Atoms and Molecules*, New York, NOVA Science Publishers, Inc., 539-580.

[70] Putz, M. V. (2012). Chemical Orthogonal Spaces. *In: Mathematical Chemistry Monographs of MATCH-Commun. Math. Comput. Chem.*, Kragujevac University Press, Kragujevac, in preparation, to be submitted, 14, http://www.pmf.kg.ac.rs/match/mcm14.html.

[71] Dirac, P. A. M. (1944). *The Principles of Quantum Mechanics*, Oxford, Oxford University Press.

[72] Randić, M. (1991). Resolution of Ambiguities in Structure-Property Studies by Use of Orthogonal Descriptors. *J. Chem. Inf. Comput. Sci.*, 31, 311-320.

[73] Randić, M. (1991). Orthogonal Molecular Descriptors. *New Journal of Chemistry*, 15, 517-525.

[74] Fadeeva, V. N. (1959). *Computational Methods of Linear Algebra*, New York, Dover Publications.

[75] Steen, L. A. (1973). Highlights in the History of Spectral Theory. *The American Mathematical Monthly*, 80, 359-381.

[76] Siegmund-Schultze, R. (1986). Der Beweis des Hilbert-Schmidt Theorem. *Archive for History of Exact Sciences*, 36, 251-270.

[77] Anderson, T. W. (1958). *An Introduction to Multivariate Statistical Methods*, New York, Wiley.

[78] Draper, N. R., & Smith, H. (1966). *Applied Regression Analysis*, New York, Wiley.

[79] Box, G. E. P., Hunter, W. G., & Hunter, J. S. (1978). *Statistics for Experimenters*, New York, John-Wiley.

[80] Green, J. R., & Margerison, D. (1978). *Statistical Treatment of Experimental Data*, New York, Elsevier.

[81] Topliss, J. (1983). *Quantitative Structure-Activity Relationships of Drugs*, New York, Academic Press.

[82] Seyfel, J. K. (1985). QSAR and Strategies in the Design of Bioactive Compounds. New York, VCH Weinheim.

[83] Chatterjee, S., Hadi, A. S., & Price, B. (2000). *Regression Analysis by Examples* (3rd Ed.), New York, John-Wiley.

[84] Amić, D., Davidović-Amić, D., & Trinajstić, N. (1995). Calculation of Retention Times of Anthocyanins with Orthogonalized Topological Indices. *Journal of Chemical Information and Computer Science*, 35, 136-139.

[85] Lučić, B., Nikolić, S., Trinajstić, N., & Juretić, D. (1995). The Structure-Property Models Can Be Improved Using The Orthogonalized Descriptors. *Journal of Chemical Information and Computer Science*, 35, 532-538.

[86] Šoškić, M., Plavšić, D., & Trinajstić, N. (1996). Link Between Orthogonal and Standard Multiple Linear Regression Models. *Journal of Chemical Information and Computer Science*, 36, 829-832.

[87] Klein, D. J., Randić, M., Babić, D., Lučić, B., Nikolić, S., & Trinajstić, N. (1997). Hierarchical Orthogonalization of Descriptors. *International Journal of Quantum Chemistry*, 63, 215-222.

[88] Putz, M. V. (2010). Cosmos, Order and Obligations: The Big CO2. *International Journal of Environmental Sciences*, 1, 1-8.

[89] Ritter, S. K. (2008). Calling all Chemists. *Chemical and Engineering News*, 18, 59-68.

[90] Anastas, P. T., Levy, I. J., & Parent, K. E. (2009). Green Chemistry Education: Changing the Course of Chemistry. *ACS Symposium Series. 1011*, Washington, DC, American Chemical Society, DOI:10.1021/bk-2009-1011.

[91] Anastas, P. T., & Zimmerman, J. B. (2003). Design through the Twelve Principles of Green Engineering. *Environmental Science and Technology*, 37, 94A-101A.

[92] OECD. (2004). *Report from the Expert Group on (Quantitative) Structure-Activity Relationships [(Q)SARs] on the Principles for the Validation of (Q)SARs, Series on Testing and Assessment* [49], 206, Paris, http://www.oecd.org/document/30/0,2340,en_2649_34365_1916638_1_1_1_1.html, accessed 3 March 2011.

[93] OECD. (2005). *Guidance Document on the Validation and International Acceptance of New or Updated Test Methods for Hazard Assessment, Series on Testing and Assessment* [34], 96, Paris, http://www.oecd.org/document/30/0,2340,en_2649_34365_1916638_1_1_1_1,00.html, accessed 3 March 2011.

[94] OECD. (2006). *Report on the Regulatory Uses and Applications in OECD Member Countries of (Quantitative) Structure-Activity Relationship [(Q)SAR] Models in the Assessment of New and Existing Chemicals, Series on Testing and Assessment* [58], 79, Paris, http://www.oecd.org/document/30/0,2340,en_2649_34365_1916638_1_1_1_1,00.html, accessed 3 March 2011.

[95] OECD. (2007). *Guidance Document on the Validation of (Quantitative) Structure-Activity Relationship [(Q)SAR] Models, Series on Testing and Assessment* [69], 154, Paris, http://www.oecd.org/document/30/0,2340,en_2649_34365_html, accessed 3 March 2011.

[96] Putz, M. V., Putz, A. M., & Barou, R. (2011). Spectral-SAR Realization of OECD-QSAR Principles. *International Journal of Chemical Modeling*, 3, 173-190.

[97] Lacrămă, A. M., Putz, M. V., & Ostafe, V. (2007). Designing a Spectral Structure-Activity Ecotoxico-Logistical Battery. *In: Putz M.V. (Ed.), Advances in Quantum Chemical Bonding Structures. Kerala: Research Signpost*, 389-419.

[98] Putz, M. V., Putz, A. M., Lazea, M., & Chiriac, A. (2009). Spectral vs. Statistic Approach of Structure-Activity Relationship. Application on Ecotoxicity of Aliphatic Amines. *Journal of Theoretical and Computational Chemistry*, 8, 1235-1251.

[99] Putz, M. V., Duda-Seiman, C., Duda-Seiman, D. M., & Putz, A.-M. (2008). Turning SPECTRAL-SAR into 3D-QSAR Analysis. Application on H+K+-ATPase Inhibitory Activity. *International Journal of Chemical Modeling*, 1, 45-62.

[100] Kaiser, K. L. E., & Palabrica, V. S. (1991). Photobacterium phosphoreum Toxicity Data Index. *Water Pollution Research Journal of Canada*, 26, 361-431.

[101] Ranke, J., Mölter, K., Stock, F., Bottin-Weber, U., Poczobutt, J., Hoffmann, J., Ondruschka, B., Filser, J., & Jastorff, B. (2004). Biological Effects of Imidazolium Ionic Liquids with Varying Chain Lengths in Acute Vibrio Fischeri and Wst-1 Cell Viability Assays. *Ecotoxicology and Environmental Safety*, 58, 396-404.

[102] Mc Queen, D. J., Post, J. R., Mills, E. L., & Fish, C. J. (1986). Trophic Relationships in Freshwater Pelagic Eco-systems. *Can. J. Fish. Aquat. Sci.*, 43, 1571-1581.

[103] Hypercube, Inc. (2002). *HyperChem 7.01. Program package, Semiempirical, AM1, Polak-Ribier optimization procedure.*

[104] Sheldon, R. A. (2005). Green Solvents for Sustainable Organic Synthesis: State of The Art. *Green Chemistry*, 7, 267-278.

[105] Freemantle, M. (2007). New Frontiers for Ionic Liquids. *Chemical Engineering and News*, 1, 23-26.

[106] Miller, J. N., & Miller, J. C. (2000). *Statistics and Chemometrics for Analytical Chemistry*, Harlow, Pretience Hall, fourth edition.

[107] Putz, M. V., & Putz, A. M. (2013). DFT Chemical Reactivity Driven by Biological Activity: Applications for the Toxicological Fate of Chlorinated PAHs. *In: Putz M.V. & Mingos D.M.P. (eds.) Applications of Density Functional Theory to Biological and Bioinorganic Chemistry. Structure & Bonding*, 150.

Chapter 5

Tribological Properties of Ionic Liquids

Yuriko Kondo, Tahahiro Koyama and Shinya Sasaki

Additional information is available at the end of the chapter

http://dx.doi.org/10.5772/ 52595

1. Introduction

An ionic liquid is a liquid salt consisting of anions and cations. Since Wilkes [1] synthesized a low melting point ionic liquid that is stable in air in 1992, there has been active fundamental research and engineering development directed at applications. In the field of tribology, where application is focused toward lubricants, the number of related papers [3]-[28] has increased every year since Liu et al. [2] published their research results in 2001. As a lubricant, ionic liquids are characterized by an extremely low vapor pressure, high thermal stability, and high ion conductivity. Based on these features, there are high expectations for the development of new ionic liquid applications as lubricants [29]-[37] in extreme environments, such as high temperatures [7]-[9] and vacuum [10]-[12], where the use of conventional lubricants is limited.

A lubricant must be used under various sliding conditions. Therefore, choosing an ionic liquid suitable for its application is necessary. Ionic liquids are also called designer's liquids because various characteristics can be created by different combinations of anions and cations. Owing to this wide variety of choices, it is important to understand the characteristics of the ionic liquid targeted for use as a lubricant. However, at present details of the lubrication mechanism are still not clearly understood, so a trial-and-error approach is inevitable for selecting ionic liquids.

This article introduces the advantages and issues related to the basic characteristics of ionic liquids as a lubricant, and it describes their future prospects, mainly for applications in vacuum.

2. Ionic Liquid as a Lubricant

2.1. Required characteristics of a lubricant

In general, the coefficient of friction is highest under dry friction conditions, where no lubricant is used. The use of a lubricant controls this type of friction conditions and may suppress damage to surfaces. Fig. 1 shows the Stribeck curve used in tribology. This figure shows the transition of the lubricating conditions of the sliding surfaces: the friction coefficient is the vertical axis, and the bearing characteristic number (fluid viscosity η × sliding speed V/average surface pressure P) is the horizontal axis. As the bearing characteristic number increases, the sliding condition transits from boundary lubrication to mixed lubrication to hydrodynamic lubrication. In dry friction, solid surfaces in the real contact area are in direct contact and adhesion; a high frictional force occurs from the shearing resistance of the adhesive parts. In boundary lubrication, an absorbed layer or tribo-chemical reaction layer lies between surfaces in the real contact area to control the adhesion between solid interfaces and to reduce the shearing resistance; this in turn decreases the coefficient of friction relative to dry friction. In mixed lubrication, the coefficient of friction further decreases because the fluid film of small shearing resistance bears a portion of the load, and the real contact area are decreased. In hydrodynamic lubrication, the coefficient of friction is at its minimum because the real contact area disappears owing to the fluid film bearing the entire load, and the frictional force is due to only the viscous resistance of the fluid. However, when the fluid viscosity and sliding speed further increase the bearing characteristic number, the coefficient of friction rises owing to the increase in viscous resistance.

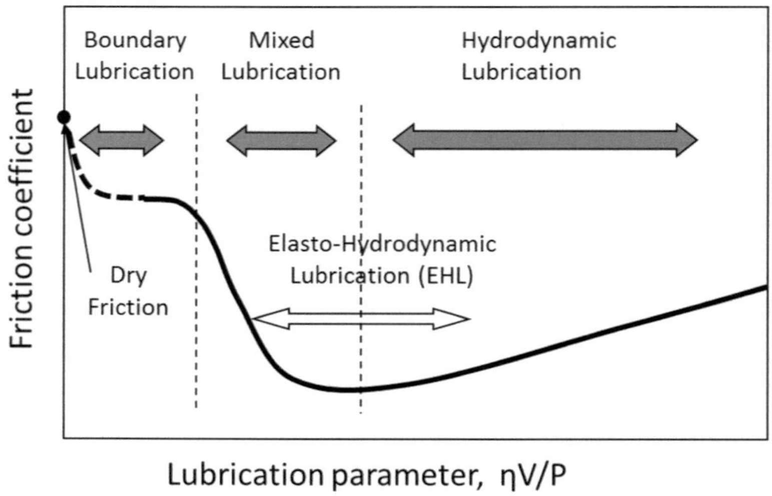

Figure 1. Schematic of the stribeck curve; the friction coefficient as a function of the lubrication parameter: ηV/P. In this formula, η is the fluid viscosity, V is the relative speed of the surfaces, and P is the load on the interface per unit bearing width.

Because most ionic liquids are in liquid state near room temperature, they can be used as a base oil for the lubricant. In addition, their application as a grease [38] by mixing the lubricant with additives [4][5][16][22] and thickeners has also been examined. The required properties of the ionic liquid differ depending upon the lubricating state. Specifically, in boundary and mixed lubrication, a firm adsorption layer must be formed on the sliding surfaces to control adhesion in the real contact area. On the other hand, in hydrodynamic lubrication, the formation of a thick liquid film must cause load burden even in the region of high surface pressure and low speed; as a result, the viscosity characteristic becomes an important factor. However, because an excessive tribo-chemical reaction can increase wear and cause corrosion—thus increasing friction coefficient—the lubricating efficiency balance must always be considered.

2.2. Hydrodynamic lubrication capacity of ionic liquid

To decrease friction and control damage to the sliding surfaces, hydrodynamic lubrication without solid contact is desired. For hydrodynamic lubrication, viscosity bears an important role with regard to the lubricant quality. An ionic liquid is in liquid state at room temperature, and its viscosity largely depends on the combination of anions and cations; it can be as thick as malt syrup or as thin as water. The viscosity of liquids decreases when the temperature rises; ionic liquids are no exception. For lubricants, the relationship between the temperature and viscosity is expressed by the viscosity index (VI). In general, this index should show a small variation in viscosity. The VI of ionic liquid is higher than that of general minerals and synthetic oils (poly-α-olefin, PAO), as shown in Table 1. Fig. 2 shows the measured results for the temperature–viscosity relation of two ionic liquid mixtures made from the same cation ([BMIM][TFSI] and [BMIM][PF6]); this verified that different mixing ratios correspond to different viscosities [39]. This indicates the possibility of arbitrarily adjusting the viscosity by mixing several types of ionic liquids. Because of the advantages of low vapor pressure and high thermal stability, ionic liquids are especially suitable for hydrodynamic lubrication applications in special environments, such as high temperatures and vacuums. Applications in bearings under dynamic and static pressure are being examined [40].

For lubricant and grease used in the elasto-hydrodynamic lubrication (EHL) state, such as for rotating bearings and gears, the dominant physical property influencing lubrication capability is the viscosity under high pressure, which can be up to several gigapascals. Ohno et al. [41][42] measured the high-pressure viscosity of a methylimidazole-type ionic liquid and reported the crystallization behavior of molecular characteristic crystals to be similar to liquid crystal under high pressure. Regarding the physical properties, such as the high-pressure viscosity index of ionic liquid, more data are expected to be reported in the future. However, behavior that causes hindrances in the EHL state, such as that reported by Ohno et al. [43], has not been reported.

Ionic liquids & Base oils		Kinematic viscosity [mm²/s]		Viscosity Index VI
		40℃	100℃	
[PP13][TFSI] *1	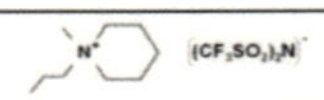	52.5	10.0	181
[P(h3)t][TFSI] *2	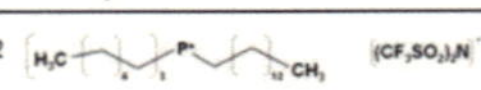	118	15.4	137
[BMIM][BF4] *3	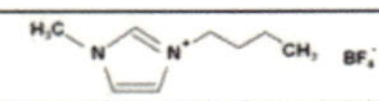	44.5	9.01	189
[BMIM][PF6] *4	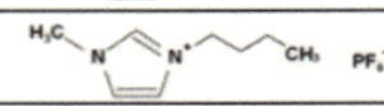	81.0	13.6	172
[BMIM][TFSI] *5	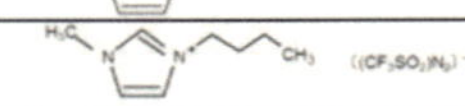	29.0	6.88	211
[BMIM][I] *6	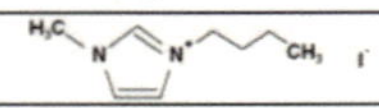	203	21.0	122
[EMIM][DCN] *7	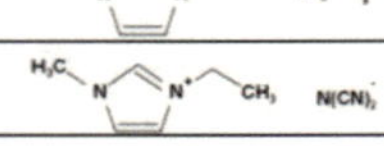	8.6	3.01	252
[BMIM][TCC] *8	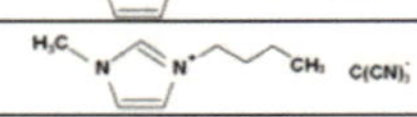	14.6	3.7	146
Synthetic oil (PAO 4)		16.8	3.90	129
Synthetic oil (PAO 6)		30.5	5.79	135
Mineral base oil		30.1	5.28	106

Table 1 Viscosity properties of ionic liquids and base oils*1: N-Methyl-N-propylpiperidinium bis(trifluoromethanesulfonyl)imide*2: Trihexyl(tetradecyl)phosphonium bis(trifluoromethylsolfonyl)imide*3: 1-Butyl-3-methylimidazolium tetrafluoroborate*4: 1-Butyl-3-methylimidazolium hexafluorophosphate*5: 1-Butyl-3-methylimidazolium bis(trifluoromethanesulfonyl)imide*6: 1-Butyl-3-methylimidazolium iodide*7: 1-Ethyl-3-methylimidazolium dicyanamide*8: 1-Butyl-3-methylimidazolium tricyanomethane

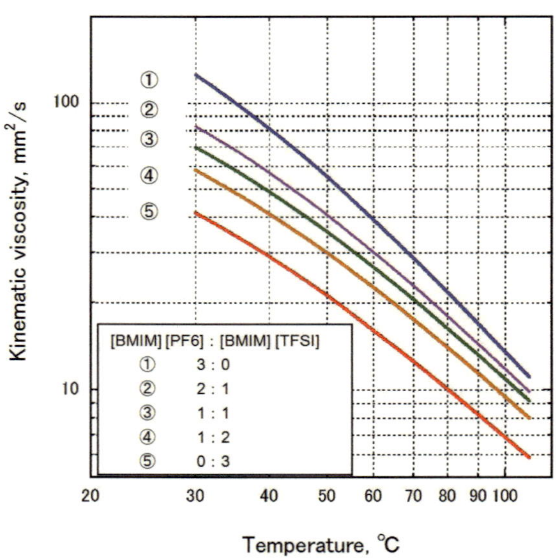

Figure 2. Viscosity property of mixed ionic liquids as a function of temperature

2.3. Boundary lubrication capacity of ionic liquid

Most studies on the special tribological qualities of ionic liquids have been on the boundary lubricating capacity. The ionic liquids examined in these studies mainly use an imidazole derivative as the cation and fluorine as the halogen element, such as tetrafluoroborate [BF4], hexafluorophosphate [PF6], and bis(trifluoromethanesulphonyl)imide [TFSI], for the anion. Fig. 3 shows the results of an investigation into the lubrication capability of different ionic liquids. Bearing steel balls and disks were used as sliding materials. The conditions for friction evaluation were as follows: temperature = 50°C, load = 50 N, reciprocating frequency = 50 Hz, amplitude = 1 mm, and friction time = 60 min. Two types of ionic liquid ([BMIM][TCC] and [EMIM][DCN]) do not include halogen. On the other hand, ionic liquids containing halogen exhibit low friction and wear and show good boundary lubrication properties. When ionic liquid containing halogen is used for lubrication, metal fluoride forms on friction surfaces by a tribochemical reaction; because this reaction product operates as a boundary lubricating layer, satisfactory lubricity is shown.

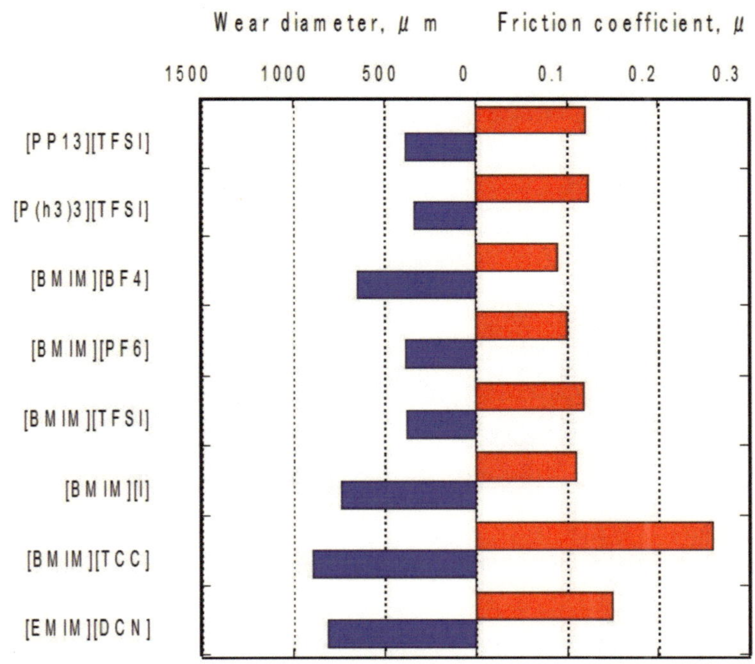

Figure 3. Lubricity of each ionic liquid for steel/steel sliding

On the other hand, ionic liquid containing a halogen such as fluorine has been known to cause corrosion in steel [16]-[22][44][45][46], aluminum alloy [17][21][22][45], bronze [45][46], and titanium alloy [9] sliding materials. The cause of corrosion has been reported to be the formation of hydrogen fluoride due to the decomposition of the ionic liquid; this is largely due to water being mixed into the ionic liquid as an impurity and participating in the reaction [21]. Decomposition and corrosion reactions of ionic liquid happen even in a static environment. However, it is more marked in a sliding environment. Because water from the surrounding atmosphere is mixed into the ionic liquid owing to enhancement by friction, metal fluoride formed on the friction surfaces is believed to further react with water by tribo-chemical reactions to generate hydrogen fluoride. Thus, corrosion occurs after friction. Fig. 4 shows the change in appearance of the friction surface for steel bearings after reciprocating sliding between the balls and disks, by using the hydrophobic ionic liquid [PP13][TFSI] as the lubricant in air at 50% relative humidity. Immediately after the rubbing test, no remarkable corrosion was seen. However, after exposure to air for 24 h, a color change was observed for all parts touched by the ionic liquid. SEM-EDX analysis verified the composition of the corrosion product, containing mainly fluorine and oxygen, in the surface marked with pit-shaped corrosion [47]. Even after the [PP13][TFSI] was applied to the bearing steel surface exposed to air for 1 week, the occurrence of corrosion could not be verified. Therefore, friction is thought to promote the corrosion reaction and the decomposition of ionic liquid.

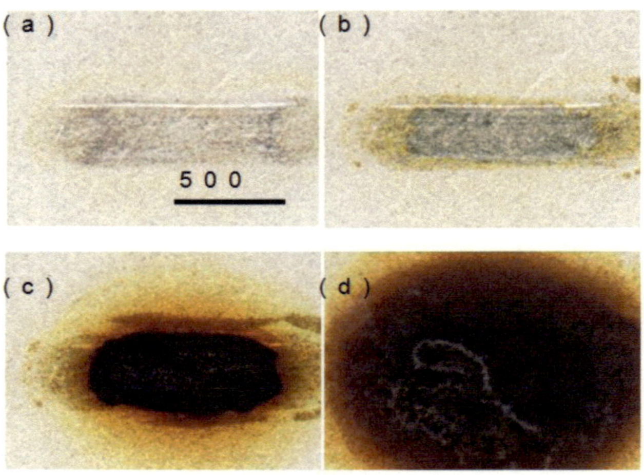

Figure 4. of corrosion on disk specimen after sliding test with [PP13][TFSI] at 50°C in air:(a)0.1 h, (b)1.0 h, (c)8.0h,(d) 24 h

2.4. Solutions to corrosion of ionic liquid

Although ionic liquids containing halogen have superior capacity in terms of boundary lubrication, they also have the problem of corrosion. There are three solutions to prevent corrosion.

2.4.1. Control of atmospheric air

Based on the discovery of the relation between corrosion and water contamination, if an ionic liquid is used in an environment where it is hard to mix water with a hydrophobic ionic http://dx.doi.org/10.5772/ 52595

liquid having a low impurity concentration, the corrosion reaction can be controlled. Fig. 5 shows the friction and wear characteristics of the halogen-containing hydrophobic ionic liquid [PP13][TFSI] in air and in dry nitrogen [47]. In a dry nitrogen atmosphere, the coefficient of friction was stable and low even at 200°C, and the wear in this atmosphere was less in comparison with that in air. In addition, further corrosion was not observed on the specimen rubbed in dry nitrogen and left in air after the rubbing test, as shown in Fig. 6. From this, it is possible to prevent the corrosion in environments such as a vacuum, where there are almost no occurrences of mixture with water.

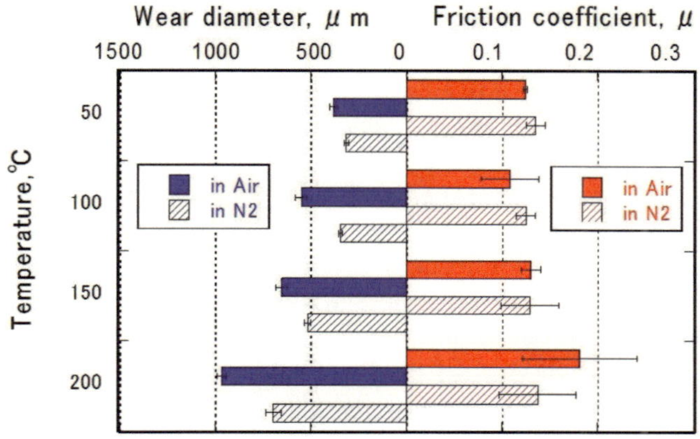

Figure 5. Comparison of friction and wear behavior of [PP13][TFSI] in air and in dry-nitrogen atmosphere

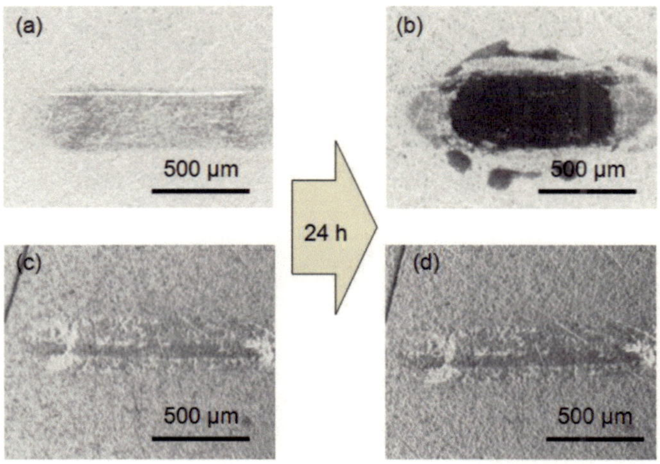

Figure 6. Optical micro-images of worn disk surfaces under lubrication with [PP13][TFSI] in air ((a),(b)) and dry-nitrogen ((c),(d)).(a) after 0.1 h, (b) after 24 h, (c) after 0.1 h and (d) after 24 h.

2.4.2. Control of metal fluoride formation reaction

To control the occurrence of hydrogen halide, which causes corrosion, methods to form a protective film over the frictional surface to prevent excessive reaction between the halogen and metal from taking place have been considered. Fig. 3 compares the effect of the presence of phosphorus on friction and wear. [BMIM][PF6] and [BMIM][BF4] have the same cation but different anions; the former has phosphorus, whereas the latter does not. [PP13][TFSI] and [P(h3)3][TFSI] have the same anion but different cations: the former has phosphorus and the latter does not. Although the ionic liquids containing phosphorus have a somewhat higher coefficient of friction, they have substantially lower wear. Phosphorus in the ionic liquid is suspected to react with the frictional surface to form a phosphoric acid compound layer that is superior in wear resistance [19][23]-[25]; also, the formation of metal fluoride, which causes the occurrence of hydrogen fluoride, was controlled. For an ionic liquid containing phosphorus as a lubricant, because the advance of corrosion in a specimen left in air after rubbing was not observed, the suppression effect is believed to be a post-rubbing phenomenon.

2.4.3. Halogen-free ionic liquid

To completely remove the corrosion reaction that originates from halogen, halogen-free ionic liquids should be selected. However, as shown in Fig. 3, their boundary lubrication ability is generally inferior to that of halogen-containing ionic liquids. However, when comparing halogen-free ionic liquids [BMIM][TCC] and [BMIM][BCN], the difference in anion results in a difference in lubricity. The discovery of a halogen-free ionic liquid with good boundary lubrication ability that does not depend on the formation of a metal halide layer is possible [48].

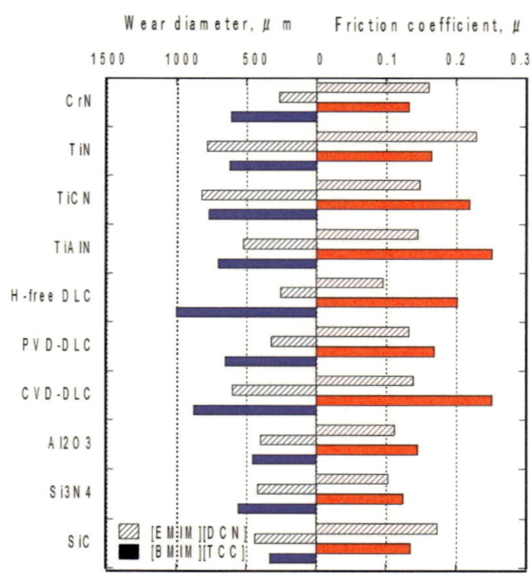

Figure 7. Friction and wear properties of hard-coatings and sintered-ceramics under lubrication with [EMIM][DCN] and [BMIM][TCC].

Fig. 7 shows the boundary lubricity of halogen-free ionic liquids corresponding to various wear resisting materials [49]. Although the lubrication ability of the two types of ionic liquid varied depending on the sliding materials, some samples showed good lubrication properties such as the combination of hydrogen-free diamond-like carbon (H-free DLC) and [BMIM][DCN]. Details on the lubricating mechanism of halogen-free ionic liquid are a future topic, but the application areas of halogen-free ionic liquid are expected to be broadened by selecting the combination of hydrogen-free diamond-like carbon with appropriate sliding materials.

3. Application as Lubricant in a Vacuum

3.1. Lubrication properties in a vacuum

In a vacuum, especially for machines in outer space, a lubricant with low vapor pressure and temperature stability is required owing to exposure to a severe temperature environment [38][43][50]. At present, perfluoropolyether (PFPE) and multiple-alkylated cycropentane (MAC) or grease that designates these as the base oil are used as the liquid lubricants in outer space because of their low vapor pressure. The lubricity of these lubricants is known to be improved by adding extreme pressure agents and the like, but issues such as evaporation of the additive and a decrease in the permanent viscosity of the base oil [43] remain to be resolved. As a non-additive base oil, ionic liquid has been reported to be superior to PFPE and MAC in terms of lubricity. In addition, as a grease, ionic-liquid-type grease has superior boundary lubrication capacity in a vacuum compared to systems of PFPE and MAC [38][50]. Regarding research on radiation for utilization in space, ionic liquid has been verified to have radiation resistance [38].

On the other hand, in industrial vacuum equipment, inevitable atmospheric release exists on sliding surfaces in most cases. Thus, extra attention must be paid to corrosion when using a halogen-containing ionic liquid. Furthermore, regarding rust prevention due to exposure to atmosphere, although the adsorption-type rust-preventive agent has been reported to be effective, these results were collected under static conditions, and the effect under sliding is unclear.

3.2. Occurrence of out-gassing

With regard to out-gassing in a vacuum, the occurrence condition, gas type, and allowed quantity differ depending on the application of the vacuum. In the case of materials for outer space equipment, measurement of the outgas (ASTM E595-93) requires a vacuum below 7 × 10-3 Pa for the sample and collector plate, a sample temperature of 125°C, and a collector temperature of 25°C (kept for 24 h) to calculate the loss mass ratio (mass change before and after test) (TML, total mass loss) and reagglutination material ratio of the collector plate (CVCM, collected volatile condensable materials) [51]. Ionic liquid based greases were developed and verified that the TML and CVCM are below 1.0% and 0.1% respectively [38]; this satisfies NASA's recommended values. In addition, the quantity of outgas that occurs during rubbing in an ionic liquid has been reported to be less than that in PFPE and MAC [50]. Thus, the outgas quality of an ionic liquid as a lubricant for space equipment applications can be concluded to be superior.

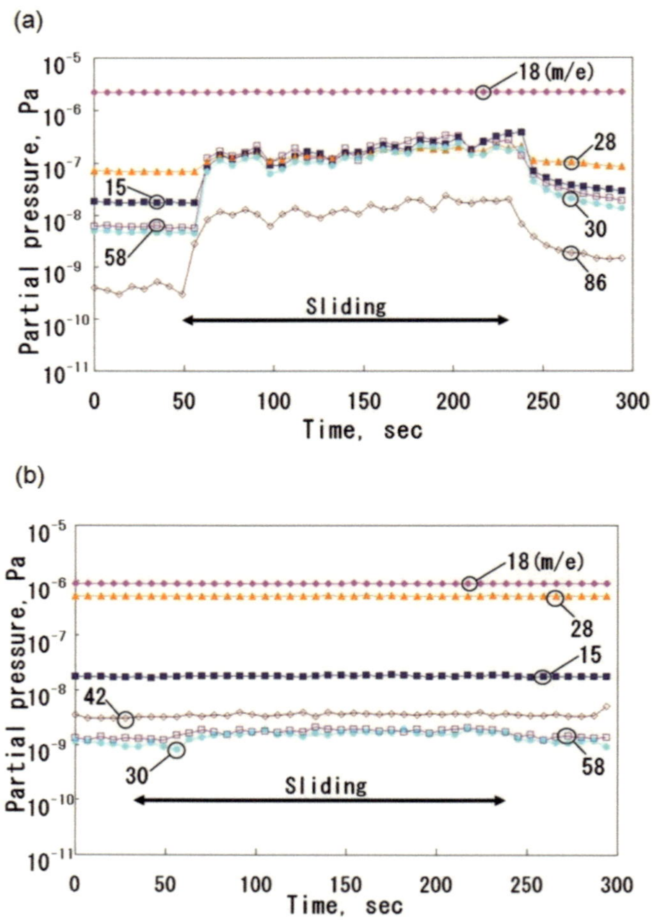

Figure 8. Partial pressure changes of generated gas species from ionic liquid during sliding under vacuum condition. (a) Titanium disk, (b) Steel disk

On the other hand, in some semiconductor processes, outgas is a contamination that can barely be tolerated. In such cases, attention must be paid to selecting the ionic liquid and sliding material. Fig. 8 shows the change in partial pressure in a vacuum during a sliding test between a SiC pin and a titanium or iron disk using N,N,N-trimethyl-N-propylammonium bis(trifluoromethanesulfonyl)imide (TMPA TFSI) as the lubricant [11]. When the iron was rubbed, a slight partial pressure rise at just m/e = 30 and 58 was verified. However, in the case of titanium, a clear rise in partial pressure was seen at m/e = 15, 30, 58, and 86 upon friction. Similar out-gassing results were seen in friction tests using aluminum. During the tribo-chemical reaction between the metal sliding surface and ionic liquid to form a metal fluoride, part of the ionic liquid decomposition is believed to release outgas into the vacuum. To suppress friction associated with out-gassing, a combination with a sliding surface having low chemical activity, such as DLC, is effective [52].

4. Conclusion

Ionic liquids have various superior qualities that traditional lubricants do not have. The development of new lubricating systems that use these features is expected. However, there are still remaining issues that must be overcome for the wide application of ionic liquids, such as disintegration and corrosion problems related to the stability and durability of ionic liquids and guidelines on optimizing the combination with sliding materials. However, these are fundamental issues that belong to the understanding of fundamental mechanisms of tribology rather than problems specific to ionic liquids. Thus, an attempt has recently been made to use an ionic liquid as a model chemical compound for understanding the action mechanism of lubricant additives. By developing an understanding of this type of fundamental lubricating mechanism, progress toward the use of ionic liquid lubricants can be expected.

Author details

Yuriko Kondo, Tahahiro Koyama and Shinya Sasaki*

Department of Mechanical Engineering, Tokyo University of Science, Japan

References

[1] Wilkes, J. S., & Zaworokto, M. J. (1992). Air and water stable 1-ethyl-3-methylimidazolium based ionic liquids. *Journal of the Chemical Society, Chemical Communications*, 13, 965-967.

[2] Ye, C., Liu, W., Chen, Y., & Yu, L. (2001). Room-temperature ionic liquids: a novel versatile lubricant. *Chemical Communications*, 21, 2244-2245.

[3] Liu, W., Ye, C., Gong, Q., Wang, H., & Wang, P. (2002). Tribological performance of room-temperature ionic liquids as lubricant. *Tribology Letters*, 13, 81-85.

[4] Kamimura, H., Kubo, T., Minami, I., & Mori, S. (2007). Effect and mechanism of additives for ionic liquids as new lubricants. *Tribology International*, 40, 620-625.

[5] Yu, B., Zhou, F., Pang, C. J., Wang, B., Liang, Y. M., & Liu, W. M. (2008). Tribological evaluation of α, $\acute{\omega}$-diimidazoliumalkylene hexafluorophosphate ionic liquid and benzotriazole as additive. *Tribology International*, 41(8), 797-801.

[6] Coerr, N., Gebseshuber, I. C., Holzer, D., Wanzenboek, H. D., Ecker, A., Pauschitz, A., & Granek, F. (2010). Evaluation of ionic liquids as lubricants. *Journal of Microengineering and Nanoelectronics*, 1, 29-34.

[7] Canter, N. (2007). Using dicationic liquids as high temperature lubricants. *Tribology & Lubrication Technology*, 63(5), 12-13.

[8] Phillips, B. S., John, G., & Zabinski, J. S. (2007). Surface chemistry of fluorine containing ionic liquids on steel substrates at elevated temperature using Mössbauer spectroscopy. *Tribology Letters*, 26, 85-91.

[9] Jimenez, A. E., & Bermudez, M. D. (2010). Ionic liquids as lubricants of titanium-steel contact part 2: friction, wear and surface interactions at high temperature. *Tribology Letters*, 37, 431-443.

[10] Suzuki, A., Shinka, Y., & Masuko, M. (2007). Tribological characteristics of imidazolium-based room temperature ionic liquids under high vacuum. *Tribology Letters*, 24, 307-313.

[11] Yagi, T., Sasaki, S., Mano, H., Miyake, K., Nakano, M., & Ishida, T. (2009). Lubricity and chemical reactivity of ionic liquid used for sliding metals under high-vacuum conditions. *Proceedings of the Institution of Mechanical Engineers, Part J: Journal of Engineering Tribology*, 223, 1083-1090.

[12] Iijima, S., Mazuko, M., Suzuki, A., Nogi, T., & Obara, S. (2011). Effect of oxide layer of metal surface on lubrication performance of liquid lubricants in high vacuum. *J. of Japanese Soc. of Tribologist*, 56(5), 320-330, (in Japanese).

[13] Kamimura, H., Chiba, T., Watanabe, N., Kubo, T., Nanao, H., Minami, I., & Mori, S. (2006). Effects of carboxylic acids on wear and friction reducing properties for alkylimidazolium-derived ionic liquids. *Tribology Online*, 1, 40-43.

[14] Xia, Y., Sasaki, S., Murakami, T., Nakano, M., Shi, L., & Wang, H. (2007). Ionic liquid lubrication of electrodeposited nickel Si_3N_4 composite coatings. *Wear*, 262, 765-771.

[15] Lu, Q., Wang, H., Ye, C., Liu, W., & Xue, Q. (2004). Room temperature ionic liquid 1-ethyl-3-hexylimidazoliumbis (trifluoromethylsulfonyl)-imide as lubricant for steel/steel contact. *Tribology International*, 37, 547-552.

[16] Qu, J., Truhan, J. J., & Dai, S. (2006). Ionic liquids with ammonium cations as lubricants or additives. *Tribology Letters*, 22, 207-214.

[17] Qu, J. (2009). Tribological characteristics of aluminum alloys sliding against steel lubricated by ammonium and imidazolium ionic liquids. *Wear*, 267, 1226-1231.

[18] Liu, X., & Zhou, F. (2006). Benzotriazole as the additive for ionic liquid lubricant: one pathway towards actual application of ionic liquids. *Tribology Letters*, 23, 191-196.

[19] Zhang, L., Feng, D., & Xu, B. (2009). Tribological characteristics of alkylimidazolium diethyl phosphates ionic liquids as lubricants for steel-steel contact. *Tribology Letters*, 34, 95-101.

[20] Zhao, Z., Shao, Y. W., Wang, T. M., Feng, D. P., & Liu, W. M. (2011). Study on corrosion property of a series of hexafluorophosphate ionic liquids on steel surface. *Corrosion Engineering Science and Technology*, 46(4), 330-333.

[21] Bermudez, M. D., Jimenez, A. E., & Martinez-Nicolas, G. (2007). Study of surface interactions of ionic liquids with aluminium alloys in corrosion and erosion-corrosion processes. *Applied Surface Science*, 253(17), 7295-7302.

[22] Jimenez, A. E., Bermudez, M. D., Carrion, F. J., & Martinez-Nicolas, G. (2006). Room temperature ionic liquids as lubricant additives in steel-aluminium contacts: influence of sliding velocity, normal load and temperature. *Wear*, 261, 347-359.

[23] Minami, I., Inada, T., Sasaki, R., & Nanao, H. (2010). Tribo-chemistry of phosphonium-derived ionic liquids. *Tribology Letters*, 40, 225-235.

[24] Fox, M. F., & Priest, M. (2008). Tribological properties of ionic liquids as lubricants and additives. Part 1: synergistic tribofilm formation between ionic liquids and tricresyl phosphate. *Proceedings of the Institution of Mechanical Engineers, Part J: Journal of Engineering Tribology*, 222, 291-303.

[25] Minami, I., Kita, M., Kubo, T., Nanao, H., & Mori, S. (2008). The tribological properties of trifluorotris (pentafluoroethyl) phosphate derived ionic liquids. *Tribology Letters*, 30, 215-223.

[26] Shah, F. U., Glavatskih, S., Mac Farlane, D. R., Somers, A., Maria, Forsyth. M., & Antzutkin, O. N. (2011). Novel halogen-free chelated orthoborate-phosphonium ionic liquids: synthesis and tribophysical properties. *Physical Chemistry Chemical. Physics*, 13, 12865-12873.

[27] Choa, S. H., Ludema, K. C., Potter, G. E., De Koven, B. M., Morgan, T. A., & Kar, K. K. (1995). A model for the boundary film formation and tribological behavior of a phosphazene lubricant on steel. *Tribology Transaction*, 38, 757-768.

[28] Zhu, J., Chu, R., & Meng, X. (2009). Chemical structure of phosphazenes in relation to the tribological properties of a steel-on-steel system. *Lubrication Science*, 21, 103-109.

[29] Bermudez, M. D., Jiménez, A. E., Sanes, J., & Carrion, F. J. (2009). Ionic Liquids as advanced lubricant fluids". *Molecules*, 14, 2888-2908.

[30] Zhou, F., Liang, Y., & Liu, W. (2009). Ionic liquid lubricants: designed chemistry for engineering applications. *Chemical Society Reviews*, 38, 2590-2599.

[31] Minami, I. (2009). Ionic liquids in tribology. *Molecules*, 14, 2286-2305.

[32] Palacio, M., & Bhushan, B. (2010). A review of ionic liquids for green molecular lubrication in nanotechnology. *Tribology Letters*, 40(2), 247-268.

[33] Schlücker, E., & Waserscheid, P. (2011). Ionic liquids in mechanical engineering. *Chemie Ingeniuer Technik*, 83(9), 1476-1484.

[34] Mori, S. (2005). Ionic liquids II. *Edited by Oono H.(CMC,Tokyo,)277 (in Japanese).*

[35] Mori, S. (2009). Ionic liquids as a candidate for lubricants. *The journal of the Surface Finishing Society of Japan*, 60(8), 502-507, in Japanese).

[36] Minami, I., & Mori, S. (2007). Tribology of ionic liquids. *Journal of The Surface Science Society of Japan*, 28(6), 311-317, (in Japanese).

[37] Minami, I. (2012). Ionic liquid lubricants. *Jyunkatsukeizai*, 2, 2-9, (in Japanese).

[38] Hayama, M., & Sasaki, S. (2012). Application of ionic liquids as space lubricants. *Proc. of JAST Tribology Conference, Tokyo*, 387-388, (in Japanese).

[39] Koyama, T., Tsuboi, R., & Sasaki, S. (2011). Research on lubricity of ionic liquid for metal alloys. Proc. of JAST Tribology Conference, Tokyo (in Japanese), 47-48.

[40] Mihashi, k., Yoshimoto, S., & Miyatake, M. (2008). Development of a small-sized hydrodynamic bearing. *Proc. of JAST Tribology Conference, Nagoya*, 491-492, (in Japanese).

[41] Tomozawa, K., & Ohno, N. (2008). Construction of phase diagram of ionic liquid and liquid crystal by diamond anvil cell. Proc. of JAST Tribology Conference, Nagoya (in Japanese), 259-260.

[42] Mia, S., Tomozawa, K., Morita, S., & Ohno, N. . (2009). High-pressure rheology of ionic liquid. *Proceedings of World Tribology Congress IV (Kyoto,)*, 572, 978-4-99001-399-8.

[43] Ohno, N. (2012). Rheological and tribological properties of space lubricant under high pressure. *J. of Japanese Soc. of Tribologist*, 57(2), 103-111, (in Japanese).

[44] Koyama, T., Kondo, Y., Tsuboi, R., & Sasaki, S. (2012). Research on lubricity of ionic liquid on bearing steel. Proc. of JAST Tribology Conference, Tokyo (in Japanese) , 365-366.

[45] Uerdingen, M., Treber, C., Balser, M., Schmittc, G., & Wernerb, C. (2005). Corrosion behaviour of ionic liquids. *Green Chemistry*, 7, 321-325.

[46] Gabler, C., Tomastik, C., Brenner, J., Pisarova, L., Doerra, N., & Allmaierb, G. (2011). Corrosion properties of ammonium based ionic liquids evaluated by SEM-EDX, XPS and ICP-OES. *Green Chemistry*, 13, 2869-2877.

[47] Kondo, Y., Yagi, S., Koyama, T., Tsuboi, R., & Sasaki, S. Lubricity and corrosiveness of ionic liquids for steel-on-steel sliding contacts. *Proceedings of the Institution of Mechanical Engineers, Part J: Journal of Engineering Tribology (in press).*

[48] Sasaki, S., Kondo, Y., Koyama, T., & Tsuboi, R. (2012). Lubricity of halogen-free ionic liquids for hard coatings. *Books of synopses, 18th International Colloquium Tribology*, (Esslingen, Germany,), 161, 3-92481-397-3.

[49] Kondo, Y., Koyama, T., Tsuboi, R., Nakano, M., Miyake, K., & Sasaki, S. (2012). Study on tribological properties if halogen-free ionic liquids for hard materials. *Proc. of JAST Tribology Conference,Tokyo*, 373-374.

[50] Nogi, T. (2011). Challenge to the ultimate lubricants for space mechanisms. *J. of Japanese Soc. of Tribologist*, 56(9), 561-566, (in Japanese).

[51] http://matdb.jaxa.jp/Outgas/Outgasdoc1j.html.

[52] Sasaki, S., Miyake, K., Nakano, M., & Ishida, T.. (2010). Lubricity of ionic liquids for hard coatings under high vacuum condition. *Books of synopses, 17th International Colloquium Tribology*, (Esslingen, Germany,), 202, 3-92481-380-9.

Chapter 6

Hydrodynamics of Ionic Liquids in Bubble Columns

Vicky Lange, Barry J. Azzopardi and Pete Licence

Additional information is available at the end of the chapter

http://dx.doi.org/10.5772/51658

1. Introduction

Over the last ten years, research into ionic liquids (ILs) and their physical/chemical and thermodynamic properties has intensified such that significant progress has been achieved in their application to a wide range of chemical processes. If one considers the patent and secondary literature however, it can be seen that there are very few IL based applications that have successfully reached commercialization. The first IL based process on a pilot scale, called Difasol, was the dimerization of olefins with a biphasic, homogeneous catalyst developed by the Institue Française du Petrole (IFP) [1]. Other examples of applications on pilot scale or even industrial scale include acid scavenging (BASIL, BASF) [2], extractive distillation (BASF) [3], compatibilizers in pigment pastes (Degussa/Evonik) [4], cooling agent (BASF) [5] and storage of gases (Air Products) [6]. It may be a fair assumption that one of the major challenges that restrict their application on an industrial scale is a lack of engineering data which are needed for the optimum scale-up, design and operation of industrial units used with ILs. Thus for the successful transition of ILs from academic labs to industry, not only must the fundamental physical/thermodynamic properties be investigated but we must also better understand the hydrodynamic or flow behaviour that govern reactions on a larger scale under real process conditions. This chapter focuses on IL based multiphase flow, a topic of specific interest to a variety of industrial platforms including biotechnology, biphasic catalysis and gas extraction involving for example H_2, CO, CO_2, H_2S and SO_2.

To apply ILs under multiphasic conditions, the simplest equipment to contact the liquid and gas is the bubble column. In these devices, the gas phase is bubbled through a column of liquid to promote close contact between the two phases. Due to their ease of construction and large applicability, these reactors find widespread use in industry. However despite its simple construction, the hydrodynamics involved inside these units is quite complex because of the very deformable nature of the gas-liquid interface. In bubble column operation, the specific interfacial area is an important criterion since it determines the rate of heat and mass transfer across the interface. The specific interfacial area is defined as the surface area of the bubbles per unit volume of the reactor space occupied by the two-phase flow mixture. To maximise this area and improve the efficiency of transport processes, small bubbles and a uniform distribution across the column cross section are desired.

This chapter examines how the physical properties of ILs, in particular its viscosity hinder achieving optimally high values of specific interfacial area. Available gas holdup data as well as bubble characteristics are analysed. Firstly factors which can influence the specific interfacial area such as the flow regime and gas holdup are discussed in Section 2. Section 3 describes the experimental approach and measurement techniques employed to study ILs in bubble columns. Bubble formation and coalescence in ILs is examined in Section 4, whilst the effect of operating conditions on the gas holdup is considered in Section 5. Lastly in Section 6, results from statistical analysis of the gas holdup data are presented to describe the frequency and velocity of the flow structures formed.

2. Influencing the specific interfacial area

For a given gas-liquid system, the transport coefficients and the interfacial area are highly dependent on the prevailing operating regime of the bubble column as this is directly related to the bubble size distribution. Because of the very deformable gas/liquid interface, there are a large number of ways in which the different phases may distribute in the column. To simplify this problem, the flow is conventionally distinguished by two main flow regimes, i.e., the homogeneous bubble regime and the heterogeneous (churn-turbulent) regime. The homogeneous regime normally exists at very low gas flow rates and is characterised by smaller, more uniformly sized bubbles which rise with similar velocity. At higher gas flow rates, a wide distribution of bubble sizes is observed due to bubble coalescence and the flow is described as heterogeneous. In this regime, the bubble swarm consists mostly of large fast-ascending bubbles and a few smaller bubbles which are trapped in the liquid recirculation flow. In cases where the column diameter is small (≤100 mm), the larger bubbles of heterogeneous flow are stabilised by the surrounding walls and can occupy the entire column cross section, forming gas plugs (called slug flow). As a result of the larger, fast-ascending bubbles, the mean gas phase residence time and the specific interfacial area in the heterogeneous regime is lower in comparison to the homogeneous regime. In reality, due to high gas throughputs the heterogeneous regime is often observed in industrial bubble columns. Although, the homogeneous regime which offers greater interfacial contacting and a low gas shear environment can be more desirable, in particular for those applications which involve sensitive media, for example biotechnology.

To estimate the specific interfacial area, the bubble size distribution is used in conjunction with the gas holdup or void fraction. The gas holdup is defined as the fraction of the gas present in the two-phase mixture in the reactor. It depends strongly on the operating conditions, physiochemical properties of the two phases, the gas distributor design (the number and size of the holes) and the column geometry (height to diameter ratio). To better under-

http://dx.doi.org/10.5772/51658

stand the effect of these parameters on the flow behaviour in the bubble column, their effect on the gas holdup and the gas flow rate has been investigated by several researchers through extensive experimentation as well as theoretical analysis in the last few decades. An extensive review was given by [7], and there have been continuous publications since.

Understanding the effect of system pressure (or gas density) has been important since most industrial applications of bubble columns operate under pressurized conditions. This is because the increased solubility of a gas with pressure is expected to enhance the mass transfer and reaction rates. Based on the findings by several workers [8-15], it has been commonly established that gas holdup increases with operating pressure. The increase of gas holdup at the elevated system pressures has been attributed to the smaller bubbler size which is caused by a reduction in the coalescence rate, the enhancement of the bubble breakup or the decreased size of bubbles formed at the gas disperser [16-18].

The effect of the liquid phase properties has also been investigated, although to a lesser degree. For convenience, most studies of bubble columns have used water as the liquid phase, even though in reality the physical properties of many of the liquids used industrially differ significantly from those of water. In particular, the ILs considered here have viscosities considerably higher than water. Several studies have looked at the effect of liquid viscosity during bubble column operation [19-26]. These show that at a given gas flow rate, the gas holdup decreases with increasing viscosity. Furthermore in a larger diameter column, the transition point from homogeneous to heterogeneous flow has been found to shift to lower gas velocities with increasing liquid viscosity.

3. Determination of gas holdup

When a gas is bubbled through a column filled with liquid, the bed of liquid begins to expand or swell as soon as gas is introduced. If the gas holdup is expressed in terms of a global or total voidage for a bubble column, it can be calculated by measuring the difference between the gassed and ungassed height of the liquid since this represents the total gas volume present in the column. The total gas holdup can also be determined from pressure gradient measurements using pressure sensors. However, in cases where more local information on the gas phase is required, the gas holdup can be measured by a selection of invasive or non-invasive techniques. A comprehensive review of methods which can be applied for closed systems (pipes, columns etc.) was presented in [27], wherein all pertinent literature until roughly 1977 was surveyed. More recent reviews have been presented by workers such as [28] and [29]. Examples of some of the options available include using needle probes, quick-closing valves (QCVs), flush mounted conductance probes, capacitance probes, wire mesh sensors, neutron and radioactive absorption, resistive and ultrasonic tomography.

To obtain the local gas holdup for an IL, the conductance probe is an appropriate choice since the IL is a very good electrical conductor, while the conductivity of the gas phase is infinitely low. This measurement technique follows the approach developed by researchers such as [30], whereby the electrical impedance of the gas-liquid region close to a system of electrodes is measured and the phases distinguished due to the difference in the electrical conductivity of the phases. The conductivity is measured between two metallic rings (ring-type conductance probe) which are mounted flush with the bubble column walls. For the experiments presented here, the metallic rings were constructed by placing two stainless steel plates between three alternating plates of an acrylic resin and machining out a cylinder through them which had an internal diameter equal to the column diameter. Figure 1(a) shows a picture of these probes, while a schematic of their configuration is presented in Figure 1(b). The configuration is characterised by the thickness of the rings s and the spacing between them D_e. The column diameter is represented by D_t. The dimensions D_e/D_t and s/D_t were 0.357and 0.075 respectively.

Each probe pair was supplied with an a.c. carrier voltage of -1 to 1 volt peak to peak, at a frequency of 20 kHz. An instrumentation amplifier, a full wave rectifier and a band pass filter were installed before the signal was sent to the data acquisition board. A cut-off frequency of 100 Hz was applied in order to eliminate the high frequency noise generated from the power supply. To account for any differences in the test fluid conductivity during experiments due to any temperature variations, all conductivities were normalized to produce a dimensionless conductance by measuring the conductivity at the start of each experimental run with the column full of liquid.

To obtain values for gas holdup from the conductance measurements, it is necessary to determine the relationship between the liquid phase conductance and gas holdup. This relationship is obtained via careful calibration. The simplest approach involves artificially creating instantaneous gas fractions between the probes, for example using plastic non-conducting beads to simulate bubbly flow inside the column. A detailed description for the calibration procedure is available in [31].

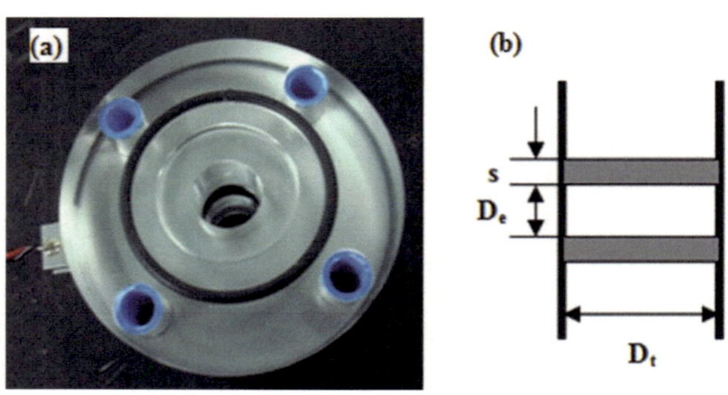

Figure 1. Conductance Probe: (a) picture; (b) schematic of ring shaped conductance probe.

For the IL 1-ethyl-3-methylimidazolium ethylsulfate, $[C_2C_1Im][EtOSO_3]$, the time-varying gas holdup in a cylindrical bubble column (inner diameter 0.038 m; height 1.1 m) has been measured by utilising two pairs of conductance probes installed flush in the column walls. The experimental setup is illustrated in Figure 2. For comparison, water, a glycerol/water solution and solution of glycerol/water with sodium chloride dissolved in it were also inves-

http://dx.doi.org/10.5772/51658

tigated. To prepare the glycerol solutions, a total of 15% of weight of water was mixed with glycerol to produce a liquid which has a viscosity similar to that of [C_2C_1Im][$EtOSO_3$]. A total of 1.3% by weight of sodium chloride was also added to one of the glycerol/water mixtures to ensure sufficient conductivity for use with the conductance probes. The physical properties of the liquid phases used in all experiments at 20°C are summarised in Table 1. To determine the viscosities and conductivities of the liquids a Brookfield viscometer and WTW KF 340 conductivity meter were used respectively. The viscosity of water and surface tensions were obtained from physical property tables. The properties of [C_2C_1Im][$EtOSO_3$] were obtained from the paper by [32].

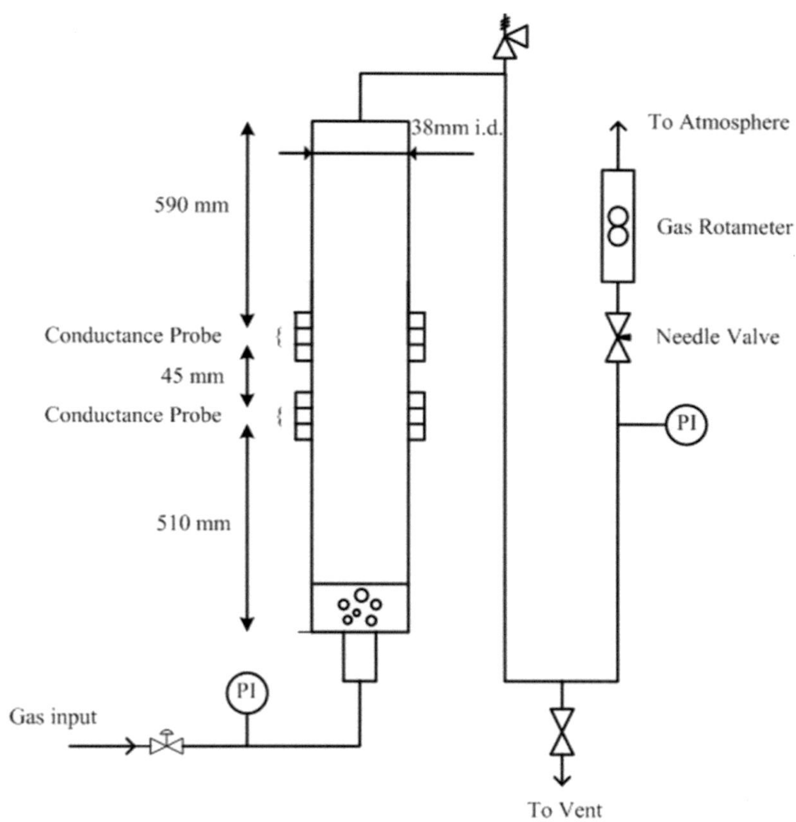

Figure 2. Experiment setup for gas holdup measurements.

Prior to experiments, [C_2C_1Im][$EtOSO_3$] was dried and degassed under vacuum at 60°C for 24 hours since it is known that ILs will absorb a couple weight percent of water when left open to the atmosphere. Here, the main concern is the effect of the absorbed water on the physical properties of [C_2C_1Im][$EtOSO_3$], in particular its viscosity as well as ion mobility which relates to its electrical conductivity. The estimated water content of [C_2C_1Im][$EtOSO_3$] after drying was approximately 0.76 wt% water, as measured by Karl Fischer test.

Property	$[C_2C_1Im][EtOSO_3]$	Glycerol (15% water)	Glycerol (15% water + 1.3% NaCl)	Water
Density (kg/m^3)	1241	1212	1224	1000
Viscosity (mPa s)	91.5	93.1	107.9	1.09
Surface tension (N/m)	0.047	0.062	0.062	0.072
Conductance (S/cm)	4160	1.4	485	42

Table 1. Physical properties of the liquid phases (T = 20°C).

Measurements for gas holdup at a given operating pressure were performed by increasing the gas superficial velocity to a maximum value of 60 mm/s, while the operating pressure was systematically varied in the range of 0.1-0.8 MPa. The column was first filled with the liquid under test at room temperature (20°C) and compressed nitrogen gas from a gas cylinder was then fed into the bottom of the column through a single nozzle gas distributor which had an inner diameter of 0.635 mm. Glass beads were also employed at the gas inlet to the column for improved phase distribution. A pressure regulating valve at the gas cylinder fixed the maximum gas inlet pressure and a pressure relief valve set at 15 MPa protected the facility against overpressure. Flow was created in the system by using a needle valve which was installed at the inlet to a gas rotameter and had an outlet open to the atmosphere. To minimise liquid carry-over from the column into the downstream instrumentation, several pieces of wire mesh were installed at the top of the column to catch stray droplets.

4. Bubble Formation and Coalescence

When a gas is passed through an orifice into a pool of liquid, certain forces act on the gas which breaks it up to form individual bubbles. If the bubble column is transparent, photographic techniques involving a high-speed camera can be quite helpful to analyse the shape and the size of the bubbles formed. Stills taken from high-speed videos at atmospheric pressure of $[C_2C_1Im][EtOSO_3]$, glycerol solutions and water are shown in Figure 3. These reveal that even at the lowest gas superficial velocity (u_{gs} = 3 mm/s), the homogeneous flow regime is essentially absent for the IL and the glycerol solutions. In a distinct contrast to water, it is seen that the bubbles formed in these high viscous liquids are bigger and less uniform in size.

In $[C_2C_1Im][EtOSO_3]$ three different sized bubbles are formed. The largest are bullet-shaped Taylor bubbles which characterize slug flow. These only occur for the largest gas flow rates studied; instead clearly defined spherical-cap shaped bubbles are formed at velocities ≥ 15 mm/s. The second, intermediate sized bubbles in the IL are essentially spherical of 0.5 – 2.5 mm, while the smallest are also spherical however with diameters ≤ 250 μm. It is believed that the smallest bubbles are formed by the bursting of Taylor bubbles at the gas/liquid interface. These tiny bubbles formed have very low rise velocities and tend to flow with the liquid. As a result, they can be distributed throughout the entire column due to backmixing of the liquid.

As the bubbles rise in the column, their size can decrease or increase through various breakup or coalesce mechanisms respectively. In both $[C_2C_1Im][EtOSO_3]$ and glycerol solutions coalescence is observed immediately at the lower gas velocities. In contrast, due to the reduced viscosity in water coalescence is only seen at velocities > 30 mm/s. The smaller viscous or drag forces in water also leads to the formation of spherical-cap bubbles which in comparison to the IL and other viscous liquids, are rather irregular and distorted in shape (Figure 3(D)).

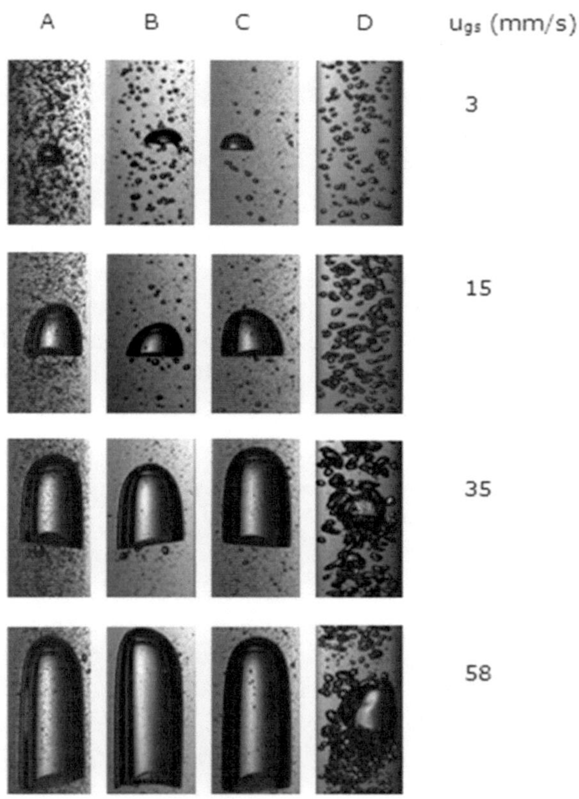

Figure 3. Stills taken from high-speed videos illustrating typical flow structures. The letters identify the liquids (A) $[C_2C_1Im][EtOSO_3]$; (B) glycerol (15% water); (C) glycerol (15% water + 1.3% NaCl); (D) water.

In the IL, coalescence of Taylor or spherical-cap bubbles is frequently observed, however coalescence between spherical-cap and intermediate bubbles is less common. In Figure 4, the coalescence between two spherical-cap bubbles is shown as an example. In the first frame the lower bubble appears to be distorted by being in the wake of the upper spherical-cap bubble. It can also be seen that the intermediate-sized bubbles in the wake of the leading spherical-cap bubble do not coalescence with the upper bubble but instead they are forced to one side by the arrival of the second spherical-cap bubble. Thereafter the intermediate-sized bubbles eventually reposition themselves in the wake of the combined spherical-cap bubble.

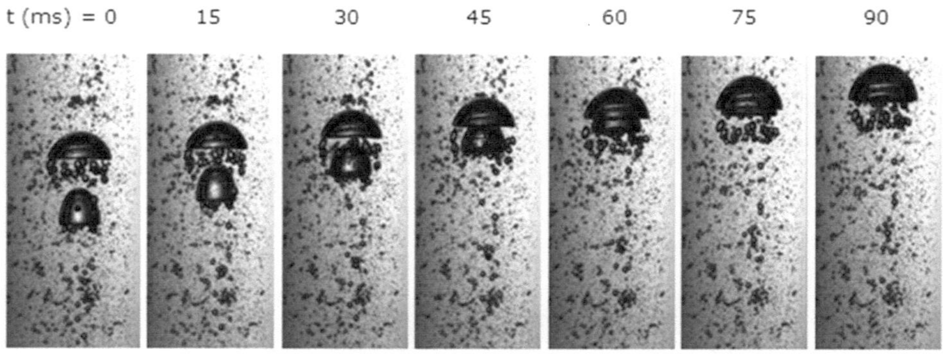

Figure 4. Sequence of stills from high-speed videos illustrating the coalescence of two spherical-cap bubbles in $[C_2C_1Im][EtOSO_3]$.

Bubble coalescence in $[C_2C_1Im][EtOSO_3]$ was also observed in high-speed videos taken of a larger diameter column (125 mm) with a gas distributor consisting of a plate with 25, 1 mm diameter holes. Trailing bubbles were seen to travel considerable distances across the column, move into the wake of a preceding one and then coalescence. A still at a gas superficial velocity of 10 mm/s is provided in Figure 5, which shows the formation of small spherical-cap bubbles even in the larger sized column.

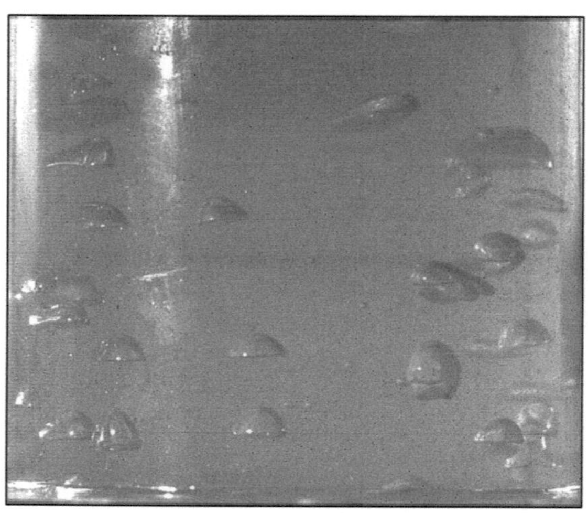

Figure 5. Still from high-speed video illustrating the formation of spherical-cap bubbles in 125 mm i.d. column filled with $[C_2C_1Im][EtOSO_3]$. (taken by D. Sreevasan)

The behaviour of single bubbles rising through liquids has been classified on the basis of three dimensionless groups, Morton, Eötvös and Reynolds numbers, defined as $\mu^4 g\Delta\rho/\varrho^2\sigma^3$, $g\Delta\rho D^2/\sigma^3$ and $\rho uD/\mu$ respectively. Here μ is the liquid viscosity, g the acceleration due to gravity, $\Delta\rho$ is the difference between the gas and the liquid, ρ is the liquid density, σ is the surface tension,

D is the column diameter and u is its rise velocity. The ranges of the dimensionless groups for which different types of bubbles exist and in which the effects of the channel walls become important has been identified by [33]. For $[C_2C_1Im][EtOSO_3]$ the Morton number is 0.014 and the Eötvös number on the order of 400. It is found that the characterization map of [33] accurately predicts the formation of spherical-cap bubbles at smaller bubble sizes.

The motion and deformation of a single bubble rising in ILs ($[C_4C_1Im][BF_4]$, $[C_8C_1Im][BF_4]$and $[C_4C_1Im][PF_6]$) were also studied by [34] through image analysis of high-speed videos. Two new empirical correlations were proposed to correlate the drag coefficient as a function of Reynolds number and the aspect ratio as a function of a new dimensionless parameter which could group experimental data for bubbles in ILs. The predicted drag coefficients agreed well with the experimental data however, further experiments are needed to verify these new correlations for a wider range of ILs with different ion pairs and physical properties.

5. Gas holdup at low and elevated pressures

To achieve a better understanding of the flow phenomenon present time traces of the gas holdup obtained from the conductance signals can be analysed by using statistical functions. Time-averaged values are usually obtained first and thereafter, the variations in the amplitude and in the frequency space can be explored. In each experiment presented here, the gas holdup was measured by sampling the data at 1000 Hz over a time period of 60 s. Examples of time traces at the same gas superficial velocity (u_{gs} = 50 mm/s) are given in Figure 6. Figure 6(a) shows the effect of increasing pressure in the IL system. At 0.1 MPa, the time series shows periods of very low gas holdup values which alternate with significant periodic increases to the gas holdup. This confirms the presence of alternating aerated liquid slugs with large gas pockets and even Taylor bubbles which were seen in the photographs. In contrast, at 0.8 MPa the time trace shows that the flow is no longer intermittent. The peaks are noticeably lower and non-uniform which indicates a bubbly heterogeneous flow of smaller sized bubbles.

In Figure 6(b), it is seen that similar time traces are obtained for $[C_2C_1Im][EtOSO_3]$ and the glycerol solutions at 0.8 MPa suggesting that similar sized bubbles were present in both viscous systems. However when compared with water at 0.8 MPa (Figure 6(c)) a more uniform time trace of smaller gas holdup values is observed for water which suggests that the 'larger' bubbles present in the heterogeneous flow in the viscous systems did not form in water.

Time-averaged gas holdup values have also been calculated and plotted against gas superficial velocity for the liquids. Figures 7(a) and 7(b) show examples at atmospheric pressure and at a pressure typical of industrial conditions (0.8 MPa) respectively. These increase monotonically with gas superficial velocity. In comparison to water, the gas holdup in $[C_2C_1Im][EtOSO_3]$ is significantly lower while similar results are obtained for $[C_2C_1Im][EtOSO_3]$ and the viscous glycerol solutions. It is therefore reasonable to conclude that the observed reduction in the gas holdup in $[C_2C_1Im][EtOSO_3]$ is due to its higher viscosity. For high viscous media in bubble columns, the reduced gas holdup can be attributed to increases in the magnitude of viscous or drag forces exerted during bubble formation so that a stable bubble diameter is attained before its detachment at the gas distributor. By stabilizing the bubble interface, bubble coalescence is promoted and bubble breakup is suppressed in the gas distributor region. It is probable that this led to the formation of larger fast-rising bubbles in the IL and glycerol solutions. These faster bubbles spend a shorter time in the column and consequently the overall gas holdup is decreased.

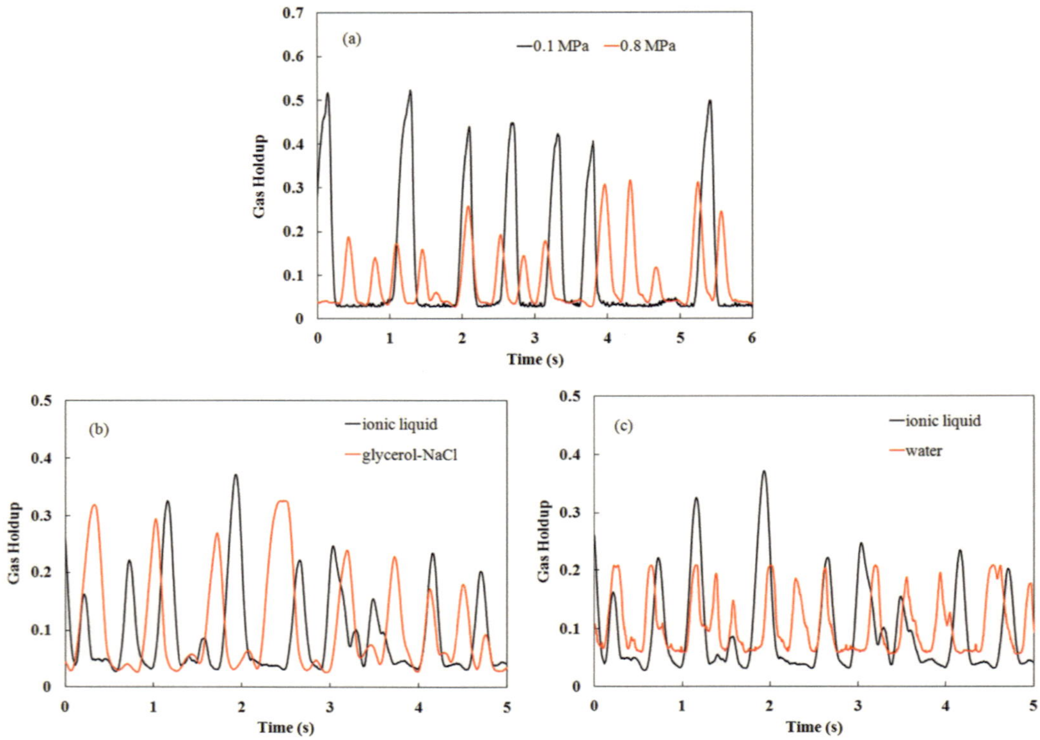

Figure 6. Gas holdup time trace: (a) $[C_2C_1Im][EtOSO_3]$ at 0.1 and 0.8 MPa; (b) $[C_2C_1Im][EtOSO_3]$ and glycerol-NaCl; (c) $[emim]EtSO_4$ and water.

For the estimation of the gas holdup, there exists a vast number of equations in the open literature, most of which are empirically based. In the review by [35], 37 published equations for gas holdup are listed. In selecting equations to predict gas holdup overall correlations based on experimental data is often used, despite the fact that it is generally better to use more physically based methods since empirical correlations are usually most suited for the experimental conditions which they were developed from. However, due to the extreme sensitivity of the gas holdup to other factors for example the gas distributor design, purity of the continuous phase and the physical properties of the phases, developing a generalised model can be complex.

The time-averaged gas holdup data for $[C_2C_1Im][EtOSO_3]$ was tested against three available models in the literature: Urseanu et al. [26], Wilkinson et al. [20] and Krishna et al.

http://dx.doi.org/10.5772/51658

[36]. It was found that the prediction by [26] gives the best agreement at both low and elevated pressures (Figure 8). This model was based on their work with high viscous liquids (0.05-0.55 Pa s) at elevated pressures (0.1-1 MPa). The model by [20] which was developed specifically for the industrial scale-up of pressurized bubble column reactors is presented in Figure 9. The model is shown to be reliable at lower gas superficial velocities while the correlation by [36] significantly overestimates the gas holdup in the homogenous regime as well as the transition point.

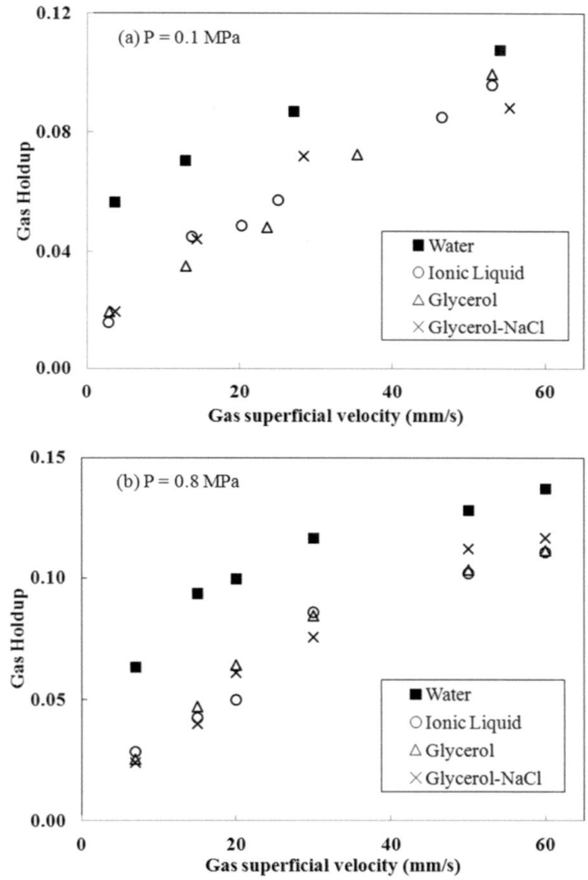

Figure 7. Effect of liquid viscosity on the average gas holdup: (a) P = 0.1 MPa; (b) P = 0.8 MPa.

The effect of system pressure (or gas density) on the gas holdup can be significant in certain circumstances. Figure 10 shows how increasing the pressure increases the time-averaged gas holdup in $[C_2C_1Im][EtOSO_3]$, water and glycerol solutions. It is believed that increased system pressure leads to enhanced local turbulence. This destabilises the larger bubbles that would otherwise form in the heterogeneous regime at atmospheric pressure. As a result bubble break-up occurs and small bubbles which have lower rise velocities are created. These bubbles have longer residence times in the column and the gas holdup is consequently increased. However it is clear that for the range of gas superficial velocities investigated, the change in gas holdup with pressure is not very significant for [C_2C_1Im][$EtOSO_3$]. Similar findings were reported by [26] in their work with high viscous oils at elevated system pressures.

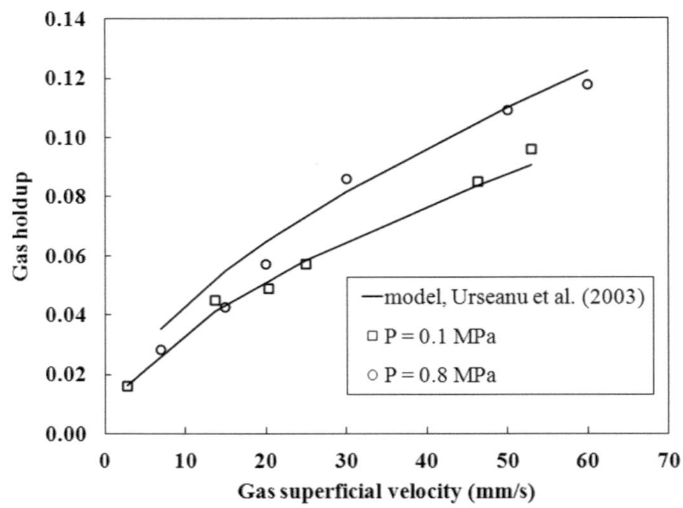

Figure 8. Comparison of predictions of the correlations of Urseanu et al. [26] with the present gas holdup data for [C_2C_1Im][$EtOSO_3$].

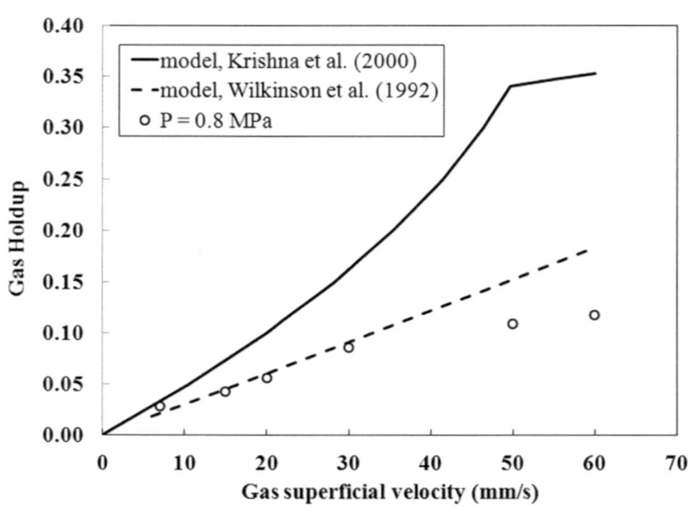

Figure 9. Comparison of predictions of the correlations of Krishna et al. [36] and Wilkinson et al. [20] with the present gas holdup data for [C_2C_1Im][$EtOSO_3$].

http://dx.doi.org/10.5772/51658

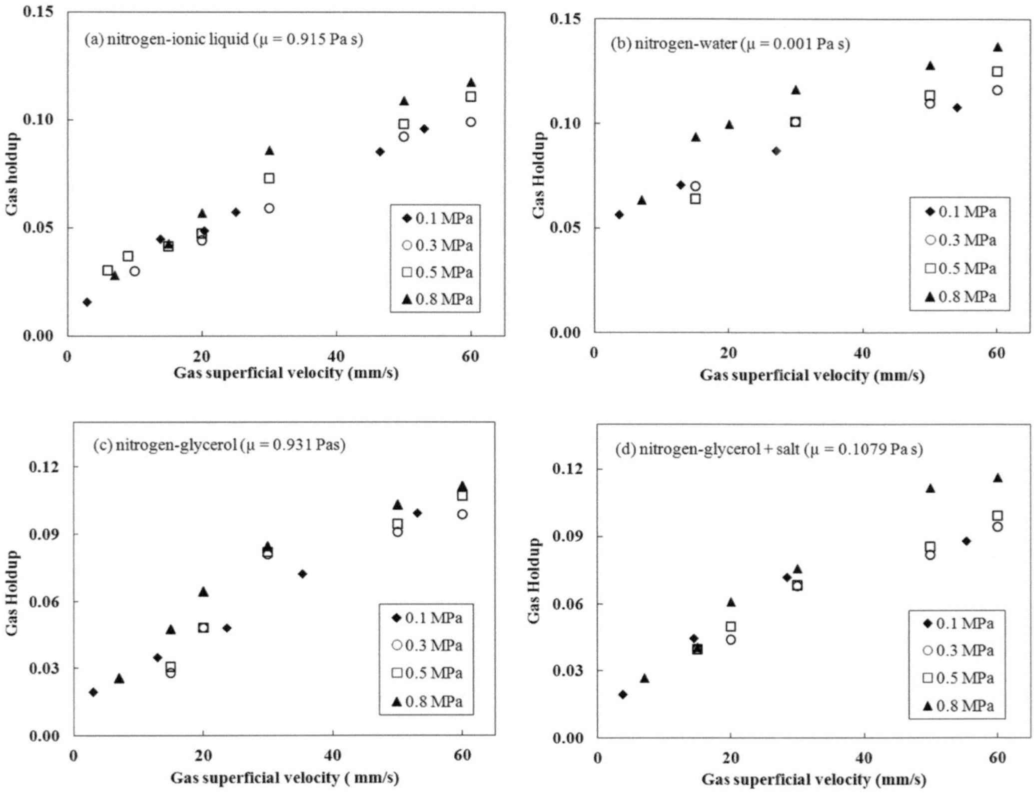

Figure 10. Effect of system pressure on gas holdup: (a) [C_2C_1Im][$EtOSO_3$]; (b) water; (c) glycerol; (d) glycerol + 1.3% NaCl.

6. Detailed Behaviour of the Flow

6.1. Flow patterns

To examine the flow in more detail, the variations in amplitude of the gas holdup time traces can be considered using the probability density functions (PDFs), i.e., the fraction or how often particular gas holdup values occur. In the approach made popular by [37], PDF plots of gas holdup time series are often used to identify different flow patterns in gas-liquid flows, based on the distinctive shape that exists for each regime. For example, the signature PDF for homogeneous bubbly flow is characterised by a narrow single peak at low gas hold-up, whereas the PDF for slug flow is double-peaked. PDFs for [C_2C_1Im][$EtOSO_3$] and glycer-ol are presented in Figures 11 and 12 respectively. It is seen that the shapes of the PDFs for both liquids are quite similar. At atmospheric pressure, there is a single peak at low gas holdup with a broadening tail which extends to a smaller second peak at higher gas holdup values. According to [37] this signature shape is indicative of spherical-cap bubbly flow or even slug flow. The small second peak for the spherical-cap bubble is marked with arrows. This is validated by what was seen in the high-speed videos (Section 3).

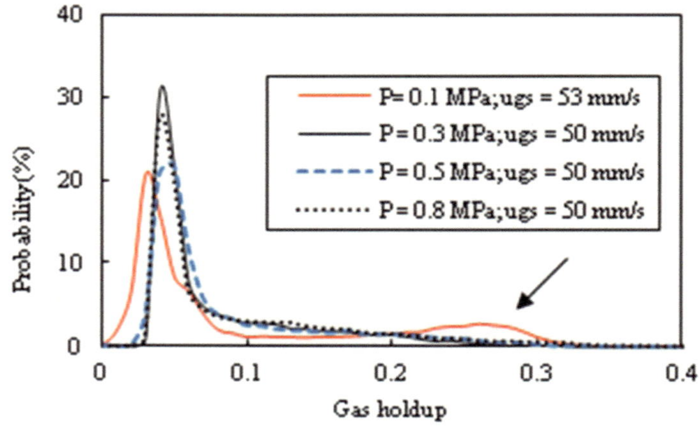

Figure 11. PDF of gas holdup: $[C_2C_1Im][EtOSO_3]$.

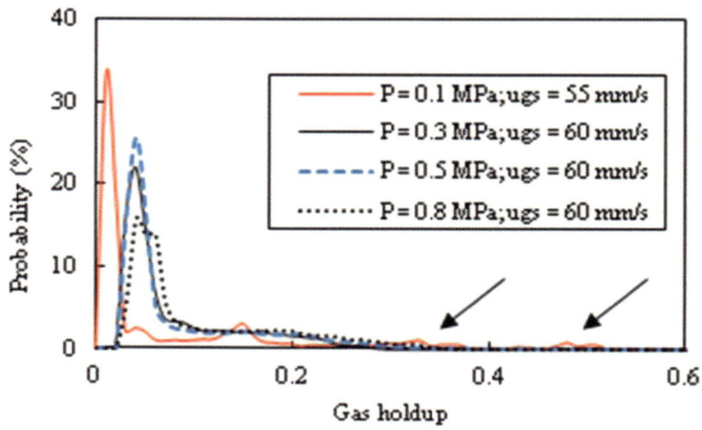

Figure 12. PDF of gas holdup: glycerol.

At higher pressures, the probabilities of higher gas holdup values decrease for both $[C_2C_1Im][EtOSO_3]$ and glycerol. This provides evidence that Taylor or larger spherical-cap bubbles are not formed in the viscous systems at system pressures exceeding atmospheric. Instead the PDF traces are single-peaked, although there is some spread in the distributions which suggests that the flow consists of unequally sized bubbles, i.e., heterogeneous. The Morton number and the Eötvös number for $[C_2C_1Im][EtOSO_3]$ were on the order of 50 and 30 respectively for the system at higher pressures. The characterization map of [33] predicts that ellipsoidal bubbles instead of spherical-cap bubbles are expected. It also indicates that the terminal velocities of the bubbles will be strongly influenced by the walls.

http://dx.doi.org/10.5772/51658

Generally for gas-liquid flows, the most common way of identifying which flow pattern occurs for a given set of flow rates is to use a flow pattern map. For bubble columns, these flow pattern maps are often plots of the gas superficial velocity against the column diameter. If the gas holdup data for $[C_2C_1Im][EtOSO_3]$ is plotted on such a graph, conditions at which heterogeneous flow were observed would be in the homogenous region of the map. The inability of these maps to accurately predict the flow regimes in high viscous media such as $[C_2C_1Im][EtOSO_3]$ is probably due to the fact that they have been developed empirically using data obtained from air-water experiments.

6.2. Structure frequencies

Much can also be learned about the flow structures in $[C_2C_1Im][EtOSO_3]$ by examining their frequencies. This can be obtained by power spectrum analysis of the gas holdup time traces. Here, power spectrum densities (PSDs) have been obtained by using the Fourier transform of the auto-covariance functions. Essentially the Fourier transform is used to transform the time series from a time domain into a frequency spectrum. Examples of PSDs for $[C_2C_1Im][EtOSO_3]$ at 0.8 MPa are shown in Figure 13. At each gas flow rate, a clear peak in the range of 2 – 2.5 Hz is seen. These peak values are the frequencies of recurrence of any periodic structures. The frequencies obtained using this method have been compared with those determined from manually counting the peaks in the time traces. Values obtained from the two methods agree within 10%.

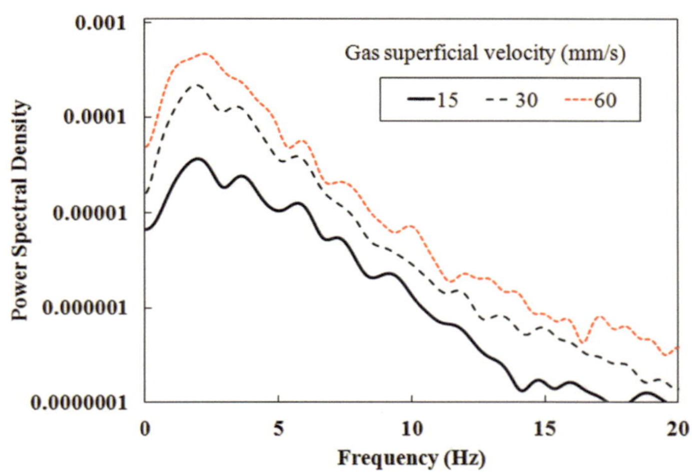

Figure 13. PSD of gas hold time series: liquid = $[C_2C_1Im][EtOSO_3]$.

The effect of increasing pressure on the structure frequencies in $[C_2C_1Im][EtOSO_3]$ is shown in Figure 14. It is seen that for the same gas flow rates, the frequencies generally increase with increasing system pressure. The increasing number of structures at higher pressures further suggests that at higher pressures bubble breakup or coalescence suppression occurs in $[C_2C_1Im][EtOSO_3]$.

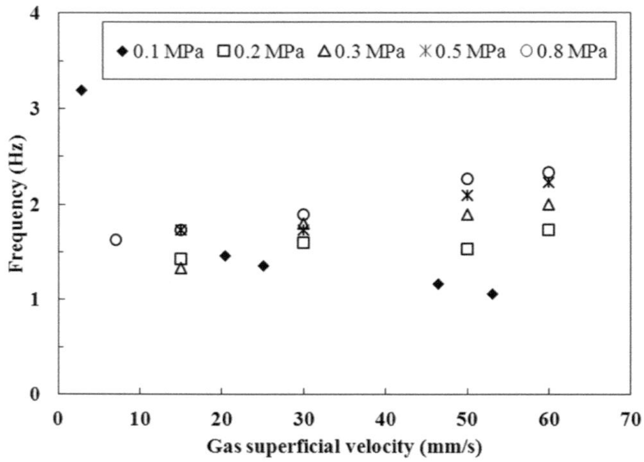

Figure 14. Effect of system pressure on the structure frequencies in $[C_2C_1Im][EtOSO_3]$.

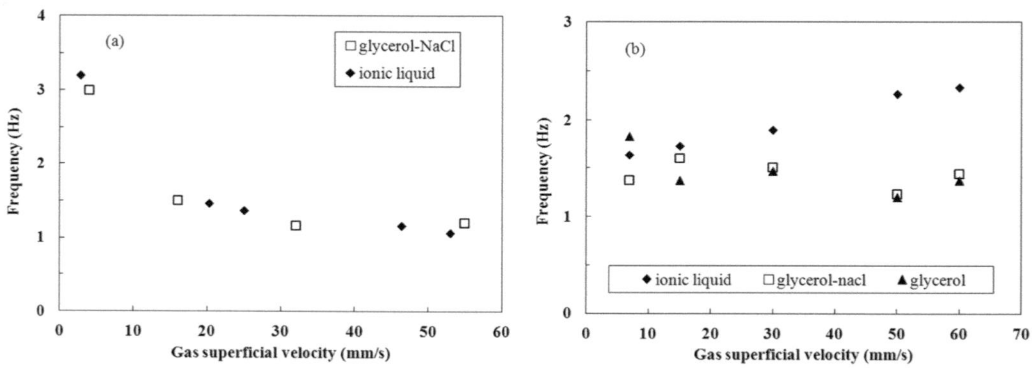

Figure 15. Effect of the liquid on the frequency: (a) 0.1 MPa; (a) 0.8 MPa.

The frequencies obtained for $[C_2C_1Im][EtOSO_3]$ at 0.1 MPa and 0.8 MPa can also be compared to the equivalent data for glycerol-NaCl. These are shown in the subplots presented in Figure 15. At atmospheric pressure it is seen that there are only small differences between the frequencies for the different liquids and they decrease with increasing gas flow rates. However at a higher system pressure, larger variation is observed particularly at the highest gas flow rates. To analyse this further aspects of the axial variations in the flow have also been examined. The flow development in $[C_2C_1Im][EtOSO_3]$ and glycerol-NaCl based on their frequency and PDF data obtained from both conductance probes are shown in Figures 16 and 17 respectively. It is seen that frequencies obtained for the probe downstream (B) in $[C_2C_1Im][EtOSO_3]$ is slightly higher than probe upstream (A) while the PDF downstream displays a taller peak indicating a larger probability of small gas holdup values there. The opposite is observed in the results for glycerol-NaCl. A possible explanation for this could be that the difference in molarity between the two liquids had an effect on the coalescence of bubbles. The glycerol-NaCl solution has a molarity of 0.17 M, while that of $[C_2C_1Im][EtOSO_3]$ is essentially infinite. In the literature, a critical value of 0.2 M has been cited for the suppression of coalescence in liquids [38-40]. Thus, it is possible that coalescence is much reduced for the IL at higher pressures.

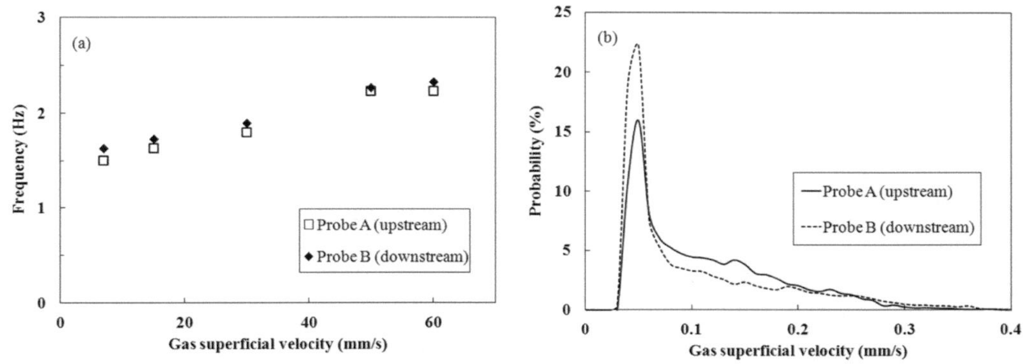

Figure 16. Effect of axial distance in $[C_2C_1Im][EtOSO_3]$: (a) frequency; (b) PDF.

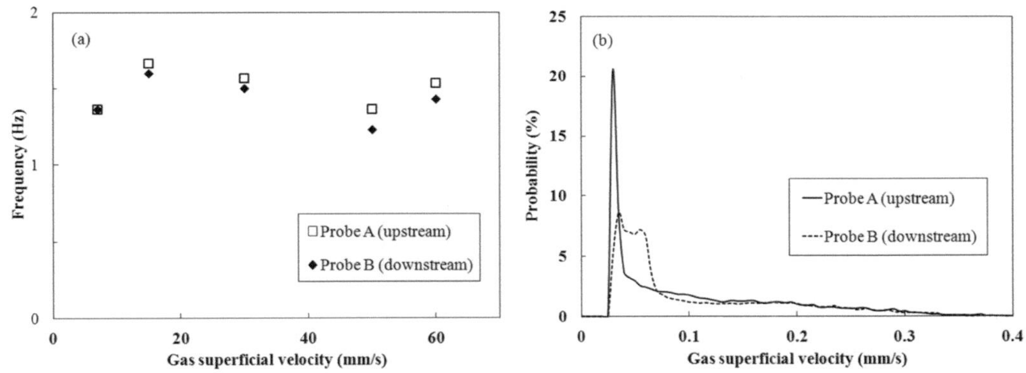

Figure 17. Effect of axial distance in glycerol + 1.3% NaCl: (a) frequency; (b) PDF.

6.3. Structure velocities

The velocities of bubbles can be determined from image analysis of high-speed videos or more objectively from cross-correlating the signals of probes placed slightly apart. For the latter, a time lag which corresponds to a peak in the cross-correlation function represents the average time required for flow structures to travel between the two probes. Since the distance between the two probes are known, the structure velocity is then easily calculated. This method of velocity measurement has been used to determine the structure velocities in $[C_2C_1Im][EtOSO_3]$. In Figure 18, the flow velocities in $[C_2C_1Im][EtOSO_3]$ at atmospheric pressure and 0.8 MPa are shown respectively. For comparison, the equivalent data for glycerol is also presented. It can be seen that there is very good agreement between both viscous systems.

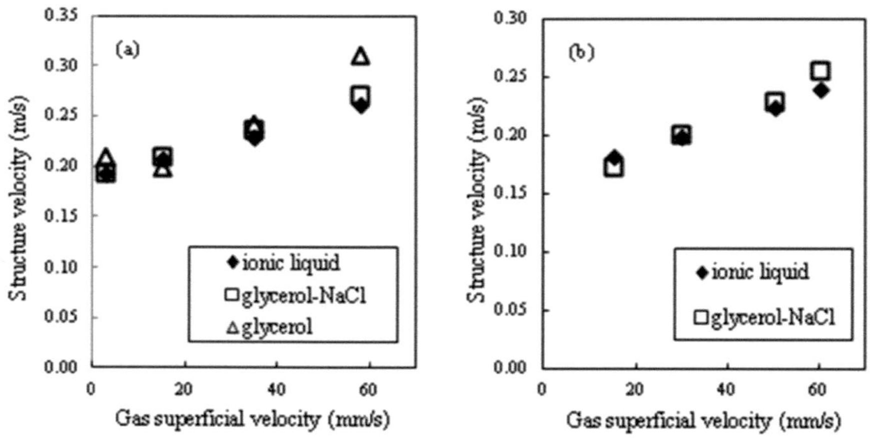

Figure 18. Effect of the liquid on the structure velocities: (a) 0.1 MPa; (b) 0.8 MPa.

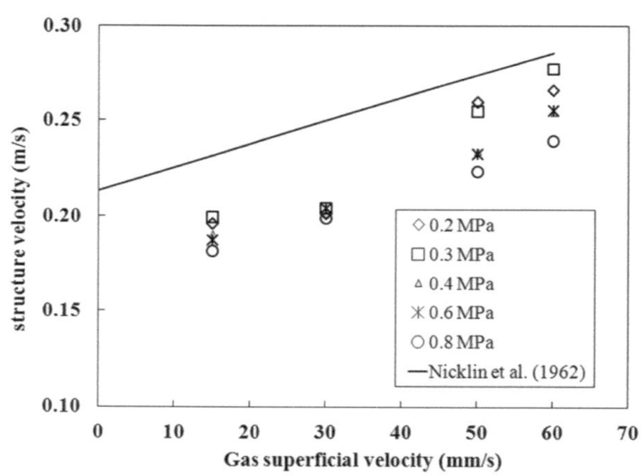

Figure 19. Effect of system pressure on structure velocities in $[C_2C_1Im][EtOSO_3]$.

The effect of increasing system pressure on the bubble rise velocities in $[C_2C_1Im][EtOSO_3]$ is illustrated in Figure 19. It is seen that in general the structure velocities decrease with increasing pressure. This is expected given the smaller bubble sizes in the system at elevated pressures. For gas-liquid flows the correlation by Nicklin et al. [41] is often used to predict the velocities of flow structures including void fraction waves and slugs. This correlation has been tested against the present IL data. It is found that the correlation predicts the correct linear trend, although with higher absolute values.

7. Conclusion and outlook

From the perspective of industrialization, research on the hydrodynamics of ionic liquids in bubble columns has been presented in this chapter. Despite being in its beginning stages, it is evident that this work is important for future industrial scale-up and process design. The experimental results reveal that the flow characteristics of the ionic liquid are similar to those of the other viscous media studied, both at atmospheric and elevated system pressures. Due to its high viscosity, the bubbles formed in the ionic liquid are larger compared to water which results in a significant reduction to the gas holdup. This creates smaller specific interfacial areas and less effective gas-liquid contacting. Thus to successfully achieve intimate contact between an ionic liquid and a gas stream, the viscosity of the ionic liquid is an important factor to be considered in the choice and design of industrial equipment.

Author details

Vicky Lange*, Barry J. Azzopardi and Pete Licence

*Address all correspondence to: enxls7@nottingham.ac.uk

School of Chemistry, University of Nottingham, UK

References

[1] Olivier, H. (1999). Recent Developments in the Use of Non-Aqueous Ionic Liquids for Two-Phase Catalysis. *Journal of Molecular Catalysis A, Chemical*, 146(1), 285-289.

[2] Maase, M., & Masonne, K. (2005). Bipasic Acid Scavenging Utilizing Ionic Liquids: The First Commercial Process with Ionic Liquids. Ionic Liquids IIIB: Fundamentals, Progress, Challenges, and Opportunities. *Transformations and Processes, ACS Symposium Series 902*, American Chemical Society, Washington DC, 126-132.

[3] Jork, C., Seiler, M., & Beste, Y. (2004). Influence of ionic liquids on the phase behaviour of aqueous azeotropic systems. *Journal of Chemical Engineering Data*, 49(4), 852-857.

[4] Weyershausen, B., & Lehmann, K. (2005). Industrial application of Ionic Liquids as Performance Additivies. *Green Communications*, 7(1), 15-19.

[5] Vagt, U. (2007, August 5-10). Ionic Liquids: Overview on Commercial Applications and First Toxicological Assessments. Yokohama, Japan. *In: Proceedings 2nd International Congress on Ionic Liquids (COIL)*, 125.

[6] Tempel, D., Henderson, P. B., & Brzozowski, J. (2006). Ionic Liquid based mixtures for gas storage and delivery. *US 2006/0060817 AI.*

[7] Deckwer, W. D. (1992). *Bubble Column Reactors*, New York, John Wiley and Sons Limited.

[8] Chiba, S., Idogawa, K., Maekawa, Y., Moritomi, H., Kato, N., & Chiba, T. (1989). Neutron radiographic observation of high pressure three-phase fluidization. *Fluidization VI*, Engineering Foundation, New York, 523.

[9] Krishna, R., Wilkinson, P. M., & van Dierendonck, L. L. (1991). A model for gas holdup in bubble columns incorporating the influence of gas density on flow regime transitions. *Chemical Engineering Science*, 46(10), 2491-2496.

[10] Wilkinson, P. M. (1991). Physical Aspects and scale-up of high pressure bubble columns. *PhD thesis*, University of Groningen, The Netherlands.

[11] Clark, K. N. (1990). The effect of high pressure and temperature on phase distributions in a bubble column. *Chemical Engineering Science*, 45(8), 2301-2307.

[12] Reilly, I., Scott, D. S., De Bruijn, T., & MacIntyre, D. (1994). The role of gas phase momentum in determining gas holdup and hydrodynamic flow regimes in bubble column operations. *Canadian Journal of Chemical Engineering*, 72(1), 3-12.

[13] Jiang, P., Bakshi, B. R., Zhong, H., & Fan, L. S. (1995). Analysis of flow in gas-liquid bubble columns using multi-resolution methods. *Transactions of the Institution of Chemical Engineers*, 73(6), 608-614.

[14] Lin, T. J., Reese, J., Hong, T., & Fan, L. S. (1995). Quantitative Analysis and Computation of Two-dimensional Bubble Columns. *AIChE Journal*, 42(2), 301-318.

[15] Letzel, H. M., Schouten, J. C., Krishna, R., & van den Bleek, C. M. (1999). Gas holdup and mass transfer in bubble column reactors operated at elevated pressure. *Chemical Engineering Science*, 54(13-14), 2237-2246.

[16] Kling, G. (1962). Uber die Dynamik der Blasenbildung Beim Begasen von Flussigkeiten Unter Druck. *International Journal of Heat and Mass Transfer*.

[17] Nauze, La, & Harris, I. J. (1974). Gas bubble formation at elevated system pressures. *Transactions of Institutions of Chemical Engineers*, 52, 337-348.

[18] Tsuge, H., & Hibino, S. (1980). Theoretical approach for the formation of gas bubble under the elevated system pressure G116. Japan. *Preprint of 14th Autumn Meeting of the Society of Chemical Engineers*.

[19] Schugerl, Eissa K. (1975). Holdup and backmixing investigation in co-current and counter-current bubble columns. *Chemical Engineering Science*, 30(10), 1251-1256.

[20] Wilkinson, P. M., Spek, A. P., & van Dierendonck, L. L. (1992). Design parameters estimation for scale-up of high-pressure bubble columns. *AIChE J.*, 38(4), 544-554.

[21] Kantak, M. V., Hesketh, R. P., & Kelkar, B. G. (1995). Effect of gas and liquid properties on gas phase dispersion in bubble columns. *Chemical Engineering Journal*, 59(2), 91-100.

[22] Kuncová, G., & Zahradnik, J. (1995). Gas holdup and bubble frequency in a bubble column reactor containing viscous saccharose solutions. *Chemical Engineering Process*, 34(1), 25-34.

[23] Ruzicka, M. C., Zahradnık, J., Drahoš, J., & Thomas, N. H. (2001). Homogeneous-heterogeneous regime transition in bubble columns. *Chemical Engineering Science*, 56(15), 4609-4626.

[24] Ruzicka, M. C., Drahoš, J., Fialová, M., & Thomas, N. H. (2001). Effect of bubble column dimensions on flow regime transition. *Chemical Engineering Science*, 6117-6124.

[25] Ruzicka, M. C., Drahos, J., Mena, P. C., & Teixeira, J. A. (2003). Effect of viscosity on homogeneous-heterogeneous flow regime transition in bubble columns. *Chemical Engineering Journal*, 96(1-3), 15-22.

[26] Urseanu, M. I., Guit, R. P. M., Stankiewicz, A., van Kranenberg, G., & Lommen, J. H. G. M. (2003). Influence of operating pressure on the gas hold-up on bubble columns of high viscous media. *Chemical Engineering Science*, 58(3), 697-704.

[27] Hewitt, G. F. (1978). *Measurements of Two Phase Flow Parameters*, London, Academic Press.

[28] Kumar, S. B., Moslemian, D., & Dudukovic, M. P. (1997). Gas-holdup measurements in bubble columns using computed tomography. *A.I.Ch.E. Journal*, 43(6), 1414-1425.

[29] Boyer, C., Duquenne, A. M., & Wild, G. (2002). Measuring techniques in gas-liquid and gas-liquid-solid reactors. *Chemical Engineering Science*, 57(16), 3185-3215.

[30] Fossa, M. (1998). Design and performance of a conductance probe for measuring the liquid fraction in two-phase gas-liquid flows. *Flow Measurement Instrumentation*, 9(2), 103-109.

[31] Kaji, R., & Azzopardi, B. J. (2009). Investigation of flow development of co-current gas-liquid vertical slug flow. *International Journal of Multiphase Flow*, 35(4), 335-348.

[32] Gomez, E., Gonzalez, B., Calvar, N., Tojo, E., & Domınguez, A. (2006). Physical properties of pure 1-Ethyl-3-methylimidazolium Ethyl Sulfate and its binary mixtures with ethanol and water at several temperatures. *Journal of Chemical Engineering Data*, 51(6), 2096.

[33] Clift, R., Grace, J. R., & Weber, M. E. (1978). *Bubbles, Drops, and Particles*, New York, Academic Press.

[34] Dong, H., Wang, X., Liu, L., Zhang, X., & Zhang, S. (2010). The rise and deformation of a single bubble in ionic liquids. *Chemical Engineering Science*, 65(10), 3240-3248.

[35] Riberio, C. P. (2007). On the estimation of the regime transition point in bubble columns. *Chemical Engineering Journal*, 140(1-3), 473-482.

[36] Krishna, R., Urseanu, M. I., & Dreher, A. J. (2000). Gas hold-up in bubble columns: influence of alcohol addition versus operation at elevated pressures. *Chemical Engineering and Processing*, 39(4), 371-378.

[37] Costigan, G., & Whalley, P. B. (1997). Slug flow regime identification from dynamic void fraction measurements in vertical air-water flows. *International Journal of Multiphase Flow*, 23(2), 263-282.

[38] Boyd, J. W. R., & Varley, J. (2004). Acoustic Emission Measurement of Low Velocity Plunging Jets to Monitor Bubble Size. *Chemical Engineering Journal*, 97(1), 11-25.

[39] Oolman, T. O., & Blanch, H. W. (1986). Bubble coalescence in air-sparged bioreactors. *Biotechnology Bioengineering*, 28(4), 578-584.

[40] Machon, V., Pacek, A. W., & Niwnow, A. W. (1997). Bubble sizes in electrolyte and alcohol solutions in a turbulent stirred vessel. *Chemical Engineering Research and Design*, 75(3), 339-348.

[41] Nicklin, D. J., Wilkes, J. O., & Davidson, J. F. (1962). Two-phase flow in vertical tubes. *Transactions of the Institution of Chemical Engineers*, 60, 61-68.

Chapter 7

Synthesis, Properties and Physical Applications of IoNanofluids

Carlos Nieto de Castro, Ana P. C. Ribeiro, Salomé I.C. Vieira, João M. P. França, Maria J.V. Lourenço, Fernando V. Santos, Sohel M.S Murshed, Peter Goodrich and Christopher Hardacre

Additional information is available at the end of the chapter

http://dx.doi.org/10.5772/52596

1. Introduction

Ionic liquids have proved to be one of the most impressive classes of fluids, due to their properties and applications to chemistry and engineering. One of the most recent applications of complex systems of ionic liquids and nanomaterials are IoNanofluids, from heat transfer to catalysis, solar absorbing panels, lubricants or luminescent materials. These novel materials belong to the class of nanofluids proposed in the last years and are a mixture of ionic liquid and nanomaterial, in the form of nanoparticle dispersion, and have already resulted in a number of publications in chemical and physical journals.

There are several reasons to study these new materials/fluids, rather complex in structure. The most important property is their enhanced thermal properties, like thermal conductivity and heat capacity, heat transfer and heat storage. The complex interactions existing between the anion/cation of the ionic liquid and the nanomaterial surface create nano-regions that can enhance reactivity and selectivity of chemical reactions. As the physiochemical properties of ionic liquids themselves can be tailored to the desired application employing an ionic liquid as a base fluid means IoNanofluids can also be designed to meet any specific application or task requirement. They also are non-flammable and non-volatile at ambient conditions and can, therefore, be considered as "green" fluids.

The possible scientific, technical and economic success of IoNanofluids open new markets, as new products for engineering fluids, heat insulators, catalysts, etc., can be envisaged. In addition, the replacement of environmentally aggressive chemicals, foams, solid composites are in daily order, imposed by regulations or social responsibility. These new challenges open new jobs for chemists, material scientists and engineers.

The precursors of the IoNanofluids are nanofluids. Their impact in the scientific/ technical literature is very big. A total of 1575 nanofluids related publications, which include journal and conference articles, patent, news, letter and other, have appeared over the past 11 years. [1] In Figure 1 it can be seen an exponential growthof the total number of these items published per year since 2002. It is also believed that there are more than 300 research groups and companies worldwide involved in nanofluids research, a hot topic in an interdisciplinary field. Several spin-off companies have originated from nanofluids research and the number continues to increase.

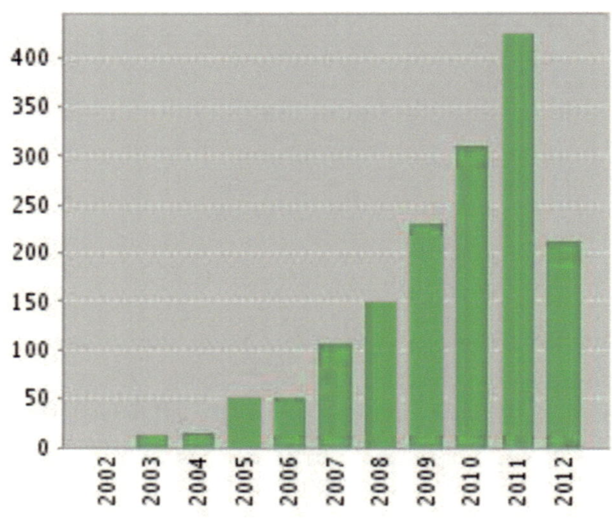

Figure 1. Web of Knowledge record of nanofluids related publications

The current review, which focuses the syntheses of ionic liquids, the preparation of IoNanofluids, their properties, experimental measurements, processes requirements and economic impact, pretends to contribute to clarify the overall question, always present in the appearance of a new field, which crosses several minds: Will IoNanofluids be useful to our society?

Comparison between the IoNanofluids properties and those of the base ionic liquids will be the key for understanding the role of the interface between the ionic liquid and the nanomaterial in the determination of physical properties. The unexpected behavior observed in

1 Web of Knowledge©, January 1, 2002 to June 20, 2012

http://dx.doi.org/10.5772/52596

some IoNanofluids and the difference observed regarding the base fluid is, in our view, the most important aspect in this review.

2. Ionic Liquids

Ionic liquids (ILs) are materials composed entirely of ions, which have melting points below 100 ºC [1]. Such liquids are an elaborate network of ions which are governed by electrostatic charges and hydrogen bonding [2]. More recently *room temperature ionic liquids* (RTILs) have moved to the forefront of IL research due to their melting temperatures being below 30°C. Generally the asymmetric form of the ions (usually the cation) reduces the Coulombic interactions which results in a lowering of the melting points, see figure 2. These liquids exhibit favourable properties for solvent use due to the formation of air and moisture stable, low-volatile liquids, low flammability [3] under ambient conditions high ionic conductivity which also maintains their thermal and electrical stability over a large temperature range [4,5]. With judicious tailoring of the cation-anion structure a range of ILs can be synthesised with various pre-chosen physiochemical properties. This has resulted in them being applied in a multitude of techniques including catalysis [6], elemental analysis [7], synthesis [8], solar absorbing panels [9], lubricants [10], luminescent materials [11] and supercritical fluids [12,13]. Ionic liquids also provide very different solvent–solute interactions which can give rise to distinct chemistries compared with molecular solvent systems [14]. In addition to this, the use of ILs acting as both the solvent system and reactant/catalyst in a reaction process [15] makes them a "hot topic" for researchers. They have also been shown to immobilize and stabilize catalytic complexes or even act as modifiers that accelerate the reaction [16]. These IL-catalyst systems can be recycled potentially reducing chemical waste and increasing the lifetime of the catalyst, further adding to the 'green' aspects of ionic liquids [17]. More recent understanding and analysis of catalytic reactions show that even so called benign ILs are now not considered chemically passive and are capable of modifying the catalyst resulting in differing chemistries [18].

Pioneering physical and chemical research in ILs where usually focused on imidazolium based cations with the corresponding tetrafluoroborate ($[BF_4]^-$) or hexafluorophosphate ($[PF_6]^-$) anions. These were initially chosen due to their ease of synthesis and purification. However, these ILs have been found to be of lower thermal stability and undergo hydrolysis reactions resulting in the production of HF and BF_3 [19]. Recently, hydrophilic fluorine based anions such as bis(trifluorosulfonyl)imide ($[NTf_2]^-$) and the tris(perfluoroalkyl)trifluorophosphate ($[FAP]^-$) have been developed. The stability of these anions is well established and has found many applications in the fields of catalysis [20]. Other halide free anions include alklysulfate, alkylphosphate and alklycarbonate molecules.

The synthesis of all ionic liquids starts with the 'neutralisation' of a Lewis base, typically alkylimidazoles, trialkylamines including pyrolles and piperidines, trialkylphopshines and pyridines are the most frequently used. For the synthesis of protic ILs, the Lewis base (represented as methylimidazole) can be neutralised directly by the addition of a Brönsted acid, see Figure 3, step 1. However, some of these salts are generally thermally unstable or exist in equilibrium with the free acid and base thus limiting their applications [21].

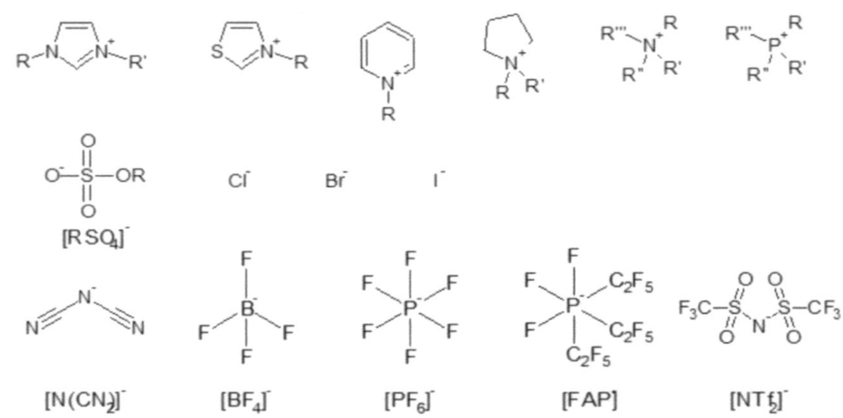

Figure 2. Common cations and anions used in air and moisture stable RTILs

For many ILs the preparation is strongly associated with the considerations about the required purity of the ILs post synthesis. For example in the synthesis of imidazolium ILs, the most common synthetic strategy involves the synthesis of the corresponding commercially available alkylimidazolium halide (Cl^- or Br^-) followed by anion metathesis to the desired IL, see Figure 3, steps 2 and 3. For hydrophobic ILs such as those containing the $[NTf_2]^-$ and $[PF_6]^-$ anions, this can be achieved by anion exchange of the imidazolium halide with the corresponding Li^+, Na^+ or $NH_4{}^+$ salt. Therein, the IL forms a separate phase which can be further purified by washing with water to remove any remaining halide. The water solubility of the ILs is very dependent on both the anion and cation present, and in general will decrease with increasing organic character (alkyl chain length) of the cation and decrease with increasing fluorinated character in the anion. For hydrophilic ILs such as those containing the, $[RSO_4]^-$ $[CF_3SO_2]^-$, $[BF_4]^-$ or $[N(CN)_2]^-$ anions the corresponding Ag^+ salt is used to precipitate out Ag halide salt which can be removed by filtration.

Impurities in ILs such as water, halides, starting materials and organic solvents not only have a profound effect on their physical properties [22] but have also resulted in significant changes in reaction chemistry [23-27]. Within both these areas significant steps have been made in the creation of cleaner and more economic routes to the preparation of ILs [28,29].

Other so halide free synthetic pathways involve direct alklylation to form the corresponding, alkylsulphate ($[RSO_4]^-$), alkylcarbonate ($[RCO_3]^-$) and trilfate ($[CF3SO_2]^-$) ILs, see Figure 3, step 4. The alkylsulphate [30] or alkylcarbonate [31] ILs which have found applications in their own right can be further reacted to form the corresponding halide free hydrophilic ILs.

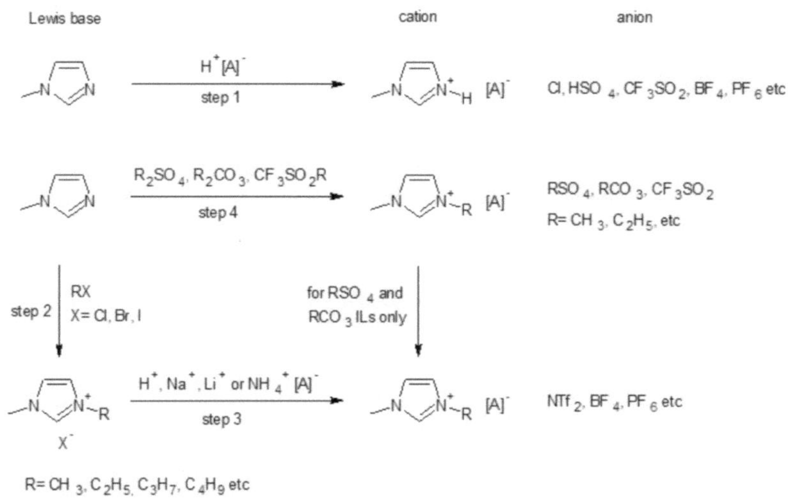

Figure 3. General synthesis of air stable, metal free Ionic liquids

3. IoNanofluids

IoNanofluids are complex systems of nanofluids with nanomaterials. Before we deal directly with these systems, a brief digression in the field of nanofluids will help to clarify the properties and these especial fluids. The origin of the nanofluids is linked to the current challenges faced by many high-tech industries and thermal management systems for cooling of smaller features of microelectronic and more power output-based devices. The conventional method to increase the cooling rate is to use extended heat transfer surfaces but this approach requires an undesirable increase in the size of the thermal management systems. This fact, added to the thermophysical properties of the traditionally heat transfer fluids (HTFs) used in industry, such as water, ethylene glycol (EG) or engine oil (EO) greatly limit the cooling performance. This situation was partially overcome by the use of materials with high thermal conductivity, such as nanomaterials suspended or dispersed in this type of base fluids. Choi [32] at Argonne National Laboratory of USA coined the concept of "nanofluids" to meet the aforementioned cooling challenges facing many advanced industries and devices. This new class of heat transfer fluids is engineered by dispersing nanometer-sized solid particles, rods or tubes in traditional heat transfer fluids and they were found to exhibit significantly higher thermophysical properties, particularly thermal conductivity and thermal diffusivity than those of base fluids (BFs) [33-38].

From practical application-based studies such as convective and boiling heat transfer characteristics [39-45], nanofluids (NFs) were also found to be even more promising as their convective heat transfer coefficient and critical heat flux were reported to be substantially higher as compared to those of their base fluids. In particular, nanofluids containing high thermal conductive materials such as carbon nanotubes (CNTs) shows anomalously enhanced thermal performance [45-47]. This is justified by the great difference between the thermal conductivity of CNTs (between 2000 and 3000 $Wm^{-1}K^{-1}$) and that of the base fluid (0.6 $Wm^{-1}K^{-1}$ for water).[2]

The concept of "IoNanofluids" was recently proposed by Nieto de Castro and co-workers [48] and it represents a very new class of heat transfer fluids where nanomaterials (particles, tubes and rods) are dispersed in ionic liquids only [49]. Since IoNanofluids are a specific type of nanofluids i.e., ionic liquids-based nanofluids, they are expected to have similar thermal properties than nanofluids, a fact that was proved recently by the authors [50]. The term IoNanofluids is therefore a new term in multidisciplinary fields such as nanoscience, nanotechnology, thermofluidity, chemical and mechanical engineering.The discovery that carbon nanotubes (CNT) and RTILs can be blended to form gels termed as "Bucky gels" which can potentially be used in many engineering or chemical processing such as making novel electronic devices, coating materials, and antistatic materials and thus, it opens a completely new field [51,52]. The "Bucky gels" are blends or emulsions of ILs with nanomaterials, mostly nanocarbons (tubes, fullerenes, and spheres) and they are actually CNT laden IoNanofluids. The possibility of using ionic liquids containing dispersed nanoparticles with specific functionalization such as functionalized single-walled nanotubes (SWCNT), multi-walled nanotubes (MWCNT) and fullerenes (C60, C80 etc.) opens the door to many applications. In recent reviews the authors have shown the properties of the IoNanofluids and nanofluids and highlighted their possible applications in different areas [9,53-55].

3.1. Manufacture of IoNanofluids

Although significant progress has been made in the last years, variability in the heat transfer characteristics of the nanofluids so far reported is presented, with very different thermal conductivity enhancements for the same systems. This variability may be the result of the various synthetic techniques employed, and the purity of the starting materials. The manufacture of nanofluids is delicate, as it does not mean necessarily a simple mixture of solid particles and a liquid, in the thermodynamic definition, and the techniques used by different authors are sometimes ill-defined [37,56].

Thus the synthesis of IoNanofluids can be a delicate operation. There are two main techniques used with normal solvents, the two-step process and the direct evaporation technique or single step. Most researchers use the two-step process, by dispersing commercial or self-produced nanoparticles in the liquid, a technique that can create large particle agglomerates, which can be destroyed by adding surfactants or using mechanical or ultrasound dispersion techniques.However the synthesis in situ seems to be the most efficient to produce very homogeneous particles, with a narrow size distribution, originating long period stability, especially for metal nanoparticles [57,58]. Aida and co-workers found that imidazolium-cation-based ionic liquids were excellent dispersants for CNT's, forming physical gels, that could be reproduced using sonication or by grinding the suspension in an agate mortar with a pes-

---

2 The thermal conductivity of molecular, organic and inorganic liquids, with the exception of molten metals, ranges from 0.1 to 0.6 Wm-1K-1.

http://dx.doi.org/10.5772/52596

tle. These techniques were followed by the current authors groups, to obtain very stable emulsions, without surfactants, with 0-3% (w/w) loading of MWCNTs (Multi-walled carbon nanotubes) in a range of imidazolium and pyrrolidinium ILs.

The importance of the purity of both starting materials and the technique used for manufacturing the dispersed nanomaterial is very important. In first place, the ionic liquid must be as pure as possible, as small quantities of water can affect its properties [59], and subsequently those of the Ionanofluid. Although the preparation of the dispersions is made open to atmosphere and therefore capable of introducing water in the samples, these were monitored by Karl-Fisher Coulometric analysis, and the amount of water in the pure ionic liquids never exceeded 400 ppm before the measurements and 800 ppm after the measurements, values that do not affect the thermal conductivity measurements [60].

The properties and characteristics of the nanomaterials are an extremely important issue. Most of the nanomaterials are spherical, rod or oblong in shape. Current manufacturing procedures can result in batch to batch variability as the degree of outer shell nanomaterial functionalization caused during manufacture and geometries of the nanomaterials can change[3]. One such example is in the case for silver nanoparticles, where 80% of the manufacturers use polymer coatings of non-disclosed thickness and properties. These coatings determine completely the heat transfer properties of the nanomaterials, and if we want to use silver nanoparticles they have to be chemically treated to eliminate them. Many publications on nanofluids systems previously reported were probably not aware of these problems and therefore those results have to be confirmed.

Our current experience in Ionanofluid manufacture is based on MWCNTs, and therefore we restrict our analysis to IoNanofluids based on these nanomaterials. However several studies are currently in progress using TiO_2 and Ag spherical particles, as well as nanomaterials delivered from nature [9,61-63].

The IoNanofluids based on MWCNTs must obey the following conditions: homogeneous dispersion, stable over a great period of time (not producing phase separation, even at a micro scale), and be free of additives, such as surfactants or salts. In addition, one of envisaged applications involves their use as heat transfer fluids, a control of viscosity is crucial, in order to create fluids with sensible heat transfer coefficients in dynamic regime (good fluidity). The preparation procedure involves weighing the nanomaterial, addition to the ionic liquid and introduction into the sonicator cell. Optimization of the time and sonication energy, visual observation for phase separation of the IL from the Ionanofluid is crucial[4]. Excessive time and high energy can result in the break-up of the Ionanofluid and breakage of tube walls. Moreover, impurities in the glass cells can lixiviate them and introduce further elements into the dispersion. The IoNanofluids dispersions produced are then allowed to settle several hours before the thermophysical properties measurements. The existence in micro-

3 In a majority, the producers of nanoparticles do not disclose the real structure of the particles, the presence of polymer coatings or oxidative type reactions

4 The dispersion is black, as shown in Figure 4. Therefore a great care as to be taken in the observation, helped by reflected light analysis through the suspensions

phase separation is controlled by measuring the properties of the dispersion, in several occasions. As an example, for thermal conductivity, dispersions in [C_2mim][dca], were measured within an interval of one year and their values did not differ by more than 2%, well within the uncertainty of the measurements [60]. Figure 4 illustrates the main steps for the preparation, starting with the ionic liquid, adding the carbon nanotubes, using the sonicator probe and finally the dispersion.

3.2. Experimental properties and their impact on process equipment design

As described above, the most important thermophysical properties for heat exchange equipment design are thermal conductivity, heat capacity, density and viscosity. They control the dynamics of heat transfer, namely the heat transfer coefficients and the heat transfer areas of the exchangers. In order to decide if any fluid can be used as an alternative to current engineering fluid, there are two essential questions to answer:

1. Are the properties of the fluid adequate to the process(es) where it will be used?

2. If the answer to the first technical question is affirmative, is the fluid economic competitive?

The answer to the first question raises another two problems: a) how accurate the thermophysical property data available is and b) what is the sensitivity of the main design parameters in heat exchange (heat transfer area, flow rates, pressure drops) are affected by the uncertainty of those properties. These factors will be discussed in detail below.

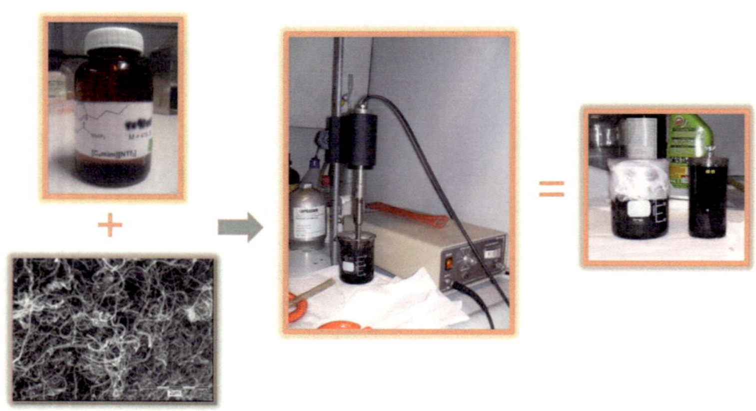

Figure 4. Main steps for the preparation of IoNanofluids with MWCNTs.

The determination of experimental values of the thermophysical properties of ionic liquids has been discussed in two recent reports, regarding the methods of measuring (existing and foreseen new developments) and how important is to characterize the samples in order to trust the measurements made with the highest accuracy available [59,64].This fact, using ionic liquids as base fluids for the nanofluids, is very important as they can absorb water

http://dx.doi.org/10.5772/52596

from the environment. As an example, Table 1 shows the results obtained for 3 ionic liquids in our laboratory, which were used for IoNanofluids preparation [60]. The ILs were obtained from Io-Li-Tec, DE, dried under vacuum for several days at approximately 60ºC and the water content was determined with a Karl-Fisher Coulometer.

The effect on the measured values of the properties depends on these, being more significant for viscosity [65,66]. For thermal conductivity an effect of 1000 ppm (0.02 in the molar fraction of the mixture), the maximum effect at 70ºC[5] is 0.2% (much smaller than measurement uncertainty). These results show that it is very important to characterize the samples, determining its water content, before and after the measurements, a fact that has not been recognized so far by many authors and journal editors.

The methods used to measure the thermophysical properties have also to be well characterized: many deviations between data obtained by different methods and different laboratories are caused by ill-defined measuring methods [59,64]. It is not the purpose of this review to describe the best experimental systems, as a full discussion was presented before [64]. However the readers should be attentive to this problem. It is very tempting for some research group that enters the field of ionic liquids to use equipments already available, using measuring cells that sometimes are not adequate. In addition the availability on the market of reliable measuring instruments also creates an opportunity for "fast" data production.

Ionic Liquid	Manufacturer Purity / %	Manufacturer H_2O / ppm	After Drying H_2O / ppm	After Measurements H_2O / ppm
$[C_2mim][dca]$	>98	1850	234.7 ± 48.6	488.9 ± 58.6
$[C_4mim][dca]$	>98	1480	324.9 ± 86.4	683.6 ± 41.7
$[C_4mpyr][dca]$	>98	1710	349.6 ± 47.9	637.0 ± 57.0

Table 1. Water content of 3 ionic liquids, before and after the thermal conductivity measurements

Other factors are known to affect the determination of experimental values, not strictly dependent of the instrumental methods used. For example the compatibility of ILs with seals, gaskets and metals contained in the measuring cell is very important. Ionic liquids anions and cations can be very different in size and to date most of the existing information has been obtained for imidazolium cations, making it difficult to generalize for other non-imidazolium ionic liquids. These ions are not mutual independent, can form aggregates and complicate structures in the liquid phase. In addition the viscosity is moderate to high, the liquids are electrical conducting and the heat capacity per unit volume is rather high. All these factors condition heat and mass transfer in the transport properties determination and must be known "a priori" to avoid systematic errors.

An example selected from reference [64] is sufficient to illustrate the point about "bad" and good" measurements. Figure 5 shows the heat capacity of $[C_4mim][BF_4]$, obtained using DSC

5 The effect decreases the value of the thermal conductivity and increases with temperature

[67-73]. A wide variation between the datasets of up to 20 % at room temperature is observed. This situation is very uncommon in calorimetry; however it is known that, apart from differences in sample purity, the DSC used must be well calibrated before use, a fact which could also help to explain the scatter of data. However, it can be seen [50] that 5 sets of data agree within 2%, within their mutual uncertainties [50,67,68,72,73], a result that would be considered excellent for DSC data, the most popular measuring method, due to its speed and excellent repeatability. Bearing in mind that handling and measuring ILs physical properties is not trivial, the authors would like to recommend a careful analysis of all these aspects prior to making measurements and reporting data.

The second point to analyze is how the sensitivity of the main design parameters in heat exchange (heat transfer area, flow rates, pressure drops) are affected by the uncertainty of those properties. This was the subject of many publications in the past, and recently we have applied this to ionic liquids. Using the same methodology [74], analyzing the effect of the uncertainty of thermophysical data of ionic liquids (density, heat capacity, thermal conductivity and viscosity) in the design of some current equipment, used in processes as solvents or heat transfer fluids. Data has been collected from IL Thermo database [75] for alkylmethylimidazolium, $[C_nmim]^-$ liquids, with $[BF_4]^-$ and $[PF_6]^-$ anions. This was justified by the fact that the thermophysical properties of ionic liquids, measured in different laboratories and by different methods do not agree within their mutual uncertainties. This was probably caused by incorrect methods of measurement and/or purity problems, as stated above. Results obtained show that the influence of actual errors in the thermophysical properties of ionic liquids can render any future design of chemical plant equipment as not working or excessively costing. Although the actual cost of ionic liquids is higher than conventional heat transfer fluids, the future production of higher quantities can make their use competitive, especially if a target price of 25US$/kg is achieved.

The heat storage capacities of $[BF_4]^-$ and $[PF_6]^-$ and other ionic liquids, containing anions like $[C_2H_5SO_4]$, $[(CF_3SO_2)_2N]$, $[CF_3SO_3]$ and $[C_8H_{17}SO_4]$, which can be considered as possible replacements of current heat transfer fluids, have been analyzed. A comparison with the properties of synthetic compounds (based on hydrocarbons, polyaromatics and siloxanes), showed that common imidazolium IL systems have higher heat capacities per unit volume than high performance commercial thermal fluids, such as Paratherm HE(a registered mark of Paratherm Corporation) and Syltherm 800™, Syltherm HF™ Dowtherm A™ and Dowtherm MX™ (trademarks of Dow Chemical Company, USA) [75]. Details of the methodology application can be found in this reference. The analysis was limited to study the effect of the uncertainty in the properties in the major design parameter, the heat transfer area, in a preselected heat transfer equipment, a shell and tube heat exchanger, as it reflects the changes in the design arising from the changes in the thermophysical properties of the ionic liquid process stream. This also permits a more facile estimation of the economic consequences of these changes in the design.

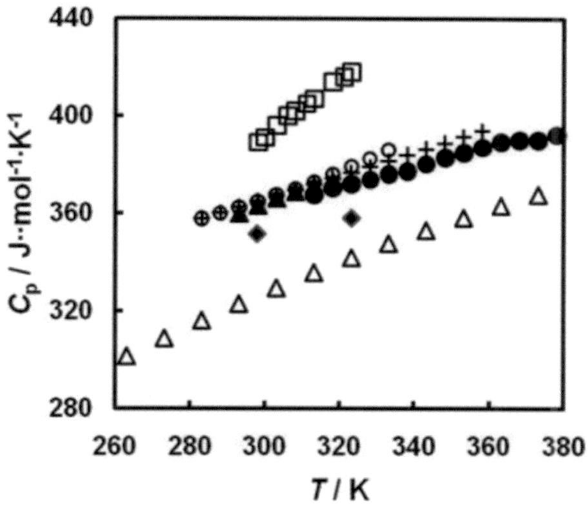

Figure 5. Existing values of the heat capacity for $[C_4mim][BF_4]$ as a function of temperature (2010). - Nieto de Castro et al. [50]; - Van Valkenburg et al. [67]; - Rebelo et al.[68]; - Kim et al.[69]; - Fredlake et al. [70]; - Waliszewski et al. [71]; - Garcia-Miaja et al. [72]; - Garcia-Miaja et al. [73].

The equipment chosen is one modern solar power plant that uses a molten salt receiver as thermal energy storage system, which captures the sun's energy and stores it in hot molten sodium nitrate or molten nitrates mixtures, so that power can be generated when needed, not just when the sun is shining. The heat transfer unit uses a molten salt through oil to salt heat exchanger [76,77]. The molten salt mixture used was replaced by a ionic liquid, one set of reference conditions for the operation were chosen based on a given set of values of the thermophysical properties and then the assigned values were perturbed about their reference values, within ranges commensurate with the actual stage of experimental uncertainty reported in IL Thermo database [75]. The effect of the uncertainties of the thermophysical properties of the ionic liquids can be calculated by calculating the new heat transfer area A_0 and its variation A as a function of the variation in percentage of the thermophysical properties, ,$\Delta\varrho$,$\Delta\eta$,$\Delta\lambda$ and ΔC_P. Figure 6 shows in a 3D plot the effect of ΔC_P and $\Delta\lambda$ in ΔA for $[C_2mim][BF_4]$, using again the viscosity uncertainty, $\Delta\eta$ as a parameter for the surfaces. The density is not shown, as this is the property known with less uncertainty. Not using extreme values, whereby the area can be overestimated by 50 %, an error of + 20 % in viscosity, - 20 % in thermal conductivity and -10 % in heat capacity generates an error of + 20 % in the area of the heat exchanger.

From the results presented we can conclude that the effects of the uncertainty in the thermophysical properties of RTIL's are high and that the heat transfer areas (see Table 2) are higher than those obtained with the used heat transfer oils. The exception to this is $[C_2mim][BF_4]$. These effects can render equipment obsolete and/or induce additional operational costs, well above of those estimated by design. When a heat exchanger is built, its cost will be weakly dependent on the size/length of the pipes, but highly dependent on the heat transfer area and the operational costs will increase significantly with size. However, to increase the capacity of the unit after having been built would represent an additional cost, possibly smaller than replacing it with a new unit. Therefore, the wise approach would be to obtain good experimental measurements of the heat transfer used, as described above.

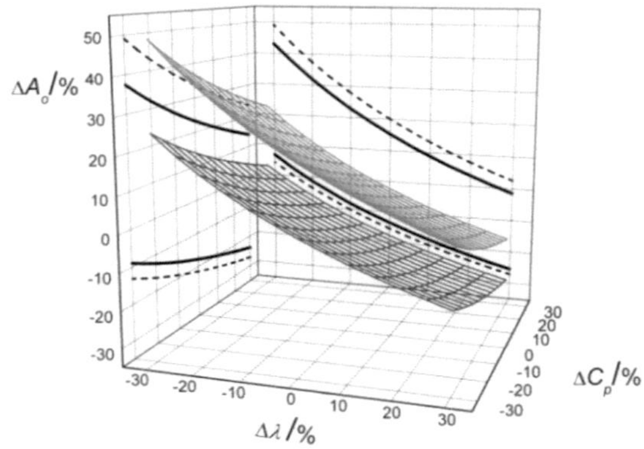

Figure 6. 3D plot of the effect of uncertainty of heat capacity (C_P) and thermal conductivity () for $[C_2mim][BF_4]$ on the area (A) of a shell and tubes heat exchanger,using viscosity uncertainty, , as a parameter for the surfaces (upper, + 20 %; lower, - 20 %); the lines in the planes represent the maximum and minimum values of the area variation for ± 20 %, , and ± 30 %, ---, in Δλ and ΔC_P. Adapted from [74].

In order to determine whether ionic liquids are economic as practical heat transfer fluids the costs of heat transfer equipment need to be examined. Details of the cost estimation of the heat exchanger can be found in [75]. The total cost required for a new design can be broken in five parts [84], the battery limits investment, the utility investment, the off-site investment, the engineering fees and the working capital. From these, we were concerned first with the battery limit investment, which is the cost of individual plant items and their installation to form the working process. The cost of the heat exchanger will be a function of its size (were the type of heat exchanger and the heat transfer area are critical), the materials of its construction (materials compatibility between metal parts and heat transfer fluids), design pressure and temperature, and it can be given by:

$$C_{\mathrm{E}} = C_{\mathrm{B}}\left(\frac{X}{X_{\mathrm{B}}}\right)^{m} f_{\mathrm{M}}\, f_{\mathrm{P}}\, f_{\mathrm{T}} \qquad (1)$$

where C_E represents the cost of the equipment with a given capacity X (here the heat transfer area, A_0), C_B the base cost of a reference equipment with a capacity X_B, and m a constant depending of the equipment type (m=0.68 for a shell and tube heat exchanger). The base cost of a carbon steel shell and tube heat exchanger, with a heat transfer area of 80 m^2 would be 3.2810^4 US$. f_M is the correction factor for materials of construction different from carbon steel, f_P the correction factor for design pressure and f_T the correction factor for design temperature. Values used for these factors can be found in references [74, 84].

The results obtained for the estimated costs for the shell and tube heat exchangers using the different heat transfer fluids are shown also in Table 2. Keeping other factors constant the heat exchangers for the $[PF_6]^-$ based RTIL's are significantly more expensive that the conventional fluids. However $[C_2mim][BF_4]$ IL is only moderately more expensive. Depending on the process, it may be possible to obtain savings in the ILs equipment, either by changing other design variables (D_e, D_i, pressure drops, materials, temperatures of the cold and hot streams, etc.,) or, by using other ionic liquids than the ones tested, especially those ILs with lower viscosity values. In addition, the cost of ionic liquids, which is currently in the order of 100-500 US$/kg for research due to high value of the personnel costs involved at this scale, are now evolving for industrial production to targeted prices of 25 US$/kg [85], a value which makes them valuable alternative heat transfer fluids, from the economic side. For example, the ionic liquid $[C_2mim][EtSO_4]$ is now available on an tonne scale, from Solvent Innovation GmbH, Cologne (www.solventinnovation.de) or BASF AG (www.basionics.de).

Heat Transfer Fluid	ρ /kgm^{-3}	C_P /Jkg^{-1}K^{-1}	η/mPas	λ mWm^{-1}K^{-1})	A_0 /m^2	C_E /k$
$[C_4mim][PF_6]$ (50°C)	1346 ± 1[83]	1493 ± 30[79]	68.8 ± 1.8[83]	146 ± 7[81]	480.75	738
$[C_6mim][PF_6]$ (50°C)	1273 ± 3[78]	1409 ± 61[79]	111.9 ± 3.2[80]	146 ± 7[81]	634.60	891
$[C_2mim][BF_4]$ (50°C)	1280 ± 2[78]	1600 ± 25[67]	15.9 ± 1.1[82]	196 ± 6[67]	217.29	430
Dowtherm A™ (50°C)	1041	1632	2.12	134	138.60	317
Dowtherm MX™ (100°C)	905	1870	2.09	114	159.08	348
Syltherm 800™ (80°C)	882	1711	3.86	124	202.86	410
Syltherm HF™ (80°C)	811	1830	0.83	92	135.47	312

Table 2. Values of Heat Transfer Liquids Properties, Reference Area $A_0$and Estimated Costs for the Shell and Tube Heat Exchanger

For IoNanofluids, there is only one study, still unpublished, performed in Lisbon laboratory [86], with $[C_4mim][NTf_2]$ and $[C_2mim][EtSO_4]$ and MWCNTs. As no data is available for the heat capacity and viscosity enhancements[6], we have assumed for the corresponding enhancements $E_\lambda = E_\eta$, E_{Cp} = 1.5; 2; 5 % and $E_\rho = 0$. Figure 7 shows the variation in the area A of

6 Measurements are under progress

the same heat exchanger as a function of the volume fraction of the Ionanofluid[7]. The data points are labeled to show the values of the areas obtained. The maximum enhancements in the thermal conductivity were found for the 3% w/w IoNanofluids, around 25% for $[C_4mim]$ $[NTf_2]$. The effect, only of the thermal conductivity enhancement can be as much as 20% decrease in the area, and therefore in the cost of the heat exchanger, saving about US$ 20 mfor a 20% volume fraction of MWCNT (3% w/w), a value very significant. Progress in the experimental measurement of density, heat capacity and viscosity of these IoNanofluids will be reported soon.

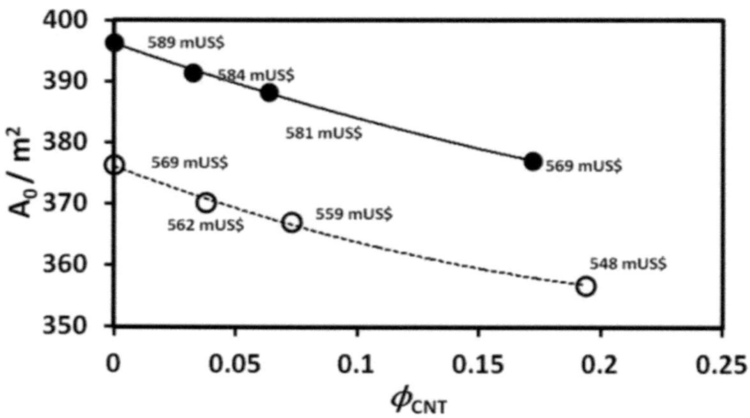

Figure 7. $A_o$vs$_{CNT}$ for ionic liquids and IoNanofluids. - fluids based on $[C_4mim][NTf_2]$; - fluids based on $[C_2mim][EtSO_4]$. Close to each data point is the cost of the heat exchanger, in US$.

3.3. Other process requirements

There are a number of factors that have to be considered to make the possible the use of these IoNanofluids in industry. We restrict the discussion associated with the analysis of thermal stability and toxicity.

3.3.1. Thermal stability

The thermal stability of the IoNanofluids has two major components: the thermal stability of the ionic liquid base fluid and the thermal stability of the dispersion, on a long term. These are facts that have to be studied. In the case of ionic liquids there is already a substantial amount of information that allow us to be confident that it is easy to target an operational value of 200ºC[8], especially for the liquids mentioned in this review. However, ionic liquids

7 As the densities of the carbon nanotubes are much smaller than those of the base fluids, the volume fraction is much greater than the weight fraction (0.5, 1, 3%)

8 In a great majority of chemical processes, temperatures for heat transfer above 200ºC are not used, for energy cost reasons

http://dx.doi.org/10.5772/52596

have the tremendous advantage of being target designed, so there will be always the possibility of synthesizing a new high temperature ionic liquid. However the thermal stability of the IoNanofluids is a totally different game, as the homogeneous system can be transformed in a heterogeneous phase, with micro phase equilibria, promoting aggregation of the nanomaterials and/or phase separation, at a micro or macroscopic level. In principle, the increase in temperature will facilitate the homogeneous system, due to the increase in entropy of the base fluid. No data is yet available for temperatures above 70ºC, so any further conclusion will be premature. New studies in this area are needed.

3.3.2. Toxicity of ionic liquids, nanomaterials and IoNanofluids

Regarding toxicity, there are not any studies for IoNanofluids, and none to date for nanofluids. In the absence of any additional effect, the toxicity of an Ionanofluid will be a sum of the toxicity of the base ionic liquid and that of the nanomaterial used. For ionic liquids, our current knowledge is still scarce, and without long term consequences, but some conclusions can already been suggested [87,88]. Following the extensive review by Petkovic et al. [87], "it is clear that the numerous formulations of ionic liquids available provide a great pool of, and impetus for, many commercial applications, but not without significant toxicological and environmental concerns. The vast majority of toxicological studies on ionic liquids, available up to the present date, have focused on imidazolium ionic liquids. In addition, frequently, the ionic liquids "selected" for study tackled under a common assay were randomly chosen. Despite the scientific weight of these studies, the lack of systematisation (e.g. monitoring the effect of defined structural alterations in a specific head group) means that it is impossible at the moment to achieve a holistic analysis, which weakens conclusions and devalues the predictive algorithms under development."(*citation*) Some problems regarding the selection of the bioassay, namely subjectivity and regional–orientation restrict generalizations. Legislation demands and standardised tests should be kept as a priority, engaging models of different complexity. The environmental persistence of any chemical should be taken as one of the most critical ecotoxicological parameters [88].

The European Community regulation on chemicals and their safe use—REACH (Registration, Evaluation, Authorisation and Restriction of CHemical substances)—[89] aims to increase the awareness of the industry on hazards and risk management. REACH registration, in force since 2007, is mandatory for any chemical produced in the quantity over one tonne per year. Although it is being criticised for itsever-increasing cost and the number of animals employed intesting, [90] REACH undoubtedly provides a meaningful, and necessary, framework to raise human and environmental safety. Currently, only the ionic liquids which have already found application on industrial scale are undergoing REACH registration. As an example, as $[C_2mim][X]$ ($X=Cl^-$, $[C_2SO_4]^-$, $[C_1SO_3]^-$,$[O_2CMe]^-$ and $[NTf_2]^-$) and $[C_4mim][Cl]$ [91], but there are nodoubts that this number will continuously increase.It appears, however that the cytotoxicity of ionic liquids cannot be systematically estimated by a summation of the independent effects of the cation and anion [92, 93]. Moreover, mixtures (binary or ternary) of ionic liquids have been rarely investigated [94].

The next components of IoNanofluids to be analysed are the nanomaterials used. However the toxicity of nanomaterials, although present in small mass content in the IoNanofluids, raises several problems and requires certain rules to be obeyed.

First, the nanomaterials dispersed in the IL have different chemical and physical properties than those of bulk materials of identical composition. It is then reasonable to expect that the biological properties of nanomaterials are different as that of bulk materials. Secondly, nanomaterials may be unique in environmental or biological systems. The properties of a nanoparticle in nonpolar solvents change when extracted into the aqueous phase; furthermore, nanoparticles in biological fluids (buffered solutions, cell culture media, or blood) may behave differently, as well. The nanoparticle surface is the part of the nanoparticle system that will have direct interactions with the biological entity (as with the ionic liquid); therefore, the surface of the nanoparticle will influence the biological response. Finally, a full characterization profile of the nanoparticles system being tested in biology must be reported [95].

The identity of the nanoparticle sample must be known in order to accurately report the positive, negative, or neutral effects of nanoparticles in vitro or in vivo. National and international standards committees (such as International Organization for Standardization, American Society for Testing and Materials, and International Council on Nanotechnology) have begun to establish recommendations for adequate nanomaterial physicochemical characterization data relevant to toxicology. These recommendations come from the literature of a variety of disciplines, including biological, environmental, and material sciences [96-98].

There are a few key points that the growing body of nanotoxiciology literature has taught us. First, morphological characterization, such as particle size and shape, should be measured in the most dispersed state achievable. Second, ideally, particle characterization data should be measured under conditions as close to the point application as possible a property should be measured using more than one method. The successful development of safe nanomaterials requires a strong collaborative effort between toxicologists, physical scientists and engineers. All characterization data should be validated using multiple techniques. Scientists from chemistry, biology, and engineering backgrounds must work together to address issues related to the potential impacts of nanomaterials, nanocomposites, and nanoparticle-containing consumer and medicinal products on the environment, human health, and even the synthesis and manufacture of nanomaterials. The ultimate goal of this collaborative effort is to determine the effects of nanomaterials in environmental and biological systems [95].

One key property essential to an understanding of the responses between nanomaterials and biological systems is the interaction between cells and the surface of the nanoparticle.

Primary characterization is performed on particles as-synthesized or as-received: in its dry native state. Secondary characterization is performed on particles in the wet phase as a solution or suspension in aqueous media. This media could be in ultrapure water, vehicle solution, or cell culture media (prokaryotic or eukaryotic). Physical and chemical characterization relevant to toxicity testing includes size and size distribution (including aggregation/agglomeration/coagulation state), concentration and purity, surface activity/reactivity, particle composition of surface coatings. Tertiary characterizations are performed on particles following interactions with cells under in vivo or in vitro conditions, and imaging the nano–bio interface. Characterization data and functionality information of nanoparticles suspended in serum, media, buffers, or other biological fluids may be different than data gathered in water. Determining the toxic effect of a nanomaterial is a challenging endeavour because each study requires a comprehensive material characterization component (which includes both physical and chemical properties) and adequate toxicological evaluation (relevant to the hypothesized route of exposure and eventual biological fate) [99].

It is difficult to produce nanomaterials on a large scale, because the mass of material produced is very small. One gram of nanoparticles is approximately equivalent to one billion particles. The immediate challenge of toxicological studies of nanoparticles is not only producing enough material for a complete in vivo study, but to also produce enough material for characterization purposes. Therefore, both new toxicological testing and characterization methods are needed when tackling this problem. New methodologies, as well as, standardization of common techniques are needed within the realm of determining the safety of nanomaterials [95,100].

Finally this problem must be tackled as a logical sequence from laboratory synthesis to industrial production. The priority aspects will be the ecological, human health, and waste elimination besides the costs of the full operation, as the functionalities of IoNanofluids are many and extremely varied. The most important parameter to be considered is the interference parameter for the mixtures (interfacial behaviour), a logic consequence of the primordial role of the nanomaterial – IL interaction [96]. A wide range of physicochemical properties are relevant to toxicology, like particle size distribution, morphology, chemical composition, solubility and surface chemistry and reactivity [95]. Not a single method can be used, but combinations of standard tests have to devise.

Some authors state that the surfactant may cause physical and/or chemical instability problems. The use of surfactants or any other additives to stabilize the microemulsions of the nanofluids can be worse than the nanomaterials and ionic liquids [99,100].

A new network of existing infrastructures to ensure a cost-effective and time-efficient examination of health, safety, and environmental aspects of nanomaterials throughout Europe and linking effectively with other international related activities is urgently needed, through European Chemical Agency, ECHA [101].

4. Theoretical Modelling of Nano and IoNanofluids

Since nanofluids were found to exhibit anomalously high thermal conductivity which cannot be predicted by the existing classical models, based on macro and nanoscale mechanisms numerous theoretical models for nanofluids have been developed over the last decade [37]. However, most of these models are neither validated with wide ranges of nanofluids systems nor accepted widely. On the other hand, no theoretical model so far is available (to the best of our knowledge) for the prediction of thermal conductivity of this newly emerged IoNanofluids. Thus, in an attempt to predict the effective thermal conductivity of IoNanofluids, representative classical model as well as recent models are used here.

The effective thermal conductivity of suspensions of milli- or micro-sized solid particles can be predicted by numerous classical models like the most popular Maxwell [102] and Hamilton-Crosser [103] models. The Maxwell model for the effective thermal conductivity (λ_{eff}) of suspensions of spherical inclusions can be expressed as [102]:

$$\lambda_{eff}=\lambda_f\left[\frac{\lambda_p+2\lambda_f+2\phi_p(\lambda_p-\lambda_f)}{\lambda_p+2\lambda_f-\phi_p(\lambda_p-\lambda_f)}\right] \tag{2}$$

where ϕ_p is the particle volume fraction, and $_f$ and $_p$ are the thermal conductivities of the base fluid and particle, respectively.

The Maxwell model [102] was later modified by Hamilton and Crosser [103] for the effective thermal conductivity of both the spherical and non-spherical particles by using a shape factor. Their model is a function of the thermal conductivities of both solid and liquid phases, volume fraction, and the shape of the disperse particles. The Hamilton-Crosser model [103] has the form:

$$\lambda_{eff}=\lambda_f\left[\frac{\lambda_p+(n-1)\lambda_f-(n-1)\phi_p(\lambda_f-\lambda_p)}{\lambda_p+(n-1)\lambda_f+\phi_p(\lambda_f-\lambda_p)}\right] \tag{3}$$

where the shape factor n = 3 for spherical particles and n = 6 for cylindrical particles. For spherical particles, Eq.(3) reduces to Eq.(2).

Most of the researchers working on nanofluids [37] found these classical models are unable to predict the anomalous thermal conductivity of nanofluids. Therefore, many theoretical studies have been carried out to understand the heat transfer mechanism and to develop models for predicting the effective thermal conductivity of nanofluids [37]. Among a handful of efforts, by taking into account the effects of particle size, concentration, and interfacial nanolayer two models for the prediction of thermal conductivity of nanofluids ($\lambda_{eff\text{-}nf}$) containing spherical and cylindrical nanoparticles were developed by Murshed et al. [36]. The model for suspensions of spherical nanoparticles is expressed as [36]:

$$\lambda_{eff\text{-}nf}=\lambda_f\frac{\phi_p\omega(\lambda_p-\omega\lambda_f)\left[2\gamma_1^3-\gamma^3+1\right]+(\lambda_p+2\omega\lambda_f)\gamma_1^3\left[\phi_p\gamma^3(\omega-1)+1\right]}{\gamma_1^3(\lambda_p+2\omega\lambda_f)-(\lambda_p-\omega\lambda_f)\phi_p\left[\gamma_1^3+\gamma^3-1\right]} \tag{4}$$

where $\omega=\lambda_{lr}/\lambda_f$, $\gamma=1+t/r_p$, $\gamma_1=1+t/(2r_p)$, r_p is the radius of the particle, t and λ_{lr} are the thickness and the thermal conductivity of interfacial nanolayer, respectively. On the other hand, model for the cylindrical nanoparticles has the form [36]:

$$\lambda_{eff\text{-}nf}=\lambda_f\frac{\phi_p\omega(\lambda_p-\omega\lambda_f)\left[\gamma_1^2-\gamma^2+1\right]+(\lambda_p+\omega\lambda_f)\gamma_1^2\left[\phi_p\gamma^2(\omega-1)+1\right]}{\gamma_1^2(\lambda_p+\omega\lambda_f)-(\lambda_p-\omega\lambda_f)\phi_p\left[\gamma_1^2+\gamma^3-1\right]} \tag{5}$$

Although the thickness of nanolayer (t)is considered to be 1 nm [36], the thermal conductivity of a nanolayer still cannot be determined by experimental or theoretical means. However, the orderness and orientation of fluid molecules absorbed on a nanoparticle surface result in an intermediate value of thermal conductivity of nanolayer i.e., $\lambda_f<\lambda_{lr}<\lambda_p$. Hence, the thermal conductivity nanolayer is given by$\lambda_{1r}=\omega\lambda_f$, where $\omega>1$ is an empirical parameter which depends on the orderness of fluid molecules in the interface as well as the nature and surface chemistry of nanoparticle. In this study, $t = 1$ nm and $\omega=1.2$ were used.

The measured and predicted thermal conductivity of nanofluids and IoNanofluids were compared in our previous study [104] and it has been depicted here. It can be seen from Figure 8that while the recent model by Murshed et al. [36] (i.e., Eq.(4)) shows fairly good prediction of the effective thermal conductivity of $[C_4mim][NTf_2]$-based Ionanofluid at low concentration of multi-wall carbon nanotubes (MWCNT), classical Hamilton-Crosser model [103] severely over-predicts the results. It is noted that the weight concentrations of MWCNT were converted to corresponding volumetric concentrations for the predictions by both models. Due to almost identical thermal conductivity values of both base ionic liquids i.e., $[C_4mim][NTf_2]$ and $[C_2mim][EtSO_4]$, both models give similar predictions for $[C_2mim][EtSO_4]$-based Ionanofluid [104]. However, at a high MWCNT concentration of 3 wt % (equivalent to 19.3 volume %) it is anticipated that the increase in thermal conductivity will not be as high as for the low concentration and any classical or recent model can easily over-predict the thermal conductivity at such a high concentration. This is mainly due to absence of dynamic mechanism for such high concentration.

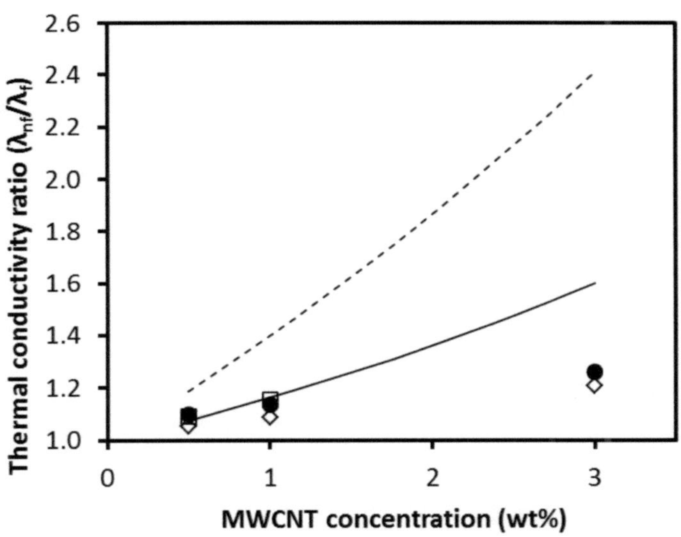

Figure 8. Comparison between predicted and measured thermal conductivity of nanofluids and IoNanofluids as a function of MWCNTconcentration [104]. - $[C_4mim][NTf_2]$ + MWCNT; - $[C_2mim][EtSO_4]$ + MWCNT; - water + MWCNT;- - - Eq. (3) for $[C_4mim][NTf_2]$ + MWCNT;Eq. (5) for $[C_4mim][NTf_2]$ + MWCNT.

Although by adjusting fitting parameters nanofluids′ thermal conductivity models [36] can be used for IoNanofluids, it is important to understand the underlying mechanisms and to develop model for thermal conductivity and other thermo-physical properties of IoNanofluids by taking into account electrochemical factors and molecular level structuring and interactions of both the base ionic liquids and the dispersed nanoparticles.

5. Conclusions and Looking Forward

IoNanofluids are shown to be a very interesting class of nanofluids for many engineering applications, namely in heat transfer and storage. Although there are still a short number of publications with IoNanofluids, the number is definitely increasing, as a very recent paper with grapheme shows [105]. Regardless of the small current industrial production of ionic liquids, the base fluids, it is expected to increase in a short term, because the properties of ionic liquids are very attractive as green solvents and reactants. The evolution of their chemistry has been tremendous and therefore, the price of their production will decrease to levels that will make them competitive with current fluids used in industry.

The manufacture, handling and characterization of IoNanofluids is still in development, but it is thought that significant progress will be made in a near future, namely in the nanoparticles and emulsions characterization. Further fundamental and applied studies are needed.

One special topic of concern is the toxicity of IoNanofluids, a reflex not only of the toxicity of the base ionic liquids but also of the constituent nanomaterials. Depending on the ionic liquids used, the systems can be toxic and all the users must be aware of it. However those IoNanofluids based on hydrophilic ionic liquids are a safe choice. In addition, there is also a risk with the nanomaterials used, and there is a wide awareness of how important is to study their effect in vitro and in vivo systems. A wide range of physicochemical properties are relevant to toxicology, like particle size distribution, morphology, chemical composition, solubility and surface chemistry and reactivity. The specific interactions of the ionic liquids with the nanomaterials can alter the individual toxicity of them. Many studies are still necessary to reach all the necessary conclusions and compliance with existing legislation, namely in Europe.

The application of existing models to predict the behaviour of the IoNanofluids, namely the enhancement in the thermal conductivity, showed that it is fundamental to understand better the mechanism of heat transfer in these systems, namely the role played by the interface ionic liquid (cation and anion)-nanoparticle, whatever shape they have. This needs theoretical developments and molecular simulation studies that will give the insight for developing new heat transfer models.

Acknowledgements

This research was partially funded by FCT- Fundação para a Ciência e a Tecnologia, Portugal, through Centro de Ciências Moleculares e Materiais, PEst-OE/QUI/UI0536/2011 and PDCT/QUE-FTT/104614/2008.

Author details

Carlos Nieto de Castro[1*], Ana P. C. Ribeiro[1], Salomé I.C. Vieira[1], João M. P. França[1], Maria J.V. Lourenço[1], Fernando V. Santos[1], Sohel M.S Murshed[1], Peter Goodrich[2] and Christopher Hardacre[2]

*Address all correspondence to: cacastro@fc.ul.pt

1 Departamento de Química e Bioquímica e Centro de Ciências Moleculares e Materiais Faculdade de Ciências, Universidade de Lisboa, Portugal

2 School of Chemistry and Chemical Engineering/QUILL, Queen's University, U. K.

References

[1] Holbrey, J. D., & Rogers, R. D. (2008). Physicochemical Properties of Ionic Liquids: Melting Points and Phase Diagrams. *In: Wasserscheid P, Welton T. (ed.) Ionic Liquids in Synthesis. Weinheim: VCH-Wiley*, 57-72.

[2] Wasserscheid, P., & Welton, T. (2008). editors. Ionic Liquids in Synthesis. *2nd Ed-completely revised and enlarged. Weinheim: VCH-Wiley.*

[3] Smiglak, M., Reichert, W. M., Holbrey, JD, Wilkes, J. S., Sun, L. Y., Thrasher, J. S., Kirichenko, K., Singh, S., Katritzky, A. R., & Rogers, R. D. (2006). Combustible Ionic Liquids by Design: Is Laboratory Safety Another Ionic Liquid Myth? *Chemical Communications*, 24, 2554-2556, 10.1039/B602086K.

[4] Ngo, H. L., Le Compte, K., Hargens, L., & Mc Ewen, A. B. (2000). Thermal Properties of Imidazolium Ionic Liquids. *Thermochimica Acta*, 357-358, 97-102.

[5] Baranyai, K. J., Deacon, G. B., Mac, Farlane. D. R., Pringle, J. M., & Scott, J. L. (2004). Thermal Degradation of Ionic Liquids at Elevated Temperatures. *Australian Journal of Chemistry*, 57(2), 145-147.

[6] Gordon, C. M. (2001). New Developments in Catalysis Using Ionic Liquids. *Applied Catalysis A: General*, 222(1-2), 101-117.

[7] Rodrigues, F., Do Nascimento, G. M., & Santos, P. S. (2007). Studies of Ionic Liquid Solutions by Soft X-ray Absorption Spectroscopy. *Journal of Electron Spectroscopy and Related Phenomena*, 155(1-3), 148-154.

[8] Crowhurst, L., Falcone, R., Lancaster, N. L., Liopis-Mestre, V., & Welton, T. (2006). Using Kamlet–Taft Solvent Descriptors to Explain the Reactivity of Anionic Nucleophiles in Ionic Liquids. *Journal of Organic chemistry*, 71(10), 8847-8853.

[9] Ribeiro, A. P. C., Vieira, S. I. C., França, J. M. P., Queirós, C. S., Langa, E., Lourenço, M. J. V., Murshed, S. M. S., & Nieto de Castro, CA. (2011). Thermal Properties of Ion-

ic Liquids and IoNanofluids. *In: Kokorin A. (ed.) Ionic Liquids: Theory, Properties, New Approaches. Rijeka: InTech*, 36-60.

[10] Kamimura, H., Kubo, T., Minami, I., & Mori, S. (2007). Effect and Mechanism of Additives for Ionic Liquids as New Lubricants. *Tribology International*, 40(4), 620-625.

[11] Mudring-V, A., Babai, A., Arenz, S., Giernoth, R., Binnemans, K., Driesen, K., & Nockemann, P. (2006). Strong Luminescence of Rare Earth Compounds in Ionic Liquids: Luminescent Properties of Lanthanide(III) Iodides iIn the Ionic Liquid 1-dodecyl-3-methylimidazolium bis(trifluoromethanesulfonyl)imide. *Journal of Alloys and Compounds*, 418(1-2), 204-208.

[12] Fredlake, C. P., Muldoon, M. J., Aki, S. N. V. K., Welton, T., & Brennecke, J. F. (2004). Solvent Strength of Ionic Liquid/CO2 Mixtures. *Physical Chemistry Chemical Physics*, 6(13), 3280-3285.

[13] Keskin, S., Kayrak-Talay, D., Akman, U., & Hortaçsu, Ö. (2007). A Review of Ionic Liquids towards Supercritical Fluid Applications. *The Journal of Supercritical Fluids*, 43(1), 150-180.

[14] Earle, M. J., Katdare, S. P., & Seddon, K. R. (2004). Paradigm Confirmed: The First Use of Ionic Liquids to Dramatically Influence the Outcome of Chemical Reactions. *Organic Letters*, 6(5), 707-710.

[15] Duan, Z., Gu, Y., & Deng, Y. (2006). Green and Moisture-Stable Lewis Acidic Ionic Liquids (choline chloride x $ZnCl_2$) Catalyzed Protection of Carbonyls at Room Temperature under Solvent-free Conditions. *Catalysis Communications*, 7(9), 651-656.

[16] Doherty, S., Goodrich, P., Hardacre, C., Knight, J. G., Nguyen, M. T., Parvulescu, V. I., & Paun, C. (2007). Recyclable Copper Catalysts Based on Imidazolium-Tagged Bis(oxazolines): A Marked Enhancement in Rate and Enantioselectivity for Diels-Alder Reactions in Ionic Liquid. *Advanced Synthesis & Catalysis*, 349(6), 951-963.

[17] Hemeon, I., Barnett, N. W., Gathergood, N., Scammells, P. J., & Singer, R. D. (2004). Manganese Dioxide Allylic and Benzylic Oxidation Reactions in Ionic Liquids. *Australian Journal of Chemistry*, 57(2), 125-128.

[18] Sowmiah, S., Srinivasadesikan, V., Tseng-C, M., & Chu-H, Y. (2009). On the Chemical Stabilities of Ionic Liquids. *Molecules*, 14(9), 3780-3813.

[19] Carda-Broch, S., Berthod, A., & Armstrong, D. W. (2003). Solvent Properties of the 1-butyl-3-methylimidazolium Hexafluorophosphate Ionic Liquid. *Analytical &Bioanalytical Chemistry*, 375(2), 191-199.

[20] Pârvulescu, V. I., & Hardacre, C. (2007). Catalysis in Ionic Liquids. *Chemical Reviews*, 107(6), 2615-2665.

[21] Greaves, T. L., Weerawardena, A., Fong, C., Krodkiewska, I., & Drummond, C. J. (2006). Protic Ionic Liquids: Solvents with Tunable Phase Behavior and Physicochemical Properties. *Journal of Physical Chemistry B*, 110(45), 22479-22487.

[22] Seddon, K. R., Stark, A., & Torres, M-J. (2000). Influence of Chloride, Water, and Organic Solvents on the Physical Properties of Ionic Liquids. *Pure Applied Chemistry*, 72(12), 275-287.

[23] Holbrey, J. D., & Seddon, K. R. (1999). Ionic Liquids. *Clean Products and Processes*, 1(4), 223-236.

[24] Wasserscheid, P., & Keim, W. (2000). Ionic Liquids-New "Solutions" for Transition Metal Catalysis. *Angewandte Chemie International Edition*, 39(21), 3772-3789.

[25] Welton, T. (2004). Ionic Liquids in Catalysis. *Coordination Chemistry Reviews*, 248(21-24), 459-477.

[26] Freemantle, M. (1998). Designer Solvents- Ionic Liquids May Boost Clean Technology Development. *Chemical and Engineering News*, 76(13), 32-37.

[27] Suarez, P. A. Z., Dullius, J. E. L., Einloft, S., de Souza, R. F., & Dupont, J. (1997). Two-Phase Catalytic Hydrogenation of Olefins by Ru(II) and Co(II) Complexes Dissolved in 1-n-butyl-3-methylimidazolium Tetrafluoroborate Ionic Liquid. *Inorganica Chimica Acta*, 255(1), 207-209.

[28] Holbrey, J. D., Reichert, W. M., Swatloski, R. P., Broker, G. A., Pitner, W. R., Seddon, K. R., & Rogers, R. D. (2002). Efficient, Halide Free Synthesis of New, Low Cost Ionic Liquids: 1,3-dialkylimidazolium Salts Containing methyl- and ethyl-sulfate Anions. *Green Chemistry*, 4(5), 407-413.

[29] Wasserscheid, P., van Hal, R., & Bosmann, A. (2002). n-Butyl-3-methylimidazolium ([bmim]) octylsulfate-An Even 'Greener' Ionic Liquid. *Green Chemistry*, 4(4), 400-404.

[30] Wasserscheid, P., van Hal, R., Bösmann, A., Eßer, J., & Jess, A. (2003). New Ionic Liquids Based on Alkylsulfate and Alkyl OligoetherSulfate Anions: Synthesis and Applications. *In: Rogers R.D., Seddon K.R. (eds.) Ionic Liquids as Green Solvents. Washington D.C.: American Chemical Society*, 57-69.

[31] Jessop, P. G., Mercer, S. M., & Heldebrant, D. J. (2012). CO2-Triggered Switchable Solvents, Surfactants, and other Materials. *Energy and Environmental Science*, 5(6), 7240-7253.

[32] Choi, S. U. S. (1995). Enhancing Thermal Conductivity of Fluids with Nanoparticles. *ASME FED*, 231, 99-105.

[33] Lee, S., Choi, S. U. S., Li, S., & Eastman, J. A. (1999). Measuring Thermal Conductivity of Fluids Containing Oxide Nanoparticles. *Journal of Heat Transfer*, 121(2), 280-289.

[34] Murshed, S. M. S., Leong, K. C., & Yang, C. (2005). Enhanced Thermal Conductivity of TiO_2-Water based Nanofluids. *International Journal of Thermal Sciences*, 44(4), 367-373.

[35] Murshed, S. M. S., Leong, K. C., & Yang, C. (2006). Determination of the Effective Thermal Diffusivity of Nanofluids by the Double Hot-Wire Technique. *Journal of Physics D: Applied Physics*, 39(24), 5316-5322.

[36] Murshed, S. M. S., Leong, K. C., & Yang, C. (2008). Investigations of Thermal Conductivity and Viscosity of Nanofluids. *International Journal of Thermal Sciences*, 47(5), 560-568.

[37] Murshed, S. M. S., Leong, K. C., & Yang, C. (2008). Thermophysical and Electrokinetic Properties of Nanofluids- A Critical Review. *Applied Thermal Engineering*, 28(17-18), 109-125.

[38] Yu, W., France, D. M., Routbort, J. L., & Choi, S. U. S. (2008). Review and Comparison of Nanofluid Thermal Conductivity and Heat Transfer Enhancements. *Heat Transfer Engineering*, 29(5), 432-460.

[39] You, S. M., Kim, J. H., & Kim, K. M. (2003). Effect of Nanoparticles on Critical Heat Flux of Water in Pool Boiling of Heat Transfer. *Applied Physics Letters*, 83(16), 3374-3376.

[40] Wen, D., & Ding, Y. (2004). Experimental Investigation into Convective Heat Transfer of Nanofluids at the Entrance Region under Laminar Flow Conditions. *International Journal of Heat and Mass Transfer*, 47(24), 5181-5188.

[41] Wen, D., & Ding, Y. (2005). Experimental Investigation into the Pool Boiling Heat Transfer of Aqueous based Alumina Nanofluids. *Journal of Nanoparticle Research*, 7(2-3), 265-274.

[42] Bang, I. C., & Chang, S. H. (2005). Boiling Heat Transfer Performance and Phenomena of Al2O3-Water Nano-Fluids from a Plain Surface in a Pool. *International Journal of Heat and Mass Transfer*, 48(12), 2407-2419.

[43] Heris, S. Z., Etemad, S. G., & Esfahany, M. S. (2006). Experimental Investigation of Oxide Nanofluids under Laminar Flow Convective Heat Transfer. *International Communications of Heat Mass Transfer*, 33(4), 529-535.

[44] Murshed, S. M. S., Leong, K. C., Yang, C., & Nguyen, N. T. (2008). Convective Heat Transfer Characteristics of Aqueous TiO_2 Nanofluids under Laminar Flow Conditions. *International Journal of Nanoscience*, 7(6), 325-331.

[45] Murshed, S. M. S., Milanova, D., & Kumar, R. (2009). An Experimental Study of Surface Tension-Dependent Pool Boiling Characteristics of Carbon Nanotubes-Nanofluids. In: Proceedings of the 7th ASME International Conferences on Nanochannels, Microchannels and Minichannels (ICNMM'09), Pohang, South Korea June., 22-24.

[46] Choi, S. U. S., Zhang, Z., Yu, W., Lockwood, F., & Grulke, E. (2001). Anomalous Thermal Conductivity Enhancement in Nanotube Suspensions. *Applied Physics Letters*, 79(14), 2252-2254.

[47] Ding, Y., Alias, H., Wen, D., & Williams, R. A. (2006). Heat Transfer of Aqueous Suspensions of Carbon Nanotubes (CNT Nanofluids). *International Journal of Heat and Mass Transfer*, 49(1-2), 240-250.

[48] Ribeiro, A.P.C, Lourenço, M.J.V, Nieto de Castro, CA, & Hardacre, C. (2008). Thermal Conductivity of "Bucky Gels". In: Proceedings of Conference on Molten Salts and Ionic Liquids (EUCHEM2008), Copenhagen, Denmark August , 24-29.

[49] Ribeiro, A. P. C., Lourenço, M. J. V., & Nieto de Castro, C. A. (2009). Thermal Conductivity of IoNanofluids. In: Proceedings of 17th Symposium on Thermophysical Properties, Boulder, USA June., 21-26.

[50] Nieto de Castro, CA, Lourenço, M. J. V., Ribeiro, A. P. C., Langa, E., Vieira, S. I. C., Goodrich, P., & Hardacre, C. (2010). Thermal Properties of Ionic Liquids and IoNanofluids of Imidazolium and Pyrrolidinium Liquids. *Journal of Chemical Engineering Data*, 55(2), 653-661.

[51] Fukushima, T., Kosaka, A., Ishimura, Y., Yamamoto, T., Takigawa, T., Ishii, N., & Aida, T. (2003). Molecular Ordering of Organic Molten Salts Triggered by Single-Walled Carbon Nanotubes. *Science*, 300(5628), 2072-2074.

[52] Fukushima, T., & Aida, T. (2007). Ionic Liquids for Soft Functional Materials with Carbon Nanotubes. *Chemistry-A European Journal*, 13(18), 5048-5058.

[53] Murshed, S. M. S., & Nieto de Castro, C.A. (2012). Nanofluids as Advanced Coolants. *In: Mohammad A., Inamuddin (eds.) Green Solvents I: Properties and Applications in Chemistry. Dordrecht: Springer*, 397-415.

[54] Nieto de Castro, CA, Murshed, S. M. S., Lourenço, M. J. V., Santos, F. J. V., Lopes, M. L. M., & França, J. M. P. (2012). IoNanofluids- New Heat Transfer Fluids for Green Process Development. *In: Mohammad A., Inamuddin (eds.) Green Solvents I: Properties and Applications in Chemistry. Dordrecht: Springer*, 233-249.

[55] Murshed, S. M. S., & Nieto de Castro, C. A. (2011). Forced Convective Heat Transfer of Nanofluids in Minichannels. *In: Ahsan A. (ed.) Two Phase Flow, Phase Change and Numerical Modeling. Rijeka: InTech.*

[56] Keblinski, P., Prasher, R., & Eapen, J. (2008). Thermal Conductance of Nanofluids: Is the Controversy Over? *Journal of Nanoparticle Research*, 10(7), 1089-1097.

[57] Patil, V. S., Krishna, S. R., Hadalwar, R. R., Gaikwad, A. B., Sathaye, S. D., & Patil, K. R. (2011). One-Step In Situ Synthesis of Nhx-Adsorbed Rhodium Nanocrystals in Liquid-Liquid Interfaces for Possible Electrocatalytic Applications. *Journal of Colloid and Interface Science*, 358(1), 238-244.

[58] Patil, V., Hadalwar, R., Sathaye, S., Joshi, M., Salavera, D., & Coronas, A. (2012). One Pot Synthesis of Aqueous Gold Nanoparticles Dispersion without External Reducing or/and Dispersing/Capping Agent. *Chemical Communications (submitted).*

[59] Nieto de Castro, C. A. (2010). Thermophysical Properties of Ionic Liquids: Do We Know How to Measure them Accurately? *Journal of Molecular Liquids*, 156(1), 10-17.

[60] França, J. M. P., Vieira, S. I. C., Murshed, S. M. S., Lourenço, M. J. V., & Nieto de Castro, C. A. (2012). Thermal Conductivity of [C2mim][dca], [C4mim][dca] and

[C4mpyr][dca] and their IoNanofluids with Nanosystems. *Fluid Phase Equilibria (to be submitted).*

[61] Vieira, S. I. C., Lourenço, M. J., & Nieto de Castro, C. A. (2012). Paints with IoNanofluids as Pigments for Improvement of Heat Transfer on Architectural and Heat Exchangers Surfaces. In: Proceedings of 18th Symposium on Thermophysical Properties, Boulder, USA June., 24-29.

[62] Queirós, C. S., Vieira, S. I. C., Lourenço, M. J. V., & Nieto de, Castro. C. A. (2012). Study of Fruit Waste Reuse as New Thermal Absorbing Materials. In: Proceedings of 18th Symposium on Thermophysical Properties, Boulder, USA June., 24-29.

[63] Queirós, C. S., Vieira, S. I., Lourenço, M. J. V., & Nieto de Castro, C. A. (2012). Natural Nano Resources to Enrich Scientific and Economical Global Needs. In: Proceedings of International Workshop on Ionic Liquids- Seeds for New Engineering Applications (WILS2012) Lisbon, Portugal February., 2-3.

[64] Nunes, V. M. B., Lourenço, M. J. V., Santos, F. J. V., Matos, M. L. M., & Nieto de, Castro. C. A. (2010). Accurate Measurements of Physico-Chemical Properties on Ionic Liquids and Molten Salts. *In: Gaune-Escard M, Seddon K.E. (eds.) Ionic Liquids and Molten Salts: Never the Twain. London: John Wiley*, 229-263.

[65] Widegren, J. A., Laesecke, A., & Magee, J. W. (2005). The Effect of Dissolved Water on the Viscosities of Hydrophobic Room-Temperature Ionic Liquids. Chemical Communications; 10.1039/B417348A , 12, 1610-1612.

[66] Widegren, J. A., & Magee, J. W. (2007). Density, Viscosity, Speed of Sound, and Electrolytic Conductivity for the Ionic Liquid 1-Hexyl-3-Methylimidazolium Bis(trifluoromethylsulfonyl)imide and its Mixtures with Water. *Journal of Chemical Engineering Data*, 52(6), 2331-2338.

[67] Van Valkenburg, ME, Vaughn, R. L., Williams, M., & Wilkes, J. S. (2005). Thermochemistry of Ionic Liquid Heat-Transfer Fluids. *ThermochimicaActa*, 425(1-2), 181-188.

[68] Rebelo, L. P. N., Najdanovic-Visak, V., Visak, Z. P., Nunes, da., Ponte, M., Szydlowski, J., Cerdeirina, CA, Troncoso, J., Romani, L., Esperanca, J. M. S. S., Guedes, H. J. R., & de Sousa, H. C. (2004). A Detailed Thermodynamic Analysis of [C_4mim][BF_4] + Water as a Case Study to Model Ionic Liquid Aqueous Solutions. *Green Chemistry*, 6(8), 369-381.

[69] Kim, K. S., Shin, B. K., & Ziegler, F. (2004). Refractive Index and Heat Capacity of 1-Butyl-3-Methylimidazolium Bromide and 1-Butyl-3-Methylimidazolium Tetrafluoroborate, and Vapor Pressure of Binary Systems for 1-Butyl-3-Methylimidazolium Bromide + Trifluoroethanol and 1-Butyl-3-Methylimidazolium Tetrafluoroborate + Trifluoroethanol. *Fluid Phase Equilibria*, 218(2), 215-220.

[70] Fredlake, C. P., Crosthwaite, J. M., Hert, D. G., Aki, S. N. V. K., & Brennecke, J. F. (2004). Thermophysical Properties of Imidazolium-based Ionic Liquids. *Journal of Chemical Engineering Data*, 49(4), 954-964.

[71] Waliszewski, D., Stepniak, I., Piekarski, H., & Lewandowski, A. (2005). Heat Capacities of Ionic Liquids and their Heats of Solution in Molecular Liquids. *ThermochimicaActa*, 433(1-2), 149-152.

[72] Garcia-Miaja, G., Troncoso, J., & Romani, L. (2008). Excess Properties for Binary Systems Ionic Liquid Plus Ethanol: Experimental Results and Theoretical Description using the ERAS Model. *Fluid Phase Equilibria*, 274(1-2), 59-67.

[73] Garcia-Miaja, G., Troncoso, J., & Romani, L. (2009). Excess Molar Properties for Binary Systems of Alkylimidazolium-Based Ionic Liquids Plus Nitromethane. Experimental Results and ERAS-Model Calculations. *Journal of Chemical Thermodynamics*, 41(3), 334-341.

[74] França, J. M. P., Nieto de, Castro. . C. A., Nunes, V. M. B., & Lopes, M. L. M. (2009). The Influence of Thermophysical Properties of Ionic Liquids in Chemical Process Design. *Journal of Chemical Engineering Data*, 54(9), 2569-2575.

[75] Ionic Liquids Database- (IL Thermo), NIST Standard Reference Database #147. (2006). *US Secretary of Commerce.*, http://ilthermo.boulder.nist.gov/ILThermo/mainmenu.uix.

[76] Pacheco, J. E., Showalter, S. K., & Kolb, W. J. (2001). Development of a Molten-Salt Thermocline Thermal Storage System for Parabolic Trough Plants. In: Proceedings of the Solar Forum Solar Energy: The Power to Choose, Washington D.C., USA April 2001., 21-25.

[77] Nunes, V. M. B., Lourenço, M. J. V., Santos, F. J. V., & Nieto de Castro, CA. (2003). The Importance of the Accurate Data on Viscosity and Thermal Conductivity in Molten Salts Applications. *Journal of Chemical Engineering Data*, 48(3), 446-450.

[78] Gardas, R. L., Freire, M. G., Carvalho, P. J., Marrucho, I. M., Fonseca, I. M. A., Ferreira, A. G., & Coutinho, J. A. P. (2007). High-Pressure Densities and Derived Thermodynamic Properties of Imidazolium-based Ionic Liquids. *Journal of Chemical Engineering Data*, 52(1), 1880-1888.

[79] Holbrey, J. D., Reichert, W. M., Reddy, R. G., & Rogers, R. D. (2003). Heat Capacities of Ionic Liquids and Their Applications as Thermal Fluids. *In: Rogers R.D., Seddon K.R. (eds.) Ionic Liquids as Green Solvents. Washington D.C.: American Chemical Society*, 121-133.

[80] Pereiro, A. B., Legido, J. L., & Rodriguez, A. (2007). Physical Properties of Ionic Liquids based on 1-alkyl-3-Methylimidazolium Cation and Hexafluorophosphate as Anion and Temperature Dependence. *The Journal of Chemical Thermodynamics*, 39(8), 1168-1175.

[81] Tomida, D., Kenmochi, S., Tsukada, T., Qiao, K., & Yokoyama, C. (2007). Thermal Conductivities of $[C_4mim][PF_6]$, $[C_6mim][PF_6]$, and $[C_8mim][PF_6]$ from 294 to 335 K at Pressures up to 20 MPa. *International Journal of Thermophysics*, 28(4), 1147-1160.

[82] Zhang, S., Li, X., Chen, H., Wang, J., Zhang, J., & Zhang, M. (2004). Determination of Physical Properties for the Binary System of 1-Ethyl-3-methylimidazolium Tetrafluoroborate + H2O. *Journal of Chemical Engineering Data*, 49(4), 760-764.

[83] Huddleston, J. G., Visser, A. E., Reichert, W. M., Willauer, H. D., Broker, G. A., & Rogers, R. D. (2001). Characterization and Comparison of Hydrophilic and Hydrophobic Room Temperature Ionic Liquids Incorporating the Imidazolium Cation. *Green Chemistry*, 3(4), 156-164.

[84] Smith, R. (2005). Chemical Process: Design and Integration. *UK: John Wiley & Sons.*

[85] Massonne, K. (2012). Ionic liquids at BASF SE: Introduction and Technical Applications. In: Proceedings of International Workshop on Ionic Liquids- Seeds for New Engineering Applications (WILS2012), Lisbon, Portugal February., 2-3.

[86] França, J. M. P., Nieto de Castro, C. A., Lopes, M. L. M., & Murshed, S. M. S. (2012). Evidence of Economic Impact of Quality Measurements. In: Proceedings of International Workshop on Ionic Liquids- Seeds for New Engineering Applications (WILS2012), Lisbon, Portugal February., 2-3.

[87] Petkovic, M., Seddon, K. R., Rebelo, L. P. N., & Pereira, C. S. (2011). Ionic liquids: a pathway to environmental acceptability. *Chemical Society Reviews*, 40-1383.

[88] Vibouda, S., Papaiconomoub, N., Cortesia, A., Chatelb, G., Drayeb, M., & Fontvieille, D. (2012). Correlating the structure and composition of ionic liquids with their toxicity on Vibrio fischeri: A systematic study. *Journal of Hazardous Materials*, 215-216(1), 40-48.

[89] REACH-Registration. Evaluation, Authorisation and Restriction of CHemicals. http://ec.europa.eu/enterprise/sectors/chemicals/reach/index_en.htm.

[90] Hartung, T., & Rovida, C. (2009). *Nature*, 460-1080.

[91] Ionic Liquids from BASF-Solutions for Your Success. http://ionmet.eu/fileadmin/ionmet/training/20090324_Munich/9_Vagt_REACH.pdf.

[92] Torrecilla, J. S., Palomar, J., Lemus, J., & Rodriguez, F. (2010). A quantum-chemical-based guide to analyze/quantify the cytotoxicity of ionic liquids. *Green Chemistry*, 12-123.

[93] Palomar, J., Torrecilla, J. S., Lemus, J., Ferro, V. R., & Rodriguez, F. (2010). A COSMO-RS based guide to analyze/quantify the polarity of ionic liquids and their mixtures with organic cosolvents. *Physical Chemistry Chemical Physics*, 12, 2991-2000.

[94] Matzke, M., Stolte, S., Böschen, A., & Filser, J. (2008). Mixture effects and predictability of combination effects of imidazolium based ionic liquids as well as imidazolium based ionic liquids and cadmium on terrestrial plants (Triticumaestivum) and limnic green algae (Scenedesmusvacuolatus). *Green Chemistry*, 10-784.

[95] Sayes, C. M., & Warheit, D. B. (2009). Characterization of nanomaterials for toxicity assessment. *Wiley Interdisciplinary Reviews Nanomedicine and Nanobiotechnology*, 1(6), 660-670.

[96] Regulatory Aspects of Nanomaterials. Summary of legislation in relation to health, 2008 safety and environment aspects of nanomaterials, regulatory research needs and related measures. *COM(2008) 366 final; CEC, Brussels*, 1-45.

[97] Riediker, M., & Katalagarianakis, G. (2012). Compendium of Projects in the European NanoSafety Cluster.

[98] Dawson, K. A., Anguissola, S., & Lynch, I. (2012). The need for in situ characterisation in nanosafety assessment: funded transnational access via the QNano research infrastructure. *Nanotoxicology*, 10.3109/17435390.2012.658096.

[99] Kunz, W., Maurer, E., Klein, R., Touraud, D., Rengstl, D., Harrar, A., Dengler, S., & Zech, O. (2011). Low Toxic Ionic Liquids, Liquid Catanionics, and Ionic Liquid Microemulsions. *Journal of Dispersion Science and Technology*, 32-1694.

[100] Carrera, G. V. S. M., Frade, R. F. M., Aires-de-Sousa, J., Afonso, C. A. M., & Branco, L. C. (2010). Synthesis and properties of new functionalized guanidinium based ionic liquids as non-toxic versatile organic materials. *Tetrahedron*, 66-8785.

[101] http://echa.europa.eu/.

[102] Maxwell, J. C. (1891). A Treatise on Electricity and Magnetism. *Oxford: Clarendon Press*.

[103] Hamilton, R. L., & Crosser, O. K. (1962). Thermal Conductivity of Heterogeneous Two Component Systems. *Industrial & Engineering Chemistry Fundamentals*, 1(3), 187-191.

[104] Nieto de Castro, C. A., Murshed, S. M. S., Lourenço, M. J. V., Santos, F. J. V., Lopes, M. L. M., & França, J. M. P. (2012). Enhanced thermal conductivity and specific heat capacity of carbon nanotubes IoNanofluids. *International Journal of Thermal Sciences (2012)*, 10.1016/j.ijthermalsci.03.010.

[105] Wang, F., Han, L., Zhang, Z., Fang, X., Shi, J., & Ma, W. (2012). Surfactant-free ionic liquid-based nanofluids with remarkable thermal conductivity enhancement at very low loading of grapheme. Nanoscale Research Letters 10.1186/1556-276X-7-314On line 19 June 2012., 7(314)

Chapter 8

The Structure of Supported Ionic Liquids at the Interface

Fatemeh Moosavi

Additional information is available at the end of the chapter

http://dx.doi.org/10.5772/53646

1. Introduction

Ionic liquids (ILs) may acceptably be defined as fluid semi-organic salts composed entirely of bulky asymmetric organic cations and organic or inorganic anions at or near room temperature. There is considerable consensus that a qualified IL must melt below 100 °C. Ionic liquids are salt, existing in the liquid phase at and around 298 K. A typical IL with a bulky organic cation (e.g., N-alkylpyridinium, N-N-dialkylimidazolium, alkylimidazolium, alkylphosphonium, and alkylammonium) is weakly coordinated to an organic or inorganic anion, such as BF_4^-, Cl^-, I^-, $AlCl_4^-$, PF_6^-, NO_3^-, CH_3COO^-, $CF_3SO_3^-$, $[(CF_3SO_2)_2N^-]$, etc. to constitute a series of low melting ILs, as shown in Figure 1.

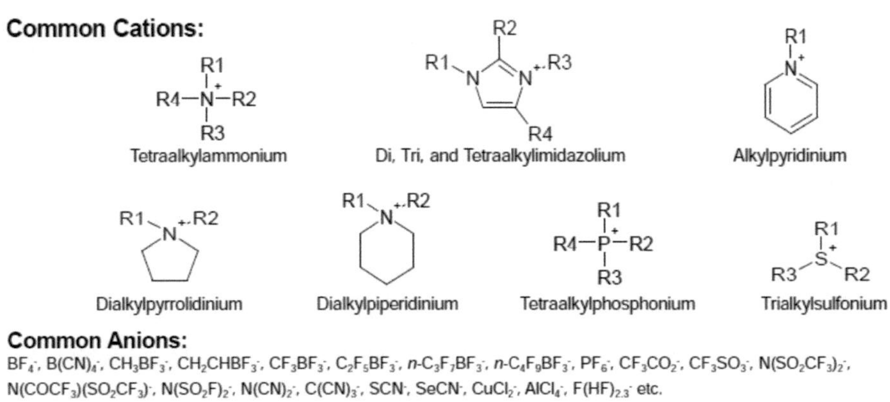

Figure 1. Common cations and anions for room temperature ionic liquids

INTECH
open science | open minds

Ionic liquids have attracted much attention as a replacement for traditional organic solvents as they possess many attractive properties. Among these properties, intrinsic ion conductivity, low volatility, high chemical and thermal stability, low vapor pressure, low combustibility, hydrophobicity, wide electrochemical windows, and high heat capacity and cohesive energy density are few. Compared with commonly used organic compounds, they have low toxicity, are non-flammable, and have negligible or nonzero volatility. Furthermore, alteration of the anions or the length of the alkyl groups allows fine-tuning of the physicochemical properties of ILs, such as viscosity, conductivity, solvation, catalytic activity, hydrophobicity, melting points, etc. Thus, ILs can be strategically designed for different applications [1,2]. The application of ILs can primarily be traced to the pioneering work in the beginning of the 1980s on pyridinium-based [3] and imidazolium-based [4] ILs for electrochemical studies. Over recent years, the concept of "green chemistry" has become well known among scientists worldwide. In particular, exploration of environmentally friendly green solvents as alternatives to volatile organic compounds in synthesis, catalysis, extraction and separation, and electrolytic processes has been persistently pursued. IL, possessing many novel properties, is the most competitive candidate that caters to all trades and professions. Properties such as nonflammability, high ionic conductivity, and electrochemical and thermal stability of ILs make them ideal electrolytes in electrochemical devices like in batteries [5-8], capacitors [9-11], fuel cells [12], photovoltaics [13-18], actuators [19], and electrochemical sensors. In addition, ILs have been widely used in electrodeposition, electrosynthesis, electrocatalysis, electrochemical capacitor, lubricants, plasticizers, solvent, lithium batteries, solvents to manufacture nanomaterials, extraction, gas absorption agents, and so forth. ILs can improve separation of complex mixtures of both polar and nonpolar compounds when used either as stationary phase or as additives in gas-liquid chromatography [20-23], liquid chromatography [22], and capillary electrophoresis [24]. They are also used in optical sensors [25,26] and also to enhance the analytical performance of the matrix-assisted laser desorption ionization mass spectrometry (MALDI-MS) [27]. The use of ILs in different applications is determined by their intrinsic properties.

To date, there have been many feature reviews dealing with different aspects of ILs, including catalysis, extraction, synthesis, nanomaterials, biosensing, energy applications, etc [28-44].

The advantages of ILs for the synthesis of conducting polymer and nanoparticle when compared to conventional media and also their electrochemical sensors and biosensors based on IL/composite modified macro-disk electrodes are the major purpose of Singh et al. in [45]. These compounds have become a novel solution to problems encountered with organic solvents and these molecules are a prospective solution to the limitations encountered in electrochemical systems [46,47]. This new chemical group can reduce the use of hazardous and polluting organic solvents due to their unique characteristics as well as taking part in various new syntheses. Due to these unique properties, ionic liquids have been widely used in different field of applications, see Figure 2.

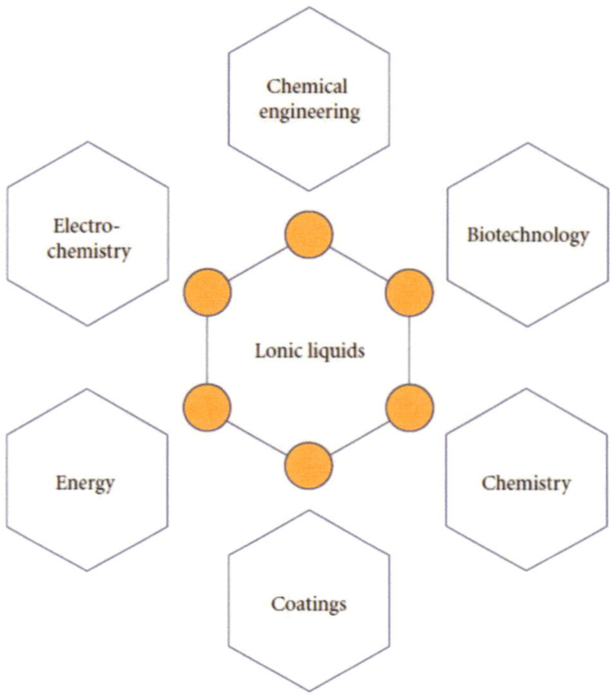

Figure 2. Applications of ILs

Considering the structural aspects of ILs, especially their surface structure on electrodes, is usually helpful for the rationalization of physical and chemical processes in ILs. The electrochemical reactions in principle are the processes at the ILs/electrode interface, including the diffusion of electroactive species, transport, capacitance at the interface, electron transfer processes etc., and which will dominate the performances in the electrochemical applications of ILs. On the other hand, the electrochemical properties in ILs are strongly dependent on the role of charge, size, polarization, and intermolecular interactions of ILs on the electrode surface. As a result, the understanding of the surface electrochemistry in ILs will benefit the performance enhancement of their application, and some novel application directions will be explored. These physical properties can be varied by selecting different combinations of ions [28,48]. Since the electrochemical window of the pure ILs depends on the electrochemical stability of the cation and/or anion, understanding the ion behavior at the electrode surface leads to improvement and implementation of the IL to the desired system [49]. The presence of an abundance of charge carriers means that when ILs are used as solvents, no supporting electrolyte is required for electrochemical experiments and this minimizes waste towards greener site [48].

2. Electrochemical supercapacitors

Supercapacitors (also called electric double-layer capacitors or ultracapacitors) are electrochemical capacitors that store energy through reversible ion adsorption onto active materials that have high specific surface area [50]. Because of their many advantageous properties, such as high power density, high capacitance, and long cycle life (>100000 cycles), these systems play an important role in electrical energy storage. To generate a high specific capacitance, the specific surface area of the electrode materials needs to be as high as possible to accommodate a large number of electrolyte ions at the electrode/electrolyte interface thereby promoting the electrical double-layer capacitance [50-52]. The latter property greatly exceeds that of conventional electrochemical energy storage devices, *e.g.*, batteries and fuel cells. Furthermore, supercapacitors can store a much greater charge per unit volume of mass than conventional dielectric-based capacitors. An electrochemical supercapacitor is based on the electrochemical double layer resulting from the electrostatic adsorption of ionic species at the electrode-solution interface, *i.e.*, no actual redox reaction is supposed to take place during the charging-discharging of these devices. To obtain the maximum possible capacitance, supercapacitor electrodes must have a high surface area; the standard material used in these devices is typically high surface area carbon. Because these devices are based on the electrosorption of ionic species, the region between the electrodes of the capacitor must contain an electrolyte with mobile ions. To obtain the maximum operating voltage without solvent decomposition, it is necessary to use aprotic solvents such as acetonitrile. Carbon-based supercapacitors based on conventional aprotic electrolytes are commercially available.

Not surprisingly, because of the numerous favorable properties described above, ionic liquids are considered promising electrolytes for electrochemical supercapacitors. $[C_2mim]BF_4$ and $[C_2mim]NTf_2$ dissolved in alkyl carbonate solvents were among the first ILs to be investigated [53]. More recently, an electrochemical supercapacitor based on *N*,*N*-diethyl-*N*-methyl-*N*-(2-methoxyethyl)ammonium tetrafluoroborate was shown to have superior properties compared to supercapacitors based on conventional aprotic electrolytes such as mixtures of Et_4NBF_4 in propylene carbonate [10].

An IL-based supercapacitor has even been prepared from carbon nanotubes. This device utilizes carbon gel electrodes fabricated by combining $[C_2mim]NTf_2$ with pulverized single-walled carbon nanotubes [54]. Hybrid supercapacitors obtain energy storage from the electrostatic double layer capacitance obtained at a high surface area carbon electrode and from a rapid, reversible charge-transfer process that occurs at a dopable conjugated polymer, *e.g.*, poly(3-methylthiophene) [55]. This charge-transfer process is designated as a pseudocapacitance. Not surprisingly, ionic liquids have also found their way into hybrid supercapacitors. In fact, a hybrid supercapacitor based on activated carbon, poly(3-methylthiophene), and *n*-$BuMePyrNTf_2$ (Pyr = pyrrolidinium) may be the first viable supercapacitor based on a IL, producing 24 Wh kg^{-1} and 14 kW kg-*1* [56].

Graphene, a monolayer of sp^2-hybridized carbon atoms arranged in a two-dimensional lattice, has attracted tremendous attention in recent years owing to its exceptional thermal, mechanical, and electrical properties. One of the most promising applications of this material is

http://dx.doi.org/10.5772/53646

in polymer nano-composites, polymer matrix composites incorporating nano-scale filler materials [57]. In addition, graphene, as an atom-thick 2D nanostructure [58,59], is a promising material for supercapacitor electrodes owing to its low mass density, excellent electronic conductivity, and high surface area (~2630 m^2/g, theoretical) [60-62]. Reduced graphene oxide, RG-O, a composition closely related to graphene, is a promising material for supercapacitor applications, as specific capacitance values of 135 and 99 F/g based on RG-O-based electrodes in aqueous and organic electrolytes, respectively, have been obtained [60]. However, dispersed graphene oxide, G-O, platelets can agglomerate during reduction by, for example, hydrazine in a solvent system such as water, resulting in the possible decrease of effective surface area, resulting in a lower specific capacitance than might be expected for an ideal graphene-based supercapacitor [60]. Moreover, current supercapacitors have energy densities well below the values required to provide power assists in various applications including hybrid electric vehicles or other high energy uses [51,63]. Hence, recent efforts have been focused on the development of supercapacitors with high energy densities, which may be achieved both by enhancing the operating voltage of the devices and by improving the accessibility of the ions from the electrolyte to the active regions of electrode materials.

Toward this goal, graphene-based electrodes combined with ionic liquid electrolytes can provide an attractive alternative for supercapacitors since such combinations result in an optimal pairing of high specific surface area electrodes and wider operating potentials that may be afforded by some IL electrolytes. Generally, ILs feature moderately high ion conductivity, nonvolatility, high decomposition temperatures, and wide electrochemical stability windows, and many ILs are being considered as electrolytes to increase supercapacitor operating voltages [10,44,64,102,65]. Despite the potential of ILs as electrolytes, further work is needed to explore their potential for supercapacitors assembled with graphene-based electrodes. One of the challenges is achieving graphene-based electrode materials capable of being well wetted by the chosen ILs [51], which may be attainable by the surface modification of graphene.

Kim *et al.* have reported their progress toward high performance supercapacitors based on poly(ionic liquid)-modified RG-O electrodes and an IL electrolyte, $[C_2mim]NTf_2$. Poly(ionic liquid), PIL, polymers formed from IL monomers can be prepared by the polymerization of unsaturated salts. Specifically, the use of poly(1-vinyl-3-ethylimidazolium) salts bearing NTf_2^- or CF_3SO_2-N-SO_2CF_3 anion has been reported to effectively stabilize hydrazine-reduced graphene oxide (RG-O) platelets *via* electrostatic and cation-π interactions, resulting in the formation of PIL-modified RG-O materials [66]. The PIL is likely physisorbed to surface of RG-O platelet and not covalently linked. These PIL-modified reduced G-O materials, PIL:RG-O, are expected to offer advantages for supercapacitor applications in that they should provide enhanced compatibility with certain IL electrolytes and improved accessibility of IL electrolyte ions into the graphene electrodes.

To investigate whether surface modification with PIL can be extended into other types of carbon electrodes, comparative experiments with the same PIL were applied to activated carbons (ACs) without RG-O platelets present. The result showed that the PIL is blocking the pores of the ACs to such an extent that the IL electrolyte cannot penetrate. For example, the PIL-modified AC electrodes exhibited far lower specific areas as measured. Given the apparent blocking of mesopores by the PIL, it would seem that good options for electrode materials include those having relatively high aspect ratios such as conductive platelets and conductive nanorods. For example, it is reasonably likely that a good configuration for SWNTs can be found where they would be PIL-coated and combined with an appropriate IL. Other options could include 3D solids with significantly larger pore size distributions than are present in the typical activated carbon currently used in supercapacitors. The use of $[C_2mim]NTf_2$ takes advantage of its larger electrochemical stability window, allowing for operation at 3.5 V, which in turn increased both the energy density and power density of the device. To further evaluate the device performance, the frequency response of the supercapacitor incorporating the PIL: RG-O electrodes and $[C_2mim]NTf_2$ electrolyte was analyzed using electrical impedance spectroscopy (EIS). The general concept of a supercapacitor design based on PIL-modified RG-O electrodes and a compatible IL electrolyte holds potential as an electrical energy storage device indicating stable electrochemical performance and a specific capacitance as high as 187 F/g. This relatively high capacitance is presumably due to improved wettability of the chosen IL electrolyte on the PIL-modified RG-O materials which, synergistically, enhanced the effective surface area of the electrode/electrolyte interfaces.

3. Lithium-ion batteries

Lithium-ion batteries are now ubiquitous in society and serve as the power sources in almost all portable electronic devices that are marketed to today's consumer. With such widespread use and in view of the safety issue of lithium-ion batteries, there are considerable ongoing efforts by battery manufacturers to improve the performance of these devices. As a result of the many attractive aspects of ILs, there is a modest but continuing interest in using them as electrolytes for these cells. Until several years ago, the IL electrolyte of choice was some variety of chloroaluminate [67], but interest in the use of nonchloroaluminate ILs has gradually increased. It is difficult to pinpoint the first instance where nonchloroaluminate ILs were used in lithium-ion batteries. An early report describes a successful $Li/LiMn_2O_4$ cell prepared with 1,2-dimethyl-4-fluoropyrazolium tetrafluoroborate + $LiBF_4$ or $LiAsF_6$ [68]. Li/ $LiCoO_2$ cells utilizing *n*-$PrMePipNTf_2$ (Pip = piperidinium) show good cyclic efficiency [69,70], and it is clear that ILs based on those anions that offer good anodic stability, *e.g.*, NTf_2- or $N(SO_2F)_2$-, give the best performance [70,71]. At the present time, the main problem is the incompatibility of the anode, *e.g.*, Li metal, and the ILs. That is, the solid electrolyte interphase film that is produced on the anode during the charge/discharge process is less stable than that obtained in conventional organic solvents. This incompatibility problem limits the cycling efficiency of the cell. MacFarlane *et al.* [72] have succeeded in elucidating the mechanism of film formation on Li in ILs based on *N*-alkyl-*N*-methylpyrrolidinium ions and NTf_2^-. Perhaps, future research of this nature will lead to resolution of this problem, enabling the practical use of ILs as electrolytes in Li batteries.

4. IL/electrode interface

By comparison of the background electrochemical behavior of an IL at both GC and platinum electrodes, one can approximately determine the amounts of protonic impurities present. The microstructure and capacitance of the electrical double layers (EDLs) at the interface of ILs and electrodes play an essential role in determining the system performance. Compared with simpler electrolytes, such as aqueous electrolytes and high-temperature molten salts, the ions in ILs are larger and often feature a complex shape. In addition, their charges are typically delocalized among many atoms [73]. Considering the solvent-free nature and the complex shape of ILs, it is expected that the classical theories for the EDLs in dilute aqueous electrolytes and high-temperature molten salts cannot accurately describe the structure and properties of the EDLs at the interface of ILs and electrified surfaces. Therefore, it is necessary to rediscover the IL/electrode interface and renew the model of IL/electrode EDLs.

Although ILs have always being investigated from the first finding in 1941 for a long period time, only a few works [74,75] concentrated on their intrinsic capacitive behaviors. A small capacitance (compared to smooth electrodes) was achieved in a practical capacitor comprising high-surface-area carbon cloth electrodes. To understand whether the observed capacitance might be due to the microporosity of the carbon cloth electrode or to the practical limitation of the device itself, the differential capacitance of 1-ethyl-3-methyl imidazolium imide IL was determined by measuring the potential of electrocapillary maximum or the point of zero charge (PZC) at an Hg electrode. The obtained capacitance at the cathodic potential is mainly determined by the cation, rather than the anion, as expected. The rate of charge, discharge, and the accessibility of electrolyte to the electrode surface ultimately determine the realizable capacitance in a practical device employing high-surface-area carbon electrodes.

The intrinsic capacitance of imidazolium-based ILs at carbon paste electrodes was investigated by using an electrochemical impedance technique [76]. The large capacitance was accounted for by intramolecular hydrogen bonding interactions that created a third charge layer between the IL and electrode EDL, which might make the interface rougher and hold more charge. The importance of ion chemistry and structure for the capacitive response of carbonaceous electrodes in ILs was stressed [77]. The double-layer capacitance of negatively charged carbon electrodes was strongly determined by the cation polarizability, which affected the dielectric constant in the double-layer, as well as double-layer thickness, which in turn also depended on the preferred orientation of the cations under the applied electric field. This suggests that the EDL constitutes a monolayer of cations up against the negatively charged carbon surface. Differential capacitance–potential curves were measured at IL/Hg, GC and Au electrode interfaces [78]. Unlike in aqueous or conventional organic solvents, capacitance–potential curves were found to vary significantly with the electrode substrates in ILs. It could be due to the absence of the inner Helmholtz layer of the molecular solvent between the electrode and ionic species, which usually works as a dominant factor in shaping the capacitance curves. The use of an electrochemical gate and ILs can reduce the

Debye ionic screen length to a few angstroms, which makes the measured capacitance nearly equal to the real quantum capacitance. The unique properties allow the direct measurement of the quantum capacitance of graphene as a function of gate potentials using a three-electrode electrochemical configuration in ILs electrolytes [79].

However, different types of ILs consist of various ions and ionic pairs structured in the bulk or IL/electrode interfaces. The bulk and interfacial behavior is governed by Columbic, van der Waals, dipole-dipole, hydrogen-bonding, and solvophobic forces [80]. The molecular organization of ILs is more complex than traditional solvents, and ILs cannot be considered as unstructured molten salts [81]. Many investigations have shown ionic liquids are nanostructured, which helps to explain their solvating power and some other unusual physical properties [82]. Spectroscopy and molecular dynamics simulation studies revealed distinct alkyl and ionic clusters in alkylimidazolium ILs [83-86]. Self-assembled IL nanostructures have also been elucidated using small-angle neutron scattering (SANS) for ILs [87], in which electrostatic contributions to solvophobicity are enhanced by the extensive hydrogen-bond network of the liquids. IL solvation layers were first detected for the ethylammonium nitrate/mica interface using surface force apparatus (SFA) [88], and subsequently detected for both protic and aprotic ILs confined between atomic force microscope (AFM) tips and mica, silica, graphite, and Au(111) [89-91]. It was found that IL nanostructures are the consequence of alkyl tail aggregation, driven by solvophobic forces inducing alkyl chains to segregate from the charged cation group and the anion, forming ionic domains. The increasing alkyl chain length leads to larger, more regular domains. The self-organized IL/substrate interfaces were observed to compose of three to seven solvation layers depending on the IL species [92]. Surface frequency generation data indicated that the interfacial cations exhibited orientational ordering and their orientation depended not only on the electrical potential of the electrode but also on the type of anions in the ILs [93]. Some experimental data suggest that the IL/electrode interface is one ion layer thick (typically 3-5 Å), which supports the idea that the EDLs in ILs are essentially Helmholtz layers [94]. Despite the lack of sufficient experimental data, two major approaches have been proposed to model the IL/electrode interface.

A molecular dynamics (MD) simulation model for an IL/metallic electrode in which the metallic electrode is maintained at a preset electrical potential is described in [95]. The model uses variable charges whose magnitudes are adjusted on the fly according to a variational procedure to maintain the constant potential condition. As such, the model also allows for the polarization of the electrode by the electrolyte, sometimes described by the introduction of image charges. The model has been implemented in a description of an electrochemical cell as a pair of parallel planar electrodes separated by the electrolyte using a two-dimensional Ewald summation method. The method has been applied to examine the interfacial structure in two ILs, consisting of binary mixtures of molten salts, chosen to exemplify the influences of dissimilar cation size and charge. The stronger coordination of the smaller and more highly charged cations by the anions prevents them from closely approaching even the negatively charged electrode. This has consequences for the capacitance of the electrode and will also have an impact on the rates of electron transfer processes. The calculated capacitances qualitatively exhibit the same dependence on the applied potential as have been observed in experimental studies.

The MD simulation study of the EDL structure near electrodes with different surface charge densities indicates that a Helmholtz-like interfacial counterion layer exists when the electrode charge density is negative or strongly positive but becomes poorly defined when the electrode charge density is weakly positive [96]. However, regardless of the presence of a distinct Helmholtz layer, the charge separation and orientational ordering of the ions persist up to a depth of 1.1 nm into the bulk ILs. The structure of the EDL is affected strongly by the liquid nature of the IL and the short-range ion-electrode and ion-ion interactions, especially at low electrode charge densities. In addition, the charge delocalization is found to affect the mean force experienced by the bulky ions near the electrode and, thus, can play an important role in shaping the EDL structure. An investigation of the electrified Au(100) surfaces in ILs by combined in situ scanning tunneling microscopy (STM) and differential capacitance measurements has revealed that the differential capacitance curves of the IL/Au interfaces have a bell-shaped feature, and ion adsorption at Au single-crystal electrodes depends critically on the structure of the surfaces [97].

The capacitance-potential (C-V) curve is virtually bell-shaped with metal electrodes in ILs composed of cations and anions of comparable size [98-101]. However, U-like C-V curves are observed at IL/nonmetallic electrode interfaces in contrast to those observed at IL/metallic electrodes [64]. The degree of curvature of the "U-like" curve measured at the GC electrode decreases in the ILs with low inherent ionic concentrations. The capacitance at the GC electrode exhibits a complex potential dependence, being different from those at highly oriented pyrolytic graphite (HOPG) and metal electrodes. However, in most real IL systems they are not hard and are not spheres. Therefore, long and challenging investigations to better understand IL/electrode interfaces are still ahead.

5. Other electrochemical applications

Besides the above application, ionic liquids are also used in some other electrochemical applications, such as electrochemical biosensing, electrochemical capacitors, lithium batteries, etc. Precisely, controlling the interface between the electrode and IL solvents allows scientists to alter the electron transfer and storage ability in the devices, and thus improve their performances. The details in this field have been well reviewed by other authors in references [102-104] and here is just glance at the use of ILs in these fields.

Enzyme electrodes are one of the most intensively researched biosensors because enzymes are highly selective and respond quickly to a specific substrate [105]. A new type of amperometric biosensor based on IL sol-gel material using the hydrolysis of tetraethyl orthosilicate in $[C_4mim]BF_4$ solution was reported [106]. The IL sol-gel enzyme electrodes retained the high activity of horseradish peroxidase (HRP) and provided long-term stability of HRP in storage. The uniform porous structure of the IL sol-gel matrix resulted in a fast mass transport, which provided a unique microenvironment around the enzyme, resulting in high sensitivity and excellent stability of the enzyme. An interesting electrochemical biosensor towards electrocatalysis of H_2O_2 was designed by entrapping HRP in IL doped DNA network [107], in which DNA with a specific double-helix structure and stacked base pairs may afford both biocompatible microenvironments around enzymes entrapped in the membranes and an effective pathway for electron transfer. Recently, IL-based sensors have been used for trace explosive detection by integrated electrochemical and colorimetric methods [108]. The explosives are pre-concentrated in ILs first.

A recent report showed that the use of [C_4mPy]TFSI in combination with a graphite and silicon electrode could maintain good lithium cycle performance in the presence of film-forming additives, such as vinylene carbonate (VC), that served as the only source of reactive chemical for an effective SEI formation at the electrolyte/electrode interface [109]. Silicon electrodes displayed higher compatibility with [C_4mPy]TFSI compared with graphite.

The latest researches of ILs in surface electrochemistry, including the IL/electrode interface and electron transfer in ILs has been highlighted in a review by Liu *et al.* [110]. As a result, the absorption, mass transport, and electron transfer at the interfaces of ILs/electrodes are complicated and quite different to that in traditional solvents. Some experimental data suggest that the IL/electrode interface is one ion layer thick (typically 3-5 Å) showing that they are essentially Helmholtz layers. Through spectroscopic techniques, *e.g.* SFA, AFM, SFG, and SANS, ILs are recently illustrated to be nanostructured as a consequence of alkyl tail aggregation driven by solvophobic forces. In the IL medium, the mobile carrier densities are shown to be able to be determined using a simplified capacitor model in a three-electrode electrochemical system, in which the electron transport characteristics are interpreted in terms of charged impurity induced scattering [79].

These strategies give ILs great promise for applications of ultrahigh frequency electronics and for studying the intrinsic transport properties of the Dirac fermions in graphene devices. It is believed that some new application directions beyond those mentioned above will be explored with the deep understanding of the structure and composition information at the ILs/electrode interface in the near future. There is no doubt that ILs will play more important roles in the electrochemical applications in both science and technology because of their unique properties.

Ionic liquids are a good choice as binder in carbon paste electrodes due to their interesting properties, mentioned previously. Recently, ionic liquids have been widely used as efficient pasting binders instead of non-conductive organic binders in preparation of carbon composite electrodes [111].

Carbon paste electrode based on MWCNTs and ILs types of electrodes show superior performance over traditional carbon paste electrodes. Solution studies shows a selective interaction between cerium acetylacetonate complex (CAA) and monohydrogen phosphate respect to a number of anions tested, therefore, the complex was used as sensing material in construction of a HPO_4^{2-} nano-composite carbon paste sensor based on MWCNTs and ILs. A new HPO_4^{2-} nano-composite carbon paste electrode was introduced [111]. From the other side of view, the purity of the nanotubes affects the gelation.

http://dx.doi.org/10.5772/53646

6. Molecular-scale insights into the mechanisms of ionic liquids interactions with carbon nanotubes and graphite surfaces

Ionic liquids have shown great promise for application in heterogeneous systems, such as lubricants, heterogeneous reactions, heterogeneous catalysis, electrochemistry, and fuel cells [15,29,31,112-116]. However, for these particular applications, a more detailed understanding of the molecular structure of the interface between IL and solid is essential.

Recently, the structures of the gas-liquid and liquid/liquid interfaces of ILs have been investigated computationally [117-121] and experimentally [122-137]. Some groups have studied the structure and dynamics of ILs by analytical techniques, such as X-ray diffraction [136-137], sum-frequency generation technique, SFG, [126,127] IR and Raman spectroscopies [134,135], direct recoil spectroscopy [122], neutron reflectometry [129], etc. They have suggested that the cations lie normal to the liquid surface and a density oscillation near the surface is also evident at the vapor-liquid interface. These studies, as well as several molecular dynamics simulations [117-121], have indicated that both cations and anions are populated at the liquid surface via a specific orientation.

Compared to gas-liquid and liquid-liquid properties of ILs, efforts to investigate solid-liquid surfaces or interface structures of ILs have so far been limited [93,132,138]. Owing to possible strong interactions between ILs and solid surfaces, a transition of IL from liquid to solid may take place. Evidence [116,139,140], has already demonstrated the coexistence of liquid and solid phases of IL [C_4mim]PF_6 on mica surfaces at room temperature by using atomic force microscopy, AFM. Although it has previously been found that some liquids, e.g., water become ordered or solid-like in the layers adjacent to the surface of crystallized solid substrates [141], a solid layer of [C_4mim]PF_6 formed on mica surfaces is much more stable. Since the melting point of [C_4mim]PF_6 is ~7 °C, it is a good model for the study of liquid/solid interactions and the possible phase transformation. Moreover, it has also been observed that, when confined to multiwalled carbon nanotubes, [C_4mim]PF_6 can be transformed into a crystal with a high melting point [142]. Some other groups have also studied the interactions between solid surface and IL. For example, by using sum-frequency vibrational spectroscopy, Fitchett and Conboy [143] have reported that the alkyl chains of the imidazolium cations are nearly normal to SiO_2 surfaces for a series of hydrophobic ILs. Romero and Baldelli [144] have studied [C_4mim]PF_6 and [C_4mim]BF_4 with hydrophobic and hydrophilic properties, respectively, and shown that the imidazolium ring lies on the plane of the quartz surface with the methyl group pointing 43 °C-47 °C from normal, displaying resonances from the alkyl chain as well as the aromatic ring by using SFG. Atkin and Warr *et al.* [89] have measured the solvation force profiles for several ILs on different solid surfaces by AFM. However, there is still a lack of detailed understanding of the ordered or solid-like molecular structures of ILs at solid/liquid interfaces.

ILs have been demonstrated to be interactive toward a number of solid materials, such as single-walled carbon nanotubes (SWNTs), graphite, silica, mica, and kaolinite, through interaction mechanisms such as H-bonding, π–π stacking, Van der waals forces, electrostatic forces, and so on [145-147]. For instance, Fukushima *et al.* [146] discovered uniform SWNTs bucky gels by grinding SWNTs in ILs, since the cation–π interaction exists between the imidazolium ion of IL and the π-electrons of the fullerene of SWNTs. In contrast, common organic solvents such as dichlorobenzene, ethanol, *N*, *N*-dimethylformamide, and 1-methylimidazole (a precursor for ionic liquids) did not form gels, even upon prolonged grinding with SWNTs. Likewise, no gelation of ionic liquids took place with other carbon allotropes such as graphite (1 to 2 μm) and C60 [148]. With all the above features in mind, rheological properties of the bucky gels of ionic liquids have been investigated in reference. SWNTs can orient imidazolium ions on their π-electronic surfaces by way of a possible "cation-π" interaction. Such a molecular ordering may trigger clustering of the surrounding imidazolium ions columbically and consequently interconnect neighboring SWNT bundles.

Lungwitz *et al.* [147] suggested that, due to the anion-silica interactions, both physisorption and chemisorptions occurred in the phase boundary of N-methylimidazolium chloride/silica. In addition, the adsorption of IL on other silicates or soils surfaces are mostly spontaneous; many adsorption mechanisms have been proposed, such as ion exchange, ion pairing, dispersive force, and so on [149]. Adsorption of IL on these solid materials indicates that these ILs could modify the surface of clay and eventually leads to significantly changed properties of both the clay and IL, such as the thermal stability, supermolecular structures [140], and crystallization behavior [142]. Halloysite nanotubes (HNTs), mined from natural deposits, are a kind of clay aluminosilicate mineral with hollow nanotubular structure and a high aspect ratio. The length of HNTs is usually in the range of 1-15 μm and the inner diameter and outer diameter of HNTs are 10-30 and 50-70 nm, respectively [150]. Halloysite, also named metahalloysite and chemically similar to kaolin, has the molecular formula $Al_2Si_2O_5(OH)_4.nH_2O$. In the pH range of 2-12, the inner and outer surfaces of HNTs are negatively charged [151] so it may absorb imidazolium cations of IL via an electrostatic effect. In addition, it was found that the outer surfaces of HNTs are mainly composed of siloxane and have only a few silanols and aluminols, which indicates that HNTs possess potential ability for the formation of hydrogen bonding with imidazolium-based ILs.

Guo and his coworkers confirmed [107], by thermogravimetric analysis, the retention of $[C_4mim]PF_6$ IL on HNTs in a tetrahydrofuran-water mixture and the formation of IL-coated HNTs (m-HNTs). In addition, the interaction was confirmed and the hydrogen bonding was proposed to as a possible mechanism. HNTs have been utilized as effective reinforcements for various polymers [152], therefore, the curing and performance of rubber compounds with m-HNTs were examined. The unique changes in the rubber compounds were correlated to the changes in filler dispersion and interaction between IL and HNTs. The IL absorbed on HNTs surface to form a mesostructure, which is different from that of the neat crystallized IL. The interaction between IL and HNTs was proposed to be hydrogen bonding and verified by the spectral results. Because of the interaction, the crystallization behavior of IL in the presence of HNTs was found to be changed. Compared with the compounds with HNTs, the uncured compounds with HNTs coated with IL showed significantly faster curing and the resulting vulcanizates showed substantially higher tensile strength and much lower hardness.

http://dx.doi.org/10.5772/53646

Interest in carbon nanotubes (CNTs) dispersed in ILs is rapidly growing [44,51]. One of the main reasons is the extraordinary electrochemical and bioelectrochemical properties of ILs [153] as well as of carbon nanotube composite materials. Such, CNTs/ILs have promising supercapacitor applications.

It has been shown in several experimental [89,91,92,117,154] and theoretical [98,155] studies that the molecular structure of IL ions strongly influences the IL interface properties at different charged and uncharged interfaces. For instance, Lockett *et al.* [154] showed that asymmetry in the size and shape of molecular ions results in unequal distribution of molecular cations and anions in direction normal to the IL-vacuum interface. These results suggest that the molecular structure of IL ions should make significant effects at the CNT/IL interface. However, still, there is a lack of molecular level information on the mechanisms of the IL interactions with nanocarbon electrodes.

Endres *et al.* [156] quoted an "undoubted" formation of at least three solvation layers of [C_2mim]TFSI on metal electrodes detected by atomic force microscopy (AFM). Recently, Hayes *et al.* [157] investigated the influence of the electric potential on the interface solvation layers in [C_2mim]FAP and [C_4mPyrr]FAP, by AFM, at the charged Au(111) electrode the support the experimental observations. They showed that IL layering is more pronounced at charged Au(111) surface compared to the neutral surface and increase of the potential leads to flattening of the tightly bound cation layer, indicating possible reorientation of cations ([C_2mim] and [Py]) to lay flat on the surface. The similar effects can be observed: increase of the potential at the CNT cathode significantly increases the tendency of [C_2mim] cation to lay flat on the surface. Applying an electric potential on the CNT electrode and/or varying the chemical structure of IL molecular ions, it is possible to change ion orientations and thus the structure of the CNT-IL interface shell.

Applying an electric potential, they found that the innermost layer changes its structure and becomes more strongly bound to the surface. At the cathode, for "the first time an interfacial (innermost) anionic layer at a solid interface has been detected by AFM". Atkin *et al.* [91] published an AFM study of the gold interface solvated in [C_2mim]TFSI and reported results which coincide well with the observed layered structure in MD simulations of neat [C_2mim]TFSI at the CNT surface performed by Frolov *et al.* [158]. They attributed the "weaker" layering pattern in simulations to the differences in temperatures: temperature in the simulations was about 70 °C, while "all force curves were acquired continuously at room temperature (22 °C)" for the AFM measurements.

There are some experimental studies on the orientation of IL molecules on the liquid-vacuum interface, see for example references [154,159,160]. Nakajima *et al.* [159] investigated the liquid-vacuum interface of different 1-alkyl-3-methylimidazolium-TFSI ionic liquids using high-resolution Rutherford backscattering spectroscopy. They showed that, due to their solvophobic nature, long alkyl chains of cations point away from the bulk liquid to vacuum and therefore stimulate the imidazolium ring to stay perpendicular to the surface. Simulations on the IL-carbon interface were observed an opposite effect: increase of alkyl chains increases the tendency for the imidazolium ring to lay parallel on the surface. The differences between IL-vacuum and IL-carbon nanotube interfaces to the strong Van der Waals attraction between the non-polar alkyl chains and carbon nanotube surface can be attributed. Contrary to the IL-vacuum interface, at the IL-CNT interface the alkyl chains of imidazolium-based cations tend to lay parallel on the CNT surface and force the imidazolium rings also to lay flat on the carbon nanotube surface.

One of the main reasons on rapidly growing interest in carbon nanotubes, CNTs, dispersed in ILs is the extraordinary electrochemical and bioelectrochemical properties of ILs as well as of carbon nanotube composite materials. Such, carbon nanotubes (nanotube forests) with ILs have promising supercapacitor applications. It has been shown in several experimental and theoretical studies that the molecular structure of IL ions strongly influences the IL interface properties at different charged and uncharged interfaces. For instance, Lockett *et al.* [161] showed that asymmetry in the size and shape of molecular ions results in unequal distribution of molecular cations and anions in a direction normal to the IL-vacuum interface. These results suggest that the molecular structure of IL ions should make significant effects at the CNT-IL interface.

Molecular simulations can provide complimentary information to the experimental data that should help to obtain a detailed picture of the interface phenomena in IL systems. Therefore, fully atomistic simulation for studying basic mechanisms of the IL interactions with the CNT surface has been applied [158]. The results have shown that ILs based on combination of imidazolium-based cations with hydrophobic anions (e.g. BF_4^- or TFSI) are moisture stable and have very promising electrochemical applications. Ion conductivity of [C_2mim]TFSI is comparable to the best of organic electrolyte solutions, and this liquid is stable up to 300-400 °C. The TFSI-based ionic liquids are practically not miscible with water but they are well miscible with several organic solvents, e.g., acetonitrile, AN [162]. The effects of the cation molecular geometry on the properties of the interface structure in the IL systems was investigated by a set of three ILs with the same anion (TFSI) but with different cations, namely, [C_2mim], [C_4mim], and [C_8mim] [158]. The cations had identical charged methylimidazolium 'heads' but different nonpolar alkyl 'tails' where the length of the tail increases from ethyl to octyl and the focus was concentrated on a set of the following questions:

- What is the interfacial structure of IL-AN mixture at the neutral CNT surface?
- How does the interfacial structure change at the positively charged CNT surface?
- How does the interfacial structure change at the negatively charged CNT surface?
- Does the length of the cation alkyl tail affect the interfacial IL-AN structure and preferred orientation of the IL ions at the CNT surface?
- What is the role of acetonitrile solvent in these interfacial effects?

Taking into account the molecular volume of the investigated ions, the probability of finding an ion inside the CNT pore was assumed to be low.

The analysis of the simulation data results in the following conclusions:

http://dx.doi.org/10.5772/53646

1. There is an enrichment of all molecular components of ILs under study at the CNT surface with formation of several distinct layers even at the non-charged CNT surface.

2. Mixing IL with acetonitrile decreases ion-counterion correlations in the electric double layer.

3. Increase of the length of the non-polar cation 'tail' increases propensity of imidazolium-based cations to lay parallel to the CNT surface.

4. At the CNT cathode TFSI anions and molecular cations are preferentially oriented parallel to the surface.

5. At the CNT anode the TFSI anions are oriented parallel to the surface, however the preferred orientations of cations depend on the length of non-polar tail: $[C_2mim]^+$ cations are oriented perpendicular to the surface, $[C_4mim]^+$ are in both parallel and perpendicular orientations, $[C_8mim]^+$ are oriented parallel to the surface.

6. By applying electric potential on the CNT electrode or/and varying the chemical structure of IL molecular ions it is possible to change the interfacial orientation of IL ions and, consequently, the structure of the CNT-IL interface shell.

As an effective computation technique, molecular dynamics program has been widely used for simulating interfacial phenomena. The interfacial molecular structure of $[C_4mim]PF_6$ in contact with the graphite surface has been studied for the first time at 2008 [163]. Maolin and his worker used $[C_4mim]PF_6$ as the model because it is hydrophobic and widely investigated. They investigated how the hydrophobic graphite surface affects the structure and orientation of hydrophobic $[C_4mim]PF_6$ at the interface. The MD calculation indicated the formation of a stable bottom layer, as well as possible single, double, or triple layer of $[C_4mim]PF_6$ on the graphite surface. The orientation calculations showed that the alkyl chains and imidazolium ring of cations both lie in the plane of the hydrophobic graphite surface.

Molecular dynamics simulations were performed to understand the microscopic structure of the IL $[C_4mim]PF_6$ on a graphite interface. In addition, MD simulations showed the existence of a solid-like IL bottom layer of about 6 Å thickness on the graphite surface. Compared to the bulk IL, the mass density peak of the bottom layer is 90% higher and its electron density peak is 80% higher. The butyl group and imidazolium ring of the cation of the IL bottom layer are parallel to the graphite surface. Due to the strong interactions between the cations and the graphite surface, ILs possessing longer alkyl tails and more imidazolium ring or aromatic ring may be applied to form more stable and well-regulated layers at graphite surfaces. This finding is important for the understanding of modification or lubrication mechanisms of ILs on solid surfaces, especially on the surfaces of carbon nanotubes and carbon black.

Several experimental and computational articles suggest that a possible layering of molecules in IL occur at the vapor-IL interfaces [119-121]. The influence of the vapor-IL interface on single, double, and triple layers of $[C_4mim]PF_6$ on the graphite surfaces was shown that the mass density profiles of different surface layered films are similar, and the vapor-IL interface appears to have little effect on the layered IL formation. Moreover, it is also shown that the thickness of the monolayer or the bottom layer away from the graphite surface is nearly identical, *i.e.*, 6.0 Å. Very early simulations of a short alkyl chain IL, dimethylimidazolium chloride, have indicated a layered surface with a clear oscillatory density profile akin to that observed for liquid metals [117]. It was also found that $[C_4mim]PF_6$ exhibit density oscillations at the liquid-vapor interface [119-121]. However, the density oscillation of dimethylimidazolium chloride is more obvious than that of $[C_4mim]PF_6$. From MD simulation, vapor-IL interface appears to have little effect on the layered IL formation. In the monolayer, the interaction energy between per cation and per anion is −175.29 kJ/mol. The interaction energy between per ion and the graphite also achieves a comparable quantity of -80.54 kJ/mol. This may be used for the explanation that the strong configurational effect and the interactions between the graphite surface and IL molecules induce the ordering of IL and stretching into a couple of IL layers. Furthermore, the above results imply that the configurational effect and the interactions between the IL layers and the graphite collectively induce different degrees of layered distribution. The vapor-IL interface may disorder to some extent the ordering of the outmost layer and weaken the interactions between the underlayers, resulting in a lower mass density peak of the outmost layer and higher mass density peak of the underlayer. This finding is important for the understanding of modification or lubrication mechanisms of ILs on solid surfaces, especially on the surfaces of carbon nanotubes and carbon black.

The behavior of a model IL, $[C_1mim]Cl$, confined between two parallel walls have been studied at various inter-wall distances, focusing on confinement effects on the structure and dynamics of the ions, and its impact on the charge-transport capacity [164]. The results focus both on structural and dynamical properties. Mass and charge density along the confinement axis reveal a structure of layers parallel to the walls that lead to an oscillatory profile in the electrostatic potential. Orientational correlation functions indicate that cations at the interface orient tilted with respect to the surface and that any other orientational order is lost thereafter. The diffusion coefficients of the ions exhibit a maximum as a function of the confinement distance, a behavior that results from a combination of the structure of the liquid as a whole and a faster molecular motion in the vicinity of the walls. Density profiles perpendicular to the walls confirmed an interfacial liquid layer twice as dense as the bulk followed by oscillations that decay toward the center of the cell; a major part of the layering is due to the distribution of anions. The number of layers changes with the interwall distance and so does the concentration of ions at the interface, showing a maximum at 28 Å. The observed structural effects result from the response of the liquid to the boundary conditions imposed and are detected also in the pressure on the walls. The pressure is determined by the number of particles in the first density layer, a result of the short-range interactions between the atoms and the walls. This effect depends on the details of the ion-wall interactions and may be significantly different in the case of charged surfaces. This is a result that, in light of recent experimental findings [143], may be independent of the type of confining walls used in simulations.

The distribution of charge perpendicular to the surface was determined by the arrangement of the ions and, in particular, by the alignment of the cations. As a result, any test charge entering the liquid encounters a first layer of positive charge followed by a second layer of opposite sign. The electrostatic potential drop between a point inside the walls and the middle of the liquid slab was around -0.5 V, although the potential oscillates strongly into the liquid.

An interesting result is that, under confinement, ionic diffusion is faster than in the bulk, at least in the presence of noncorrugated walls. Ions close to the surfaces diffuse faster than those in the middle of the slab and the diffusion coefficients reflect the changes in the density and proportion of ions near the walls. The reasons for the faster diffusion near the walls may be related to their lack of corrugation, their solvophobic nature, or a combination of both. However, it must also be kept in mind that the local mobility depends on the local density and orientation of the ions, which are the result of collective structural effects.

If the observed trends in ionic mobility apply also to more realistic surfaces, then the higher diffusivity of the ions will surely have an impact on the electrical conductivity and response of the IL to internal electric field changes. In fact, it has been proposed that the ability of the liquid to screen, fast and efficiently, an external field controls the rate of charge-carrier percolation across the nanocrystalline film.

There are several aspects of ILs that need further attention in connection with their uses in solar cells. First, a more realistic modeling of the walls would be necessary as well as the consideration of more sophisticated ILs. Second, it is important to understand the dynamical response of the semiconductor/IL interface to changes in the surface charge.

A recent comprehensive computer simulation of ILs at the SiO_2 surface clearly revealed that the interfacial structure is sensitive to polar or apolar surface as well as hydrophobic or hydrophilic IL components. The study nicely corroborates the experimental sum frequency generation vibrational spectroscopy (SFG) studies of ILs at SiO_2 substrate. Despite the detailed interfacial structure depending on the nature of substrate and ILs, a common feature shared by these simulations is the well-ordered structure at the IL/solid interface. Such ordered interfacial structure was also found in both simulation and experimental SFG study of the IL/TiO_2 interface, as well as the simulations of the IL/graphite interface by Wu *et al.* [163] and by Kislenko *et al.* [165]. Wu's latter study also highlights the solidation of ILs at graphite, as analogue to the confined ILs in carbon nanotube. Recently, Reichert and co-workers [166] investigated ILs interfacial ordering mechanism at the charged Al2O3 substrate via high-energy X-ray reflectivity and observed strong interfacial layering which decays exponentially into the bulk region. Such interfacial layering was expected to be a generic trait of ILs at charged walls. Understanding the interfacial structure, especially the electric double layer (EDL), is crucial in exploring the applications of ILs in electrochemical devices. Recently, Kornyshev [167] proposed a mean-field theory in which a compressibility parameter γ is incorporated. It was shown that the differential capacitance (DC) is bell-shaped when $\gamma >$ 1/3, and it is came l-shaped otherwise, while the U-shaped DC, predicted by the classical Couy-Chapman theory, is recovered for the γ>0 limit for the infinite dilute electrolyte solution. The bell-shaped DC is supported by Oldham's modification of the Couy-Chapman model for ionic liquid interface in a specific case with $\gamma = 1$. The bell-shaped DC was observed experimentally on an IL/metal electrode (platinum and gold), with similar IL ion sizes, by Ohsaka and co-workers [101]. The study highlights the compressibility of ILs and the different sizes of cations and anions. The focus was on liquid-solid systems and reported molecular dynamics studies at the liquid-solid interface between imidazolium-based ILs, $[C_4mim]PF_6$ and $[C_8mim]PF_6$, and an apolar uncharged graphite surface. The main aim of the study was to investigate the influence of different alkyl-side chain lengths of imidazolium on the interfacial structures. The density of IL was much enhanced at the interfacial region and the density oscillation extended to ~15 Å into the bulk with three layers. The results also demonstrated the polar groups tend to aggregate forming a polar network while the nonpolar groups fill up the rest of the vacancy. The imidazolium rings and the side chains preferentially lie flat at the graphite surface with the alkyl side chains of the cations elongated at the interfacial region, and the cations are closer to the graphite surface (ca. 3.6-3.7 Å) than the anions. The surface potential drop across the interface is more profound for $[C_8mim]PF_6$ than for $[C_4mim]PF_6$ due to relatively larger local density of the anions for $[C_8mim]PF_6$ near the graphite surface.

7. Liquid-to-solid phase transition of ionic liquids monolayer confined between graphite walls

While receiving much attention due to ILs importance in a broad range of applications, yet little is understood about their microstructure and phase transition in confined systems. Understanding the microstructure and freezing processes of ILs in confined systems is of practical importance in lubrication, adhesion, and the fabrication of solar cells or IL/nanomaterial composites, in which ILs are in contact with solid surfaces or under confinement. In general, the reduction of the liquid film thickness to fewer than 4-6 molecular layers will promote solidification. This results from the characteristic transverse density profile of thin films, which can induce lateral ordering and lead to freezing. Some evidence has indicated a possible liquid-solid phase transition for ILs in confined systems. Several reports have also revealed the astonishing property of melting point depression of 1,3-dialkylimidazolium-based ILs confined to nanospaces, which was discovered utilizing differential scanning thermal calorimetry. However, the phase behavior of ILs confined to nanospaces remains largely unexplored. The first simulation results of a liquid-solid freezing transition of $[C_1mim]Cl$ IL between two parallel graphite walls has been reported by Sha *et al.* [168]. Their result is of importance to understand the interfacial interactions between ILs and carbon nanotubes because of dispersing uniformly on the surface of single-walled carbon nanotubes via simple mulling. The molecular dynamics simulations were utilized to investigate the freezing of a $[C_1mim]Cl$ monolayer. The simulations predicted a first-order freezing transition from a liquid monolayer to a solid monolayer induced by varying the distance between the parallel graphite walls. The resulting monolayer solid consisting of a hydrogen-bonded network structure is very different from bulk crystalline $[C_1mim]Cl$. The phase transition can be induced only at a molecular surface density of $\rho=1.9/nm^2$. It is important to note that $[C_1mim]Cl$ IL confined between hydrophobic walls indicated no "hardening" or transition to a solid like phase structure, but only down to a wall distance of 2.5 nm. The wall distance plays a crucial role in the phase transition. Like an ice monolayer confined between walls, the molecule area density ($\varrho=N/l_x l_y$) of [C_1mim]Cl is another important factor. During the simulations, the molecule number of the system was decreased in steps and simulated for time periods equal to the first set of simulations. The change in N reflects the change in surface density since the area of the graphite wall remains unchanged in the simulation. As the molecule number is decreased, the lateral pressure of the monolayer liquid decreases to a lower value. If the surface density exceeds this value, a solid monolayer will not be formed. Meanwhile, the interaction energy between the IL and the walls is about -62 kJ/mol at a distance of 0.7 nm. Hence, the results imply that the configurational effect and the attraction interaction between the IL and the walls collectively induce the solid monolayer. The distance dependences of the diffusion coefficient and the potential energy indicate a strong first-order phase transition in confined [C_1mim]Cl. These results are helpful for the understanding of microstructures of ILs in nanostructured confinement as well as the thermodynamic mechanism of liquid-to-solid phase transition.

For the applications of electrolyte membranes and catalysts, there is a need to immobilize ILs on solid supports or within a solid matrix.

The microstructure of the IL bilayer was studied by varying the graphite wall distance. A liquid-to-solid phase transition of bilayer [C_1mim]Cl was observed at 425 K in this confined system, whereas the melting point of the bulk [C_1mim]Cl crystal is 399 K. Further calculation indicates a high melting point of the confined IL: melting point ~825-850 K. The imidazolium ring of the solid bilayer forms a strong $\pi \ldots \pi$ stacking structure in which each cation is surrounded by the three nearest-neighbor anions. The bilayer solid is a new phase of [C_1mim]Cl differing from the monolayer solid [168] under identical confinement conditions or the bulk crystal.

The simulation on the [C_4mim]NO_3/rutile (110) system shows that the adsorbed NO_3 anions at the interface organized themselves into a highly ordered manner, while changing the anion to PF_6^- does not present such an ordered interfacial structure. On the other hand, Grimes and coworkers have conducted experiments to study the high rate photocatalytic conversion of CO_2 to hydrocarbon production on the high surface area TiO_2 nanotube arrays.

An operating efficiency of electrochemical devices is greatly influenced by the molecular structure of the electrode/electrolyte interface. For example, the electron lifetime and open-circuit voltage in dye-sensitized solar cells depend on the structure of the electrical double layer (EDL). The specific capacitance of supercapacitors also depends on the EDL structure and is capable of being varied on change of the ions or even molecular ion fragments in IL. However, a double layer in ILs has not yet been adequately explored.

8. The structure of ionic liquid [C_4mim]PF_6/rutile (110) interface

The dye-sensitized solar cells (DSSCs) have been extensively investigated since their applications in the production of high light-to-electricity conversion efficiency. Traditionally, the devices are immersed in electrolytes, which are usually composed of an I^-/I_3^- redox couple and organic solvents such as acetonitrile. IL crystal system ($C_{12}mim^+/I^-/I_2$) has been used as electrolyte of dye-sensitized TiO_2 solar cells; it has been found that though large viscosity of the IL, the diffusion coefficient is still high since the exchange reactions facilitate the charge transport process. To acquire detailed information of DSSCs, Aliaga and Baldelli have performed a sum frequency generation study on one component of the solar devices, the liquid/solid interface between [C_4mim]DCA and [C_4mim]MS and TiO_2. The results pointed out that both the ions are present at the interface for [C_4mim]DCA while only the cations are detected for [C_4mim]MS. The spectra also emphasized on a stronger charge adsorption of the DCA anions than the MS anions. In addition to experimental researches, it has been previously shown that the interface between [C_4mim]NO_3 and rutile (110) surface by means of molecular dynamics simulation. The results indicated that the NO_3 anions prefer to aggregate at the interface and arrange themselves in a highly ordered manner. As for the [C_4mim]$^+$ cations, occupying the region adjacent to the NO_3^- layer, they tend to stand vertically on the TiO_2 (110) surface.

Shu *et al.* [169] presented a simulation on the interface between [C_4mim]PF_6 and rutile TiO_2 (110) surface, considering the other polymorphs of TiO_2 are less stable at room temperature. The main objective of their work was to model the IL/semiconductor interface, where the recombination process occurs, and study the structural behaviors of the confined IL. The results revealed that both ions are gathered on the surface, forming ionic double layers with regularity. It is also interesting to see that the cations lie flat on the surface, with alkyls stretched out in the bottom region. In addition, the ions are assembled at the interface, forming several enhanced layers at each side of the slab and are self-organized into alternate double ionic layers. The adsorbed cations are inclined to lie flat on the rutile (110) surface- This MD simulation casts new light on the microscopic structure features on the IL/solid interface and provides insights on the development of DSSCs. They have performed a molecular dynamics simulation on the [C_4mim]PF_6/ rutile (110) system.

Thanks to their low vapor pressure, ILs are ideal extraction solvents or reaction media because simple evaporation methods can be used to separate solutes from ILs. In addition, ILs can be custom-made with targeted functions. Because of these advantages, ionic liquids have been engineered as extraction solvents, reaction media and drug delivery materials. In most IL applications – such as extraction, lubrication, and IL super capacitors – the core function of the IL occurs at the ionic liquid–solid interfaces. Ionic liquids are different from conventional molecular liquids because no individual molecule exists in the liquid. Moreover, they are not diluted electrolyte solutions either. Hence, no existing theory and model can precisely describe the behavior of ILs, especially at the IL interfaces. Therefore, studies of the IL interfacial properties are necessary for further developments of IL-based applications. Furthermore, new applications – such as IL reactor, IL-circuit, and surface pattern visualization – require the precise control over the position of the IL drop on surface. The chemical pattern-directed assembly of IL on surface has been under investigation by Zhang *et al.* [170]. The chemical patterns can control the shape, size, and position of the IL on surface. Furthermore, IL drops on surface can be coated with a layer of silane film, forming an IL capsule.

However, experimental data are still lack, and none of the existing theories can completely explain the much diversity. Thus, first, the challenge for surface chemists, electrochemists, and theoreticians is to understand the detailed interface structure, including the adsorption, configuration, distribution, and orientation of ILs on the electrode surface. Some new models, new experimental techniques, and the details of this important system should be explored in the future to further realize the IL/electrode interface structures. Moreover, up to now, only quite a few number of ILs species were studied as models. However, the interface properties, structure and functions of the ILs with different cations and anions on the electrodes varies distinguishably. Thus, there is a need to have systematic studies on a wide range of ionic liquids so that more meaningful and comparable results can be obtained.

In an IL system, the heterogeneous electron transfer kinetics is highly dependent on the nature of both cation and anion in ILs. In addition, it should be pointed out that the physicochemical properties of ILs are largely dependent on the temperature and impurities such as water and some organic solvents, which will remarkably affect the transport and heterogeneous electron transfer ability. Though vast works have been carried out in this aspect, there is a huge data gap with respect to the thermodynamic and transport properties over a wide range of temperatures, as well as the effect of solutes and impurities on these properties. It is worth noting that the classical Marcus theory for outer sphere electron transfer, which is based on the reorganization of solvent dipoles, is not suitable to the entire ion systems of ILs. Developing a novel theory and electron transfer model system for the entire ion systems to explain the electron transfer process in ILs is also an important task in the future.

To date, the applications of ILs have involved nearly all aspects of electrochemistry, such as electrodeposition, electrosynthesis, electrochemical capacitors, electrochemical biosensors, and lithium batteries. Besides those stated above, it was recently shown that ILs can be used as a simple and effective way to reduce the scattering of carriers by charged impurities in graphene transistors supported on solid substrates [171].

Confinement can induce unusual behavior in the properties of matter. Using molecular dynamics simulations, a liquid-to-solid transition of a bilayer of [C_1mim]Cl confined between graphite walls was studied in order to mimic the phase transition of an IL confined to hydrophobic nanospace. It was found that the IL bilayer undergoes a clear and drastic phase transition at a wall distance of about 1.1 nm, forming a new high-melting-point solid phase with different hydrogen bonding networks. In the new phase, each cation is surrounded by the three nearestneighbor anions, and each anion is also encircled by the three nearest-neighbor cations. Strong π-π stacking interactions are found between the cations of the bilayer solid. The anions can be formed into a hexagonal ring around the cations. The new bilayer solid is a high-melting-point crystal possessing a melting point of 825-850 K, which is higher than that of the bulk crystal by more than 400 K.

For imidazolium-based ILs, the effect of the alkyl-chain length on the differential capacitance of EDL, electrocapillary curves, and the potential of zero charge (PZC) were experimentally investigated [8,9]. Based on the results obtained, a qualitative pattern of the Hg/[C_8mim]BF_4 interface at the electrode potentials close to PZC was suggested [9]. A quantitative information on the EDL structure is given in a number of papers wherein the imidazolium cations' orientation at the platinum and quartz surfaces has been determined using the sum frequency generation vibrational spectroscopy [10–13]. Since the electrode/electrolyte interface affects appreciably the relevant electrochemical processes, the development of adequate theoretical models of EDL in ILs takes on great significance. Owing to high ion concentration in ILs, the Gouy-Chapman classical theory is no longer valid and the double layer model must take into account a finite size of ions [14]. Investigating the double layer in IL $[C_4mim]PF_6$ at the graphite surface by use of the molecular dynamics simulation was the major goal for Kislenko, Samoylov, and Amirov [165].

Account was also taken of a large body of experimental and theoretical data available for comparison with the simulation results. The basal plane of graphite emulates the surface of an activated carbon used as an electrode material for supercapacitors. The calculations were performed both for an uncharged graphite surface and for positively and negatively charged ones. It is found that near an uncharged surface the IL structure differs from its bulk structure and represents a well-ordered region, extending over 20 Å from the surface. Three dense layers of ca 5 Å thick are clearly observed at the interface, composed of negative ions and positively charged rings. It is established that in the first adsorption layer the imidazolium ring in the $[C_4mim]^+$ cation tends to be arranged in parallel to the graphite surface at a distance of 3.5 Å. The PF_6^- anion is oriented in such a way that the phosphorus atom is at a distance of 4.1 Å from the surface and triplets of fluorine atoms form two planes parallel to the graphite surface. Ions adsorbed at the uncharged surface are arranged in a highly defective 2D hexagonal lattice and the corresponding lattice spacing is approximately four times larger than that of the graphene substrate. The influence of the electrode potential on the distribution of electrolyte ions and their orientation has also been investigated. Increase in the electrode potential induces broadening of the angle distribution of adsorbed rings and a shift of the most probable tilt angle towards bigger values. It was shown that there are no adsorbed anions on the negatively charged surface, but the surface concentration of adsorbed cations on the positively charged surface has a non-zero value. In addition, the influence of the surface charge on the volume charge density and electric potential profiles in an electrolyte was studied. The differences in the cation and anion structure result in the fact that the integral capacitance of the electrical double layer depends on the electrode polarity. The phenomena confirmed experimentally such interfacial layers formation and increase of an average tilt angle of adsorbed rings with increasing the electrode potential. In addition, the calculated values of the EDL capacitance coincide in the order of magnitude with the experimentally measured specific capacitance at the electrode/IL interface for activated carbons as electrode material and imidazolium-based ILs.

Author details

Fatemeh Moosavi

Address all correspondence to: moosavibaigi@um.ac.ir

Department of Chemistry, Ferdowsi University of Mashhad, Mashhad, Iran

References

[1] Patil ML, Rao CVL, Yonezawa K, Takizawa S, Onitsuka K, Sasai H. Design and Synthesis of Novel Chiral Spiro Ionic Liquids. Organic Letters 2006; 8(2) 227–230.

[2] Newington I, Perez-Arlandis JM, Welton T. Ionic Liquids as Designer Solvents for Nucleophilic Aromatic Substitutions. Organic Letters 2007; 9(25) 5247–5250.

[3] Gale RJ, Osteryoung RA. Electrochemical Reduction of Pyridinium Ions in ionic Aluminum Chloride: Alkylpyridinium Halide Ambient Temperature Liquids. Journal of the Electrochemical Society 1980; 127(10) 2167–2172.

[4] Wilkes JS, Levisky JA, Wilson RA, Hussey CL. Dialkylimidazolium Chloroaluminate Melts: A New Class of Room-Temperature Ionic Liquids for Electrochemistry, Spectroscopy and Synthesis. Inorganic Chemistry 1982; 21(3), 1263–1264.

[5] Fung YS, Zhu DR. Electrodeposited Tin Coating as Negative Electrode Material for Lithium-Ion Battery in Room Temperature Molten Salt. Journal of the Electrochemical Society 2002; 149(3) A319–A324.

[6] Shobukawa H, Tokuda H, Susan MABH, Watanabe M. Ion Transport Properties of Lithium Ionic Liquids and Their Ion Gels. Electrochimica Acta 2005; 50(19) 3872–3877.

[7] Shobukawa H, Tokuda H, Tabata SI, Watanabe M. Preparation and Transport Properties of Novel Lithium Ionic Liquids. Electrochimica Acta 2004; 50(2-3) 305–309.

[8] Seki S, Kobayashi Y, Miyashiro H, Ohno Y, Usami A, Mita Y, Watanabe M, Terada N. Highly Reversible Lithium Metal Secondary Battery Using a Room Temperature Ionic Liquid/Lithium Salt Mixture and a Surface-Coated Cathode Active Material. Chemical Communications 2006; (5) 544–545.

[9] Stenger-Smith JD, Webber CK, Anderson N, Chafin AP, Zong K, Reynolds JR. Poly(3,4-alkylenedioxythiophene)-Based Supercapacitors Using Ionic Liquids as Supporting Electrolytes. Journal of the Electrochemical Society 2002; 149(8) A973–A977.

[10] Sato T, Masuda G, Takagi K. Electrochemical Properties of Novel Ionic Liquids for Electric Double Layer Capacitor Applications. Electrochimica Acta 2004; 49(21) 3603–3611.

[11] Liu H, He P, Li Z, Liu Y, Li J. A Novel Nickel-Based Mixed Rare-Earth Oxide/Activated Carbon Supercapacitor Using Room Temperature Ionic Liquid Electrolyte, Electrochimica Acta, 2006; 51(10) 1925–1931.

[12] Noda A, Susan MABH, Kudo K, Mitsushima S, Hayamizu K, Watanabe M. Brønsted Acid-Base Ionic Liquids as Proton-Conducting Nonaqueous Electrolytes. Journal of Physical Chemistry B 2003; 107(17) 4024–4033.

[13] Kubo W, Murakoshi K, Kitamura T, Quasi-Solid-State Dye-Sensitized TiO_2 Solar Cells: Effective Charge Transport in Mesoporous Space Filled with Gel Electrolytes

Containing Iodide and Iodine. Journal of Physical Chemistry B 2001; 105(51) 12809–12815.

[14] Kubo W, Kitamura T, Hanabusa K, Wada Y, Yanagida S. Quasi-Solid-State Dye-Sensitized Solar Cells using Room Temperature Molten Salts and a Low Molecular Weight Gelator. Chemical Communications 2002; (4) 374–375.

[15] Wang P, Zakeeruddin SM, Moser JE, Grätzel M. A New Ionic Liquid Electrolyte Enhances the Conversion Efficiency of Dye-Sensitized Solar Cells. Journal of Physical Chemistry B 2003; 107(48) 13280–13285.

[16] Xia J, Masaki N, Jiang K, Yanagida S. Deposition of a Thin Film of TiO_x from a Titanium Metal Target as Novel Blocking Layers at Conducting Glass/TiO_2 Interfaces in Ionic Liquid Mesoscopic TiO_2 Dye-Sensitized Solar Cells. Journal of Physical Chemistry B 2006; 110(50) 25222–25228.

[17] Wang P, Zakeeruddin SM, Moser JE, Humphry-Baker R, Grätzel M. A Solvent-Free, $SeCN^-/(SeCN)_3^-$ Based Ionic Liquid Electrolyte for High-Efficiency Dye-Sensitized Nanocrystalline Solar Cells. Journal of the American Chemical Society 2004; 126(23) 7164–7165.

[18] Wang P, Zakeeruddin SM, Comte P, Exnar I, Grätzel M. Gelation of Ionic Liquid-Based Electrolytes with Silica Nanoparticles for Quasi-Solid-State Dye-Sensitized Solar Cells. Journal of the American Chemical Society 2003; 125(5) 1166–1167.

[19] Zhou D, Spinks GM, Wallace GG, Tiyapiboonchaiya C, MacFarlane DR, Forsyth M, Sun J. Solid State Actuators Based on Polypyrrole and Polymer-in-Ionic Liquid Electrolytes. Electrochimica Acta, 2003; 48(14–16) 2355–2359.

[20] Anderson JL, Armstrong DW. High-Stability Ionic Liquids. A New Class of Stationary Phases for Gas Chromatography. Analytical Chemistry 2003; 75(18) 4851–4858.

[21] Xiaohua X, Liang Z, Xia L, Shengxiang J. Ionic Liquids as Additives in High Performance Liquid Chromatography: Analysis of Amines and the Interaction Mechanism of Ionic Liquids. Analytica Chimica Acta 2004; 519(2) 207–211.

[22] Peng JF, Liu JF, Hu XL, Jiang GB. Direct Determination of Chlorophenols in Environmental Water Samples by Hollow Fiber Supported Ionic Liquid Membrane Extraction Coupled with High-Performance Liquid Chromatography. Journal of Chromatography A 2007; 1139(2) 165–170.

[23] Qi S, Cui S, Cheng Y, Chen X, Hu Z. Rapid Separation and Determination of Aconitine Alkaloids in Traditional Chinese Herbs by Capillary Electrophoresis using 1-Butyl-3-Methylimidazoium-Based Ionic Liquid as Running Electrolyte. Biomedical Chromatography 2006; 20(3) 294–300.

[24] Tian K, Qi S, Cheng Y, Chen X, Hu Z. Separation and Determination of Lignans from Seeds of Schisandra Species by Micellar Electrokinetic Capillary Chromatography using Ionic Liquid as Modifier. Journal of Chromatography A 2005; 1078(1-2) 181–187.

[25] Oter O, Ertekin K, Topkaya D, Alp S. Room Temperature Ionic Liquids as Optical Sensor Matrix Materials for Gaseous and Dissolved CO_2. Sensors and Actuators B: Chemical 2006; 117(1) 295–301.

[26] Fletcher KA, Pandey S, Storey IK, Hendricks AE, Pandey S. Selective Fluorescence Quenching of Polycyclic Aromatic Hydrocarbons by Nitromethane within Room Temperature Ionic Liquid 1-Butyl-3-Methylimidazolium Hexafluorophosphate. Analytica Chimica Acta 2002; 453(1) 89–96.

[27] Mank M, Stahl B, Boehm G. 2,5-Dihydroxybenzoic Acid Butylamine and other Ionic Liquid Matrixes for Enhanced MALDI-MS Analysis of Biomolecules. Analytical Chemistry 2004; 76(10) 2938–2950.

[28] Welton T. Room Temperature Ionic Liquids. Solvents for Synthesis and Catalysis. Chemical Reviews 1999; 99(8) 2071–2083.

[29] Wasserscheid P, Keim W. Ionic Liquids-New "Solutions" for Transition Metal Catalysis. Angewandte Chemie International Edition 2000; 39(21) 3772–3789.

[30] Huddleston JG, Visser AE, Reichert WM, Willauer HD, Broke GA, Rogers RD. Characterization and Comparison of Hydrophilic and Hydrophobic Room Temperature Ionic Liquids Incorporating the Imidazolium Cation. Green Chemistry 2001; 3(4) 156–164.

[31] Dupont J, de Souza RF, Suarez PAZ, Ionic Liquid (Molten Salt) Phase Organometallic Catalysis, Chemical Reviews 2002; 102(10) 3667–3692.

[32] Wasserscheid P, Welton T. Ionic Liquids in Synthesis, John Wiley & Sons: New York; 2002.

[33] Welton T. Ionic liquids in Catalysis. Coordination Chemistry Reviews 2004; 248(21-24) 2459–2477.

[34] Buzzeo MC, Evans RG, Compton RG. Non-Haloaluminate Room-Temperature Ionic Liquids in Electrochemistry. ChemPhysChem 2004; 5(8) 1106–1120.

[35] Baker GA, Baker SN, Pandey S, Bright FV. An Analytical View of Ionic Liquids. Analyst 2005; 130(6) 800–808.

[36] Weyershausen B, Lehmann K. Industrial Application of Ionic Liquids as Performance Additives, Green Chemistry 2005; 7(1) 15–19.

[37] Zhang S, Sun N, He X, Lu X, Zhang X. Physical Properties of Ionic Liquids: Database and Evaluation. Journal of Physical and Chemical Reference Data 2006; 35(4) 1475–1517.

[38] Anderson JL, Armstrong DW, Wei GT. Ionic Liquids in Analytical Chemistry. Analytical Chemistry 2006; 78(9) 2892–2902.

[39] Galiński M, Lewandowski A, Stępniak I. Ionic liquids as Electrolytes. Electrochimica Acta 2006; 51(26) 5567–5580.

[40] Pârvulescu VI, Hardacre C. Catalysis in Ionic Liquids. Chemical Review 2007; 107(6) 2615–2665.

[41] Greaves TL, Drummond CJ. Ionic liquids as Amphiphile Self-Assembly Media. Chemical Society Reviews 2008; 37(8) 1709–1726.

[42] Hapiot P, Lagrost C. Electrochemical Reactivity in Room-Temperature Ionic Liquids. Chemical Review 2008; 108(7) 2238–2264.

[43] Greaves TL, Drummond CJ. Protic Ionic Liquids: Properties and Applications, Chemical Review 2008; 108(1) 206–237.

[44] Armand M, Endres F, MacFarlane DR, Ohno H, Scrosati B. Ionic-Liquid Materials for the Electrochemical Challenges of the Future. Nature Materials 2009; 8(6) 621–629.

[45] Singh VV, Nigam AK, Batra A, Boopathi M, Singh B, Vijayaraghavan R. Applications of Ionic Liquids in Electrochemical Sensors and Biosensors. International Journal of Electrochemistry 2012; 2012, Article ID 165683, 19 pages.

[46] Fuller J, Carlin RT, Osteryoung RA. The Room Temperature Ionic Liquid 1-Ethyl-3-Methylimidazolium Tetrafluoroborate: Electrochemical Couples and Physical Properties. Journal of the Electrochemical Society 1997; 144(11) 3881–3886.

[47] Quinn BM, Ding Z, Moulton R, Bard AJ. Novel Electrochemical Studies of Ionic Liquids. Langmuir 2002; 18(5) 1734–1742.

[48] Holbrey JD, Seddon KR, Ionic liquids. Clean Products and Processes 1999; 1(4) 223–236.

[49] Trulove PC, Mantz RA. Electrochemical Properties of Ionic Liquids, in Ionic Liquids in Synthesis, Welton T, Wasserscheld P. Eds., p. 368, John Wiley & Sons, Morlenbach, Germany, 2003.

[50] Conway BE. Electrochemical Supercapacitors: Scientific Fundamentals and Technological Applications; Kluwer Academics and Plenum: New York, 1999.

[51] Simon P, Gogotsi Y. Materials for Electrochemical Capacitors. Nature Materials 2008; 7(11) 845–854.

[52] Miller JR, Simon P. Electrochemical Capacitors for Energy Management. Science 2008; 321(5889) 651–652.

[53] McEwen AB, McDevitt SF, Koch VR, Nonaqueous Electrolytes for Electrochemical Capacitors: Imidazolium Cations and Inorganic Fluorides with Organic Carbonates, Journal of the Electrochemical Society 1997; 144(4) L84–L86; McEwen AB, Ngo HL, LeCompte K, Goldman JL. Electrochemical Properties of Imidazolium Salt Electrolytes for Electrochemical Capacitor Applications, Journal of the Electrochemical Society 1999; 146(5) 1687–1695.

[54] Katakabe T, Kaneko T, Watanabe M, Fukushima T, Aida T. Electric Double-Layer Capacitors Using "Bucky Gels" Consisting of an Ionic Liquid and Carbon Nanotubes. Journal of the Electrochemical Society 2005; 152(10) A1913–A1916.

[55] Laforgue A, Simon P, Fauvarque JF, Mastragostino M, Soavi F, Sarrau JF, Lailler P, Conte M, Rossi E, Saguatti S. Activated Carbon/Conducting Polymer Hybrid Supercapacitors BATTERIES AND ENERGY CONVERSION. Journal of the Electrochemical Society 2003; 150(5) A645–A651.

[56] Balducci A, Henderson WA, Mastragostino M, Passerini S, Simon P, Soavi F. Cycling Stability of a Hybrid Activated Carbon//Poly(3-Methylthiophene) Supercapacitor with N-Butyl-N-Methylpyrrolidinium Bis(Trifluoromethanesulfonyl)Imide Ionic Liquid as Electrolyte. Electrochimica Acta 2005; 50(11) 2233–2237.

[57] Potts JR, Dreyer DR, Bielawski CW, Ruoff RS. Graphene-Based Polymer Nanocomposites. Polymer 2011; 52(1) 5–25.

[58] Geim AK, Novoselov KS. The Rise of Graphene. Nature Materials 2007; 6(3) 183–191.

[59] Novoselov KS, Geim AK, Morozov SV, Jiang D, Zhang Y, Dubonos SV, Grigorieva IV, Firsov AA. Electric Field Effect in Atomically Thin Carbon Films. Science 2004; 306(5696) 666–669.

[60] Stoller MD, Park S, Zhu Y, An J, Ruoff RS. Graphene-Based Ultracapacitors. Nano Letters 2008; 8(10) 3498–3502.

[61] Wang Y, Shi Z, Huang Y, Ma Y, Wang C, Chen M, Chen Y. Supercapacitor Devices Based on Graphene Materials. Journal of Physical Chemistry C 2009; 113(30) 13103–13107.

[62] Stankovich S, Dikin DA, Dommett GHB, Kohlhaas KM, Zimney EJ, Stach EA, Piner RD, Nguyen ST, Ruoff RS. Graphene-Based Composite Materials. Nature 2006; 442(7100) 282–286.

[63] Kötz R, Carlen M. Principles and Applications of Electrochemical Capacitors. Electrochimica Acta 2000; 45(15-16) 2483–2498.

[64] Islam MM, Alam MT, Okajima T, Ohsaka T. Electrical Double Layer Structure in Ionic Liquids: An Understanding of the Unusual Capacitance–Potential Curve at a Nonmetallic Electrode. Journal of Physical Chemistry C 2009; 113(9) 3386–3389.

[65] Largeot C, Portet C, Chmiola J, Taberna P, Gogotsi Y, Simon P. Relation between the Ion Size and Pore Size for an Electric Double–Layer Capacitor. Journal of the American Chemical Society 2008; 130(9) 2730–2731.

[66] Kim T, Lee H, Kim J, Suh KS. Synthesis of Phase Transferable Graphene Sheets Using Ionic Liquid Polymers. ACS Nano 2010; 4(3) 1612–1618.

[67] Webber A, Blomgren GE. Ionic Liquids for Lithium Ion and Related Batteries, in Advances in Lithium-Ion Batteries, W. A. van Schalkwijk and B. Scrosati, Editors, p. 185,

Kluwer Academic / Plenum Publishers, New York (2002); Wilkes JS. The Past, Present and Future of Ionic Liquids as Battery Electrolytes, in Green Industrial Applications of Ionic Liquids, Rogers RD, Seddon KR, Volkov S. Editors, p. 295, NATO Science Series, Vol. 92, Kluwer Academic Publishers, Dordrecht, The Netherlands (2002).

[68] Caja J, Dunstan TDJ, Ryan DM, Katovic V, in Molten Salts XII, P. C. Trulove, H. C. De Long, G. R. Stafford, and S. Deki, Editors, PV 99-41, p. 150, The Electrochemical Society Proceedings Series, Pennington, NJ (1999).

[69] Matsumoto H, Sakaebe H, Tatsumi K, Kikuta M, Ishiko E, Kono M. Fast Cycling of Li/$LiCoO_2$ Cell with Low-Viscosity Ionic Liquids Based on Bis(Fluorosulfonyl)Imide [FSI]. Journal of Power Sources 2006; 160(2) 1308–1313.

[70] Ishikawa M, Sugimoto T, Kikuta M, Ishiko E, Kono M. Pure Ionic Liquid Electrolytes Compatible with a Graphitized Carbon Negative Electrode in Rechargeable Lithium-Ion Batteries. Journal of Power Sources 2006; 162(1) 658–662.

[71] Garcia B, Lavallee S, Perron G, Michot C, Armand M. Room Temperature Molten Salts As Lithium Battery Electrolyte, Electrochimica Acta 2004; 49(26) 4583–4588.

[72] Howlett PC, Brack N, Hollenkamp AF, Forsyth M, MacFarlane DR. Characterization of the Lithium Surface in N-Methyl-N-alkylpyrrolidinium Bis(trifluoromethanesulfonyl)amide Room-Temperature Ionic Liquid Electrolytes Batteries, Fuel Cells, and Energy Conversion. Journal of Electrochemical Society 2006; 153(3) A595–A606.

[73] de Andrade J, Boes ES, Stassen H. Computational Study of Room Temperature Molten Salts Composed by 1-Alkyl-3-methylimidazolium Cations-Force-Field Proposal and Validation. Journal of Physical Chemistry B 2002; 106(51), 13344–13351.

[74] Gale RJ, Osteryoung RA. The Electrical Double Layer at Mercury in Room Temperature Aluminum Chloride: 1-Butylpyridinium Chloride Ionic Liquids. Electrochimica Acta 1980; 25(11) 1527–1529.

[75] Nanjundiah C, Mcdevitt SF, Koch VR. Differential Capacitance Measurements in Solvent-Free Ionic Liquids at Hg and C Interfaces. Journal of Electrochemical Society 1997; 144(10) 3392–3397.

[76] Liu H, He P, Li Z, Liu Y, Li J, Zheng L, Li J. The Inherent Capacitive Behavior of Imidazolium-based Room-Temperature Ionic Liquids at Carbon Paste Electrode. Electrochemical and Solid-State Letters 2005; 8(7) J17–J19.

[77] Lazzari M, Mastragostino M, Soavi F. Capacitance Response of Carbons in Solvent-Free Ionic Liquid Electrolytes. Electrochemistry Communications 2007; 9(7) 1567–1572.

[78] Alam MT, Islam MM, Okajima T, Ohsaka T. Measurements of Differential Capacitance in Room Temperature Ionic Liquid at Mercury, Glassy Carbon and Gold Electrode Interfaces. Electrochemistry Communications 2007; 9(9) 2370–2374.

[79] Xia J, Chen F, Li J, Tao N. Measurement of the quantum capacitance of graphene, Nature Nanotechnology 2009; 4(8) 505–509.

[80] Crowhurst L, Lancaster NL, Arlandis JMP, Welton T. Manipulating Solute Nucleophilicity with Room Temperature Ionic Liquids. Journal of the American Chemical Society 2004; 126(37) 11549–11555.

[81] Wishart JF, Castner EWJ. The Physical Chemistry of Ionic Liquids. Journal of Physical Chemistry B 2007; 111(18) 4639–4640.

[82] Seddon KR, Ionic Liquids for Clean Technology. Journal of Chemical Technology & Biotechnology 1997; 68(4) 351–356.

[83] Triolo A, Russina O, Bleif HJ, Di Cola E. Nanoscale Segregation in Room Temperature Ionic Liquids. Journal of Physical Chemistry B 2007; 111(18) 4641–4644.

[84] Triolo A, Russina O, Fazio B, Triolo R, Di Cola E. Morphology of 1-Alkyl-3-Methylimidazolium Hexafluorophosphate Room Temperature Ionic Liquids. Chemical Physics Letters 2008; 457(4-6) 362–365.

[85] Xiao D, Rajian JR, Hines LG, Li JS, Bartsch RA, Quitevis EL. Nanostructural Organization and Anion Effects in the Optical Kerr Effect Spectra of Binary Ionic Liquid Mixtures. Journal of Physical Chemistry B 2008; 112(42) 13316–13325.

[86] Bhargava BL, Balasubramanian S, Klein ML. Modelling Room Temperature Ionic Liquids. Chemical Communications 2008; (29) 3339–3351.

[87] Atkin R, Warr GG. The Smallest Amphiphiles: Nanostructure in Protic Room-Temperature Ionic Liquids with Short Alkyl Groups. Journal of Physical Chemistry B 2008; 112(14) 4164–4166.

[88] Horn RG, Evans DF, Ninham BW. Double-Layer and Solvation Forces Measured in a Molten Salt and its Mixtures with Water. Journal of Physical Chemistry 1988; 92 (12) 3531–3537.

[89] Atkin R, Warr GG. Structure in Confined Room-Temperature Ionic Liquids. Journal of Physical Chemistry C 2007; 111(13) 5162–5168.

[90] Wakeham D, Hayes R, Warr GG, Atkin R. Influence of Temperature and Molecular Structure on Ionic Liquid Solvation Layers. Journal of Physical Chemistry B 2009; 113(17) 5961–5966.

[91] Atkin R, Abedin SZE, Hayes R, Gasparotto LHS, Borisenko N, Endres F. AFM and STM Studies on the Surface Interaction of [BMP]TFSA and [EMIm]TFSA Ionic Liquids with Au(111). Journal of Physical Chemistry C 2009; 113(30) 13266–13272.

[92] Hayes R, Abedin SZE, Atkin R. Pronounced Structure in Confined Aprotic Room-Temperature Ionic Liquids. Journal of Physical Chemistry B 2009; 113(20) 7049–7052.

[93] Baldelli S. Probing Electric Fields at the Ionic Liquid–Electrode Interface Using Sum Frequency Generation Spectroscopy and Electrochemistry. Journal of Physical Chemistry B 2005; 109(27) 13049–13051.

[94] Baldelli S. Surface Structure at the Ionic Liquid-Electrified Metal Interface. Accounts of Chemical Research 2008; 41(3) 421–431.

[95] Reed SK, Lanning OJ, Madden PA. Electrochemical Interface Between an Ionic Liquid and a Model Metallic Electrode. The Journal of Chemical Physics 2007; 126(8) 084704(1–13).

[96] Feng G, Zhang JS, Qiao R. Microstructure and Capacitance of the Electrical Double Layers at the Interface of Ionic Liquids and Planar Electrodes. Journal of Physical Chemistry C 2009; 113(11) 4549–4559.

[97] Su YZ, Fu YC, Yan JW, Chen ZB, Mao BW. Double Layer of Au(100)/Ionic Liquid Interface and Its Stability in Imidazolium-Based Ionic Liquids. Angewandte Chemie International Edition 2009; 48(28) 5148–5151.

[98] Fedorov MV, Kornyshev AA. Ionic Liquid near a Charged Wall: Structure and Capacitance of Electrical Double Layer. Journal of Physical Chemistry B 2008; 112(38) 11868–11872.

[99] Fedorov MV, Kornyshev AA. Towards understanding the structure and capacitance of electrical double layer in ionic liquids. Electrochimica Acta 2008; 53(23) 6835–6840.

[100] Oldham KB. A Gouy-Chapman-Stern Model of the Double Layer at a (Metal)/(Ionic Liquid) Interface. Journal of Electroanalytical Chemistry 2008; 613(2) 131–138.

[101] Islam MM, Alam MT, Ohsaka T. Electrical Double-Layer Structure in Ionic Liquids: A Corroboration of the Theoretical Model by Experimental Results. Journal of Physical Chemistry C 2008; 112(42) 16568–16574.

[102] Tsuda T, Hussey CL. Electrochemical Applications of Room-Temperature Ionic Liquids. he Electrochemical Society Interface 2007; 16(1) 42–49.

[103] Wei D, Ivaska A. Applications of Ionic Liquids in Electrochemical Sensors, Analytica Chimica Acta 2008; 607(2) 126–135.

[104] Shvedene NV, Chernyshov DV, Pletnev IV. Ionic liquids in Electrochemical Sensors. Russian Journal of General Chemistry 2008; 78(12) 2507–2520.

[105] Gomes SASS, Nogueira JMF, Rebelo MJF. An Amperometric Biosensor for Polyphenolic Compounds in Red Wine. Biosensors and Bioelectronics 2004; 20(6) 1211–1216.

[106] Liu Y, Shi L, Wang M, Li Z, Liu H, Li J. A Novel Room Temperature Ionic Liquid Sol–Gel Matrix for Amperometric Biosensor Application. Green Chemistry 2005; 7(9) 655–658.

[107] Guo C, Song Y, Wei H, Li P, Wang L, Sun L, Sun Y, Li Z. Room Temperature Ionic Liquid Doped DNA Network Immobilized Horseradish Peroxidase Biosensor for

Amperometric Determination of Hydrogen Peroxide. Analytical and Bioanalytical Chemistry 2007; 389(2) 527–532.

[108] Forzani ES, Lu DL, Leright M, Aguilar AD, Tsow F, Iglesias R, Zhang Q, Lu J, Li JH, Tao NJ. A Hybrid Electrochemical–Colorimetric Sensing Platform for Detection of Explosives, Journal of the American Chemical Society 2009; 131(4) 1390–1391.

[109] Lux SF, Schmuck M, Appetecchi GB, Passerini S, Winter M, Balducci A. Lithium Insertion in Graphite from Ternary Ionic Liquid - Lithium Salt Electrolyte. II. Specific Capacity, Cycling Stability and Cycling Efficiency. Journal of Power Sources 2009; 192(2) 606–661.

[110] Liu H, Liu Y, Li J. Ionic Liquids in Surface Electrochemistry. Physical Chemistry Chemical Physics 2010; 12(8) 1685–1697.

[111] Norouzi P, Ganjali MR, Faridbod F, Shahtaheri SJ, Zamani HA. Electrochemical Anion Sensor for Monohydrogen Phosphate Based on Nano-composite Carbon Paste. International Journal of Electrochemical Science 2012; 7(3) 2633–2642.

[112] Sheldon R. Catalytic Reactions in Ionic Liquids. Chemical Communications 2001; (23) 2399–2407.

[113] Huddleston JG, Rogers RD. Room Temperature Ionic Liquids as Novel Media for 'Clean' Liquid–Liquid Extraction. Chemical Communications 1998; (16) 1765–1766.

[114] Abraham MH, Zissimos AM, Huddleson JG, Willauer HD, Rogers RD, Acree WE. Some Novel Liquid Partitioning System: Water-Ionic Liquids and Aqueous Biphasic Systems. Industrial and Engineering Chemistry Research 2003; 42(3) 413–418.

[115] Papageorgiou N, Athanassov Y, Armand M, Bonhote P, Petterson H, Azam A, Grätzel M. The Performance and Stability of Ambient Temperature Molten Salts for Solar Cell Applications. Journal of Electrochemical Society 1996; 143(10) 3099–3108.

[116] Nainaparampil JJ, Phillips BS, Eapen KC, Zabinski JS. Micro–nano Behaviour of DMBI-PF_6 Ionic Liquid Nanocrystals: Large and Small-Scale Interfaces. Nanotechnology 2005; 16(11) 2474–2481.

[117] Mezger M, Schramm S, Schröder H, Reichert H, Deutsch M, De Souza EJ, Okasinski JS, Ocko BM, Honkimäki V, Dosch H. Layering of [BMIM]+-Based Ionic Liquids at a Charged Sapphire Interface. The Journal of Chemical Physics 2009; 131(9) 094701 (1–13).

[118] Lynden-Bell RM. Gas-Liquid Interfaces of Room Temperature Ionic Liquids, Molecular Physics 2003; 101(16) 2625–2633.

[119] Yan T, Li S, Jiang W, Gao X, Xiang B, Voth GA, Structure of the Liquid–Vacuum Interface of Room-Temperature Ionic Liquids: A Molecular Dynamics Study. Journal of Physical Chemistry B 2006; 110(4) 1800–1806.

[120] Bhargava BL, Balasubramanian S. Layering at an Ionic Liquid-Vapor Interface: A Molecular Dynamics Simulation Study of [bmim][PF_6]. Journal of the American Chemical Society 2006; 128(31) 10073–10078.

[121] Sloutskin E, Lynden-Bell RM, Balasubramanian S, Deutsch M. The Surface Structure of Ionic Liquids: Comparing Simulations with X-ray Measurements. The Journal of Chemical Physics 2006; 125(17) 174715(1–7).

[122] Gannon TJ, Law G, Watson PR. First Observation of Molecular Composition and Orientation at the Surface of a Room-Temperature Ionic Liquid. Langmuir 1999; 15(24) 8429–8434.

[123] Law G, Watson PR. Surface Tension Measurements of N-Alkylimidazolium Ionic Liquids. Langmuir 2001; 17(20) 6138–6141.

[124] Law G, Watson PR. Surface Orientation in Ionic Liquids. Chemical Physics Letters 2001; 345(1-2) 1–4.

[125] Law G, Watson PR, Carmichael AJ, Seddon KR. Molecular Composition and Orientation at the Surface of Room-Temperature Ionic Liquids: Effect of Molecular Structure. Physical Chemistry Chemical Physics 2001; 3(14) 2879–2885.

[126] Baldelli S. Influence of Water on the Orientation of Cations at the Surface of a Room-Temperature Ionic Liquid: A Sum Frequency Generation Vibrational Spectroscopic Study. Journal of Physical Chemistry B 2003; 107(25) 6148–6152.

[127] Rivera-Rubero S, Baldelli S. Influence of Water on the Surface of Hydrophilic and Hydrophobic Room-Temperature Ionic Liquids. Journal of the American Chemical Society 2004; 126(38) 11788–11789.

[128] Fletcher KA, Pandey S. Surfactant Aggregation within Room-Temperature Ionic Liquid 1-Ethyl-3-methylimidazolium Bis(trifluoromethylsulfonyl)imide. Langmuir 2004; 20(1) 33–36.

[129] Bowers J, Vergara-Gutierrez MC, Webster JRP. Surface Ordering of Amphiphilic Ionic Liquids. Langmuir 2004; 20(2) 309–312.

[130] Bowers J, Butts CP, Martin PJ, Vergara-Gutierrez MC, Heenan RK. Aggregation Behavior of Aqueous Solutions of Ionic Liquids. Langmuir 2004; 20(6) 2191–2198.

[131] Iimori T, Iwahashi T, Ishii H, Seki K, Ouchi Y, Ozawa R, Hamaguchi H, Kim D. Orientational Ordering of Alkyl Chain at the Air/Liquid Interface of Ionic Liquids Studied by Sum Frequency Vibrational Spectroscopy. Chemical Physics Letters 2004; 389(4-6) 321–326.

[132] Sloutskin E, Ocko BM, Tamam L, Kuzmenko I, Gog T, Deutsch M. Surface Layering in Ionic Liquids: An X-ray Reflectivity Study. Journal of the American Chemical Society 2005; 127(21) 7796–7804.

[133] Halka V, Tsekov R, Freyland W. Peculiarity of the Liquid/Vapour Interface of an Ionic Liquid: Study of Surface Tension and Viscoelasticity of Liquid $BmimPF_6$ at Various Temperatures. Physical Chemistry Chemical Physics 2005; 7(9) 2038–2043.

[134] Hayashi S, Ozawa R, Hamaguchi H. Raman Spectra, Crystal Polymorphism, and Structure of a Prototype Ionic-liquid [bmim]Cl. Chemistry Letters 2003; 32(6) 498–499.

[135] Ozawa R, Hayashi S, Saha S, Kobayashi A, Hamaguchi H. Rotational Isomerism and Structure of the 1-Butyl-3-methylimidazolium Cation in the Ionic Liquid State. Chemistry Letters 2003; 32(10) 948–949.

[136] Neilson GW, Adya AK, Ansell S. Neutron and X-Ray Diffraction Studies on Complex Liquids. Annual Reports Section "C" (Physical Chemistry) 2002; 98 273–322.

[137] Abdul-Sada AK, Greenway AM, Hitchcock PB, Mohammed TJ, Seddon KR, Zora JA. Upon the Structure of Room Temperature Halogenoaluminate Ionic Liquids. Journal of the Chemical Society, Chemical Communications. 1986; 1753–1754.

[138] Lee BS, Chi YS, Lee JK, Choi IS, Song CE, Namgoong SK, Lee S. Imidazolium Ion-Terminated Self-Assembled Monolayers on Au: Effects of Counteranions on Surface Wettability. Journal of the American Chemical Society 2004; 126(2) 480–481.

[139] Wong DSH, Chen JP, Chang JM, Chou CH. Phase Equilibria of Water and Ionic Liquids [Emim][PF_6] and [Bmim][PF_6]. Fluid Phase Equilibria 2002; 194-197 1089–1095.

[140] Liu Y, Zhang Y, Wu G, Hu J. Coexistence of Liquid and Solid Phases of Bmim-PF_6 Ionic Liquid on Mica Surfaces at Room Temperature. Journal of the American Chemical Society 2006; 128(23) 7456–7457.

[141] Oh SH, Kauffmann Y, Scheu C, Kaplan WD, Rühle M. Ordered Liquid Aluminum at the Interface with Sapphire. Science 2005; 310(5748) 661–663; Hu J, Xiao XD, Ogletree DF, Salmeron M. Imaging the Condensation and Evaporation of Molecularly Thin Films of Water with Nanometer Resolution. Science 1995; 268(5208) 267–269.

[142] Chen S, Wu G, Sha M, Huang S. Transition of Ionic Liquid [Bmim][PF_6] from Liquid to High-Melting-Point Crystal when Confined in Multiwalled Carbon Nanotubes. Journal of the American Chemical Society 2007; 129(9) 2416–2417.

[143] Fitchett BD, Conboy JC. Structure of the Room-Temperature Ionic Liquid/SiO_2 Interface Studied by Sum-Frequency Vibrational Spectroscopy, Journal of Physical Chemistry B 2004; 108(52) 20255–20262.

[144] Romero C, Baldelli S. Sum Frequency Generation Study of the Room-Temperature Ionic Liquids/Quartz Interface, Journal of Physical Chemistry B 2006; 110(12) 6213–6223.

[145] Dupont J. On the Solid, Liquid and Solution Structural Organization of Imidazolium Ionic Liquids. Journal of the Brazilian Chemical Society 2004; 15(3) 341–350.

[146] Fukushima T, Kosaka A, Ishimura Y, Yamamoto T, Takigawa T, Ishii N, Aida T. Molecular Ordering of Organic Molten Salts Triggered by Single-Walled Carbon Nanotubes. Science 2003; 300(5628) 2072–2074.

[147] Lungwitz R, Spange S. Structure and Polarity of the Phase Boundary of N-Methylimidazolium Chloride/Silica. Journal of Physical Chemistry C 2008; 112(49) 19443–19448.

[148] Fukushima T, Kosaka A, Ishimura Y, Yamamoto T, Takigawa T, Ishii N, Aida T. Molecular Ordering of Organic Molten Salts Triggered by Single-Walled Carbon Nanotubes. Science 2003; 300(5628) 2072–2074.

[149] Mrozik W, Jungnickel C, Skup M, Urbaszek P, Stepnowski P. Determination of the Adsorption Mechanism of Imidazolium-Type Ionic Liquids onto Kaolinite: Implications for their Fate and Transport in the Soil Environment. Environmental Chemistry 2008; 5(4) 299–306.

[150] Joussein E, Petit S, Churchman J, Theng B, Righi D, Delvaux B. Halloysite Clay Minerals. Clay Minerals 2005; 40(4) 383–426.

[151] Levis SR, Deasy PB. Characterisation of Halloysite for Use as a Microtubular Drug Delivery System. International Journal of Pharmaceutics 2002; 243(1-2) 125–134.

[152] Guo B, Chen F, Lei Y, Liu X, Wan J, Jia D. Styrene-Butadiene Rubber/Halloysite Nanotubes Nanocomposites Modified by Sorbic Acid. Applied Surface Science 2009; 255(16) 7329–7336.

[153] El Abedin SZ, Polleth M, Meiss SA, Janek J, Endres F. Ionic Liquids as Green Electrolytes for the Electrodeposition of Nanomaterials. Green Chemistry 2007; 9(6) 549–553.

[154] Lockett V, Sedev R, Harmer S, Ralston J, Horne M, Rodopoulos T. Orientation and Mutual Location of Ions at the Surface of Ionic Liquids, Physical Chemistry Chemical Physics 2010; 12(41) 13816–13827.

[155] Sarangi SS, Raju SG, Balasubramanian S. Molecular Dynamics Simulations of Ionic Liquid–Vapour Interfaces: Effect of Cation Symmetry on Structure at the Interface. Physical Chemistry Chemical Physics 2011; 13(7) 2714–2722.

[156] Endres F, Höfft O, Borisenko N, Gasparotto LH, Prowald A, Al-Salman R, Carstens T, Atkin R, Bund A, El Abedin SZ. Do Solvation Layers of Ionic Liquids Influence Electrochemical Reactions? Physical Chemistry Chemical Physics 2010; 12(8) 1724–1732.

[157] Hayes R, Borisenko N, Tam MK, Howlett PC, Endres F. Atkin R. Double Layer Structure of Ionic Liquids at the Au(111) Electrode Interface: An Atomic Force Microscopy Investigation. Journal of Physical Chemistry C 2011; 115(14) 6855–6863.

[158] Frolov AI, Kirchner K, Kirchner T, Fedorov MV. Molecular-Scale Insights into the Mechanisms of Ionic Liquids Interactions with Carbon Nanotubes. Faraday Discussions. 2012; 154 235–247.

[159] Nakajima K, Ohno A, Hashimoto H, Suzuki M, Kimura K. Observation of Surface Structure of 1-Alkyl-3-Methylimidazolium Bis(Trifluoromethanesulfonyl)Imide Using High-Resolution Rutherford Backscattering Spectroscopy. The Journal of Chemical Physics 2010; 133(4) 044702 (1–7).

[160] Hashimoto H, Ohno A, Nakajima K, Suzuki M, Tsuji H, Kimura K. Surface Characterization of Imidazolium Ionic Liquids by High-Resolution Rutherford Backscattering Spectroscopy and X-ray Photoelectron Spectroscopy. Surface Science 2010; 604(3-4) 464–469.

[161] Lockett V, Horne M. Sedev R, Rodopoulos T, Ralston J. Differential Capacitance of the Double Layer at the Electrode/Ionic Liquids Interface. Physical Chemistry Chemical Physics 2010; 12(39) 12499–12512.

[162] Lin D, Liu N, Yang K, Zhu L, Xu Y, Xing B. The Effect of Ionic Strength and pH on the Stability of Tannic Acid-Facilitated Carbon Nanotube Suspensions. Carbon 2009; 47(12) 2875–2882.

[163] Maolin S, Fuchun Z, Guozhong W, Haiping F, Chunlei W, Shimou C, Yi Z, Hu J. Ordering Layers of [Bmim]PF_6 Ionic Liquid on Graphite Surfaces: Molecular Dynamics Simulation. The Journal of Chemical Physics 2008; 128(13) 134504 (1–7).

[164] Pinilla C, Del Popolo MG, Lynden-Bell RM, Kohanoff J. Structure and Dynamics of a Confined Ionic Liquid: Topics of Relevance to Dye-Sensitized Solar Cells. Journal of Physical Chemistry B 2005; 109(38) 17922–17927.

[165] Kislenko SA, Samoylov IS, Amirov RH. Molecular Dynamics Simulation of the Electrochemical Interface Between a Graphite Surface and the Ionic Liquid [BMIM][PF_6]. Physical Chemistry Chemical Physics. 2009; 11(27) 5584–5590.

[166] Mezger M, Schröder H, Reichert H, Schramm S, Okasinski JS, Schröder S, Honkimäki V, Deutsch M, Ocko BM, Ralston J, Rohwerder M, Stratmann M, Dosch H. Molecular Layering of Fluorinated Ionic Liquids at a Charged Sapphire (0001) Surface. Science 2008; 322(5900) 424–428.

[167] Kornyshev AA. Double-Layer in Ionic Liquids: Paradigm Change? Journal of Physical Chemistry B 2007; 111(20) 5545–5557.

[168] Sha M, Wu G, Fang H, Zhu G, Liu Y. Liquid-to-Solid Phase Transition of a 1,3-Dimethylimidazolium Chloride Ionic Liquid Monolayer Confined between Graphite Walls. Journal of Physical Chemistry C 2008; 112(47) 18584–18587.

[169] Shu W, Zhen C, Shu L, TianYing Y. A Molecular Dynamics Simulation of the Structure of Ionic Liquid [BMIM]PF_6/Rutile (110) Interface. Science in China Series B-Chemistry 2009; 52(9) 1434–1437.

[170] Zhang X, Cai Y. Octadecyltrichlorosilane (OTS)-Coated Ionic Liquid Drops: Micro-Reactors for Homogenous Catalytic Reactions at Designated Interfaces. Beilstein Journal of Nanotechnology. 2012; 3 33–39.

[171] Chen F, Qing Q, Xia J, Li J, Tao N. Electrochemical Gate-Controlled Charge Transport in Graphene in Ionic Liquid and Aqueous Solution. Journal of the American Chemical Society 2009; 131(29) 9908–9909.

Section 2

Energies, Fuels, and Biomass Conversions

Chapter 9

Ionic Liquids for Green Energy Applications - Local Structure and Dynamics by Advanced Spectroscopic Techniques

Anna Martinelli

Additional information is available at the end of the chapter

http://dx.doi.org/10.5772/52863

1. Introduction

Although ionic liquids (ILs) have been known since the beginning of the last century (1814) [1], they are materials of recent advent in the field of green energy applications where their recognition as solvents or electrolytes has ever increased during the last decades.

By definition ILs are molecular salts that melt at low temperatures, *i.e.* around room temperature or below 100 °C [2]. ILs also combine a remarkable set of properties such as high ion density (only ions!), wide temperature ranges of chemical and thermal stability, inflammability and negligible vapor pressure. Hence ILs are particularly suitable for applications where loss to the vapor phase limits the performance or constitutes an hazard, and have therefore been highlighted as the solvents for "green chemistry" in industrial chemical processes [3]. More recently (and as we will see in this chapter), the application of ILs has extended to other fields like electrochemistry, food science and pharmacy [4, 5]. Yet another field of application for ILs is in the metal extraction process from waste water using emulsion ionic liquid membranes, where the IL contributes significantly to the membrane's stability [6].

Typically, an IL consists of an organic cation, such as an imidazolium, pyridinium, pyrrolidinium, or ammonium derivative [7] combined with an organic or an inorganic anion, such as BF_4^-, PF_6^-, $CF_3SO_3^-$ and $(CF_3SO_2)_2N^-$ [8], see Figure 1 for the corresponding molecular structures. Since the discovery of the first IL ($EtNH_3^+$:NO_3^-, with melting point at 12 °C [1]), the number of cation–anion combinations has ever increased. In fact, organic cations can essentially be designed with any molecular structure resulting in a huge diversity of possible cation-anion pairs that enables tuning the properties of ILs to suit a particular application.

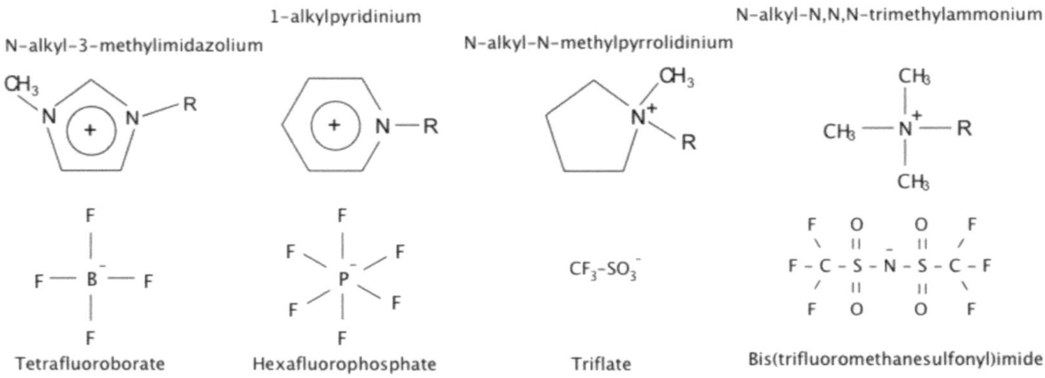

Figure 1. Examples of cation and anion structures typically found in ionic liquids. In these structures, R is the alkylated chain that can be methyl, ethyl, propyl, butyl, *etc.* In protic ILs R is a proton (H).

Roughly, ILs can be divided into two subgroups: protic (suitable for fuel cells) and aprotic (suitable for lithum batteries and supercapacitors). The former can be obtained through the proton transfer from a Brönsted acid to a Brönsted base (also called neutralization procedure) that, by being a one-step synthesis, results in very pure compounds [8]. The latter can be obtained by the quaternization procedure or the metathesis reaction, which are two-step syntheses involving the exchange of an intermediate anion. Since the intermediate reagents are not always easily eliminated, high-purity can become an issue. The two synthesis procedures to obtain protic and aprotic ILs are also schematically shown in Figure 2. There is a difference between the two types of ILs in the reversibility of the reactions. In protic ILs, for instance, the reaction is theoretically reversible and the completeness of proton transfer can a priori only be predicted by the acid-base pair strength, expressed by the ΔpK_a value.[1] Aprotic ILs, on the other hand, are irreversible compounds. However, apart from this aspect, there is no general difference in the chemico-physical properties between protic and aprotic ILs.

It has been empirically observed that small structural variations on the constituting ions of the ILs have an important influence on the macroscopically observed properties. Understanding these structure-property relationship is therefore key to design new ILs for specific applications. For example, provided a fixed anion, the viscosity and the ionic association degree both increase with the length of the aliphatic chain attached to the cation, which has been related to stronger van der Waals interactions [10]. On the other hand, provided a fixed cation, larger anions result in lower glass transition temperatures, lower melting points, lower viscosities and higher ionic conductivities, which is attributed to a more effective delocalization of the negative charge, hence more loosely coordinating ions [11]. These properties are also strongly affected by the symmetry of the cation and the position of the alkylated substituents on the cationic ring [8]. In protic ILs, it has also been found that provided a fixed anion as for instance the bis(trifluoromethanesulfonyl)imide ($(CF_3SO_2)_2N^-$), smaller cations result in lower glass transition temperatures.

1 The ΔpK_a is defined as $pK_a^{base} - pK_a^{acid}$ [9].

2. Ionic liquids for energy conversion devices

Because ILs can provide high ionic density, intrinsic ionic conductivity, non-volatility and non- flammability, as well as wide windows of electrochemical stability (up to 5–6 V for certain cation-anion combinations), they represent very interesting materials for applications where transport of ionic species and structural stability are key properties. Concrete examples are electrochemical conversion devices like fuel cells, Li-ion batteries, solar cells and capacitors [12]. In these fields, the ionic conductivity represents a measure of how easily ionic species are transported through the electrolyte. ILs typically display conductivities in the range 10^{-3}–10^{-2} Scm^{-1} at room temperature and stay liquid in wide temperature ranges extending to several hundreds degrees.[2]

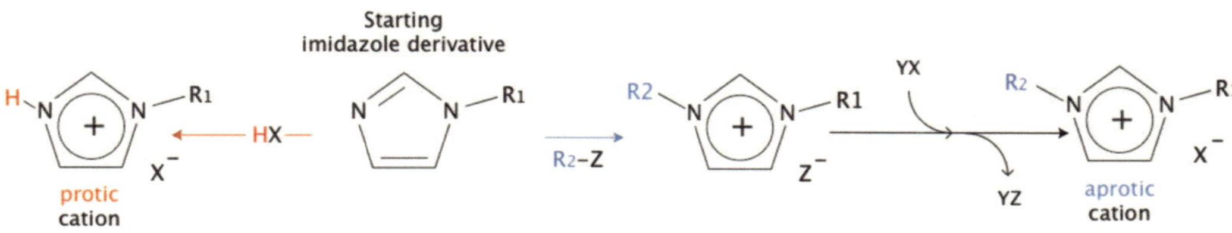

Figure 2. Synthesis procedures for protic ILs by neutralization of a base with an acid (left route, red) and for aprotic ILs by the quaternization method (right route, blue).

As we will see in more detail below (section 5.1) the ionic conductivity in ILs follows a non-Arrhenius dependence on temperature. Compared to the electrolytes conventionally used in for instance commercially available Li-ion batteries, ILs are both safer and greener, which represents an advantage with respect to both environmental and societal issues.

Despite these advantages, the use of ILs in electrochemical devices is limited by their melted state, since leakage can constitute a serious hazard.[3] Considerable research efforts are therefore being devoted to find proper ways of confining the ILs into solid-like matrices without, however, sacrificing the ionic conductivity. Proposed confining systems include both polymer based membranes and silica gels, as we will discuss more thoroughly in the next sections.

2.1. Ionic liquids for PEM fuel cells

The operational principles of a low-temperature fuel cell are schematically shown in Figure 3A. The main components in a proton exchange membrane (PEM) fuel cell are the anode, the PEM, and the cathode. The fuel (like H_2 or methanol) is fed at the anode where it is electrochemically split by platinum nano-particles into protons (H^+) and electrons (e^-). The latter follow an external circuit whereas the former diffuse through the PEM towards the cathode.

2 Indeed, many ILs decompose before evaporation occurs.

3 Loss of the liquid electrolyte can lead to short circuit and dangerous chemical reactions.

Here, electrons, protons and oxygen recombine to produce water and heat solely. Since the fuel cell does not produce pollutant or in other way hazardous elements, it is considered one of the most promising future devices for clean energy supply.

The archetypical proton conducting material used in low-temperature fuel cells is Nafion, a perfluorinated polymer membrane containing sulfonic acid pending groups ($-SO_3H$). Nafion has outstanding chemical and mechanical properties and, upon hydration, separates into hydrophilic and hydrophobic domains of the nano-meter size (see e.g. figure 2 in reference [13]), a structural property that results in well defined channels facilitating the transport of the protonic species (H^+ and H_3O^+). A drawback of Nafion, however, is that at temperatures higher than 80 °C the membrane dehydrates (due to water evaporation) and drastically looses its conducting properties, see also the conductivity of hydrated Nafion in Figure 3B. Meanwhile, for a realistic implementation of the fuel cell into the transport sector (in *e.g.* cars and buses) the goal has been set to operate fuel cells at above 120 °C.[4] At these higher temperatures, electrochemical reactions are faster and less platinum loading is needed at the electrodes, hence the costs of production are also reduced. The challenge today is to find proton conducting materials that are both solid-like (non-leaching) and can provide high enough ionic conductivities (*i.e.* greater than 10^{-1} Scm^{-1}) at temperatures above 120 °C.

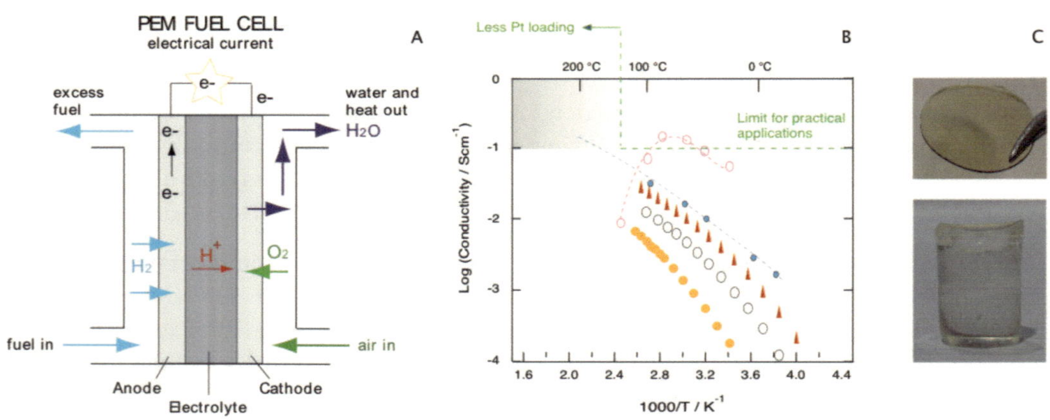

Figure 3. A: schematic of the operational principles in a PEM fuel cell. B: Arrhenius plot of ionic conductivities for diverse PEM fuel cell electrolytes: 80EIM:20PVdF (in wt%,∘), 80PMP/HTFSI:20PVdF (in wt%, △), and 40TFTEA:20Nafion (in wt%, •); an ion gel (•); and hydrated Nafion (∘). Here, EIM is ethylimidazoliumbis(trifluoromethanesulfonyl)imide, PMP is N-propyl-3-methylpyridinium bis(trifluoromethanesulfonyl)imide, and MSTEA is methane sulfonatetriethylammonium. Data points have been reproduced with permission from references [14, 15, 16]. The shadowed area shows the target set by the U.S. DOE for next-generation proton conducting materials (see also footnote 4). C: a photo of an IL swelled PVdF based membrane (top) and of an ionogel of the C_6mimTFSI (bottom).

4 In the Multi-Years Development Program of the U.S. Department of Energy (DOE) for the Fuel Cell Technology the requirement for next-generation proton exchange membrane (PEM) electrolytes is (≥10^{-1} Scm^{-1} at temperatures above 120 °C. Achieving this goal will facilitate the implementation of the fuel cell into the transport sector (buses, cars, scooters, etc).

Because ILs have a high ionic density, an intrinsically high ionic conductivity and are non-volatile, they are considered suitable materials to replace the concept of hydrated Nafion membranes in fuel cell applications. One investigated approach in this direction has been the swelling of PVdF-based polymer membranes by aprotic ILs of the methylpyridinium or propylimidazoliumcation and the bis(trifluoromethanesulfonyl)imide (TFSI) anion [14], see also Figure 3C (top). However, to provide the protonic species a chemically compatible acid had to be included, more specifically HTFSI. Swelling the membrane with a protic IL instead resulted in a simpler system since the protic species were intrinsically provided by the (protonated) cation of the IL [15]. The IL-swelled polymer membrane concept results in reasonable ionic conductivities, however at the expense of poor mechanical stability. Nafion membranes swelled with protic ILs of the triethylammoniumcation and the methane sulfonate ($CH_3SO_3^-$) or triflate ($CF_3SO_3^-$) anion is an alternative and more recent concept of proton exchange membranes [16]. This type of electrolytes show a smaller loss in conductivity with respect to the bulk IL and also better mechanical resistance than PVdF based membranes. In addition, these ammonium based ILs display a range of different degrees of cation:anion dissociation and an ability for proton exchange that also contribute to higher conductivities [17].

The ionic conductivities of these IL swelled polymer membranes are compared on a common Arrhenius plot in Figure 3B, together with the conductivity of hydrated Nafion. The reader may note that the shadowed area, representing the target set by the U.S. DOE (see footnote 4), is still not hit, implying that further scientific efforts are needed to develop new materials able to meet the set requirements.

2.2. Ionic liquids for Li-ion batteries

The most modern Li-ion battery is based on the rocking chair electrode model and the intercalation of Li ions into and from the electrodes [18]. In this battery concept the anode is typically carbonaceous (*e.g.* graphite) while the cathode consists in a layered oxide (*e.g.* $LiCoO_2$), a polyanion (*e.g.* $LiFePO_4$) or a spinel (*e.g.* $LiMnO_2$), and the electrolyte is typically a Li-salt (*e.g.* $LiPF_6$ or $LiBF_4$) dissolved in an organic solvent like ethylene carbonate (EC) or propylene carbonate (PC). The operational principles of a conventional Li-ion battery are shown in Figure 4A. The main issue with this battery concept concerns safety, which slows down the implementation of large-scale cells for energy storage and vehicle applications. Unpredictable events like short circuits or local overheating can trigger undesired reaction between battery components and the liquid organic electrolyte that can develop in drastic local heating and, eventually, into fire or explosion. Therefore, the replacement of conventional organic solvents with ILs, represents a huge improvement in terms of safety, the major advantages with this respect being non-volatility and non-flammability. In addition, the solubility of lithium salts (*e.g.* $LiBF_4$ or LiTFSI) in ILs is very high, and comparable to that of organic polar solvents. Different IL structures for Li-ion battery applications are under investigation in many industrial and academic laboratories, the electrochemical stability towards the electrodes being one main issue to solve. However, for an improved battery performance the growing of the surface electrode interface (SEI) must also be given further attention.

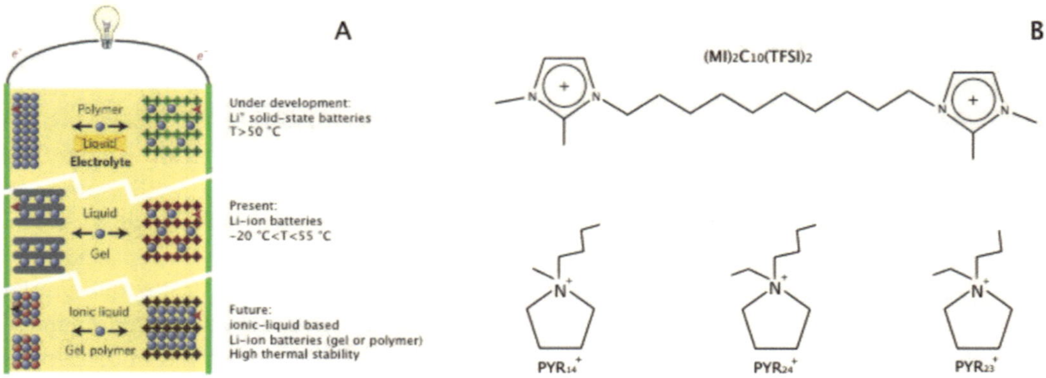

Figure 4. A: schematic of under development, present, and future lithium battery concepts, including those based on ILs as electrolytes (reproduced with permission from reference [5]). B: some cationic structures of interest for use in Li-ion batteries, a dicationic imidazolium based (top) and three pyrrolidinium derivatives (bottom), respectively.

The most investigated family of ILs for Li-ion battery application is that of the pyrrolidinium cation (Pyr), which has shown a better stability in time towards lithium metal electrodes than the previously investigated imidazolium based systems, and a wider electrochemical stability window [19]. Also, $TFSI^-$ results in low-melting ILs and high conductivities in combinations with many cationic structures due to the high charge delocalization and is therefore also the most used and investigated anion. To achieve chemical compatibility, the lithium salt most commonly dissolved in these ILs is consequently the LiTFSI. Thus, the typical IL based electrolyte for Li-ion battery applications is LiTFSI-Pyr_{xy}TFSI, where x and y are alkyl chains of variable length, see also Figure 4B. Nevertheless, as we will discuss in more details below, even small structural variations on the Pyr cation can result in different properties, like different phase behavior and dependence on temperature of the ionic conductivity.

2.3. Ionogels

Ionogels constitute a very recent material concept based on the nano-confinement of ILs into silica; see Figure 3C (bottom) for a photo of an ionogel prepared in our laboratories. Compared to IL swelled polymer membranes, ionogels can incorporate a considerably higher volume of liquid (up to 98%) without loosing in mechanical resistance [20, 21]. Ionogels can be prepared through a sol-gel synthesis that follows a non-aqueous route proposed by K. G. Sharp [22], consisting in the reaction of tetramethylorthosilicate (TMOS) with formic acid (FA). During this reaction, nano-sized particles of silica (SiO_2) are formed that first undergo aggregation and then gelation. If the sol-gel reaction is let occurring inside an IL as a co-solvent, the final gel will be three-dimensionally interpenetrated by the IL.

Recent results from magic angle spinning ^{1}H NMR experiments have shown that even at high degrees of nano-confinement[5] the dynamical properties of the IL are only marginally slowed down [23, 24]. Thus, ionogels represent an elegant route to robust and non-leaching

5 Or low ionic liquid contents in the ionogel.

http://dx.doi.org/10.5772/52863

nano-structured electrolytes. So far, the ILs used in the synthesis of ionogels have primarily included 1-ethyl-3-methyl imidazolium (C_2mim) or 1-butyl-3-methyl imidazolium (C_4mim) cations, whose ionic conductivity is shown in Figure 3B together with those of hydrated Nafion and IL swelled polymer membranes. From this plot it is evident that the ionogel concept has the potentiality to meet the requirements in terms of high conductivity at temperatures higher than 120 °C, as set by the U.S. DOE.

In our laboratories we are currently investigating the use of the 1-hexyl-3-methyl imidazolium (C_6mim) cation, which has a longer alkyl side chain and may thus induce interesting structural features in the gel. We have indeed found by time resolved Raman and ^{1}H NMR spectroscopy that a structural reorientation of the cation may occur in concomitance with the sol-gel transition [25], which may also be accompanied by a local cation-anion reorganization.[6] We are undertaking detailed structural investigations at the molecular level since this understanding will guide the design of next-generation ionogels tailor made for specific applications. To be suitable for fuel cell applications, for instance, ionogels must be prepared with protic ILs, whereas to be used in Li-ion battery applications, a suitable Li salt-IL should be used in the sol-gel synthesis. To the best of our knowledge, none of these approaches have been reported so far and, in our laboratories, efforts will be primarily spent on the former.

3. Structural investigations

The local structure is one hot issue in the field of ILs, where recent progresses have mainly concerned the understanding of the mesoscopic segregation in the liquid state. With local in this context we mean the nm or molecular scale range, where effects of conformational evolution, crystalline-to-amorphous transitions, ionic clustering *etc* occur. Experimental techniques that have been successfully used to resolve this size domain are vibrational (Raman and infrared) spectroscopy, solid state Nuclear Magnetic Resonance (NMR) spectroscopy and Small Angle X-ray Scattering (SAXS), which will also be discussed in the sections that follow. Knowing the local organization of cations and anions in ILs is very important, in particular because the resulting ion-ion association can in turn have an impact on the ionic transport properties. The reader should recall that ILs are only comprised of ions, that the inter-ionic electrostatic forces are very important and that, as a consequence, anions and cations should be considered as structurally and dynamically associated pairs.

A particular structural feature can also affect the electrochemical performance of the IL. For instance, it has been found that in aprotic ILs based on the 1-ethyl-3-methylimidazolium (C_2mim) cation the substitution of the acidic proton on the ring (position C_2) by an alkyl chain decreases the reduction potential [27], which is otherwise too large for practical applications in Li-ion batteries. Also, the addition of a Li-salt to ILs results in ionic clustering and consequently a reduced number of free charges, as also discussed in section 4.1 and reference [28].

6 From ongoing analysis of confocal μ-Raman/x-ray (small angle x-ray scattering, SAXS) data collected at the ID13 beam line of the ESRF facility in Grenoble [26].

3.1. Raman spectroscopy: Conformational isomerism

Vibrational spectroscopy (including Raman, infrared, neutron and luminescence spectroscopy) is a powerful technique to investigate the structure in materials on a molecular or smaller level. More specifically, issues like dissociation, inter-molecular interactions, crystallinity and conformations can be investigated in both the liquid and solid state. In this section however, only Raman spectroscopy will be treated.

The basic principle of Raman spectroscopy[7] is the excitation of the material by monochromatic light with wavelength usually in the visible range[8] and the collection of the inelastically scattered light. The latter contains detailed information on the molecular vibrations characteristic of the investigated material. If the energy of the incident light is $h\omega_0$, a minor part of this ($\sim 10^{-6}$) will be inelastically scattered (or Raman scattered) with a gain or loss in energy. The inelastically scattered light has energy $h\omega_0 \pm h\omega_{vib}$, where $h\omega_{vib}$ is the vibrational energy exchanged in the scattering process. The plus/minus signs correspond to anti-Stokes and Stokes scattering respectively, see Figure 5A. In the simplest classical treatment, where atoms are treated as particles bound by weightless springs and vibrate around their equilibrium position, the vibrational frequency can be described as

$$\omega_{vib} \propto \sqrt{\frac{k}{M}} \tag{1}$$

where k is the bond strength and M the reduced mass of the oscillating system.[9] From this expression it follows that molecules with atoms of different masses and bond strengths will have distinct fingerprints in a Raman spectrum, where Raman intensity is plotted versus frequency of vibration (expressed in cm^{-1}). As a consequence, changes in the dissociation state, in the closest chemical environment, or in the internal bond rotations can be studied analyzing intensity and frequency shifts of the vibrational modes.

The conformation adopted by cations and anions in ILs has been one such investigated structural feature. This is of interest since the relative orientation of the cation-anion pair can in turn affect the association degree of the ions and thus the dynamical properties of the IL. The TFSI anion, for instance, can adopt two different conformations, the *cisoid*(or C_1) and the *transoid* (or C_2) that differ in the orientation of the $-CF_3$ groups with respect to the internal S-N-S bond, see Figure 5B. The fingerprints of these two conformations were first theoretically predicted to be found in the low-frequency spectral range 260–370 cm^{-1} [31] and then also experimentally distinguished for protic and aprotic ILs of the imidazolium, pyridinium and pyrrolidinium cations [29]. This study shows that as the temperature is increased the population of *cisoid* conformers increases[10] (see also Figure 5C) and that the enthalpy of conforma-

7 A more thorough description of the Raman spectroscopy technique can be found in reference [30].

8 Nevertheless, laser light in the UV and near IR range can also be used.

9 In a two-atoms system with masses $m_1$and $m_2$the reduced mass M is defined as $(m_1 m_2)/(m_1+m_2)$.

10 The *transoid* is indeed the conformation most stable at low temperatures and commonly found in the crystalline phase.

tional change, ΔH, also varies with the structure of the associated cation. Moreover, it seems from recent investigation of ionogels prepared with the IL C_1C_6TFSI, that nano-confinement in silica also can increase the population of the C_1 conformers [25]. So, it is clear that even very small changes on the cationic structure can induce important effects on the anion's conformation. In this respect, J.D. Holbrey*et al.* have reported the unusual *cisoid* conformation for $TFSI^-$ in the solid-state structure for the case of the 1,3-dimethylimidazolium cation [32], as opposed to what is found with the 1,2,3- triethylimidazolium cation and most other investigated imidazolium-based ILs (that adopt the *transoid* form in the solid-state).

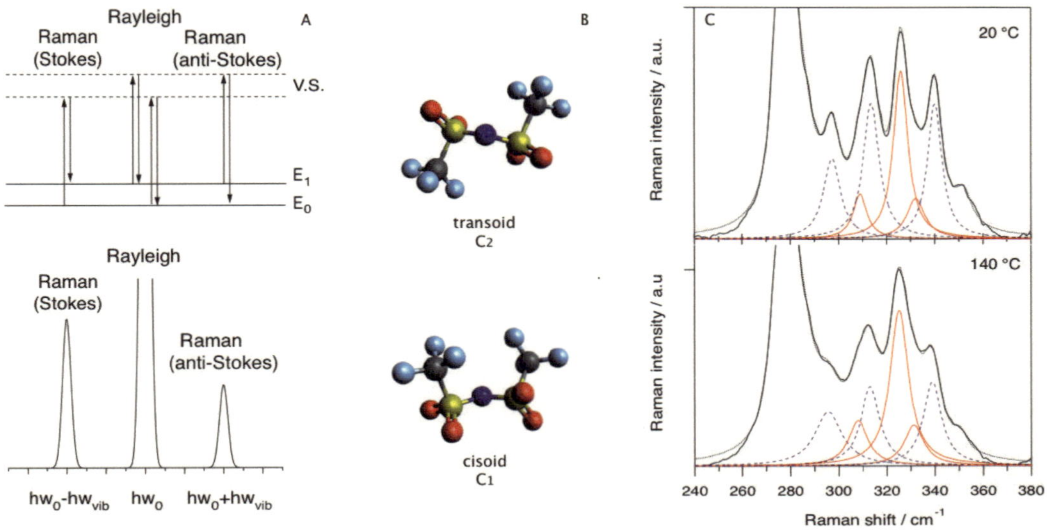

Figure 5. A: the inelastic scattering process during a Raman experiment with Stokes lines corresponding to energy loss and anti-Stokes to energy gain, respectively. The Rayleigh lines correspond to no energy exchange. B: the two conformations in which the TFSI anion can be found in, *i.e.* the *cisoid* (or C_1, bottom) and the *transoid* (or C_2, top). C: Deconvolution of Raman spectra recorded at different temperatures for the protic IL EIMTFSI (ethylimidazolium bis(trifluoromethanesulfonyl)imide). The fitting (Lorentzian) components corresponding to the *cisoid* and *transoid* contributions are shown in red (solid line) and blue (dashed line), respectively. These Raman spectra are reproduced with permission from reference [29].

Also the conformational isomerism of cations can be investigated by vibrational spectroscopy, as for instance demonstrated in reference [33]. The number of conformations in cations can be significantly increased when long alkyl side chains are attached, due to a larger degree of rotational freedom around the C-C bonds and the orientation of the chain with respect to the cationic head. For the case of 1-butyl-3-methylimidazolium tetrafluoroborate (C_4mimBF_4) the coexistence of at least four conformers was found, GG, GA, TA and AA, with the population of the most stable GA and AA increasing as temperature is decreased [33].

Raman spectroscopy has also greatly contributed to understand the formation of larger ionic aggregates, or $[Li(TFSI)_n]^{-(n-1)}$ clusters, upon addition of LiTFSI to ILs of the pyrrolydinium cation. In reference [28], the authors discuss possible types of $[Li(TFSI)_n]^{-(n-1)}$ aggregates in ILs of different cations, *i.e.* pyrrolidinium and imidazolium, the latter both in the mono- and di-cationic form, analyzing the TFSI-characteristic Raman mode at ~740 cm^{-1}. This study shows that for all ILs the number of TFSI anions strongly coordinating to one Li-ion drastically increases at very low salt concentrations, where values of *n* in the range 4–6 dominate. For intermediate LiTFSI concentrations, triplets of the type $[Li(TFSI)_2]^-$ are found instead. It is interesting to note that these different coordination regimes also correspond to different strengths of TFSI:Li-ion interaction, which in turn affect the macroscopically observed ionic conductivity (see also Figure 3 in reference [28] and the discussion teherein).

3.2. NMR spectroscopy: Heteronuclear coupling

Nuclear magnetic resonance (NMR) is a phenomenon based on the exchange of electromagnetic radiation when magnetic nuclei are exposed to a magnetic field. A requisite for this phenomenon is that the nuclei have non-zero spins, which applies for all isotopes with an odd number of protons and/or neutrons. A key feature in NMR spectroscopy is that the resonance frequency of a particular nucleus is directly proportional to the strength of the applied field and to the magnetic properties of the nucleus itself. The basic relation is thus:

$$\omega_0 = \gamma \cdot B_0 \tag{2}$$

where ω_0 is the precession frequency of the nucleus, B_0 is the externally applied magnetic field, and γ_0 is the gyromagnetic ratio characteristic of the investigated nucleus. It might appear from this relation that all nuclei having the same γ_0 would resonate at the same frequency. This is not the case since the most important perturbation of the NMR frequency is the 'shielding' effect of the surrounding electrons. The rotation of these electrons produces a spin, which results in a magnetic field that counteracts the magnetic field of the nucleus. In general, this electronic shielding reduces the resonance frequency, whereby same nuclei found in different molecular structures can be resolved by their characteristic chemical shift (∂, expressed in ppm).

By using different types of pulse sequences, where the pulses vary in shape, frequency and duration, dynamical or structural information can be extracted from an NMR experiment. Multi-dimensional NMR spectroscopy is a kind of Fourier Transformed (FT) NMR that allows detecting nuclear-nuclear interactions through magnetization transfer. Through-bond and through-space interactions can be detected, the latter in particular allowing to establish distances between atoms (*e.g.* by 2D-FT NMR). Although widely used to investigate proteins and crystalline materials, NMR spectroscopy has scarcely been used to elucidate the local structure in ILs. Of great potentiality in this context are heteronuclear and cross-polarization (CP) MAS NMR experiments, the latter being particularly suitable where the IL phase is present in a solid-like matrix, as in a gel or in a polymer.

One of the exceptions is the study reported in reference [17], where the structure and local organization in protic ILs of the triethylammonium (TEA) cation have been elucidated by combining 1- and 2D heteronuclear NMR experiments The authors show that the choice of different TEA$^-$ anionic species pairs strongly affects the dissociation scheme:

$$TEA\cdots H^{+}\cdots A^{-} \leftrightarrow TEAH^{+} + A^{-} \leftrightarrow TEAH^{+},\ A^{-} \qquad (3)$$

where A− denotes the anion. These experiments also evidence the coupling between the dissociation state of the proton in the IL and the diffusive behaviors of the individual ionic species (see also Table 1 in reference [17]). For instance, even though in TEA-TFSI the ions are fully dissociated the proton diffuses with the cation and faster than the TFSI anion, whereas in the IL TEA-acetate the proton is fully dissociated from the cation, and is also the fastest diffusing species. These observations have obvious implications for practical use in fuel cells where the transport of the protic species through the electrolyte and its reactivity at the cathode are key properties.

3.3. SAXS: Nano-segregation

Along with a peculiar set of physico-chemical properties, ILs also show a complex local organization with self-aggregating polar and non-polar domains of the nanometer size. This mesoscopic separation was first predicted by molecular dynamic (MD) simulations and later also experimentally confirmed by small (and wide) angle x-ray scattering measurements (SAXS (and WAXS)). Before discussing in detail these results, the basic principles of an x-ray scattering experiment will be briefly explained.

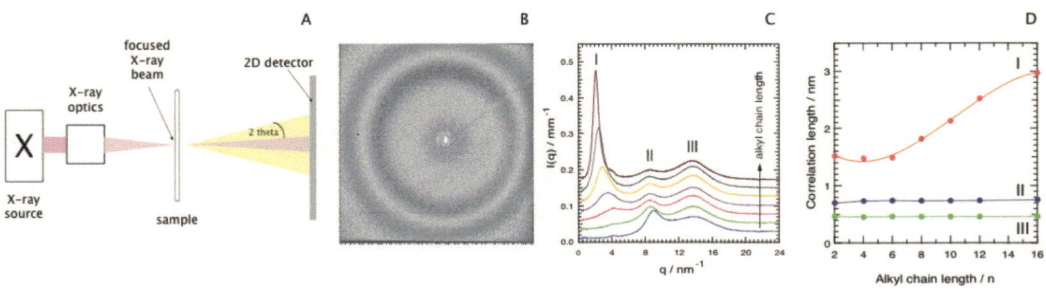

Figure 6. A: Schematic of a SAXS experimental set up. B: a typical 2D diffraction pattern recorded for IL containing samples [26]. C: the 2D diffraction pattern recorded for 1-alkyl-3-methylimidazoliumTFSI ILs transformed into intensity, I, as function of scattering vector, q [37]. D: correlation lengths derived from the diffraction pattern in C using d=2π/q.

In a SAXS experiment the sample is exposed to x-ray radiation with wavelength in the range of a few Å, and the elastically scattered x-rays are recorded at low angles, typically close to 0°, Figure 6A. In the presence of structural inhomogeneities in the nm range, a diffraction pattern is recorded if the following condition is fulfilled (Bragg's law):

$$n\lambda = 2dsin(\theta) \tag{4}$$

where λ is the wavelength of the incoming x-ray, d is the repeat distance of the local structure, and θ the angle at which diffraction peaks are collected. Unoriented samples yield a centro-symmetric pattern on a 2D detector (Figure 6B), which is radially averaged to give the typical plots of diffracted intensity versus scattering vector q (Figure 6C). Since the scattering vector q is defined as:

$$q = (4\pi / \lambda) \sin\theta \tag{5}$$

the real space correlation length d can be experimentally estimated as $d=2\pi/q$, see also Figure 6D.

The intriguing property of ILs is that even in the liquid state they can display clear scattering peaks in the x-ray diffraction pattern. This feature has been repeatedly reported by several authors and for diverse cation-anion combinations. The first studies focused primarily on ILs of the imidazolium cations [34], but more recent investigations have extended to those of ammonium and pyrrolidinium derivatives also, associated with PF_6^-, BF_4^-, Cl^- or $TFSI^-$ [35]. SAXS (and WAXS) diffraction patterns show that the low-q feature (corresponding to long real space distances) increases in intensity and shifts to lower values as the length of the alkyl side chain on the cation increases. This behavior is now rationalized as the separation of non-polar domains (consisting of aggregated alkyl side chains) from the polar matrix composed of anions and cationic head groups. This model can be compared to the micelle-like structuring occurring in n-alcohols [36].

As shown in Figure 6C, a typical SAXS diffraction pattern recorded for 1-alkyl-3-methylimidazolium TFSI ILs, and covering the wide q-range 0.2–30 nm^{-1}, displays three peaks attributed to cation-cation (0–5 nm^{-1}), anion-anion (5–11 nm^{-1}) and intramolecular (12–16 nm^{-1}) correlations. This IL series includes cations with the alkyl chain varying from ethyl (n=2) to hexadodecyl (n=16) [37]. In this context the reader should know that the central peak is relatively strong for the TFSI anions but can be much less intense for smaller anions like Cl^- or Br^-. In agreement with previous results we also record an increasing intensity for the cation-cation correlation length upon longer alkyl chains; however, by including very long chains in the investigated series we also evidence a non-strictly linear dependence of the corresponding correlation length d, Figure 6D [37]. For ILs where the alkyl chain is an octyl or shorter ($n \leq 8$) a linear dependence has previously been assumed, whereas our results point to a deviation for longer chains. Since we have independently also found a non-linear dependence on n for the cationc hydrodynamic radius extrapolated from self-diffusion meas-

http://dx.doi.org/10.5772/52863

urements ($r_+(n)=kT/c\pi\eta D_+(n)$), we believe that this non-linearity reflects a certain degree of inter-digitation of the alkyl chains, previously not believed to occur for shorter alkyl chains.

This finding fits well into the vivid debate currently ongoing on the true interpretation of the SAXS patterns: some researchers believing in a real mesoscopic separation as several times evidenced by SAXS and WAXS experiments [38], and other claiming that the observed diffraction pattern only reflects the internal structural inhomogeneity of the cation with no implications of a long-range ordering [39]. The main point of our study [37], however, is the correlation experimentally found between the transport properties (from pfg-NMR measurements) and the local ordering (from SAXS data) of the individual ionic species. In particular, we have found that there is a correlation between the dispersion curves of the anion-anion and cation-cation correlations and the measured self-diffusion constants independently measured for anions (D_-) and cations (D_+) in the ILs (see also Figure 8B and the thorough discussion in reference [37]).

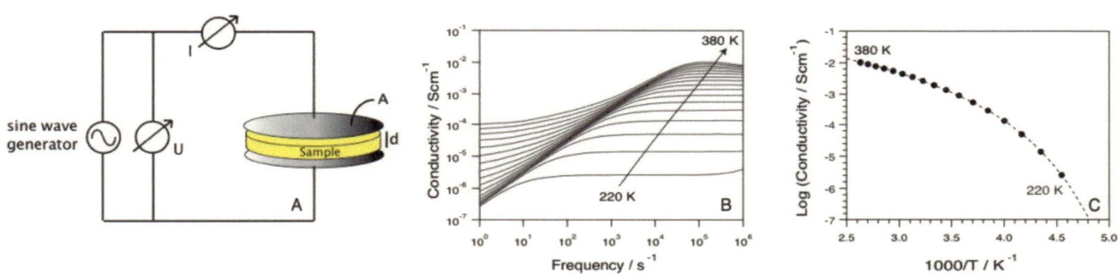

Figure 7. A: Schematic picture of the experimental set up for a dielectric spectroscopic measurement. Sample thickness, *d*, and surface area, *A*, are indicated in the figure. B: Conductivity–frequency plot for the protic ionic liquid N-ethylimidazolium bis(trifluoromethanesulfonyl)imide (EIMTFSI). C: Temperature dependence of conductivity, extrapolated from the constant plateau in B.

4. Dynamical investigations

When ILs are used as electrolytes in energy conversion devices like fuel cells or Li-ion batteries, they have the two-fold functionality to separate the electrodes (and thus prevent from short circuit) and to be the conducting medium for the electro-active ionic species (H^+ or protic species in PEM fuel cells, and Li^+ in Li-ion batteries). It is therefore of great interest to investigate and understand the transport properties in ILs, in particular the ionic conductivity and the self-diffusion of ionic species, but also transport phenomena under operational conditions that could lead to concentration gradients of the electrolyte (*e.g.* due to electro-osmotic drag). The most common techniques to investigate these phenomena, as well as recent important results, are presented and discussed in the following sections.

4.1. Dielectric spectroscopy: Ionic conductivity

Dielectric spectroscopy can be used to measure the ionic conductivity in diverse materials, including liquids. In such a dielectric experiment, the sample is sandwiched between two electrodes of surface area *A* and separation *d*, as shown in Figure 7A. This corresponds to a parallel plate capacitor, with capacitance *C* defined by

$$C = \varepsilon_0 \varepsilon \frac{A}{d} \tag{6}$$

where ε_0 is the dielectric constant in vacuum and ε is the frequency-dependent complex dielectric function of the material, $\varepsilon(f)$. In the dielectric experiment, an alternating voltage (*U*) is applied to the electrodes and the resulting alternating current (*I*) is measured. These quantities are related through the complex impedance (Z) of the material, *i.e.* $Z=U/I$. The complex impedance is in turn related to the dielectric function of the material through the relation

$$\varepsilon = \varepsilon'(f) - i\varepsilon''(f) = -\frac{i}{2\pi f C_0 Z} \tag{7}$$

where C_0 is the empty cell capacitance. In Eq. 7, $\varepsilon'(f)$ and $\varepsilon''(f)$ are the real and imaginary parts of the dielectric function $\varepsilon(f)$. The dependence of ε on frequency, *f*, and temperature, *T*, is typically investigated in the experiments. The presence of mobile charges in a material results in conductivity, which can be obtained in a dielectric experiment from the dielectric function through the relation:

$$\varepsilon''(f) = \frac{\sigma'(f)}{2\pi f C_0 \varepsilon_0} \tag{8}$$

This equation shows that the real part of conductivity contributes to the imaginary part of the dielectric constant. Thus, because of the f^{-1} dependence in Eq. 8, in the plot of $\varepsilon''(f)$ the dc conductivity will contribute as a rapidly increasing signal at low frequencies. The dc-conductivity can be directly investigated in a frequency plot ($\sigma(f)$), as shown in Figure 7B, and is extrapolated from the constant plateau, which moves to lower frequencies as temperature is decreased. At lower frequencies, the decay in conductivity is due to polarization effects at the electrodes. This occurs when the conductivity of the material is appreciable and a high charge density is created at the surface of the electrodes [40, 41]. In Figure 7C, the temperature dependence of the dc conductivity of the IL ethylimidazolium-bis(trifluoromethanesulfonly)imide (EIMTFSI) is shown. This plot shows the typical non-Arrhenius dependence of conductivity on temperature observed for all ILs in their melted state. Instead, the conductivity data are very well described by the Vogel-Fulcher-Tamman (VFT) equation

http://dx.doi.org/10.5772/52863

$$\sigma = \sigma_0 e^{-\frac{B}{T-T_0}} \tag{9}$$

where σ_0 is the ionic conductivity at very high temperatures ($T>>T_0$), B is a constant related to fragility (commonly discussed in glass-forming liquids) and T_0 is the ideal glass transition temperature. If compared on the same plot, ILs with larger values of B will display less curved conductivity data sets, similarly to what is observed when plotting viscosity as a function of inverse temperature. In our recent investigation of a large series of 1-alkyl-3-methylimidazolium ILs of the TFSI anions [37], we have extrapolated B values by fitting the conductivity data with a VFT equation. These values are also given in Table 1, indicating that for longer alkyl chains attached to the cation the IL becomes progressively less fragile (or stronger, larger B values). This is in agreement with that the cation-anion association also becomes more important, as also shown by the trend of $\Lambda_{imp}/\Lambda_{NMR}$ (also given in Table 1).

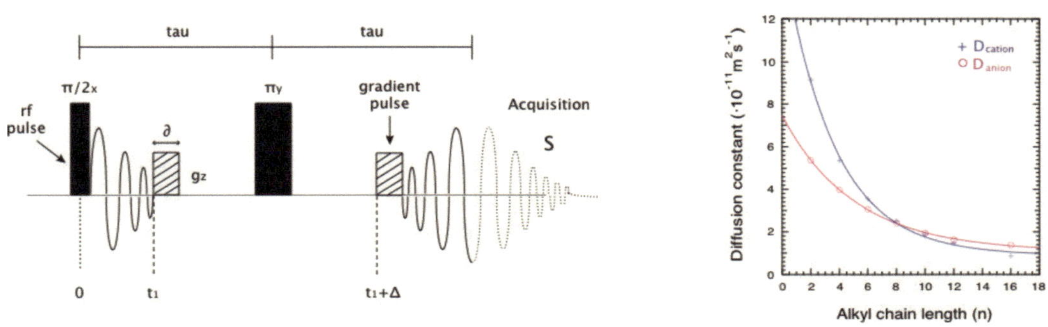

Figure 8. A: Schematic of the most basic pfg-NMR pulse sequence for diffusion experiments (reproduced with permission from reference [44]). B: Self-diffusion constants of the anions (pink) and cations (blue) in 1-alkyl-3-methyimidazolium TFSI ILs, for increasing alkyl chain lengths (see also reference [37]).

In our laboratories we have found that if the ionic conductivity of ILs is plotted as a function of T_g/T, where T_g is the experimentally found glass transition temperature, data fall onto master curves [42, 43]. This universal behavior resulting from T_g-scaling is a strong indication that conductivity is dominated by the viscous properties in the whole temperature range investigated.[11] Further, we have found that ILs of the same cationic structure but doped with different amounts of Li-TFSI also fall onto master plots through this scaling. However, even small structural changes on the cation in terms of, for instance, alkyl chain length can result in different B values, and therefore different curvatures, also in a T_g-scaled plot (compare for example the case of Pyr_{14} with Pyr_{24} in reference [42]).

11 Since viscosity is a quantity strongly related to the glass transition temperature T_g.

4.2. NMR spectroscopy: Self-diffusion

A very powerful tool to study the dynamics of ionic species in liquid materials is by pulsed field gradient nuclear magnetic resonance (pfg-NMR) spectroscopy. This technique measures the translational diffusion of molecules in time scales larger than milliseconds.

The most basic pfg (or Stejskal and Tanner 1967) pulse sequence used to estimate self-diffusion constants is schematically shown in Figure 8A. This consists of a spin-echo experiment with the 180° pulse in between two equal gradient pulses of magnitude g_z and duration δ. The first field gradient pulse introduces a dephasing in the precession frequency and the second pulse partially refocuses the phases. The phase difference due to the diffusing spins that cannot recover the initial phase leads to the attenuation of the NMR echo, while the signal loss in the case of unrestricted diffusion is proportional to the average root mean square displacement occurring between the two gradient pulses.

Typically, the magnetic field gradient g_z is imposed along the *z*-direction (that is also the direction of the static field, see B_0 in Eq. (2)) and, as a consequence, the Larmor frequency of a spin becomes a position label (from the NMR basic relation $\omega = \gamma B$).[12] If the duration between the leading edges of the gradient pulses is denoted Δ, the duration and the strength of the gradient field δ and g, the attenuation of the echo signal is described by

$$I = I_0 e^{-(\gamma^2 g^2 D \delta^2)(\Delta - \delta/3)} \tag{10}$$

From this relation the self-diffusion constant D of the diffusing species can be extrapolated. Typical self-diffusion constants at room temperature in ILs are found in the range 10^{-10} - 10^{-11} m^2s^{-1}, as also given in Table 1 for ILs of the imidazolium cation and the TFSI anion [45]. These values show that D tend to decrease with the length of the alkyl chain on the imidazolium, which is a direct effect of increased viscosity (η) and, in theory, also of the ionic size (r_s). Indeed, these quantities are closely related through the Stokes-Einstein relation

$$D = kT / 6\pi\eta r_2 \tag{11}$$

Where k is the Boltzmann's constant, T is the temperature, and r_s the hydrodynamic (or Stokes) radius. Currently a vivid debate is ongoing on whether, and to what extent, Eq. 11 is valid for ILs. In fact, not always do larger molecules display the lowest D values [46], and in some IL systems the fractional form of the $D(\eta)$ dependence has found to be more appropriate:

$$D \propto (T/\eta)^{\beta} \tag{12}$$

12 Where ω is the Larmor frequency (radians s^{-1}), γ the gyromagnetic ratio (rad $T^{-1}s^{-1}$) and B is the strength of the magnetic field (T).

For the few ILs investigated through the fractional form of the Stokes-Einstein equation, β has found to be less than unity and in the range 0.92–0.95 [47, 48]. These observations indicate that ILs cannot always be treated as classical hydrodynamic systems, and that a complex combination of solvating and electrostatic forces govern the self-diffusion of the molecular species.

Another interesting aspect of pfg-NMR measurements is that the molar conductivity (Λ_{NMR}) can be calculated from the self-diffusion constants using the Nernst-Einstein equation[13]

$$\Lambda_{NMR} = N_A e^2 \left(D_+ + D_- \right) \ / \ kT \tag{13}$$

In electrolytic systems this quantity becomes interesting if compared to the molar conductivity directly measured by impedance spectroscopy (Λ_{imp}, see section 5.1). The molar conductivity ratio $\Lambda_{imp}/\Lambda_{NMR}$ illustrates well the degree of cation-anion aggregation in ILs at equilibrium and represents a measure of the tendency to form higher ionic complexes, as opposed to completely dissociated systems. Indeed, while impedance measurements record the displacement of charged species only, pfg-NMR measurements record the movement of all probed molecules regardless their charged state (or ionic complexation). As a representative case, values for the $\Lambda_{imp}/\Lambda_{NMR}$ ratio in ILs of the imidazolium cation are given in Table 1. These are all smaller than one indicating that not all the diffusing species in the IL contribute to the ionic conduction, *i.e.* ionic aggregates and/or clusters are formed [45]. In addition, the trend for $\Lambda_{imp}/\Lambda_{NMR}$ indicates that for longer alkyl chains the cation-anion association also becomes stronger. Table 1 also illustrates that cations and anions do not always have the same diffusivity, even though this discrepancy becomes smaller for longer alkyl chains. We have recently addressed this specific issue by combining SAXS and pfg-NMR experiments on a large series of 1-alkyl-3-methylimidazoliumTFSI ILs [37].

Structure of cation	**$\Lambda_{imp}/\Lambda_{NMR}$ [45]**	**B_{imp} [37]**	**D_{NMR+} [45] ($\cdot 10^{-11}$ $m^2 s^{-1}$)**	**D_{NMR-} [45] ($\cdot 10^{-11}$ $m^2 s^{-1}$)**
C_1mim	0.76	n.a.	5.8	3.3
C_2mim	0.75	627	6.2	3.7
C_4mim	0.61	772	3.4	2.6
C_6mim	0.57	841	2.2	1.9
C_8mim	0.54	887	1.5	1.5

Table 1. Experimental values of the molar conductivity $\Lambda_{imp}/\Lambda_{NMR}$ ratio, the *B*-parameter from the VFT conductivity dependence, as well as the cationic and anionic self-diffusion constants D_{NMR+} and D_{NMR-}, for ILs of the 1-alkyl-3-methylimidazolium cation and TFSI anion.

13 Where NA is the Avogadro number, e is the electric charge on each ionic carrier, k is the Boltzmann constant and T is the temperature.

4.3. Confocal μ-Raman spectroscopy: *In-situ* fuel cell diagnostic

A very informative way to investigate the transport properties within a PEM is by *in situ* confocal Raman spectroscopy. With this technique the state of the electrolyte can be resolved in both space and time and correlated to the overall performance of the PEM fuel cell in which it is operated. This technique was first demonstrated for an H_2/H_2 cell [49] and later further developed for a H_2/O_2 PEM fuel cell [50, 51], see also Figure 9A. Using the confocal set up, good quality Raman spectra can be recorded across the whole thickness of the PEM, from anode to cathode (or *vice versa*), with spatial intervals of a few μm. Provided transparency of the material to the visible and small refraction index mismatch (with respect to air), the loss in Raman intensity and spatial resolution at depth are negligible. Under these conditions, the user can diagnostic the state of both the membrane and the liquid phase (water in hydrated Nafion and IL in IL swelled membranes) varying the operational conditions such as temperature, relative humidity, gas pressure, *etc.* Thus, with this *in situ* technique, both the structural robustness of the membrane and transport phenomena can be investigated *operando.*

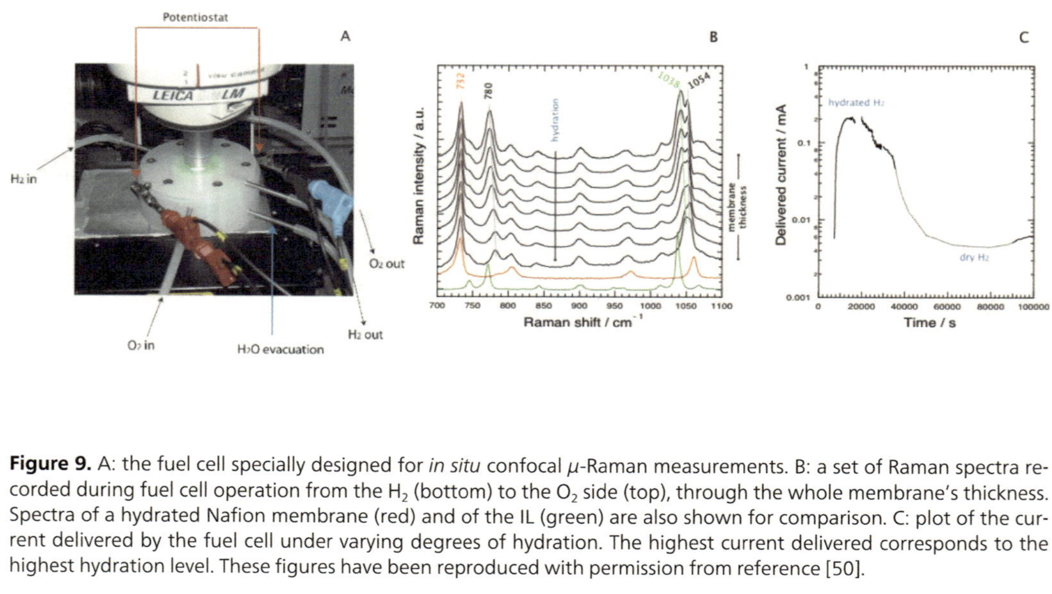

Figure 9. A: the fuel cell specially designed for *in situ* confocal μ-Raman measurements. B: a set of Raman spectra recorded during fuel cell operation from the H_2 (bottom) to the O_2 side (top), through the whole membrane's thickness. Spectra of a hydrated Nafion membrane (red) and of the IL (green) are also shown for comparison. C: plot of the current delivered by the fuel cell under varying degrees of hydration. The highest current delivered corresponds to the highest hydration level. These figures have been reproduced with permission from reference [50].

In reference [50] the authors demonstrate the potentiality of this technique through the study of a Nafion membrane swelled with protic ILs of the triethylammonium cation. In this particular case, the humidification of the H_2 gas was varied during fuel cell operation, whereby also the membrane became more or less hydrated. This *in situ* investigation showed that while the vibrational modes associated to the backbone of Nafion (ν_sCF_2 at 732 cm^{-1}) remained unchanged at different hydration degrees, those due to $-SO_3^-$ groups (νSO_3

http://dx.doi.org/10.5772/52863

at 1038–1049 cm^{-1} for the IL's anion (in this specific case $CH_3SO_3^-$) and at 1050–1061 cm^{-1} (for the sulfonic acid group of Nafion) varied in frequency, Figure 9B.

The blue-shift of the νSO_3 (IL's anion) in the more hydrated state indicates a stronger association state, most probably to water molecules that are thought to find interstitial positions between cations and anions in H_2O/IL mixtures, thus disrupting the original local organization. These features are of high relevance for fuel cell applications and deserve further investigations. The employment of *in situ* μ-Raman spectroscopy on operating fuel cell is one appropriate tool to understand the local dynamics and interactions around the protic species, putting these in relation to the overall performance of the PEM.

The authors also demonstrate that if the Raman spectra are correctly interpreted,[14] possible con- centration gradients of the IL induced by electro-osmotic drag across the membrane can also be resolved *operando*. From the quantitative analysis of Raman spectra, the authors also found that the membrane does not loose significant amounts of IL even after hydration cycles and four days of fuel cell operation. This is a very important result that supports the concept of IL swelled Nafion membranes for real fuel cell applications.

5. Conclusions

ILs are materials with an incredible variety of application fields. In this chapter, we have given examples of IL structures that can be used in fuel cells and Li-ion batteries, but use in super-capacitors, solar cells and green chemistry must not be forgotten. Very recently, ILs have also shown to have an important role in the extraction process of heavy metals from waste water. In all these applications, dynamical and structural properties jointly govern the functionality of ILs; yet in the field of ILs local structure and dynamics are rarely investigated in strict relation to each other. The combination of complementary experimental techniques like SAXS and pfg-NMR and the use of *in situ* spectroscopic methods are highly recommended for a better understanding of the real functionality of IL-derived materials, in particular those of interest for green energy applications. In this context, I foresee that *in situ* μ-Raman, *in situ* μ-NMR imaging and CP MAS NMR spectroscopy will be of increased importance in the next coming years.

Acknowledgements

The author acknowledges the financial support from the Chalmers' Areas of Advance Energy and Materials Science, as well as the Swedish Foundation for Strategic Research (SSF). Dr. M. Maréchal is also acknowledged for fruitful scientific discussions on the use (and interpretation) of SAXS in ionic liquids. A special thank goes also to all my previous and current collaborators.

14 Through calibrated relative intensity curves.

Author details

Anna Martinelli

Address all correspondence to: anna.martinelli@chalmers.se

Applied Surface Chemistry, Department of Chemical and Biological Engineering, Chalmers University of Technology, Gothenburg, Sweden

References

[1] Walden P., Bull. Acad. Imper. Sci. (St. Petersburg) 1914; 8: 405–422.

[2] Wasserscheid P., Keim W. Ionic liquids – New 'solutions' for transition metal catalysis. Angewandte Chemie - International Edition 2000; 39(21): 3773–3789.

[3] Wilkes J.S. A short history of ionic liquids - from molten salts to neoteric solvents. Green Chemistry 2002; 4: 73–80.

[4] Ohno H., editor, Ionic Liquids: The Front and Future of Material Developments. CMC, Tokyo; 2003.

[5] Armand M. Ionic-liquid materials for the electrochemical challenges of the future. Nature Materials 2009; 8(8): 621–629.

[6] Goyal R.K., Jayakumar N.S., Hashim M.A. Emulsion stabilization using ionic liquid [BMIM]+[NTf2]- and performance evaluation on the extraction of chromium. Journal of Hazardous Materials 2011; 195: 55–61.

[7] Tokuda H., Ishii K., Susan M.A.B.H., Tsuzuki S., Hayamizu K., Watanabe M.. Physicochemical properties and structures of room-temperature ionic liquids. 3. Variation of cationic structures. Journal of Physical Chemistry B 2006; 110(6): 2833–2839.

[8] Hirao M., Sugimoto H., Ohno H. Preparation of novel room-temperature molten salts by neutralization of amines. Journal of The Electrochemical Society 2000; 147(11): 4168–4172.

[9] Yoshizawa M., Xu M., Angell C.A. Ionic Liquids by Proton Transfer: Vapor Pressure, Conductivity, and the Relevance of ΔpKa from Aqueous Solutions. Journal of the American Chemical Society 2003; 125(50): 15411–15419.

[10] Tokuda H., Hayamizu K., Ishii K., Susan M.A.B.H., Watanabe M. Physicochemical properties and structures of room temperature ionic liquids. 2. Variation of alkyl chain length in imidazolium cation. Journal of Physical Chemistry B 2005; 109(13): 6103–6110.

[11] Tokuda H., Hayamizu K., Ishii K., Susan Md.A.B.H., Watanabe M. Physicochemical properties and structures of room temperature ionic liquids. 1. Variation of anionic species. Journal of Physical Chemistry B 2004; 108(42): 16593–16600.

[12] Hagiwara R., Lee J.S. Ionic liquids for electrochemical devices. Electrochemistry 2007; 75(1): 23–34.

[13] Kreuer K. D. On the development of proton conducting polymer membranes for hydrogen and methanol fuel cells. Journal of Membrane Science 2001; 185(1): 29–39.

[14] Martinelli A. Matic A., Jacobsson P., Börjesson L., Navarra M.A., Panero S., Scrosati B. A structural study on ionic-liquid-based polymer electrolyte membranes. Journal of the Electrochemical Society 2007; 154(8): G183–G187.

[15] Martinelli A., Matic A., Jacobsson P., Börjesson L., Fernicola A., Panero S., Scrosati B., Ohno H., Physical properties of proton conducting membranes based on a protic ionic liquid. Journal of Physical Chemistry B 2007; 111 (43): 12462–12467.

[16] Iojoiu C., Martinez M., Hanna M., Molmeret Y., Cointeaux L., Leprêtre J.-C., El Kissi N., Sanchez J.-Y. PILs-based Nafion membranes: A route to high-temperature PEFMCs dedicated to electric and hybrid vehicles. Polymers for Advanced Technologies 2008; 19(10): 1406–1414.

[17] Judeinstein P., Iojoiu C., Sanchez J.-Y., Ancian B. Proton conducting ionic liquid organization as probed by NMR: self-diffusion coefficients and heteronuclear correlations. Journal of Physical Chemistry B 2008; 112(12): 3680–3683.

[18] Lazzari M., Scrosati B. A Cyclable Lithium Organic Cell Based on Two Intercalation Electrodes. Journal of the Electrochemical Society 1980; 127(3): 773–774.

[19] Fernicola A., Croce F., Scrosati B., Watanabe M., Ohno H. LiTFSI-BEPyTFSI as an improved ionic liquid electrolyte for rechargeable lithium batteries. Journal of Power Sources 2007; 174(1): 342–348.

[20] Shimano S., Zhou H., Honma I. Preparation of nanohybrid solid-state electrolytes with liquid like mobilities by solidifying ionic liquids with silica particles. Chemistry of Materials 2007; 19(22): 5216–5221.

[21] Ueno K., Hata K., Katakabe T., Kondoh M., Watanabe M. Nanocomposite ion gels based on silica nanoparticles and an ionic liquid: ionic transport, viscoelastic properties, and microstructure. Journal of Physical Chemistry B 2008; 112(30): 9013–9019.

[22] Sharp K.G. A two-component, non-aqueous route to silica gel. Journal of Sol-Gel Science and Technology 1994; 2(1-3): 35–41.

[23] Néouze M.-A., Le Bideau J., Gaveau P., Bellayer S., Vioux A. Ionogels, new materials arising from the confinement of ionic liquids within silica-derived networks. Chemistry of Materials 2006; 18(17): 3931–3936.

[24] Le Bideau J., Gaveau P., Bellayer S., Néouze M.-A., Vioux A. Effect of confinement on ionic liquids dynamics in monolithic silica ionogels: 1H NMR study. Physical Chemistry Chemical Physics 2007; 9(40): 5419–5422.

[25] Martinelli A., Nordstierna L. An investigation of the sol-gel process in ionic liquid-silica gels by time resolved Raman and 1H NMR spectroscopy. Physical Chemistry Chemical Physics 2012; 14(38): 13216-13223.

[26] Nayeri M., Martinelli A. Simultaneous -Raman and SAXS investigation of the sol-gel process in ionogels of the 1-hexyl-3-methylimidazolium bis(trifluoromethanesulfonyl)imide ionic liquid. In manuscript; 2012.

[27] Hayashi K. Journal of The Electrochemical Society, 2002; 202nd Meeting, Abstracts, MA 2002-2, No. 205.

[28] Pitawala J., Kim J.-K., Jacobsson P., Koch V., Croce F., Matic A. Phase behaviour, transport properties, and interactions in Li-salt doped ionic liquids. Faraday Discussions 2012; 154: 71–78.

[29] Martinelli A., Matic A., Johansson P., Jacobsson P., Börjesson L., Fernicola A., Panero S., Scrosati B., Ohno H. Conformational evolution of TFSI- in protic and aprotic ionic liquids. Journal of Raman Spectroscopy 2011; 42(3): 522–528.

[30] Chalmers J.M. and Griffiths P.R. Handbook of Vibrational Spectroscopy; Theory and Instrumentation (John Wiley and Sons, 2002).

[31] Herstedt M., Smirnov M., Johansson P., Chami M., Grondin J., Servant L., Lassègues J.C. Spectroscopic characterization of the conformational states of the bis(trifluoromethanesulfonyl)imide anion (TFSI-). Journal of Raman Spectroscopy 2005; 36(8): 762–770.

[32] Holbrey J.D., Reichert W.M., Rogers R.D. Crystal structures of imidazolium bis(trifluoromethanesulfonyl)imide 'ionic liquid' salts: The first organic salt with a cis-TFSI anion conformation. Dalton Transactions 2004; (15): 2267–2271.

[33] Holomb R., Martinelli A., Albinsson I., Lassègues J.C., Johansson P., Jacobsson P. Ionic liquid structure: The conformational isomerism in 1-butyl-3-methyl-imidazolium tetrafluoroborate ([bmim][BF4]). Journal of Raman Spectroscopy 2008; 39(7): 793-805.

[34] Triolo A., Russina O., Bleiff H.-J., Di Cola E. Nanoscale segregation in room temperature ionic liquids. Journal of Physical Chemistry B 2007; 111(18): 4641–4644.

[35] Pott T., Méléard P. New insight into the nanostructure of ionic liquids: A small angle X-ray scattering (SAXS) study on liquid tri-alkyl-methyl-ammonium bis(trifluoromethanesulfonyl)amides and their mixtures. Physical Chemistry Chemical Physics 2009; 11(26): 5469–5475.

[36] Triolo A., Russina O., Fazio B., Triolo R., Di Cola E. Morphology of 1-alkyl-3-methylimidazolium hexafluorophosphate room temperature ionic liquids. Chemical Physics Letters 2008; 457(4-6): 362–365.

[37] Martinelli A., Maréchal M., Åsa Östlund. Correlation between molecular structure and transport properties in 1-alkyl-3-methylimidazolium ionic liquids: a combined pfg-NMR and X-ray scattering study. In manuscript 2012.

[38] Russina O., Triolo A. New experimental evidence supporting the mesoscopic segregation model in room temperature ionic liquids. Faraday Discussions 2012; 154: 97–109.

[39] Hardacre C., Holbrey J.D., Mullan C.L., Youngs T.G.A., Bowron D.T. Small angle neutron scattering from 1-alkyl-3-methylimidazolium hexafluorophosphate ionic liquids ([Cnmim] [PF6], n=4, 6, and 8). Journal of Chemical Physics 2010; 133(7): 074510-074517.

[40] Kremer F. and Schonhals A., editors. Broadband Dielectric Spectroscopy. Springer-Verlag; 2003.

[41] Hohne G., Hemminger W. and Flammersheim H.-J., editors. Differential Scanning Calorimetry. Springer-Verlag; 1996.

[42] Martinelli A., Matic A., Jacobsson P., Börjesson L., Fernicola A., Scrosati B. Phase behavior and ionic conductivity in lithium bis(trifluoromethanesulfonyl)imide-doped ionic liquids of the pyrrolidinium cation and bis(trifluoromethanesulfonyl)imide anion. Journal of Physical Chemistry B 2009; 113(32): 11247–11251.

[43] Pitawala J., Matic A., Martinelli A., Jacobsson P., Koch V., Croce F. Thermal properties and ionic conductivity of imidazolium bis(trifluoromethanesulfonyl)imide dicationic ionic liquids. Journal of Physical Chemistry B 2009; 113(32): 10607–10610.

[44] Price W.S. Pulsed-field gradient nuclear magnetic resonance as a tool for studying translational diffusion: Part 1. Basic theory. Concepts in Magnetic Resonance 1997; 9(5): 299–335.

[45] Tokuda H., Tsuzuki S., Susan M.A.B.H., Hayamizu K., Watanabe M. How ionic are ionic liquids? An indicator of the physicochemical properties. Journal of Physical Chemistry B 2006; 110(39): 19593–19600.

[46] Noda A., Hayamizu K., Watanabe M. Pulsed-gradient spin-echo 1H and 19F NMR ionic diffusion coefficient, viscosity, and ionic conductivity of non-chloroaluminate room-temperature ionic liquids. Journal of Physical Chemistry B 2001; 105(20): 4603–4610.

[47] Kanakubo M., Harris K.R., Tsuchihashi N, Ibuki K., Ueno M. Effect of pressure on transport properties of the ionic liquid 1-butyl-3-methylimidazolium hexafluorophosphate. Journal of Physical Chemistry B 2007; 111(8): 2062–2069.

[48] Liu H., Maginn E. A molecular dynamics investigation of the structural and dynamic properties of the ionic liquid 1-n-butyl-3-methylimidazolium bis(trifluoromethanesulfonyl) imide. Journal of Chemical Physics 2011; 135(12): 124507–16.

[49] Matic H., Lundblad A., Lindbergh G., Jacobsson P. In situ micro-Raman on the membrane in a working PEM cell. Electrochemical and Solid-State Letters 2005; 8(1): A5–A7.

[50] Martinelli A., Iojoiu C., Sergent N. A H2/O2 fuel cell for in situ -Raman measurements. In-depth characterization of an ionic liquid filled Nafion membrane. Fuel Cells 2011; 12(2): 169–178.

[51] Huguet P., Morin A., Gebel G., Deabate S., Sutor A.K., Peng Z. In situ analysis of water management in operating fuel cells by confocal Raman spectroscopy. Electrochemistry Communications 2011; 13(5): 418–422.

Chapter 10

Solid-State Ionic Liquid Based Electrolytes for Dye-Sensitized Solar Cells

Chuan-Pei Lee, Te-Chun Chu, Ling-Yu Chang,
Jiang-Jen Lin and Kuo-Chuan Ho

Additional information is available at the end of the chapter

http://dx.doi.org/10.5772/53647

1. Introduction

1.1. Dye-sensitized solar cells (DSSCs)

The increasing global need for energy coupled with the depletion of easily accessible, hence cheap, fossil fuel reserves, poses a serious threat to the human global economy in the near future [1]. Considering in addition the harmful ecological impact of conventional energy sources, it becomes obvious that development of clean alternative energy sources is a necessity [2, 3]. Best renewable energy options must rely on a reliable input of energy onto the earth. Since the sun is our only external energy source, harnessing its energy, which is clean, non-hazardous and infinite, satisfies the main objectives of all alternative energy strategies. Mastering the conversion of sunlight to electricity or to a nonfossil fuel like hydrogen is without any doubt the most promising solution to the energy challenge. It is remarkable that a mere 10 min of solar irradiation onto the Earth's surface is equal to the total yearly human energy consumption [4]. Therefore, solar power is considered to be one of the best sustainable energies for future generations. To date photovoltaics has been dominated by solid-state junction devices, usually in silicon, crystalline or amorphous, and profiting from the experience and materials availability resulting from the semiconductor industry. However, the expensive and energy-intensive high-temperature and high-vacuum processes is needed for the silicon based solar cells. Therefore, the dominance of the photovoltatic field by such kind of inorganic solid-state junction devices is now being challenged by the emergence of a third generation solar cell based on interpenetrating network structures, such as dye-sensitized solar cells (DSSCs) [5].

Since Professor M. Grätzel in EPFL introduced the nanoporous films into dye-derived wide-band semiconductor research and made the breakthrough in the photoelectric conversion efficiency of DSSCs, academic and commercial interests have been focused on DSSCs for their high efficiency, potential low-cost and simple assembly technology. This became especially noticeable when the first cell with a certified efficiency of greater than 10% was demonstrated [6-10]. By incorporating the novel YD2-o-C8 dye and cosensitizing with Y123 dye, the DSSC with a traditional liquid electrolyte has achieved a 12.3% efficiency record [11], encouraging the surge to explore new organic materials for the conversion of solar to electric power.

The DSSC device is composed of three adjacent thin layers such as a high band-gap nanocrystalline semiconductor-based mesoporous film adsorbed with a dye sensitizer on the working electrode for the absorption in the visible region, a platinized counter electrode for the collection of electrons and a redox electrolyte, sandwiched between the two electrodes. The usual choice for the semiconductor material is titanium dioxide (TiO_2), whereas ruthenium bipyridyl derivatives (N3, N719, Z907 and black dye etc.) are for the dye sensitizer. The electrolyte mostly contains I^-/I_3^- redox couple, which was obtained by the mixing of iodine (I_2) and inorganic or organic iodides in suitable non-aqueous solvents. Upon absorption of light, an electron is injected from a metal-to-ligand charge transfer excited state of the dye into the conduction band of the metal oxide. The rate of this electron injection reaction is ultrafast, typically occurring on the order of hundreds of femtoseconds to tens of picoseconds. The injected electron percolates through the TiO_2 film, and is thought to move by a "hopping" mechanism and is driven by a chemical diffusion gradient (rather than an electric field), and is collected at a transparent conductive substrate of fluorine doped tin oxide glass (SnO_2: F), on which the TiO_2 film is formed. After passing through an external circuit, the electron is reintroduced into the solar cell at the platinum counter electrode, where triiodide is reduced to iodide. The iodide then regenerates the oxidized dye, thereby completing the circuit with no net chemical change.

1.2. Ionic liquids (ILs)

Ionic liquids (ILs) are low-temperature molten salts with melting points below 100 °C, that is, liquids composed of ions only. The salts are characterized by weak interactions, owing to the combination of a large cation and a charge-delocalized anion. This results in a low tendency to crystallize due to flexibility (anion) and dissymmetry (cation). ILs are basically composed of organic ions that may undergo almost unlimited structural variations because of the easy preparation of a large variety of their components. Thus, various kinds of salts can be used to design the ionic liquid that has the desired properties for a given application. These include, among others, imidazolium, pyrrolidinium and quaternary ammonium salts as cations and bis(trifluoromethanesulphonyl)imide, bis(fluorosulphonyl) imide and hexafluorophosphate as anions.

An IL, triethylammonium nitrate (a pure low-melting salt), was firstly identified more than a century ago. In the 1930s, a patent application described cellulose dissolution using a molten pyridinium salt above 130 °C. It was the need for a sturdy medium for nuclear fuel reprocessing that prompted the study of low-melting-point chloroaluminates. Among the onium cations with positive nitrogen(s), those derived from the imidazolium ring proved to be the best choice in terms of melting points and electrochemical stability [12]. At the same time, the need for new anions for organic polymer electrolytes based on polyethylene oxide led to the concept of a plasticizing anion, that is, an anion having a delocalized charge and multiple conformations differing only marginally in energy. The archetype of such anions is the bis(trifluoromethylsulphonyl) amide (CF_3SO_2-N-SO_2CF_3) ion, also known as NTf_2, in which the extremely electron-withdrawing CF_3SO_2-groups are conjugated and linked by flexible S-N-S bonds. When combined with an imidazolium cation, such as the ethylmethylimidazolium cation, this anion produces a fluid IL (melting point: -15 °C) with an ion conductivity comparable to that of the best organic electrolyte solutions; it shows no vapour pressure or no decomposition up to ~300-400 °C [13]. It is not miscible with water (~1,000 p.p.m. in equilibrium with liquid H_2O), and thus defies the conventional wisdom that states polarity is synonymous with hydrophilicity. ILs then developed rapidly, with a reinvestigation of ions, for example quaternary ammonium cations, that had been avoided previously by organic chemists because of unsymmetrical shapes that hindered easy purification through crystallization. The organic chemistry community had earlier engaged in research of media with controllable Lewis acidity (chloroaluminate ILs), but the modern era of ILs has produced numerous neutral ILs, that is, those based on ions which are unreactive towards acids or bases, be they Lewis or Bronsted. As a result, it is now difficult to name an organic reaction that has not been performed successfully in these potentially green solvents, which can be recycled almost indefinitely with no or minimal use of volatile organic compounds. Most products made in ILs can be distilled off, in the case of small molecules, or extracted with water or hydrocarbon solvents, at least one of which is usually immiscible with the ionic liquid.

It is this unique solvent potential that makes ILs key materials for the development of a range of emerging technologies. The advent of ILs has made viable processes that fail, or are even impossible, with conventional solvents. Water sensitive metals or semiconductors that previously could not be deposited from conventional water baths can now, by turning to ILs, be directly electroplated. Energy devices, such as the quasi-solid/all-solid-state DSSCs, polymer-electrolyte-membrane fuel cells, lithium batteries and supercapacitors presently under development to address the challenges of increasing energy costs and global warming, may greatly benefit from a switch to low-vapour-pressure, non-flammable, ILs-based electrolytes.

1.3. ILs as the electrolyte for DSSCs

The role of the electrolyte in DSSCs is very important as it provides the necessary ionic conductivity in the bulk of the solution and sets the potential barrier necessary for the energy conversion. In addition, it offers a reduction reaction at the counter electrode and helps for the dye regeneration by the charge-transfer reaction with the dye molecule [14]. Usually, the conventional inorganic and organic iodide electrolyte salts are lithium and tetra-alkyl ammonium iodides, respectively. Besides this, several molten salts, particularly ionic liquid based imidazolinium salts, have also been used for improving the performance of the DSSCs [15-19]. In these works, cations play an important role in determining the conversion efficiency of the DSSCs. For example, the interaction of Li^+ with TiO_2 enhances the electron transfer from the sensitized dye to the TiO_2 and from I^- to the oxidized dye, leading to high photocurrent [20-22]. In the case of imidazolinium cations, the increase in the concentration of imidazolinium cations leads to the decrease of recombination at the working electrode due to the multilayer adsorption; thus improving the DSSC performance [23]. Kubo *et al.* [15] further studied the effect of the alkyl chain length of some imidazolium cations on the electron recombination lifetime. They found that the chain length does affect the electron recombination lifetime. The lifetime increased with the increase in the alkyl chain length. This is because hydrophobic alkyl chains may impede I_3^- from reaching TiO_2. The result reveals that bigger cations may enhance photocurrent due to the lower probability of electron recombination.

The highest efficiency record of DSSC was obtained based on the highly volatile organic solvent electrolyte due to the efficient infiltration of organic electrolyte in nanocrystalline films. However, commercialization of the cells with organic liquid electrolytes was impeded owing to technological problems related to hermetic sealing, precipitation of salts at low temperature and evaporation of liquids at high temperature; long-term stability is thus a major problem for these types of cells. Therefore, p-type inorganic semiconductors [24-26], organic hole conducting materials [27-30], ionic gel electrolytes having a polymer or a gelator [15, 31-33], and ionic liquid (IL) based electrolytes (or IL based electrolyte containing dispersed nano-components) [18, 34-40] were recently investigated for preparing the electrolytes. In these cases, imperfect filling of the dye-coated porous TiO_2 film by p-type inorganic semiconductors or polymers has resulted in poor efficiency for the cells. Another weakness of the inorganic p-type materials is the decided chemical structure, resulting in the limitedly adjustable chemical/physical properties in the application of solid-state electrolytes. Meanwhile, the inorganic p-type material derived all-solid-state DSSCs shows no stability. This should be ascribed to inorganic p-type materials tending to be oxidized under continuous illumination and the worsening of the interfacial contact between dye-sensitized TiO_2 and electrolyte along with the growth of age [41]. Moreover, the carrier diffusion length was limited in the case of conducting polymers due to their low conductivity. Thereby, ILs based electrolytes were considered to be most attractive for replacing the organic solvents; they are preferred because of their negligible vapor pressure, high thermal stability, wide electrochemical window, and high ionic conductivity [42-46]. However, most ILs based electrolytes are liquid at room temperature [36, 43, 45-48]. Therefore, the fluidity and potential leakage of ILs based electrolytes during long-term operation is still unavoidable, which limits their wide application in DSSCs. To overcome this problem, the solid-state ILs have been applied as solid-state electrolytes for DSSCs recently.

This chapter mainly reviewed the recent researches on the topic of solid-state ILs-based electrolytes for DSSCs. Here the solid-state ILs employed in the electrolytes of DSSCs can be classified as follows: (a) ILs crystals (system A), (b) ILs polymers (system B), and (c) ILs conductors (system C).

http://dx.doi.org/10.5772/53647

2. Solid-state ILs-based electrolytes for DSSCs

2.1. ILs crystals (system A)

In year 2005, Yamanaka *et al.* [49] reported a new strategy for enhancing the conductivity of ILs based electrolytes; employing an IL crystal as a constituent of an electrolyte, which forms a self-assembled structure and promotes the exchange reaction by the locally increased concentrations of I_2 and I_3^- [15, 50]. They selected 1-dodecyl-3-methylimidazolium iodide (C12MImI) as the IL crystal. This provides a self-assembled structure of the imidazolium cations like a solid, while maintaining the molecular dynamics like a liquid. The IL crystal (C12MImI) with the smectic A phase (SA) has a bilayer structure of interdigitated alkyl chains of the imidazolium cations, and I_2 and I_3^- would be localized between the SA layers. The locally high concentration would promote the exchange reaction. So, the IL crystal with the SA phase would be suitable for the electrolyte of DSSC. A few examples of the IL crystal with the SA phase, such as imidazolium salts consisting of cations with alkyl chains of C12-C18 and anions of hexafluorophosphate or bromide, have been reported [51, 52]. Before their study [49], an imidazolium salt with iodide as the counter anion has not been reported to be an IL crystal with a SA phase. Therefore, they showed for the first time that imidazolium iodides with alkyl chains longer than C12 exhibit a SA phase and that the liquid crystalline nature is preferable in terms of the hole transport layer in DSSC. The DSSC with C12MImI/I_2 electrolyte achieved a cell efficiency (η) of 2.30% under AM 1.5 irradiation. Zhao *et al.* [53] reported a solid-state DSSC utilizing imidazolium-type ionic crystal (1-methyl-3-hydroxyethyl-imidazolium iodide, MH-II) as the charge transfer layer, and obtained a good cell efficiency of 3.10% under one sun irradiation by adopting 1-methyl-3-propylimidazolium tetra-fluoroborate (MP-BF_4) as a crystal growth inhibitor, lithium bis-trifluoromethanesulfonylimideas ($Li[(CF_3SO_2)_2N]$) a charge transport enhancer, and 4-tert-butylpyridine (tBP) as a carrier recombination inhibitor. As shown in their report, the cell efficiency remained 60% of the initial value after 30 days at room temperature (R. T.) without any sealing and protection from ambient condition. Lee *et al.* [54] have fabricated all-solid-state DSSCs with a hybrid SWCNT-binary charge transfer intermediate (CTI), consisting of single wall carbon nanotubes (SWCNT), 1-ethyl-3-methylimidazolium iodide (EMII) and 1-methyl-3-propyl imidazolium iodide (PMII), without the addition of I_2 and tBP. A solid organic ionic crystal, EMII, was employed as CTI to fabricate all-solid-state DSSCs. In addition, SWCNTs were incorporated into the CTI as the extended electron transfer materials (EETM), which can reduce charge diffusion length and serve simultaneously as catalyst for the electrochemical reduction of I_3^-. An all-solid-state DSSC with this hybrid SWCNT-EMII achieved the higher cell efficiency (1.88%), as compared to that containing bare EMII (0.41%). To further improve the cell efficiency, they utilized PMII, which acts simultaneously as a co-charge transfer intermediate and crystal growth inhibitor. The highest cell efficiency (3.49%) was obtained using a hybrid SWCNT-binary CTI. In their studies, the durability of the solid-state DSSCs were studied at R. T. and was found to be far superior to that of a cell with an organic solvent electrolyte. In their further study, Lee *et al.* [55] also developed a solid-state composite electrolyte, comprising two ionic liquids and a carbon material, to fabricate a solid-state DSSC; the ILs were EMII and 1-ethyl-3-methylimidazolium tetrafluoroborate ($EMIBF_4$), and the carbon materials were carbon black (CB), multi-wall carbon nanotubes (MWCNT), and single-wall carbon nanotubes (SWCNT). A cell efficiency of 0.41% was achieved by using the bare EMII as the CTI; an efficiency of 2.52% was achieved for a solid-state DSSC by the incorporation of carbon black (CB) in the EMII. To further improve the cell efficiency, they utilized $EMIBF_4$, a crystal growth inhibitor, as an additive to the electrolyte. A cell efficiency of 3.09% was obtained using an electrolyte conaining the CB and the binary CTI (EMII plus $EMIBF_4$). When the CB was replaced with MWCNT and SWCNT, the cell efficiency could be improved to 3.53% and 4.01%, respectively. Long-term durability of the DSSC with SWCNT-binary CTI was found to be far superior to that of the cell with an organic solvent electrolyte, and in fact the durability was uninterrupted for at least 1,000 h. Armel *et al.* [56] have develpoed the organic ionic plastic crystals as a new class of solid-state electrolyte for DSSCs. The DSSC with their solid-state electrolyte, containing N,N-dimethylpyrrolidinium dicyanamide ($C_1mpyrN(CN)_2$), EMII, lithium iodide (LiI), I_2 and N-methylbenzimidazole (NMB), achieved a good cell efficiency of 5.10% under the illumination of 1 sun (AM 1.5G). Among the system A, Chen *et al.* [57] have developed a most efficient solid-state electrolyte employing an ionic liquid (1, 2-dimethyl-3-propylimidazolium iodide, DMPII) as CTI for DSSCs. Simultaneously, potassium iodide (KI) and polyethylene oxide (PEO, MW = 100,000) were incorporated into the CTI as the charge transfer auxiliary agent and the crystal growth inhibitor, respectively. Where, the strong interactions between the potassium cations and PEO can prevent the crystallization of the CTI and then enhance its ionic conductivity. As shown in Figure 1, an optimal cell efficiency of 5.87% can be obtained for the DSSC fabricated with the solid-state electrolyte of DMPII/KI/PEO. The corresponding photovoltaic parameters of the DSSCs with their solid-state electrolyte containing different KI contents are listed in Table 1. Recently, Li *et al.* [58] also reported the development of organic ionic plastic crystals, instead of molecular plastic crystals (such as succinonitrile), as the electrolytes for solid-state DSSCs. Compared with molecular plastic crystals, the advantages of ionic plastic crystal materials are their higher conductivity and very low volatility [59]. In their work, 1-ethyl-1-methylpyrrolidinium bis(tri-fluoromethane sulfonyl)imide ($P_{12}TFSI$), a well-known example of an organic ionic plastic crystal [40, 60-62], which shows a broad solid–solid transformation at 14 °C (entropy of transition 3.1 J K^{-1} mol^{-1}) and a melting point at 86 °C, was applied as the electrolyte for solid-state DSSCs. A room-temperature (R.T.) IL, PMII, was used as the iodide source for the electrolyte. The DSSCs with $P_{12}TFSI$/PMII electrolyte yielded a cell efficiency of 3.92% under the illumination of 100 mW cm^{-2} (AM 1.5G). Further addition of LiI and N-butylbenzimidazole (NBB) to the $P_{12}TFSI$/PMII electrolyte improved the cell efficiency to 4.78%. Their DSSCs, using $P_{12}TFSI$/PMII/LiI/NBB electrolyte, displayed better long-term stability compared to conventional liquid electrolytes, and remained 90% of the initial value after 50 days under ambient conditions without further sealing. Cao-Cen *et al.* [63] have synthesized the organic ionic crystal (N-4-(4-cyanobiphenyl-40-oxy)-butyl-3-butylimidazolium bromide, C_4BImBr) carrying 4-cyano-4′-hydroxybiphenyl and imidazolium units and applied them as the electrolytes for DSSCs. It has been demonstrated that biphenyl and 4-cyanobiphenyl are important core units for mesogenic molecules [64-67]. Cyanobiphenyl-functionalized compounds can act as a kind of low light-transmitting and high light-scattering material and obtained a high light-harvesting efficiency when they were employed as an electrolyte placed at the back of the working electrode of DSSC. The fabricated all-solid-state DSSCs with EMII/I_2/C_4BImBr electrolyte achieved a good cell efficiency of 4.45% under the illumination of 100 mW cm^{-2} (AM 1.5G) because of the enhanced light harvesting capability of the electrolyte containing C_4BImBr. To further improved the cell efficiency, PMII, was added into the EMII/I_2/C_4BImBr electrolyte as a crystal growth inhibitor. This fabricated DSSC with EMII/I_2/C_4BImBr/PMII electrolyte showed an enhanced cell efficiency of 5.07% under the illumination of 100 mW cm^{-2} (AM 1.5G), and it also possessed a good long-term stability (decay 5%) for 1,000 h during the accelerated aging test (1sun light soaking) at 25 °C.

Table 2 is a partial list of the all-solid-state DSSCs with ILs crystals-based electrolytes, which were obtained from the literatures.

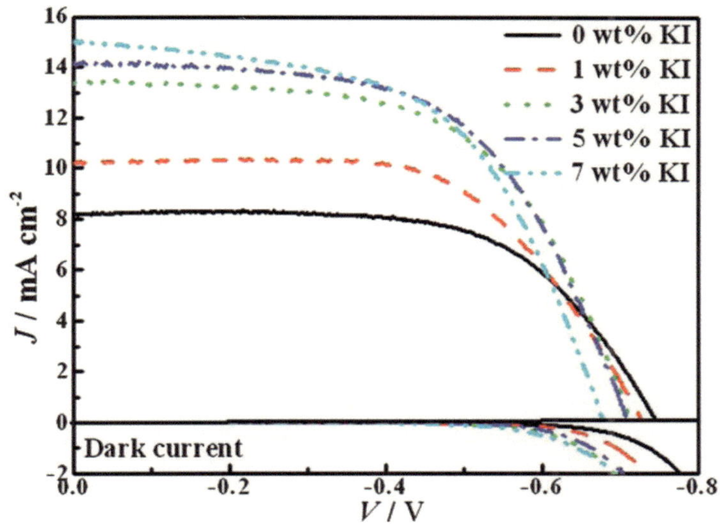

Figure 1. Photovoltaic characteristics of DSSCs assembled with the quasi solid-state electrolyte containing different KI contents under 100 mW cm^{-2} and in the dark [57].

KI (wt%)	J_{SC} (mA cm^{-2})	V_{OC} (mV)	FF	η (%)
0	8.22	750	0.65	4.05
1	10.21	730	0.61	4.56
3	13.44	710	0.59	5.66
5	14.11	710	0.59	5.87
7	15.00	680	0.56	5.72

Table 1. Photovoltaic characteristics of DSSCs assembled with the quasi solid-state electrolytes containing different KI contents under 100 mW cm^{-2}.

References	ILs crystals	Structures	Electrolytes	Dye	J_{SC} (mA cm^{-2})	V_{OC} (mV)	FF	η (%)	Durability
Yamanaka *et al.*, (2005)	C12MImI	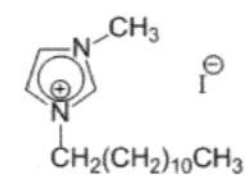	C12MImI; I_2	N. A.	~7.00	~520	~0.63	~2.30	N. A.
Zhao *et al.*, (2008)	MH-II	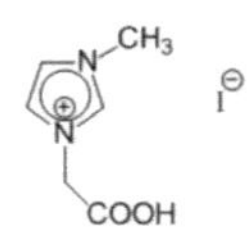	MH-II; MP-BF$_4$; Li[$(CF_3SO_2)_2N$] ; tBP	N3	7.01	646	0.69	3.10	30 days at-rest at 25 °C, decay 40%
Lee *et al.*, (2010)	EMII	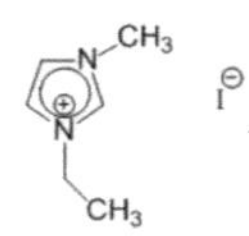	EMII; PMII; SWCNTs	N719	8.07	716	0.61	3.49	1,000 h at-rest at 25 °C, no decay
Lee *et al.*, (2011)	EMII		EMII; EMIBF$_4$; SWCNTs	N719	9.74	620	0.66	4.01	1,000 h at-rest at 25 °C, no decay
Armel *et al.*, (2011)	C_1mpyrN(CN)$_2$	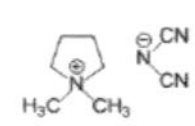	C_1mpyrNCN)$_2$; EMII; LiI; I_2; NMB	N719	8.60	775	0.77	5.10	N. A.
	EMII	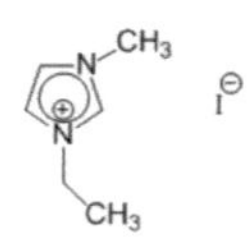							
Chen *et al.*, (2011)	DMPII	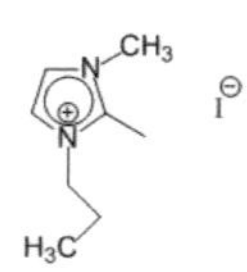	DMPII; KI; PEO	N719	14.11	710	0.59	5.87	N. A.
Li *et al.*, (2012)	P_{12}TFSI	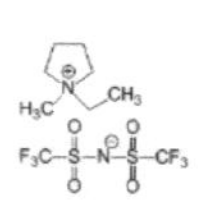	P_{12}TFSI; PMII; LiI; NBB	Z907	12.45	588	0.65	4.78	50 days at-rest at 25 °C, decay 10%
Cao-Cen *et al.*, (2012)	C_4BImBr	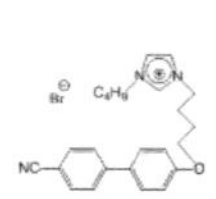	EMII; I_2; C_4BImBr; PMII	Z907	12.39	609	0.67	5.07	1,000 h light soaking at 25 °C, decay 5%

Table 2. Partial literatures reported on the solid-state DSSCs with ILs crystals.

2.2. ILs polymers (system B)

Wang *et al.* [68, 69] have synthesized the IL polymers, poly (1-alkyl-3-(acryloyloxy)hexylimidazolium iodide) (PAAII) and employed it as an iodine-free electrolyte for all-solid-state DSSCs. PAAII is an ionic liquid polymer formed from the polymerization of monomers containing IL moieties [70, 71]. IL polymers contain both the IL structure and the polymer main chain. They show good ionic conductivity without liquid components due to their specific functions such as transport of target ions, specific polar environment and mechanical strength [72]. They used poly (1-ethyl-3-(acryloyloxy)hexylimidazolium iodide) (PEAII) as the electrolyte without the addition of iodine, a cell efficiency of 5.29% has been achieved in solid-state DSSCs under illumination of 100 mW cm^{-2} (AM 1.5G). PEAII also exhibited high thermal stability, and the DSSC assembled with PEAII electrolyte maintained about 85% of their initial efficiency after 1,000 h without sealing. An acidic ionic liquid polymer P [((3-(4-vinylpyridine) propanesulfonic acid) iodide)-co-(acrylonitrile)], which is named as P-HI for short, has been synthesized and employed in ILs electrolyte for DSSCs by Fang *et al.* [73]. The polymer P-HI contains sulfonic acid group, which has electrostatic forces with ILs to form the homogeneous and continuous framework for enhancing transportationof redox couples in the electrolyte. The DSSC with the novel ILs electrolyte, containing P-HI, NMB, guanidinium thiocyanate (GuNCS), 1-hexyl-3-methylimidazolium iodide (HMII) and 1-allyl-3-methylimidazoliumiodide (AMII), achieved the highest cell efficiency of 6.95% under AM 1.5G illumination at 100mW cm^{-2} in the system B. They also investigated the effects of the concentration of iodine on the performance of DSSCs with their electrolyte system. As shown in Figure 2, they found that the addition of iodine mainly reduces the open-circuit voltage (V_{OC}) and slightly decreases the short-circuit current density (J_{SC}) of their cells due to the increase in dark current and the serious visible light by I_3^-, respectively. Therefore, they demonstrated that their system works best without the addition of iodine. Chi *et al.* [74] have synthesized a polymerized ionic liquid of poly((1-(4-ethenylphenyl)methyl)-3-butyl-imidazolium iodide) (PEBII) and employed it as a solid electrolyte for I_2-free solid-state DSSCs. In their study, the photoanode/electrolyte interfaces were significantly improved using a graft copolymer-directed and organized mesoporous TiO_2 thin film. The cell efficiency of the DSSC with PEBII has reached 5.93% at 100 mW cm^{-2}. In their further study [75], they utilized a novel TiO_2 photoanode with double layer structures containing mesoporous TiO_2 beads and PEBII electrolyte for constructing an solid-state DSSC; the cell efficiency was then enhanced up to 6.70%. Bis-imidazolium based poly(ionic liquid), poly(1-butyl-3-(1-vinylimidazolium-3-hexyl)-imidazolium bis(trifluoromethanesulfonyl)imide) (Poly[BVIm] [HIm] [TFSI]), was synthesized by Chen *et al.* [76], and it was dissolved in the ILs electrolyte (EMII/PMII/1-ethyl-3-methylimidazolium thiocyanate (EMISCN)/I_2/GuSCN/N-butylbenzimidazole (NBB)) to form solid-state electrolytes for DSSCs, without using any volatile organic solvent. They found that the bis-imidazolium based Poly[BVIm][HIm][TFSI] electrolyte possessed good thermal stability and conductivity due to the charge transport networks formed in the electrolyte via the π-π stacked imidazolium rings. The DSSCs based on Poly[BVIm][HIm] [TFSI] electrolyte yielded the cell efficiency of 5.92% under the simulated AM 1.5G solar spectrum illumination at 100 mW cm^{-2}. The Poly[BVIm][HIm][TFSI] based DSSC also showed a good long-term stability during accelerated aging test under 1 sun light-soaking at

60 °C, and it retains about 96% of the initial efficiency even after about 1,200 h test. In the recent report from Chang *et al.* [77], multi-walled carbon nanotubes (MWCNT) coated with a thin layer of 1-(2-acryloyloxy-ethyl)-3-methyl-benzoimidazol-1-ium iodide (AMBImI) were successfully fabricated by physical adsorption. They were then incorporated into poly(1-(2-acryloyloxy-ethyl)-3-methyl-imidazol-1-ium iodide (poly(AMImI))-based electrolytes to fabricate an all-solid state DSSC. The DSSC with the solid-state electrolyte,containing, achieved a cell efficiency of 3.55% at 100 mW cm^{-2}.

Table 3 is a partial list of the all-solid-state DSSCs with ILs polymers-based electrolytes, which were obtained from the literatures.

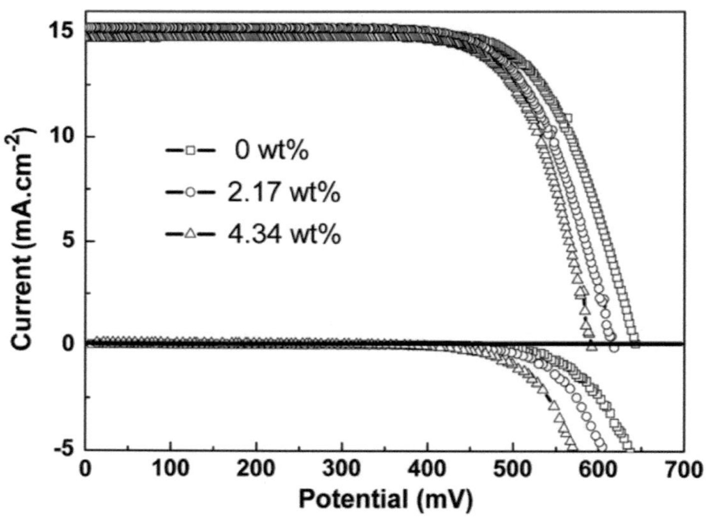

Figure 2. Photovoltaic characteristics of DSSCs using PH-I based electrolyte containing different contents of iodine measured under 100 mW cm^{-2} and in the dark [73].

References	ILs crystals	Structures	Electrolytes	Dye	J_{SC} *(mA cm⁻²)*	V_{OC} *(mV)*	*FF*	*η (%)*	Durability
Wang *et al.* (2011&2012)	PEAII	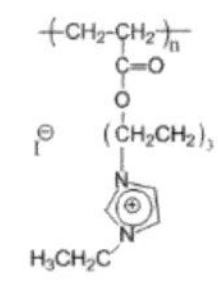	PEAII	N3	9.75	838	0.65	5.29	1,000 h at-rest at 25 °C, decay 15%
Fang *et al.*, (2011)	P-HI	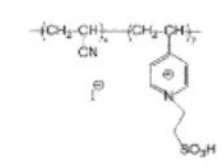	P-HI; HMII; AMII; NMB; GuNCS	N3	15.10	643	0.72	6.95	N. A.
Roh *et al.*, (2012)	PEBII	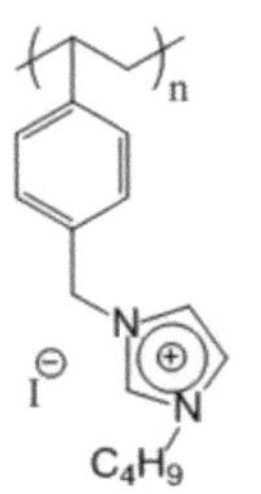	PEBII	N719	16.60	760	0.53	6.70	N. A.
Chen *et al.*, (2012)	Poly[BVIm][HIm][TFSI]	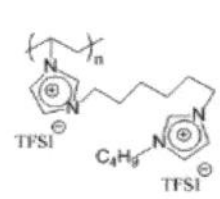	Poly[BVIm][HIm][TFSI]; EMII; PMII; EMISCN); I_2; GuSCN; NBB	N719	12.92	676	0.68	5.92	1,200 h light soaking at 60 °C, decay 4%
Chang *et al.*, (2012)	Poly(AMImI)	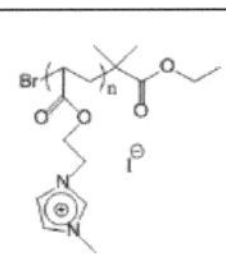	Poly(AMImI); NMB; GuSCN; I_2; MWCNT-poly(AMImI)	N3	8.51	646	0.64	3.55	N. A.

Table 3. Partial literatures reported on the solid-state DSSCs with ILs polymers.

2.3. ILs conductors (system C)

Recently, Midya *et al.* [78] have designed and synthesized a new class of solid-state ionic conductors (Table 4) based on a carbazole-imidazolium ionic salt as electrolytes for solid-state DSSCs. Carbazole is chosen as the hole conductor because polyvinyl carbazole has already been employed successfully in a solid-state DSSC as a hole conductor [79]. In their report [78], the solid-state IL conductors with SCN^- anions (SD1) or I^- anions (SD2) were synthesized and applied in an all-solid-state DSSC. The solid-state electrolyte containing SD2 and I_2 can provide dual channels for hole/triiodide transportation (Figure 3). In their systm (system C), the DSSC with a solid-state electrolyte, containing SD2, I_2, $Li[(CF_3SO_2)_2N]$, tBP and 1-ethyl-3-methyl-imidazolium tetracyanoborate ($EMIB(CN)_4$), achieved the highest cell efficiency of 2.85% under one sun irradiation, and the efficiency of the DSSC with SD1 is 1.43%.

References	ILs crystals	Structures	Electrolytes	Dye	J_{SC} *(mA cm^{-2})*	V_{OC} *(mV)*	*FF*	*η (%)*	Durability
Midya *et al.*, (2010)	SD2	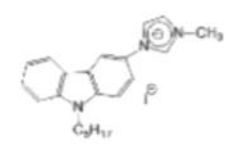	SD2; I_2; $Li[(CF_3SO_2)_2N]$; tBP; $EMIB(CN)_4$	N719	6.23	718	0.64	2.85	N. A.

Table 4. Partial literature reported on the solid-state DSSCs with ILs conductors.

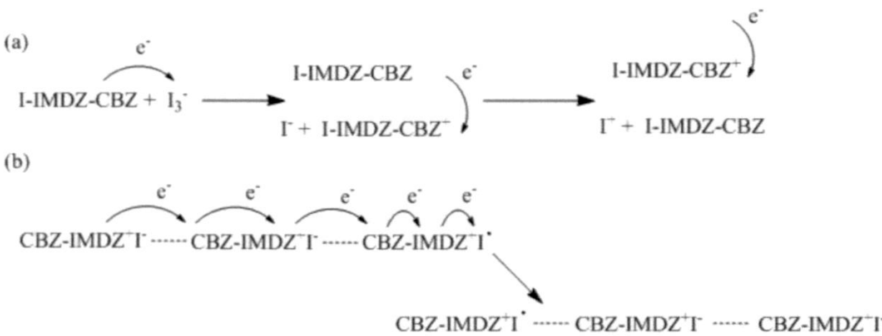

Figure 3. Schematic illustration of the mechanisms of (a) hole hopping and (b) iodine radical transport through CBZ-IMDZ-I solid-state ionic conductors [78].

3. Summary and future prospects

ILs are organic salts, composed mostly of organic ions that may undergo almost unlimited structural variations. Recently, the ILs act as useful electrolyte materials in DSSCs due to their negligible vapor pressure, high thermal stability, high ionic conductivity, and wide electrochemical window properties. This chapter mainly deals with the topic of novel ILs based electrolytes for solid-state DSSCs. The novel ILs based electrolytes include the IL-crystals (system A), IL-polymers (system B) and IL-conductors (system C).

Among system A, Chen *et al.* [57] have developed a novel solid-state electrolyte employing an ionic liquid crystal (DMPII) as CTI for DSSCs. An optimal cell efficiency of 5.87% can be obtained for the DSSC fabricated with the solid-state electrolyte of DMPII/KI/PEO. In system B, an acidic IL polymer, P-HI, has been synthesized and employed in ILs electrolyte for DSSCs by Fang *et al.* [73]. The DSSC with the novel ILs electrolyte, containing P-HI, NMB, GuNCS, HMII and AMII, achieved a cell efficiency of 6.95%. In the last system, Midya *et al.* [78] have designed and synthesized a new class of solid-state ionic conductor (SD2) based on a carbazole-imidazolium ionic salt as electrolytes for solid-state DSSCs. The DSSC with a solid-state electrolyte, containing SD2, I_2, $Li[(CF_3SO_2)_2N]$, tBP and $EMIB(CN)_4$, achieved a cell efficiency of 2.85%. A literature survey on the solid-state DSSCs with IL-crystals, IL-polymers, and IL-conductors based electrolytes have shown that these systems possessed superior long-term durability over the traditional organic solvent based electrolytes.

Recently, the stable organic radical, 2,2,6,6-tetramethyl-1-piperidinyloxy (TEMPO), was demonstrated to be a promising redox system for DSSCs [80, 81], offering an alternative to the widely used iodide/triiodide couple. In the future, we can synthesize a novel IL with TEMPO-imidazole complex [82-84] for an iodine (I_2)-free mediator system, and make application on solid-state DSSCs. This kind of TEMPO-imidazole complex containing TEMPO-redox radical and iodide-redox anion could potentially provide dual channels for charge transportation within the DSSCs.

Acknowledgments

This work was supported in part by the National Science Council of Taiwan under grant numbers NSC 100-2923-E-002-004-MY3 and NSC 100-2221-E-002-242-MY2.

Author details

Chuan-Pei Lee, Te-Chun Chu, Ling-Yu Chang, Jiang-Jen Lin and Kuo-Chuan Ho

National Taiwan University, Taipei, Taiwan

References

[1] Armaroli N, Balzani V. The future of energy supply: Challenges and opportunities. *Angewandte Chemie - International Edition* 2007,46:52-66.

[2] Gust D, Moore TA, Moore AL. Solar fuels via artificial photosynthesis. *Accounts of Chemical Research* 2009,42:1890-1898.

[3] Nocera DG. Chemistry of personalized solar energy. *Inorganic Chemistry* 2009,48:10001-10017.

[4] Yum JH, Baranoff E, Wenger S, Nazeeruddin MK, Grätzel M. Panchromatic engineering for dye-sensitized solar cells. *Energy and Environmental Science* 2011,4:842-857.

[5] O'Regan B, Grätzel M. A low-cost, high-efficiency solar cell based on dye-sensitized colloidal TiO_2 films. *Nature* 1991,353:737-740.

[6] Barbé CJ, Arendse F, Comte P, Jirousek M, Lenzmann F, Shklover V, *et al.* Nanocrystalline titanium oxide electrodes for photovoltaic applications. *Journal of the American Ceramic Society* 1997,80:3157-3171.

[7] Hagfeldt A, Grätzel M. Molecular photovoltaics. *Accounts of Chemical Research* 2000,33:269-277.

[8] Nazeeruddin MK, De Angelis F, Fantacci S, Selloni A, Viscardi G, Liska P, *et al.* Combined experimental and DFT-TDDFT computational study of photoelectrochemical cell ruthenium sensitizers. *Journal of the American Chemical Society* 2005,127:16835-16847.

[9] Gao F, Wang Y, Shi D, Zhang J, Wang M, Jing X, *et al.* Enhance the optical absorptivity of nanocrystalline TiO_2 film with high molar extinction coefficient ruthenium sensitizers for high performance dye-sensitized solar cells. *Journal of the American Chemical Society* 2008,130:10720-10728.

[10] Chen CY, Wang M, Li JY, Pootrakulchote N, Alibabaei L, Ngoc-Le CH, *et al.* Highly efficient light-harvesting ruthenium sensitizer for thin-film dye-sensitized solar cells. *ACS Nano* 2009,3:3103-3109.

[11] Yella A, Lee HW, Tsao HN, Yi C, Chandiran AK, Nazeeruddin MK, *et al.* Porphyrin-sensitized solar cells with cobalt (II/III)-based redox electrolyte exceed 12 percent efficiency. *Science* 2011,334:629-634.

[12] Armand M, Endres F, MacFarlane DR, Ohno H, Scrosati B. Ionic-liquid materials for the electrochemical challenges of the future. *Nature Materials* 2009,8:621-629.

[13] Earle MJ, Esperanca JMSS, Gilea MA, Canongia Lopes JN, Rebelo LPN, Magee JW, *et al.* The distillation and volatility of ionic liquids. *Nature* 2006,439:831-834.

[14] Stathatos E, Lianos P, Zakeeruddin SM, Liska P, Grätzel M. A quasi-solid-state dye-sensitized solar cell based on a sol-gel nanocomposite electrolyte containing ionic liquid. *Chemistry of Materials* 2003,15:1825-1829.

[15] Kubo W, Murakoshi K, Kitamura T, Yoshida S, Haruki M, Hanabusa K, *et al.* Quasi-solid-state dye-sensitized TiO 2 solar cells: Effective charge transport in mesoporous space filled with gel electrolytes containing iodide and iodine. *Journal of Physical Chemistry B* 2001,105:12809-12815.

[16] Mikoshiba S, Murai S, Sumino H, Hayase S. Anomalous increase in photocurrent density for quasi-solid dye sensitized solar cells by addition of tetra(bromomethyl)benzene. *Chemistry Letters* 2002:918-919.

[17] Wang P, Zakeeruddin SM, Exnar I, Grätzel M. High efficiency dye-sensitized nanocrystalline solar cells based on ionic liquid polymer gel electrolyte. *Chemical Communications* 2002,8:2972-2973.

[18] Wang P, Zakeeruddin SM, Comte P, Exnar I, Grätzel M. Gelation of ionic liquid-based electrolytes with silica nanoparticles for quasi-solid-state dye-sensitized solar cells. *Journal of the American Chemical Society* 2003,125:1166-1167.

[19] Kubo W, Kambe S, Nakade S, Kitamura T, Hanabusa K, Wada Y, Yanagida, S. Photocurrent-determining processes in quasi-solid-state dye-sensitized solar cells using ionic gel electrolytes. *Journal of Physical Chemistry B* 2003,107:4374-4381.

[20] Haque SA, Tachibana Y, Willis RL, Moser JE, Grätzel M, Klug DR, Durrant, J. R. Parameters Influencing Charge Recombination Kinetics in Dye-Sensitized Nanocrystalline Titanium Dioxide Films. *Journal of Physical Chemistry B* 2000,104:538-547.

[21] Hara K, Horiguchi T, Kinoshita T, Sayama K, Arakawa H. Influence of electrolytes on the photovoltaic performance of organic dye-sensitized nanocrystalline TiO_2 solar cells. *Solar Energy Materials and Solar Cells* 2001,70:151-161.

[22] Wagemaker M, Kentjens APM, Mulder FM. Equilibrium lithium transport between nanocrystalline phases in intercalated TiO_2 anatase. *Nature* 2002,418:397-399.

[23] Kambe S, Nakade S, Kitamura T, Wada Y, Yanagida S. Influence of the electrolytes on electron transport in mesoporous TiO_2-electrolyte systems. *Journal of Physical Chemistry B* 2002,106:2967-2972.

[24] Kumara GRA, Konno A, Shiratsuchi K, Tsukahara J, Tennakone K. Dye-sensitized solid-state solar cells: Use of crystal growth inhibitors for deposition of the hole collector. *Chemistry of Materials* 2002,14:954-955.

[25] Perera VPS, Pitigala PKDDP, Jayaweera PVV, Bandaranayake KMP, Tennakone K. Dye-Sensitized Solid-State Photovoltaic Cells Based on Dye Multilayer-Semiconductor Nanostructures. *Journal of Physical Chemistry B* 2003,107:13758-13761.

[26] Tennakone K, Kumara GRRA, Kottegoda IRM, Wijayantha KGU, Perera VPS. A solid-state photovoltaic cell sensitized with a ruthenium bipyridyl complex. *Journal of Physics D: Applied Physics* 1998,31:1492-1496.

[27] Bach U, Lupo D, Comte P, Moser JE, Weissörtel F, Salbeck J, Spreitzer, H., Grätzel, M. Solid-state dye-sensitized mesoporous TiO_2 solar cells with high photon-to-electron conversion efficiencies. *Nature* 1998,395:583-585.

[28] Krüger J, Plass R, Cevey L, Piccirelli M, Grätzel M, Bach U. High efficiency solid-state photovoltaic device due to inhibition of interface charge recombination. *Applied Physics Letters* 2001,79:2085-2087.

[29] Krüger J, Plass R, Grätzel M, Matthieu HJ. Improvement of the photovoltaic performance of solid-state dye-sensitized device by silver complexation of the sensitizer cis-bis(4,4′- dicarboxy-2,2′bipyridine)-bis(isothiocyanato) ruthenium(II). *Applied Physics Letters* 2002,81:367-369.

[30] Peter K, Wietasch H, Peng B, Thelakkat M. Dual-functional materials for interface modifications in solid-state dye-sensitised TiO_2 solar cells. *Applied Physics A: Materials Science and Processing* 2004,79:65-71.

[31] Wu J, Hao S, Lan Z, Lin J, Huang M, Huang Y, Fang, L. Q., Yin, S., Sato, T. A thermoplastic gel electrolyte for stable quasi-solid-state dye-sensitized solar cells. *Advanced Functional Materials* 2007,17:2645-2652.

[32] Yang Y, Zhou Ch, Xu S, Hu H, Chen Bl, Zhang J, Wu, S. J., Liu, W., Zhao, X. Z. Improved stability of quasi-solid-state dye-sensitized solar cell based on poly (ethylene oxide)-poly (vinylidene fluoride) polymer-blend electrolytes. *Journal of Power Sources* 2008,185:1492-1498.

[33] Kubo W, Kitamura T, Hanabusa K, Wada Y, Yanagida S. Quasi-solid-state dye-sensitized solar cells using room temperature molten salts and a low molecular weight gelator. *Chemical Communications* 2002:374-375.

[34] Usui H, Matsui H, Tanabe N, Yanagida S. Improved dye-sensitized solar cells using ionic nanocomposite gel electrolytes. *Journal of Photochemistry and Photobiology A: Chemistry* 2004,164:97-101.

[35] Katakabe T, Kawano R, Watanabe M. Acceleration of redox diffusion and charge-transfer rates in an ionic liquid with nanoparticle addition. *Electrochemical and Solid-State Letters* 2007,10:23-25.

[36] Lee KM, Chen PY, Lee CP, Ho KC. Binary room-temperature ionic liquids based electrolytes solidified with SiO2 nanoparticles for dye-sensitized solar cells. *Journal of Power Sources* 2009,190:573-577.

[37] Lee CP, Lee KM, Chen PY, Ho KC. On the addition of conducting ceramic nanoparticles in solvent-free ionic liquid electrolyte for dye-sensitized solar cells. *Solar Energy Materials and Solar Cells* 2009,93:1411-1416.

[38] Lee CP, Chen PY, Vittal R, Ho KC. Iodine-free high efficient quasi solid-state dye-sensitized solar cell containing ionic liquid and polyaniline-loaded carbon black. *Journal of Materials Chemistry* 2010,20:2356-2361.

[39] Chen PY, Lee CP, Vittal R, Ho KC. A quasi solid-state dye-sensitized solar cell containing binary ionic liquid and polyaniline-loaded carbon black. *Journal of Power Sources* 2010,195:3933-3938.

[40] Chen Z, Yang H, Li X, Li F, Yi T, Huang C. Thermostable succinonitrile-based gel electrolyte for efficient, long-life dye-sensitized solar cells. *Journal of Materials Chemistry* 2007,17:1602-1607.

[41] Taguchi T, Zhang XT, Sutanto I, Tokuhiro KI, Rao TN, Watanabe H, Nakamori T, Uragami M, Fujishima A. Improving the performance of solid-state dye-sensitized solar cell using MgO-coated TiO_2 nanoporous film. *Chemical Communications* 2003,9:2480-2481.

[42] Pringle JM, Golding J, Forsyth CM, Deacon GB, Forsyth M, MacFarlane DR. Physical trends and structural features in organic salts of the thiocyanate anion. *Journal of Materials Chemistry* 2002,12:3475-3480.

[43] Wang P, Zakeeruddin SM, Moser JE, Grätzel M. A new ionic liquid electrolyte enhances the conversion efficiency of dye-sensitized solar cells. *Journal of Physical Chemistry B* 2003,107:13280-13285.

[44] Fredlake CP, Crosthwaite JM, Hert DG, Aki SNVK, Brennecke JF. Thermophysical properties of imidazolium-based ionic liquids. *Journal of Chemical and Engineering Data* 2004,49:954-964.

[45] Wang P, Zakeeruddin SM, Humphry-Baker R, Grätzel M. A binary ionic liquid electrolyte to achieve ≥7% power conversion efficiencies in dye-sensitized solar cells. *Chemistry of Materials* 2004,16:2694-2696.

[46] Wang P, Zakeeruddin SM, Moser JE, Humphry-Baker R, Grätzel M. A solvent-free, $SeCN^-/(SeCN)_3^-$ based ionic liquid electrolyte for high-efficiency dye-sensitized nanocrystalline solar cells. *Journal of the American Chemical Society* 2004,126:7164-7165.

[47] Kuang D, Klein C, Zhang Z, Ito S, Moser JE, Zakeeruddin SM, Grätzel, M. Stable, high-efficiency ionic-liquid-based mesoscopic dye-sensitized solar cells. *Small* 2007,3:2094-2102.

[48] Jhong HR, Wong DSH, Wan CC, Wang YY, Wei TC. A novel deep eutectic solvent-based ionic liquid used as electrolyte for dye-sensitized solar cells. *Electrochemistry Communications* 2009,11:209-211.

[49] Yamanaka N, Kawano R, Kubo W, Kitamura T, Wada Y, Watanabe M, Yanagida, S. Ionic liquid crystal as a hole transport layer of dye-sensitized solar cells. *Chemical Communications* 2005:740-742.

[50] Kawano R, Watanabe M. Equilibrium potentials and charge transport of an I^-/I_3^- redox couple in an ionic liquid. *Chemical Communications* 2003,9:330-331.

[51] Gordon CM, Holbrey JD, Kennedy AR, Seddon KR. Ionic liquid crystals: Hexafluorophosphate salts. *Journal of Materials Chemistry* 1998,8:2627-2636.

[52] Bradley AE, Hardacre C, Holbrey JD, Johnston S, McMath SEJ, Nieuwenhuyzent M. Small-angle x-ray scattering studies of liquid crystalline 1-alkyl-3-methylimidazolium salts. *Chemistry of Materials* 2002,14:629-635.

[53] Zhao Y, Zhai J, He J, Chen X, Chen L, Zhang L, Tian, Y., Jiang, L., Zhu, D. High-performance all-solid-state dye-sensitized solar cells utilizing imidazolium-type ionic crystal as charge transfer layer. *Chemistry of Materials* 2008,20:6022-6028.

[54] Lee CP, Lin LY, Chen PY, Vittal R, Ho KC. All-solid-state dye-sensitized solar cells incorporating SWCNTs and crystal growth inhibitor. *Journal of Materials Chemistry* 2010,20:3619-3625.

[55] Lee C-P, Yeh M-H, Vittal R, Ho K-C. Solid-state dye-sensitized solar cell with a charge transfer layer comprising two ionic liquids and a carbon material. *Journal of Materials Chemistry* 2011,21:15471-15478.

[56] Armel V, Forsyth M, MacFarlane DR, Pringle JM. Organic ionic plastic crystal electrolytes; a new class of electrolyte for high efficiency solid state dye-sensitized solar cells. *Energy & Environmental Science* 2011,4:2234-2239.

[57] Chen J, Peng T, Fan K, Xia J. Iodine-free quasi solid-state dye-sensitized solar cells based on ionic liquid and alkali salt. *Journal of Materials Chemistry* 2011,21:16448-16452.

[58] Li Q, Chen X, Zhao J, Qiu L, Zhang Y, Sun B, Yan, F. Organic ionic plastic crystal-based electrolytes for solid-state dye-sensitized solar cells. *Journal of Materials Chemistry* 2012,22:6674-6679.

[59] Pringle JM, Howlett PC, MacFarlane DR, Forsyth M. Organic ionic plastic crystals: Recent advances. *Journal of Materials Chemistry* 2010,20:2056-2062.

[60] MacFarlane DR, Forsyth M. Plastic crystal electrolyte materials: New perspectives on solid state ionics. *Advanced Materials* 2001,13:957-966.

[61] Li Q, Zhao J, Sun B, Lin B, Qiu L, Zhang Y, Chen, X., Lu, J., Yan, F. High-temperature solid-state dye-sensitized solar cells based on organic ionic plastic crystal electrolytes. *Advanced Materials* 2012,24:945-950.

[62] Wang P, Dai Q, Zakeeruddin SM, Forsyth M, MacFarlane DR, Grätzel M. Ambient temperature plastic crystal electrolyte for efficient, all-solid-state dye-sensitized solar cell. *Journal of the American Chemical Society* 2004,126:13590-13591.

[63] Cao-Cen H, Zhao J, Qiu L, Xu D, Li Q, Chen X, Yan, F. High performance all-solid-state dye-sensitized solar cells based on cyanobiphenyl-functionalized imidazolium-type ionic crystals. *Journal of Materials Chemistry* 2012,22:12842-12850.

[64] Sebastián N, De La Fuente MR, López DO, Pérez-Jubindo MA, Salud J, Diez-Berart S, Ros, M. B. Dielectric and thermodynamic study on the liquid crystal dimer α-(4-cyanobiphenyl-4′-oxy)-ω-(1-pyreniminebenzylidene-4′-oxy)undecane (CBO11O Py). *Journal of Physical Chemistry B* 2011,115:9766-9775.

[65] Vijayakumar G, Lee MJ, Song M, Jin SH, Lee JW, Lee CW, Gal, Y. S., Shim, H. J., Kang, Y., Lee, G. W., Kim, K., Park, N. G., Kim, S. New liquid crystal-embedded PVdF-co-HFP-based polymer electrolytes for dye-sensitized solar cell applications. *Macromolecular Research* 2009,17:963-968.

[66] Binnemans K. Ionic liquid crystals. *Chemical Reviews* 2005,105:4148-4204.

[67] Lammi RK, Fritz KP, Scholes GD, Barbara PF. Ordering of Single Conjugated Polymers in a Nematic Liquid Crystal Host. *Journal of Physical Chemistry B* 2004,108:4593-4596.

[68] Wang G, Zhuo S, Wang L, Fang S, Lin Y. Mono-ion transport electrolyte based on ionic liquid polymer for all-solid-state dye-sensitized solar cells. *Solar Energy* 2012,86:1546-1551.

[69] Wang G, Wang L, Zhuo S, Fang S, Lin Y. An iodine-free electrolyte based on ionic liquid polymers for all-solid-state dye-sensitized solar cells. *Chemical Communications* 2011,47:2700-2702.

[70] Ohno H. Molten salt type polymer electrolytes. *Electrochimica Acta* 2001,46:1407-1411.

[71] Yoshizawa M, Ohno H. Synthesis of molten salt-type polymer brush and effect of brush structure on the ionic conductivity. *Electrochimica Acta* 2001,46:1723-1728.

[72] Ogihara W, Washiro S, Nakajima H, Ohno H. Effect of cation structure on the electrochemical and thermal properties of ion conductive polymers obtained from polymerizable ionic liquids. *Electrochimica Acta* 2006,51:2614-2619.

[73] Fang Y, Xiang W, Zhou X, Lin Y, Fang S. High-performance novel acidic ionic liquid polymer/ionic liquid composite polymer electrolyte for dye-sensitized solar cells. *Electrochemistry Communications* 2011,13:60-63.

[74] Chi WS, Koh JK, Ahn SH, Shin JS, Ahn H, Ryu DY, Kim, J. H. Highly efficient I_2-free solid-state dye-sensitized solar cells fabricated with polymerized ionic liquid and graft copolymer-directed mesoporous film. *Electrochemistry Communications* 2011,13:1349-1352.

[75] Roh DK, Seo JA, Chi WS, Koh JK, Kim JH. Facile synthesis of size-tunable mesoporous anatase TiO_2 beads using a graft copolymer for quasi-solid and all-solid dye-sensitized solar cells. *Journal of Materials Chemistry* 2012,22:11079-11085.

[76] Chen X, Zhao J, Zhang J, Qiu L, Xu D, Zhang H, Han, X., Sun, B., Fu, G., Zhang, Y., Yan, F. Bis-imidazolium Based Poly(ionic liquid) Electrolytes for Quasi-Solid-State Dye-Sensitized Solar Cells. *Journal of Materials Chemistry* 2012.

[77] Chang Y-H, Lin P-Y, Huang S-R, Liu K-Y, Lin K-F. Enhancing photovoltaic performance of all-solid-state dye-sensitized solar cells by incorporating ionic liquid-physisorbed MWCNT. *Journal of Materials Chemistry* 2012,22:15592-15598.

[78] Midya A, Xie Z, Yang JX, Chen ZK, Blackwood DJ, Wang J, Adams, S., Loh, K. P. A new class of solid state ionic conductors for application in all solid state dye sensitized solar cells. *Chemical Communications* 2010,46:2091-2093.

[79] Ikeda N, Miyasaka T. A solid-state dye-sensitized photovoltaic cell with a poly(N-vinyl- carbazole) hole transporter mediated by an alkali iodide. *Chemical Communications* 2005:1886-1888.

[80] Nakahara K, Oyaizu K, Nishide H. Electrolyte anion-assisted charge transportation in poly(oxoammonium cation/nitroxyl radical) redox gels. *Journal of Materials Chemistry* 2012,22:13669-13673.

[81] Zhang Z, Chen P, Murakami TN, Zakeeruddin SM, Grätzel M. The 2,2,6,6-Tetramethyl-1-piperidinyloxy Radical: An Efficient, Iodine- Free Redox Mediator for Dye-Sensitized Solar Cells. *Advanced Functional Materials* 2008,18:341-346.

[82] Suga T, Takeuchi S, Nishide H. Morphology-driven modulation of charge transport in radical/ion-containing, self-assembled block copolymer platform. *Advanced Materials* 2011,23:5545-5549.

[83] Strehmel V, Rexhausen H, Strauch P. 2,2,6,6-Tetramethylpiperidine-1-yloxyl bound to the imidazolium ion by an acetamido group for investigation of ionic liquids. *Tetrahedron Letters* 2010,51:747-750.

[84] Lee SH, Kim JK, Cheruvally G, Choi JW, Ahn JH, Chauhan GS, Song, C. E. Electrochemical properties of new organic radical materials for lithium secondary batteries. *Journal of Power Sources* 2008,184:503-507.

Chapter 11

Recent Advances in the Science and Technology of Desulfurization of Diesel Fuel Using Ionic Liquids

Elaheh Kowsari

Additional information is available at the end of the chapter

http://dx.doi.org/10.5772/51651

1. Introduction

Sulfur-containing compounds in transportation fuels are converted by combustion to SO_x, which is a major source of acid rain and air pollution [1]. For environmental protection purposes, many countries have mandated a reduction in fuels sulfur level down to 10 ppm by 2009 [2,3], and with more and more stringent regulatory constraints, it is a trend to achieve little-to-no sulfur fuels in the next several years.

In the petroleum industry, low-sulfur fuels are often obtained from hydrocracking processes or hydrotreating processes [4]. Although hydrotreating processes have been highly effective for the reduction of sulfur levels, further improvement of the hydrodesulfurization efficiency is limited to increasingly severe operational conditions at escalated cost. Moreover, when the deep hydrodesulfurization of motor fuels is needed, not only the energy and hydrogen consumption will be evidently increased, but undesired side reactions (such as the saturation of more olefins) also will be induced. Such side reactions result in a decrease in the octane number of the gasoline.

Ionic liquids, a new class of green solvents, have recently been undergoing intensive research on the removal of thiophenic sulfur species (e.g., dibenzothiophene) from fuels because of the limitation of the traditional hydrodesulfurization method in removing these species. Ionic liquids have the ability of extracting aromatic sulfur-containing compounds at ambient conditions without H_2 consumption. In addition Ionic liquids are immiscible with fuel, and the used Ionic liquids can be regenerated and recycled by solvent washing or distillation [5-9]. The desulfurization using ionic liquids has received growing attention [10-26].

In 2003, Lo et al. first reported chemical oxidation in conjunction with Ionic liquid extraction for oxidative desulfurization. Using the IL $[BMIm]PF_6$ as extractant, acetic acid as catalyst, and H_2O_2 as oxidant, the sulfur removal of DBT in model oil was increased significantly to 85%. The oxidation of organosulfur compounds to their corresponding sulfones catalyzed by polyoxometallic acids and their salts in Ionic liquids was also reported [27, 28]. Furthermore, it is well-known that homogeneous catalysts are difficult to separate from their reaction products limiting their recyclability. Currently, more efficient ODS systems solely containing H_2O_2 as oxidant, acidic ionic liquid, [HMIm]BF_4 or [Hnmp]BF_4 as extractant and catalyst have been reported by Lu et al. [29] and Zhao et al.[30] respectively.

Different types of ionic liquids, Imidazolium, pyridinium, and ammonium based Ionic liquids with different anion were demonstrated to be potentially applicable for sulfur removal from transportation fuels. Holbrey et al. [31] investigated the DBT extraction power from dodecan. This study ranks ionic liquids desulfurization ability by cation following the sequence methyl pyridinium ≥ pyridinium≈ imidazolium ≈ pyrrolidinium with much less significant variation with anion type. Recently, several pyridinium ionic liquids have been investigated for desulfurization-oriented purposes [32, 33]. In this chapter, the removal of sulfur compounds (S-compounds) from diesel fuels with different Ionic liquids are systematically reviewed.

2. Current Desulfurization Technologies

2.1. General Overview

Diesel fuel is a multi-purpose petroleum fuel used in trucks, trains, boats, buses, planes, heavy machinery and off-road vehicles. It also remains one of the largest sources of fine particle air pollution, which has serious health impacts. Besides fine particles or soot, Diesel-fueled engines also emit nitrogen oxides that can form ground level ozone.

Beginning in 2001, the U.S. Environmental Protection Agency (EPA) passed rules requiring use of ultra-low sulfur diesel (ULSD) fuel in diesel engines like trucks and buses, construction equipment, and more recently, stationary sources. Ultra-low sulfur diesel (ULSD) has only 15 parts per million (ppm) of sulfur. Low sulfur fuel has 500 ppm sulfur and uncontrolled sulfur diesel may have levels much higher [34-37].

The use of ULSD fuel in conjunction with re-designed advanced emission-control devices lowers the levels of released hydrocarbons, sulfur and nitrogen compounds, along with harmful particulate matter, to almost zero. Nitrogen oxides chemically react to form a lower-atmosphere ozone layer and contribute to acid rain. Burning ULSD fuel greatly cuts the amount of sulfur dioxide, a major contributor to acid rain. The oxides rise high into the atmosphere, lowering the pH of rain drops.

2.2. Description of Hydrodesulfurization (HDS) Process

Hydrodesulfurization (HDS) also know as a hydrotreating process, is one of the most common desulfurization methods that have been used in refinery processes, since the

1950s. HDS is a catalytic chemical process widely used to remove sulfur (S) from natural gas and from refined petroleum products such as gasoline or petrol, jet fuel, kerosene, diesel fuel, and fuel oils. The purpose of removing the sulfur is to reduce the sulfur dioxide (SO_2) emissions that result from using those fuels in automotive vehicles, aircraft, railroad locomotives, ships, gas or oil burning power plants, residential and industrial furnaces, and other forms of fuel combustion.

The industrial hydrodesulfurization processes include facilities for the capture and removal of the resulting hydrogen sulfide (H_2S) gas. In petroleum refineries, the hydrogen sulfide gas is then subsequently converted into byproduct elemental sulfur or sulfuric acid (H_2SO4). In fact, the vast majority of the 64,000,000 metric tons of sulfur produced worldwide in 2005 was byproduct sulfur from refineries and other hydrocarbon processing plants. Sulfur contents in crude oil may be categorized to the following groups [38]:

1- Free Elemental Sulfur

2- Mercaptans & Tiols (R-SH)

3- Hydrogen Sulfide

4- Sulfides

5- Disulfides (R-S-S-R')

6- Poly Sulfides ($R-S_n-R'$)

7- Thiophenes and their derivatives such as BT and DBT

In a typical catalytic hydrodesulfurization unit, the feedstock is deaerated and mixed with hydrogen, preheated in a fired heater (600°-800° F) and then charged under pressure (up to 1,000 psi) through a fixed-bed catalytic reactor

Although HDS, a high-pressure, high-temperature catalytic process that converts organic sulfur to hydrogen sulfide gas, can remove various types of sulfur compounds, some types of heterocyclic sulfur compounds existing in petroleum cannot be removed [39].

2.3. Biodesulfurization (BDS)

Biodesulfurization (BDS), based on the application of microorganisms that selectively remove sulfur atoms from organosulfur compounds, appears as a viable technology to complement the traditional hydrodesulfurization of fuels.

Enzymes in the bacteria selectively oxidize the sulfur, then cleave carbon-sulfur bonds. BDS will operate at ambient temperatures and atmospheric pressure and thus will require substantially less energy than conventional HDS methods to achieve sulfur levels below those required by current regulatory standards. BDS generates a fraction of the CO_2 that is generated in association with HDS, and it does not require hydrogen. Additionally, BDS can effectively remove some key sulfur-containing compounds that are among the most difficult for HDS to treat. BDS can be used instead of, or complementary with, HDS [40-52].

2.4. Oxidative Desulfurization (ODS)

Oxidative desulfurization is considered as the latest unconventional desulfurization process which involves chemical oxidation of divalent organic sulfur compounds to the corresponding hexavalent sulfur, also known as sulfone[53-61].

2.5. Adsorptive Desulfurization (ADS)

In adsorptive desulfurization process, OSCs are adsorbed into a specified solid adsorbent so as to produce none- or low-sulfur fuel. Depending onthe interaction between OSCs and the adsorbent, adsorptive desulfurization can be classified into direct adsorption desulfurization and reactive adsorption desulfurization [62].

3. Desulfurization of diesel fuels by extraction with ionic liquids

3.1. Desulfurization of diesel fuels by extraction with N-alkyl-pyridinium-based ionic liquids

3-Methylpyridinium-based ionic liquids were demonstrated to be effective for the selective removal of aromatic heterocyclic sulfur compounds from diesel at room temperature by Gao and coworkers [32]. The results indicated that the extractive performance using 3-methylpyridinium-based ionic liquids followed the order of 1-octyl-3-methylpyridinium tetrafluoroborate ($[C_8\ ^3MPy][BF_4]$) > 1-hexyl-3-methylpyridinium tetrafluoroborate ($[C_6\ ^3MPy][BF_4]$) > 1-butyl-3-methylpyridinium tetrafluoroborate ($[C_4\ ^3MPy][BF_4]$). For a given IL, the sulfur removal selectivity of sulfur compounds followed the order of dibenzothiophene (DBT) > benzothiophene (BT) > thiophene (TS) > 4,6-dibenzothiophene (4,6-DMDBT) under the same conditions, except for $[C8^3MPy][BF_4]$ ionic liquid, which followed the order of DBT > BT > 4,6-DMDBT > TS. The 3-methylpyridinium-based ionic liquids are insoluble in diesel while diesel has a certain solubility in 3-methylpyridinium-based ionic liquids, with the content varying from 6.1 wt %for $[C_4\ ^3MPy][BF_4]$ to 9.5 wt % for $[C_8\ ^3MPy][BF_4]$. The spent ionic liquid saturated sulfur compounds could be regenerated by a water dilution process. Considering these results, ionic liquids studied in this work are more competitive and feasible for extractive desulfurization applications. Moreover, the extractive desulfurization using 3-methylpyridinium-based ionic liquids could be used at least as a complementary process to hydrodesulfurization (HDS).

The pyridinium-based ionic liquids are employed as phase-transfer catalysts (PTCs) for phase-transfer catalytic oxidation of dibenzothiophene (DBT) dissolved in n-octane by D. Zhao and coworkers [63]. The partition coefficients of DBT between ionic liquids and n-octane are investigated. Then H_2O_2–formic acid is used as an oxidant and ionic liquids are used as PTCs. The reaction turns to be heterogeneous and desulfurization rate of DBT increased apparently. When IL ($[BPy]HSO_4$) is used as PTC, and the condition are: temperature is 60 °C, time is 60 min, H_2O_2/sulfur molar ratio (O/S) is 4, the desulfurization rate reaches the maximum (93.3%), and the desulfurization of the real gasoline is also investigated, 87.7% of sulfur contents are removed under optima reaction conditions. The PTC [BPy]HSO_4 can be recycled for five times without significant decrease in activity. The ability of the ionic liquids to extract DBT from oil phase follows the order below: [BPy HSO_4>[BPy]H_2PO_4>[BPy]SCN>[BPy]BF_4. The ionic liquid [BPy]HSO_4 has the highest KN value 1.42, followed by the ionic liquid [BPy]H_2PO_4,the KN value is 1.24. figure 1 is shown oxidization mechanism of DBT using ionic liquid as phase-transfer catalyst.

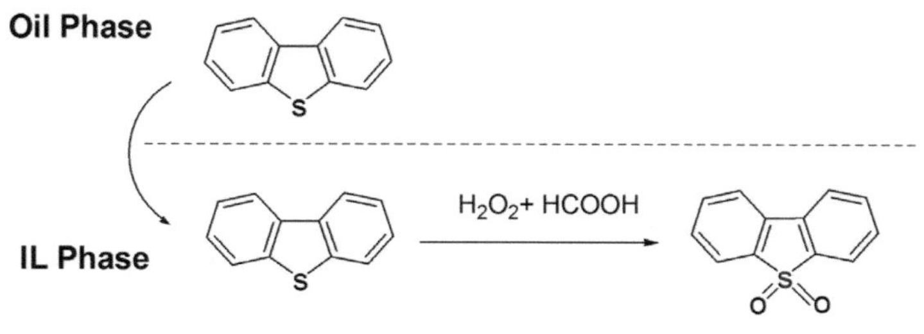

Figure 1. Oxidization mechanism of DBT using ionic liquid as phase-transfer scatalyst.

[Reproduced from Zhao D, Wang Y, Duan E, Zhang J. Oxidation Desulfurization of Fuel using Pyridinium-Based Ionic Liquids as Phase-Transfer Catalysts. Fuel Proces Tech. 2010; 91(12):1803–1806, Copyright (2010), with permeation from Elsevier]

Suitability of a pyridinium ionic liquid as a solvent in desulfurization has been analyzed by Arce and coworkers [64]. (Liquid + liquid)equilibria for ternary systems composed by 1-hexyl-3,5-dimethyl pyridinium {bis[trifluoromethylsulfonyl] imide, thiophene, and three hydrocarbons representative of fuel (n-heptane, 2,2,4 trimethylpentane, and toluene) have been determined at T = 298.15 K and atmospheric pressure. High solubility of thiophene in the ionic liquid and also of toluene have been found, being this solvent practically immiscible with 2,2,4 trimethylpentane and heptane. The chemical structure of [hmmpy] [NTf2] was shown in figure 2.

Six N-alkyl-pyridinium-based ILs, N-butyl-pyridinium nitrate ([BPy]NO_3), N-ethyl-pyridinium nitrate ([EPy]NO_3), *N*-butyl-pyridinium tetrafluoroborate ([BPy]BF_4), N-ethyl-pyridinium tetrafluoroborate ([EPy]BF_4), N-ethyl-pyridinium acetate ([EPy]Ac), and N-butyl-pyridinium acetate ([BPy]Ac), were prepared and tested in the extraction desulfurization of gasoline ny wang and coworkers [65]. It is found that [BPy]BF_4 has the best effect on the selective removal of sulfur-containing compounds from gasoline at room temperature among these ionic liquids. The extraction rate of [BPy]BF_4 is 45.5%. The desulfurization effect of [EPy]BF_4 is the lowest. The used ILs can be regenerated by rotary evaporation or re-extraction using tetrachloromethane. Thermosolvatochromism has been studied in three series of ionic liquids, pyridinium-based 1-butylpyridinium, 1-hexylpyridinium, and 1-octylpyridinium with bis(trifluoromethylsulfonyl) imide [NTf2] and tetrafluoroborate [BF4] anions, pyrrolidinium-based ionic liquids 1-methyl-1-butylpyrrolidium, 1-methyl-1-hexylpyrrolidium, 1-methyl-1-octylpyrrolidium with bis(trifluoromethylsulfonyl) imide anion and phosphonium-based ionic liquids tetrabutylphosphonium with alanate and valinate anions by Khupse and coworkers [66]. The effect of increase in alkyl chain length and temperature on normalized polarity ET N, Kamlet-Taft parameters, hydrogen bond donor ability (R), hydrogen bond acceptor ability (), and polarizability (π*) was investigated in the temperature range of 298-353 K.

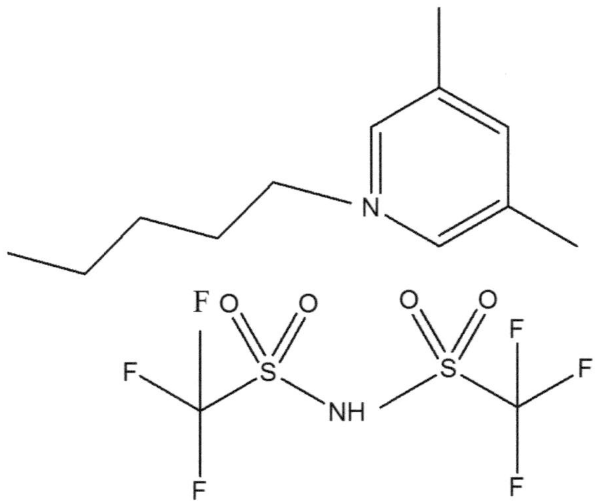

Figure 2. Chemical structure of ionic liquid [hmmpy][NTf$_2$]. No permitiobn.

[Reproduced from Arce A, Francisco M, Soto A, Evaluation of the Polysubstituted Pyridinium Ionic Liquid [hmmpy][Ntf2] as a Suitable Solvent for Desulfurization: Phase Equilibria. J. Chem. Thermodynamics 2010; 42(6): 712–718, Copyright (2010), with permeation from Elsevier]

Interestingly, the polarity decreases with temperature in the case of pyridinium- and pyrrolidinium-based ionic liquids, and it increases with temperature in the case of phosphonium-based ionic liquids.

3.2. Desulfurization of of fuel Using Imidazolium-based ILs

N-butylimidazole-derived dialkylphosphate ionic liquids are demonstrated to be effective for extractive removal of aromatic sulfur compounds (S-compounds) from fuel oils by Nie and coworkers [67], and show strong preferential extraction for aromatic S-compound versus toluene. Sulfur partition coefficients (KN) between ionic liquid and fuel oil at 298.15 K are determined experimentally over a wide range of sulfur content. The results show that the sulfur removal selectivity for a specific ionic liquid is dependent on the molecular structure of the S-compounds and follows the order dibenzothiophene >benzothiophene>thiophene >3-methylthiophene, and the efficiency of the ionic liquids for removal of aromatic S-compounds is dependent on the size and structure of both cations and anions of the ionic liquids. For the dialkylphosphate ionic liquids studied with the same anion, the longer the alkyl substitute to the imidazolium ring is the higher the KN value for that ionic liquid, and a similar trend is found for the ionic liquids with same cation. The chemical structures of ionic liquids were shown in figure 3.

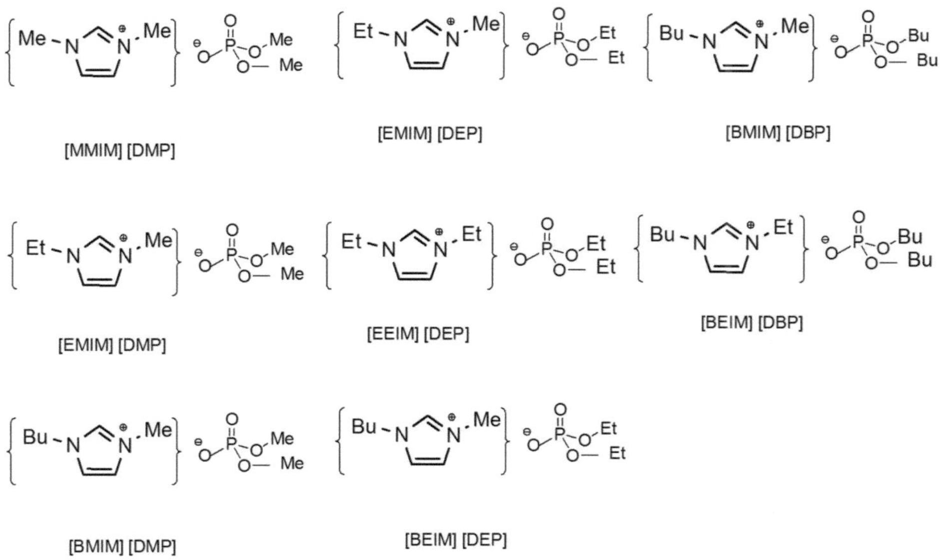

Figure 3. Structures of *N,N*-dialkylimidazolium dialkylphosphate ionic liquids.

[Reproduced from Nie Y, Li C, Meng H, Wang Z. N,N-Dialkylimidazolium Dialkylphosphate Ionic Liquids: Their Extractive Performance for Thiophene Series Compounds from Fuel Oils Versus the Length of Alkyl Group. Fuel Proces Tech. 2008; 8 9 (10): 978 –983, Copyright (2008), with permeation from Elsevier]

Two types of ionic liquids, 1-alkyl-3-methylimidazolium [AMIM] tetrafluoroborate and hexafluorophosphate and trimethylamine hydrochloride ($AlCl_3$-TMAC), were demonstrated to be potentially applicable for sulfur removal from transportation fuels by Zhang and Coworkers [68]. $EMIMBF_4$ (E) ethyl), $BMIMPF_6$ (B) butyl), $BMIMBF_4$, and heavier $AMIMPF_6$ showed high selectivity, particularly toward aromatic sulfur and nitrogen compounds, for extractive desulfurization and denitrogenation.

The used ionic liquids were readily regenerated either by distillation or by water displacement of absorbed molecules. The absorbed aromatic S-containing compounds were quantitatively recovered. Organic compounds with higher aromatic ð-electron density were favorably absorbed. Alkyl substitution on the aromatic rings was found to significantly reduce the absorption capacity, as a result of a steric effect. The cation and anion structure and size in the ionic liquids are important parameters affecting the absorption capacity for aromatic compounds. At low concentrations, the N- and S-containing compounds were extracted from fuels without mutual hindrance. $AlCl_3$-TMAC ionic liquids were found to have remarkably high absorption capacities for aromatics.

To develop an advanced desulfurization process that can be carried out under mild conditions without pressurized hydrogen or catalysis that has been evaluated for the extraction of thiophenic sulfur from a model fuel using the ionic liquids, 1-alkyl-3-alkyl imidazolium alkyl sulfate at room temperature was investigated by Mochizuki [69].

Six types of halogen-free ionic liquids with different alkyl chain lengths were prepared. The extraction yield of dibenzothiophene was higher than that of diphenylsulfide and diphenyldisulfide. The extraction yield of dibenzothiophene increased linearly with an increase in the length of alkyl chains and the mass ratio of the ionic liquid to the model fuel. The effect because of the change in the type of solvent was not appreciable, and dibenzothiophene was efficiently removed regardless of whether tetralin, benzene, or n-dodecane was used as the solvent.

The extractive and oxidative deep desulfurizations of model fuel oils using a low-viscosity ionic liquids, i.e.,1-ethyl-3-methylimidazolium dicyanamide ([C_2mim][$N(CN)_2$]), are investigated by Yu and coworkers [70]. [C_2mim][$N(CN)_2$] is capable of effectively extracting thiophene (TS) and dibenzothiophene (DBT) from oils. The sulfur content in the raffinate phases is only ~10 ppm after a few extraction steps. A short extraction equilibrium time of <5 min is observed.

The extraction operation is insensitive to temperature, and it can be effectively performed at or around room temperature. Unexpectedly, the oxidative removal of DBT by such a dicyanamide-based ionic liquid is not effective and is not as good as the corresponding extraction operation. Such an undesirable oxidative desulfurization is understood at a molecular level from ab initio calculations, and it may be ascribed to the strong intermolecular interaction between CH_3COOH or CH_3COOOH and [C_2mim][$N(CN)_2$] phase. Therefore, such a dicyanamide-based ionic liquid is efficient for direct extractive desulfurization, while it is less efficient for oxidative desulfurization.

An extraction and catalytic oxidation desulfurization (ECODS) system composed of V_2O_5, 30 wt%H_2O_2 and 1-butyl-3-methylimidazolium tetrafluoroborate ([Bmim]BF_4) as the basic experiment was used for the removal of DBT from the model oil at moderate temperature (30 °C) by Xu and coworkers [71]. In the reaction process, V_2O_5 was oxidized by H_2O_2 into peroxovanadium compounds. Meanwhile, the sulfur-containing compounds, such as benzothiophene (BT), dibenzothiophene (DBT), and 4,6-dimethyldibenzothiophene (4,6-DMDBT), were extracted into ionic liquid from the model oil and oxidized into their corresponding sulfones by peroxovanadium compounds. The reactivity of sulfur-containing compounds in the ECODS followed this order: DBT>BT>4,6-DMDBT. In the case of ECODS, the sulfur removal of DBT can reach 98.7%, which was superior to that of the simple extraction with IL (16.5%) or the catalytic oxidation (2.8%). This ECODS system could be recycled seven times without a significant decrease in activity. Supposed Mechanism of the Deep ECODS Using V_2O_5 Catalyst in ionic liquid shows in figure 4.

In order to obtain the ultra low-sulfur diesel, deep desulfurization of diesel oil has become a vital subject of environmental catalysis studies. Extraction and catalytic oxidation desulfurization (ECODS) system is one of the most promising desulfurization processes. A series of Keggin-type POM-based ionic liquids hybrid materials [MIMPS]$_3PW_{12}O_{40}$ $2H_2O$ (1-(3-sulfonic group) propyl-3-methyl imidazolium phosphotungstate), [Bmim]$_3PW_{12}O_{40}$ (1-butyl 3-methyl imidazolium phosphotungstate), [Bmim]$_3PMo_{12}O_{40}$ (1-butyl 3-methyl imidazolium phosphomolybdate) and [Bmim]$_4SiW_{12}O_{40}$ (1-butyl-3-methyl imidazolium silicotungstate) have been developed in this study, and the reaction has performed using the POM-ILs materials as catalysts, H_2O_2 as oxidant, and ionic liquid (IL) as solvent by Zhu and coworkers [72]. Through experimental evaluations, $[MIMPS]_3PW_{12}O_{40}\ 2H_2O$ was found to be the best catalyst, with an S-removal of 100% at 30 °C for 1 h. The main factors affecting the process including temperature, catalyst dosage, and O/S (H_2O_2/DBT) molar ratio were investigated in detail. Under the optimal conditions, DBT (dibenzothiophene) and 4,6-DMDBT (4,6-dimethyl-dibenzothiophene)could achieve high desulfurization efficiency. Moreover, the reaction system also exhibited high activity in actual diesel oil, which could be reduced from 1113 ppm to 198 ppm. The reaction system could recycle 8-times with a slight decrease in activity.

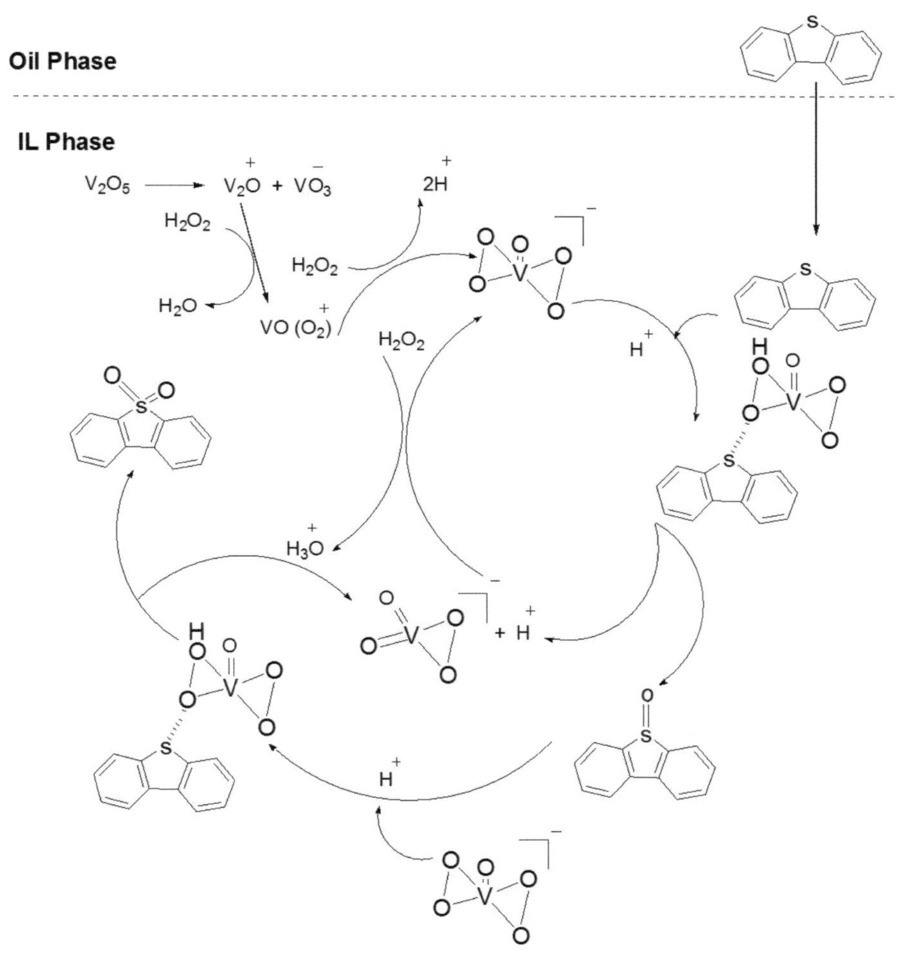

Figure 4. Supposed Mechanism of the Deep ECODS Using V_2O_5 Catalyst in ionic liquid.

[Reproduced from Xu D, Zhu W, Li H, Zhang J, Zou F, Shi H, Yan Y. Oxidative Desulfurization of Fuels Catalyzed by V_2O_5 in Ionic Liquids at Room Temperature. Energy Fuels 2009; 23(12): 5929–5933, Copyright (2009), with permeation from American Chemical Society]

Ozone associated with hydrogen peroxide as an advanced oxidation process had been employed to remove dibenzothiophene (DBT) in model oil in an ionic liquid system by Wang and coworkers [73]. DBT was oxidized availably by ozone and hydroxyl radicals that were generated by ozone and hydrogen peroxide. The oxidative productions of DBT were extracted to the IL phase because of their high polarity. The IL can be recycled 5 times without a significant decrease in desulfurization activity.

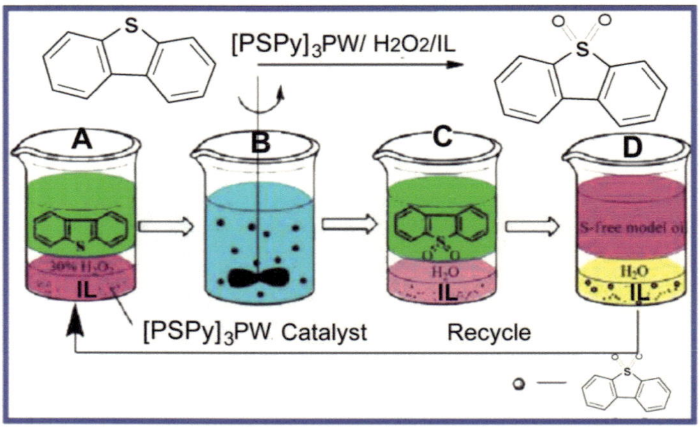

Figure 5. Catalytic Oxidation and Extraction of Sulfur Content Present in Model Oil: (A) before Oxidation; (B) during Oxidation; (C) after Oxidation; (D) with Extraction of Oxidative.

[Reproduced from Huang W, Zhu W, Li H,, Shi H, Zhu G, Liu H, Chen G. Heteropolyanion-Based Ionic Liquid for Deep Desulfurization of Fuels in Ionic Liquids, Ind. Eng. Chem. Res. 2010, 49(19):8998–9003.Copyright (2010), with permeation from American Chemical ociety]

Peroxotungsten and peroxomolybdenum complexes such as [$WO(O_2)_2$. Phen. H_2O] and [$MoO(O_2)_2$.Phen] (Phen: 1,10-phenanthroline) have been synthesized and characterized and were immobilized in 1-butyl-3-methylimidazolium tetrafluoroborate ([Bmim]BF_4), 1-n-octyl-3-methylimidazolium tetrafluoroborate ([Omim]-BF_4), 1-butyl-3-methyl-imidazolium hexafluorophosphate ([Bmim]PF_6), and 1-n-octyl-3-methylimidazolium hexafluorophosphate ([Omim]PF_6) for extraction and catalytic oxidation of dibenzothiophene (DBT) remaining in n-octane by Zhu and coworkers [27]. The results demonstrated that ionic liquid was only used as an extractant for DBT-containing model oil and the removal of sulfur was only about 12.2-22.0%. After addition of 30 wt % H_2O_2 in IL, model oil with 30.0-63.0% sulfur removal was given via chemical oxidation. While H_2O_2 and catalyst were introduced together, the removal of sulfur increased sharply. In the case of the system containing H_2O_2, $WO(O_2)_2$. Phen.H_2O and [Bmim]BF_4, extraction and catalytic oxidation increased the sulfur removal to 98.6%. However, the oxidative desulfurization systems containing $WO(O_2)_2$.Phen.H_2O and H_2O_2 only led to 50.3% sulfur removal in the absence of ionic liquid. This experiment demonstrated that a combination of catalytic oxidation and extraction in ionic liquid can deeply remove DBT from model oil. This result

http://dx.doi.org/10.5772/51651

also indicated the remarkable advantage of this process over the desulfurization by mere solvent extraction with ionic liquid or catalytic oxidation without ionic liquid.

A heteropolyanion-based ionic liquid, [(3-sulfonic acid) propylpyridine]$_3$PW$_{12}$O$_{40}$ 2H$_2$O, [PSPy]$_3$PW$_{12}$O$_{40}$ 2H$_2$O, abbreviated [PSPy]$_3$PW, was synthesized and approved as an effective catalyst for desulfurization of fuels in [omim]PF$_6$ by using aqueous H_2O_2 as oxidant by Huang and coworkers [74]. The catalysis was fulfilled with advantages of high activity, simplified workup, and flexible recyclability. The catalytic oxidation reactivity of sulfur-containing compounds was in the order dibenzothiophene (DBT) > 4,6-dimethyldibenzothiophene (4,6-DMDBT) > benzothiophene (BT). The effects of the amount of [PSPy]$_3$PW, H_2O_2, and reaction time and temperature were investigated in detail. Under the optimal conditions, the removal of DBT achieved 99.4%. Especially, we found that the removal of 4,6-DMDBT could be up to 98.8%, and the system could be recycled at least 9 times without significant decrease in activity. The sulfur level of FCC gasoline could be reduced from 360 to 70 ppm in the extraction and catalytic oxidation system. Catalytic Oxidation and Extraction of Sulfur Content Present in Model Oil shows in figure 5.

3.3. Desulfurization of fuel using quaternary ammonium -based ionic liquids

With the aim of deep desulfurization of the gasoline, an amphiphilic catalyst, which is composed of lacunary anion $[PW_{11}O_{39}]^{7-}$ and quaternary ammonium cation $[C_{18}H_{37}(CH_3)_3]N^+$, assembled in hydrophobic ionic liquid emulsions, can oxidize the sulfur compounds present in oil into their corresponding sulfones under ambient reaction conditions by Ge and coworkers [75].

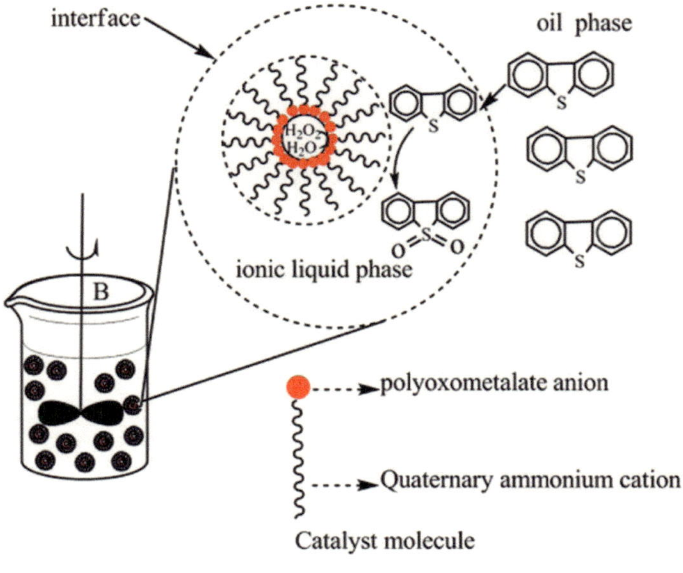

Figure 6. Catalytic Oxidation of DBT in Ionic Liquid Emulsion System.

In this process, catalytic oxidation of sulfur-containing molecules in model oil was investigated in detail under di ▪ erent reaction conditions (including di ▪ erent desulfurization systems, H_2O_2/DBT molar ratio, temperature, and various sulfur compounds). Moreover; this ionic liquid emulsion systemcould be recycled five times with an unnoticeable decrease in catalytic activity, and, fromthe kinetics study, it can be shown that the catalytic oxidative reaction is a pseudofirst-order reaction and the half-life is 30.4 min. Furthermore, the mechanism of catalytic oxidation desulfurization was elaborated, and the total sulfur level of real gasoline can be decreased from 1236 to 65 ppm after catalytic oxidation using an ionic liquid emulsion system.

In this emulsion reaction system, the catalystmolecule acts as an emulsifying agent, could be uniformly distributed in the interface of H_2O_2 ionic liquid, and forms a film around the dispersed ionic liquid droplets (Scheme 1)

3.4. Extractive Desulfurization Using Fe-Containing Ionic Liquids

Fe^{III} -containing ionic liquids, prepared from the reaction of anhydrous $FeCl_3$ and imidazolium chloride ([imidazolium]Cl), were used as effective extractants for the desulfurization of a model oil containing dibenzothiophene (DBT) by Ko and coworkers [76]. The amount of DBT extracted increased with an increasing molar ratio of $FeCl_3$/imidazolium]Cl. The ability of the ionic liquids to extract DBT seems to be attributed to the combined effects of Lewis acidity and fluidity of ionic liquids.

xEt_3NHCl_3 $FeCl_3$ (x=1.4-1.8) ionic liquids were synthesized by mixing Et_3NHCl and anhydrous FeC13 at 80 °C by Li and coworkers [77]. These were liquid at room temperature, with low viscosities, and exhibited remarkable abilities in effective desulfurization of thiophene in n-octane and fluid catalytic cracking (FCC) gasoline. Among them, $1.6Et_3NHCl_3$ $FeCl_3$ showed the highest sulfur removal. The anionic species $FeCl_4$-existed in $1.6Et_3NHCl_3$ $FeCl_3$ ionic liquid, as detected by electrospray ionization-mass spectrometry (ESI-MS), and the ionic liquidwas stable in air andmoisture. Sulfur-free (<10mg/L) gasoline could be obtained after extraction twice using an ionic liquid/oil volume ratio of 1. The ionic liquid could be recycled 10 times by distillation with a slight decrease in activity.The influence of the $Et_3NHCl/FeCl_3$ molar ratio on the sulfur removal of thiophene is shown in Table 1.

Sulfur Removal (%)	ionic liquids	Sulfur Removal (%)	ionic liquids
41.2	$[BMIm][OcSO_4]^b$	80.2	$1.4\ Et_3NHCl.\ Fe\ Cl_3$
40.0	$[C_8MIm]BF_4{}^c$	84.3	$1.5\ Et_3NHCl.\ Fe\ Cl_3$
51.7	$[C_8MPy]BF_4{}^d$	87.6	$1.6\ Et_3NHCl.\ Fe\ Cl_3$

44.1	$[OPy]BF_4$ [e]	86.5	1.7 Et_3NHCl. Fe Cl_3
		85.2	1.8 Et_3NHCl. Fe Cl_3

Table 1. Sulfur Removal of Thiophene Using ILs.

[a]Extraction conditions:model oil, 10mL; IL, 10mL; the mixture was stirred at 20 °C for 10 min. [b] Results from Esser et al. [c]Results from Alonso et al.5 [d]Results from Liu et al.,in which the extraction conditions were mass ratio oil/IL, 1:1; extraction time, 15 min; room temperature.[e] Results from Liu et al.

[Reproduced from Li FT, Liu Y, Sun ZM, Chen LJ, Zhao DS, Liu RH, Kou CG, Deep Extractive Desulfurization of Gasoline with xEt_3NHCl_3 $FeCl_3$ Ionic Liquids. Energy Fuels, 2010; 24(8): 4285–4289, Copyright (2010), with permeation from American Chemical Society]

A series of polymer-supported metal chlorides imidazolium ionic liquid moieties, M/CMPS-Im(Cl) (M =CuCl, $ZnCl_2$ and $FeCl_3$), were synthesized by grafted method using chloromethylated polystyrene (CMPS) resin as support by Wang and coworkers [78].

The results showed that the surface of CMPS resin was covered with a thin layer of extraction activity components. Then, the synthesized CMPS-supported imidazolium-based ionic liquids were investigated to extract thiophene and its derivatives from model gasoline (n-octane/thiophene) under certain conditions. For a given imidazolium-based ionic liquid: first, the order of extraction capacity of extractant was CuCl/CMPS-Im(Cl) > ZnCl2/CMPS-Im(Cl) > $FeCl_3$/CMPS-Im(Cl); the reason for this was that the π-complexation capability between Cu+ and thiophene was stronger than those of $Fe^{3+}$and Zn^{2+}. Second, the sulfur removal selectivity of sulfur compound followed the order of TS < BT < DBT under the same conditions; it indicated that the extraction was favored for those aromatic heterocyclic sulfur compounds with higher density aromatic π-electrons density. Meanwhile, the effect of mass ratio of model gasoline to M/CMPS-Im(Cl) ionic liquids, different initial sulfur concentrations, and extraction time on desulfurization rates of M/CMPS-Im (Cl) ionic liquids was performed, respectively.

3.5. Optimization of oxidative desulfurization of dibenzothiophene using acidic ionic liquid

The oxidative desulfurization of dibenzothiophene (DBT) in n-octane as model oil with Brönsted acidic ionic liquids *N*-methyl-pyrrolidonium phosphate ($[Hnmp]H_2PO_4$) as catalytic solvent and H_2O_2 as oxidant was optimized by orthogonal experiments. 99.8% of DBT in the model oil was removed under the optimal conditions of molar ratio of H_2O_2 to sulfur of 16:1, reaction temperature of 60°C, reaction time of 5 h, and volume ratio of model oil to ionic liquids of 1:1 by ZHAO and coworkers [79]. The desulfurization efficiency of actual diesel was 64.3% under the optimized conditions. The influences of the desulfurization efficiency of DBT decreased in the following order: oxidation temperature > oxidation time> molar ratio of H_2O_2/sulfur (O/S) > volume ratio of $[Hnmp]H_2PO_4$ to model oil (VIL/Vmodel oil), according to extreme analysis of orthogonal test. The ionic liquid $[Hnmp]H_2PO_4$ can be recycled six times without a significant decrease in activity.

An acidic ionic liquid N-butyl-N-methylimidazolium hydrogen sulfate ([BMIm]HSO_4) was applied as extractant and catalyst for the oxidative desulfurization of dibenzothiophenes (DBT) in the presence of H_2O_2 in model oil by Zhang and coworkers [80]. Several parameters, e.g., catalyst amount, hydrogen peroxide quantity, reaction time, and temperature, were investigated in detail. The catalytic oxidative desulfurization rate can reach 100% for DBT in model oil. The ionic liquid [BMIm]HSO_4 can be recycled 5 times with only a slight reduction in activity.

Ionic liquids, a new class of green solvents, have recently been undergoing intensive research on the removal of thiophenic sulfur species (e.g., dibenzothiophene) from fuels because of the limitation of the traditional hydrodesulfurization method in removing these species. In this work, deep oxidative desulfurization of diesel fuels by six functional acidic ionic liquids are studied, in which ionic liquids are used as both extractant and catalyst, and 30 wt % H_2O_2 solution as oxidant by Yu and coworkers [81]. These ionic liquids include both Lewis acidic species such as 1-butyl-3-methylimidazolium chloride/2$ZnCl_2$ ([C_4mim]Cl/2$ZnCl_2$ and [C_4mim]Cl/$ZnCl_2$) and Brønsted acidic species such as 1-methyl-3-ethylcarboxylic acid imidazolium hydrogen sulfate ([CH_2COOHmim]HSO_4), 1-methyl-3-(butyl-4-sulfinate) imidazolium hydrogen sulfate ([SO3HC4mim]HSO_4), [Hmim]HSO_4, and [C_4mim]HSO_4 where di ▪ erent acidic groups such as H, COOH, and SO_3H are appended to the cations. Except for [CH_2COOHmim]HSO_4, both Brønsted and Lewis acidic ILs are capable of e ▪ ectively removing dibenzothiophene from model diesel fuels, where 100% sulfur removal is obtained for [C_4mim]Cl/2$ZnCl_2$ and [SO_3HC_4mim]HSO_4.Thee ▪ ects of temperature, molar ratio of O/S, mass ratio of ionic liquid /oil, and ionic liquid regeneration on desulfurization are investigated systematically for [C_4mim]Cl/2$ZnCl_2$ and [SO_3HC_4mim]HSO_4. The desulfurization ability is not sensitive to themass ratio of IL/oil, which is desired for reducing ionic liquid dosage in industrial application; the ionic liquids can be recycled six times with merely a negligible loss in activity. [C_4mim]Cl/2$ZnCl_2$ can reduce the sulfur content in real commercial diesel fuel from64 to 7.9 ppmwith a sulfur removal of 87.7%; however, it is not too e ▪ ective for coke diesel fuel with high initial sulfur content of 5380 ppm. This work tends to show that diesel fuels can be purified to sulfur-free or ultralow sulfur fuels by further deep oxidative desulfurization by using ionic liquids after hydrodesulfurization. Lewis and Brønsted acidic ionic liquids used in this work show in figure 7.

http://dx.doi.org/10.5772/51651

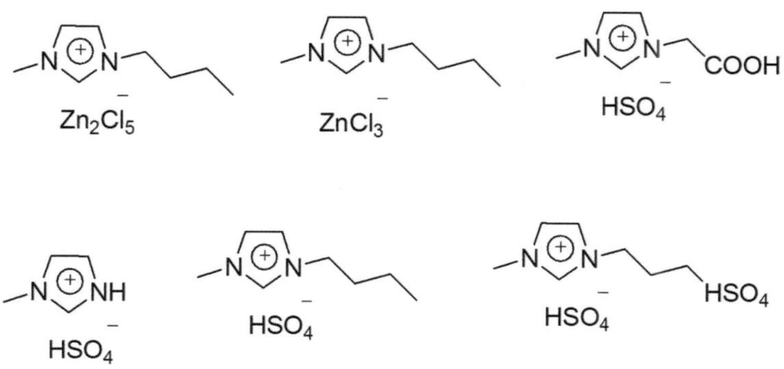

Figure 7. Lewis and Brønsted acidic ionic liquids used in this work.

[Reproduced from Yu G, Zhao J, Song D, Asumana C, Zhang X, Chen X. Deep Oxidative Desulfurization of Diesel Fuels by Acidic Ionic Liquids. Ind Eng Chem Res 2011; 50(20): 11690–11697, Copyright (2011), with permeation from American Chemical Society]

4. Conclusion

In view of stringent environmental regulations, utilization of sulfur-containing fuel oils has severe limitations regarding emission of sulfur dioxide. Technology for reduction of sulfur in diesel fuel to 15 ppm is currently available and new technologies are under development that could reduce the cost of desulfurization. Chemical oxidation in conjunction with ionic liquid extraction can increase the removal of sulfur sharply. Ionic liquids have the ability of extracting aromatic sulfur-containing compounds at ambient conditions without H_2 consumption. The cations, anions structure, and size of ionic liquids are important parameters affecting the extracting ability. In addition ionic liquids are immiscible with fuel, and the used ionic liquids can be regenerated and recycled by solvent washing or distillation.

Author details

Elaheh Kowsari*

Address all correspondence to: kowsarie@aut.ac.ir

Department of Chemistry, Amirkabir University of Technology, Tehran, Iran

References

[1] Li, H. M., He, L. N., Lu, J. D., Zhu, W. S., Jiang, X., Wang, Y., & Yan, Y. S. (2009). Deep Oxidative Desulfurization of Fuels Catalyzed by Phosphotungstic Acid in Ionic Liquids at Room Temperature. *Energy Fuels*, 23(3), 354-357.

[2] Song, CS. (2003). An Overview of New Approaches to Deep Desulfurization for Ultra-Clean Gasoline. *Diesel Fuel and Jet Fuel. Catal. Today*, 86(1-4), 211-263.

[3] Babich, I. V., & Moulijn, J. A. (2003). External and Internal Factors Influencing Modern Refineries. *Fuel*, 82(6), 607-631.

[4] Kwak, C., Lee, J. J., Bae, J. S., Choi, K., & Moonm, S. H. (2000). Hydrodesulfurization of DBT, 4-MDBT, and 4,6-DMDBT on Fluorinated CoMoS/Al_2O_3 Catalysts. *Appl. Catal. A.*, 200(1), 233-242.

[5] Zhang, SG, & Zhang, ZC. (2002). Novel Properties of Ionic Liquids in Selective Sulfur Removal from Fuels at Room Temperature. *Green Chem.*, 4, 376-379.

[6] Esser, J., Wasserscheid, P., & Jess, A. (2004). Deep Desulfurization of Oil Refinery Streams by Extraction with Ionic Liquids. *Green Chem.*, 6, 316-322.

[7] Planeta, J., Karásek, P., & Roth, M. (2006). Distribution of Sulfur-Containing Aromatics Between [hmim][Tf_2N] and supercritical CO_2: A Case Study for Deep Desulfurization of Oil Refinery Streams by Extraction WithIonic Liquids. *Green Chem.*, 8, 70-77.

[8] Nie, Y., Li, C. X., & Wang, Z. H. (2007). Extractive Desulfurization of Fuel Oil Using Alkylimidazole and Its Mixture with Dialkylphosphate Ionic Liquids. *Ind. Eng. Chem. Res.*, 46(15), 5108-5112.

[9] Li, F. T., Liu, R. H., Wen, J. H., Zhao, Sun Z. M., & Liu, Y. (2009). Desulfurization of Dibenzothiophene by Chemical Oxidation and Solvent Extraction with $Me_3NCH_2C_6H_5Cl\bullet 2ZnCl_2$ Ionic Liquid. *Green Chem.*, 11, 883-888.

[10] Zhang, J., Wang, A. J., Li, X., & Ma, X. H. (2011). Oxidative Desulfurization of Dibenzothiophene and Diesel Over [Bmim]$_3PMo_{12}O_{40}$. *J. Catal.*, 279(2), 269-275.

[11] Kulkarni, P. S., & Afonso, C. A. M. (2010). Deep Desulfurization of Diesel Fuel Using Ionic Liquids: Current Status and Future Challenges. *Green Chem.*, 2, 139-149.

[12] Bösmann, A., Datsevich, L., Jess, A., Lauter, A., Schmitz, C., & Wasserscheid, P. (2001). Deep Desulfurization of Diesel Fuel by Extraction with Ionic Liquids. *Chem Commun.*, 2494.

[13] Liu, Z. C., Hu, J. R., & Gao, J. S. (2006). FCC naphtha desulfurization via alkylation process over ionic liquid catalyst. *Petroleum Proc. Petrochem.*, 37(10), 22-26.

[14] Alonso, L., Arce, A., Francisco, M., & Soto, A. (2008). Solvent Extraction of Thiophene from n-Alkanes (C_7, C_{12}, and C_{16}) Using the Ionic Liquid [C_8mim][BF_4]. *J Chem Thermodyn.*, 40(6), 966-972.

[15] Qiu, J., Wang, G., Zeng, , Tang, D. Y., Wang, M., & Li, Y. (2009). Oxidative desulfurization of diesel fuel using amphiphilic quaternary ammonium phosphomolybdate catalysts. *Fuel Process. Technol.*, 90(12), 538-1542.

[16] Zhang, J., Huang, C. P., Chen, B. H., Li, Y. X., & Qiao, C. Z. (2007). Extractive Desulfurization from Gasoline by [BMIM][Cu_2Cl_3]. *J Fuel Chem Tech.*, 33(4), 431-434.

[17] Lu, H. Y., Gao, J. B., Jiang, Z. X., Jing, F., Yang, Y. X., Wang, G., & Li, C. (2006). Ultra-Deep Desulfurization of Diesel by Selective Oxidation With $[C_{18}H_{37}N(CH_3)_3]_4[H_2NaPW_{10}O_{36}]$ Catalyst Assembled in Emulsion Droplets. *J Catal*, 239(2), 369-375.

[18] Zhao, D. S., Liu, R., Wang, J. L., & Liu, B. Y. (2008). Photochemical Oxidation ionic Liquid Extraction Coupling Technique in Deep Desulphurization of Light Oil. *Energy Fuel*, 22(2), 1100.

[19] Jiang, X. C., Nie, Y., Li, C. X., & Wang, Z. H. (2008). Imidazolium-Based Alkylphosphate Ionic Liquids: A Potential Solvent for Extractive Desulfurization of Fuel. *Fuel*, 87(1), 79-84.

[20] Yansheng, C., Changping, L., Qingzhu, J., Qingshan, L., Peifang, Y., Xiumei, L., & Welz-Biermann, U. (2011). Desulfurization by Oxidation Combined with Extraction Using Acidic Room-Temperature Ionic Liquids. *Green Chem.*, 13, 1224-129.

[21] Zhu, W. S., Li, H. M., Gu, Q. Q., Wu, P. W., Zhu, G. P., Yan, S., & Chen, Y. (2011). Kinetics and mechanism for oxidative desulfurization of fuels catalyzed by peroxomolybdenum amino acid complexes in water-immiscible ionic liquids. *J. Mol. Catal. A: Chem.*, 336(1-2), 16-22.

[22] Nie, Y., Li, C. X., Sun, A. J., Meng, H., & Wang, Z. (2006). Extractive Desulfurization of Gasoline Using Imidazolium-Based Phosphoric Ionic Liquids. *Energy Fuels*, 20(5), 2083.

[23] Zhu, H. P., Yang, F., Tang, J., & He, M. Y. (2003). Brønsted Acidic Ionic Liquid 1-Methylimidazolium Tetrafluoroborate: A Green Catalyst and Recyclable Medium for Esterification. *Green Chem*, 5, 38-39.

[24] Zhang, S., & Zhang, Z. (2002). Novel Properties of Ionic Liquids in Selective Sulfur Removal from Fuels at Room Temperature. *Green Chem*, 4, 376-379.

[25] Holbrey, J. D., Reichert, W. M., Nieuwenhuyzen, M., Sheppard, O., Hardacre, C., & Rogers, R. D. (2003). Liquid Clathrate Formation in Ionic Liquid-Aromatic Mixtures. *Chem Commun*, 4, 476-477.

[26] Zhao, DS. , Sun, Z. M., Li, F. T., Liu, R., & Shan, H. D. (2008). Oxidative Desulfurization of Thiophene Catalyzed by $(C_4H_9)_4NBr\ 2C_6H_{11}NO$ Coordinated Ionic Liquid. *Energy Fuels*, 22(5), 3065-3069.

[27] Zhu, W. S., Li, H. M., Jiang, X., Yan, Y. S., Lu, J. D., & Xia, J. X. (2007). Oxidative Desulfurization of Fuels Catalyzed by Peroxotungsten and Peroxomolybdenum Complexes in Ionic Liquids. *Energy Fuels*, 21(5), 2514.

[28] He, L. N., Li, H. M., Zhu, W. S., Guo, J. X., Jiang, X., Lu, J. D., & Yan, Y. S. (2008). Deep Oxidative Desulfurization of Fuels Using Peroxophosphomolybdate Catalysts in Ionic Liquids. *Ind. Eng. Chem. Res.*, 47(18), 6890-6895.

[29] Lu, L., Cheng, S. F., Gao, J. B., Gao, G. H., & He, M. Y. (2007). Deep Oxidative Desulfurization of Fuels Catalyzed by Ionic Liquid in the Presence of H_2O_2 . *Energy Fuels*, 21(1), 383-384.

[30] Zhao, D. S., Wang, J. L., & Zhou, E. P. (2007). Oxidative desulfurization of diesel fuel using a Brønsted acid room temperature ionic liquid in the presence of H_2O_2 . *Green Chem.*, 9, 1219.

[31] Holbrey, J. D., López-Martin, I., Rothenberg, G., Silveiro, G., & Zheng, X. (2008). Desulfurisation of Oils Using Ionic Liquids: Selection of Cationic and Anionic Components to Enhance Extraction Efficiency. *Green Chem.*, 10, 87-92.

[32] Gao, H., Li, Y., Wu, Y., Luo, M., Li, Q., & Xing, J. (2009). Extractive Desulfurization of Fuel Using 3-Methylpyridinium-Based Ionic Liquids. *Energy Fuels*, 23(5), 2690-2694.

[33] Gao, H. S., Luo, M. F., Xing, J. M., Wu, Y., Li, Y. G., Li, W. L., Liu, Q. F., & Liu, H. Z. (2008). Desulfurization of Fuel by Extraction with Pyridinium-Based Ionic Liquids. *Ind. Eng. Chem. Res.*, 47(21), 8384-8388.

[34] Ristovski, Z. D., Jayaratne, E. R., Lim, M., Ayoko, G. A., & Morawska, L. (2006). Influence of diesel fuel sulfur on nanoparticle emissions from city buses. *Environ Sci.Technol.*, 40(4), 1314.

[35] Oh, S. K., Baik, D. S., & Han, Y. C. (2003). Emission characteristics in ultra low sulfur diesel. *Int. J. Automot. Technol.*, 4(2), 95-100.

[36] Demonstration of Advanced Emission Control Technologies Enabling Diesel powered Heavy-duty Engines to Achieve Low Emission levels. (1999). *Final Report*, Manufacturers of Emission Controls Association, Washington, DC 20036.

[37] California Air Resources Board. (2000, October). *Fuels Report: Appendix to the Diesel Risk Reduction Plan, Appendix IV*, http://www.arb.ca.gov/diesel/documents/rrpfinal.pdf.

[38] Iranian Research Institute of Petroleum, Industry. (2000). http://www.ripi.ir/index.php?option=content&task=view&id=61, accessed July 2000.

[39] Gates, B. C., & Topsoe, H. (1997). Reactivities in Deep Catalytic Hydrodesulfurization: Challenges, opportunities, and the importance of 4-methyldibenzothiophene and 4,6-Dimethyldibenzothiophene. *Polyhedron*, 16(18), 3213-3217.

[40] Gray, K. A., Mrachko, C. T., & Squires, C. H. (2003). Biodesulfurization of fossil fuels. *Curr Opin Microbiol.*, 6(3), 229-235.

[41] Gray, K. A., Pogrebinsky, O. S., Mrachko, G. T., Xi, L., Monticello, D. J., & Squires, C. H. (1996). Molecular Mechanisms of Biocatalytic Desulfurization of Fossil fuels. *Nat. Biotechnol.*, 14(13), 1705.

[42] Gupta, N., Roychoudhury, P. K., & Deb, J. K. (2005). Biotechnology of Desulfurization of Diesel: Prospects and Challenges. *Appl. Microbiol. Biotechnol.*, 66(4), 356-366.

[43] Kilbane, J. J. (2006). Microbial Biocatalyst Developments to Upgrade Fossil Fuels. *Curr Opin. Microbiol.*, 17(3), 1-10.

[44] Lee, W. C., Ohshiro, T., Matsubara, T., Izumi, Y., & Tanokura, M. (2006). Crystal-structure and Desulfurization Mechanism of 2-hydroxybiphenyl-2-Sulfinic Acid Desulfinase. *J Biol Chem.*, 281(43), 32534-32539.

[45] Li, G., Ma, T., Li, J., Liang, F., & Liu, R. (2006). Desulfurization of Dibenzothiophene by Bacillus Subtilis Recombinants Carrying dszABC and dszD Genes. *Biotechnol Lett.*, 28(14), 1095-1100.

[46] Martín, A. B., Alcón, A., Santos, V. E., & Garcia-Ochoa, F. (2004). Production of a Biocatalyst of Pseudomonas Putida CECT5279 for Dibenzothiophene (DBT Biodesulfurization for Different media Composition. *Energy and Fuels*, 18(3), 851-857.

[47] Martín, A. B., Alcón, A., Santos, V. E., & Garcia-Ochoa, F. (2005). Production of a Biocatalyst of Pseudomonas Putida CECT5279 for Dibenzothiophene (DBT): influence of the Operational Conditions. *Energy and Fuels*, 19, 775-782.

[48] Monticello, D. J. (2000). Biodesulfurization and the Upgrading of Petroleum Distillates. *Curr. Opin. Biotechnol.*, 11(6), 540-546.

[49] Noda, K. I., Watanabe, K., & Maruhashi, K. (2003). Recombinant Pseudomonas Putida Carrying Both the dsz and hcu Genes Can Desulfurize Dibenzothiophene in n-Tetradecane. *Biotechnol. Lett.*, 25(14), 147-150.

[50] Piddington, C. S., Kovacevich, B. R., & Rambosek, J. (1995). Sequence and Molecular Characterization of a DNA Region Encoding the Dibenzothiophene Desulfurization Operon of Rhodococcus Sp. Strain IGTS8. *Appl. Environ. Microbiol.*, 61(2), 468-475.

[51] Soleimani, M., Bassi, A., & Margaritis, A. (2007). Biodesulfurization of Refractory Organic Sulfur Compounds in Fossil Fuels. *Biotechnol. Adv.*, 25(6), 570-596.

[52] Xu, P., Yu, B., Li, F. L., Cai, X. F., & , C. Q. (2006). Microbial degradation of sulfur, nitrogen and oxygen heterocycles. *Trends Microbiol.*, 14, 398-405.

[53] Te, M., Fairbridge, C., & Ring, Z. (2001). Oxidation Reactivities of Dibenzothiophenes in Polyoxometalate/H_2O_2 and Formic acid/H_2O_2 Systems. *Appl. Catal. A: Gen.*, 219(1-2), 267-280.

[54] Mondal, S., Hangun-Balkir, Y., Alexandrova, L., Link, D., Howard, B., Zandhuis, P., Cugini, A., Horwitz, B., & Collins, T. J. (2006). Oxidation of Sulfur Components in

Diesel Fuel Using Fe-TAML® Catalysts and Hydrogen Peroxide. *Catal. Today*, 116(4), 554-561.

[55] García-Gutiérrez, J. L., Fuentes, G. A., Hernández-Terán, M. E., García, P., Murrieta-Guevara, F., & Jiménez-Cruz, F. (2008). Ultra-Deep Oxidative Desulfurization of Diesel Fuel by the Mo/AlO3-H_2O_2 System: The Effect of System Parameters on Catalytic Activity. *Appl Catal A: Gen*, 334(1-2), 366-373.

[56] Hulea, V., Fajula, F., & Bousquet, J. (2001). Mild Oxidation With H_2O_2 over Ti-Containing Molecular Sieves-a Very Efficientmethod for Removing Aromatic Sulfur Compounds from Fuels. *J. Catal.*, 198(2), 179-186.

[57] Anisimov, A. V., Fedorova, E. V., Lesnugin, A. Z., Senyavin, V. M., Aslanov, L. A., Rybakov, V. B., & Tarakanova, A. V. (2003). Vanadiumperoxocomplexes as Oxidation Catalysts of Sulfur Organic Compounds by Hydrogen Peroxide in bi-phase Systems. *Catal. Today*, 78(1-4), 319-325.

[58] Palomeque, J., Clacens, J. M., & Figueras, F. (2002). Oxidation of Dibenzothiophene by Hydrogen Peroxide Catalyzed by Solid Bases. *J. Catal.*, 211(1), 103-108.

[59] Yazu, K., Yamamoto, Y., Furuya, T., Mild, K., & Ukegawa, K. (2001). Oxidation of Dibenzothiophenes in an Organic Biphasic System and its Application to Oxidative Desulfurization of Light Oil. *Energy Fuels*, 15(6), 535-536.

[60] Djangkung, S., Murti, S., Yang, H., Choi, K., Kora, Y., & Mochida, I. (2003). Influences of Nitrogen Species on the Hydrodesulfurization Reactivity of a Gas Oil Over Sulfide Catalysts of Variable Activity. *Appl. Catal. A: Gen.*, 252(2), 331-346.

[61] Shiraishi, Y., Naito, T., & Hirai, T. (2003). Vanadosilicate Molecular Sieve as a Catalyst for Oxidative Desulfurization of Light Oil. *Ind Eng Chem Res.*, 42(24), 6034-6039.

[62] Song, C., & Ma, X. (2003). New Design Approaches to Ultra-Clean Diesel Fuels by Deep Desulfurization and Deep Dearomatization. *Appl. Catal. B.*, 41(1-2), 207-238.

[63] Zhao, D., Wang, Y., Duan, E., & Zhang, J. (2010). Oxidation Desulfurization of Fuel using Pyridinium-Based Ionic Liquids as Phase-Transfer Catalysts. *Fuel Proces Tech.*, 91(12), 1803.

[64] Arce, A., Francisco, M., & Soto, A. (2010). Evaluation of the Polysubstituted Pyridinium Ionic Liquid [hmmpy][Ntf_2] as a Suitable Solvent for Desulfurization: Phase Equilibria. *J. Chem. Thermodynamics*, 42(6), 712-718.

[65] Wang, J. L., Zhao, D. S., Zhou, E. P., & Dong, Z. (2007). Desulfurization of Gasoline by Extraction with N-alkyl-Pyridinium-Based Ionic Liquids. *Journal of Fuel Chemistry and Technology*, 35(3), 293-296.

[66] Khupse, N. D., & Kumar, A. (2010). Contrasting Thermosolvatochromic Trends in Pyridinium-, Pyrrolidinium-, and Phosphonium-Based Ionic Liquids. *J. Phys. Chem. B*, 114(1), 376-381.

http://dx.doi.org/10.5772/51651

[67] Nie, Y., Li, C., Meng, H., & Wang, Z. N. (2008). N-Dialkylimidazolium Dialkylphosphate Ionic Liquids: Their Extractive Performance for Thiophene Series Compounds from Fuel Oils Versus the Length of Alkyl Group. *Fuel Proces Tech.*, 8(9), 10-978.

[68] Zhang, S., Zhang, Q., & Zhang, Z. C. (2004). Extractive Desulfurization and Denitrogenation of Fuels Using Ionic Liquids. *Ind Eng Chem Res.*, 43(2), 614-622.

[69] Yuuki, M., & Katsuyasu, S. (2008). Removal of Organic Sulfur from Hydrocarbon Resources Using Ionic Liquids. *Energy & Fuels*, 22(5), 3303-3307.

[70] Yu, G., Li, X., Liu, X., Asumana, C., & Chen, X. (2011). Deep Desulfurization of Fuel Oils Using Low-Viscosity-Ethyl-3-methylimidazolium Dicyanamide Ionic Liquid. *Ind Eng Chem Res*, 50(4), 2236-2244.

[71] Xu, D., Zhu, W., Li, H., Zhang, J., Zou, F., Shi, H., & Yan, Y. (2009). Oxidative Desulfurization of Fuels Catalyzed by V2O5 in Ionic Liquids at Room Temperature. *Energy Fuels*, 23(12), 5929-5933.

[72] Zhu, W., Huang, W., Li, H., Zhang, M., Jiang, W., Chen, G., & Han, C. (2011). Polyoxometalate-Based Ionic Liquids as Catalysts for Deep Desulfurization of Fuels. *Fuel Processing Technology*, 92(10), 1842-1848.

[73] Wang, J., Zhao, D., & Li, K. (2010). Oxidative Desulfurization of Dibenzothiophene Using Ozone and Hydrogen Peroxide in Ionic Liquid. *Energy Fuels.*, 24(4), 527-529.

[74] Huang, W., Zhu, W., Li, H., Shi, H., Zhu, G., Liu, H., & Chen, G. (2010). Heteropolyanion-Based Ionic Liquid for Deep Desulfurization of Fuels in Ionic Liquids. *Ind. Eng. Chem. Res.*, 49(19), 8998-9003.

[75] Ge, J., Zhou, Y., Yang, Y., & Xue, M. (2011). Catalytic Oxidative Desulfurization of Gasoline Using Ionic Liquid Emulsion System. *Ind. Eng. Chem. Res.*, 50(24), 13686-13692.

[76] Ko, N. H., Lee, J. S., Huh, E. S., Lee, H., Jung, K. D., Kim, H. S., & Cheong, M. (2008). Extractive Desulfurization Using Fe-Containing Ionic Liquids. *Energy & Fuels*, 22(3), 687-690.

[77] Li, F. T., Liu, Y., Sun, Z. M., Chen, L. J., Zhao, D. S., Liu, R. H., & Kou, C. G. (2010). Deep Extractive Desulfurization of Gasoline with xEt_3NHCl_3 $FeCl_3$ Ionic Liquids. *Energy Fuels*, 24(8), 4285-4289.

[78] Wang, X., Wan, H., Han, M., Gao, L., & Guan, G. (2012). Removal of Thiophene and Its Derivatives from Model Gasoline Using Polymer-Supported Metal Chlorides Ionic Liquid Moieties. *Ind. Eng. Chem. Res.*, 51(8), 3418-3424.

[79] Zhao, D. S., Sun, Z. M., Li, F. T., & Shan, H. D. (2009). Optimization of oxidative desulfurization of dibenzothiophene using acidic ionic liquid as catalytic solvent. *J Fuel Chem Technol*, 37(2), 194-198.

[80] Zhang, W., Ke, X., Qian, Z., Daliang, L., Shuyao, W., Francis, V., & Xi-Ming, S. (2010). Oxidative Desulfurization of Dibenzothiophene Catalyzed by Ionic Liquid [BMIm]HSO_4 . *Ind Eng Chem Res.*, 49(22), 11760-11763.

[81] Yu, G., Zhao, J., Song, D., Asumana, C., Zhang, X., & Chen, X. (2011). Deep Oxidative Desulfurization of Diesel Fuels by Acidic Ionic Liquids. *Ind Eng Chem Res*, 50(20), 11690-11697.

Chapter 12

Liquefaction of Wood by Ionic Liquid Treatment

Hisashi Miyafuji

Additional information is available at the end of the chapter

http://dx.doi.org/10.5772/51798

1. Introduction

Energy and environmental issues such as the exhaustion of fossil resources and global warming are major concerns. There is increasing interest in biomass resources as alternatives to fossil resources owing to their renewable and environmentally friendly properties. Among the various types of biomass, wood is a promising resource, because of its huge stocks and because it is not an edible crop. However, effective conversion technologies are required to use wood for the production of bioenergy or bio-based products. There have been many studies on various conversion technologies, including acid hydrolysis [1-3], enzymatic saccharification [4-5], hot-compressed water treatment [6], supercritical fluid treatment [7-9] and pyrolysis [10-12].

Recently, treatment of wood with ionic liquids has been reported as one of the most promising new conversion technologies for biomass. Ionic liquids are organic salts that have melting points close to ambient temperature. These liquids have many notable characteristics, such as negligible vapor pressure, thermal stability, recyclability, and non-flammability. Some ionic liquids can dissolve cellulose [13-22], and there have been several reports on applications of ionic liquids to liquefy wood [23-30].

In this chapter, recent progress on the liquefaction of wood in an ionic liquid, 1-ethyl-3-methylimidazolium chloride ([C2mim][Cl]), which has a chemical structure as shown in Figure 1, is presented. [C2mim][Cl] is well known as an ionic liquid that can dissolve cellulose. The difference in reactivity of lignin and polysaccharides, such as cellulose and hemicellulose, is described. Swelling behavior and the distortion of cell walls during the liquefaction of wood by [C2mim][Cl] is also considered. Additionally, the liquefaction of cellulose in [C2mim][Cl] with sulfuric acid is presented.

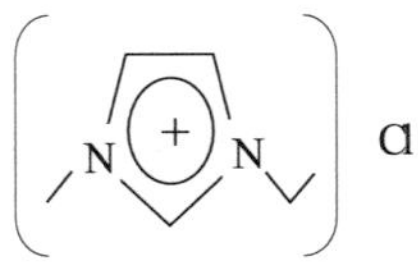

Figure 1. Chemical structure of 1-ethyl-3-methylimidazolium chloride ([C2mim][Cl])

2. Liquefaction behavior of wood

2.1. Changes in chemical components

Reaction of Japanese beech (*Fagus crenata*) in [C2mim][Cl] was investigated. Figure 2 shows the changes in residual weight of Japanese beech after [C2mim][Cl] treatment at various temperatures. Although 85 % of the residue remains after 24 h treatment at 90 °C, little residue is recovered after 24 h treatment at 120 °C. These results indicate that a significant amount of the components of wood can be liquefied in [C2mim][Cl] with higher temperatures and longer treatment times.

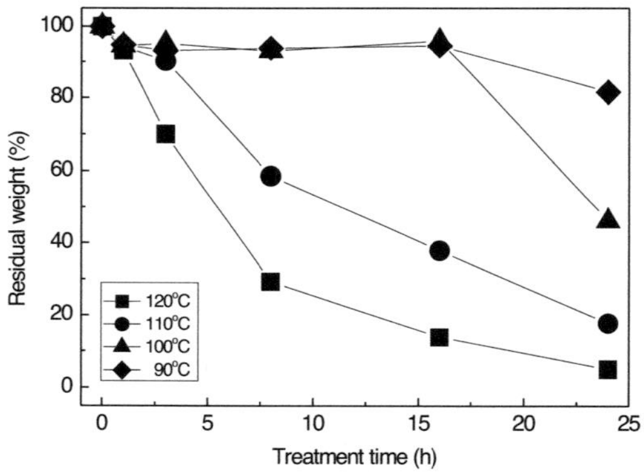

Figure 2. Changes in residual weight of Japanese beech after [C2mim][Cl] treatment at various temperatures

Figure 3 shows the changes in the chemical composition of cellulose, hemicellulose and lignin in the residue of western red cedar (*Thuja plicata*) as softwood and Japanese beech as hardwood treated with [C2mim][Cl] at 120 °C. The residue after 8 h treatment can be reduced to 30 % of western red cedar and 35 % of Japanese beech. In both species, the reduction of lignin was small. Therefore, the decrease of residue until 8 h was caused mainly by the decrease in the cellulose and hemicellulose in wood. These results indicate that although both lignin and polysaccharides such as cellulose and hemicelluloses can be liquefied, the liquefaction of the latter occurs mainly at the beginning of the reaction with [C2mim][Cl].

Japanese beech was found to be liquefied at a slightly faster rate than western red cedar up to 8 h. However, a significant difference between them was observed at 24 h. While western red cedar remains at a 17 % level, Japanese beech remains at only 5 %.

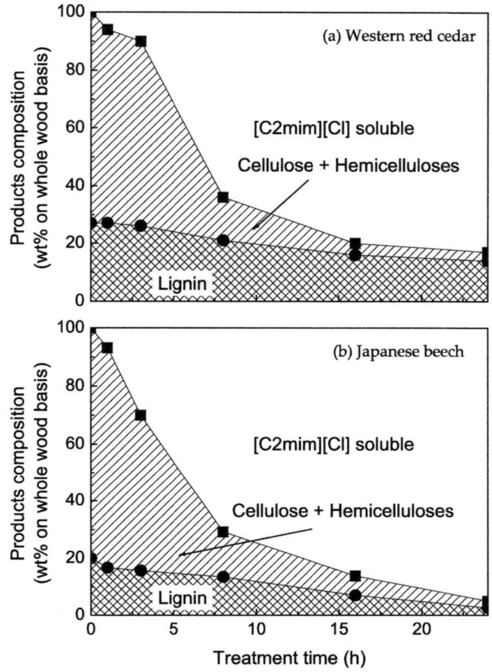

Figure 3. Changes in the chemical composition of cellulose, hemicellulose and lignin in the residue of (a) western red cedar and (b) Japanese beech treated with [C2mim][Cl] at 120 °C

For more detailed analysis of the changes in chemical components, Figure 4 shows the changes in the percentages of polysaccharide (cellulose and hemicellulose) and lignin following treatment by [C2mim][Cl]. Each plot was calculated as a percentage of the original amount of each component. The similar trend in the decrease of polysaccharide can be seen in both species. However, lignin in Japanese beech was removed much faster than that in western red cedar. After 24 h treatment, lignin decreases to 13% in the former and 52% in the latter. These results indicate that the reactivity of lignin to [C2mim][Cl] is very different between Japanese beech and western red cedar, and this is due to the difference in the chemical structure of between Japanese beech and western red cedar [27].

The solubilized compounds in [C2mim][Cl] during the liquefaction of wood were analyzed by gel permeation chromatography (GPC). Figure 5 shows the GPC chromatograms obtained at various treatment times. Analyses were conducted using a refractive index detector (RID), which can detect all solubilized compounds, and a photodiode array detector (PDA), which can detect those solubilized compounds that exhibit UV absorption. Pullulan was used as a standard for the molecular weight (MW) distribution. At 0 h in RID, both species have broad peaks with a MW around a few hundred thousand. However, neither of the two species have any peaks at 0 h in PDA. These results indicate that wood components solubilized in [C2mim][Cl] are not lignin but cellulose and hemicellulose at the early stages of reaction. They decrease to MW values of a few tens of thousands after 3 h in RID. After 8 h and 24 h, the peaks around 180 MW appear, indicating that the depolymerization of cellulose and hemicellulose occurred in [C2mim][Cl] as the treatment was extended. A peak observed at around 180 MW in RID is equivalent to that of a hexose such as glucose, mannose and galactose, which are components of cellulose and hemicellulose. Moreover, the low molecular compounds observed below 180 MW in the PDA chromatograms could possibly include the lignin derived compounds. No significant differences in this depolymerization behavior were found between western red cedar and Japanese beech.

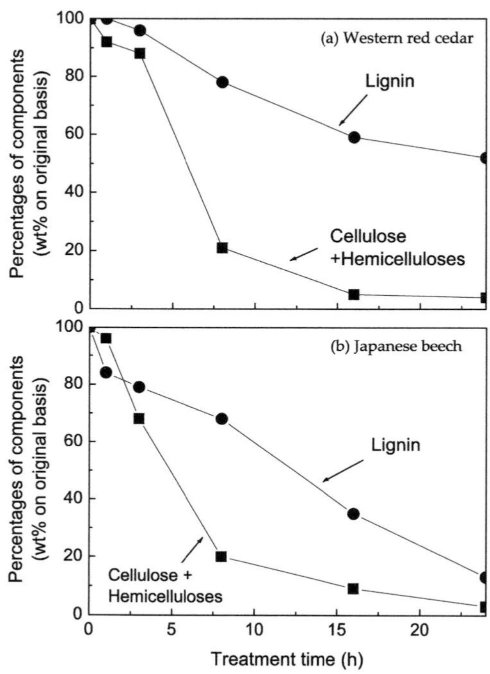

Figure 4. Changes in the percentages of cellulose, hemicellulose and lignin as treated by [C2mim][Cl], based on the original amounts present

In order to identify the low molecular compounds that were found by GPC analysis, gas chromatography coupled with mass spectrometry (GC-MS) analysis was carried out as shown in Figure 5. Products were identified by comparing the retention times and mass fragmentation patterns of the samples to those of pure compounds. The chromatograms of both species show few peaks at 0 h, 1 h and 3 h. However, various major monosugars such as glucose, arabinose, xylose, mannose and galactose are confirmed to be produced after 8 h and 24 h treatment. This is strong evidence that the polysaccharides are hydrolyzed to monosugars by the [C2mim][Cl] treatment.

http://dx.doi.org/10.5772/51798

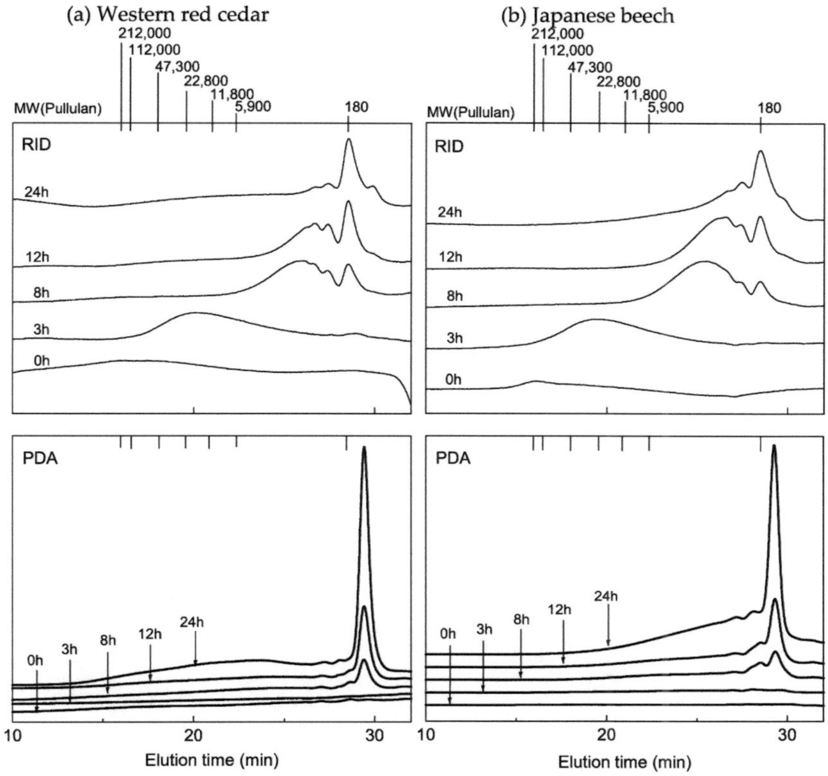

Figure 5. GPC chromatograms for the solubilized compounds in [C2mim][Cl] from (a) western red cedar and (b) Japanese beech

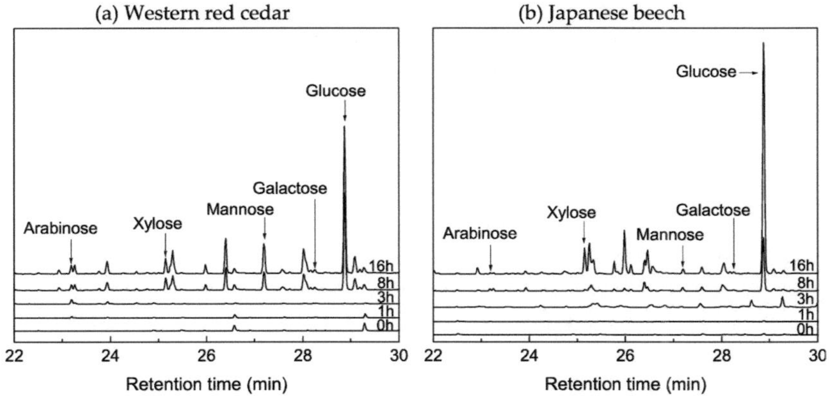

Figure 6. GC-MS chromatograms of the solubilized compounds in [C2mim][Cl] from (a) western red cedar and (b) Japanese beech

2.2. Morphological changes in wood tissue

The liquefaction behavior of wood in [C2mim][Cl] was studied not only from the changes of chemical components but also from morphological changes taking place in the wood tissues. Figure 7 shows light microscopy images of sugi (*Cryptomeria japonica*) after treatment with [C2mim][Cl] at 120 °C for 0 h and 24 h. In the transverse sections, we observed that cell walls in latewood were well ordered at 0 h (Figure 7a) but disordered and distorted after 24 h of treatment (Figure 7b). In contrast, no significant morphological changes were seen in earlywood. Although the cell walls in earlywood swelled as a result of [C2mim][Cl] treatment, the cells retained a similar form to that seen before treatment. At the boundary regions of latewood and earlywood in the radial sections, dissociation of tracheids was found after 24 h treatment (Figure 7d).

To analyze in detail the dissociations and distortions in latewood resulting from [C2mim][Cl] treatment, the swelling behavior of cell walls in latewood and earlywood in transverse sections was studied.

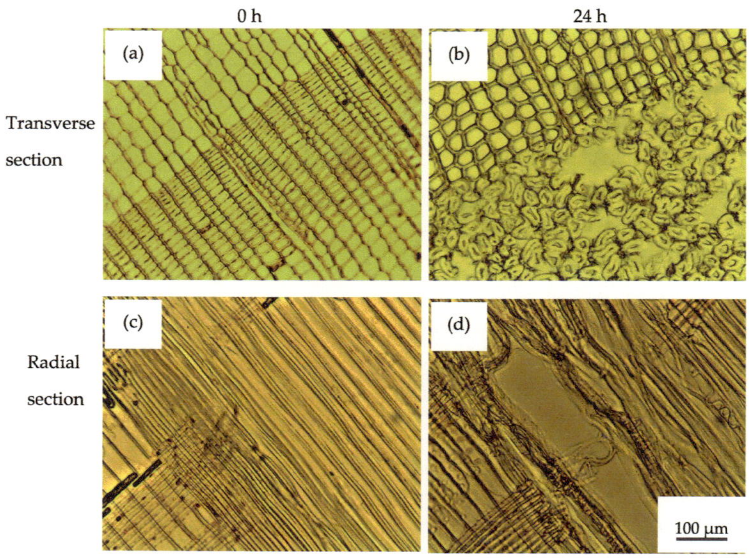

Figure 7. Light microscopy images of transverse sections (a,b) and radial sections (c,d) after treatment with [C2mim][Cl] at 120°C for 24 h

The results are shown in Figure 8. In earlywood, the cell wall area increased slightly at an early stage of [C2mim][Cl] treatment. After these initial changes in cell wall area, the cell wall area remained stable and did not show further changes during the [C2mim][Cl] treatment. These results indicate that cell walls of tracheids in earlywood did not swell significantly. In latewood, on the other hand, there were marked increases in the cell wall area at an early stage of [C2mim][Cl] treatment. At 48 h of treatment, the cell wall area had increased by five times. This swelling is likely to have caused the dissociation and distortions in latewood. Once the tracheids have dissociated, their cell walls can swell freely because there are no longer the physical restraints of neighboring cell walls. These results indicate differences in the morphological changes between earlywood and latewood (Figure 7)

Figure 9 shows scanning electron microscopy (SEM) images of transverse sections after treatment by [C2mim][Cl] for 24 h. The dissociation and distortions of cell walls are found in latewood after 24 h treatment (Figure 9b) as observed in the light micrograph images in Figure 8. Magnified SEM image (Figure 9d) reveals the dissociation between the secondary cell wall and the intercellular layer (indicated by arrows). [C2mim][Cl] is known to liquefy cellulose and hemicelluloses much more than lignin as shown in Figure 3. Thus, the reaction behavior of secondary cell walls and the intercellular layer is thought to be different from each other because the chemical components in those tissues are different. It is speculated that such differences in reaction behavior cause differences in their swelling behavior, and dissociation between secondary cell walls and the intercellular layer occurs.

Figure 10 shows SEM images of radial sections after treatment by [C2mim][Cl] for 24 h. The dissociation of tracheids with flaking and distortion in latewood is found (Figure 10b). The magnified image (Figure 10d) shows that the ray tracheids are segmented and the segments can be clearly observed on the tracheids (indicated by arrows). These results indicate that swelling of the tracheids in the radial direction is much greater than that in the axial direction.

Figure 11 shows SEM images of bordered pit-pair at earlywood in tangential sections. At 0 h treatment (Figure 11a), a torus is found in bordered pit-pair as shown by the arrow. However, it disappears after 48 h treatment (Figure 11b) while bordered pit-pair can be observed without any morphological changes. As shown in Figs. 7, 9 and 10, significant morphological changes were not found in tracheids in earlywood. Pit membrane is built up mainly by accumulation of cellulose microfibrils [31]. In the previous section, it is mentioned that cellulose is easily liquefied compared with lignin. Thus, many pit membranes are thought to be destroyed by [C2mim][Cl] treatment.

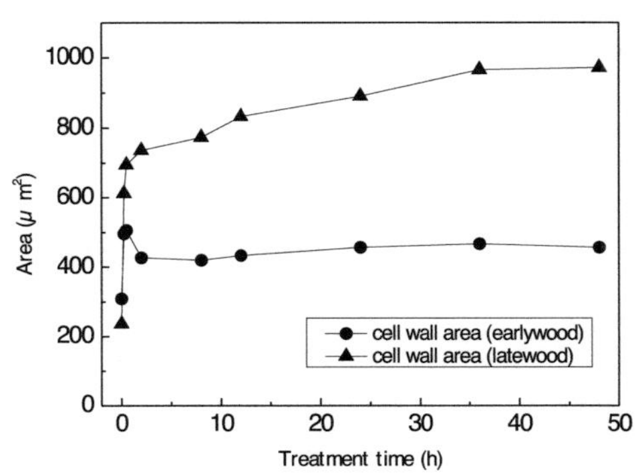

Figure 8. Changes of cell wall area in earlywood and latewood during [C2mim][Cl] treatment at 120°C

Consequently, these results indicate that [C2mim][Cl] is an effective solvent and reagent for the liquefaction of wood components and subsequent depolymerization of them. However, the reaction of wood liquefaction by [C2mim][Cl] treatment is not homogeneous, from either chemical or morphological viewpoints.

2.3. Influence of reaction atmosphere

The influence of moisture and reaction atmosphere on the liquefaction of wood (western red cedar) was studied. The changes in the residue and the composition after treatment by [C2mim][Cl] for 24 h under various atmospheres are shown in Figure 12. The samples treated under humidified oxygen (O_2+H_2O) and oxygen (O_2) drop to 4 % and 6 % respectively, while those under carbon dioxide (CO_2), nitrogen (N_2) and vacuum remain above 35 %. The samples treated under humidified pseudo-air (air+H_2O) and pseudo-air (air), which contain 21 % oxygen and 79 % nitrogen respectively, drop to 18 % and 19 % respectively. These results indicate that oxygen considerably accelerates the liquefaction of wood in [C2mim][Cl]. In addition, a few percentage points difference can also be observed between the samples treated under gas and those with moisture. The presence of water slightly affects the liquefaction of wood in [C2mim][Cl].

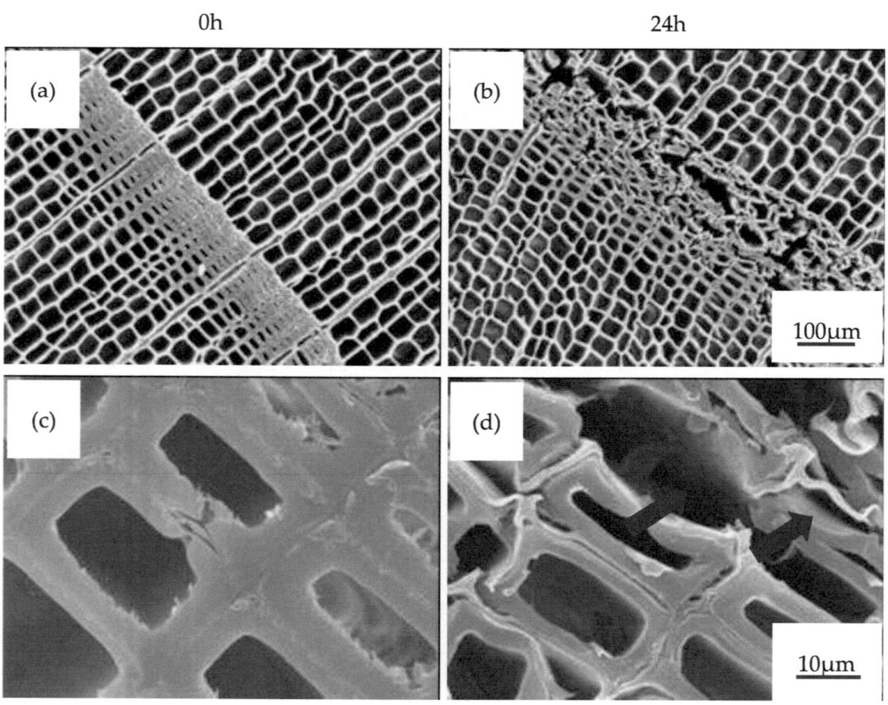

Figure 9. SEM images of transverse sections treated with [C2mim][Cl] at 120°C for 24 h. (a,c) 0 h treatment, (b,d) 24 h treatment.

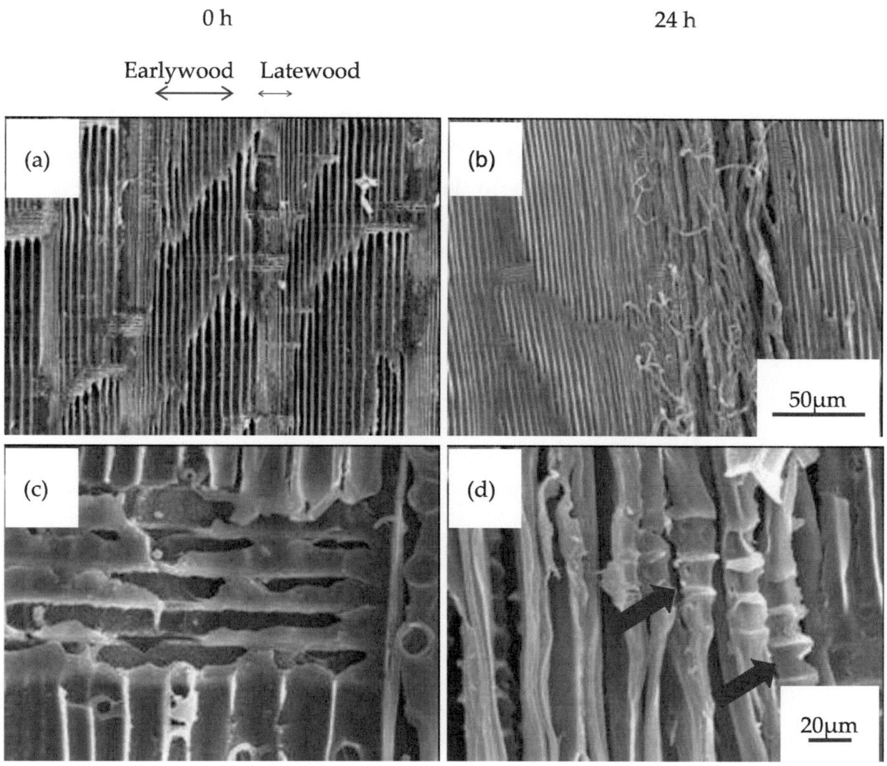

Figure 10. SEM images of radial sections after treatment with [C2mim][Cl] at 120 °C for 24 h. (a,c) 0 h treatment, (b,d) 24 h treatment.

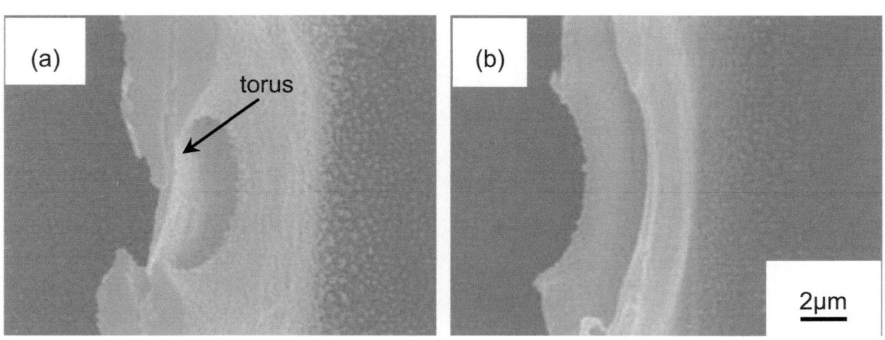

Figure 11. SEM images of bordered pit-pair in earlywood in tangential sections after treatment by [C2mim][Cl] at 120 °C for 0 h (a) and 48 h (b)

For further investigation of low molecular compounds solubilized in [C2mim][Cl], high performance liquid chromatography analysis was carried out as shown in Figure 13. Both samples treated under Air+H_2O and Air show peaks at around 10.5 min and 12.5 min in retention time, which are cellobiose and glucose, respectively. The complex peaks observed between 7.5 min and 9.5 min in retention time are thought to be oligomers. The samples treated under inactive gases are degraded to oligomers and those treated under Air+H_2O and air are degraded to glucose by the [C2mim][Cl] treatment. Under O_2+H_2O and O_2, there are no clear peaks in the chromatograms. This is due to the fact that glucose is quickly degraded to other lower molecular compounds such as 5-hydroxymethylfurfural because of the high activity of O_2.

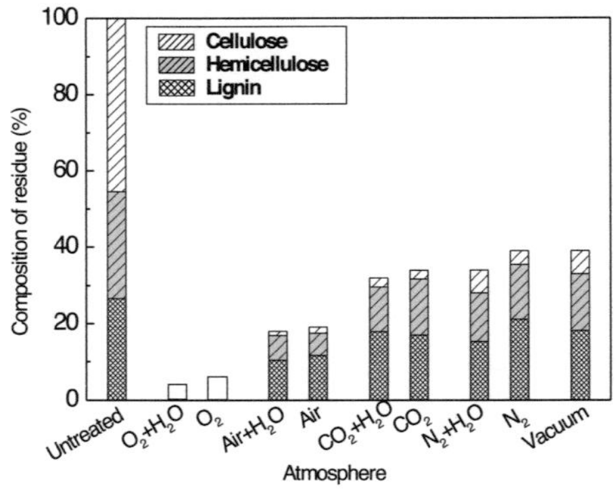

Figure 12. Changes in the residue and composition as treated by [C2mim][Cl] for 24 h under various atmospheres

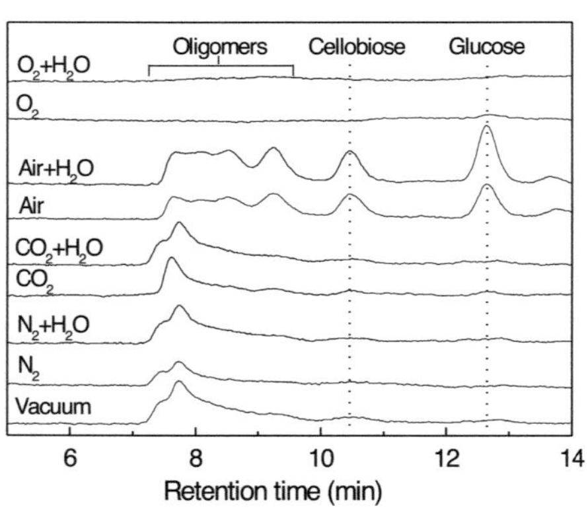

Figure 13. High performance liquid chromatograms of the solubilized compounds in [C2mim][Cl] obtained by 24 h treatment under various reaction atmospheres

In general, it is shown that active gases such as O_2 and air considerably accelerate wood liquefaction in [C2mim][Cl], and even with inactive gases such as N_2 and CO_2, liquefaction proceeds, which means [C2mim][Cl] itself has the ability to liquefy wood.

3. Liquefaction of cellulose in [C2mim][Cl] with sulfuric acid

In the previous section, it is clear that glucose can be obtained by the liquefaction of wood. This is mainly due to the depolymerization of cellulose. Glucose is one of the most important compounds among various compounds derived from wood because it can be converted to a range of valuable chemicals. For producing glucose, therefore, the liquefaction of cellulose in [C2mim][Cl] with sulfuric acid was also studied.

Figure 14 shows the changes in glucose yield at various reaction temperatures. The concentration of sulfuric acid in the reaction system was set at 1.5 wt%. In any reaction temperature except for 90 °C, glucose yield shows the optimum around 30 % to 40 %. Although the significant difference in maximal yield at each reaction temperature was not observed, the highest yield was 40.9 % at 90 °C. The maximal yield could be attained at shorter reaction times at increasing reaction temperature.

Figure 15 shows the changes of the glucose yield at various reaction temperatures with 0.5 wt% of sulfuric acid. Maximal yields were found at 120 °C and 60 min and at 110 °C and 120 min, respectively, although the glucose yield could not reach the optimum yet at 100 °C or 90 °C, even after 360 min. Compared with the results in 1.5 wt% of sulfuric acid as shown in Figure 14, a longer reaction time is necessary to attain the maximal yield. However, the maximum value at 120 °C or 110 °C shows the same levels as in 1.5 wt% of sulfuric acid. These results indicate that the reaction of cellulose in [C2mim][Cl] can be controlled by various reaction conditions such as the concentration of sulfuric acid, reaction time and reaction temperature.

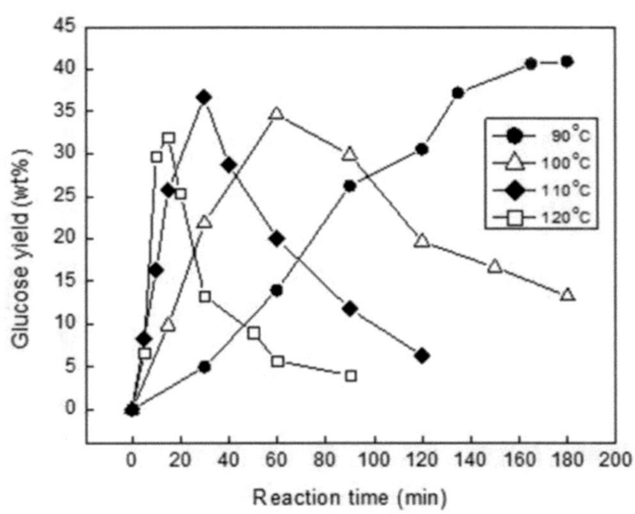

Figure 14. Changes in glucose yield from cellulose treated in [C2mim][Cl] with 1.5wt% of sulfuric acid at various reaction temperatures.

Figure 16 shows the comparisons of the glucose yield in the reaction system of [C2mim][Cl] and water as solvent at 90 °C (Figure 16a) or 120 °C (Figure 16b) reaction temperature. Sulfuric acid was added at 1.5 wt%. At both reaction temperatures, it reveals that much higher yields can be achieved in the reaction system of [C2mim][Cl] although they are at a negligible level in the reaction system of water. From these results, glucose productivity was calculated by the equation shown below.

$$\text{Glucose productivity} = \frac{\text{Maximal glucose yield (\%)}}{\text{Reaction time at maximal glucose yield (min)}}$$

The obtained glucose productivity is shown in Table 1. The reaction system of [C2mim][Cl] at 90 °C and 120 °C showed, respectively, 10 and 100 times the glucose productivity of the reaction system of water at 120 °C. These results indicate that cellulose can be converted to glucose much more effectively in [C2mim][Cl] than in water. By dissolving cellulose in [C2mim][Cl], the rigid crystalline structure of cellulose can be destroyed. This is a reason for the higher reactivity of cellulose in the reaction system of [C2mim][Cl].

It is revealed that much higher glucose yield can be achieved in the reaction system of [C2mim][Cl] around 100 °C, compared with that obtained in water, which cannot dissolve cellulose. Therefore, it can be concluded that the ionic liquid that can dissolve cellulose is a promising solvent for producing glucose.

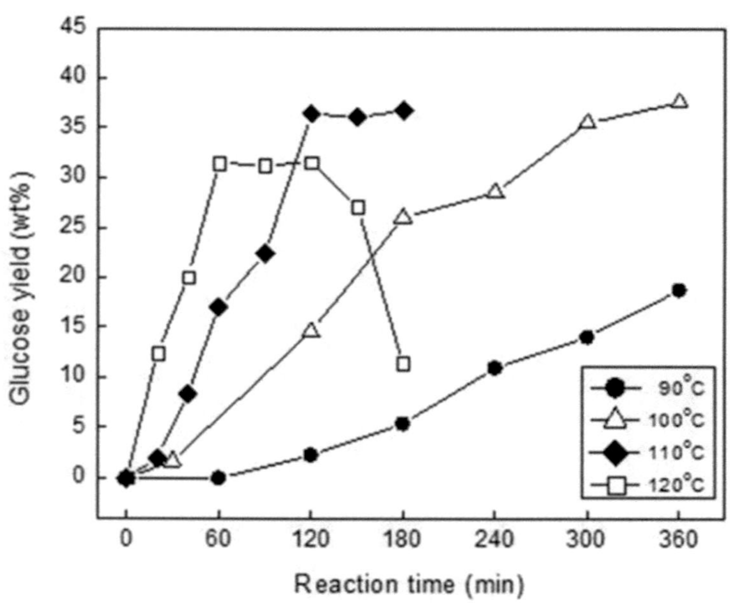

Figure 15. Changes in glucose yield from cellulose treated in [C2mim][Cl] with 0.5 wt% sulfuric acid at various reaction temperatures

http://dx.doi.org/10.5772/51798

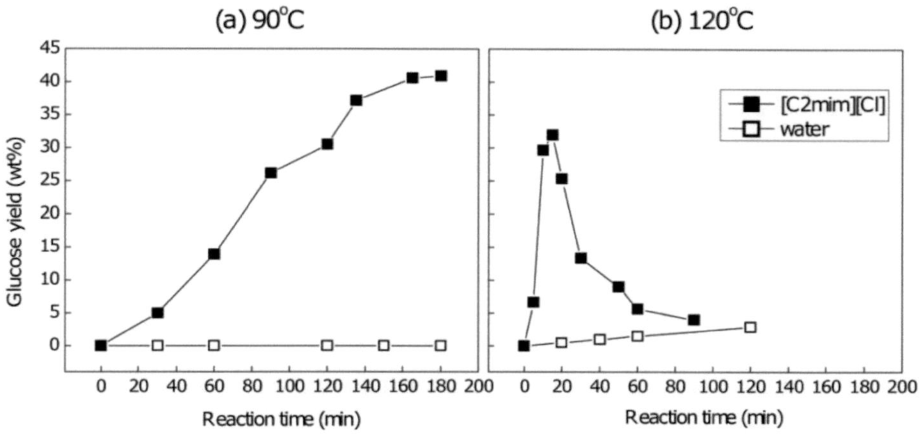

Figure 16. Comparisons of glucose yield from cellulose treated in the reaction system of [C2mim][Cl] and water as solvent at (a) 90 °C and (b) 120 °C with 1.5 wt% sulfuric acid

Reaction system	Reaction temperature (°C)	Glucose productivity*
Water	90	0.00
	120	0.02
[C2mim][Cl]	90	0.23
	120	2.13

*Glucose productivity = Maximal glucose yield (%) / Reaction time at maximal glucose yield (min)

Table 1. Glucose productivity from cellulose in water and the [C2mim][Cl] reaction system.

4. Conclusion

This chapter demonstrated that [C2mim][Cl], which can dissolve cellulose, liquefies wood components with the depolymerization of these substances. Cellulose can be effectively converted to glucose in [C2mim][Cl] with sulfuric acid. It is concluded from these results that [C2mim][Cl] can work not only as a solvent for wood or cellulose but also as a reagent for converting them to low MW compounds. These findings suggest that [C2mim][Cl] is applicable to the chemical conversion of wood or cellulose to useful chemicals. This achievement opens the way for an effective utilization of wood or cellulose.

However, the liquefaction of wood by [C2mim][Cl] treatment is not homogeneous, from either a chemical or morphological viewpoint. Additionally, with wood species as raw materials, the reaction atmosphere significantly influences the liquefaction reaction. Thus, further specific research is necessary for industrialization to maximize the target product from wood.

Acknowledgments

This work was partly supported by JFE 21st Century Foundation, the Sumitomo Foundation, a Grant-in-Aid for Exploratory Research (20658041) and the TOSTEM Foundation for Construction Materials Industry Promotion for which authors are grateful.

Author details

Hisashi Miyafuji

Division of Environmental Sciences, Graduate School of Life and Environmental Sciences Kyoto Prefectural University, Kyoto, Japan

References

[1] Goldstein IS. The hydrolysis of wood. TAPPI 1980;63 141-143.

[2] Kim, J. S., & Lee, Y. Y. (2000). Fundamental aspects of dilute acid hydrolysis/fractionation kinetics of hardwood carbohydrates. 1. Cellulose hydrolysis. Industrial & Engineering Chemistry Research 2000;39 2817-2825.

[3] Iranmahboob J, Nadim F, Monemi S. Optimizing acid-hydrolysis: a critical step for production of ethanol from mixed wood chips. Biomass and Bioenergy 2002;22 401-404.

[4] Chang V. S., & Holzapple, M. T. (2000). Fundamental factors affecting biomass enzymatic reactivity. Applied Biochemistry and Biotechnology 2000;38 53-87.

[5] Ortega N, Busto, M. D., & Perez-Mateos, M. Kinetics of cellulose saccharification by *Trichoderma reesei* cellulases. International Biodeterioration & Biodegradation 2001;47 7-14.

[6] Mok, W. S., & Antal Jr., M. J. Uncatalyzed solvolysis of whole biomass hemicelluloses by hot compressed liquid water. Industrial & Engineering Chemistry Research 1992;31 1157-1161.

[7] Saka S, Konishi R. Chemical conversion of biomass resources to useful chemicals and fuels by supercritical water treatment. In: Bridgewater AV (ed) Progress in thermochemical biomass conversion, Blackwell, Oxford; 2001. p1338-1348.

[8] Yamazaki J, Minami E, Saka S. Liquefaction of beech wood in various supercritical alcohols. Journal of Wood Science 2006;52 527-532.

[9] Mishra G, Saka S. Kinetic behavior of liquefaction of Japanese beech subcritical phenol. Bioresource Technology 2011;102 10946-10950.

[10] Kwon G J, Kuga S, Hori K, Yatagai M, Ando K, Hattori N. Saccharification of cellulose by dry pyrolysis. Journal of Wood Science 2006;52 461-465.

[11] Hosoya T, Kawamoto H, Saka S. Influence of inorganic matter on wood pyrolysis at gasification temperature. Journal of Wood Science 2007;53 351-357.

[12] Asmadi M, Kawamoto H, Saka S. Pyrolysis reaction of Japanese cedar and Japanese beech woods in a closed ampoule reactor. Journal of Wood Science 2010;56 319-330.

[13] Pinkert A, Marsh K N, Pang S, Staiger MP. Ionic liquids and their interaction with cellulose. Chemical Reviews 2009;109 6712-6728.

[14] Moulthrop JS, Swatroski RP, Moyna G, Rogers RD. High-resolution ^{13}C NMR studies of cellulose and cellulose oligomers in ionic liquid solutions. Chemical Communications 2005; 2005-1557.

[15] Remsing RC, Swatroski RP, Rogers RD, Moyna G. Mechanism of cellulose dissolution in the ionic liquid 1-*n*-butyl-3-methylimidazolium chloride: a ^{13}C and $^{35/37}Cl$ NMR relaxation study on model systems. Chemical Communications 2006; 2006-1271.

[16] Remsing RC, Harnandez G, Swatroski RP, Massefski WW, Rogers RD, Moyna G. Solvation of carbohydrates in N,N′-dialkylimidazolium ionic liquids: A multinuclear NMR spectroscopy study. Journal of Physical Chemistry B 2008;112 11071-11078.

[17] Youngs TGA, Holbrey JD, Deetlefts M, Nieuwenhuyzen M, Gomes MFC, Hardacre C. A molecular dynamics study of glucose solvation in the ionic liquid 1,3-dimethylimidazolium chloride. ChemPhysChem 2006, 7-2279.

[18] Youngs TGA, Hardacre C , Holbrey JD. Glucose solvation by the ionic liquid 1,3-dimethylimidazolium chloride: A simulation study. Journal of Physical Chemistry B 2007, 111-13765.

[19] Liu H, Sale KL, Holmes BM, Simmons BA, Singh S. Understanding the interactions of cellulose with ionic liquids: A molecular dynamics study. Journal of Physical Chemistry B 2010;114 4293-4301.

[20] Zhang H, Wu J, Zhang J, He J. 1-Allyl-3-methylimidazolium chloride room temperature ionic liquid: A new and powerful nonderivatizing solvent for cellulose. Macromolecules 2005, 38-8272.

[21] Fukaya Y, Hayashi K, Wada M, Ohno H. Cellulose dissolution with polar ionic liquids under mild conditions: required factors for anions. Green Chemistry 2008;10 44-46.

[22] Nakamura A, Miyafuji H, Saka S, Mori M, Takahashi H. (2010) Recovery of cellulose and xylan liquefied in ionic liquids by precipitation in anti-solvents. Holzforschung 2010;64 77-79.

[23] Xie H, Shi T. Wood liquefaction by ionic liquid. Holzforschung 2006;60 509-512.

[24] Kilpeläinen I, Xie H, King A, Granstrom M, Heikkinen S, Argyropoulos DS. Dissolution of wood in ionic liquids. Journal of Agricultural and Food Chemistry 2007;55 9142-9148.

[25] Pu Y, Jiang N, Ragauskas AJ. Ionic liquid as a green solvent for lignin. Journal of Wood Chemistry and Technology 2007;27 23-33.

[26] Miyafuji H, Miyata K, Saka S, Ueda F, Mori M. (2009) Reaction behavior of wood in an ionic liquid, 1-ethyl-3-methylimidazolium chloride. Journal of Wood Science 2009;55 215–219.

[27] Nakamura A, Miyafuji H, Saka S. Liquefaction behaviour of Western red cedar and Japanese beech in the ionic liquid 1-ethyl-3-methylimidazolium chloride. Holzforschung 2010, 64-289.

[28] Nakamura A, Miyafuji H, Saka S. Influence of reaction atmosphere on the liquefaction and depolymerisation of wood in an ionic liquid, 1-ethyl-3-methylimidazolium chloride. Journal of Wood Science 2010, 56-256.

[29] Miyafuji H, Suzuki N. Observation by light microscope of sugi (*Cryptomeria japonica*) treated with the ionic liquid, 1-ethyl-3-methylimidazolium chloride. Journal of Wood Science 2011, 57-459.

[30] Miyafuji H, Suzuki N. Morphological changes in sugi (*Cryptomeria japonica*) wood after treatment with the ionic liquid, 1-ethyl-3-methylimidazolium chloride, Journal of Wood Science 2012, 58-222.

[31] Tsoumis G. Chemical composition and ultrastructure of wood. In: Tsoumis G (ed) Science and technology of Wood. Van Nostrand Reinhold, New York; 1991. p42–46.

Chapter 13

Applications of Ionic Liquids in Lignin Chemistry

Zhu Yinghuai, Karen Tang Yuanting and
Narayan S. Hosmane

Additional information is available at the end of the chapter

http://dx.doi.org/10.5772/51161

1. Introduction

1.1. Lignin

Lignin is a naturally occurring aromatic cross-linked polymer with molecular weight of more than 10,000 Daltons (Da). It is estimated that about 30% of the organic carbon in the earth's biosphere are from lignin [1,2]. It is made up of about 20-40% of wood and annual plants, depending on the species [1-3]. Together with cellulose and hemicellulose, they formed the lignocellulose, which is an important source of biomass. Lignocellulose is found largely in the cell walls with lignin acting as a linker between sets of cellulose and hemicellulose as shown in Figure 1. It is covalently bonded to hemicellulose, thus increasing the mechanical strength of the cell walls [1,2,4]. Due to its hydrophobicity, lignin prevents water from entering the cells; thus it provides an efficient way to transport water and nutrients by repelling them away from the cells. Hence, transportation cells such as tracheid, sclereid and xylem cells have more lignin in them [2].

1.1.1. Monolignols

As a natural polymer, lignin is made up of three main, but not limited to, basic monomers. These monomers are a variation of phenylpropane species collectively known as 'monolignols' which primarily includes *p*-coumaryl alcohol, coniferyl alcohol and sinapyl alcohol, shown in Figure 2 [3,5]. As observed from Figure 2, the structure of these monolignols differs only by the number of substituted methoxy group at the phenyl ring. Figure 2(a) has also shown the numbering system, based on lignin nomenclature rules, that is different and must not be confused with IUPAC nomenclature rules.

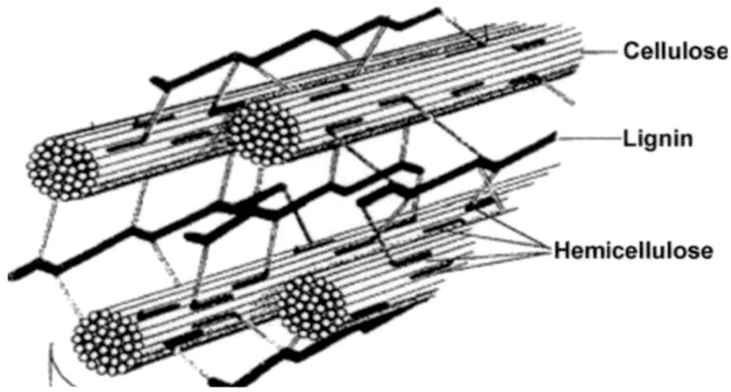

Figure 1. Lignocellulose network

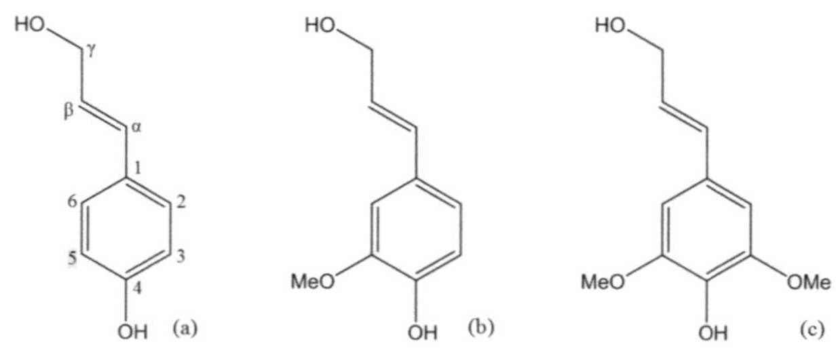

Figure 2. Lignin monomers: (a) *p*-coumaryl, (b) coniferyl and (c) sinapyl alcohols

Additionally, the composition of each lignin monomer differs significantly with the type of plant material, shown in Table 1. Softwood lignin, found in coniferous trees such as pine tree, composes of mainly of coniferyl alcohol and only a trace amount of sinapyl alcohol as their repeating units. On the other hand, hardwood lignin, found in tropical and subtropical trees such as oak and teak, contains both coniferyl and sinapyl alcohols in considerable amounts. It should also be noted that neither hardwood nor softwood lignin contains significant proportion of *p*-coumaryl alcohol which can be found in grass lignin along with other two monolignols.

	p-Coumaryl alcohol	Coniferyl alcohol	Sinapyl alcohol
Softwood	<5	<95	Trace amount
Hardwood	0-8	25-50	46-75
Grasses	5-33	33-80	20-54

Table 1. Compositions in monolignols in plant

1.1.2. Formation of Lignin

From the monolignols, discussed above, it is apparent that lignin can be formed through phenylpropanoid, *p*-hydroxyphenyl, guaiacyl, and syringyl moieties by *p*-coumaryl, coniferyl and sinapyl alcohols, respectively.

The first step of the polymerization process involves oxidation of the monolignols forming a radical to initiate the reaction. Figure 3 shows the reaction and the various resonance structures of coniferyl alcohol radical formed [6]. Due to the various resonance structures, there is a randomness of bonding when radical coupling takes place. Table 2 shows the various common types of linkages that can be possibly formed. Out of the all linkages, the *β*-O-4 linkage is the most common which make up 50-60% of the total linkages depending on the type of wood materials [2]. A simplified overall mechanism of the radical coupling of two coniferyl alcohols forming the *β*-O-4 linkage is shown in Figure 4 [6].

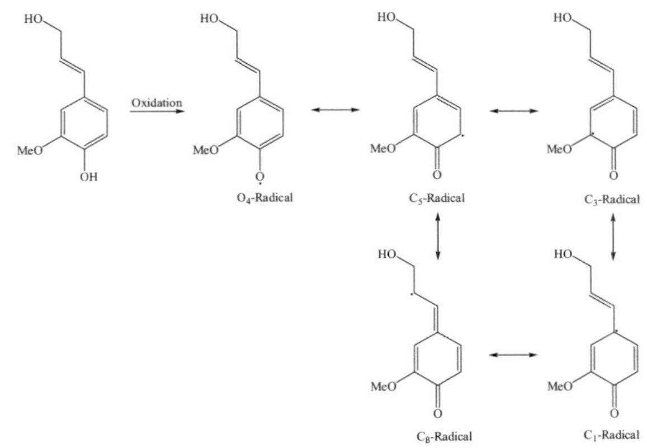

Figure 3. Oxidation of coniferyl alcohol

Name	Linkage	Chemical Structure
β-aryl ether	*β*-O-4	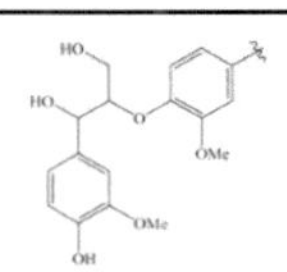
Phenylcoumaran	*β*-5(*α*-O-4)	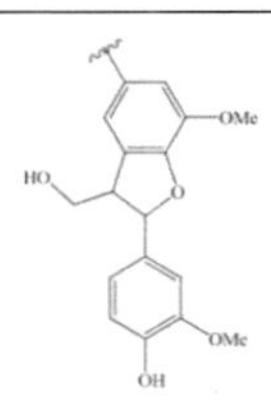

Name	Linkage	Chemical Structure
Resinol	β-β(α-O-α)	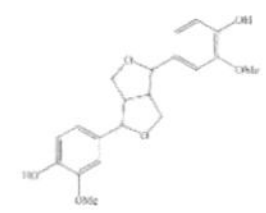
Biphenyl	5-5	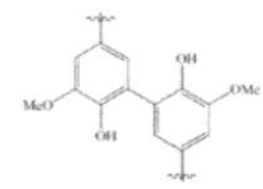
Biphenyl ether	4-O-5	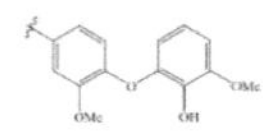
Benzodioxane	β-O-4(α-O-5)	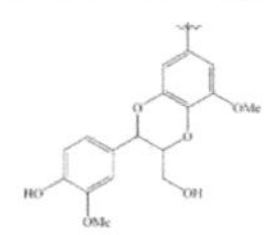
β-C1	β-1	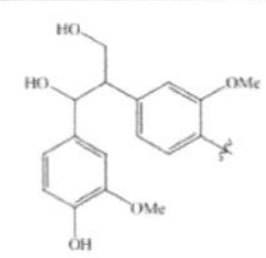

Table 2. Common types of linkages

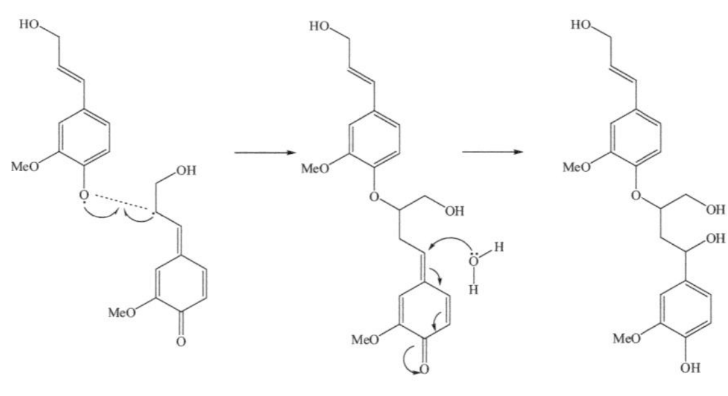

Figure 4. β-O-4 linkage formation via radical coupling of 2 coniferyl alcohols

It can be observed from Table 2 that the common types of linkages utilize most of the radicals formed in Figure 3 except for C_3-radical. Radical coupling at the C_3 site is just as likely to occur, but no stable product could be obtained (see Figure 5). Since the methoxyl group at the C_3 site is a poor leaving group, the aromatic ring could not be regenerated. Thus, the coupling reaction will move backwards and the initial radicals can be obtained to form more stable linkages [6].

Due to the randomness in the bonding nature, the overall structure of the macromolecular lignin has not been accurately predicted. The overall 3D structure of lignin is also unknown as isolating them without modifications are still difficult even though better isolating methods are found [6].

1.2. Ionic Liquids: Classifications and Synthetic Methods

Ionic Liquids (ILs) are salts that are in liquid state. For example, sodium chloride (NaCl) is an IL when it melts at 801°C, forming sodium cations (Na^+) and chloride anions (Cl^-). However, ILs of such high temperature cannot be used.Hence, in literature [7,8], ILs are usually referred to salts that are in liquid state at ambient temperatures (<100 °C). They possess unique characteristics that distinguish them from molecular solvents. Many ILs are excellent prerequisities for efficient IL recycling as they have negligible vapour pressure and good thermal stability. Some of the ILs are very good solvents for large biomolecules, such as cellulose and lignin [9].

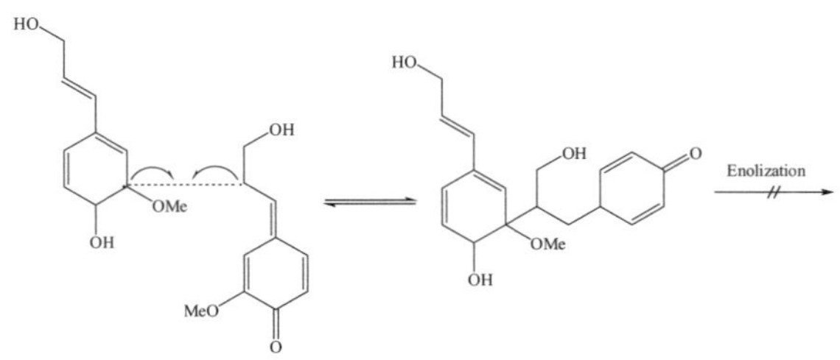

Figure 5. Possible but disallowed C_3-C_β linkage

1.2.1. *Classifications of ILs*

Ionic liquids can be classified into many categories. Figure 6 shows one of the methods in which they are being classified.

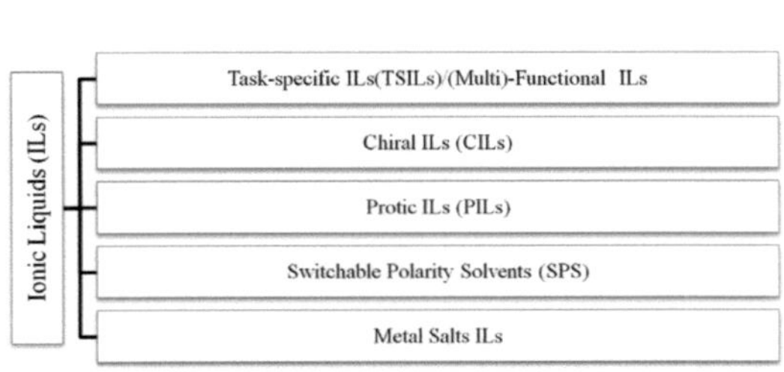

Figure 6. Various classes of ionic liquids

1.2.1.1. *Task-Specific ILs/ (Multi)-Functional ILs*

In the past decades, a vast amount of research has been performed to incorporate additional functional groups into functional ILs. This incorporation can introduce certain properties to ILs including enhancing capacity of catalyst reusability. In addition, when specific functionalities are added, the resulting IL will be task-specific, hence they are sometimes also referred as task-specific ILs (TSILs). Many of these TSILs are produced by introducing functional groups into a branch appended to the cation, especially imidazoliumcation (Figure 7). The imidazolium salts are only defined as TSILs when functional group is covalently bonded to the cation/anion of the salt, which behaves as a reaction medium and reagent/catalyst [10].

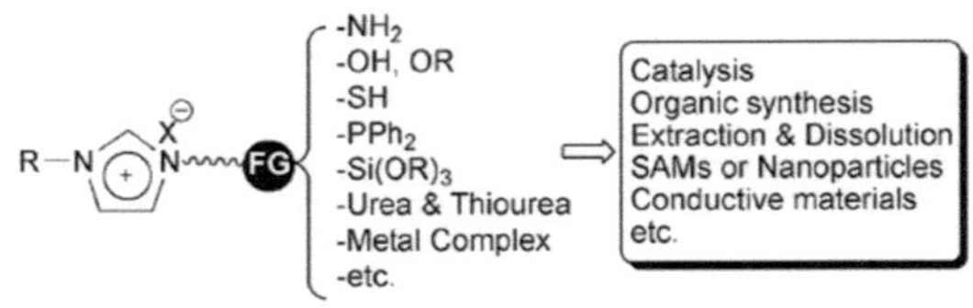

Figure 7. Functionalized imidazolium salts for task-specific ILs

1.2.1.2. Chiral ILs

Chiral ILs (CILs) are a special class of TSILs [11]. They have a chiral center either on the cation, anion or both. Due to their ease to synthesis, they had gained its popularity as a chiral solvent in asymmetric synthesis [12]. Figure 8 shows some examples of CILs.

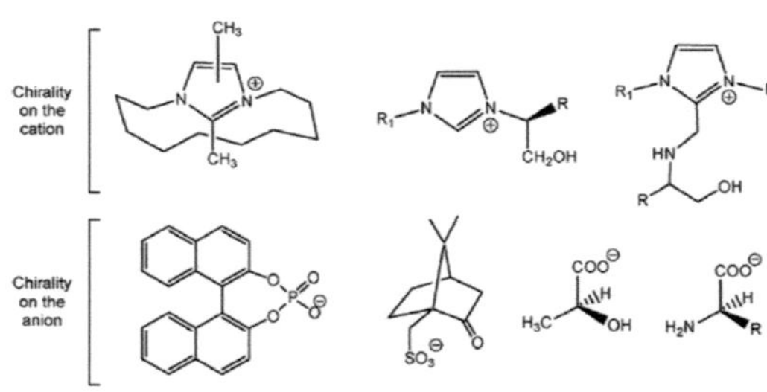

Figure 8. Examples of CILs.

1.2.1.3. Protic ILs

The major difference between Protic ILs (PILs) and other ILs is the presence of exchangeable proton. There is a resurrection of interest in these PILs recently due to their great capabilities for proton transfer in fuel cell chemistry. Some examples of PILs are shown in Figure 9. Many of these PILs are very strong acid, thus, the equilibrium is shifted heavily to the right. This, in terms, results in fully ionized ILs.

1.2.1.4. Switchable Polarity Solvent (SPS)

Switchable polarity solvents (SPS) equilibrate between a higher polarity and a lower polarity when trigger is applied. These solvents are particularly useful when two different polarities of the solvent are needed for two different steps. Recently, secondary amines have been used as SPS, with carbon dioxide as the trigger, forming carbamatesalts [13] (Figure 10).

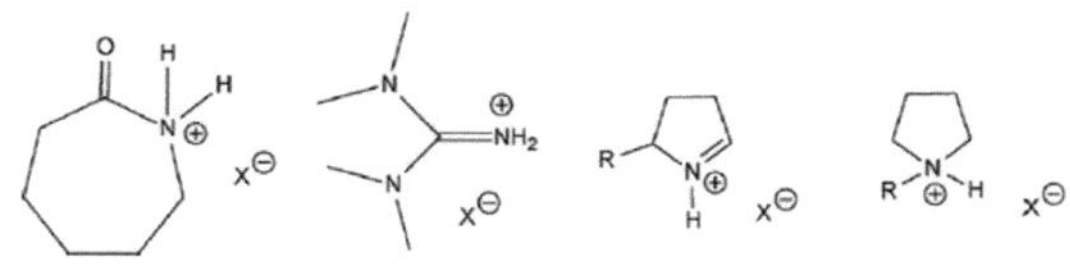

Figure 9. Some examples of PILs.

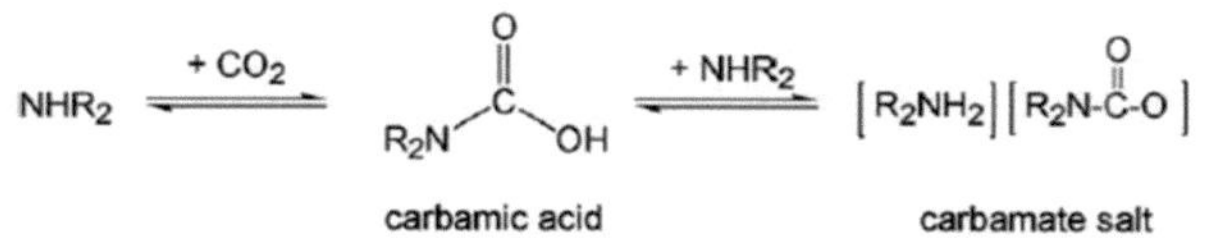

Figure 10. New generation of SPS.

1.2.1.5. Metal Salts ILs

Many metal salts ILs have been developed [15]. Chloroaluminates associating with imidazolium or pyridiumcations were part of a focused investigation, but was recently extended to include other chlorometalate salts (e.g. $[CuX_3]^-$, $[NiCl_4]^{2-}$, $[Co(CO)_4]^-$). The recently developed chlorometalate salts are not water sensitive, unlike the chloroaluminates, but are generally more viscous. The introduction of the metal ions inside the ILs is able to immobilize catalysts while it is being the integral part of the potentially ordered structure of ILs.

1.2.2. Preparations of ILs

There are 4 main synthetic routes to prepare ILs. [7]. Figure 11 shows the overview of the synthetic methods.

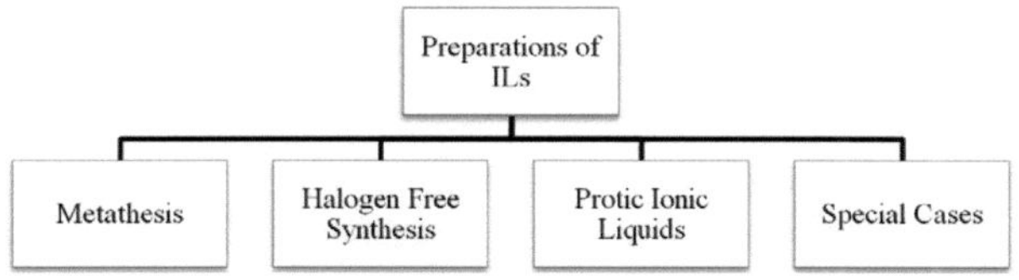

Figure 11. Synthetic routes of preparations of ILs

1.2.2.1. Metathesis

Many ILs are prepared through a metathesis reaction from a halide or similar salt of the desired cation. This process can be subdivided into two classes depending on the water solubility of the targeted ILs. Figure 12 shows the overall reaction scheme of the metathesis reaction.

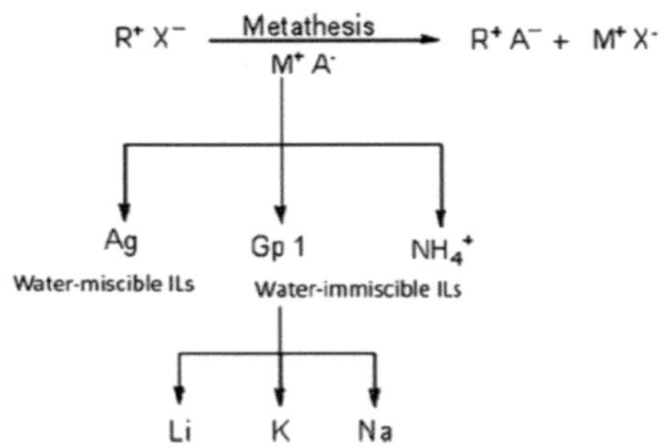

Figure 12. Overall metathesis routes to ILs

There are two main methods to prepare the water-immiscible ILs. The more commonly used approach is the metathesis reaction of the corresponding halide salt, using either the free anions of an acid or its salt. The use of free anions of an acid is favored as the hydrogen halide produced can be easily removed though washing with water. Alternatively, the metathesis reaction can also be conducted in organic solvents such as dichloromethane or acetone. However, the starting materials are not completely soluble and the reaction is carried out as a suspension. The organic layer must be washed several times with water to remove the unwanted halides.

The preparation of water-miscible ILs is proven to be more difficult as it requires purification steps before the desired ILs are obtained. This can be done easily by a metathesis reaction of the corresponding halide with a silver salt of the anion such as $AgBF_4$. Unfortunately, this approach is not cost effective as large amount of silver halide is produced as a by-product. The ILs might also be silver-contaminated as complete precipitation of silver halides from organic solvents tends to be slow.

Ion exchange materials are also used to obtain ILs via the metathesis method although little information is available openly.

1.2.2.2. Halogen Free Synthesis

Production of high purity ILs via the metathesis method can be problematic due to contamination of the residual halide. The physical properties of the ILs can be drastically changed with the presence of the halides and may even result in catalyst poisoning and deactivation; hence, halide free synthetic methods had been devised.

http://dx.doi.org/10.5772/51161

Imidazoles can be produced by using N-heterocyclic carbene (NHC) as its intermediate. Carbenes have a lone pair of electrons on a carbon atom which makes them very reactive. The synthesis of ILs via carbenes can be achieved by reacting with acids using either NHC adducts (Figure 13) or NHC-organometallic intermediates (Figure 14).

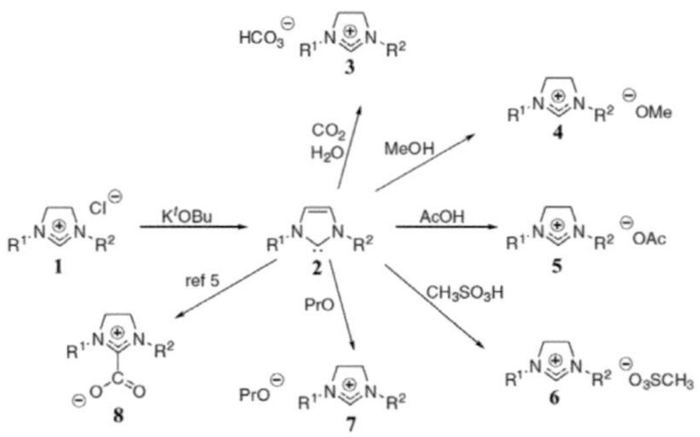

Figure 13. Synthetic methods of various ILs via NHC adducts

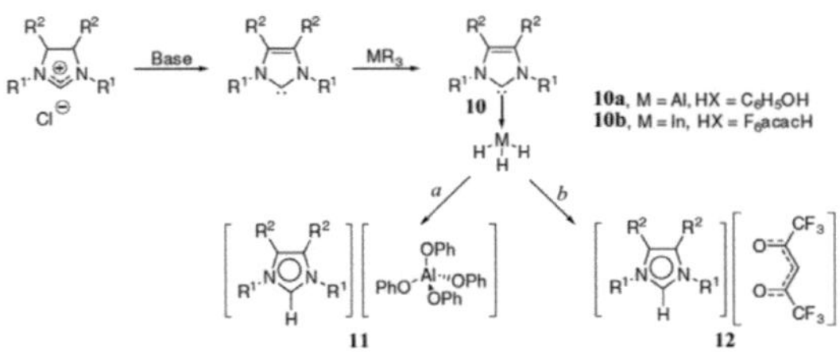

Figure 14. Reaction of NHC-organometallic intermediates with acids

Halogen-free phosphorus and sulfur based ILs can be produced by using the synthetic routes shown in Figure 15 and Figure 16 respectively.

1.2.2.3. Protic Ionic Liquids (PILs)

PILs can be synthesized by proton transfer between an equimolar mixture of a Brønsted acid and Brønsted base. This presents the advantage of being cost-effective and easily prepared as synthesis does not form residual products. PILs can produce hydrogen bonding between the acid and base or even hydrogen-bonded extended network.

1.2.2.4. Special Cases

Metal Based ILs

Transition metal complexes containing anions ILs were among the earliest developed room temperature ILs (RTILs). These RTILs can be synthesized by reaction of phosphonium/

imidazolium halides with metal halides or metathesis reaction with alkali salts of metal based anions. These metal based salts can be subdivided into three groups: transition metal, p-block and f-block metal salts.

Functionalized ILs

Till recently, functionalized ILs are prepared by displacement of halide from the organic halide containing the functional group by a parent imidazole, phosphine, etc., shown in Figure 17.

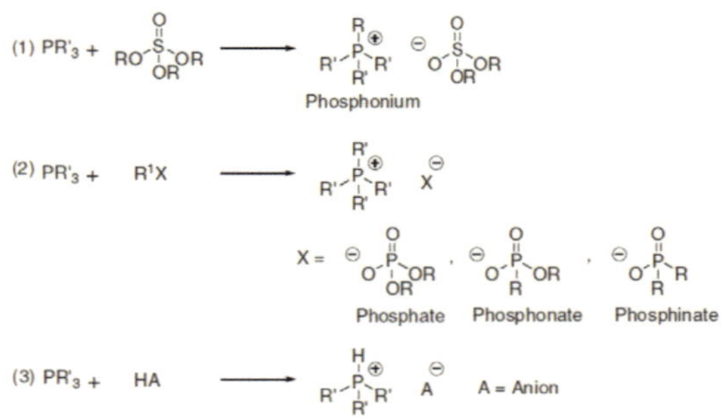

Figure 15. Halogen free synthetic routes for phosphonium ILs: (1) phosphines with sulphates; (2) 3°phosphines or imidazoles with alkylating agents; (3) phosphines with acid.

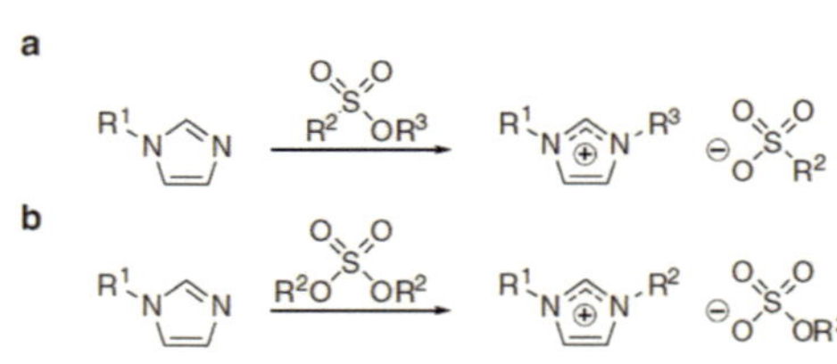

Figure 16. Halogen free synthetic routes for (a) sulphate and (b) sulphonate ILs.

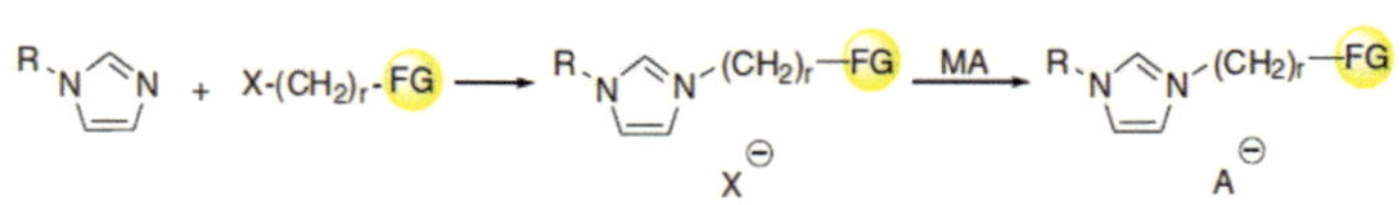

Figure 17. Classical method for functionalized ILs

Recently, novel functionalized ILs have been designed, synthesized and characterized. Some of the examples are:

- ILs with two Brønsted acid sites with COOH, HSO_4 or H_2PO_4 groups are synthesized by using methylimidazole with Brønsted acid moieties.

http://dx.doi.org/10.5772/51161

- ILs with amino acids as anions are synthesized by neutralizing between [Emin][OH] and amino acid
- ILs with ether or alcohol functional groups are synthesized though alkylation of methylimidazole with alkyl halide, followed by halogen exchange with slight excess of KPF_6 to reduce the remaining halogen content.

Microwave-Assisted

Using classical heating methods in reflux solvent, several hours of reaction time is needed to obtain reasonable yields. It also uses in large excess of alkylhalides and organic solvents. By using the microwave-assisted method, reaction time can be shortened with a cleaner work-up procedure. However, ILs may decompose under such conditions.

1.3. Ionic liquids in Lignin Chemistry: How compatible?

As seen above, ILs are beneficial to a wide range of applications. These applications include, but not limited to, solvation, catalysis and as a reagent. In order to obtain lignin from plant, the ideal IL should possess properties as followed: (1) high dissolution capacity for lignin; (2) low melting point; (3) good thermal stability; (4) non-volatile; (5) non-toxic; (6) chemically stable; (7) no lignin decomposition; (8) easy lignin regeneration and (9) low cost and simple process [8]. For surface modifications and conversion of lignin to better materials, ILs can be used as a catalyst or solvent. In the next few sections, we will fully discuss in details of extracting lignin (Section 2), surface modification of lignin (Section 3) and conversion of lignin to value-added chemicals (Section 4) using ILs.

2. Extracting Lignin with Ionic Liquid

One of the most traditional methods to lignin extraction used in the industry is by the kraft process. Even though most of the lignin is extracted, it has several major disadvantages: (1) high temperature and pressure; (2) pollution; (3) odour problem (due to the use of sulphite); (4) high water usage and (5) large plant size [14].

Several other methods to extract lignin from lignocellulose have been designed and developed in the past. These include physical (limited pyrolysis and mechanical disruption/comminution [15]), physiochemical (steam explosion, ammonia fiber explosion [16,17]), chemical (acid hydrolysis, alkaline hydrolysis, high temperature organic solvent pretreatment, oxidative delignification [18-20]) methods.These extraction methods have one main disadvantage: the lignin starts to degrade after a certain amount of lignin is extracted. This often leads to loss of fermentable sugars in the cellulose and hemicellulose, which can be used to produce other value-added products like the levulinic acid [21] and 5-hydroxymethylfurfural [22].

In the past decade, ILs have been used as solvents for natural polymers, including cellulose and starch [23-28]. This leads to an interest in designing and developing ILs that are able to dissolve lignin and thus, extracting them from the lignocellulosic biomass.

2.1. Dissolution of Lignin

Pu and co-workers [29] used the imidazoilum-based ILs for the studies of dissolution of residual softwood lignin isolated from a southern pine kraft pulp. Table 3 shows the solubility of the residual softwood lignin that they have obtained.

Ionic Liquid	Temperature (°C)	Solubility (g L^{-1})
[Mmim][$MeSO_4$]	50	344
	25	74.2
[Hmim][CF_3SO_4]	70	275
	50	<10
[Bmim][$MeSO_4$]	50	312
	25	61.8
[Bmim][Cl]	75	13.9
[Bmim][Br]	75	17.5
[Bmim][PF_6]	70-120	Insoluble
[Bm_2im][BF_4]	70-100	14.5
[Bmpy][PF_6]	70-120	Insoluble

Table 3. Solubility of residual softwood kraft pulp lignin in ionic liquids

They show that the lignin solubility can be influenced by the nature of anion as the solubility of the lignin differs when different $[Bmim]^+$-containing ILs are used. They also concluded that ILs containing large, non-coordinating anions like $[PF_6]^-$ and $[BF_4]^-$ are not suitable for dissolving lignin; and that methylsulfateimidazolium-based ILs are effective for dissolution of residual softwood lignin isolated from a southern pine kraft pulp.

2.2. Lignin Extraction without Dissolution of Biomass

Lee and co-workers [30] had done lignin extraction from maple wood. They had also used Indulin AT (kraft lignin) as standards for solubility test. The results they had obtained are shown as Table 4. Using Indulin AT, the solubility of lignin shows similar results as Pu and co-workers as discuss earlier. However, when maple wood flour is used, there are difficulties dissolving and extracting lignin using the same ILs. Instead, Cl^--containing ILs, [Amim][Cl] and [Bmim][Cl], show better capabilities of extracting lignin from maple wood flour. This might be as a result of high solubility of the wood flour as a whole. Cl^- ions are good hydrogen acceptors and are able to interact with the hydroxyl groups of the sugars, causing dissolution of cellulose too. They had also found an IL, [Emim][Ac], that provides a balance between good lignin extraction and low wood flour solubility. Hence, lignin extraction can be done using [Emim][Ac] without much disruption of the cellulose and hemicellulose structures.

ILs	Lignin Solubility (g/kg)[a]	Wood Flour Solubility (g/kg)[b]	Extracted Lignin Content (g/kg)[c]
[Mmin][$MeSO_4$]	<500	ND	0.8
[Bmim][CF_3SO_3]	<500	ND	0.5
[Emim][Ac]	<300	<5	4.4
[Amim][Cl]	<300	<30	5.2
[Bmim][Cl]	<100	<30	3.2
[Bzmim][Cl]	<100	<10	1.9
[Bmim][BF_4]	40	ND	ND
[Bmim][PF_6]	~1	ND	ND

[a]Solubility of Indulin AT (kraft lignin) at 90 °C after 24 hours incubation[b]Solubility of maple wood flour after 24 hours incubation at 80 °C under N_2, ND indicates <1 g/kg[c]0.5 g maple wood flour was incubated in 10 g ILs for 24 hours at 80 °C under N_2. Lignin content was determined with Indulin AT standard. ND indicates <0.1 g/kg

Table 4. Solubility and extraction efficiency of lignin in various ILs.

Another group, Tan, S. S. Y. et. al. [31], had reported of using the ethyl-methylimidazoliumalkylbenzenesulfonate IL, [Emim][ABS],for extraction of lignin from sugarcane bagasse. [Emim][ABS] was used for the study as sodium xylenesulfonate was used in a lesser known pulping process, *hydrotropic* pulping. Table 5 shows the lignin extraction increases with elevating temperature from 170 °C to 190 °C, and when the extraction time increases from 30 minutes to 120 minutes. It should be noted that the mass recovered for Entry 3 is higher than the original lignin content. This might be due to the incorporation of the xylenesulfonate anion as sulphur is detected from the elemental analysis, which is not present in the original bagasse lignocelluloses. It might be also due to the reactions between lignin and hemicelluloses products. As a whole, lignin extraction with more than 93% yield was successfully attained at atmospheric pressure with [Emim][ABS]. Although ILs proved to be effective in the dissolution of lignin, several problems such as the simplification of IL recovery and designing an IL to create lignin with desirable adducts.

Fu, Mazza and Tamaki [32] also reported that [Emim][Ac] is an effective solvent for lignin extraction from triticale straw, flax shives and wheat straw. Five other ILs ([Bmim][Cl], DMEAF, DMEAA, DMEAG, DMEAS) had been examined but the extraction yields are much lower than that of [Emim][Ac]. They had also investigated the extraction capabilities at a range of temperature (70-150 °C) and time taken (0.5-24 h). Results show that within this range, higher temperature and longer extraction time produces higher extraction yield.

Entry No.	Conditions	Mass of Recovered Lignin[a] (% ±2)
1	170 °C, 120 min	67
2	180 °C, 120 min	78

Entry No.	Conditions	Mass of Recovered Lignin[a] (% ±2)
3	190 °C, 120 min	118
4	190 °C, 90 min	97
5	190 °C, 60 min	96
6	190 °C, 30 min	67

[a]As percentage of original lignin content, corrected for ash content

Table 5. Mass of recovered lignin

Last year, Pinket and co-worker [33] published a paper that uses food additive-derived ILs, imidazoliumacesulfamate ILs, for lignin extraction from *Pinus radiate* and *Eucalyptus nitens*wood flour. They show promising results without disrupting the cellulose crystallinity. Among all, [Emim][Ace] is desirable for industrial processing due to its physical properties. The extracted lignins have a larger average molar mass as well as a more uniform molar mass distribution compared to that obtained from the Kraft process. This adds to another advantage of using imidazoliumacesulfamate ILs.

In their paper, they had also examined various extraction conditions: (1) Extraction temperature and time; (2) Water content; (3) Wood load, particle size and species; (4) Types of IL cation; (5) Effect of IL recycling; (6) Multi-step treatment and; (7) Use of co-solvents. Interestingly, the use of DMSO as a co-solvent (w_{IL}:w_{DMSO}=9:1) increases the extraction efficiency by almost 50% (Figure 18). It is believed that the penetration and interaction of the ligncellulosic biomass with IL is enhanced due to two effects caused by DMSO: (a) the tight hydrogen bond network of the cellulose is loosened and; (b) the overall viscosity of the mixture is decreased.

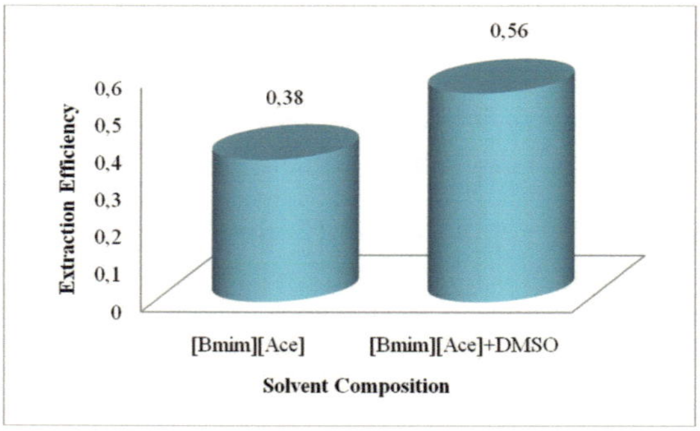

Figure 18. The extraction efficiency for extracting wood lignin with either pure [Bmim][Ace] or a mixture of [Bmim][Ace] and DMSO (w_{IL}:w_{DMSO}=9:1). *Pinusradiata*wood flour (w_T=0.05), with particle size of 100 μm, was treated at 373 K for 2 hours in an open atmosphere.

Kim, J. et. al. [34] compared the structural features of poplar wood lignin extracted using IL, [Emim][Ac], and dioxane-water (classical method). Table 6 shows the summarized data obtained by them. It can be seen that a higher yield is obtained through IL extraction. Even though the lignins obtained from both extractions have relatively similar methoxy and phenolic hydroxyl contents, the molecular weight of that obtained from classical method is higher. However, the polydispersity index (PDI) of IL extraction is lower and thus, indicating that lignin from IL extraction is of rather uniform size. This suggests that some form of depolymerisation had occurred (will discuss further in Section 4.1) On top of that, thermal behaviour of the lignins were also analysed. Lignin obtained from classical method has a higher maximum decomposition rate and temperature, indicating that it is thermally more stable.

	[Emim][Ac] Extraction	Dioxane-water Extraction
Yield (%)	5.8±0.3	4.4±0.4
Amount of OMe group (%)	15.5±1.5	14.4±0.1
Amount of phenolic OH group (%)	6.7±0.2	6.3±0.2
Molecular Weight (Da)	6347	10,002
Polydispersity Index (PDI)	1.62	2.46
Max. decomposition rate (%/°C)	0.25	0.30
Max. decomposition temperature (°C)	308.2	381.3

Table 6. Summarized data obtained by Kim, J. et. al. [34]

2.3. Extraction of Lignin through Dissolution of Biomass

Lignin extraction can also be done after complete dissolution of the whole cellulosic biomass. Lateef's group [35] demonstrated that cellulose-lignin mixture system can be separated and recovered from ILs, [Pmim][Br] and [Bmim][Cl]. Cellulose precipitates when water is added to the solution (cellulose-lignin mixture in IL). Lignin is then recovered from the precipitate formed when the filtrate is treated with ethanol. The IL can be regenerated by evaporating the ethanol from the second filtrate with more than 95% yield. On the other hand, lignin yields of 69% and 49% were isolated from [Pmim][Br] and [Bmim][Cl], respectively.

Muhammad and co-workers [36] used amino acid-based IL, [Emim][Gly], to dissolve bamboo biomass at 120 °C in 8 hours. Other than [Emim][Gly], [Emim][TFA] and choline propionate were also investigated. [Emim][Gly] and choline propionate were used as it is reported that they have high hydrogen bond basicity. On the other hand, [Emim][TFA] was chosen as it contains an acetate-based anion which was proven for its dissolution of lignocelluloses. However, both [Emim][TFA] and choline propionate could not dissolve the bamboo biomass after 24 hours. Hence, only [Emim][Gly] was effective towards dissolution of bamboo biomass.

The lignin was then extracted from the solution using acetone/water mixture (7:3 ratio), obtaining about 85.3% of the total lignin content of the bamboo biomass. Acetone/water mixture acts as an anti-solvent of cellulosic materials and thus, precipitates cellulose and cellulose-rich materials. The acetone/water mixture ratio was determined using a plot of dissolved lignin (wt %) against acetone/water ratio (v/v). It is shown that the amount of dissolved lignin is at the maximum when the acetone/water mixture ratio is 7:3.

Lan, Liu and Sun [37] performed a fractionation of bagasse to obtain cellulose (47.17%), hemicellulose (33.85%) and lignin (54.62%), shown in Figure 19. They dissolved ball milled bagasse in [Bmim][Cl] at 110°C under inert environment with agitation to obtain a clear solution. Residue 1 is regenerated using acetone/water (9:1, v/v) added to the clear solution. Filtrate 1 was then acidified to obtain Residue 2 (acetone soluble lignin). Residue 3 (cellulose) was extracted with 3% sodium hydroxide aqueous solution from Residue 1. Filtrate 3 was then treated with 4 M hydrochloric acid and precipitated using 95% ethanol to attain Residue 4 (hemicellulose). Filtrate 4 was acidified to obtain Residue 5 (Alkaline lignin).

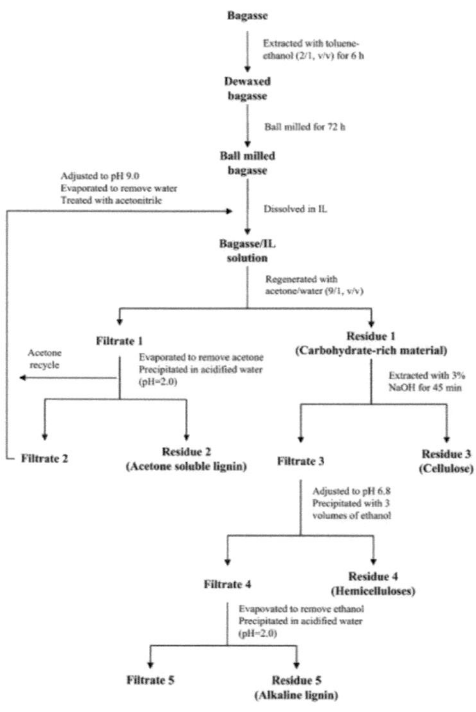

Figure 19. Schematic process of bagasse fractionation based on complete dissolution in [Bmim][Cl] followed by precipitation in acetone/water (9:1,v/v) and extraction with 3% NaOH.

Sievers, C. et. al. [38] had recovered lignin fraction as a solid residue when IL phase hydrolysis of pine wood was done. They first dissolved pine wood in [Bmim][Cl]. Acid catalyst, trifluoroacetic acid, was subsequently added to convert the carbohydrate fraction to water-soluble products. These water-soluble products, which include furfural and 5-HMF, were then extracted using aqueous phase, leaving lignin as a solid. However, this method does not extract pure lignin and the lignin may be susceptible to modifications due to the acidic condition used.

Sun's group [39] dissolved both softwood and hardwood in [Emim][OAc] and lignin was extracted using acetone/water (1:1 v/v). Figure 20 shows the ideal process flow of the dissolution and regeneration of wood in IL. However, recycling of IL was not investigated and evaluated in the published report. Although variable processes for lignin, similar to Pinket's [33], have been evaluated, but no report on dissolution of softwood and hardwood can be found. Sun's group evaluated the effects of (1) IL, (2) particle size, (3) wood species, (4) initial wood concentration and, (5) pretreatment. Using the same method as Muhammad's [36], they extracted lignin from the solution using acetone/water anti-solvent.

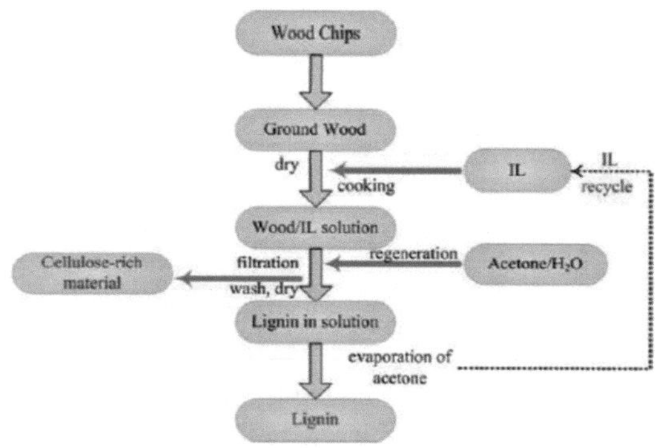

Figure 20. Flowchart for the process of dissolution and regeneration of wood in IL proposed by Sun's group

3. Surface Modification of Lignin in Ionic Liquid Media to form Functional Materials

Carbon fibres, activated carbon, polymer alloys, polyelectrolites, substituted lignins, thermosets, composites, wood preservatives, neutraceuticals/drugs, adhesives and resins are some macromolecules that can be obtained from lignin [40]. Some of them can be obtained from the extracted lignin directly, but most of them needed modifications. Surface modification of lignin allows a change in functional groups on the side-chains while maintaining the aromatic backbone of a lignin structure. Chemical modification of lignins (and also other lignocellulosic materials) leads to different structural characteristics [41]. In this section, we will discuss three types of surface modifications: oxidation, esterification and others.

3.1. Oxidation

Oxidation plays an important role in surface modification of lignin as hydroxyl groups are abundant. When these hydroxyl groups are oxidized, carbonyl groups are formed, which are more susceptible to reactions as compared to hydroxyl groups. In addition, it increases the hydrophilicty of lignin as well.

Zakzeski and co-workers [42] effectively performed oxidation of lignin using [Emim][DEP], cobalt catalyst and molecular oxygen. Although they had hypothesized that there will be an oxidative cleavage of β-O-4 linkage (will be discussed in details in Section 4) and other linkages, it did not materialize. Instead, the benzyl and other alcohol functionalities (i.e. hydroxyl groups at the side-chains) were oxidized to form aldehydes/acids. Similar results were obtained when the same reaction was done on various lignin model compounds, thus providing further evidences that the β-O-4 linkage remained intact. Figure 21 shows the proposed scheme and various reaction sites of lignin when oxidation occurs. Mild conditions were used for the oxidation process due to the special properties of IL, which includes high oxygen solubility. The abundance of diethyl phosphate, [DEP], anions from the IL also allows simple metal salts to give exceptionally high activity.

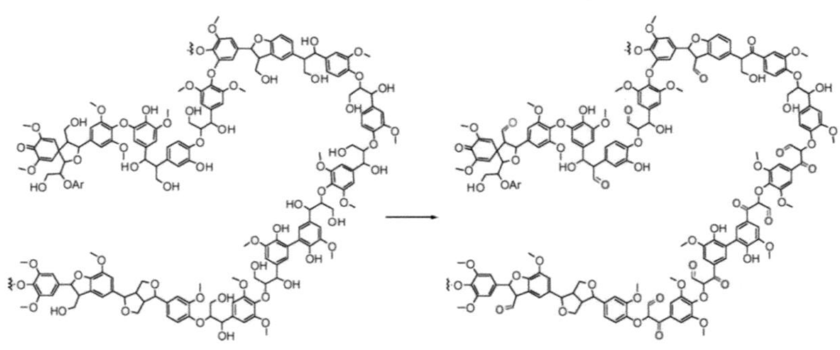

Figure 21. Proposed structure of lignin after oxidation

This Co/[Emim][DEP] catalyst system is able to increase the functionality and hydrophilicity of lignin, thus, Zakzeski had further investigated on the catalyst system [43]. Figure 22 shows the catalytic system proposed by them. As it can be seen, the hydroxyl group being oxidized will be coordinated to the cobalt catalyst forming a complex before oxidation takes place. Hence, the oxidation site must be acidic enough for the coordination to occur. This, in tern, explains why phenolic hydroxyl groups and the linkages were left untouched.

Similar oxidation reaction is done by Kumar, Jain and Chauhan [44] using water soluble iron (III) porphyrins and horseradish peroxidase (HRP) as catalyst, with hydrogen peroxide in IL. Figure 23 shows the reaction scheme that they had reported in their paper. Veratraldehydes are obtained as a major product for both catalysts with more than 70% yields when optimized. The non-coordinating nature and weak nucleophilicity of IL significantly enhances the activity of water soluble iron(III) porphyrins as compared to aqueous solutions. The stability of HRP is also improved when IL is used. Additionally, both catalysts can be recycled with appreciable activity for up to five runs.

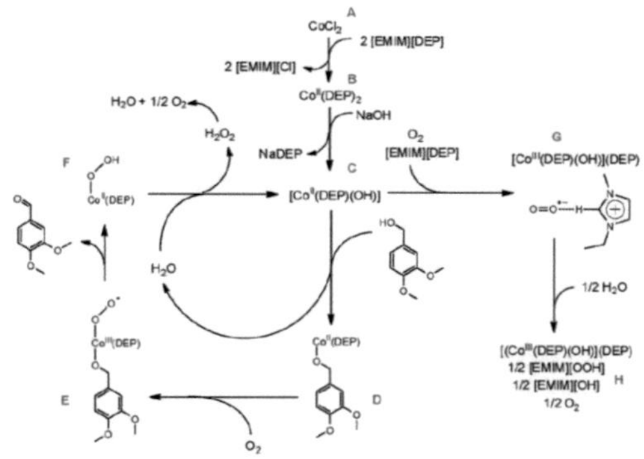

Figure 22. Proposed catalytic system using veratryl alcohol (lignin model compound) as substrate.

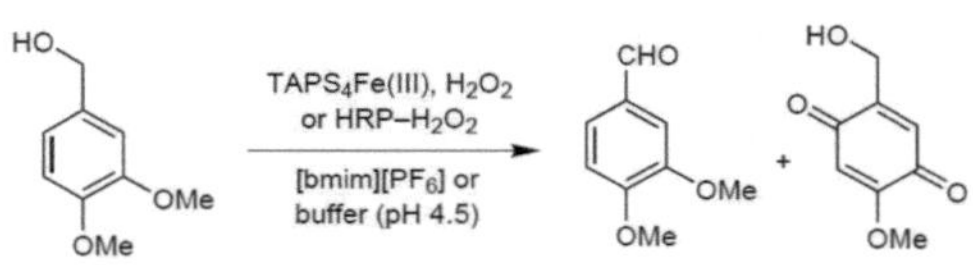

Figure 23. Reaction scheme reported by Kumar, Jain and Chauhan [44]

3.2. Dehydration

A stereospecific dehydration reaction was achieved by Kubo's group [45] by heat treatment in IL. A glycerol type enol-ether (EE), 3-(4-hydroxy-3-methoxyphenyl)-2-(2-methoxyphenoxy)-2-propenol, was produced from guaiacylglycerol-β-guaiacyl ether (GG) when temperature is increases to 120 °C, shown in Figure 24. The EE produced is previously believed to be an unstable intermediate in lignin degradation, but is obtainable as a stable compound in IL through this reaction. Furthermore, it is analyzed that the [Z] isomer of EE is formed as a major product.

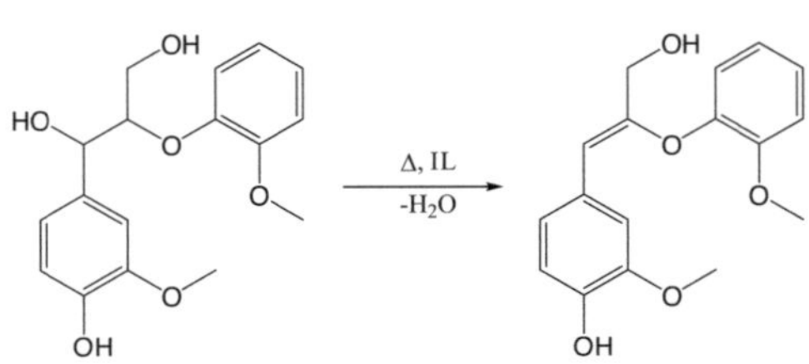

Figure 24. Stereospecific dehydration reaction achieved by Kubo's group [45]

3.3. Esterification

Esterified wood displays admirable plastic properties such as highly substituted degree and excellent hydrophobicity. Hence, it is important to investigate possible esterification methods on wood.

Wen, J. et. al. [41] successfully carried out homogeneous lauroylation of ball-milled bamboo in IL. The ball-milled bamboo was first dissolved in IL, [Bmim][Cl], to enable separation of cellulose, lignin and hemicellulose. After complete dissolution, triethylamine and lauroyl chloride were added. Triethylamine was used to neutralize the hydrochloric acid generated during the esterification process. The bamboo ester was then regenerated. The whole reaction process is shown in Figure 25. Yuan and co-workers [46] had also achieved homogeneous lauroylation and butyrylation of poplar wood with a similar process. Xie's group [47] had also done a similar process, using pyridine instead of triethylamine as neutralizer, to attain homogeneous acetylation, benzoylation and carbanilation on thermomechanical pulp fibers.

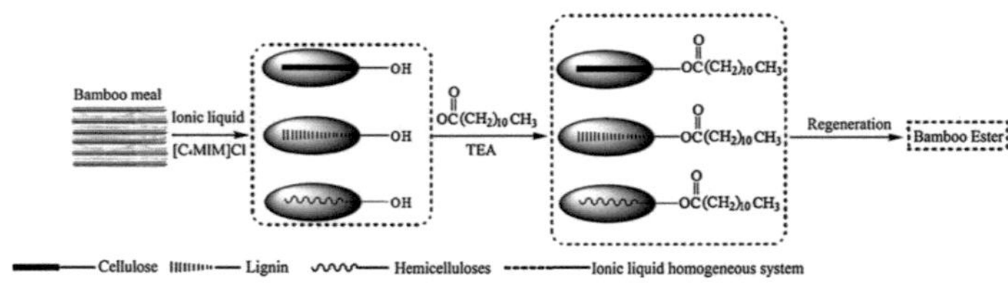

Figure 25. Schematic diagram of the dissolution and esterification process proposed by Wen, J. et. al. [41]

Cerrutti and co-workers [48] synthesized carboxymethyl lignin from organosolv lignin using monochloroacetic acid. An alkalinization of lignin, using sodium hydroxide, is performed before the carboxymethylation reaction to generate stronger nucleophiles for the reaction. The overall reaction is shown in Figure 26. One of the uses for this carboxymethyl lignin produced is as a stabilizing agent in aqueous ceramic suspension.

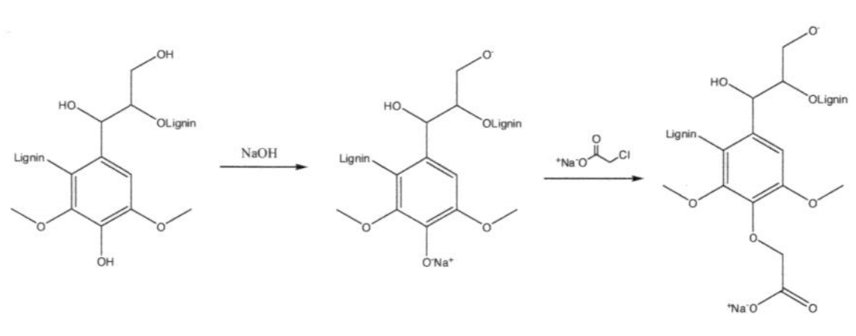

Figure 26. Overall reaction of carboxymethylation reaction

Using organosolv lignin as macro-initiator, Zhu, et al. successfully prepared polylactide-lignin hybrids with tin complex-based catalyst as detailed in Figure 27 [49]. High molecular weight copolymers with an Mw of ca. 5.3×10^4 (DPI 2.56) were produced in 77% yield. Results obtained from tensile strength testing showed that the tensile strength (σ_M, MPa) was

http://dx.doi.org/10.5772/51161

approximately doubled upon blending lignin with PLLA (PLLA:Lignin = 5:1; σ_M = 11.3 MPa) and the ratio of 10:1 (σ_M = 22.8 MPa). There was also a significant increase in tensile modulus (E_t) from 289 to 340 MPa for blending ratios of 5:1 and 10:1, respectively. Tensile strength is the maximum stress that a material can withstand while being stretched before necking. Therefore, a increase in tensile strength indicates a tougher material.

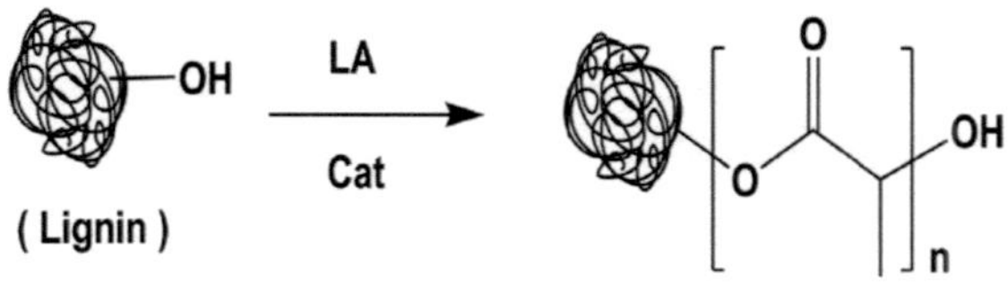

Figure 27. Synthesis of lignin-PLLA hybrid.

4. Conversion of Lignin to Value-Added Chemicals in Ionic Liquid

The lignin obtained from the biomass had be used mainly as a low value fuel in the past. [50-54] There are two main reasons for this main application. Firstly, harsh reaction conditions are required for depolymerization as the polyphenolic structure is chemically very stable. Adding on, the depolymerized products cannot be used as a substitute or additive for conventional liquid fuels due to its high oxygen content [50]. Secondly, biomass-derived feedstock are less readily available as compared to petroleum-derived feedstock. [50,51] However, due to rising fossil fuel prices and energy demand worldwide, research had turned towards obtaining value-added products from biomass-derived feedstock [51].

As mentioned in the introduction, lignin accounts for approximately 30% of organic carbon in the biosphere. Hence, it provides a promising platform for generation of value-added products from lignin [40], which is illustrated in Table 7.

Lignin	Syngas	
	Syngas Products	Methanol/Dimethyl ether, Ethanol, Mixed liquid fuels
	Hydrocarbons	Cyclohexanes, higher alkylates
	Phenols	Cresols, Eugenol, Coniferols, Syringols
	Oxidized Products	Vanillin, vanillic acid, DMSO, aldehydes, Quinones, aromatic and aliphatic acids

Table 7. Value-added chemicals potentially derived from lignin.

Current strategies to produce these value-added chemicals from lignin are typically based on a two-step process. Firstly, lignin is depolymerized into simpler aromatic compounds. After depolymerization, some of the value-added chemicals can be obtained, e.g. phenols and oxidized products. Other value-added products are obtainable by transforming the resultant aromatic compounds. [55]

4.1. Depolymerization of Lignin

The macromolecular lignin contains various types of linkages, with β-O-4 ether linkage (50-60% of total linkages) being the most common. Hence, one of the best depolymerization strategies is to target on cleavage of β-O-4 ether linkage while preserving the aromatic character of the fragments [56]. Since it is difficult for researchers to determine and do mechanistic studies when macromolecular lignin is used, most of the papers uses lignin model compounds instead.

At the earlier stages of the research, Binder J. B. et. al. [57] shows that ILs provide a suitable medium for reactions of lignin model compounds. Through their various reactions done with several lignin model compounds, they suggested that not all model lignins are able to be used for depolymerization studies. Alkene-substituted aromatics and simple ethers are not suitable, being more reactive than natural lignin. While models like eugenol and 2-phenylethyl phenyl ether could be used, having similar reactivity trends as lignin, although they react under milder conditions than lignin.

Reichert and co-workers [58] successfully depolymerized lignin through electro-catalytic oxidative cleavage. They performed the depolymerization in the PIL, triethylammoniummethanesulfonate, using ruthenium-vanadium-titanium mixed oxide coated electrodes. The PIL offers a suitable medium for lignin dissolution, ensures higher potential electrolysis as well as promotes the oxidative cleavage mechanism as shown in Figure 28. They have also demonstrated that smaller molecular weight molecules are obtained when a higher applied potential is used.

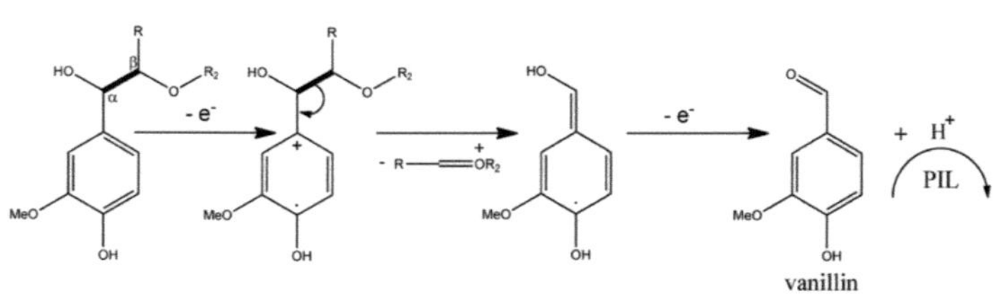

Figure 28. Mechanism of Oxidative cleavage of lignin at C_α-C_β bond

George, A. et. al. [59] observed that depolymerization occurs when various ILs are used to treat three types of technical lignins, organosolv, alkali and ALS lignins. They noted that the cation of ILs does not significantly affect the process. On the other hand, the size of the depolymerized lignin is decreased when sulfates>lactate>acetate>chlorides>phosphates are used as the anionic counterpart of ILs. Although no mechanistic studies have been done, there are indications that different anions cause cleavage at different linkages. Organosolv lignin, which is deemed closest to natural lignin, breaks down to smallest molecules, fol-

http://dx.doi.org/10.5772/51161

lowed by ALS lignin and lastly alkali lignin. Organosolv lignin became more conjugated after treatment and thus, indicates a deconstruction mechanism consistent with that of an attack on strong nucleophile.

Jia's group [56] had demonstrated a method for the β-O-4 bond cleavage of two lignin model compound, guaiacylglycerol-β-guaiacyl ether (GG) and veratrylglycerol-β-guaiacyl ether (VG), in IL ([Bmim][Cl]) with metal chlorides. The simplified reaction scheme is shown in Figure 29. Iron(III) chloride, copper(II) chloride and aluminum(III) chloride are more effective in cleaving β-O-4 bond of GG. In this process, hydrochloric acid is formed. It was concluded that more β-O-4 bond cleavage of GG occurs when there is an increase in available water. On the other hand, only aluminum(III) chloride is effective towards β-O-4 bond cleaving of VG.

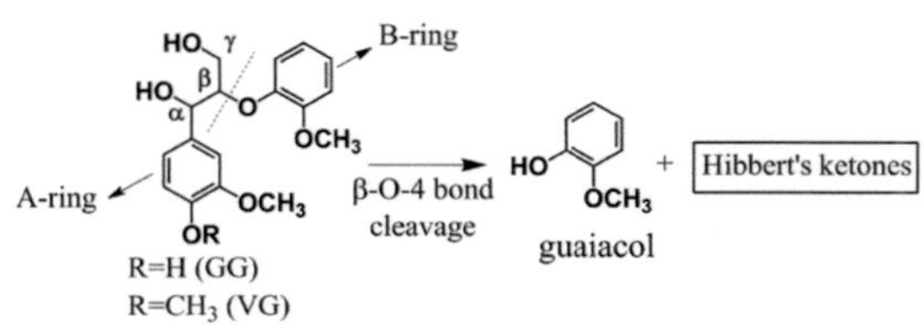

Figure 29. β-O-4 bond cleavage of GG and VG

Another group, Cox and co-workers [52], studied the Hammett acidity and anion effects of catalytic depolymerization of GG and VG in acidic imidazolium based IL. They had successfully used acidic ILs to hydrolyze the β-O-4 linkage. Although the acidic nature of ILs catalyzes the hydrolysis reaction, the Hammett acidity of ILs does not correlate with their reactivity toward GG and VG. The reactivity, however, is heavily dependent on the anion of the ILs. Hence, they postulated that the anion forms a hydrogen bond with GG and VG, making it more susceptible for hydrolysis (Figure 31).

4.2. Value-added small molecular compounds

After depolymerization, the products obtained can either be used directly or further processed to other value-added chemicals. Further processing in ILs have proven difficult even though they still serve as good solvents. For example, the colloidal catalyst, used by Bonilla [60], was inactivated for hydrogenation when IL, [Bmim][BF_4], was used. It was postulated that the IL itself caused the inactivation or the trace amount of chloride ions from the preparation of the IL is responsible.

Yan and co-workers [55] achieved the transformation of lignin-derived phenolic compounds to alkanes in ILs. The reaction system consists of metal (ruthenium, rhodium or platinum) nanoparticles and a SO_3-functionalised Brønsted acid IL, which forms a catalytic cycle, in a non-functionalized IL, [Bmim][BF_4] or [Bmim][TF_2N], as solvent. This system allows hydrogenation and dehydration processes to occur in tadem, shown in Figure 32. Metal nanoparticles and the SO_3-functionalised Brønsted acid IL were investigated while all others are kept constant. Rhodium, being the most active metal in benzene hydrogenation, is able to attain high alkane yields, even for branched phenols. It is also notable that methanol is produced when substrates with methoxy group are used. Various SO_3-functionalized Brønstedacid ILs were examined. It was noticed that, in general, the stronger the acidity of the IL, the higher the obtainable yields. The exception being 1-(4-sulfobutyl)-3-methylimidazolium hydrogen sulphate, that has a Hammett acidity of around 1.75, resulting in a yield of over 80%. It was predicted that the result is due to the dehydration power and poor nucleophilicity of the hydrogen sulphate anion.

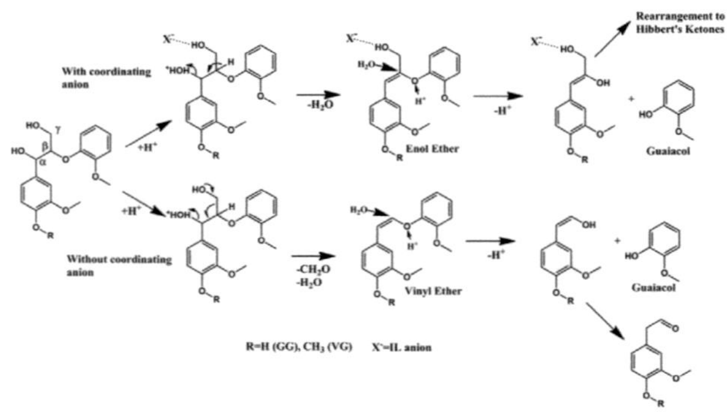

Figure 30. Mechanistic pathways of GG and VG degradation in acidic ILs.

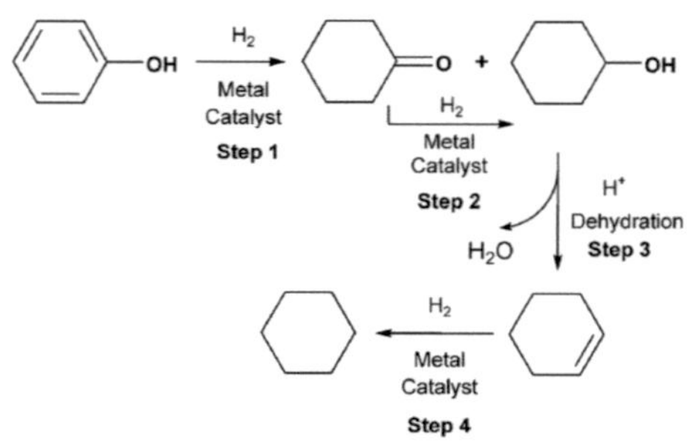

Figure 31. Reaction Scheme of cyclohexane from phenol

On the other hand, transition metal nanoparticle-based catalysts have been found to exhibit attractive catalytic activities relative to their corresponding bulk materials [61]. In our lab, ionic liquid stabilized metal nanoparticles have been found to be robust and recyclable catalyst composites for organic transformations [62]. Supported nano-Pd catalysts have been used as recyclable catalysts for alcohol oxidations [63-65]. Recently, the use of ionic liquid stabilized metal nanoparticles as catalysts have been employed in our group to conduct the oxidation conversion of lignin to value-added chemicals such as aromatic aldehydes (Figure 33) [66]. With a co-catalyst of pyridinium salt of iron bis(dicarbollide) (Figure 33), ionic liquid stabilized nano-Pd was found to be efficient for the oxidation of benzyl alcohol and lignin to produce aromatic aldehydes. The new Pd(0) catalytic system was found to be efficient, robust and recyclable with high product selectivity [66].

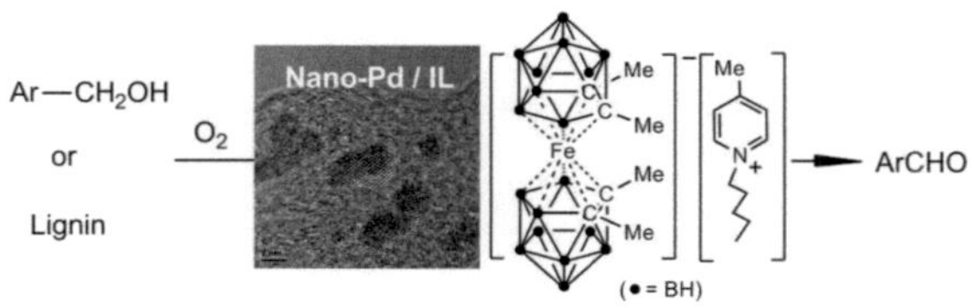

Figure 32. Application of IL stabilized nano-Pd in lignin oxidation reaction.

Also using rhodium nanoparticles, Denicourt-Nowicki and partners [67] successfully did hydrogenation on oxygen-containing arenes. Their rhodium nanoparticles are stabilized using polynitrogen ligands, like bipyridines. IL plays an interesting role in the reaction as it acts as both a solvent and a stabilizing agent. Hence, it has the ability to further stabilize the nanoparticles. One particularly interesting product obtained from the hydrogenation is the cyclohexanone from anisole (Figure 34). Cyclohexanone is used for synthesizing caprolactam and adipic acid, both of which are utilized in polyamides manufacturing. As shown in Figure 26, there are two plausible reaction pathways. The first pathway shows 1-methoxycyclohexene formed through the partial hydrogenation of anisole, then a hydrogenolysis into cyclohexen-1-ol, which results in the thermodynamically stable cyclohexanone. The second pathway demethylates anisole to phenol, followed by partial hydrogenation to obtain cyclohexen-1-ol, transforming to the thermodynamically stable cyclohexanone.

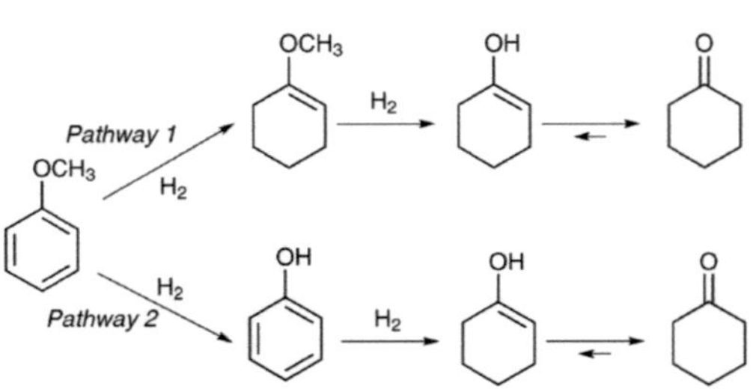

Figure 33. Potential pathways of anisole hydrogenation

Hydrogenation of acetophenone leads to several products and byproducts. This is primarily due to the competitive and consecutive hydrogenation of the carbonyl and aryl groups. Interesting selectivities were achieved when different ligand is used. Usage of 2,2'-bipyridine as a ligand produces an additional product, phenylethanol, which is not present when TPTZ (2,4,6-tris(2-pyridyl)-s-triazine) was used as the ligand. Hydrogenation of *o*- and *m*-cresols leads to a mixture of methylcyclohexanols and methylcyclohexanone as products. It was noticed that the conversion of *o*-cresol is much slower than *m*-cresol. This can be explained by a kinetically less reactive disubstituted enol intermediate and also an increase in steric hindrance. Another point of notice is that the major products obtained are *cis*-isomers, just like the products obtained from normal heterogeneous catalytic systems.

5. Conclusions and perspectives

Lignin, which is estimated to be around 30% of the biosphere carbon, is a naturally occurring aromatic cross-linked polymer. It is one of the three components of the lignocellulose. As lignin is able to dissolve in selective ionic liquids (ILs), it can be easily extracted from lignocellulose. There are mainly two methods of extraction and the main difference of the two methods is whether the whole lignocellulose is dissolved. The extracted lignin can then be chemically modified to form value-added chemicals. A surface modification of lignin is sufficient when macromolecular lignins are desired. Three types of surface modifications are discussed in this chapter which includes oxidation, dehydration and esterification. When smaller value-added molecules are desired, a depolymerization process should take place. Further transformation of the depolymerized lignin can also take place when the depolymerization step could not yield the desired product. As lignins, in any form, are able to dissolve in ILs, ILs become very good candidates as solvents. On top of having superior dissolution capabilities, ILs are considered as green solvents due to their non-volatility and low flammability. ILs are not only used as solvents but also play an important part in the catalytic cycles in some reactions discussed above. Hence, there is a surge in research in this area in the past five years.

As described above, the research on the applications of ILs in lignin chemistry is still at its budding stage, keen interest in developing this area will increase. As seen in this chapter, ILs are currently used in a small area of lignin chemistry. Since ILs are proven to be good solvents in lignin chemistry, they should be applied and used in other areas of lignin chemistry to create a greener and environmentally-friendly chemistry. However, ILs have a major shortcoming as they are much more expensive when compared to common and traditional solvents. Hence, recoverability of ILs should be explored and emphasized. Due to the π-π interaction between ILs and lignin, removal of lignin from ILs are proven a complex process and, therefore, requires multiple steps [54]. This makes the recycling and regeneration of ILs, particularly in extremely large volumes, as cost inefficient. Therefore, this must also be addressed in future investigations in this area [68].

http://dx.doi.org/10.5772/51161

Acknowledgements

Special thanks are expressed to all researchers whose works have contributed in one way or the other to this chapter. Special gratitude is also given to the editor of the book, Dr Jun-ichi Kadokawa, for giving us the opportunity to contribute to the book. We thank ICES in Singapore for support of this work. NSH thanks the support by grants from the National Science Foundation (CHE-0906179 and CHE-0840504), Alexander von Humboldt Foundation, and the NIU Inaugural Board of Trustees Professorship Award.

Author details

Zhu Yinghuai[1*], Karen Tang Yuanting[1] and Narayan S. Hosmane[2]

*Address all correspondence to: zhu_yinghuai@ices.a-star.edu.sg

1 Institute of Chemical and Engineering Sciences, No. 1 Pesek Road, Jurong Island, Singapore

2 Department of Chemistry and Chemical Biology, Northern Illinois University, DeKalb, Illinois 60115-2862, USA

References

[1] Boerjan, W., Ralph, J., & Baucher, M. (2003). Lignin Biosynthesis. *Annu. Rev. Plant Biol*, 54, 519-546.

[2] Notley, S. M., & Norgren, M. (2009). Lignin: Function Biomaterial with Potential in Surface Chemistry and Nanoscience. *Nanoscience and Technology of Renewable Biomaterials; Lucia, L. A., Rojas O. J., Eds.*, Wiley & Sons, United Kingdom, 173-206.

[3] Wang, J. S., Manley, R. S., & Feldman, F. (1992). Synthetic-polymer Lignin Copolymer and Blends. *Prog. Polym. Sci*, 28, 271-282.

[4] Chabannes, M., Ruel, K., Yoshinaga, A., Chabbert, B., Jauneau, A., Joseleau, J., & Boudet, A. (2001). Situ analysis of lignins in transgenic tobacco reveals a differential impact of individual transformations on the spatial patterns of lignin deposition at the cellular and subcellular levels. *Plant J.*

[5] Davin, L. B., Jourdes, M., Patten, A. M., Kim, K. W., Vassao, D. G., & Lewis, N. G. (2008). Dissection of Lignin Macromolecular Configuration and Assembly: Comparison to Related Biochemical Processes. *Allyl/Propenyl Phenol and Lignan Biosynthesis, Nat. Prod. Rep.*, 25, 1015-1090.

[6] Dimmel, D. (2010). Overview. *Lignin and Lignans: Advances in Chemistry; Heither, C., Dimmel, D. R., Schmidt, J. A., Eds.*, CRC Press, Boca Raton, 1-10.

[7] Clare, B., Sirwardanam, A., & Mac, Farlane. D. R. (2009). Synthesis, Purification and Characterization of Ionic Liquids. *Top. Curr. Chem.*, 209, 1-40.

[8] Olivier-Bourbigou, H., Magna, L., & Morvan, D. (2010). Ionic liquids and catalysis: Recent progress from knowledge to applications. *Appl. Catal. A,*, 373, 1-56.

[9] Lee, S. (2006). Functionalized imidazolium salts for task-specific ionic liquids and their applications. *Chem. Commun.*, 1049-1063.

[10] Winkel, A., Reddy, P. V. G., & Wilhelm, R. (2008). Recent Advances in the Synthesis and Application of Chiral Ionic Liquids. *Synthesis*, 7, 999-1016.

[11] Baudequinm, C., Baudoux, J., Levillian, J., Cahard, D., Gaumont, A., & Plaquevent, J. Ionic Liquids and chirality: opportunities and challenges. Tetrahedron: Asymmetry (2003). (14), 3081-1093.

[12] Phan, L., Andreatta, J. R., Horvey, L. K., Edie, C. F., Luco, A., Mirchandani, A., Darensbourg, D. J., & Jessop, P. G. (2008). Switchable-Polarity Solvents Prepared with a Single Liquid Component. *J. Org. Chem.*, 73, 127-132.

[13] Lin, I. J. B., & Vasam, C. S. (2005). Metal-containing ionic liquids and ionic liquid crystals based on imidazolium moiety. *J. Organomet. Chem.* [690], 3498-3512.

[14] Wegener, G. (1992). Pulping innovations in Germany. *Ind. Crops Prod.* [1], 113-117.

[15] Mosier, N., Wyman, C., Dale, B., Elander, R., Lee, Y. Y., Holtzapple, M., & Ladisch, M. (2005). Features of promising technologies for pretreatment of lignocellulosic biomass. *Bioresour. Technol.* [96], 673-686.

[16] Grous, W. R., Converse, A. O., & Grethlein, H. E. (1986). Effect of steam explosion pretreatment on pore size and enzymatic hydrolysis of poplar. *Enzyme Microb. Technol.*, 8, 274-280.

[17] Mes-Hartree, M., Dale, B. E., & Craig, W. K. (1988). Comparison of steam and ammonia pretreatment for enzymatic hydrolysis of cellulose. Appl. Microbiol. Biotechnol. , 29, 462-468.

[18] Chum, H. L., Johnson, D. K., Black, S., Baker, J., Grohmann, K., Sarkanen, K. V., Wallace, K., & Schroeder, H. A. (1988). Organosolv pretreatment for enzymatic hydrolysis of poplars: I. Enzyme hydrolysis of cellulosic residues. *Biotechnol. Bioeng.*, 31, 643-649.

[19] Zhang, Y. H. P., Ding, S. Y., Mielenz, J. R., Cui, J. B., Elander, R. T., Laser, M., Himmel, M. E., Mc Millan, J. R., & Lynd, L. R. (2007). Fractionating recalcitrant lignocellulose at modest reaction conditions. Biotechnol. *Bioeng.*, 97, 214-223.

[20] Gierer, J., & Norén, I. (1982). Oxidative Pretreatment of Pine Wood to Facilitate Delignification during Kraft Pulping. *Holzforschung*, 36, 123-130.

[21] Vyver, S. V., Thomas, J., Geboers, J., Keyzer, S., Smet, M., Dehaen, W., Jacobs, P. A., & Sels, B. F. (2011). Catalytic production of levulinic acid from cellulose and other biomass-derived carbohydrates with sulfonatedhyperbranched poly(aryleneoxindole)s. *Energy Environ. Sci.* [4], 3601-3610.

[22] Román-Leshkov, Y., Chheda, J. N., & Dumesic, J. A. (2006). Phase Modifiers Promote Efficient Production of Hydroxymethylfurfural from Fructose. *Science* [312], 1933-1937.

[23] Swatloski, R. P., Spear, S. K., Holbrey, J. D., & Rogers, R. D. (2002). Dissolution of cellulose with ionic liquids. *J. Am. Chem. Soc.* [124], 4974-4975.

[24] Liu, Q., Janssen, M. H. A., Van Rantwijk, F., & Sheldon, R. A. (2005). Room-temperature ionic liquids that dissolve carbohydrates in high concentrations. *Green Chem.* [7], 39-42.

[25] Moulthrop, J. S., Swatloski, R. P., Moyna, G., & Rogers, R. D. High-resolution 13C NMR studies of cellulose and cellulose oligomers in ionic liquid solutions. *Chem. Commun.*, 1557-1559.

[26] Fort, D. A., Swatloski, R. P., Moyna, P., Rogers, R. D., & Moyna, G. (2006). Use of ionic liquids in the study of fruit ripening by high-resolution 13C NMR spectroscopy: "green" solvents meet green bananas. *Chem. Commun.*, 714-716.

[27] Zhu, S., Wu, Y., Chen, Q., Yu, Z., Wang, C., Jin, S., Ding, Y., & Wu, G. (2006). Dissolution of cellulose with ionic liquids and its application: A mini-review. *Green Chem.* [8], 325-327.

[28] Fort, D.A., Remsing, R.C., Swatloski, R.P., Moyna, P., Moyna, G., & Rogers, R.D. (2007). Can ionic liquids dissolve wood? *Processing and analysis of lignocellulosic materials with 1-n-butyl-3-methylimidazolium chloride. Green Chem.*, 9, 63-69.

[29] Pu, Y., Jiang, N., & Ragauskas, A. J. (2007). *Ionic Liquid as a Green Solvent for Lignin. J. Wood Chem.Technol.*, 27, 23-33.

[30] Lee, S. H., Doherty, T. V., Linhardt, R. J., & Dordick, J. S. (2009). Ionic liquid-mediated selective extraction of lignin from wood leading to enhanced enzymatic cellulose hydrolysis. *Biotechnol. Bioeng.*, 102, 1368-1376.

[31] Tan, S. S. Y., Mac, Farlane. D. R., Upfal, J., Edye, L. A., Doherty, W. O. S., Patti, A. F., Pringle, J. M., & Scott, J. L. (2009). Extraction of lignin from lignocellulose at atmospheric pressure using alkylbenzenesulfonate ionic liquid. *Green Chem.*, 11, 339-345.

[32] Fu, D., Mazza, G., & Tamaki, Y. (2010). Lignin Extraction from Straw by Ionic Liquids and Enzymatic Hydrolysis of the Cellulosic Residues. *J. Agric. Food Chem.* [58], 2915-2922.

[33] Pinkert, A., Goeke, D. F., Marsh, K. N., & Pang, S. (2011). Extracting wood lignin without dissolving or degrading cellulose: investigations on the use of food additives-derived ionic liquids. *Green Chem.* [13], 3124-3136.

[34] Kim, J., Shin, E., Eom, I., Won, K., Kim, Y. H., et al. (2011). Structural features of lignin macromolecules extracted with ionic liquid from poplar wood. *Bioresour. Technol.* [102], 9020-9025.

[35] Lateef, H., Grimes, S., Kewcharoenwong, P., & Feinberg, B. (2009). Separation and recovery of cellulose and lignin using ionic liquids: a process for recovery from paper-based waste. *J. Chem. Technol. Biotechnol.* [84], 1818-1827.

[36] Muhammad, N., Man, Z., Bustam, M. A., Mutalib, M. I. A., Wilfred, C. D., & Rafiq, S. (2001). Dissolution and Delignification of Bamboo Biomass Using Amino Acid-Based Ionic Liquid. *Appl. Biochem. Biotechnol.* [165], 998-1009.

[37] Lan, W., Liu, C., & Sun, R. (2011). Fractionation of Bagasse into Cellulose, Hemicelluloses, and Lignin with Ionic Liquid Treatment Followed by Alkaline Extraction. *J. Agric. Food Chem.* [59], 8691-8701.

[38] Sievers, C., Valenzula-Olarte, M., Marzialetti, T., Musin, I., Agrawal, P. K., & Jones, C. W. (2009). Ionic-Liquid-Phase Hydrolysis of Pine Wood. *Ind. Eng. Chem. Res.*, 48, 1277-1286.

[39] Sun, N., Rahman, M., Qin, Y., Maxim, M. L., Rodríguez, H., & Rogers, R. D. (2009). Complete dissolution and partial delignification of wood in the ionic liquid 1-ethyl-3-methylimidazolium acetate. *Green Chem.* [11], 646-655.

[40] Menon, V., & Rao, M. (2012). Trends in bioconversion of lignocellulose: Biofuels, platform chemicals &biorefinery concept. *Prog. Energy Combust. Sci.*, 1-29.

[41] Wen, J., Sun, Y., Meng, L., Yuan, T., Xu, F., & Sun, R. (2011). Homogeneous lauroylation of ball-milled bamboo in ionic liquid for bio-based composites production Part I; Modification and characterization. *Ind. Crops Prod.* [34], 1491-1501.

[42] Zakeski, J., Jongerius, A. L., & Weckhuysen, B. M. Transition metal catalyzed oxidation of Alcell lignin, soda lignin, and lignin model compounds in ionic liquid. Green Chem. (2010). , 2010(12), 1225-1236.

[43] Zakeski, J., Bruijninex, P. C. A., & Weckhuysen, B. M. (2011). In situ spectroscopic investigation of the cobalt-catalyzed oxidation of lignin model compounds in ionic liquids. *Green Chem.* [13], 671-680.

[44] Kumar, A., Jain, N., & Chauhan, S. M. S. (2007). Biomimetic Oxidation of Veratryl Alcohol with H_2O_2 Catalyzed by Iron(III) Porphyrins and Horseradish Peroxidase in Ionic Liquid. *Synlett* [3], 411-414.

[45] Kubo, S., Hashida, K., Yamada, T., Hishiyama, S., Magara, K., et al. (2008). A Characteristic Reaction of Lignin in Ionic Liquids; Glycelol Type Enol-Ether as the Primary Decomposition Product of β-O-4 Model Compound. *J. Wood Chem. Technol.*, 28, 84-96.

[46] Yuan, T., Sun, S., Xu, F., & Sun, R. (2011). Homogeneous butyrylation and lauroylation of poplar wood in the ionic liquid 1-butyl-3methylimidazolium chloride. *Bioresour. Technol.* [102], 4590-4593.

[47] Xie, H., King, A., Kilpelainen, I., Granstrom, M., & Argyropoulos, D. S. (2007). Thorough Chemical Modification of Wood-Based Lignocellulosic Materials in Ionic Liquids. *Biomacromolecules* [8], 3740-3748.

[48] Cerrutti, B.M., de Souza, C.S., Castellan, A., Ruggiero, R., & Frollini, E. (2012). Carboxymethyl lignin as stabilizing agent in aqueous ceramic suspensions. *Ind. Crops Prod.*, 36, 108-115.

[49] Zhu, Y., Yang, S.H., & Tham, Y.H. Polymer composites cored with lignin. *3rd EuCheMS Chemistry Congress, 29 August-2 September, Numberg, Germany.*

[50] Kleinert, M., & Barth, T. (2008). Phenol from Lignin. *Chem. Eng. Technol.*, 31(5), 736-745.

[51] Clark, J. H., Deswarte, F. E. I., & Farmer, T. J. (2009). The integration of green chemistry. *Biofuels, Bioprod. Biorefin.* [3], 72-90.

[52] Cox, B. J., Jia, S., Zhang, Z. C., & Ekerdt, J. G. (2011). Catalytic degradation of lignin model compounds in acidic imidazolium based ionic liquids: Hammett acidity and anion effects. *Polym. Degrad. Stab.* [96], 426-431.

[53] Holladay, J. E., White, J. F., Bozell, J. J., & Johnson, D. Top Value-Added Chemicals from Biomass- Volume II-Results of Screening for Potential Candidates from Biorefinery Lignin. Pacific Northwest National Laboratory Richland, WA. (2007).

[54] Zakzeski, J., Bruijnincx, P. C., Jongerius, A. L., & Weckhuysen, B. M. (2010). The catalytic valorization of lignin for the productionof renewable chemicals. *Chem. Rev.*, 3552-3599.

[55] Yan, N., Yuan, Y., Dykeman, R., Kou, Y., & Dyson, P. J. (2010). Hydrodeoxygenation of Lignin-Derived Pehnols into Alkanes by Using Nanoparticles Catalysts Combined with Brønsted Acidic Ionic Liquids. *Angew. Chem. Int. Ed.* [49], 5549-5553.

[56] Jia, S., Cox, B. J., Guo, X., Zhang, Z. C., & Ekerdt, J. G. (2011). Hydrolytic cleavage of β-O-4 ether bonds of lignin model compounds in an ionic liquid with metal chlorides. *Ind. Eng. Chem. Res.* [50], 849-855.

[57] Binder, J. B., Gray, M. J., White, J. F., Zhang, Z. C., & Halloday, J. E. (2009). Reactions of lignin model compounds in ionic liquids. *BiomassBioenergy* [33], 1122-1130.

[58] Reichert, E., Wintringer, R., Volmer, D. A., & Hempelmann, R. (2012). Electro-catalytic oxidative cleavage of lignin in a protic ionic liquid. *Phys. Chem. Chem. Phys.* [14], 5214-5221.

[59] George, A., Tran, K., Morgan, T. J., Benke, P. I., Berrueco, C., et al. (2011). The effect of ionic cation and anion combinations on the macromolecular structure of lignins. *Green Chem.* [13], 3375-3385.

[60] Bonilla, R. J., James, B. R., & Jessop, P. G. (2000). Colloid-catalysedarene hydrogenation in aqueous/supercritical fluid biphasic media. *Chem. Commun.*, 941-942.

[61] Johnson, B. F. G. (1999). *Coord. Chem. Rev.*, 1269-1285.

[62] Zhu, Y., Widjaja, E., Shirley, L. P. S., Wang, Z., Carpenter, K., Maguire, J. A., Hosmane, N. S., & Hawthorne, M. F. (2007). *J. Am. Chem. Soc.*, 129, 6507-6512.

[63] Karimi, B., Abedi, S., Clark, J. H., & Budarin, V. (2006). *Angew. Chem. Int. Ed.*, 45, 4776-4779.

[64] Parlett, C. M. A., Bruce, D. W., Hondow, N. S., Lee, A. F., & Wilson, K. (2011). *ACS Catal.*, 1, 636-640.

[65] Chen, Y., Zheng, H., Guo, Z., Zhou, C., Wang, C., Borgna, A., & Yang, J. (2011). *Catal.*, 283, 34-44.

[66] Zhu, Y., Li, C., Meriska, S., Ng, H. M., Algin, O. B., Maguire, J. A., & Hosmane, N. S. (2012). *Chem. OPEN*, 1, 67-70.

[67] Denicourt-Nowicki, A., Léger, B., & Roucoux, A. N. (2011). N-Donor ligands based on bipyridine and ionic liquids: an efficient partnership to stabilize rhodium colloids. Focus on oxygen-containing compounds hydrogenation. *Phys. Chem. Chem. Phys.* [13], 13510-13517.

[68] Mora-Pale, M., Meli, L., Doherty, T. V., Linhardt, R. J., & Dordick, J. S. (2011). Room Temperature Ionic Liquids as Emerging Solvents for the Pretreatment of Lignocellulosic Biomass. *Biotechnol. Bioeng.* [108], 1229-1245.

Section 3

Organic Reactions and Biological Applications

Chapter 14

Ionic Liquids: "Green" Solvent for Catalytic Oxidations with Hydrogen Peroxide

Liangfang Zhu and Changwei Hu

Additional information is available at the end of the chapter

http://dx.doi.org/10.5772/51936

1. Introduction

Over the past decade, ionic liquids (ILs) have received great deal of attention as possible "green" replacement for volatile organic solvent mainly due to their nonmeasurable vapor pressure and good dissolubility for other salts. [1-5] Reaction types successfully performed in ILs include Diels–Alder, [6] Friedel–Crafts, [7] olefin hydrogenation, [8] hydroformylation, [9] , [10] oligomerization, [11] and Heck and Suzuki coupling reactions. [12] , [13] In addition to solvent, ILs may have multiple functions in catalytic reactions. They may act as catalyst, co-catalyst, support, or ligands for the catalytic process. [14] In particular, some "unexpected" effects have been observed in affecting the catalytic reaction pathway. For example, the cations/anions in ILs may be involved in the formation of the active species changing the reaction mechanism. [15] - [20] Understanding the functions of ILs in the catalytic reaction is of critical importance for deliberately modifying existed reaction system and exploiting new types of synthetic route by using this "green" solvent.

Catalytic oxidation is a class of commercially important reaction. As an environmentally benign oxidant, hydrogen peroxide (H_2O_2) has been used for several catalytic oxidation. So far, significant improvements on the catalytic performance, in terms of yield and selectivity, have been observed using ILs as the solvent for the H_2O_2 oxidation reaction. [21] - [23] Actually, in many cases, ILs are active participant because the formation of radical species, stabilization of the charged reactive intermediate, and immobilization of the actual catalyst can be strongly affected by the presence of an ionic environment. In comparison with traditional organic solvent, the use of ILs in catalytic oxidation has been regarded as a new means for recycling the catalyst and enhancing the yield and selectivity of the product. Though a great number of catalytic oxidation have been performed in ILs, there are still rare examples which demonstrate how the ILs affect the reaction pathway and the reactivity.

This chapter aims at summarizing the examples that concern the H_2O_2 oxidation reactions in ILs, in particular the benzene hydroxylation, alcohol oxidation, and olefin oxidation. The effects of ILs on the reaction pathway and the selectivity are discussed, drawing to the conclusion that ILs are offering unique properties as solvent by recycling the catalyst and enhancing the yield and selectivity of the product. What should be pointed out is that the examples are limited as far as possible to those that inform the readers' understanding of the role of ILs in the H_2O_2 oxidation reaction. We apologize that some fine work is not covered, and we hope to stimulate more discussions in the future.

2. Benzene hydroxylation

Direct hydroxylation of benzene to phenol with H_2O_2 has been extensively investigated owing to the reduced reaction steps and environmentally benign byproduct of water when comparing to the commercial cumene process for phenol production. One of the fundamental targets in this intriguing investigation is to enhance the utilization efficiency of H_2O_2 and the selectivity of phenol. The low efficiency of H_2O_2 always derives from the fast decomposition of H_2O_2, and the low product selectivity is mainly originated from the over-oxidation of phenol. Studies have shown that solvents used in the hydroxylation play an important role on enhancing both the H_2O_2 efficiency and the product selectivity. For example, water was the solvent in the traditional Fenton's reagent (Fe^{II}-H_2O_2) catalyzed hydroxylation, [24] whereas, the decomposition of H_2O_2 was very fast. [25] The selectivity to phenol was rather poor in the aqueous solution since phenol is more reactive toward oxidation than benzene itself. Acetonitrile and acetic acid were then used as the solvents for most of the catalyzed hydroxylation of benzene, [26] - [28] and a biphasic water-acetonitrile (1:1) system was developed to decrease the over-oxidation of phenol. [29] In Bianchi *et al.*'s work, [30] sulfolane was believed to form complexes with phenolic compounds inducing increased selectivity to phenol.

In addition to organic solvents, Peng *et al.* [31] introduced a biphasic aqueous–imidazolium-based IL system for benzene hydroxylation in the presence of ferric tri(dodecanesulfonate) catalyst (Figure 1). In this aqueous- [$C_n$mim] [X] (n=4, 8, 10; X=PF_6, BF_4) IL biphasic system, both the catalyst and benzene were dissolved in the IL, whereas, H_2O_2 was mainly dissolved in aqueous phase. The produced phenol was extracted into water phase, minimizing the over-oxidation. A highest yield of 54% and selectivity of 100% to phenol was obtained in the aqueous-IL biphasic system. However, as H_2O_2 was existed in the IL-free aqueous phase, the active oxidizing agent were deemed to be radical species. Therefore, this biphasic system was not operative for the hydroxylation of toluene: only 1% of toluene was converted to benzaldehyde although its selectivity reached 100%.

In our work, [25] , [32] a benzene–triethylammonium acetate ([Et_3NH] [CH_3COO]) IL biphasic system was constructed for the benzene hydroxylation (Figure 2). The Fenton-like reagent (Fe^{III}–H_2O_2) existed in the IL phase and most of the phenol was extracted to the benzene layer. The [Et_3NH] [CH_3COO] IL was found to be stable in the water- and oxygen-rich environment. Benzene acted as both the substrate and the extractant in the hydroxylation reaction. In comparison with the aqueous-IL biphasic system, the continuous extraction of phenol by benzene from [Et_3NH] [CH_3COO] IL protected phenol from further oxidation by directly avoiding the contact of phenol with the catalyst and oxidant. As a result, moderate yield (20%, based on benzene converted, excluding evaporated) and high selectivity (> 99.5%) of phenol were obtained in the IL-benzene biphasic system.

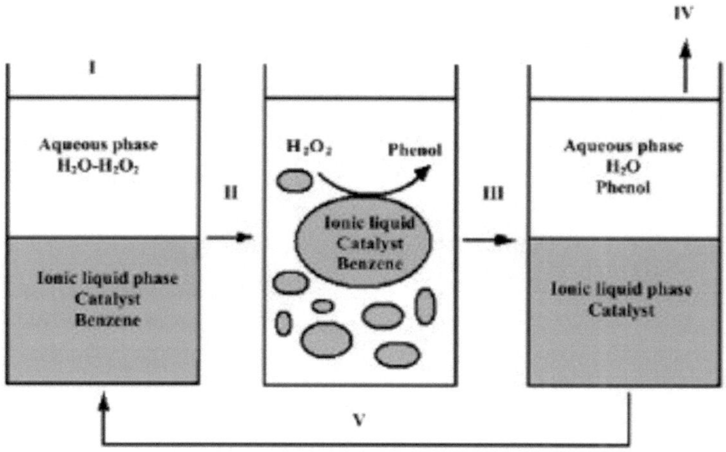

Figure 1. A schematic representation of the aqueous–IL biphasiccatalytic reaction system for benzene hydroxylation to phenol with H_2O_2.Step I charging; Step II reaction; Step III still; Step IV recovery of phenol*via* extraction; Step V the IL and catalyst are reused for anotherreaction cycle. IL = [$C_n$mim] [X] (n = 4,8,10; X = PF_6, BF_4). [31] Reprinted with permission from ref. 31. Copyright ©2003, Royal Society of Chemistry.

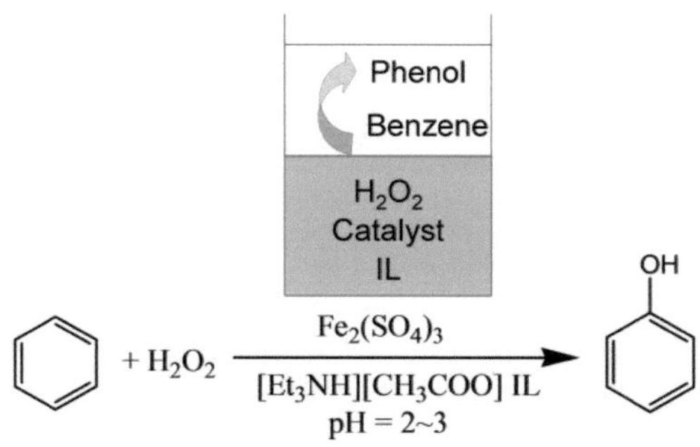

Figure 2. Schematic illustration of the benzene- [Et_3NH] [CH_3COO] IL biphasic system for benzene hydroxylation to phenol with H_2O_2. [25]

Moreover, the [Et_3NH] [CH_3COO] IL exhibited retardation performance for the decomposition of H_2O_2 and protection performance for the over-oxidation of phenol. From a molecular aspect, the CH_3COO- anions of [Et_3NH] [CH_3COO] IL were found to be coordinated with the Fe ions, forming Fe complexes, *by virtue of* the solution of Fenton-like reagent in the IL phase (Figure 2). [25] Such coordination "anchored" the catalyst in the IL, affecting the subsequent activation of benzene and H_2O_2. A higher electrophilicity of the Fe-complexes was favorable for the interaction of H_2O_2 with the Fe center, which may be the origin of the retardation role of IL on the decomposition of H_2O_2. High-valent Fe^{IV}-oxo species, formed from the O-O bond homolysis of a Fe^{III}-OOH intermediate, was found to be the main active oxidizing species in the ionic environment rather than the widely accepted oxidizing species of hydroxyl radical ($OH^{\bullet}$) in an aqueous Fenton system. The mechanism for hydroxylation of benzene in the [Et_3NH] [CH_3COO] IL was thus different from that occurred in aqueous solution (Figure 3) [32] : the activation of benzene was mainly achieved *via* the electrophilic attack by the Fe^{IV}-oxo species, rather than *via* hydrogen abstraction to form phenyl radical. Over-oxidation of phenol through H-abstraction from O-H of phenol by the Fe^{IV}-oxo species was partly prohibited by the hydrogen-bond interaction between phenol and the CH_3COO- anion. The electrophilic character of the Fe^{IV}-oxo species made the [Et_3NH] [CH_3COO] IL suitable for hydroxylation of other alkyl-benzenes. As an instance, the oxidation of toluene in [Et_3NH] [CH_3COO] IL resulted in the selective activation of benzene ring with the selectivity to methylphenols of about 62%. [33]

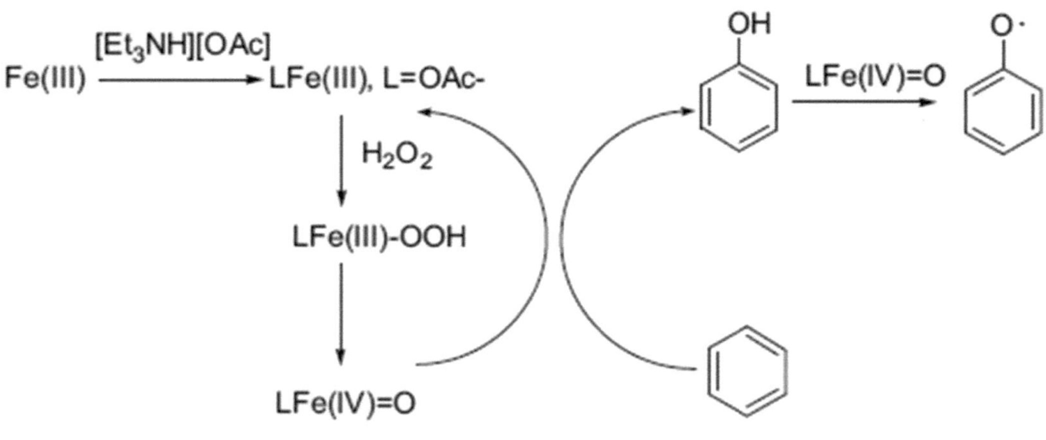

Figure 3. Mechanism for hydroxylation of benzene with H_2O_2 in [Et_3NH] [OAc] system. [32] Reprinted with permission from ref. 32. Copyright © 2011,Elsevier Science Ltd.

In the benzene hydroxylation, both hydrophobic and hydrophilic ILs have shown the feasibility of acting as solvent for enhancing both the yield and selectivity of phenol when compared with that in aqueous solution (*vide supra*). The nature of the active oxidizing species is mostly depended on the dispersion of both H_2O_2 and the catalyst in the ILs because the generation of active species is thus influenced by the ionic environment. Furthermore, the combination of ILs with a second solvent (co-solvent), either traditional organic solvent or water, may offer opportunity for breaking the thermodynamic equilibrium by mass-transfer of the product from phase to phase. Most importantly, the ILs-solvent biphasic system provides opportunity for the stabilization of charged reactive intermediate and the protection of unstable product from over-oxidation. Therefore, the ILs-co-solvent biphasic system can be tentatively developed as solvent for a wide range of H_2O_2 oxidation reaction to meet specific requirements, for example, yield, selectivity, or solubility, *etc.*

3. Alcohol oxidation

The partial oxidation of alcohols to aldehydes, or secondary alcohols to corresponding ketones is a fundamental synthetic transformation in organic chemistry and is industrially important. [34] - [36] However, this transformation always suffers from drawbacks such as poor conversion and selectivity due to over-oxidation. Stable-free nitroxyl radicals such as TEMPO(2,2,6,6-tetramethylpiperidine-1-oxyl) has recently emerged as a catalyst or co-catalyst to promote the formation of the catalytically active species for selective oxidation of alcohols to aldehydes or ketones where volatile organic solvents such as CH_2Cl_2 are frequently used. [37] - [44] However, the recycling of the quite expensive TEMPO is problematic due to the homogeneous character of the classic organic media. Replacement of organic solvents with ILs or immobilization of TEMPO on ILs provide alternative strategies for solving the above-mentioned problems. On the one hand, TEMPO can be anchored on ILs allowing the recycling of the catalyst; on the other hand, ILs provide advantages for increasing the selectivity of the product by promoting oxidation of alcohols to aldehydes but suppressing over-oxidation of these aldehydes to acids.

In Wang *et al.*'s work, [45] oxidation of alcohols with H_2O_2 was performed in pyridiniumtetrafluoroborate([Bpy] BF_4) IL. The catalyst (vanadate) and co-catalyst (TEMPO and sulfonic acid) were both grafted on [Bpy] BF_4 IL. The functionalized IL showed good dissolubility in the [Bpy] BF_4 solvent. The as-formed homogeneous mixture of N-n-dodecyl pyridinium vanadate, N-(propyl-1-sulfonic acid) pyridiniumtetrafluoroborate, and 4-(propanoate-TEMPO) pyridiniumtetrafluoroborate exhibited good activity and recyclability for alcohols oxidation.Jiang *et al.* [46] used the acetamido-modified TEMPO as catalyst for the selective oxidation of benzylic alcohols to aldehydes in the 1-n-butyl-3-methylimidazolium hexafluorophosphate ([bmim] [PF_6]) IL (Figure 4). The [bmim] [PF_6] ILwas immiscible with H_2O_2, favoring the partial oxidation of alcohols to aldehydes but inhibiting the over-oxidation of aldehydes to acids by reducing the contact of the product with the oxidant. The miscibility of acetamido-TEMPO in [bmim] [PF_6] IL ensured good catalytic activity and efficient recycling of the catalyst. In comparison to the common organic solvents (ethyl acetate or chloroform), the yield of aldehydes was enhanced by about three times.

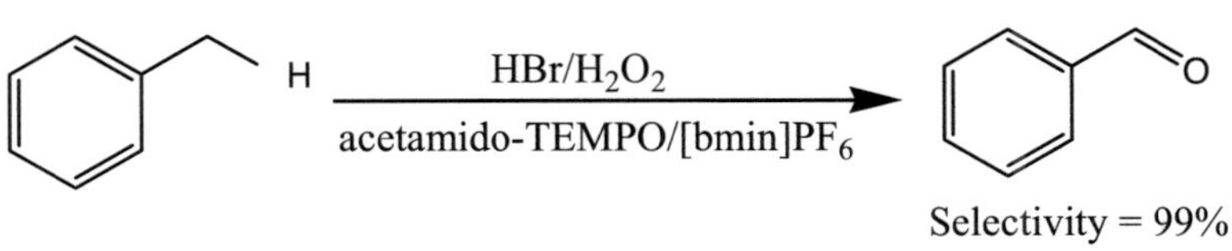

Figure 4. Highly selective oxidation of benzyl alcohol to benzaldehyde with acetamido-TEMPO/HBr//H_2O_2 in [bmim] [PF_6] IL. [46]

The strategy of anchoring TEMPO on ILs has also been applied in other catalytic oxidation reaction. In Fall *et al.*'s work, [39] TEMPO was supported on ILs through click chemistry reaction (Figure 5a). The IL-supported catalyst exhibited good solubility in [HMIM] [BF_4] IL and high activity for alcohol oxidation using bis(acetoxy)iodobenzene (BAIB) as the terminal oxidant. The catalyst can be recycled together with the IL without loss of the efficiency for several cycles. Karimi *et al.* [47] grafted TEMPO on SBA-15 solid support and then synthesized IL@SBA-15-TEMPO catalyst by physically confining 1-methyl-3-butylimidazolium ([Bmim] Br) IL within the mesopores of the TEMPO-modified SBA-15 (Figure 5b). The catalyst showed high activity, improved selectivity, and good recyclability for the oxidation of alcohols to aldehydesand ketones with *t*-butylnitrite (TBN) as oxidant in AcOH. Although the catalytic performance of the TEMPO-ILs and ILs@support-TEMPO catalysts in the H_2O_2 oxidation reaction is not investigated, we may expect the strategy of fixing TEMPO onto ILs or solid support applicable in recycling the expensive catalyst in a wide range of catalytic reactions.

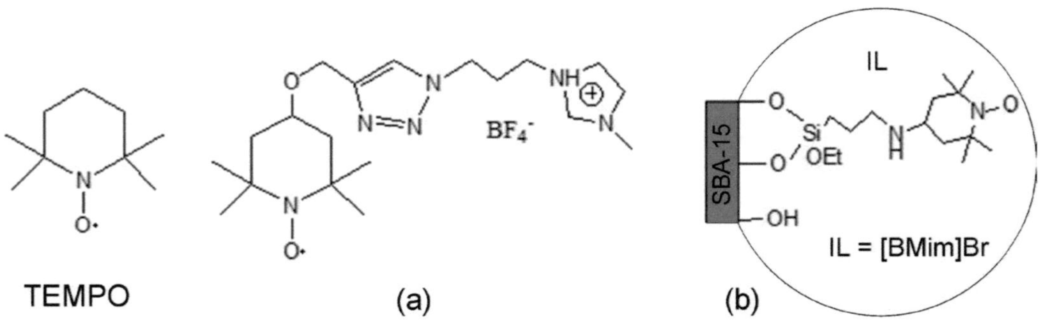

Figure 5. Strategies for immobilizing TEMPO on ILs. (a) IL supported TEMPO, [39] and (b) IL@SBA-15-TEMPO. [47]

In addition to TEMPO, the combination of various kinds of catalysts (*i.e.*, inorganic salts, transition metal complexes, and oxides) with ILs has shown prospect as effective catalytic system for the H_2O_2 oxidation reaction. Chhikara *et al.* [48] synthesized imidazolium-based phosphotungstate catalyst (Figure 6, 1) by grafting phosphotungstate onto imidazolium-based IL, which showed good catalytic performance in the homogeneous oxidation of secondary alcohols in the 1-butyl-3-methylimidazolium tetrafluoroborate ([bmim] [BF_4]) IL (Figure 6a). All of the secondary alcohols were converted to corresponding ketones in good to excellent yields. Oxidation of primary alcohol, *i.e.* benzyl alcohol, produced benzaldehydein good yield (78%), or benzoic acid in high yield (96%) after increasing the H_2O_2 amount (Figure 6b). The IL and catalyst after the extraction of the products could be reused for further catalytic oxidation.

Bianchini *et al.* [49] described the oxidation of some secondary alcohols to their ketones with H_2O_2 catalyzed by methyltrioxorhenium(VII)(MTO) or polymer supported-MTO in [bmim] [PF_6] or 1-ethyl-3-methylimidazolium bis-triflic amide ([emim] [Tf_2N]) ILs. The supported catalyst was dispersed in the IL layer, which allowed the recycling of the catalyst and solvent system that could not be realized in a wide range of organic solvent. Moreover, in comparison with ethanol or acetic acid, the activity of the catalyst in the ILs was obviously improved.

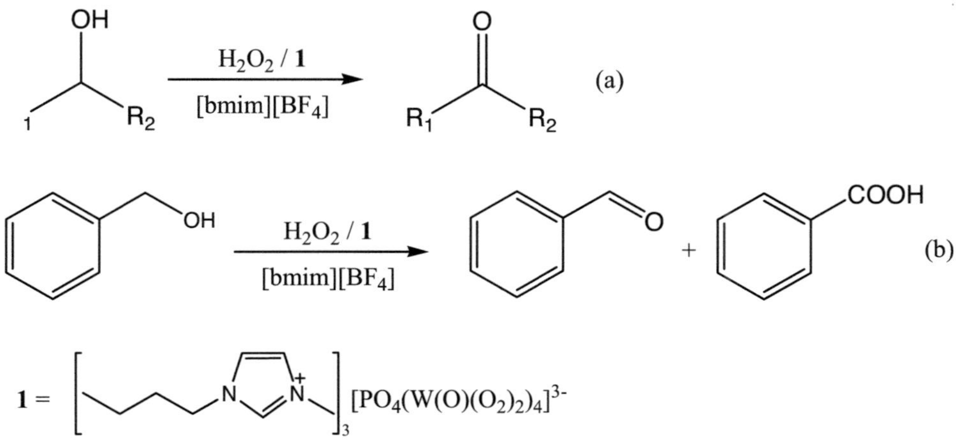

Figure 6. Oxidation of (a) secondary alcohols to ketones, and (b) benzyl alcohol to benzaldehyde and benzoic acid with imidazolium-based phosphotungstate (**1**) and H_2O_2 in [bmim] [BF_4] IL. [48]

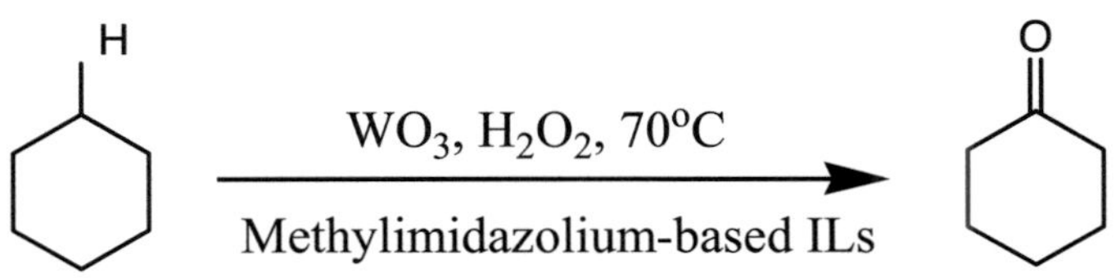

Figure 7. Selective oxidation of cyclohexanol to cyclohexanone with H_2O_2 in hydrophobic methylimidazolium-based ILs. [50]

Chen *et al.* [50] applied several hydrophobic methylimidazolium-based ILs in the oxidation of cyclohexanol to cyclohexanone with H_2O_2 using WO_3 as the catalyst (Figure 7). The oxidation of cyclohexanol in the absence of ILs produced cyclohexanone with a moderate yield (42%) or adipic acid at high cyclohexanol conversion. [51] In the biphasic cyclohexanol-ILs system, however, the ILs were found to effectively intensify cyclohexanol oxidation and resulted in 100% conversion of cyclohexanol with 100% selectivity to cyclohexanone. The high production of cyclohexanone can be explained by the fact that the oxidation of cyclohexanol occurred in aqueous phase contained H_2O_2 and WO_3, whereas, the produced cyclohexanone was abstracted into the organic phase, minimizing the further oxidation of the product. Among the three kinds of methylimidazolium-based ILs (1-hydroxyethyl-3-methylimidazoliumchloride ([HOemim] Cl); 1-hexyl-3-methylimidazoliumchloride ([Hmim] Cl); and 1-octyl-3-methylimidazolium chloride ([Omim] Cl,)) investigated, the [Omim] Cl IL exhibited the best solvent performance for enhancing the conversion of cyclohexanol when comparing with the traditional organic solvents (methanol, *n*-propanol, or acetone). Furthermore, a higher concentration of the [Omim] ClIL favored the transformation of cyclohexanol, giving evidence that the IL may be involved in the stabilization of the reaction intermediates in the catalytic process. Detailed investigation revealed that a longer alkyl chain of the IL increased the interaction between the solvent and the hydrophobic substrate, and a larger polarity of the IL increased the strength of coulombic forces arising in the solvation process. Both of these two factors improved the conversion of cyclohexanol considerably.

In this part, the anchoring of the actual catalyst on ILs offers opportunity for immobilizing catalyst with the solvent, allowing the recycling of the catalyst, especially some expensive reagent. This "anchoring" can be either chemical coupling or physical confinement. Chemical coupling requires special tailoring or functionalization of the ILs. The functionalized ILs show prospect as both catalyst and solvent for the H_2O_2 oxidation reaction. The physical confinement of ILs within some solid porous materials has dual effects on the catalytic oxidation: on the one hand, the ILs supply special microenvironment on affecting the reaction pathway; on the other hand, the micropores of the porous material allow the controlling of the selectivity of the product. In addition, we may expect structural modification of the functionalized ILs by deliberately varying the cations/anions to meet the distribution requirement of the substrate or product, which will be of great importance for increasing the yield and selectivity of the product. Beyond that, more synthetic method should be developed to support the actual catalyst in order to shed light on the effective utilization of the expensive reagent in future catalytic oxidation.

4. Olefin oxidation

Recently, significant improvements on the catalytic performance in some transition metal-catalyzed reactions have been observed using ILs as the solvent. [2] , [19] , [20] The room-temperature ILs have emerged as environmentally benign reaction media as well as new vehicles for the immobilization of transition metal-based catalysts. Singh *et al.* [52] reported the H_2O_2 epoxidation of substrates containing both electron rich and deficient olefins catalyzed with *meso*-tetraarylporphyrin iron(III) chlorides ([TAPFe(III)Cl]) in imidazolium ILs. The active oxidizing species depended on the substrate used (Figure 8): the ferric peroxy anions (TAP-Fe^{III}-OO^-) were effective intermediates in the epoxidation of electron deficient olefins, whereas the high valent oxoferrylporphyrin π-cation radicals (TAP-Fe^{IV}=$O^{\bullet +}$) were involved in the epoxidation of electron rich olefins. The ILs provide special microenvironment *via* the interactions between their cations and anions, where the active intermediates could be fast generated from TAPFe^{III}Cl/ [Bmim] [PF_6] and H_2O_2 and well stabilized in the [Bmim] [PF_6] IL. The nature of the anions in the ILs played an important role on the activity of the catalyst.

Han and coworkers [53] synthesized novel Ni^{2+}-containing 1-methyl-3- [(triethoxysilyl)propyl] imidazolium chloride (TMICl) IL immobilized on silica to catalytic oxidation of styrene to benzaldehyde with H_2O_2 under solvent-free condition (Figure 9). With the aid of the IL, both hydrophobic reactant and the hydrophilic reactant were accessible to the active sites of the catalyst: styrene and H_2O_2 are miscible with the IL, and the Ni^{2+} was coordinated by the immobilized IL that allowed both reactants to access to active sites of the catalyst effectively. Under solvent-free condition, the conversion of styrene reached 18.5%and the selectivity to benzaldehyde was as high as 95.9% on the IMM-TMICl-Ni^{2+} catalyst.

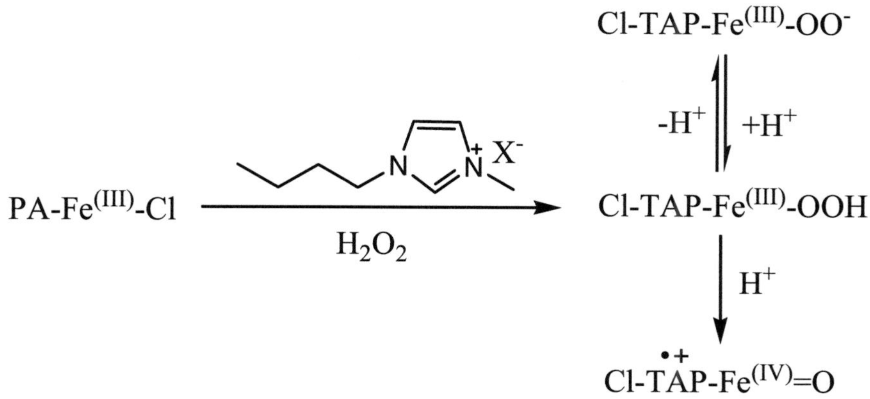

Figure 8. Mechanism for the generation of the reactive intermediate from TAPFe^{III}Cl/ [Bmim] [PF_6] and H_2O_2. [52]

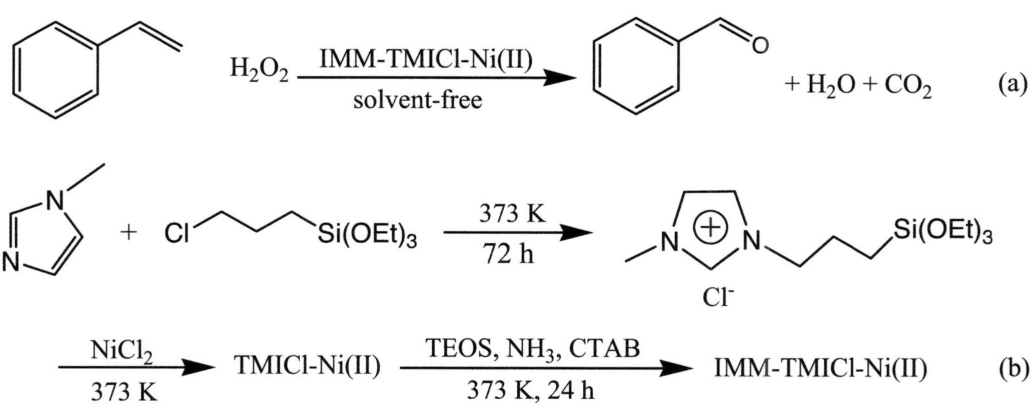

Figure 9. Schematic illustrations of (a) oxidation of styrene to benzaldehyde with H_2O_2 under solvent-free condition, (b) synthesis of IMM-TMICl-Ni(II) catalyst by grafting Ni^{2+} on IL-Silica. [53]

Some more examples are given by combining ILs with metal peroxides or polyoxometalates in the catalytic epoxidation of olefins with H_2O_2. In Yamaguchi *et al.*'s work, [54] peroxotungstate was immobilized on dihydroimidazolium-based IL-modified SiO_2 (Figure 10). The as-prepared catalyst showed high activity and selectivity for epoxidation of a wide range of olefins. Radical mechanism was excluded for the IL-involved epoxidation in CH_3CN solvent. Berardi *et al.* [55] embedded the catalytically active $[\gamma\text{-SiW}_{10}O_{36}(PhPO)_2]^{4-}$polyanions in the hydrophobic IL ([bmim] [[PF$_6$] or hydrophilic IL [bmim] [(CF$_3$SO$_2$)$_2$N]). The catalyst gave out high yield and selectivity for epoxidation of olefins under microwave irradiation in the hydrophilic IL (Figure 11). The catalyst can be recycled with the catalytic IL phase. Liu *et al.* [56] demonstrated the role of [bmim] [PF$_6$] IL as an activator for efficient olefin epoxidation with H_2O_2catalyzed by Keggin polyoxometalate [bmim] $_3PW_{12}O_{40}$ (Figure 12). In the IL, the interaction between the anions and the cations supplied a special microenvironment, accelerating the generation of the active peroxotungstate [PO$_4${W(O)(O)$_2$}$_4$] $^{3-}$ species from [bmim] $_3PW_{12}O_{40}$ and H_2O_2. In some sense, the [bmim] [PF$_6$] IL could be considered as a co-catalyst to promote the formation of active species for epoxidation. Both of the TOF and selectivity for olefin epoxidation were significantly enhanced in the IL compared to that of traditional organic solvents, *e.g.*, 289 times TOF and 1.3 times selectivity as found in CH_2Cl_2 for the epoxidation of *cis*-cyclooctene. The utilization efficiency of H_2O_2 reached as high as 87%.

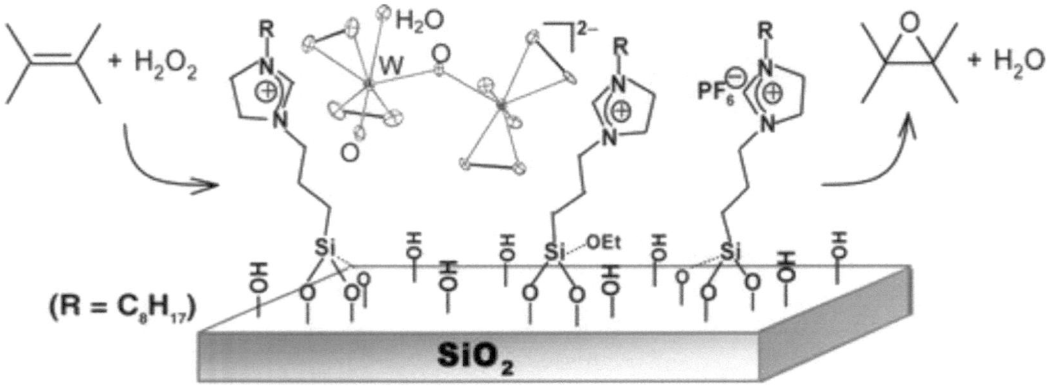

Figure 10. Epoxidation of Olefins with H_2O2 catalyzed by peroxotungstate immobilized on IL-modified SiO_2. [54] Reprinted with permission from ref.54. Copyright © 2005,AmericanChemical Society.

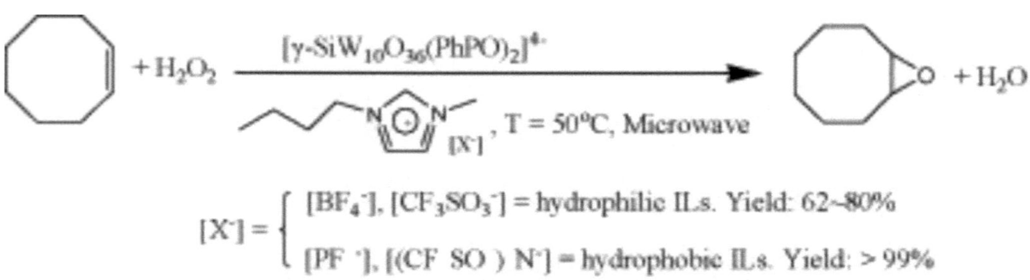

Figure 11. Epoxidaton of *cis*-Cyclooctene with H_2O_2 and polyoxometalates in both hydrophilic and hydrophobic ILs. [55]

Numerous examples have shown that ILs are offering unique properties in the transition metal-, metal peroxide- or polyoxometalates-catalyzed oxidation of olefins with H_2O_2. The

ILs supply special environment for the generation and stabilization of the active intermediate, or act as support for immobilizing and recycling the actual catalyst, both of which are necessary for performing effective catalytic oxidation. By delicately designing the combination of catalyst, support and ILs, the interactions between the hydrophobic substrate, hydrophilic oxidant, and the active site could be reinforced, intensifying the catalytic efficiency of the oxidation reaction.

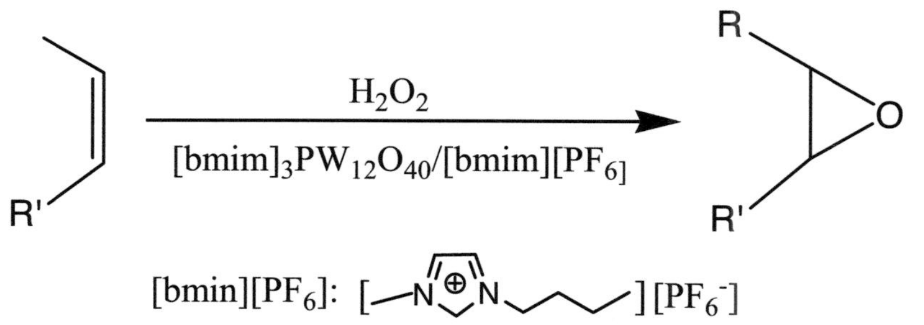

Figure 12. Epoxidation of olefins with H_2O_2 and polyoxometalates in [bmim] [PF_6] IL. [56]

5. Conclusion

Catalytic oxidations have been widely studied in ionic liquids, and much of this interest is centered on the possible use as "green" alternatives to traditionally used volatile organic solvents. This chapter summarizes limited examples that illustrate the applications of ILs in the catalytic oxidation using H_2O_2 as the oxidant, in particular benzene hydroxylation, alcohol oxidation, and olefin oxidation. We focus our discussion on understanding how the unusual solvent environment provide solute species that affect the reactions occurred in them.

As innocent and non-vaporized solvents, ILs provide good solubility to salts and most of the hydrophobic substrate, endowing them good solvent for the transition metal complexes-, peroxides-, oxides-, polyoxometalates-, or organic molecules-catalyzed oxidation. The miscibility of ILs with water and organic molecules can be elaborately tuned by varying the cations/anions (*i.e.*, length of alkyl chain, polarity, *etc.*). The interactions between ILs and the substrate, catalyst, oxidant, even reaction intermediate, make the ILs act as multi-functional solvent for the H_2O_2 involved catalytic oxidation. The cations/anions in ILs may influence the reaction pathway by stabilization of the charged transition state, active species or ligands which are necessary for many oxidation reaction. The structure of ILs can also be specially tailored to support actual catalyst and/or co-catalyst for effective recycling. The functionalized ILs have prospect as both catalyst and solvent for the H_2O_2 oxidation reaction. The grafting of ILs on some solid porous materials may improve the microenvironment of the reactive site, affecting the reaction pathway by increasing the selectivity of the product. Most importantly, the combination of ILs and a co-solvent (water or organic solvent) allows the mass transfer of unstable product from oxidative environment, protecting the product from over-oxidation by H_2O_2. As a conclusion, the integration of the multiple benefits from ILs will provide a greener scenario for recycling the catalyst and solvent, as well as improving the yield and selectivity of the product. We may expect novel synthetic strategy for functionalized ILs and their elaborate combination with prevailing catalytic materials for applications in a wide range of catalytic oxidation in future.

Acknowledgments

The financial support from the National Natural Science Foundationof China (No. 20901053 and 20872102) and PCSIRT (No. IRT0846) are greatly appreciated.

Author details

Liangfang Zhu and Changwei Hu*

*Address all correspondence to: changweihu@scu.edu.cn or chwehu@mail.sc.cninfo.net

Key Laboratory of Green Chemistry and Technology, Ministry of Education, College of Chemistry, Sichuan University, Chengdu, P.R. China

References

[1] Anastas, P. T.; Warner, J. C., *Green Chemistry: Theory and Practice*, Oxford University Press: New York, 1998.

[2] Welton, T., Room-Temperature Ionic Liquids. Solvents for Synthesis and Catalysis. *Chem. Rev.* 1999, 99, 2071-2084.

[3] Dupont, J.; De Souza, R. F.; Suarez, P. A. Z., Ionic Liquid (Molten Salt) Phase Organometallic Catalysis. *Chem. Rev.* 2002, 102, 3667-3692.

[4] Hernandez, O. R., ToTreat or Not to Treat? Applying Chemical Engineering Tools and a Life Cycle Approach to Assessing the Level of Sustainability of a Clean-up Technology. *Green Chem.* 2004, 6, 395-400.

[5] Clift, R., Sustainable Development and Its Implications for Chemical Engineering. *Chem. Eng. Sci.* 2006, 61, 4179-4187.

[6] Fischer, T.; Sethi, A.; Welton, T.; Woolf, J., Diels-Alder Reactions in Room-temperature Ionic Liquids. *Tetrahedron Lett.* 1999, 40, 793-796.

[7] Boon, J. A.; Levisky, J. A.; L., P. J.; Wilkes, J. S., Friedel-Crafts Reactions in Ambient-Temperature Molten Salts. *J. Org. Chem.* 1986, 51, 480-483.

[8] Brown, R. A.; Pollet, P.; McKoon, E.; Eckert, C. A.; Liotta, C. L.; Jessop, P. G., Asymmetric Hydrogenation and Catalyst Recycling Using Ionic Liquid and Supercritical Carbon Dioxide. *J. Am. Chem. Soc.* 2001, 123, 1254-1255.

[9] Chauvin, Y.; Mussmann, L.; Olivier, H., A Novel Class of Versatile Solvents for Two-Phase Catalysis: Hydrogenation, Isomerization, and Hydroformylation of Alkenes Catalyzed by Rhodium Complexes in Liquid 1,3-Dialkylimidazolium Salts. *Angew. Chem., Int. Ed. Engl.* 1996, 34, 2698-2700.

[10] Favre, F.; Olivier-Bourbigou, H.; Commereuc, D.; Saussine, L., Hydroformylation of 1-Hexene with Rhodium in Non-aqueous Ionic Liquids : How to Design the Solvent and the Ligand to the Reaction. *Chem. Commun.* 2001, 1360-1361.

[11] Dullius, J. E. L.; Suarez, P. A. Z.; Einloft, S.; de Souza, R. F.; Dupont, J., Selective Catalytic Hydrodimerization of 1,3-Butadiene by Palladium Compounds Dissolved in Ionic Liquids. *Organometallics* 1998, 17 (5), 815-819.

[12] Kaufmann, D. E.; Nouroozian, M.; Henze, H., Molten Salts as an Efficient Medium for Palladium Catalyzed C-C Coupling Reactions. *Synlett.* 1996, 1091-1092.

[13] Matthews, C. J.; Smith, P. J.; Welton, T., Palladium Catalysed Suzuki Cross-coupling Reactions in Ambient Temperature Ionic Liquids. *Chem. Commun.* 2000, 1249-1250.

[14] Parvulescu, V. I.; Hardacre, C., Catalysis in Ionic Liquids. *Chem. Rev.* 2007, 107, 2615-2665.

[15] Hallett, J. P.; Welton, T., Room-Temperature Ionic Liquids: Solvents for Synthesis and Catalysis. 2. *Chem. Rev.* 2011, 111, 3508-3576.

[16] Olivier-Bourbigou, H.; Magna, L.; Morvan, D., Ionic Liquids and Catalysis: Recent Progress from Knowledge to Applications. *Appl. Catal. A* 2010, 373, 1-56.

[17] Lee, J. W.; Shin, J. Y.; Chun, Y. S.; Jang, H. B.; Song, C. E.; Lee, S., Toward Understanding the Origin of Positive Effects of Ionic Liquids on Catalysis: Formation of More Reactive Catalysts and Stabilization of Reactive Intermediates and Transition States in Ionic Liquids. *Acc. Chem. Res.* 2010, 43 (7), 985-994.

[18] Stark, A., Ionic Liquid Structure-Induced Effects on Organic Reactions. *Top Curr. Chem.* 2009, 290, 41-81.

[19] Wasserscheid, P.; Keim, W., Ionic Liquids-New "solution" for Transition Metal Catalysis. *Angew Chem, Int. Ed.* 2000, 39, 3772-3789.

[20] Sheldon, R., Catalytic Reactions in Ionic Liquids. *Chem. Commun.* 2001, 2399-2407.

[21] Gharnati, L.; Doering, M.; Arnold, U., Catalytic Oxidation with Hydrogen Peroxide in Ionic Liquids. *Curr. Org. Synth.* 2009, *6* (4), 342-361.

[22] Betz, D.; Altmann, P.; Cokoja, M.; Herrmann, W. A.; Kuehn, F. E., Recent Advances in Oxidation Catalysis Using Ionic Liquids as Solvents. *Coord. Chem. Rev.* 2011, 255 (13-14), 1518-1540.

[23] Muzart, J., Ionic Liquids as Solvents for Catalyzed Oxidations of Organic Compounds. *Adv. Synth. Catal.* 2006, 348, 275-295.

[24] Walling, C., Intermediates in the Reactions of Fenton Type Reagents. *Acc. Chem. Res.* 1998, 31, 155-157.

[25] Hu, X. K.; Zhu, L. F.; Guo, B.; Liu, Q. Y.; Li, G. Y.; Hu, C. W., Hydroxylation of Benzene to Phenol via Hydrogen Peroxide in Hydrophilic Triethylammonium Acetate Ionic Liquid. *Chem. Res. Chin. Univ.* 2011, 27 (3), 503-507.

[26] Zhang, J.; Tang, Y.; Li, G. Y.; Hu, C. W., Room Temperature Direct Oxidation of Benzene to Phenol Using Hydrogen Peroxide in the Presence of Vanadium-substituted Heteropolymolybdates. *Appl. Catal. A* 2005, 278, 251-261.

[27] Zhong, Y. K.; Li, G. Y.; Zhu, L. F.; Yan, Y.; Wu, G.; Hu, C. W., Low Temperature Hydroxylation of Benzene to Phenol by Hydrogen Peroxide over Fe/activated Carbon Catalyst. *J. Mol. Catal. A* 2007, 272, 169-173.

[28] Jian, M.; Zhu, L. F.; Wang, J. Y.; Zhang, J.; Li, G. Y.; Hu, C. W., Sodium Metavanadate Catalyzed Direct Hydroxylation of Benzene to Phenol with Hydrogen Peroxide in Acetonitrile Medium. *J. Mol. Catal. A* 2006, 253, 1-7.

[29] Bianchi, D.; Bertoli, M.; Tassinari, R.; Ricci, M.; Vignola, R., Ligand Effect on the Iron-catalysed Biphasic Oxidation of Aromatic Hydrocarbons by Hydrogen Peroxide *J. Mol. Catal. A* 2003, 204, 419-424.

[30] Bianchi, D.; Balducci, L.; Bortolo, R.; D' Aloisio, R.; Ricci, M.; Span, G.; Tassinari, R.; Tonini, C.; Ungarellia, R., Oxidation of Benzene to Phenol with Hydrogen Peroxide Catalyzed by a Modified Titanium Silicalite (TS-1B). *Adv. Synth. Catal.* 2007, 349, 979-986.

[31] Peng, J. J.; Shi, F.; Gu, Y. L.; Deng, Y. Q., Highly Selective and Green Aqueous-Ionic Liquid Biphasic Hydroxylation of Benzene to Phenol with Hydrogen Peroxide. *Green Chem.* 2003, 5 (2), 224-226.

[32] Hu, X. K.; Zhu, L. F.; Wang, X. Q.; Guo, B.; Xu, J. Q.; Li, G. Y.; Hu, C. W., Active Species Formed in a Fenton-Like System in the Medium of Triethylammonium Acetate Ionic Liquid for Hydroxylation of Benzene to Phenol. *J. Mol. Catal. A* 2011, 342-343, 41-49.

[33] Hu, X. K., Study on Hydroxylation of Benzene in Triethylammonium Acetate Ionic Liquid. *Chinese Doctoral Dissertation* 2011.

[34] Ley, S. V.; Madin, A., *in: Trost, B. M.; Flemming, I. (Eds.), Comprehensive Organic Synthesis*, Vol. 7, 305-327, Pergamon Press, Oxford, 1991.

[35] Hudlick, M., *Oxidations in Organic Chemistry*, American Chemical Society: Washington, DC, 1990.

[36] Sheldon, R. A.; Kochi, J. K., *Metal Catalyzed Oxidationd of Organic Compounds*, Academic Press, New York, 1984.

[37] Bobbitt, J. M.; Br °uckner, C., *Organic Reactions*, John-Wiley & Sons, New York, 2009.

[38] Anelli, P. L.; Biffi, C.; Montanari, F.; Quici, S., *J. Org. Chem.* 1987, 52, 2559-2562.

[39] Fall, A.; Sene, M.; Gaye, M.; Gomez, G.; Fall, Y., Ionic Liquid-Supported TEMPO as Catalyst in the Oxidation of Alcohols to Aldehydes and Ketones. *Tetrahedron Lett.* 2010, 51 (34), 4501-4504.

[40] Hoover, J. M.; Stahl, S. S., Highly Practical Copper(I)/TEMPO Catalyst System for Chemoselective Aerobic Oxidation of Primary Alcohols. *J. Am. Chem. Soc.* 2011, 133 (42), 16901-16910.

[41] Ma, S. M.; Liu, J. X.; Li, S. H.; Chen, B.; Cheng, J. J.; Kuang, J. Q.; Liu, Y.; Wan, B. Q.; Wang, Y. L.; Ye, J. T.; Yu, Q.; Yuan, W. M.; Yu, S. C., Development of a General and Practical Iron Nitrate/TEMPO-Catalyzed Aerobic Oxidation of Alcohols to Aldehydes/Ketones: Catalysis with Table Salt. *Adv. Synth. Catal.* 2011, 353 (6), 1005-1017.

[42] Hoover, J. M.; Steves, J. E.; Stahl, S. S., Copper(I)/TEMPO-catalyzed aerobic oxidation of primary alcohols to aldehydes with ambient air. *Nature Protoc.* 2012, 7 (6), 1161-1166.

[43] Gheorghe, A.; Chinnusamy, T.; Cuevas-Yanez, E.; Hilgers, P.; Reiser, O., Combination of Perfluoroalkyl and Triazole Moieties: A New Recovery Strategy for TEMPO. *Org. Lett.* 2008, 10 (19), 4171-4174.

[44] Liu, R.; Liang, X.; Dong, C.; Hu, X., Transition-Metal-Free: A Highly Efficient Catalytic Aerobic Alcohol Oxidation Process. *J. Am. Chem. Soc.* 2004, 126, 4112-4113.

[45] Wang, S. S.; Popovic, Z.; Wu, H. H.; Liu, Y., A Homogeneous Mixture Composed of Vanadate, Acid, and TEMPO Functionalized Ionic Liquids for Alcohol Oxidation by H2O2. *ChemCatChem*2011, 3 (7), 1208-1213.

[46] Jiang, N.; Ragauskas, A. J., TEMPO-Catalyzed Oxidation of Benzylic Alcohols to Aldehydes with the H2O2/HBr/Ionic Liquid [bmim] PF6 System. *Tetrahedron Lett.* 2005, 46 (19), 3323-3326.

[47] Karimi, B.; Badreh, E., SBA-15-Functionalized TEMPO Confined Ionic Liquid: an Efficient Catalyst System for Transition-Metal-Free Aerobic Oxidation of Alcohols with Improved Selectivity. *Org. Biomol. Chem.* 2011, 9 (11), 4194-4198.

[48] Chhikara, B. S.; Chandra, R.; Tandon, V., Oxidation of Alcohols with Hydrogen Peroxide Catalyzed by a New Imidazolium Ion Based Phosphotungstate Complex in Ionic Liquid. *J. Catal.* 2005, 230 (2), 436-439.

[49] Bianchini, G.; Crucianelli, M.; De Angelis, F.; Neri, V.; Saladino, R., Highly Efficient C-H Insertion Reactions of Hydrogen Peroxide Catalyzed by Homogeneous and Het-

erogeneous Methyltrioxorhenium Systems in Ionic Liquids. *Tetrahedron Lett.* 2005, 46 (14), 2427-2432.

[50] Chen, L.; Zhou, T.; Chen, L.; Ye, Y.; Qi, Z.; Freund, H.; Sundmacher, K., Selective Oxidation of Cyclohexanol to Cyclohexanone in the Ionic Liquid 1-Octyl-3-Methylimidazolium Chloride. *Chem. Commun.* 2011, 47 (33), 9354-9356.

[51] Usui, Y.; Sato, K., A Green Method of Adipic Acid Synthesis: Organic Solvent- and Halide-Free Oxidation of Cycloalkanones with 30% Hydrogen Peroxide. *Green Chem.* 2003, 5, 373-375.

[52] Singh, P. P.; Ambika; Chauhan, S. M. S., Chemoselective Epoxidation of Electron Rich and Electron Deficient Olefins Catalyzed by Meso-Tetraarylporphyrin Iron(III) Chlorides in Imidazolium Ionic Liquids. *New J. Chem.* 2012, 36 (3), 650-655.

[53] Liu, G.; Hou, M. Q.; Song, J. Y.; Zhang, Z. F.; Wu, T. B.; Han, B. X., Ni2+-Containing Ionic Liquid Immobilized on Silica: Effective Catalyst for Styrene Oxidation with H2O2 at Solvent-Free Condition. *J. Mol. Catal. A* 2010, 316 (1-2), 90-94.

[54] Yamaguchi, K.; Yoshida, C.; Uchida, S.; Mizuno, N., Peroxotungstate immobilized on ionic liquid-modified silica as a heterogeneous epoxidation catalyst with hydrogen peroxide. *J. Am. Chem. Soc.* 2005, 127 (2), 530-531.

[55] Berardi, S.; Bonchio, M.; Carraro, M.; Conte, V.; Sartorel, A.; Scorrano, G., Fast Catalytic Epoxidation with H2O2 and [gamma-SiW10O36(PhPO)2]4 in Ionic Liquids under Microwave Irradiation. *J. Org. Chem.* 2007, 72 (23), 8954-8957.

[56] Liu, L. L.; Chen, C. C.; Hu, X. F.; Mohamood, T.; Ma, W. H.; Lin, J.; Zhao, J. C., A Role of Ionic Liquid as an Activator for Efficient Olefin Epoxidation Catalyzed by Polyoxometalate. *New J. Chem.* 2008, 32 (2), 283-289.

Chapter 15

New Brønsted Ionic Liquids: Synthesis, Thermodinamics and Catalytic Activity in Aldol Condensation Reactions

I. Cota, R. Gonzalez-Olmos, M. Iglesias and
F. Medina

Additional information is available at the end of the chapter

http://dx.doi.org/10.5772/51163

1. Introduction

It is a continuous challenge to find new catalysts able to perform with good activities and selectivity condensation reactions for the synthesis of pharmaceutical and fine chemicals. In the last years room temperature ionic liquids (ILs) have received a lot of interest as environmental friendly or "green" alternatives to conventional molecular solvents. They differ from molecular solvents by their unique ionic character and their "structure and organization" which can lead to specific effects [1].

Room-temperature ILs have been used as clean solvents and catalysts for green chemistry, stabilizing agents for the catalysts or intermediates, electrolytes for batteries, in photochemistry and electrosynthesis etc [2-6]. Their success as environmental benign solvents or catalysts is described in numerous reactions [7-11], such as Diels-Alder reactions [12, 13], the Friedel-Crafts reaction [14-17], esterification [18-20], cracking rections [21], and so on. The link between ionic ILs and green chemistry is related to the solvent properties of ILs. Some of the properties that make ILs attractive media for catalysis are: they have no significant vapour pressure and thus create no volatile organic pollution during manipulation; ILs have good chemical and thermal stability, most ILs having liquid ranges for more than 300^{0}C; they are immiscible with some organic solvents and therefore can be used in two-phase systems; ILs polarity can be adjusted by a suitable choice of cation/anion; they are able to dissolve a wide range of organic, inorganic and organometallic compounds; ILs are often composed of weakly coordinating anions and therefore have the potential to be highly polar.

The number of ILs has increased exponentially in the recent years. Many of them are based on the imidazolium cation and in a lesser proportion, alkyl pyridiniums and trialkylamines (Scheme 1). By changing the anion or the alkyl chain of the cation, a wide variety of ILs may be designed for specific applications. They can be of hydrophobic or hydrophilic nature depending on the chemical structures involved.

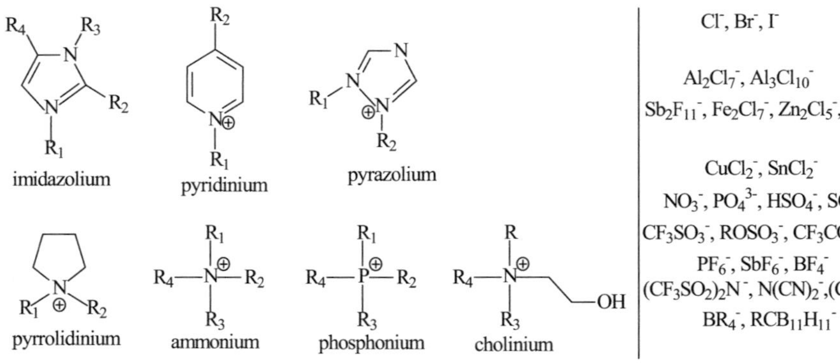

Scheme 1. Main cations and anions described in literature [1].

ILs can be divided into two broad categories: aprotic ionic liquids (AILs) and protic ionic liquids (PILs).

AILs largely dominate the open literature due to their relative inertness to organometallic compounds and their potential of applications, particularly in catalysis. They are synthesized by transferring an alkyl group to the basic nitrogen site through S_N2 reactions [1].

PILs are formed through proton transfer from a Brønsted acid to a Brønsted base. Recently there has been an increasing interest in PILs due to their greater potential as environmental friendly solvents and promising applications. Moreover, they present the advantage of being cost-effective and easily prepared as their formation does not involve the formation of residual by-products. A specific feature of the PILs is that they are capable of developing a certain hydrogen bonding potency, including proton acceptance and proton donation and they are highly tolerant to hydroxylic media [22-23].

The application of new policies on terms of environment, health and safety deals towards minimizing or substituting organic volatile solvents by green alternatives, placing a renewed emphasis on research and development of lesser harmful compounds as ILs. On the other hand, recently the interest in the use of PILs to tailor the water properties for cleaning applications in processes of minimization of CO_2/SO_2 emissions has increased [24-26].

In the last years numerous studies report the use of ILs as selective catalysts for different reactions, like aldol condensation reactions where several ILs have been successfully applied as homogeneous and heterogeneous catalysts [27-30]. Abelló et al. [28] described the use of choline hydroxide as basic catalyst for aldol condensation reactions between several ketones and aldehydes. Better conversions and selectivities were obtained when compared to other well-known catalysts, such as rehydrated hydrotalcites, MgO and NaOH. In addition, higher performance was obtained when choline was immobilized on MgO.

Zhu et al. [27] described the use of 1,1,3,3-tetramethylguanidine lactate ([TMG] [Lac]) as recyclable catalyst for direct aldol condensation reactions at room temperature without any solvent. It was demonstrated that for each reaction only the aldol adduct was produced when the molar ratio of the IL and substrate was smaller than 1. Moreover, after the reaction the IL was easily recovered and recycled without considerably decrease of activity.

Kryshtal et al. [29] described the application of tetraalkylammonium and 1,3-dialkylimidazolium perfluoro-borates and perfluoro-phosphates as recoverable phase-transfer catalysts in multiphase reactions of CH-acids, in particular in solid base-promoted cross-aldol condensations. The catalysts retained their catalytic activity over several reaction cycles.

In the study of Lombardo et al. [30] two onium ion-tagged prolines, imidazolium bis (trifluoromethylsulfonyl)imide-substituted proline and butyldimethylammonium bis (trifluoromethylsulfonyl) imide-substituted proline, were synthesized and their catalytic activity in the direct asymmetric aldol condensation was studied. The catalytic protocol developed by this group makes use of a 6-fold lower amount of catalyst with respect to the preceding reports [31, 32] and affords greater chemical yields and higher enantioselectivity.

The main objective of this chapter is to develop and study the applications of a new family of ILs based on substituted amine cations of the form RNH_3^+ combined with organic anions of the form $R'COO^-$ (being of different nature R and R'). The variations in the anion alkyl chain, in conjunction with the cations, lead to a large matrix of materials.

This kind of compounds show interesting properties for industrial use of ILs: low cost of preparation, simple synthesis and purification methods. Moreover, the very low toxicity and the degradability of this kind of ILs have been verified. Thus, sustainable processes can be originated from their use.

Recently, many studies dealing with the application of ILs in organic synthesis and catalysis have been published, pointing out the vast interest in this type of compounds [33-36]. With these facts in mind, we studied their catalytic potential for two condensation reactions of carbonyl compounds. The products obtained from these reactions are applied in pharmacological, flavor and fragrance industry.

2. Experimental

2.1. Preparation of ILs and supported ILs

The ILs synthesized in this work are: 2-hydroxy ethylammonium formate (2-HEAF), 2-hydroxy ethylammonium acetate (2-HEAA), 2-hydroxy ethylammonium propionate (2-HEAP), 2-hydroxi ethylammonium butanoate (2-HEAB), 2-hydroxi ethylammonium isobutanoate (2-HEAiB) and 2-hydroxi ethylammonium pentanoate (2-HEAPE).

The amine (Merck Synthesis, better than 99%) was placed in a three necked flask all-made-in-glass equipped with a reflux condenser, a PT-100 temperature sensor for controlling temperature and a dropping funnel. The flask was mounted in a thermal bath. A slight heating is necessary for increasing miscibility between reactants and then allow reaction. The organic acid (Merck Synthesis, better than 99%) was added drop wise to the flask under stirring with a magnetic bar. Stirring was continued for 24 h at laboratory temperature, in order to obtain a final viscous liquid. Lower viscosity was observed in the final product by decreasing molecular weight of reactants. No solid crystals or precipitation was noticed when the liquid sample was purified or stored at freeze temperature for a few months after synthesis. The reaction is a simple acid–base neutralization creating the formiate, acetate, propionate, butanoate, isobutanoate or pentanoate salt of ethanolamine that in a general form should be expressed as follows:

$$(HOCH_2CH_2)NH_2 + HOOCR \rightarrow (HOCH_2CH_2)NH_3^{+}(^{-}OOCR) \qquad (1)$$

For example, when formic acid is used this equation shows the chemical reaction for the reactants ethanolamine + formic acid, with 2-HEAF as neutralization product.

Because these chemical reactions are highly exothermic, an adequate control of temperature is essential throughout the chemical reaction; otherwise heat evolution may produce the dehydration of the salt to the corresponding amide, as in the case for nylon salts (salts of diamines with dicarboxy acids).

As observed in our laboratory during IL synthesis, dehydration begins around 423.15 K for the lightest ILs. The color varied in each case from transparent to dark yellow when the reaction process and purification (strong agitation and slight heating for the vaporization of residual non-reacted acid for at least for 24 h) were completed.

There was no detectable decomposition for the ILs studied here when left for over 12 months at laboratory temperature. Less than 1% amide was detected after this period of time. On the basis of these results it appears obvious that the probability of amide formation is low for this kind of structures.

In order to obtain the supported ILs, 1 g of IL was dissolved in 7 ml of ethanol and after stirring at room temperature for 30 min, 1 g of alanine (Fluka, better than 99%) was added. The mixture was stirred for 2 h and then heated at 348 K under vacuum to remove ethanol. The supported ILs thus obtained were labelled hereafter as a-ILs.

2.2. Spectroscopy test

FT-IR spectrum was taken by a Jasco FT/IR 680 plus model IR spectrometer, using a NaCl disk.

2.3. Physical properties equipment

During the course of the experiments, the purity of ILs was monitored by different physical properties measurements. The pure ILs were stored in sun light protected form, constant humidity and low temperature. Usual manipulation and purification in our experimental work was applied [22].

The densities and ultrasonic velocities of pure components were measured with an Anton Paar DSA-5000 vibrational tube densimeter and sound analyzer, with a resolution of 10^{-5} g cm^{-3} and 1 m s^{-1}. Apparatus calibration was performed periodically in accordance with provider's instructions using a double reference (millipore quality water and ambient air at each temperature). Accuracy in the temperature of measurement was better than $\pm 10^{-2}$ K by means of a temperature control device that apply the Peltier principle to maintain isothermal conditions during the measurements.

The ion conductivity was measured by a Jenway Model 4150 Conductivity/TDS Meter with resolution of 0.01μS to 1 mS and accuracy of ±0.5% at the range temperature. The accuracy of temperature into the measurement cell was ±0.5 °C.

2.4. Catalytic studies

The studied reactions were the condensation between citral and acetone and between benzaldehyde and acetone. The reactions were performed in liquid phase using a 100 mL batch reactor equipped with a condenser system. To a stirred solution of substrate and ketone (molar ratio ketone/substrate = 4.4) was added 1 g of IL, and the flask was maintained at 333 K using an oil bath. Samples were taken at regular time periods and analyzed by gas chromatography using a flame ionization detector and an AG Ultra 2 column (15 m x 0.32 mm x 0.25 μm). Tetradecane was used as the internal standard. Reagents were purchase from Aldrich and used without further purification.

In order to separate the ILs from the reaction mixture, at the end of the reaction 6 mL of H_2O were added. The mixture was stirred for 2 h and then left 15 h to repose. Two phases were separated: the organic phase which contains the reaction products and the aqueous phase which contains the IL. In order to separate the IL, the aqueous phase was heated up to 393 K under vacuum.

3. Results and discussion

As Figure 1 shows, the broad band in the 3500-2400 cm^{-1} range exhibits characteristic ammonium structure for all the neutralization products. The OH stretching vibration is embedded in this band. The broad band centered at 1600 cm^{-1} is a combined band of the carbonyl stretching and N-H plane bending vibrations. FT-IR results clearly demonstrate the IL characteristics of compounds synthesized in this work.

Due to space considerations, we will present the thermodynamic properties only for two of the studied ILs: 2-HEAF and 2-HEAPE.

The molar mass and experimental results at standard condition for 2-HEAF and 2-HEAPE are shown in Table 1.

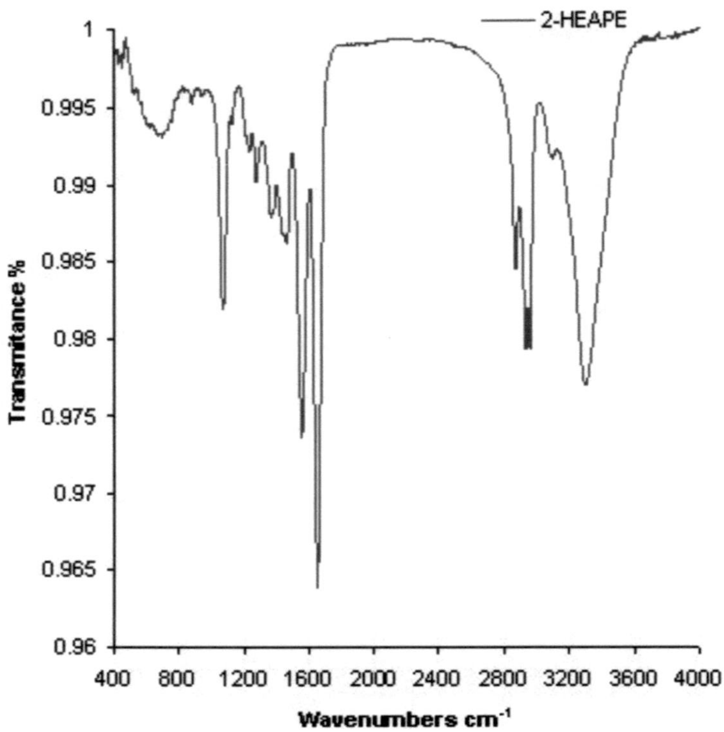

Figure 1. FT-IR spectrum for 2-HEAPE.

IL	Molecular Weight ($g \bullet mol^{-1}$)	Exp. Density ($g \bullet cm^{-3}$)	Exp. Ultrasonic Velocity (ms^{-1})	Exp. Conductivity ($\mu S \bullet cm^{-1}$)
2-HEAF	107.11	1.176489	1709.00	4197.6
2-HEAPE	163.21	1.045479	1591.59	239.6

[a]Other experimental data for comparison are not available from the literature.

Table 1. Experimental data for pure ionic liquids at 298.15 K and other relevant information[a]

The densities, ultrasonic velocities and isobaric expansibility of 2-HEAF and 2-HEAPE are given in Table 2, and the ionic conductivities are given in Table 3. From the results obtained it can be observed that an increase in temperature diminishes the interaction among ions, lower values of density and ultrasonic velocity being gathered for rising temperatures in each case.

2-hydroxy ethylammonium formate (2-HEAF)									
T (K)	ρ (gcm^{-3})	u (ms^{-1})	κ_S (TPa^{-1})	$10^3 \cdot \alpha$ (K^{-1})	T (K)	ρ (gcm^{-3})	u (ms^{-1})	κ_S (TPa^{-1})	$10^3 \cdot \alpha$ (K^{-1})
338.15	1.148091	1613.59	334.53	0.6188	327.16	1.155890	1639.38	321.90	0.6148
337.90	1.148254	1614.14	334.26	0.6187	326.91	1.156069	1639.97	321.62	0.6147
337.66	1.148433	1614.71	333.97	0.6186	326.66	1.156247	1640.57	321.34	0.6146
337.40	1.148608	1615.30	333.67	0.6185	326.41	1.156426	1641.16	321.06	0.6145
337.15	1.148785	1615.87	333.39	0.6184	326.16	1.156603	1641.75	320.78	0.6144
336.91	1.148963	1616.46	333.09	0.6183	325.91	1.156780	1642.34	320.50	0.6143
336.66	1.149139	1617.04	332.80	0.6182	325.65	1.156957	1642.94	320.21	0.6142
336.41	1.149316	1617.63	332.51	0.6182	325.40	1.157136	1643.53	319.93	0.6141
336.16	1.149494	1618.22	332.21	0.6181	325.16	1.157314	1644.12	319.66	0.6140
335.90	1.149669	1618.81	331.92	0.6180	324.90	1.157490	1644.72	319.37	0.6139
335.65	1.149848	1619.38	331.64	0.6179	324.65	1.157669	1645.32	319.09	0.6138
335.40	1.150027	1619.96	331.35	0.6178	324.40	1.157846	1645.91	318.81	0.6137
335.16	1.150205	1620.55	331.05	0.6177	324.15	1.158023	1646.50	318.54	0.6136
334.90	1.150384	1621.13	330.77	0.6176	323.90	1.158201	1647.09	318.26	0.6135
334.66	1.150560	1621.71	330.48	0.6175	323.65	1.158378	1647.68	317.98	0.6134
334.40	1.150740	1622.30	330.19	0.6174	323.40	1.158556	1648.28	317.70	0.6133
334.16	1.150916	1622.89	329.90	0.6173	323.15	1.158734	1648.90	317.42	0.6132
333.90	1.151094	1623.48	329.61	0.6173	322.90	1.158910	1649.47	317.15	0.6131
333.65	1.151271	1624.06	329.32	0.6172	322.66	1.159088	1650.06	316.87	0.6130
333.41	1.151449	1624.64	329.03	0.6171	322.41	1.159265	1650.66	316.59	0.6129
333.16	1.151625	1625.23	328.75	0.6170	322.16	1.159442	1651.25	316.32	0.6128
332.90	1.151804	1625.82	328.46	0.6169	321.91	1.159620	1651.85	316.04	0.6127
332.65	1.151981	1626.41	328.17	0.6168	321.65	1.159797	1652.43	315.77	0.6126
332.41	1.152159	1626.99	327.88	0.6167	321.40	1.159976	1653.03	315.49	0.6125
332.15	1.152338	1627.58	327.59	0.6166	321.15	1.160154	1653.63	315.22	0.6124
331.90	1.152514	1628.16	327.31	0.6165	320.91	1.160330	1654.22	314.94	0.6124
331.65	1.152694	1628.75	327.02	0.6164	320.66	1.160509	1654.81	314.67	0.6123
331.40	1.152871	1629.34	326.74	0.6163	320.40	1.160688	1655.41	314.39	0.6122
331.16	1.153048	1629.93	326.45	0.6162	320.15	1.160863	1656.01	314.12	0.6121
330.90	1.153225	1630.52	326.16	0.6162	319.90	1.161042	1656.60	313.85	0.6120
330.65	1.153405	1631.11	325.88	0.6161	319.65	1.161218	1657.19	313.58	0.6119
330.41	1.153582	1631.69	325.59	0.6160	319.40	1.161398	1657.79	313.30	0.6118

330.15	1.153761	1632.29	325.30	0.6159	319.15	1.161574	1658.39	313.03	0.6117
329.90	1.153939	1632.88	325.02	0.6158	318.91	1.161750	1658.98	312.76	0.6116
329.65	1.154114	1633.47	324.73	0.6157	318.65	1.161930	1659.58	312.48	0.6115
329.41	1.154294	1634.06	324.45	0.6156	318.40	1.162110	1660.18	312.21	0.6114
329.15	1.154469	1634.65	324.17	0.6155	318.16	1.162286	1660.78	311.93	0.6113
328.91	1.154648	1635.24	323.88	0.6154	317.90	1.162462	1661.37	311.67	0.6112
328.65	1.154826	1635.84	323.59	0.6153	317.65	1.162643	1661.97	311.39	0.6111
328.40	1.155003	1636.43	323.31	0.6152	317.41	1.162820	1662.56	311.12	0.6110
328.15	1.155181	1637.02	323.03	0.6151	317.15	1.162998	1663.16	310.85	0.6109
327.90	1.155360	1637.61	322.75	0.6150	316.91	1.163174	1663.75	310.58	0.6108
327.66	1.155535	1638.20	322.47	0.6149	316.65	1.163352	1664.35	310.31	0.6107
316.15	1.163706	1665.55	309.77	0.6105	303.90	1.172408	1695.01	296.88	0.6054
315.90	1.163885	1666.15	309.50	0.6104	303.65	1.172587	1695.62	296.62	0.6053
315.65	1.164062	1666.74	309.23	0.6103	303.40	1.172764	1696.23	296.36	0.6052
315.40	1.164240	1667.34	308.96	0.6102	303.15	1.172937	1696.81	296.11	0.6051
315.15	1.164417	1667.94	308.70	0.6101	302.90	1.173120	1697.43	295.85	0.6050
314.90	1.164597	1668.54	308.43	0.6100	302.65	1.173295	1698.04	295.59	0.6049
314.65	1.164774	1669.14	308.16	0.6099	302.40	1.173473	1698.64	295.34	0.6048
314.40	1.164951	1669.73	307.89	0.6098	302.15	1.173648	1699.25	295.09	0.6047
314.15	1.165128	1670.33	307.63	0.6097	301.90	1.173826	1699.86	294.83	0.6046
313.90	1.165305	1670.94	307.35	0.6096	301.65	1.174003	1700.47	294.57	0.6045
313.65	1.165485	1671.54	307.09	0.6095	301.40	1.174180	1701.07	294.32	0.6043
313.40	1.165661	1672.13	306.82	0.6094	301.15	1.174361	1701.68	294.06	0.6042
313.15	1.165839	1672.72	306.56	0.6093	300.90	1.174535	1702.29	293.81	0.6041
312.90	1.166018	1673.34	306.29	0.6092	300.65	1.174714	1702.90	293.56	0.6040
312.65	1.166194	1673.94	306.02	0.6091	300.40	1.174891	1703.50	293.30	0.6039
312.40	1.166372	1674.54	305.75	0.6090	300.15	1.175070	1704.12	293.05	0.6038
312.15	1.166549	1675.14	305.49	0.6089	299.90	1.175247	1704.73	292.79	0.6037
311.90	1.166726	1675.74	305.22	0.6088	299.65	1.175425	1705.33	292.54	0.6036
311.65	1.166903	1676.34	304.96	0.6086	299.40	1.175602	1705.95	292.29	0.6035
311.40	1.167085	1676.95	304.69	0.6085	299.15	1.175780	1706.55	292.04	0.6034
311.15	1.167260	1677.55	304.43	0.6084	298.90	1.175955	1707.16	291.78	0.6033
310.90	1.167437	1678.14	304.17	0.6083	298.65	1.176133	1707.77	291.53	0.6032
310.65	1.167617	1678.74	303.90	0.6082	298.40	1.176311	1708.39	291.28	0.6030

http://dx.doi.org/10.5772/51163

310.40	1.167794	1679.35	303.63	0.6081	298.15	1.176489	1709.00	291.02	0.6029
310.15	1.167970	1679.94	303.38	0.6080	297.90	1.176666	1709.61	290.77	0.6028
309.90	1.168149	1680.55	303.11	0.6079	297.65	1.176842	1710.22	290.52	0.6027
309.65	1.168325	1681.15	302.85	0.6078	297.40	1.177019	1710.84	290.27	0.6026
309.40	1.168502	1681.75	302.59	0.6077	297.15	1.177201	1711.45	290.02	0.6025
309.15	1.168680	1682.35	302.32	0.6076	296.90	1.177373	1712.06	289.77	0.6024
308.90	1.168859	1682.96	302.06	0.6075	296.65	1.177553	1712.67	289.52	0.6023
308.65	1.169036	1683.55	301.80	0.6074	296.40	1.177729	1713.28	289.27	0.6022
308.40	1.169213	1684.16	301.54	0.6073	296.15	1.177905	1713.90	289.01	0.6021
308.15	1.169391	1684.76	301.28	0.6072	295.90	1.178085	1714.52	288.76	0.6019
307.90	1.169567	1685.36	301.02	0.6071	295.65	1.178265	1715.13	288.51	0.6018
307.65	1.169742	1685.96	300.76	0.6070	295.40	1.178438	1715.75	288.26	0.6017
307.40	1.169922	1686.56	300.50	0.6069	295.15	1.178617	1716.36	288.01	0.6016
307.15	1.170102	1687.17	300.23	0.6068	294.90	1.178798	1716.97	287.76	0.6015
306.90	1.170276	1687.77	299.98	0.6067	294.65	1.178971	1717.58	287.52	0.6014
306.65	1.170454	1688.37	299.72	0.6066	294.40	1.179148	1718.20	287.27	0.6013
306.40	1.170632	1688.98	299.45	0.6065	294.15	1.179325	1718.81	287.02	0.6012
306.15	1.170810	1689.58	299.20	0.6064	293.90	1.179505	1719.42	286.77	0.6011
305.90	1.170986	1690.18	298.94	0.6063	293.65	1.179682	1720.04	286.52	0.6009
305.65	1.171165	1690.79	298.68	0.6062	293.40	1.179858	1720.66	286.27	0.6008
305.40	1.171343	1691.39	298.42	0.6060	293.15	1.180037	1721.27	286.03	0.6007
305.15	1.171518	1691.99	298.16	0.6059	292.90	1.180210	1721.88	285.78	0.6006
304.90	1.171699	1692.60	297.90	0.6058	292.65	1.180390	1722.50	285.53	0.6005
304.40	1.172053	1693.80	297.39	0.6056	292.15	1.180744	1723.72	285.04	0.6003
304.15	1.172230	1694.41	297.13	0.6055	291.90	1.180923	1724.34	284.80	0.6002
291.65	1.181104	1724.95	284.55	0.6000	279.40	1.189760	1755.38	272.77	0.5944
291.40	1.181278	1725.57	284.30	0.5999	279.15	1.189935	1756.03	272.53	0.5943
291.15	1.181453	1726.18	284.06	0.5998	278.90	1.190108	1756.62	272.31	0.5941
290.90	1.181631	1726.80	283.81	0.5997	278.65	1.190288	1757.23	272.08	0.5940
290.65	1.181809	1727.43	283.56	0.5996	278.40	1.190464	1757.88	271.83	0.5939
290.40	1.181990	1728.05	283.32	0.5995	278.15	1.190632	1758.50	271.60	0.5938
290.15	1.182162	1728.67	283.07	0.5994					
289.90	1.182339	1729.29	282.83	0.5993					
289.65	1.182515	1729.91	282.58	0.5991					

289.39	1.182700	1730.84	282.24	0.5990
289.15	1.182877	1731.59	281.95	0.5989
288.89	1.183052	1732.13	281.73	0.5988
288.64	1.183228	1732.78	281.48	0.5987
288.39	1.183407	1733.34	281.25	0.5986
288.15	1.183574	1733.91	281.03	0.5985
287.90	1.183753	1734.51	280.79	0.5983
287.64	1.183941	1735.04	280.58	0.5982
287.40	1.184107	1735.67	280.33	0.5981
287.15	1.184289	1736.27	280.10	0.5980
286.90	1.184462	1736.82	279.88	0.5979
286.65	1.184637	1737.45	279.63	0.5978
286.40	1.184815	1738.07	279.39	0.5977
286.15	1.184986	1738.68	279.16	0.5975
285.90	1.185168	1739.24	278.93	0.5974
285.65	1.185344	1739.86	278.69	0.5973
285.40	1.185519	1740.47	278.46	0.5972
285.15	1.185700	1741.08	278.22	0.5971
284.90	1.185886	1741.82	277.94	0.5970
284.64	1.186059	1742.42	277.71	0.5968
284.40	1.186228	1742.99	277.49	0.5967
284.15	1.186403	1743.61	277.25	0.5966
283.90	1.186582	1744.21	277.02	0.5965
283.65	1.186756	1744.84	276.78	0.5964
283.40	1.186933	1745.46	276.54	0.5963
283.15	1.187110	1746.08	276.30	0.5961
282.90	1.187288	1746.70	276.06	0.5960
282.65	1.187467	1747.32	275.82	0.5959
282.40	1.187641	1747.95	275.59	0.5958
282.15	1.187817	1748.57	275.35	0.5957
281.90	1.187991	1749.20	275.11	0.5956
281.65	1.188172	1749.83	274.87	0.5954
281.40	1.188344	1750.39	274.66	0.5953
281.15	1.188523	1751.00	274.42	0.5952

280.90	1.188699	1751.60	274.19	0.5951
280.40	1.189050	1752.86	273.72	0.5948
280.15	1.189231	1753.49	273.48	0.5947
279.90	1.189407	1754.12	273.24	0.5946
279.65	1.189580	1754.75	273.01	0.5945

2-hydroxy ethylammonium pentanoate (2-HEAPE)

T (K)	ρ (gcm^{-3})	u (ms^{-1})	κ_S (TPa^{-1})	$10^3 \cdot \alpha$ (K^{-1})	T (K)	ρ (gcm^{-3})	u (ms^{-1})	κ_S (TPa^{-1})	$10^3 \cdot \alpha$ (K^{-1})
338.15	1.020672	1468.15	454.54	-3.6736	307.90	1.039467	1558.18	396.24	-3.8607
337.90	1.020820	1468.77	454.09	-3.6729	307.65	1.039618	1558.99	395.77	-3.8646
337.65	1.020969	1469.46	453.60	-3.6723	307.40	1.039772	1559.78	395.31	-3.8684
337.40	1.021126	1470.18	453.08	-3.6716	307.15	1.039925	1560.61	394.83	-3.8723
337.15	1.021280	1470.87	452.59	-3.6710	306.90	1.040077	1561.44	394.35	-3.8763
336.90	1.021436	1471.58	452.09	-3.6705	306.65	1.040230	1562.25	393.89	-3.8803
336.65	1.021593	1472.29	451.58	-3.6700	306.40	1.040384	1563.08	393.41	-3.8843
336.40	1.021745	1473.00	451.08	-3.6695	306.15	1.040533	1563.89	392.94	-3.8883
336.15	1.021898	1473.73	450.56	-3.6690	305.90	1.040687	1564.73	392.47	-3.8924
335.65	1.022205	1475.12	449.58	-3.6683	305.40	1.040991	1566.38	391.52	-3.9007
335.40	1.022364	1475.83	449.08	-3.6679	305.15	1.041143	1567.19	391.06	-3.9050
335.15	1.022520	1476.54	448.58	-3.6677	304.90	1.041297	1568.03	390.59	-3.9092
334.90	1.022671	1477.24	448.09	-3.6674	304.65	1.041450	1568.87	390.11	-3.9135
334.65	1.022828	1477.95	447.59	-3.6672	304.40	1.041602	1569.72	389.63	-3.9178
334.40	1.022986	1478.66	447.09	-3.6670	304.15	1.041753	1570.56	389.16	-3.9222
334.15	1.023146	1479.37	446.59	-3.6669	303.90	1.041907	1571.39	388.69	-3.9266
333.90	1.023305	1480.07	446.10	-3.6668	303.65	1.042059	1572.25	388.21	-3.9310
333.65	1.023463	1480.78	445.60	-3.6667	303.40	1.042209	1573.09	387.74	-3.9355
333.40	1.023622	1481.49	445.11	-3.6667	303.15	1.042363	1573.94	387.26	-3.9400
333.15	1.023780	1482.20	444.61	-3.6667	302.90	1.042516	1574.79	386.79	-3.9445
332.90	1.023940	1482.92	444.11	-3.6667	302.65	1.042668	1575.65	386.31	-3.9491
332.65	1.024100	1483.63	443.62	-3.6668	302.40	1.042820	1576.51	385.83	-3.9537
332.40	1.024257	1484.34	443.12	-3.6669	302.15	1.042972	1577.39	385.34	-3.9584
332.15	1.024414	1485.06	442.63	-3.6671	301.90	1.043124	1578.23	384.88	-3.9631
331.90	1.024574	1485.77	442.13	-3.6673	301.65	1.043277	1579.11	384.39	-3.9678
331.65	1.024732	1486.48	441.64	-3.6675	301.40	1.043429	1579.97	383.92	-3.9726
331.40	1.024890	1487.19	441.15	-3.6678	301.15	1.043579	1580.82	383.45	-3.9774

331.15	1.025050	1487.90	440.66	-3.6681	300.90	1.043732	1581.71	382.96	-3.9822
330.90	1.025207	1488.62	440.17	-3.6684	300.65	1.043883	1582.58	382.49	-3.9871
330.65	1.025363	1489.35	439.67	-3.6688	300.40	1.044037	1583.48	382.00	-3.9920
330.40	1.025523	1490.05	439.19	-3.6692	300.15	1.044188	1584.38	381.51	-3.9970
330.15	1.025679	1490.79	438.69	-3.6697	299.90	1.044340	1585.27	381.02	-4.0020
329.90	1.025838	1491.51	438.20	-3.6702	299.65	1.044492	1586.16	380.54	-4.0070
329.65	1.025997	1492.23	437.71	-3.6707	299.40	1.044644	1587.08	380.04	-4.0121
329.15	1.026310	1493.70	436.71	-3.6719	298.90	1.044973	1588.87	379.07	-4.0224
328.90	1.026467	1494.41	436.23	-3.6726	298.65	1.045148	1589.78	378.57	-4.0275
328.65	1.026627	1495.14	435.74	-3.6732	298.40	1.045311	1590.70	378.07	-4.0328
327.90	1.027097	1497.32	434.27	-3.6755	297.65	1.045807	1593.44	376.60	-4.0487
327.65	1.027255	1498.06	433.77	-3.6764	297.40	1.045975	1594.39	376.09	-4.0540
327.40	1.027411	1498.78	433.29	-3.6772	297.15	1.046142	1595.32	375.59	-4.0594
327.15	1.027568	1499.51	432.80	-3.6781	296.90	1.046304	1596.24	375.10	-4.0649
326.90	1.027725	1500.24	432.32	-3.6791	296.65	1.046470	1597.18	374.60	-4.0704
326.65	1.027883	1500.98	431.82	-3.6801	296.40	1.046642	1598.12	374.10	-4.0759
326.40	1.028039	1501.70	431.34	-3.6811	296.15	1.046804	1599.08	373.59	-4.0814
326.15	1.028194	1502.44	430.85	-3.6821	295.90	1.046975	1600.00	373.10	-4.0870
325.90	1.028352	1503.16	430.38	-3.6832	295.65	1.047135	1600.95	372.60	-4.0927
325.65	1.028508	1503.88	429.90	-3.6844	295.40	1.047303	1601.93	372.08	-4.0983
325.40	1.028665	1504.64	429.40	-3.6855	295.15	1.047465	1602.89	371.58	-4.1041
325.15	1.028822	1505.36	428.92	-3.6868	294.90	1.047628	1603.86	371.07	-4.1098
324.90	1.028976	1506.11	428.43	-3.6880	294.65	1.047795	1604.81	370.58	-4.1156
324.65	1.029135	1506.84	427.95	-3.6893	294.40	1.047960	1605.78	370.07	-4.1214
324.40	1.029289	1507.58	427.47	-3.6906	294.15	1.048125	1606.77	369.56	-4.1273
324.15	1.029445	1508.32	426.98	-3.6920	293.90	1.048288	1607.74	369.05	-4.1332
323.90	1.029602	1509.05	426.50	-3.6934	293.65	1.048451	1608.73	368.54	-4.1391
323.65	1.029757	1509.79	426.02	-3.6948	293.40	1.048614	1609.75	368.02	-4.1451
323.15	1.030071	1511.28	425.05	-3.6978	292.90	1.048944	1611.75	366.99	-4.1571
322.90	1.030226	1512.02	424.57	-3.6993	292.65	1.049105	1612.77	366.47	-4.1632
322.65	1.030381	1512.75	424.10	-3.7009	292.40	1.049271	1613.76	365.96	-4.1693
322.40	1.030537	1513.50	423.62	-3.7025	292.15	1.049433	1614.77	365.45	-4.1755
322.15	1.030693	1514.23	423.14	-3.7042	291.90	1.049593	1615.76	364.94	-4.1817
321.90	1.030846	1514.98	422.66	-3.7059	291.65	1.049759	1616.79	364.42	-4.1879

321.65	1.031002	1515.72	422.18	-3.7076	291.40	1.049921	1617.83	363.90	-4.1942
321.40	1.031159	1516.46	421.71	-3.7094	291.15	1.050082	1618.87	363.37	-4.2005
321.15	1.031314	1517.21	421.23	-3.7112	290.90	1.050244	1619.95	362.83	-4.2068
320.90	1.031468	1517.96	420.75	-3.7130	290.65	1.050407	1620.99	362.31	-4.2132
320.65	1.031625	1518.71	420.27	-3.7149	290.40	1.050566	1622.02	361.80	-4.2196
320.40	1.031780	1519.46	419.79	-3.7168	290.15	1.050730	1623.16	361.23	-4.2261
320.15	1.031934	1520.22	419.31	-3.7188	289.90	1.050889	1624.19	360.72	-4.2326
319.90	1.032088	1520.97	418.83	-3.7208	289.65	1.051050	1625.29	360.18	-4.2391
319.65	1.032243	1521.73	418.35	-3.7228	289.40	1.051211	1626.38	359.64	-4.2457
319.40	1.032399	1522.49	417.87	-3.7249	289.15	1.051372	1627.47	359.10	-4.2523
319.15	1.032553	1523.24	417.40	-3.7270	288.90	1.051531	1628.60	358.55	-4.2590
318.90	1.032709	1524.00	416.92	-3.7292	288.65	1.051691	1629.70	358.01	-4.2656
318.65	1.032862	1524.77	416.44	-3.7313	288.40	1.051853	1630.82	357.46	-4.2724
318.40	1.033016	1525.53	415.96	-3.7336	288.15	1.052010	1631.92	356.93	-4.2791
318.15	1.033171	1526.28	415.49	-3.7358	287.90	1.052170	1633.05	356.38	-4.2859
317.90	1.033327	1527.05	415.01	-3.7381	287.65	1.052330	1634.18	355.83	-4.2928
317.40	1.033635	1528.57	414.06	-3.7428	287.15	1.052647	1636.52	354.71	-4.3065
317.15	1.033790	1529.33	413.59	-3.7452	286.90	1.052803	1637.66	354.16	-4.3135
316.65	1.034098	1530.86	412.64	-3.7502	286.40	1.053121	1639.97	353.06	-4.3275
316.40	1.034253	1531.63	412.16	-3.7527	286.15	1.053282	1641.17	352.49	-4.3345
316.15	1.034406	1532.39	411.69	-3.7552	285.90	1.053440	1642.36	351.93	-4.3416
315.90	1.034559	1533.16	411.22	-3.7578	285.65	1.053595	1643.59	351.35	-4.3488
315.40	1.034867	1534.71	410.26	-3.7631	285.15	1.053914	1645.91	350.25	-4.3632
315.15	1.035022	1535.47	409.80	-3.7659	284.90	1.054069	1647.20	349.65	-4.3704
314.90	1.035175	1536.22	409.34	-3.7686	284.65	1.054227	1648.38	349.10	-4.3777
314.65	1.035330	1536.99	408.86	-3.7714	284.40	1.054384	1649.68	348.50	-4.3850
314.40	1.035483	1537.77	408.39	-3.7742	284.15	1.054542	1650.96	347.91	-4.3924
314.15	1.035638	1538.53	407.92	-3.7771	283.90	1.054697	1652.23	347.32	-4.3998
313.90	1.035792	1539.30	407.46	-3.7800	283.65	1.054853	1653.49	346.74	-4.4072
313.65	1.035945	1540.06	406.99	-3.7829	283.40	1.055012	1654.78	346.15	-4.4147
313.40	1.036100	1540.83	406.53	-3.7859	283.15	1.055166	1656.17	345.52	-4.4222
313.15	1.036252	1541.60	406.06	-3.7889	282.90	1.055325	1657.46	344.93	-4.4297
312.90	1.036406	1542.37	405.60	-3.7919	282.65	1.055479	1658.73	344.35	-4.4373
312.65	1.036558	1543.14	405.13	-3.7950	282.40	1.055637	1660.17	343.70	-4.4449

312.40	1.036711	1543.91	404.67	-3.7981	282.15	1.055795	1661.49	343.10	-4.4526
312.15	1.036865	1544.69	404.20	-3.8013	281.90	1.055948	1662.83	342.50	-4.4603
311.90	1.037019	1545.47	403.73	-3.8045	281.65	1.056104	1664.24	341.87	-4.4680
311.65	1.037171	1546.25	403.26	-3.8077	281.40	1.056260	1665.61	341.26	-4.4758
311.40	1.037325	1547.02	402.80	-3.8110	281.15	1.056416	1667.01	340.63	-4.4836
311.15	1.037479	1547.82	402.33	-3.8143	280.90	1.056572	1668.41	340.01	-4.4914
310.65	1.037785	1549.39	401.39	-3.8210	280.40	1.056883	1671.29	338.74	-4.5072
310.40	1.037938	1550.17	400.93	-3.8244	280.15	1.057038	1672.76	338.10	-4.5152
310.15	1.038089	1550.96	400.46	-3.8279	279.90	1.057192	1674.21	337.46	-4.5232
309.90	1.038244	1551.75	400.00	-3.8314	279.65	1.057349	1675.59	336.86	-4.5312
309.65	1.038396	1552.56	399.52	-3.8349	279.40	1.057504	1677.18	336.17	-4.5393
309.40	1.038550	1553.36	399.05	-3.8385	279.15	1.057659	1678.69	335.52	-4.5474
309.15	1.038704	1554.16	398.58	-3.8421	278.90	1.057816	1680.20	334.86	-4.5556
308.90	1.038856	1554.95	398.12	-3.8458	278.65	1.057971	1681.62	334.25	-4.5637
308.65	1.039008	1555.77	397.64	-3.8494	278.40	1.058124	1683.11	333.61	-4.5720
308.40	1.039161	1556.55	397.18	-3.8532	278.15	1.058279	1684.75	332.91	-4.5802
308.15	1.039313	1557.36	396.71	-3.8569					

Table 2. Densities (ρ), ultrasonic velocity (u), isentropic compressibilities (κ_S), isobaric expansibilities (α), 278.15-338.15K

The contrary effect is observed for conductivity. At the same temperature, higher viscosity was observed when the salt was of higher molecular weight. The effect of the temperature is similar for all salts.

A frequently applied derived property for industrial mixtures is the isobaric expansibility or thermal expansion coefficient (α), expressed as the temperature dependence of density. Thermal expansion coefficients are calculated by means of $(-\Delta\rho\,/\,\rho)$ as a function of temperature and assuming that α remains constant in any thermal range. As in the case of pure chemicals it can be computed by way of the expression:

$$\alpha = -\left(\frac{\partial \ln\rho}{\partial T}\right)_{P,x} \tag{2}$$

taking into account the temperature dependence of density. The results gathered in Table 2 showed that a minimum of isobaric expansibility is obtained (in terms of negative values) at approximately the same temperature for all ILs. The smaller the size of the cation (mono-ethylene cation), the lower the value of isobaric expansibility was obtained.

Temperature (K)	2-HEAF	2-HEAPE
278.15	2158.20	83.6
288.15	3069.00	143.3
298.15	4197.60	239.6
308.15	5623.20	453.4
318.15	6959.70	632.6
328.15	8563.50	910.8
338.15	10404.90	1202.9

Table 3. Values of ionic conductivity (μS•cm^{-1}) of the 2-HEAF and 2-HEAPE in the range 278.15 – 338.15 K

The values of ionic conductivity are gathered in Table 3. These results show an increasing trend for higher temperatures in each case. This fact may be ascribed to the increasing mobility of the ions for increased temperatures. At the same time, the ionic conductivity values decrease when molecular weight increases, thus 2-HEAPE has a lower ionic conductivity than 2-HEAF, the shortest member of this IL family [23].

The factor studied in this work is the chain length of the anion. The influence of anion residue is higher in terms of steric hindrance, due to its longer structure [2, 23]. This factor produces a higher disturbation on ion package. This fact may be observed in terms of higher values of densities and ultrasonic velocities for those salts of the lighter anion [37].

The ILs studied in this work showed interesting properties for industrial use: low cost of preparation, simple synthesis and purification methods. Moreover, the very low toxicity and the degradability have been verified [38]. Thus, sustainable processes can be originated from their use.

With this in mind, we decided to test their catalytic potential for several aldol condensation reactions with interest for fine chemicals synthesis. At industrial level aldol condensations are catalyzed by homogeneous alkaline bases (KOH or NaOH) [39,40] but with this kind of catalysts numerous disadvantages arise such as loss of catalysts due to separation difficulties, corrosion problems in the equipment and generation of large amounts of residual effluents which must be subsequently treated to minimize their environmental impact. Consequently, new technological solutions have to be developed in order to generate new and more environmental friendly processes.

The condensation reaction between citral and acetone leads to the formation of pseudoionones which are precursors in the commercial production of vitamin A. In the last years, the aldol condensation between citral and acetone has been studied by several groups employing different types of catalysts: rehydrated hydrotalcites [41], mixed oxides derived from hydrotalcites [42, 43], organic molecules [44], ionic liquids [28] etc.

Using the mixed oxides derived from hydrotalcites Climent et al. [42, 43] obtained a conversion of 83% and selectivity to pseudoionones of 82% in 1 h. Abello et al. obtained a citral conversion of 81% in only 5 min employing rehydrated hydrotalcites as catalysts [41] highlighting that Brønsted basic sites are more active than Lewis sites for aldol condensation reactions. In the study of Cota et al. [44] it was shown that 1,8-diazabicyclo[5.4.0]undec-7-ene

(DBU) which has Lewis basic properties, is inactive for aldol condensation reactions; however when it reacts with equimolar amounts of water, this molecule transforms towards a complex that shows Brønsted basic properties and becomes active giving a conversion of 89.17% and a selectivity of 89.6% in 6 h. When choline hydroxide (ionic liquid) was used as catalyst a citral conversion of 93% and selectivity of 98.2% were obtained in 1 h [28].

Among the ILs studied in this work, for citral and acetone condensation (entry 1, Table 4) the most active IL is 2-HEAA, which gives a conversion of 52%, the less active is 2-HEAiB which gives a conversion of 10%. The selectivity obtained in this reaction ranges between 49-83%. No traces of diacetone alcohol derived from the self-condensation of acetone were found but other secondary products coming from the self-condensation of citral and oligomers derived from citral are detected in small quantities in the reaction mixture.

Entry	Substrate	Ketone	Product	Catalyst	Time	Conversion	Selectivity
					(h)	(%)	(%)
1	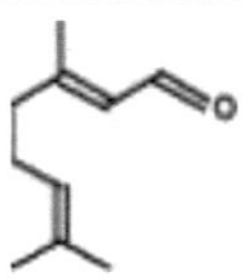	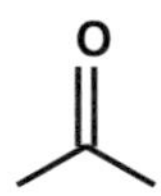	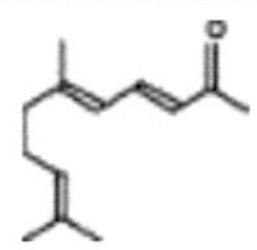	2-HEAF	7	35	83
				2-HEAP		40	63
				2-HEAA		52	74
				2-HEAB		33	60
				2-HEAiB		10	53
				2-HEAPE		38	49
2	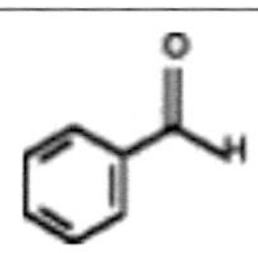	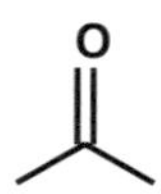	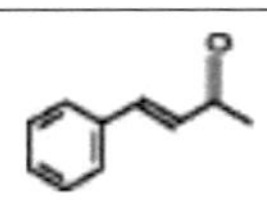	2-HEAF	4	94	82
				2-HEAP	3	100	86
				2-HEAA	4	99	85
				2-HEAB	2	99	85
				2-HEAiB	2	93	85
				2-HEAPE	2	98	77

Table 4. Condensation reactions catalyzed by the studied ILs.

For the production of benzylideneacetone from the aldol condensation between acetone and benzaldehyde, Cota et al. [44] obtained a conversion of 99.9% and 93.97 selectivity in 2 h. When choline hydroxide was employed as catalyst [28] the total conversion was obtained in 0.1 hours but due to the production of dibenzylidenacetone the selectivity to benzylidenacetone decreased around 77%.

When ILs presented in this study were employed for this reaction (entry 2, Table 4), in 2 h of reaction, a conversion of 99% and a selectivity of 85% are obtained when using 2-HEAB as catalyst. Good conversion was also obtained with 2-HEAiB (93%) and 2-HEAPE (98%) with selectivity of 85% and 77% respectively. The decrease in the selectivity to benzylidenacetone is due to the formation of secondary products which include products of aldolisation of benzylidenacetone, like dibenzylidenacetone and other oligomers. The other studied ILs reached the maximum conversion in 3h (2-HEAP) and 4h (2-HEAF and 2-HEAA) and provided high selectivities between 82-86%.

For the repeated runs experiments, we used 2-HEAB in the condensation reaction between acetone and benzaldehyde. The catalyst was recycled 3 times, and in all runs a very good conversion was obtained. The results are presented in Figure 2.

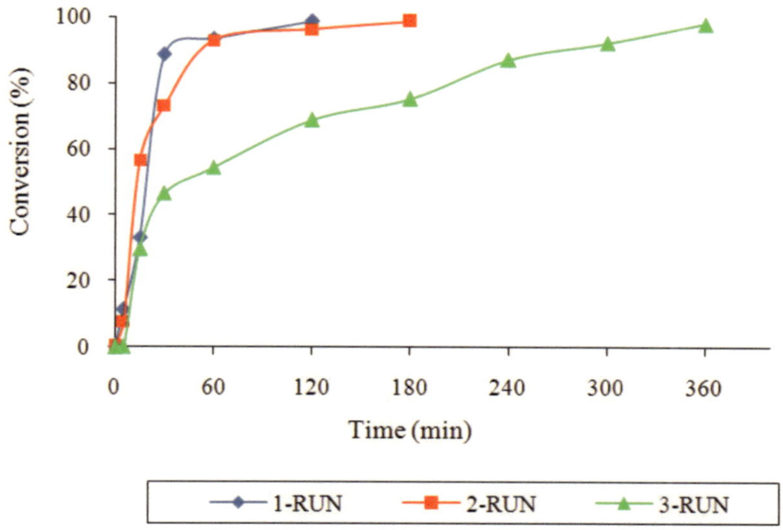

Figure 2. Repeated runs experiments using 2-HEAB in benzylideneacetone synthesis.

The loss of activity noticed in the second and third run can be attributed, on one hand to the loss of IL during the separation process and on the other hand due to the absorption of reaction products on the active sites of the catalyst. IL is partially soluble in the reaction product therefore during the separation procedure small quantities of IL can be dissolved in the organic phase and therefore lost during the separation process. This hypothesis is sustained by the evolution of the specific bands of the ILs which appear in the range 3500-2400 cm^{-1}, almost disappearing in the re-used sample as Figure 3 shows.

A weak band around 1591 cm^{-1} is present in the re-used sample accounting for the carbonyl stretching and N-H plane bending vibrations. On the other hand, deactivation of the catalyst, moreover exhibiting a dark yellow color, is probably due to the adsorption of oligomers and other secondary products on the surface of the catalyst during the reaction. This hypothesis is supported by the appearance of new bands in the re-used IL spectrum. The bands detected in the 1700-1200 cm^{-1} region corresponding to the symmetric and stretching vibrations of CH modes can be assigned to oligomeric species adsorbed on the surface. On the other hand in the 1260-700 cm^{-1} region bands which are normally weak appear and can be assigned to the C-C skeletal vibrations.

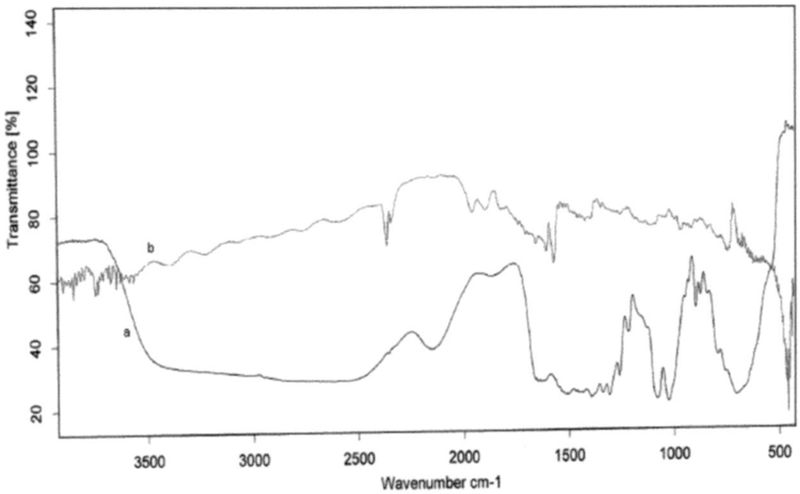

Figure 3. FT-IR spectra for (a) 2-HEAB before reaction, (b) 2-HEAB after reaction (3 consecutive runs).

In order to facilitate the recovery and re-use of the ILs we decided to immobilize them on a solid support. Immobilization and supporting of ILs can be achieved by simple impregnation, covalent linking of the cation or the anion, polymerization etc [45-47]. Compared to pure ILs, immobilized ILs facilitate the recovery and re-use of the catalyst. Previous reports describe the immobilization of ILs by adsorption or grafting onto silica surface and their use as catalysts for reactions like Friedel-Crafts acylation [45], hydrogenation [48] and hydroformilation [49]. Organic polymers [30], natural polymers [50] and zeolites [51] have been also used as supports for ILs.

For this purpose, the ILs were supported on alanine, a cheap readily available aminoacid. Their catalytic activity was tested in the same reactions as the pure ILs.

The catalytic activity results of the a-ILs for the citral-acetone condensation are presented in Table 5. After 6 h of reaction, the two isomers of citral can be converted into the corresponding pseudoionone with conversion between 30-56% except for a-HEAiB for which a conversion of 9% was obtained. The most active IL for this reaction is a-2-HEAA which provides a conversion of 56%. The selectivity obtained in this reaction ranges between 48-80%. No traces of diacetone alcohol derived from the self-condensation of acetone were found, but other secondary products coming from the self condensation of citral and oligomers derived from citral are detected in the reaction mixture. The support (entry 1) is not catalytically active.

In the condensation reaction of benzaldehyde and acetone the first step is the deprotonation of an acetone molecule to give the enolate anion whose nucleophilic attack on the C=O group of benzaldehyde leads to the β-aldol. This latter is easily dehydrated on weak acid sites and benzylidenacetone is obtained.

Entry	Catalyst	Conversion	Selectivity
		(%)	(%)
1	alanine	0	0
2	a-2-HEAF	30	61
3	a-2-HEAA	56	74
4	a-2-HEAP	49	80
5	a-2-HEAB	35	63
6	a-2-HEAiB	9	52
7	a-2-HEAPE	33	48

Table 5. Conversion at 6 h for citral-acetone condensation catalyzed by a-ILs

Entry	Catalyst	Conversion	Selectivity
		(%)	(%)
1	alanine	0	0
2	a-2-HEAF	99	83
3	a-2-HEAA	99	82
4	a-2-HEAP	99	85
5	a-2-HEAB	99	84
6	a-2-HEAiB	78	82
7	a-2-HEAPE	98	80

Table 6. Conversion at 2 h for benzaldehyde-acetone condensation catalyzed by a-ILs

In 2 hours of reaction a conversion of 98-99% is achieved for the majority of a-ILs, while a lower conversion (78%) is obtained for a-2-HEAiB (Table 6). The selectivity toward benzylidenacetone is around 80-86% due to the formation of dibenzylidenacetone as secondary product. The support, alanine (entry 1) is not active for citral acetone condensation.

It is noteworthy that, for both studied reactions, the conversions obtained with the a-ILs are in the same range as the ones obtained with free ILs (Figure 4 and 5).

The a-ILs are easily separated from the reaction mixture and reused. For the consecutive runs experiments we chose condensation between benzaldehyde and acetone as model reaction. The catalysts were recycled for 3 consecutive runs and in all runs a very good conversion was obtained. The results are presented in Figure 6.

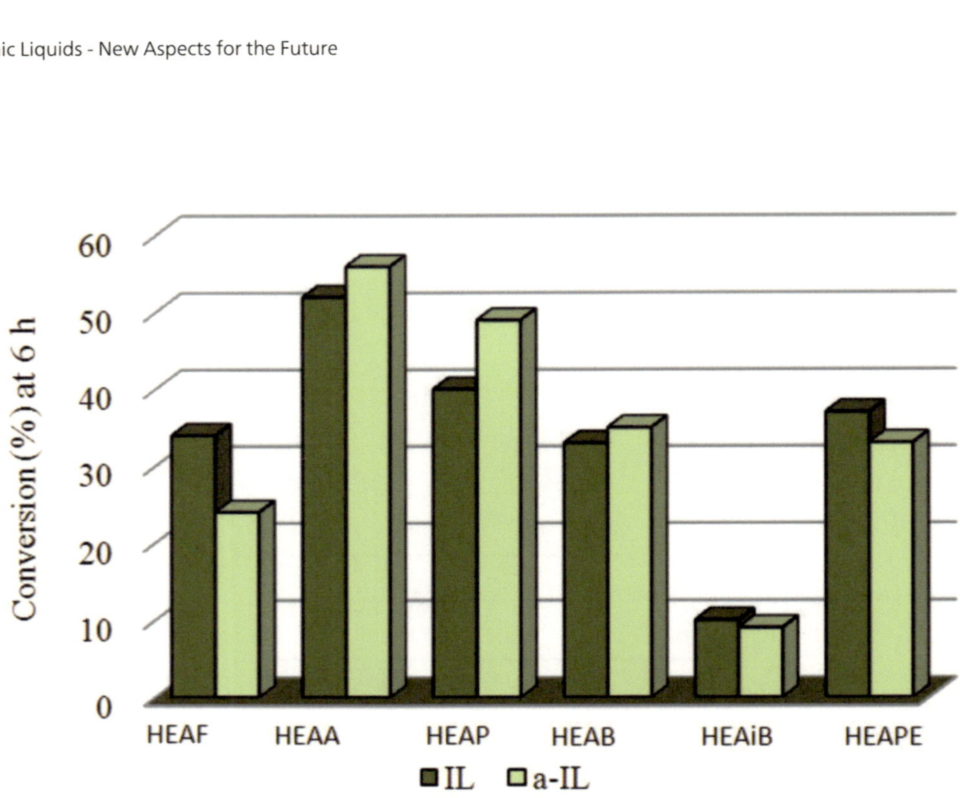

Figure 4. Conversion at 6 h for citral-acetone condensation for free ILs and a-ILs.

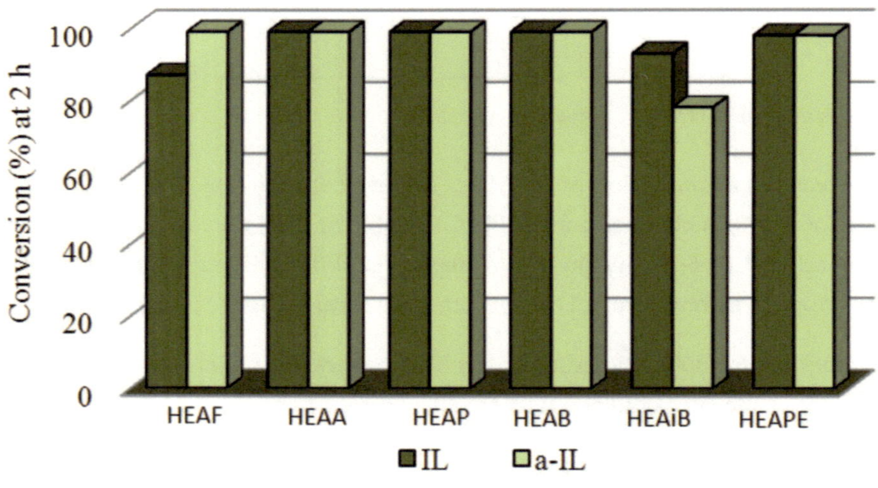

Figure 5. Conversion at 2 h for benzaldehyde-acetone condensation for free ILs and a-ILs.

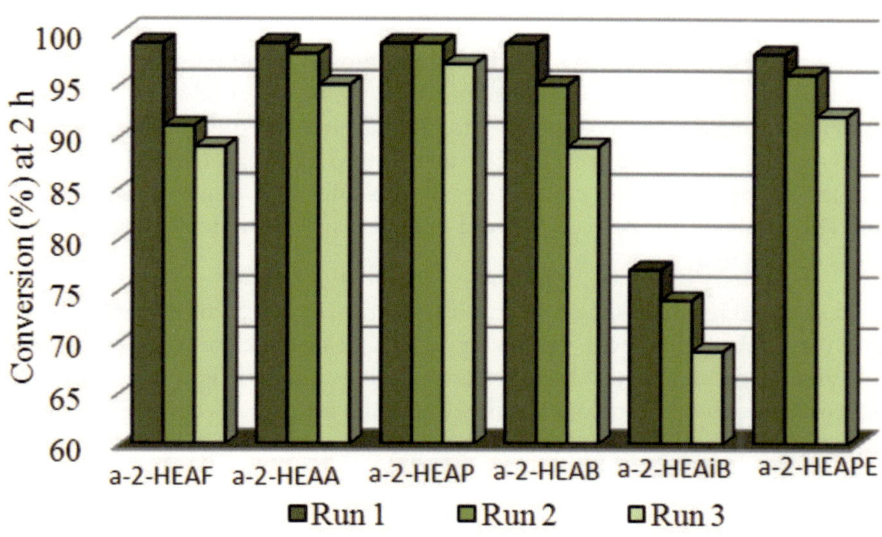

Figure 6. Consecutive runs experiments in benzaldehyde acetone condensation.

In the case of each IL, only a negligible loss of activity is detected in the second and third run which can be attributed to the possible adsorption of reactants or reaction products to the active sites of the catalyst.

From the comparison made with the aforementioned basic catalysts employed for these two aldol condensation reactions we can conclude that the ILs presented in this study are not the most active catalysts for these reactions but due to their green character and easy separation from the reaction media represent a convenient and environmental friendly alternative for the traditional homogeneous catalysts.

4. Conclusions

In this work, we present a simple and efficient synthesis protocol for protic ionic liquids and the experimental data for density, ultrasonic velocity and ionic conductivity of these liquid salts. It was found that increased temperature diminishes the interaction among ions and therefore lower values of density, ultrasonic velocity, viscosity, surface tension and refractive index are obtained for increased temperatures in each case. The contrary effect is observed for conductivity.

The influence of chain length of the anion on the physicochemical properties of the ILs has been also studied. The effect of the anion residue is higher in terms of steric hindrance, due to its longer structure. This factor produces a higher disturbation on ion package. The physicochemical data of ILs are important for both, designing cleaner technological processes and understanding the interactions in this kind of compounds

The catalytic potential of these new ILs was tested for two aldol condensation reactions with interest for fine chemistry industry. Conversions ranging from 35 to 52% and selectivities up to 83% are obtained for the condensation of citral with acetone. In the synthesis of benzilidenacetone, conversions above 93% with selectivities around 85% are obtained. We also studied the optimization of the recovery process of the ILs and their reuse in repeated runs of experiments. The catalysts can be recycled and reused for three consecutive cycles without significant loss of activity.

In addition, in order to improve the recovery process, the ILs were immobilized on alanine, a cheap readily available aminoacid. The catalytic activity of the alanine supported ILs was tested for citral-acetone and benzaldyde-acetone condensations. It is noteworthy that, for both studied reactions, the conversions obtained with the a-ILs are in the same range as the ones obtained with free ILs; moreover the catalysts can be recycled and reused for three consecutive cycles without significant loss of activity.

The ILs studied in this work showed interesting properties for industrial use: low cost of preparation, simple synthesis and purification methods. Moreover, the very low toxicity and the degradability have been verified. Thus, sustainable processes can be originated from their use.

Acknowledgements

This work has been financed by the MEC of Spain and the Generalitat of Catalunya (ICREA ACADEMIA AWARD).

Author details

I. Cota[1], R. Gonzalez-Olmos[2], M. Iglesias[3] and F. Medina[1]

1 Departament d'Enginyeria Química, Escola Tècnica Superior d'Enginyeria Química, Universitat Rovira i Virgili, Avinguda Països Catalans 26, Campus Sescelades, 43007 Tarragona, Spain

2 Laboratory of Chemical and Environmental Engineering (LEQUiA), Institute of the Environment, University of Girona, Campus Montilivi s/n, Faculty of Sciences, E-17071 Girona, Spain

3 Departamento de Engenharia Química, Escola Politécnica, Universidade Federal da Bahia, 40210-630 Salvador-Bahia, Brazil

References

[1] Olivier-Bourbigou, H., Magna, L., & Morvan, D. (2010). Ionic liquids and catalysis: Recent progress from knowledge to applications. *Applied Catalysis A: General*, 1-56.

[2] Sheldon, R. (2001). Catalytic reactions in ionic liquids. *Chemical Communications*, 2399-2407.

[3] Bates, E., D., , Mayton, R. D., Ntai, I., & Davis, J. H. (2002). CO2 Capture by a Task-Specific Ionic Liquid. *Journal of the American Chemical Society*, 926-927.

[4] Huddleston, J. G., Willauer, H. D., Swatloski, R. P., Visser, A. E., & Rogers, R. D. (1998). Room temperature ionic liquids a novel media for'clean' liquid-liquid extraction. *Chemical Communications*, 1765-1766.

[5] Zhang, S., Zhang, Q., & Zhang, Z. C. (2004). Extractive Desulfurization and Denitrogenation of Fuels Using Ionic Liquids. *Industrial & Engineering Chemistry Research*, 614-622.

[6] Fuller, J., Carlin, R. T., & Osteryoung, R. A. (1997). The Room Temperature Ionic Liquid 1 Ethyl-3-methylimidazolium Tetrafluoroborate: Electrochemical Couples and Physical Properties. *Journal of the Electrochemical Society*, 3881-3886.

[7] Welton, T. (1999). Room-Temperature Ionic Liquids. Solvents for Synthesis and Catalysis. *Chemical Reviews*, 2071-2084.

[8] Dupont, J., de Souza, R. F., & Suarez, P. A. Z. (2002). Ionic Liquid (Molten Salt) Phase Organometallic Catalysis. *Chemical Reviews*, 3667-3692.

[9] Chauvin, Y. L., Mussmann, L., & Olivier, H. (1996). A Novel Class of Versatile Solvents for Two-Phase Catalysis: Hydrogenation, Isomerization, and Hydroformylation of Alkenes Catalyzed by Rhodium Complexes in Liquid 1, 3 Dialkylimidazolium Salts. *Angewandte Chemie. International Edition in English*, 34, 2698-2700.

[10] Brausch, N., Metlen, A., & Wasserscheid, P. (2004). New, highly acidic ionic liquid systems and their application in the carbonylation of toluene. *Chemical Communications*, 1552-1553.

[11] Jiang, T., Ma, X., Zhou, Y., Liang, S., Zhang, J., & Han, B. (2008). Solvent-free synthesis of substituted ureas from CO_2 and amines with a functional ionic liquid as the catalyst. *Green Chemistry*, 465-469.

[12] Earle, M. J., Mc Cormac, P. R., & Sheldon, K. R. (1999). Diels-Alder reactions in ionic liquids. A safe recyclable alternative to lithium perchlorate-diethyl ether mixtures. *Green Chemistry*, 1, 23-25.

[13] Doherty, S., Goodrich, P., Hardacre, C., Luo, H. K., Rooney, D. W., Seddon, K. R., & Styring, P. (2004). Marked enantioselectivity enhancements for Diels-Alder reactions in ionic liquids catalysed by platinum diphosphine complexes. *Green Chemistry*, 63-67.

[14] Wasserscheid, P., Sesing, M., & Korth, W. (2002). Hydrogensulfate and tetrakis(hydrogensulfato)borate ionic liquids: synthesis and catalytic application in highly Brønsted-acidic systems for Friedel-Crafts alkylation. *Green Chemistry*, 134-138.

[15] Adams, C. J., Earle, M. J., Roberts, G., & Seddon, K. R. (1998). Friedel-Crafts reactions in room temperature ionic liquids. *Chemical Communications*, 185-190.

[16] Song, C. E., Oh, C. R., Roh, E. J., & Choo, D. J. (2000). Cr(salen) catalysed asymmetric ring opening reactions of epoxides in room temperature ionic liquids. *Chemical Communications*, 1743-1744.

[17] Song, C. E., Shim, W. H., Roh, E. J., & Choo, J. H. (2000). Scandium (III) triflate immobilised in ionic liquids: a novel and recyclable catalytic system for Friedel-Crafts alkylation of aromatic compounds with alkenes. *Chemical Communications*, 1695-1696.

[18] Fraga-Dubreuil, J., Bourahla, K., Rahmouni, M., Bazureau, J. P., & Hamelin, J. (2002). Catalysed esterifications in room temperature ionic liquids with acidic counteranion as recyclable reaction media. *Catalysis Communications*, 185-190.

[19] Alleti, R., Oh, W. S., Perambuduru, M., Afrasiabi, Z., Simm, E., & Reddy, V. P. (2005). Gadolinium triflate immobilized in imidazolium based ionic liquids: a recyclable catalyst and green solvent for acetylation of alcohols and amines. *Green Chemistry*, 203-206.

[20] Bradaric, C. J., Downard, A., Kennedy, C., Rovertson, A. J., & Zhou, Y. H. (2003). Industrial preparation of phosphonium ionic liquids. *Green Chemistry*, 143-152.

[21] Wang, Y., Li, H., Wang, C., & Jiang, H. (2004). Ionic liquids as catalytic green solvents for cracking reactions. *Chemical Communications*, 1938-1939.

[22] Cota, I., Gonzalez-Olmos, R., Iglesias, M., & Medina, F. (2007). New Short Aliphatic Chain Ionic Liquids: Synthesis, Physical Properties, and Catalytic Activity in Aldol Condensations. *Journal of Physical Chemistry B*, 12468-21477.

[23] Iglesias, M., Torres, A., Gonzalez-Olmos, R., & Salvatierra, D. (2008). Effect of temperature on mixing thermodynamics of a new ionic liquid: {2Hydroxy ethylammonium formate (2-HEAF) + short hydroxylic solvents}. *Journal of Chemical Thermodynamics*, 119-133.

[24] Yuan, X. L., Zhang, S. J., & Lu, X. M. (2007). Hydroxyl Ammonium Ionic Liquids: Synthesis, Properties, and Solubility of SO_2. *Journal of Chemical & Engineering Data*, 596-599.

[25] Kurnia, K. A., Harris, F., Wilfred, C. D., Mutalib, M. I. A., & Murugesan, T. (2009). Thermodynamic properties of CO_2 absorption in hydroxyl ammonium ionic liquids at pressures of (100-1600) kPa. *Journal of Chemical Thermodynamics*, 1069-1073.

[26] Li, X. Y., Hou, M. Q., Zhang, Z. F., Han, B. X., Yang, G. Y., Wang, X. L., & Zou, L. Z. (2008). Absorption of CO2 by ionic liquid/polyethylene glycol mixture and the thermodynamic parameters. *Green Chemistry*, 879-884.

[27] Zhu, A., Jiang, T., Wang, D., Han, B., Liu, L., Huang, J., Zhang, J., & Sun, D. (2005). Direct aldol reactions catalyzed by 1,1,3,3-tetramethylguanidine lactate without solvent. *Green Chemistry*, 514-517.

[28] Abello, S., Medina, F., Rodriguez, X., Cesteros, Y., Salagre, P., Sueiras, J., Tichit, D., & Coq, B. (2004). Supported choline hydroxide (ionic liquid) as heterogeneous catalyst for aldol condensation reactions. *Chemical Communications*, 1096-1097.

[29] Kryshtal, G. V., Zhdankina, G. M., & Zlotin, S. G. (2005). Tetraalkylammonium and 1, 3Dialkylimidazolium Salts with Fluorinated Anions as Recoverable Phase-Transfer Catalysts in Solid Base-Promoted Cross-Aldol Condensations. *European Journal of Organic Chemistry*, 2822-2827.

[30] Lombardo, M., Pasi, F., Easwar, S., & Trombini, C. (2007). An Improved Protocol for the Direct Asymmetric Aldol Reaction in Ionic Liquids, Catalysed by Onium Ion-Tagged Prolines. *Advanced Synthis & Catalysis*, 2061-2065.

[31] Kotrusz, P., Kmentova, I., Gotov, B., Toma, S., & Solcaniova, E. (2002). Proline-catalysed asymmetric aldol reaction in the room temperature ionic liquid [bmim]PF_6. *Chemical Communications*, 2510-2511.

[32] Loh, T. P., Feng, L. C., Yang, H. Y., & Yiang, J. Y. l. (2002). l-Proline in an ionic liquid as an efficient and reusable catalyst for direct asymmetric aldol reactions. *Tetrahedron Letters*, 8741-8743.

[33] Wang, C., Zhao, W., Li, H., & Guo, L. (2009). Solvent-free synthesis of unsaturated ketones by the Saucy-Marbet reaction using simple ammonium ionic liquid as a catalyst. *Green Chemistry*, 843-847.

[34] Gu, Y., Zhang, J., Duan, Z., & Deng, Y. (2005). Pechmann Reaction in Non-Chloroaluminate Acidic Ionic Liquids under Solvent-Free Conditions. *Advanced Synthesis & Catalysis*, 512-516.

[35] Mallakpour, S., & Seyedjamali, H. (2009). Ionic liquid catalyzed synthesis of organosoluble wholly aromatic optically active polyamides. *Polymer Bulletin*, 605-614.

[36] Kim, D. W., & Chi, D. Y. (2004). Polymer-Supported Ionic Liquids: Imidazolium Salts as Catalysts for Nucleophilic Substitution Reactions Including Fluorinations. *Angewandte Chemie*, 483-485.

[37] Iglesias, M., Garcia-Muñoz, R., Gonzalez-Olmos, R., Salvatierra, D., & Mattedi, S. (2007). Analysis of methanol extraction from aqueous solution by n-hexane: Equilibrium diagrams as a function of temperatura. *Journal of Molecular Liquids*, 52-58.

[38] Peric, B., Marti, E., Sierra, J., Cruañas, R., Iglesias, M., & Garau, M. A. (2011). Terrestrial ecotoxicity of short aliphatic protic ionic liquids. *Environmental Toxicology and Chemistry*, 2802-2809.

[39] Gradeff, P. S. (1974). US Patent 3,840,601, to Rhodia Inc.

[40] Mitchell, P. W. D. (1989). US Patent 4,874,900, to Union Camp Corporation.

[41] Climent, M. J., Corma, A., Iborra, S., & Velty, A. (2002). Synthesis of pseudoionones by acid and base solid catalysts. *Catalysis Letters*, 157-163.

[42] Climent, M. J., Corma, A., Iborra, S., Epping, K., & Velty, A. (2004). Increasing the basicity and catalytic activity of hydrotalcites by different synthesis procedures. *Journal of Catalysis*, 316-326.

[43] Abello, S., Medina, F., Tichit, D., Perez-Ramirez, J., Groen, J. C., Sueiras, J., Salagre, P., & Cesteros, Y. (2005). Aldol Condensations Over Reconstructed Mg-Al Hydrotalcites: Structure-Activity Relationships Related to the Rehydration Method. *Chemistry a European Journal*, 728-739.

[44] Cota, I., Chimentao, R., Sueiras, J. E., & Medina, F. (2008). The DBU-H2O complex as a new catalyst for aldol condensation reactions. *Catalysis Communications*, 2090-2094.

[45] Valkenberg, M. H., de Castro, C. W., & Hölderich, F. (2002). Immobilisation of ionic liquids on solid supports. *Green Chemistry*, 88-93.

[46] Gadenne, B., Hesemann, P. J., & Moreau, J. E. (2004). Supported ionic liquids: ordered mesoporous silicas containing covalently linked ionic species. *Chemical Communications*, 1768-1769.

[47] Mehnert, C. P. (2005). Supported Ionic Liquid Catalysis. *Chemistry a European Journal*, 50-56.

[48] Mehnert, C. P., Mozeleski, E. J., & Cook, R. A. (2002). Supported ionic liquid catalysis investigated for hydrogenation reactions. *Chemical Communications*, 3010-3011.

[49] Mehnert, C. P., Cook, R. A., Dispenziere, N. C., & Afeworki, M. (2002). Supported Ionic Liquid Catalysis – A New Concept for Homogeneous Hydroformylation Catalysis. *Journal of the American Chemical Society*, 12932-12933.

[50] Baudoux, J., Perrigaud, K., Madec-J, P., Gaumont-C, A., & Dez, I. (2007). Development of new SILP catalysts using chitosan as support. *Green Chemistry*, 1346-1351.

[51] Hu, Y. Q., Wang, J. Y., Zhao, R. H., Liu, Y., Liu, R., & Li, Y. (2009). Catalytic Oxidation of Cyclohexane over ZSM-5 Catalyst in N-alkyl-N-methylimidazolium Ionic Liquids. *Chinese Journal of Chemical Engineering*, 407-411.

Chapter 16

Protic and Nonprotic Ionic Liquids in Polar Diels-Alder Reactions Using Properly Substituted Heterocycles and Carbocycles as Dienophiles. A DFT study

Pedro M. E. Mancini, Carla M. Ormachea,
Claudia D. Della Rosa, María N. Kneeteman and
Luis R. Domingo

Additional information is available at the end of the chapter

http://dx.doi.org/10.5772/51656

1. Introduction

The Diels-Alder (D-A) reaction is one of the most useful processes in preparative organic chemistry. Its potential in heterocyclic chemistry and natural products synthesis is very well known. It provides the chemist with one of his best tool for the preparation of cyclic compounds having a six-membered ring. The process is in one step inter or intramolecular from a diene and dienophile bearing an almost unlimited number of variants. It worth noting that these variants exist not only in the substitution of the reaction component but also in the electronic nature of these dienes and dienophiles. (Carruthers W, 1990; Fringelli F. et al 2002)

The D-A reaction has remained as one of the most powerful organic transformations in chemical synthesis, particularly in obtaining polycyclic rings. With the potential of forming carbon-carbon, carbon-heteroatom and heteroatom-heteroatom bonds, the reaction underlies the synthesis of diverse carbo- and heterocycle compounds. (Corey, 2002)

The design and discovery of ionic liquids (ILs) displaying a melting point lower than 100 ºC, mainly room temperature ionic liquids (RTILs), have been the subject of considerable research efforts over the past decade RTILs have attracted considerable attention because these are expected to be ideal solvents to provide novel reactions in green chemistry (Hitchcock, et al, 1986; Welton, 1999). The interest in this class of molecules arises from their use as liquid media for a variety of chemical transformations specially D-A reactions, as substitutes of organic molecular solvents. ILs has importance due to their unique properties. Thus, this class of molecules is increasingly employed in organic chemistry, material sciences and physical chemistry (Wasserscheid & Keim, 2000; Welton, 1999). An IL is a salt -substance composed exclusively of cations and anions-, and this fact differentiates them from simple ionic solutions, in which ions are dissolved in a molecular medium. They are also different from inorganic molten salts because their melting points are lower than 100 ºC (most of them exist in liquid form at or near room temperature).

RTILs exhibit a variety of desirable properties, such as negligible vapor pressure, which makes them interesting for various applications. In particular, the option of fine-tuning chemical and physical properties by an appropriate choice of cations and anions has stimulated much of the current excitement with respect to these compounds and has led to the term "design solvents". ILs have high solvation ability for a wide variety of polar, non polar, organic and inorganic molecules as well as organometallic compounds. Moreover, the possibility of changing their properties allowing the selective salvation of solvents and thus control the mutual miscibility of particular organic compounds such as for instance alcohols and water. As a consequence, the characterization of the properties of different classes of ILs used as solvent for specific applications and for chemical reactions and catalysis, have been intensively investigated (Mancini, P.M.E., et al, 2012).

In base to their ionic nature, the structure of ILs incorporates different level of complexity. First and in order to maintain local electro neutrality, the high-charge density parts of cations and anions must create a three-dimensional network where the nearest neighbors of a given ion are of opposite sign. Second, the low-charge density residues that are often presents in the ions (generally as alkyl side chain) are segregated from the polar network, forming non polar domains. This nano-segregation/structuration between polar and non polar regions, first predicted by molecular dynamics simulation studies and later corroborated by diffraction techniques, implies the existence of differentiated and complex interactions not only in pure ionic liquid but also in their mixtures with molecular solutes or even other ionic liquids.

The imidazolium ILs were used to investigate the influence of the alkyl chain length and the presence of functional group on the cation on the polarity and the three Kamlet and Taft parameters (π^*, α, and β) measured the solvent dipolarity/polarizability, hydrogen-bond donating (HBD) acidity and hydrogen-bond accepting (HBA) basicity. The results shown that the length of the alkyl chain on the cation has a significant influence on polarizability and on HBD, but only a small influence on HBA and π^*, indicating that in these type of IL's, HBD is a major contributor to polarity.

In the last years we reported the electrophilic behavior of different aromatic heterocyclopentadienes properly mono and disubstituted with an electron-withdrawing group such as nitro or carboxylate in their exposure to different dienes under thermal conditions, using molecular solvents and ionic liquids, respectively. Moreover, we use as dienophile in this type of polar D-A reactions (P-DA) with normal electron demand, nitrotosylindoles, nitrobenzofuranes and nitrobenzothiophenes, properly substituted.

In general, these reactions are domino processes which are initialized by a P-DA reaction to give the formal [4+2] cycloadduct, which undergo an irreversible elimination of nitrous

http://dx.doi.org/10.5772/51656

acid; this elimination is the factor responsible for the feasibility of the overall process. An alternative way is the hetero D-A reaction which takes place when some thiophene derivatives act as electrophiles. This last behavior is probably due to the improved aromatic character of these heterocycles.

For P-DA reactions one of the most interesting aspects is its solvent dependence. Moreover, in recent years, these reactions have been subject of several studies in order to enhance the reactivity. For specific P-DA reactions was demonstrated that the aqueous solutions have remarkable increase in reactivity and selectivity, and these results were discussed in terms of hydrogen-bond (HB) formation. Protic ionic liquids (PILs) with similar properties to water, such as being highly ordered media and good hydrogen bonding donor, have also been shown to have potential influence the outcome of P-DA reactions. Also, in this direction it is interesting discuss which is the influence on these reactions of non protic ionic liquids.

Due to our interest in the cycloaddition chemistry of substituted aromatic heterocycles with electron-withdrawing groups, we have reported that 2- and 3-nitrotosylpyrroles, 2-nitrofurane, 2- and 3-nitrothiophenes, 1-tosyl-3-nitroindole, 2-nitrobenzofurane, 5- y 8-nitroquinolines and 1-nitronaphtalene, react as electrophiles in normal electron demand D-A reaction (Biolatto, B., et al, 2001) (Della Rosa, C., et al, 2004, 2005, 2007,2010, 2011; Brasca, R., et al, 2010, 2011; Cancian, S., et al, 2010; Paredes, E., et al, 2007). These dienophiles were exposed to different dienes strongly, moderately and poorly activated under thermal conditions using molecular solvents as reaction media and in the same cases using PILs. In these reactions, the best results were obtained with the PILs and with chloroform as molecular solvent, due to its potential character HBD which could be influence the reactivity of the reaction systems. The participation of *N*-tosyl-nitropyrroles in cycloaddition reactions made possible a one spot simple indole synthesis.

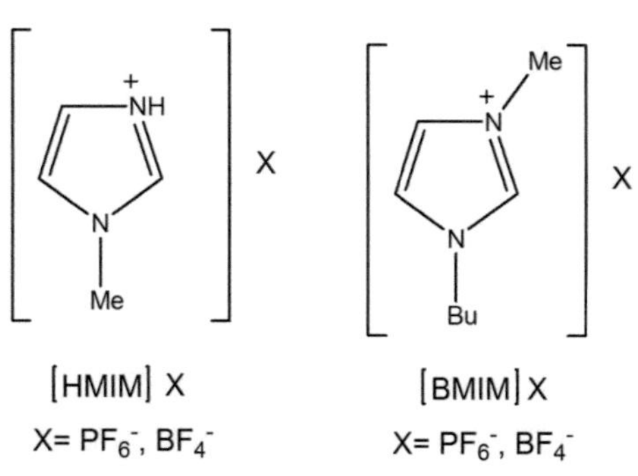

For to analyze the influence of room temperature ionic liquids (RTILs) in this type of polar cycloaddition reactions in which the dienophiles are relatively poor, ethylammoniun nitrate (EAN), 1-methylimidazolium tetrafluoroborate ([HMIM][BF_4]), and 1-methylimidazolium hexafluorophosphate ([HMIM][PF_6]), 1-*n*-butyl-3-methylimidazolium tetrafluoroborate

([BMIM][BF_4]) and 1-*n*-butyl-3-methylimidazolium hexafluorophosphate ([BMIM][PF_6]) were selected as reaction media. To explore the normal electron demand D-A dienophilicity of the proposed dienophiles we choose isoprene, 1-trimethylsilyloxy-1,3-butadiene, and 1-methoxy-3-trimethylsilyloxy-1,3-butadiene (Danishefsky diene) as dienes.

In general, in all cases studied the presences of ILs have two effects. On one hand improved the yields in reaction conditions softer than those when we used molecular solvents. Moreover manifest a clear tendency to the aromatization of adducts. In particular, when 2- and 3-nitrothiophene reacts in thermal conditions with isoprene in a molecular solvent we observed the corresponding hetero D-A adduct, however if an IL is the reaction media the reactions follow the normal D-A curse.

Considering that microwave irradiation has been used to enhance organic reactions in which an ionic liquid is used as the solvent, we have realized some experiences using a combination of microwave and PILs. In this case we noted that the microwave plus PILs constituted a synergetic mixture with strong effects on the reaction yields. ILs absorbed microwave irradiation extremely well and transfer energy rapidly by ionic conduction. At this point of the study we suppose that microwave irradiation has a major effect in a special range of energy activation barriers (ΔE). When the reaction ΔE is too low, the presence of microwave radiation is not especially important, and if the value is extremely high, this irradiation would not take effect.

Part of this work is specifically concerned with theoretically studies using DFT methods. We try to obtain information about the factors affecting reactivity and selectivity. Previous studies have been developed in this type of calculation including one molecule of the ILs corresponding cation coordinated with the dienofile. In this new generation of theoretical studies the aim is to get a better solvatation model including in the system some molecules (in this case, IL's cations and anions) obtaining a "solvent box" which involves the reactive molecule like a 3D electrostactic network.

2. Main Objetives

The aim of the present review is twofold. The first purpose is to analyze the influence of RTILs protic and nonprotic in polar cycloaddition reactions in which the dienophiles are relatively poor -aromatic carbocycles and heterocycles compounds substituted with electron withdrawing groups-. For this purpose alkylimidazolium and dialkylimmidazolium-based ILs have been selected because the differences in their HBD acidity. The second purpose concerned with theoretically studies using DFT methods. We try to obtain information about reaction mechanisms which would be affect reactivity and selectivity. In general, it would be possible demonstrated that the ILs solvent effect in these reactions is in general determined by the solvent hydrogen bond donation ability.

3. Results and Discussion

With the purpose of comparison and reference, we have taken in account the results obtained when the aromatic substrates heterocycles and carbocycles proposed, adequately substituted by electron with-drawing groups, take part in cycloaddition reactions with diverse dienes of nucleophilicity variable, development in molecular solvents or in PILs. These results will be compared with those reached when the cycloaddition reactions are developed in nonprotic ILs. If it is necessary the cited compilation will be included in the corresponding tables of results, looking for an appropriate comprehension of the influence of the reaction media in this type of Diels-Alder reactions.

3.1. Monocyclic five membered and benzofused five membered nitroheteroaromatic compounds, as dienophiles

The study of dienophilic character of substituted nitroaromatic heterocycles in the presence of ILs was carried out employing 2-nitrofuran (1a), 2,5-dimethyl-3-nitrofuran (1b), 1-tosyl-2-nitropyrrole (2a), 1-tosyl-3-nitropyrrole (2b), 2- and 3- nitrothiophene (3a, 3b), 1-tosyl-3-nitroindol (4) and 2-nitrobenzofurane (5). In addition, isoprene (6), 1- trimethylsiloxy-1,3-butadiene (7), 1-methoxy-3-trimethylsiloxy-1,3-butadiene (Danishefsky's diene) (8) were selected as dienes, covering an attractive spectre of nucleophilic character (Figure 1).

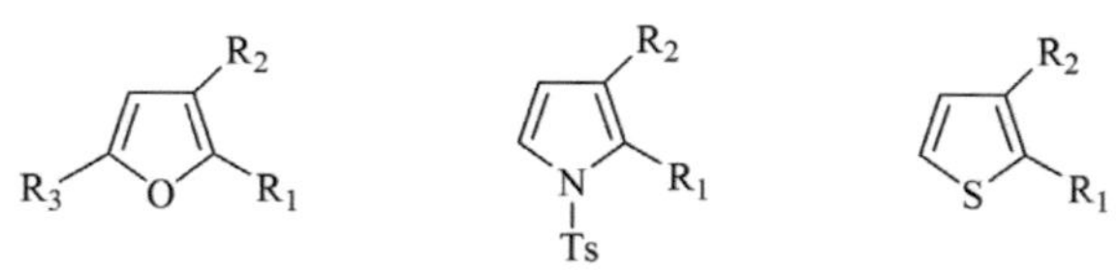

1a R 1= NO_2, R_2=R_3=H
b R_2= NO_2, R_1=R_3=Me

2a R_1= NO_2 , R_2 = H
b R_1= H, R_2= NO_2

3a R_1= NO_2, R_2=H
b R_1 = H, R_2= NO_2

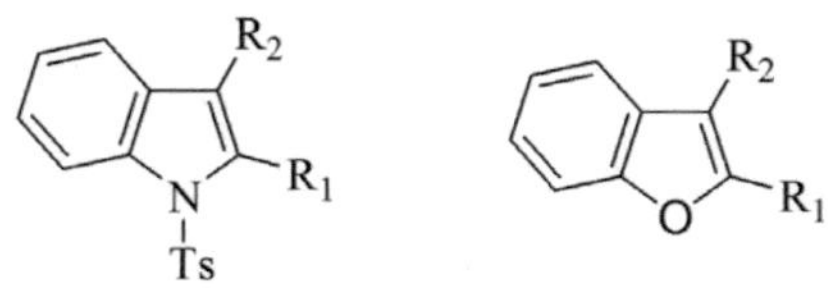

4 R_1= NO_2 , R_2 = H

5 R_1= NO_2 , R_2 = H

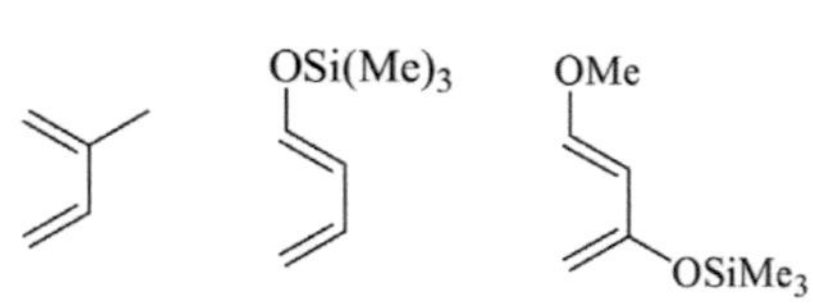

6 **7** **8**

Figure 1. Dienophiles and dienes used in the different experiences.

When 1a was reacted with the less reactive isoprene 6 in a sealed ampoule at 60ºC for 24 h using [BMIM][BF_4] or [BMIM][PF_6] as solvent, respectively, the reactions proceeded to produce a mixture of isomeric benzofurans 10a and 10b (1:1) as the principal products with reasonable yield and dihydrobenzofurans 9a and 9b (1:1) (Figure 2). If the time of the reaction increased to 48 h we observe a 1:1 mixture of isomeric benzofurans 10a and 10b in reasonable yield and traces of the isomeric dihydrobenzofurans 9a and 9b. Similar results were observed when the reaction was development in EAN and [HMIM][BF_4] although with these solvents the yields were majors (Table 1).

In the same manner, in the case of 1-tosyl-2-nitropyrrole 2a, it reacted with isoprene in [BMIM][BF_4] or [BMIM][PF_6] (60ºC, 24 h) furnishing indole isomers 12a,b as the principal products in moderate yield, and dihydroindole isomers 11a,b. All addition products showed extrusion of the nitro group as nitrous acid. (Della Rosa, et al, 2007) (Table 1)

In contrast with the above mentioned behavior, when 2-nitrothiophene 3a was tested with 6, it gave traces of pyrrolyl-thiophene 13 formed by a heterocycloaddition followed by thermal rearrangement. (Della Rosa, et al, 2004) (Figure 2, Table 1). This unexpected behavior was also found with other compounds with stronger aromatic character. The observed low yield in this reaction would be attributed to the interaction between the nitro group and the ILs (Table 1).

Entry	Dienophile[a]	Conditions[b]	Products	Yield (%)[c]
1	**1a**	EAN	**9a,b; 10a,b**	40; 17
2		[HMIM][BF_4]	**9a,b; 10a,b**	30: 15
3		[HMIM][PF_6]	**9a,b; 10a,b**	28; 14
4		[BMIM][BF_4]	**9a,b; 10a,b**	25; 05
5		[BMIM][PF_6]	**9a,b; 10a,b**	26; 05
6	**2a**	EAN	**11a,b;12a,b**	40; 12
7		[HMIM][BF_4]	**11a,b;12a,b**	30; 10
8		[HMIM][PF_6]	**11a,b;12a,b**	28; 09
9		[BMIM][BF_4]	**11a,b;12a,b**	20; 04
10		[BMIM][PF_6]	**11a,b;12a,b**	21; 05
11	**3a**	EAN	**13**	18
12		[HMIM][BF_4]	**13**	12
13		[HMIM][PF_6]	**13**	12
14		[BMIM][BF_4]	**13**	Traces
15		[BMIM][PF_6]	**13**	Traces

[a] Diene/dienophile ratio 12:1
[b] Reaction´s time 24 h, reaction´s temperature 60 ºC.
[c] Based on consumed dienophile.

Table 1. Diels-Alder reactions of 2-nitroheterocycles with isoprene.

http://dx.doi.org/10.5772/51656

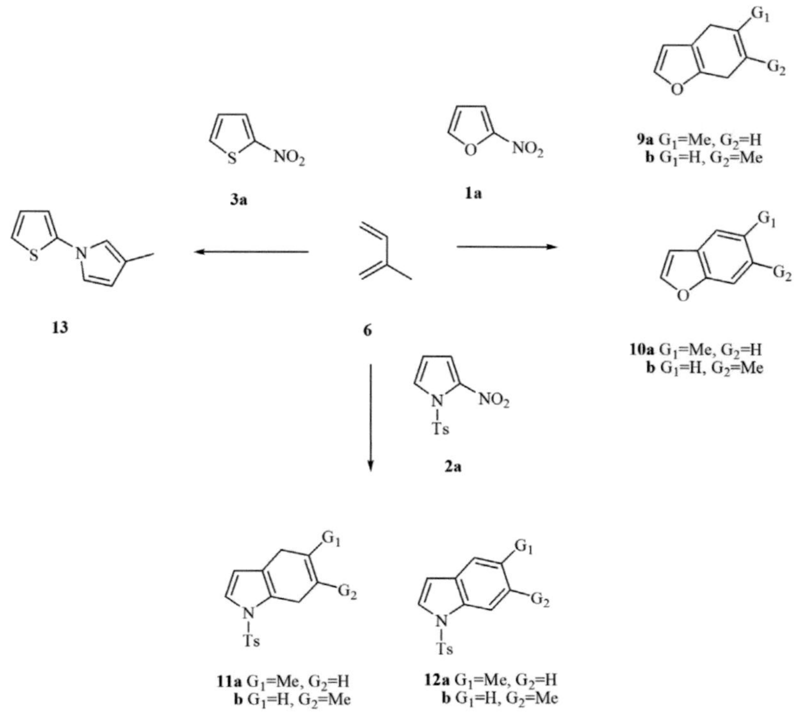

Figure 2. Diels-Alder reactions of aromatic 2- nitroheterocycles with isoprene.

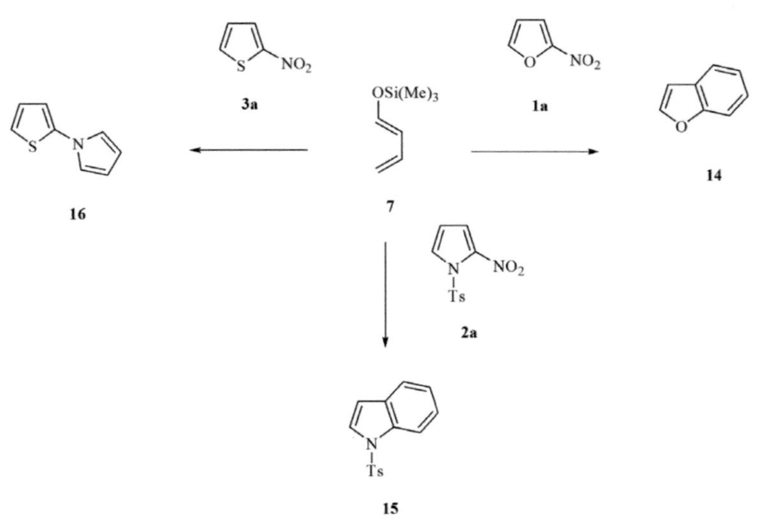

Figure 3. Diels-Alder reactions of aromatic 2- nitroheterocycles with diene 7.

The reactions of 1a with 1-trimethylsilyloxy-1,3-butadiene using [BMIM][BF_4] and [BMIM][PF_6], respectively, as solvents, in seals ampoule (60ºC, 24 h), offered in all cases good yield in benzofuran 14. The yields were lower that those obtained with EAN, [HMIM][BF_4] and [HMIM][PF_6], as solvent. On the other hand, the reaction of 2a with this diene produced *N*-tosylindole 15, with reasonable yield little lower when PILs were used. In the reaction of 3a with 7 we observed again traces of the corresponding hetero Diels-Alder product 16 followed the same trend that the reaction with isoprene (Figure 3)(Table 2).

Entry	Dienophile[a]	Conditions[b]	Products	Yield (%)[c]
1	**1a**	EAN	**14**	62
2		[HMIM][BF_4]	**14**	58
3		[HMIM][PF_6]	**14**	57
4		[BMIM][BF_4]	**14**	42
5		[BMIM][PF_6]	**14**	43
6	**2a**	EAN	**15**	63
7		[HMIM][BF_4]	**15**	51
8		[HMIM][PF_6]	**15**	52
9		[BMIM][BF_4]	**15**	39
10		[BMIM][PF_6]	**15**	38
11	**3a**	EAN	**16**	15
12		[HMIM][BF_4]	**16**	12
13		[HMIM][PF_6]	**16**	12
14		[BMIM][BF_4]	**16**	Traces
15		[BMIM][PF_6]	**16**	Traces

[a] Diene/dienophile ratio 3:1
[b] Reaction´s time 24 h, reaction´s temperature 60 ºC.
[c] Based on consumed dienophile.

Table 2. Diels-Alder reactions of 2- nitroheterocycles with 1-trimethylsilyloxy-1,3-butadiene.

The reaction of Danishefsky´s diene with 1a using the nonprotic ILs cited in the before paragraphs yielded 5-hydroxyibenzofuran 17 in reasonable yield. Similarly to the reactions with isoprene and 1-trimethylsilyloxy-1,3-butadiene, the best yield was observed with EAN (Mancini, P.M.E., et al, 2011).In turn the reactions of 2a with diene 8 developed in these nonprotic ILs offered 1-tosyl-5-hydroxyindole in good yield. However, in the reactions of 3a with this diene and the neoteric solvents, we observed only traces of the aromatic product 19. The results derived from the aromatization of the nitro-adducts promoted by the loss of the nitro and methoxyl groups as nitrous acid and methanol, respectively. The intermediate that suffered nitrous acid extrusion and retained the methoxy group was not detected in any

http://dx.doi.org/10.5772/51656

case. In all cases the presence of PIL´s as reaction media, improve the yields respect to the use of nonpolar ILs (Figure 4) (Table 3).

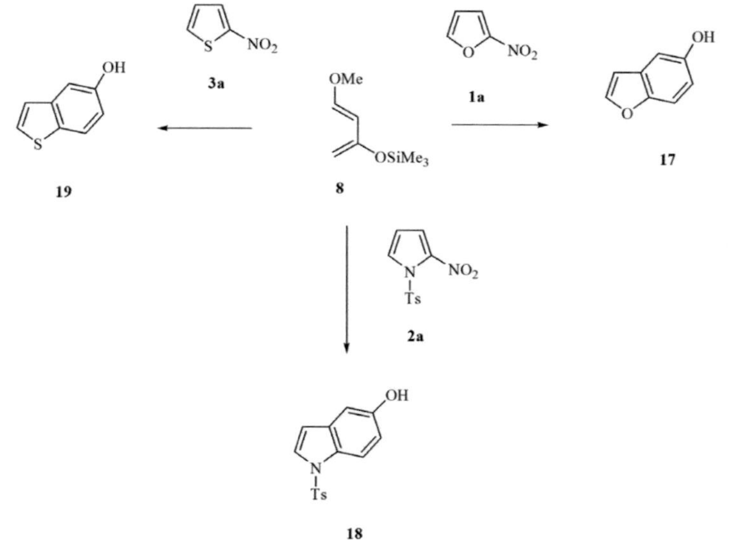

Figure 4. Diels-Alder reactions of aromatic 2- nitroheterocycles with Danishefsky´s diene.

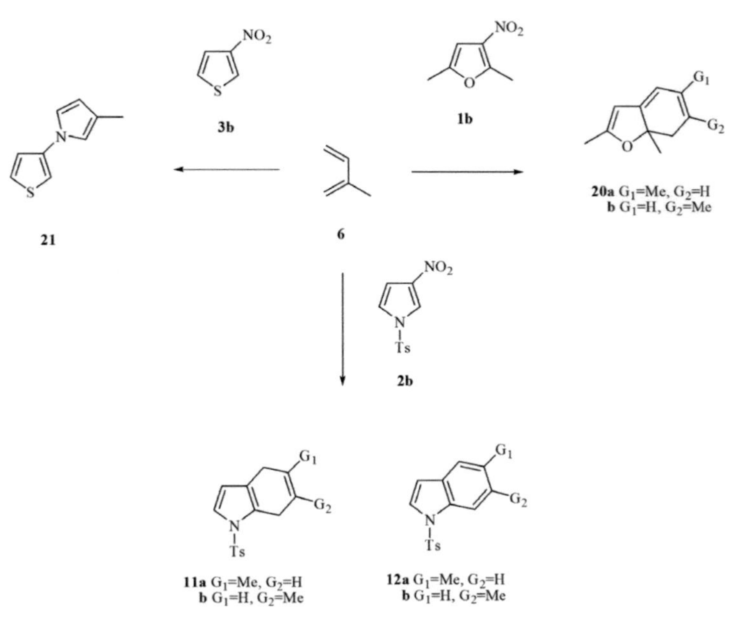

Figure 5. Diels-Alder reactions of aromatic 3- nitroheterocycles with isoprene.

Entry	Dienophile[a]	Conditions[b]	Products	Yield (%)[c]
1	**1a**	EAN	**17**	65
2		[HMIM][BF_4]	**17**	59
3		[HMIM][PF_6]	**17**	60
4		[BMIM][BF_4]	**17**	40
5		[BMIM][PF_6]	**17**	38
6	**2a**	EAN	**18**	57
7		[HMIM][BF_4]	**18**	55
8		[HMIM][PF_6]	**18**	54
9		[BMIM][BF_4]	**18**	42
10		[BMIM][PF_6]	**18**	41
11	**3a**	EAN	**19**	42
12		[HMIM][BF_4]	**19**	36
13		[HMIM][PF_6]	**19**	35
14		[BMIM][BF_4]	**19**	22
15		[BMIM][PF_6]	**19**	21

[a] Diene/dienophile ratio 2:1
[b] Reaction´s time 24 h, reaction´s temperature 60 ºC.
[c] Based on consumed dienophile.

Table 3. Diels-Alder reactions of 2-nitroheterocycles with Danishefsky´s diene.

When 1-tosyl-3-nitropyrrole 2b was tested with isoprene as diene and [BMIM][BF_4], [BMIM][PF_6] as solvent (60ºC, 24 h), the reactions afforded a mixture of regioisomeric cycloadducts previously informed: 11 a,b; 12 a,b. In the same direction and due to the impossibility of obtaining 3-nitrofuran, when 2,5-dimethyl-3-nitrofuran 1b was exposed to 6, the D-A reaction (60ºC/24 h) in the nonprotic ILs proceeded to furnish the mixture of regioisomers 20a and 20b. Once again the reaction of 3b, and this diene development in ILs yielded only traces of the pyrrole derivative 21 in the same manner that 3a, formed by a heterocycloaddition followed by thermal rearrangement (Figure 5) (Table 4). In all cases the yields of these reactions in presence of the PILs are better

Entry	Dienophile[a]	Conditions[b]	Products	Yield (%)[c]
1	**1b**	EAN	**20a,b**	38; 15
2		[HMIM][BF_4]	**20a,b**	30: 15
3		[HMIM][PF_6]	**20a,b**	28; 14
4		[BMIM][BF_4]	**20a,b**	25; 05
5		[BMIM][PF_6]	**20a,b**	26; 05
6	**2b**	EAN	**11a,b;12a,b**	41; 12

http://dx.doi.org/10.5772/51656

Entry	Dienophile[a]	Conditions[b]	Products	Yield (%)[c]
7		[HMIM][BF_4]	**11a,b;12a,b**	30; 10
8		[HMIM][PF_6]	**11a,b;12a,b**	29; 09
9		[BMIM][BF_4]	**11a,b;12a,b**	20; 04
10		[BMIM][PF_6]	**11a,b;12a,b**	21; 05
11	**3b**	EAN	**21**	15
12		[HMIM][BF_4]	**21**	11
13		[HMIM][PF_6]	**21**	11
14		[BMIM][BF_4]	**21**	Traces
15		[BMIM][PF_6]	**21**	Traces

[a] Diene/dienophile ratio 3:1
[b] Reaction´s time 24 h, reaction´s temperature 60 °C.
[c] Based on consumed dienophile.

Table 4. Diels-Alder reactions of 3- nitroheterocycles with isoprene.

Exposure of 1-tosyl-3-nitropyrrole 2b to dienes 7 and 8 in the presence of [BMIM][BF_4], [BMIM][PF_6] yielded 1-tosyl-indole 15 and 1-tosyl-indole-6-ol 23, respectively with moderate yield. At the same time, mononitrated substrate 3b in its reaction with diene 7 afforded traces of the pyrrolyl derivative 22. However, 3b did not undergo cycloaddition with diene 8. Probably this behavior is a consequence of the special reactivity of this substrate connected with its aromaticity (Figures 6 and 7) (Tables 5 and 6).

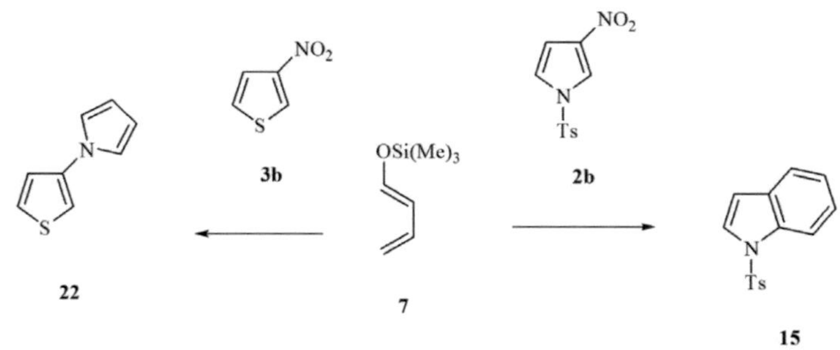

Figure 6. Diels-Alder reactions of aromatic 3- nitroheterocycles with diene 7.

Entry	Dienophile[a]	Conditions[b]	Products	Yield (%)[c]
1	**2b**	EAN	**15**	61
2		[HMIM][BF_4]	**15**	51
3		[HMIM][PF_6]	**15**	50
4		[BMIM][BF_4]	**15**	35

Entry	Dienophile[a]	Conditions[b]	Products	Yield (%)[c]
5		[BMIM][PF_6]	**15**	36
6	**3b**	EAN	**22**	12
7		[HMIM][BF_4]	**22**	11
8		[HMIM][PF_6]	**22**	11
9		[BMIM][BF_4]	**22**	Traces
10		[BMIM][PF_6]	**22**	Traces

[a] Diene/dienophile ratio 3:1
[b] Reaction´s time 24 h, reaction´s temperature 60 °C.
[c] Based on consumed dienophile.

Table 5. Diels-Alder reactions of 3- nitroheterocycles with 7.

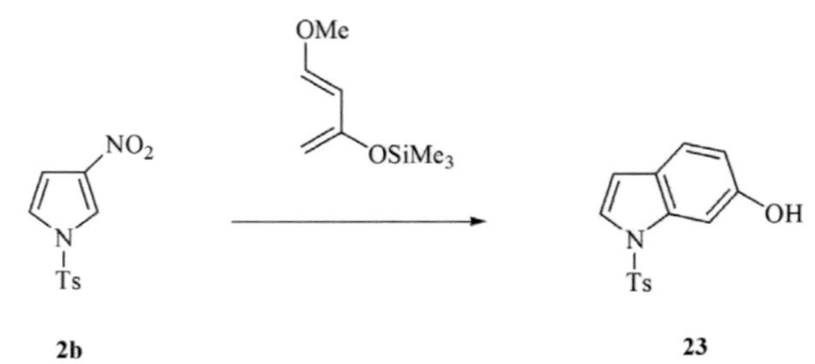

Figure 7. Diels-Alder reactions of aromatic 3- nitroheterocycles with Danishefsky´s diene.

Entry	Dienophile[a]	Conditions[b]	Products	Yield (%)[c]
1	**2b**	EAN	**23**	62
2		[HMIM][BF_4]	**23**	55
3		[HMIM][PF_6]	**23**	54
4		[BMIM][BF_4]	**23**	43
5		[BMIM][PF_6]	**23**	41

[a] Diene/dienophile ratio 3:1
[b] Reaction´s time 24 h, reaction´s temperature 60 °C.
[c] Based on consumed dienophile.

Table 6. Diels-Alder reactions of 3- nitroheterocycles with 9.

3.1.1. Nitroindole as dienophiles

When 1-tosyl-3-nitroindole 4 was tested with isoprene as diene and [BMIM][BF_4] or [BMIM][PF_6] as solvent (60ºC, 24 h), the reactions afforded a mixture of regioisomeric cycloadducts 25a,b, as principal products in reasonable yield, and traces of the regioisomers 24a,b. On the other hand the reactions of 4 with the dienes 7 and 8 in these ILs produced N-tosylcarbazole 26 and the hydroxycarbazole 27 in reasonable good yield. In all cases the yields were lower than those obtained with PILs (EAN, [HMIM][BF_4] and [HMIM][PF_6]) (Figure 8) (Table 7).

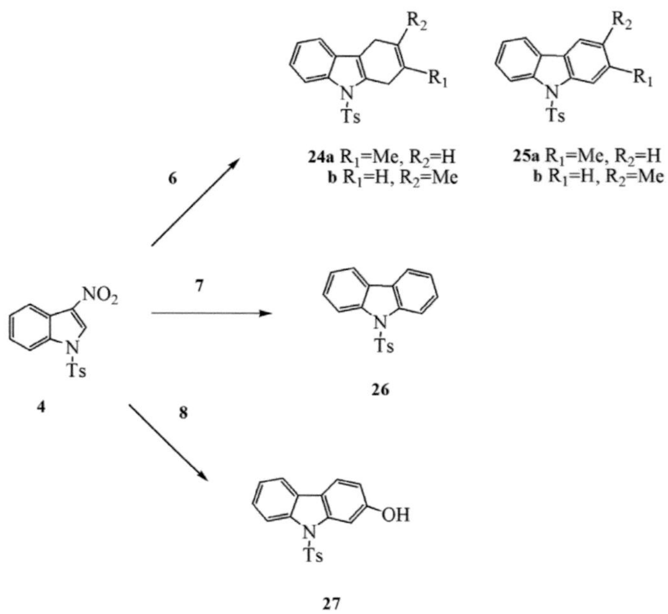

Figure 8. Diels-Alder reactions of 1-tosyl-3-nitroindol with dienes 6, 7 and 8.

Entry	Dienophile[a]	Conditions[b]	Products	Yield (%)[c]
1	**6**	EAN	**24a,b; 25a,b**	45; 05
2		[HMIM][BF_4]	**24a,b; 25a,b**	42: 03
3		[HMIM][PF_6]	**24a,b; 25a,b**	42; 04
4		[BMIM][BF_4]	**24a,b, 25a,b**	30; 02
5		[BMIM][PF_6]	**24a,b; 25a,b**	31; 02
6	**7**	EAN	**26**	62
7		[HMIM][BF_4]	**26**	55
8		[HMIM][PF_6]	**26**	56
9		[BMIM][BF_4]	**26**	40
10		[BMIM][PF_6]	**26**	41
11	**8**	EAN	**27**	72
12		[HMIM][BF_4]	**27**	65
13		[HMIM][PF_6]	**27**	63

Entry	Dienophile[a]	Conditions[b]	Products	Yield (%)[c]
14		[BMIM][BF$_4$]	**27**	51
15		[BMIM][PF$_6$]	**27**	48

[a] Diene/dienophile ratio 3:1
[b] Reaction´s time 24 h, reaction's temperature 60 ºC.
[c] Based on consumed dienophile.

Table 7. Diels-Alder reactions of dienophile 4 with different dienes.

3.1.2. 2-NitroBenzofuran as dienophile

The reactions of 5 with isoprene employing [BMIM][BF$_4$] or [BMIM][PF$_6$] as solvent (60 ºC, 24 h) afforded the mixture of aromatic regioisomeric cycloadducts 28a,b as principal products and traces of dihydrodibenzofurans 29a and 29b. On the other hand, reactions of 5 with 1-trimethylsilyloxy-1,3-butadiene (60ªc, 24 h, [BMIM][BF$_4$] and [BMIM][PF$_6$]) yielded dibenzofuran 30 with loss of trimethylsilyloxy and nitro groups. The yield is good. In the same way, in the reaction of 2-nitrobenzofuran with the Danishefsky`s diene hidroxy aromatic cycloadduct 31 was obtained with very good yield and complete regioselectivity (Figure 9, Table 8). With these solvents the yields were lower than using PILs (EAN, [HMIM][BF$_4$] and [HMIM][PF$_6$])

Entry	Dieno[a]	Conditions[b]	Products	Yield (%)[c]
1	**6**	EAN	**28a,b; 29a,b**	42; 04
2		[HMIM][BF$_4$]	**28a,b; 29a,b**	38: 03
3		[HMIM][PF$_6$]	**28a,b; 29a,b**	38; 04
4		[BMIM][BF$_4$]	**28a,b; 29a,b**	27; 02
5		[BMIM][PF$_6$]	**28a,b; 29a,b**	27; 02
6	**7**	EAN	**30**	58
7		[HMIM][BF$_4$]	**30**	52
8		[HMIM][PF$_6$]	**30**	51
9		[BMIM][BF$_4$]	**30**	36
10		[BMIM][PF$_6$]	**30**	35
11	**8**	EAN	**31**	66
12		[HMIM][BF$_4$]	**31**	59
13		[HMIM][PF$_6$]	**31**	58
14		[BMIM][BF$_4$]	**31**	45
15		[BMIM][PF$_6$]	**31**	42

[a] Diene/dienophile ratio 3:1
[b] Reaction´s time 24 h, reaction´s temperature 60 ºC.
[c] Based on consumed dienophile.

Table 8. Diels-Alder reactions of 2-nitrobenzofuran with different dienes.

http://dx.doi.org/10.5772/51656

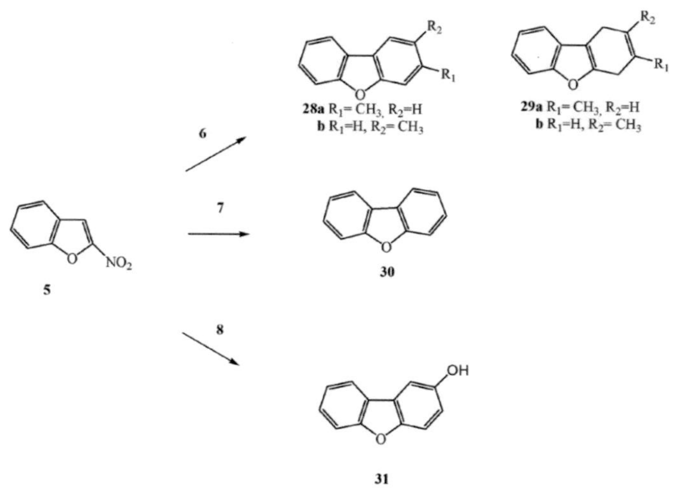

Figure 9. Diels-Alder reactions of 2-nitrobenzofuran with dienes 6, 7 and 8.

3.2. Azanitronaphathalenes as dienophiles

It was explored the cycloaddition reactions between 5-nitro and 8-nitroquinolines (32a and 32b) with the dienes 6, 7 y 8, respectively, in presence of [BMIM][BF_4] and [BMIM][PF_6] (60ºC, 24h).

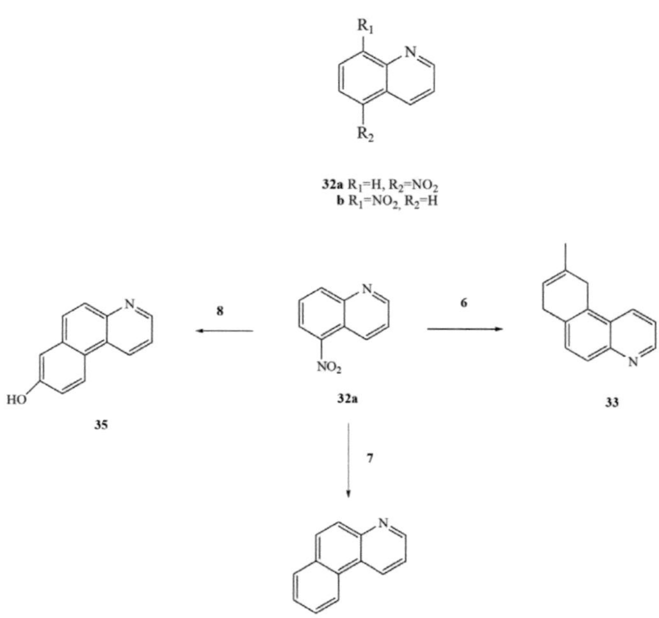

Figure 10. Diels-Alder reactions of 5-nitroquinoline with 6,7 and 8.

The reactions of 5-nitroquinoline with these dienes, yield the same products that using [HMIM][BF_4] as solvent but with lower yield. In the cases of dienes 7 and 8 the normal addition products 34 and 35, respectively, show complete aromatization due to the loss of the nitro and trimethylsyliloxy groups. The product 35 was obtained with complete regioselectivity. However, when the diene 6 was used the observed cycloaddition products was (9-Methyl-7,10-dihydro-benzo [f]quinoline) 33 with traces of its regioisomers (Cancian, et al, 2010) (Figures 10) (Tables 9).

Entry	Dieno[a]	Conditions[b]	Products	Yield (%)[c]
1	**6**	EAN	**33**	18
2		[HMIM][BF_4]	**33**	16
3		[HMIM][PF_6]	**33**	16
4		[BMIM][BF_4]	**33**	12
5		[BMIM][PF_6]	**33**	12
6	**7**	EAN	**34**	22
7		[HMIM][BF_4]	**34**	20
8		[HMIM][PF_6]	**34**	19
9		[BMIM][BF_4]	**34**	14
10		[BMIM][PF_6]	**34**	14
11	**8**	EAN	**35**	25
12		[HMIM][BF_4]	**35**	20
13		[HMIM][PF_6]	**35**	20
14		[BMIM][BF_4]	**35**	15
15		[BMIM][PF_6]	**35**	15

[a] Diene/dienophile ratio 3:1
[b] Reaction´s time 24 h, reaction´s temperature 60 °C.
[c] Based on consumed dienophile.

Table 9. Diels-Alder reactions of dienophile 32a with different dienes.

Entry	Dieno[a]	Conditions[b]	Products	Yield (%)[c]
1	**6**	EAN	**36**	17
2		[HMIM][BF_4]	**36**	15
3		[HMIM][PF_6]	**36**	15
4		[BMIM][BF_4]	**36**	12
5		[BMIM][PF_6]	**36**	12
6	**7**	EAN	**37**	20
7		[HMIM][BF_4]	**37**	18
8		[HMIM][PF_6]	**37**	19

Entry	Dieno[a]	Conditions[b]	Products	Yield (%)[c]
9		[BMIM][BF_4]	**37**	13
10		[BMIM][PF_6]	**37**	13
11	**8**	EAN	**38**	24
12		[HMIM][BF_4]	**38**	20
13		[HMIM][PF_6]	**38**	20
14		[BMIM][BF_4]	**38**	15
15		[BMIM][PF_6]	**38**	15

[a] Diene/dienophile ratio 3:1
[b] Reaction´s time 24 h, reaction´s temperature 60 °C.
[c] Based on consumed dienophile.

Table 10. Diels-Alder reactions of dienophile 32b with different dienes.

In the same way, the reactions of 8-nitroquinoline with these dienes, yield the same products that using [HMIM][BF_4] as solvent but with lower yield. When the were dienes 7 and 8 the normal addition products 37 and 38, respectively, show complete aromatization due to the loss of the nitro and trimethylsyliloxy groups. The product 37 was obtained with complete regioselectivity. However, when the diene 6 was used the observed cycloaddition products was (9-Methyl-7,10-dihydro-benzo[h]quinoline) 36 with traces of its regioisomers (Cancian, et al, 2010) (Figures 11) (Tables 10).

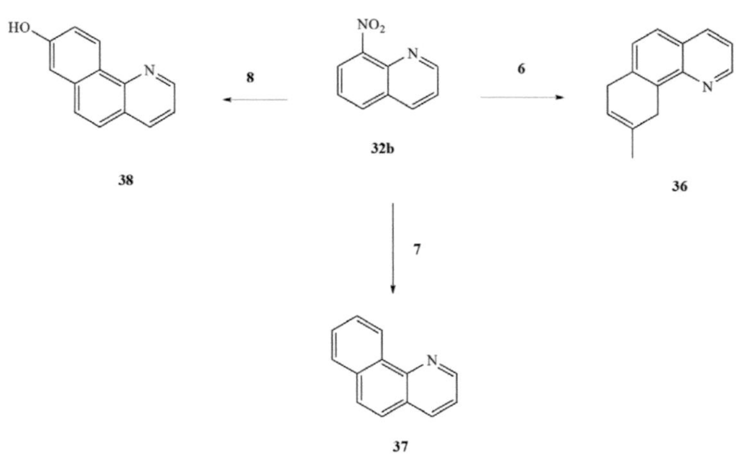

Figure 11. Diels-Alder reactions of 8-nitroquinoline with 6, 7 and 8.

3.3. Nitronaphthalenes as dienophiles

To explore the normal electron-demand D-A dienophilicity of nitronaphtahlenes in presence of no-protic ILs ([BMIM][BF_4] and [BMIM][PF_6]) we selected 1-nitronaphthalene 39 as electrophile and 8 as diene.

When 39 and 8 were heated in a sealed ampoule (60°C, 24 h) using [BMIM][BF_4] and [BMIM] [PF_6], respectively, as solvents, in both cases ca de 50% of 2-hydroxy-phenanthrene 40, was regioselectively produced. The regioselectivity of these reactions was controlled by both the nitro group of the dienophile and the methoxyl group of Danishefsky´s diene. This product was obtained when a PILs was used (e.g. [HMIM][BF_4]), however in this case with major yield The preference for the normal D-A products in the presence of ILs respect the use of molecular solvent, probably is due to the increase of the electrophilicity of the dinenophile (Figure 12).

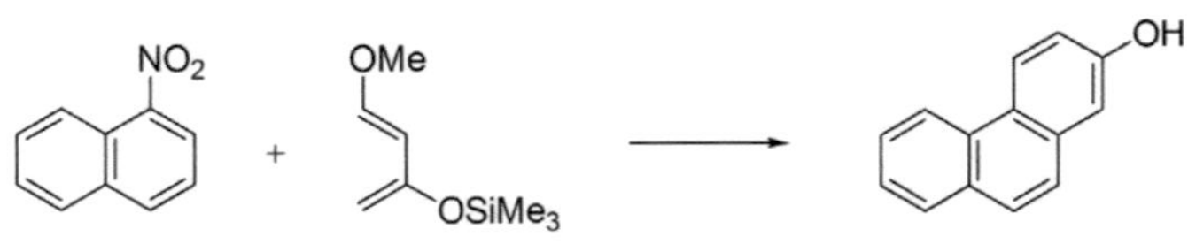

39 8 40

Figure 12. Diels- Alder reaction of 1-nitronaphthalene with Danishefsky's diene.

4. Diels-Alder reactions employing ionic liquids and microwave irradiation

Ionic liquids are becoming promising and useful substitutes for standard organic solvents. Not only they are environmentally benign, they also possess unique chemical and physical properties. Moreover, microwave irradiation has been used to enhance organic reactions in which an ionic liquid is used as the solvent. Ionic liquids absorb microwave irradiation extremely well and transfer energy rapidly by ionic conduction.

The D-A transformations usually require harsh conditions (high temperatures and pressures) and long reaction times. These cycloadditions were the first reaction type to be examined in conjunction with microwave irradiation. Microwave irradiation has also been used to enhance organic reaction in which an ionic liquid is used as the solvent.

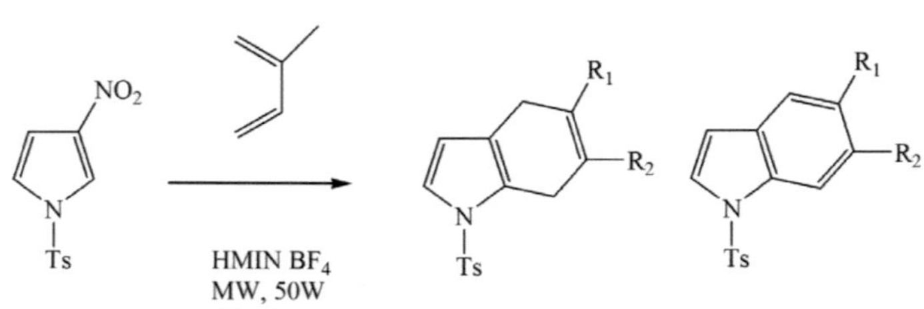

2b

11a R_1=Me, R_2=H
b R_1=H, R_2=Me

12a R_1=Me, R_2=H
b R_1=H, R_2=Me

Figure 13. Diels-Alder reaction of 1-tosyl-3-nitropyrrole with isoprene.

http://dx.doi.org/10.5772/51656

With microwave irradiation (50 W, 30 min.) and [HMIM] [BF_4] as reaction media, 1-tosyl-3-nitropyrrole reacts with isoprene yielding the mixture of isomeric dihydroindoles 11a and 11b and indoles 12a and 12b as the principal.products (global yield 95%). (Figure 13)

In turn, nitrobenzene react with isoprene using benzene as solvent to offer pyrroliybenzene as product in reasonable yield. If the diene is 8 we do not observe reaction probably due to the strong aromatic character of this substrate. It call our attention the absence of reactivity when nitrobenzene reacted with 6 in presence of [HMIM] [BF_4] and microwave irradiation. This result would be consequence of the strong interaction between the nitro group and the PIL.

5. Theoretical studies

5.1. General

The Density Functional Theory (DFT) is a model of quantum mechanics used to obtain electronic structure of different systems, in this case, molecules involved in the Diels-Alder reaction. Considering this theory, there are some parameters (or indexes) that can be used to explain the reactivity and regioselectivity of the cycloaddition reactions. The most significant ones are the chemical hardness (η),that describes the resistance of the chemical potential to a change in the number of electrons, and the electronic chemical potential (μ), which is usually associated with the charge-transfer (CT) ability of the system in its ground state geometry. Both quantities can be approximated in terms of the energies of the HOMO and LUMO frontier molecular orbitals (Eqs. 1 and 2) (Domingo, et al, 2002; Domingo & Aurell, 2002).

$$\eta = (\varepsilon_{LUMO} - \varepsilon_{HOMO}) \tag{1}$$

$$\mu = \frac{(\varepsilon_{LUMO} + \varepsilon_{HOMO})}{2} \tag{2}$$

Based on these parameters Parr (Parr, et al, 1999) introduced the global electrophilicity index (ω), an useful descriptor of the reactivity that allows a quantitative classification of the electrophilicity character of the whole molecule in an unique scale. This index is defined as:

$$\omega = \frac{\mu^2}{2\eta} \tag{3}$$

Current studies based on the DFT have shown that this classification is a powerful tool to predict and justify the feasibility of the D-A process and the type of mechanism involved. The electrophilicity scale describes the effects of electron-donor and electron-withdrawing groups in the diene/dienofile pair. Reactants can be classified for the D-A cycloaddition as strong (>1.50 eV), moderate (1.49 - 0.90 eV) and poor (<0.90 eV) electrophiles.

The difference in electrophilicity between the diene/dienofile pair can be related to the electronic pattern expected in the transition state (TS) of a D-A process and, in consequence, its has been proposed as a measure of the polar character of the reaction.

Additionally, there are local reactivity indexes that are associated with site selectivity in a chemical reaction. They can be calculated from the Fukui function (Parr, R. G. *et al* 1984). Eq. (4) provides a simple and direct formalism to obtain it from an approach based on a relationship between the FMOs.

$$f_k^{\alpha} = \sum_{\mu \in k} |c_{\mu\alpha}|^2 + \sum_{\upsilon \neq \mu} c_{\mu\alpha} c_{\upsilon\alpha} S_{\mu\upsilon} \tag{4}$$

The condensed Fukui function for electrophilic attacks involves the HOMO FMO coefficients (c) and the atomic overlap matrix elements (S).

In this direction, to analyze at which atomic site of a molecule the maximum electrophilicity value is reached, Domingo (Domingo, *et al*, 2002) has introduced Eq. (5)

$$\omega_k = \omega f_k^{+} \tag{5}$$

On the other hand, the first approach toward a quantitative description of nucleophilicity has also been reported by Domingo (Domingo, *et al*, 2008). The global nucleophilicity index, *N*, is defined in Eq. (6)

$$N = (\varepsilon_{HOMO,Nu} - \varepsilon_{HOMO,TCE}) \tag{6}$$

Where $\varepsilon_{HOMO,TCE}$ is the HOMO energy of tetracyanoethylene (TCE) (taken as a reference molecule due to the fact that it exhibits the lowest HOMO energy in a large series of molecules previously considered in D-A cycloadditions).

Its local counterpart, N_k, -Eq. (7)- has been developed with the purpose of identifying the most nucleophilic site of a molecule. In this case Eq (4) is considered for a nucleophilic attack and involves LUMO FMO coefficients.

$$N_k = N f_k^{-} \tag{7}$$

This nucleophilicity index has been useful to explain the nucleophilic reactivity of some dienes with electrophiles in cycloaddition as well as substitution reactions (Domingo, *et al*, 2008).

In general, the polarity of the normal electron demand D-A process is studied trough global electrophilicity indexes difference between reactants. And the regioselectivity of the normal electron demand D-A reaction, using the local electrophilicity index for dienophiles (electrophiles in the reaction) and the local nucleophilicity index for dienes (nucleophiles in the reaction).

http://dx.doi.org/10.5772/51656

In this stage, we show different theoretical studies related to the polar D-A reactions experimentally described, where the dienophiles are aromatic heterocycles or carbocycles. In the same way, the mechanism of these reactions, specially respect to regio-, site- and stereochemistry have been analyzed in detail.

5.2. Dienes

Global electronic properties of the dienes experimentally used in the cycloadditions previously described are exposed in Table 11.

Molecule	Global properties			
	μ (a.u.)	η (a.u.)	ω (eV)	*N (eV)*
Isoprene	-0.1209	0.1962	1.01	2.93
1-trimethylsilyloxy-1,3-butadiene	-0.0911	0.1977	0.75	3.59
Danishefsky's diene	-0.0945	0.1851	0.66	3.55

Table 11. Dienes.

As the substitution of the dienes with electron-donor groups increases, its global electrophilicity decrease. This indicates that the diene became a better nucleophile for D-A reaction.

The effect of this kind of substitutive groups is also reflected in the difference between local nucleophilicity indexes of the extreme carbon atoms. Considering these values, regioselectivity of D-A reactions is evaluated.

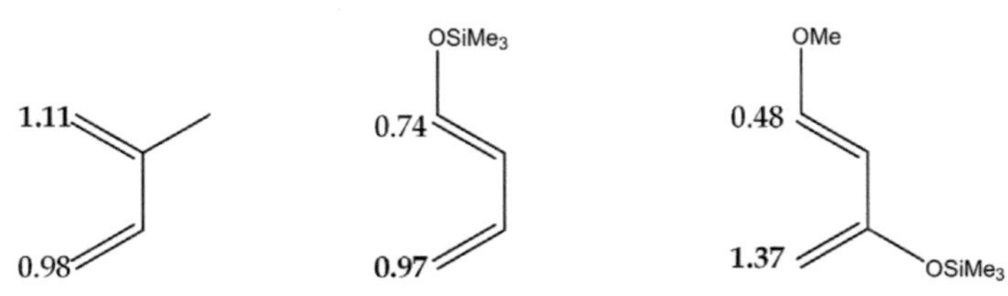

Figure 14. local nucleophilicity index (N_k) in e*V*.

In Figure 14, local nucleophilicity indexes for carbon atoms that would react in this type of cycloadditions (C1 and C4), are shown. A higher value of nucleophilicity is observed in C4 for 1-trimethylsilyloxy-1,3-butadiene and 1-methoxy-3-trimethylsilyloxy-1,3-butadiene (Danishefsky diene). For isoprene, C1 is the most nucleophilic atom.

The electrophilicity of isoprene falls in the range of moderate electrophiles within the electrophilicity scale proposed by Domingo *et al.* When electron-donating substituents, $-OCH_3$ and $-OSi(CH_3)_3$, are incorporated into the structure of butadiene, a decrease in the electrophilicity power is observed. Therefore, the electrophilicity of Danishefsky´s diene falls in the range of marginal electrophiles, good nucleophiles, within the electrophilicity scale. This behavior indicates that the nucleophilic activation in Danishefsky's diene is better than in isoprene, in clear agreement with the high nucleophilicity index of the diene. 1-trimethylsilyloxy-1,3-butadiene have a intermediate behavior.

5.3. Dienophiles

The electrophilicity power (ω) of the different dienophiles is shown in Tables 12, 13, 14, 15, 16 and 17. In the tables we also included some global properties such as the chemical potential and the chemical hardness. A good electrophile is characterized by a high absolute value of μ and a low value of η.

We also analyze its local counterpart (ω_k), the values are exposed in the Figures 15, 16, 17, 18, 19 and 20 below each table.

5.3.1. Five-membered heterocycles

- Furan and derivatives.

Molecule	Global properties			
	μ (a.u.)	η (a.u.)	ω (eV)	*N* (eV)
Furan	-0.1024	0.2441	0.99	3.23
2-Nitrofuran	-0.1810	0.1775	2.51	1.99
3-Nitrofuran	-0.1767	0.1808	2.35	2.07

Table 12. global electronic properties.

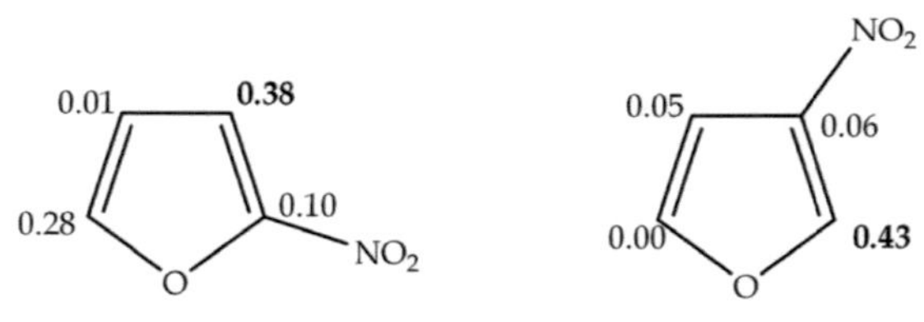

Figure 15. local electrophilicity indexes (ω_k) in *eV*.

- Tiophene and derivatives.

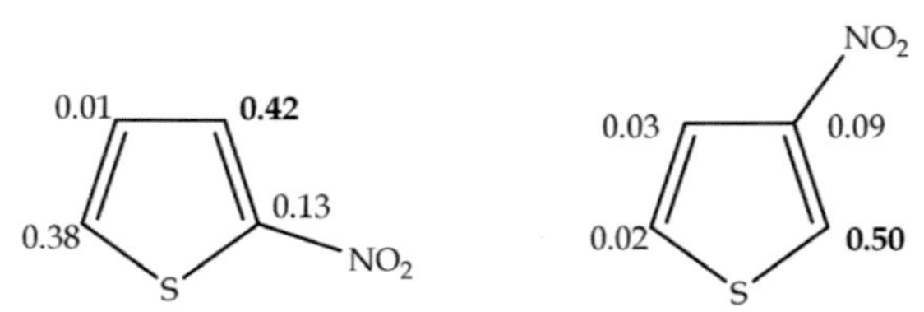

Figure 16. local electrophilicity indexes (ω_k) in *eV*.

Molecule	Global properties			
	μ (a.u.)	η (a.u.)	ω (eV)	*N (eV)*
Thiophene	-0.1545	0.1566	0.87	3.01
2-Nitrothiophene	-0.1845	0.1738	2.66	1.95
3-nitrothiophene	-0.1794	0.1821	2.40	1.98

Table 13. *global electronic properties*

- Selenophene and derivatives.

Molecule	Global properties			
	μ (a.u.)	η (a.u.)	ω (eV)	*N (eV)*
Selenophene	-0.1220	0.2195	0.49	3.16
2-Nitroselenophene	-0.1829	0.1695	2.68	2.05
3-Nitroselenophene	-0.1776	0.1803	2.38	2.05

Table 14. *global electronic properties*

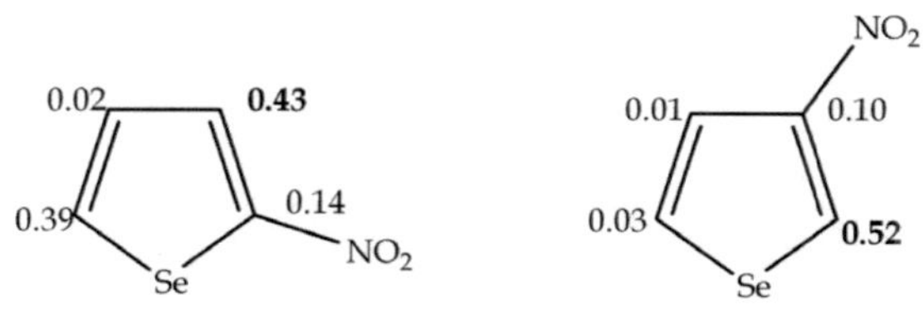

Figure 17. local electrophilicity indexes (ω_k) in *eV*.

- 1-Tosyl-pyrrole and derivatives.

Molecule	Global properties			
	μ (a.u.)	η (a.u.)	ω (eV)	*N (eV)*
1-Tosylpyrrole	-0.1348	0.1752	1.41	3.28
1-Tosyl-2-nitropyrrole	-0.1655	0.1739	2.31	2.43
1-Tosyl-3-nitropyrrole	-0.1668	0.1765	2.14	2.39

Table 15. global electronic properties.

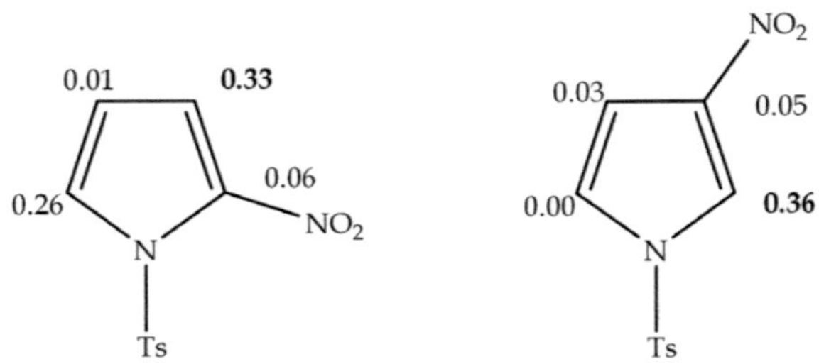

Figure 18. local electrophilicity indexes (ω_k) in *eV*.

5.3.2. Benzofused heterocycles

- Benzofuran and derivatives.

Molecule	Global properties			
	μ (a.u.)	η (a.u.)	ω (eV)	*N (eV)*
Benzofuran	-0.1180	0.2032	0.93	3.36
2- nitrobenzofuran	-0.1757	0.1536	2.73	2.46
3-nitrobenzofuran	-0.1694	0.1613	2.42	2.53

Table 16. global electronic properties.

Figure 19. local electrophilicity indexes (ω_k) in *eV*.

- *N*-Tosyl-indole and derivatives.

Molecule	Global properties			
	μ (a.u.)	η (a.u.)	ω (eV)	*N (eV)*
1-Tosyl-indole	-0.1316	0.1683	1.40	3.47
1-Tosyl-2-nitroindole	-0.1657	0.1455	2.57	2.84
1-Tosyl-3-nitroindole	-0.1606	0.1562	2.25	2.84

Table 17. global electronic properties.

http://dx.doi.org/10.5772/51656

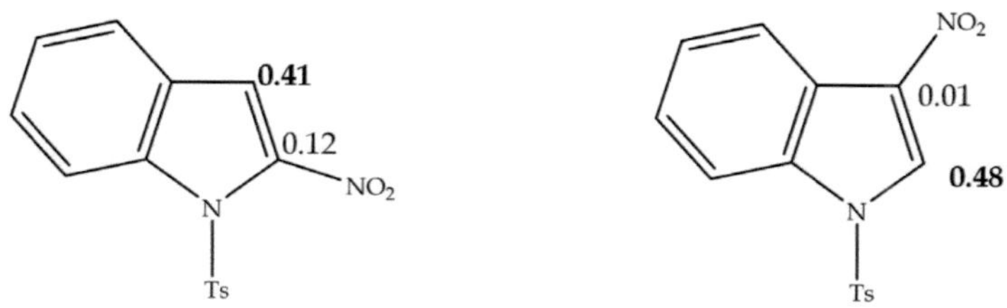

Figure 20. local electrophilicity indexes (ω_k) in *eV*.

We can assume that high nucleophilicity and high electrophilicity corresponds to opposite extremes of this scale (Della Rosa, et al, 2011; Brasca, et al, 2009).

The substitution of one hydrogen atom in all the dienophiles by the nitro group, one of the most powerful electron-withdrawing groups, produces an increment in the electrophilicity character and therefore an increase in the reaction rate is expected. The 2-nitro-substituted heterocycles show high electrophilicity power respect to the 3-nitro-substituted ones. Experimentally we obtained higher yields when the nitro group is place in the position 2 of the thiophene's ring than when it is in the 3-position. So these last results support the tendency observed in the tables.

The differences in the global electrophilicity power between the dienophile/diene pair ($\Delta\omega$) are higher for the Danishefsky's diene than for isoprene. Therefore, we can expect a high reactivity for the pair nitrosubstituted-dienophile/Danishefsky's diene.

As a consequence of the high electrophilic character of these substituted dienophiles and the high nucleophilic character of the dienes, it is expected that these D-A reactions proceed with polar character. The polarity of the process is assessed comparing the electrophilicity index of the interacting pairs. Evidently, the differences in the global electrophilicity power ($\Delta\omega$) are higher for the Danishefsky's diene than 1-trimethylsilyloxy-1,3-butadiene and isoprene.

On the other hand, a high regioselectivity is expected for Danishefsky's diene, due to the fact that the difference between local nucleophilicity indexes of C1 and C4 (ΔN_k) presents the highest value in this diene (ΔN_k=0.89). For isoprene and 1-trimethylsilyloxy-1,3-butadiene, as this difference is low (ΔN_k=0.13 and ΔN_k=0.23 respectively), both isomers are expected as D-A products. For dienophiles, the carbon atom adjacent to the nitro-substituted one is the most electrophile site in all the cases ($\Delta\omega_{k=}$ 0.49-0.27).

When asymmetric reactants participate in this polar D-A cycloaddition, the most favorable interaction will take place between the most nucleophile site of the diene and the most electrophile site of the dienophile. This fact is also consistent with the experimental researches in ionic liquids as solvents.

When 2-nitrobenzofuran and 3-nitrobenzofuran reacted with isoprene, 1-trimethylsilyloxy-1,3-butadiene and the Danishesfky's diene, under different reaction conditions they showed their dienophilic character taking part in a normal demand polar D-A cycloaddition reactions.

Finally, the flux of the electron-density in these polar cycloaddition reactions is also supported by means of a DFT analysis besed on the electronic chemical potentials of the reagents. The electronic chemical potentials of the substituted heterocyclic dienophiles, nearly -5 eV, are higher than those of the dienes, nearly -3 eV, thereby suggesting that the net charge transfer will take place from these electron-rich dienes towards the aromatic dienophiles.

According to the global electrophilicity index ω showed that the dienes will act as nucleophiles and the dienophiles as electrophiles. To study the regioselectivity we used the local electrophilicity and nucleophilicity indexes for dienophiles and dienes respectively. The more favored adducts are the ones where the most electrophilic and nucleophilic sites interact first. In the reactions in which it is possible discussed the regioselectivity of the experimental data agree with the computational results. (Della Rosa, el al, 2011)

The 2-nitrosubstituted benzofuran show higher electrophilicity power than the 3-nitrosubstituted benzofuran probably due to the proximity of the nitro group with the heteroatom.

5.4. Ionic Liquids effect

The influence of ILs has also been considered in this theoretical study. For the calculation of electronic properties of the "complexes" dienophile-IL we use a model that includes the dienophile molecule interacting with the anion-cation IL's system. In Table 18 we show an example of the different global electronic properties (ω and *N*) for these five-membered heterocycle-IL systems in comparison with the same study in gas phase.

	ω(eV)	*N(eV)*			ω(eV)	*N(eV)*	
	2.51	1.78	Gas Phase		2.35	1.85	Gas Phase
	4.31	1.83	[HMIM][BF4]		4.02	2.00	[HMIM][BF4]
2-nitrofuran	4,18	1.26	[HMIM][PF6]	3-nitrofuran	3.87	1.34	[HMIM][PF6]
	3.02	1.52	[BMIM][BF4]		2.78	1.64	[BMIM][BF4]
	2.95	1.60	[BMIM][PF6]		2.63	1.49	[BMIM][PF6]

Table 18. Global electronic properties for 2- and 3- nitrofuranes.

It can be observed that for nitrofuran dienophiles the electrophilicity index reach higher values in presence of IL ([HMIM][BF_4], [HMIM][PF_6], [BMIM][BF_4], [BMIM][PF_6]). The strongest hydrogen-bond donation ability of [HMIM] cation is reflected in the significant increment of ω in about 2 eV for all the heterocycles in study. The $[PF_6]^-$ anion present little lower values than $[BF_4]^-$.

Taking into account that the reactivity of a D-A reaction depends on the HOMO-LUMO energy separation of the reactants, and that in a normal electron demand D-A reaction the strongest interaction takes place between the HOMO of the diene and the LUMO of the dienophile, we compared the corresponding energies of the reacting partners in order to explain the experimental tendency observed.

http://dx.doi.org/10.5772/51656

	HOMO (a.u.)
Isoprene	-0.2272
1-trimethylsilyloxy-1,3-butadiene	-0.2046
Danishefsky's diene	-0.2045

Table 19. HOMO energy values.

	LUMO (a.u.)
2-nitrofuran	-0.0922
2-nitrofuran+[HMIM][BF_4]	-0.1379
2-nitrofuran+[HMIM][PF_6]	-0.1358
2-nitrofuran+[BMIM][BF_4]	-0.1016
2-nitrofuran+[BMIM][PF_6]	-0.1004

Table 20. LUMO energy values.

When the FMO of the reacting pairs are closer in energies, the interaction is higher. Thus, the FMO energies of the reactants were evaluated.

Therefore, in Tables 8 and 9 it can be observed that the expected higher reactivities for the dienophile-IL complexes is due to the fact that LUMO's energy of the dienophile gets closer to the HOMO's energy of the diene, which is consistent with the experimental results. This effect is also revealed by an increase in the yield of the D-A reaction. The energy difference between the FMOs is lower for [HMIM] based IL, what we attribute to the formation of the hydrogen-bond between the nitro group and the IL cation. The tendency is the same in all the cases of nitroheterocycle compounds.

5.5. Theoretical Mechanistic Approach

These D-A reactions could be considered domino processes that are initialized by a polar cycloadittion, and the latter concerted elimination of nitrous acid from the [4+2] cycloadduct yields the corresponding products (Della Rosa, et al, 2011).

5.5.1. Monocyclic five membered nitroheteroaromatic compounds as dienophiles.

Specifically the reactions of nitro-substituted five-membered heterocycles with Danishefsky ′s diene were studied using the hardness, the polarizability and the electrophilicity of the corresponding D-A primary adducts as global reactivity indexes. The experimentally observed products for these D-A reactions using different conditions were indicated in the Figure 21 and related experiments. It has been demonstrated that both the hardness as well as the electrophilicity power of the adducts are appropriate descriptors to predict the major product of the reactions at least in the cases in study. (Brasca, et al, 2011)

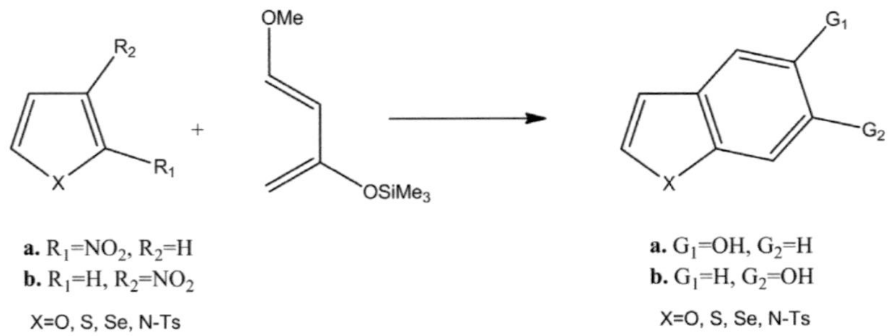

Figure 21. Reactions of nitrofuran whit Danishefsky diene.

For each reaction four channels, which lead to the regioisomers I, I', II and II' are feasible (Figure 9). As we can observe, depending on the orientation of the nitro group, two stereoisomers can be obtained in each channel (i.e. *endo* and *exo* adducts).

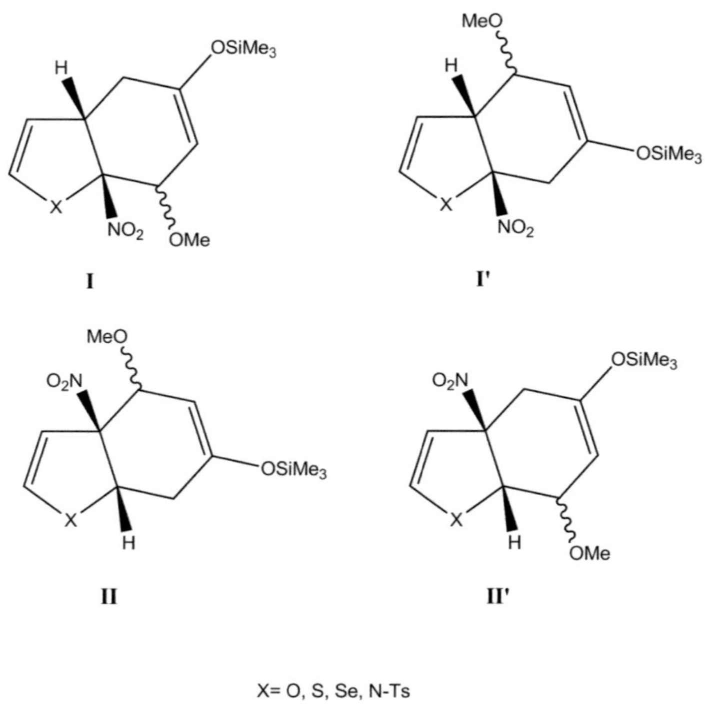

Figure 22. Possible regioisomeric D-A adducts.

The regioisomer that have the higher value of η and the lower values of α and ω, should correspond to the major product. The calculated hardness and electrophilicity power correctly predict the regioisomers I.a-I.c as the main adducts of the D-A reactions.

http://dx.doi.org/10.5772/51656

The results obtained in gas phase revealed the same tendency as in ionic liquids as solvents.

We can conclude that the predominant regioisomeric adduct of the reactions between five-membered heterocycles derivatives and Danishefsky's diene have always the less electrophilicity and high hardness values. Moreover, the regioselectivity experimentally observed can be confirmed by this approach.

The obtained energies show that the I.*endo* isomer is more stable than the I.*exo* one. Moreover, the more stable I.*endo* isomer has lower electrophilicity value than the I.*exo* isomer in all cases.

The investigation by DFT theory, in which we include solvent effects (considering cation and anion of the ionic liquids), show that these cycloadditions proceeded by a concerted but asynchronus reaction mechanism. The lowest activation energies for concerted reactions are obtained. However, the stepwise additions have significantly lower activation energy lead to substantially less stable products. Moreover, the primary cycloadducts could never be isolated but were converted into 5-hydroxybenzofused heterocycles by subsequent extrusion of nitrous acid, hydrolysis of the silyl enol ether, and elimination of methanol. Elimination of nitrous acid is calculated to have lower overall barriers than cycloaddition reactions and is strongly exothermic, thus explaining the preferred reaction channel.

5.5.2. Nitronaphthalenes as dienophiles

The reactions of 1-nitronaphthalene with a serie of dienes were evaluated with the the frontier molecular orbitals (FMO) theory which provide qualitative information about the feasibility of this D-A reaction. Besides, the global electrophilicity index (ω) is employed to estimate the electrophilic character of the dienophiles used in the cycloaddition reactions.

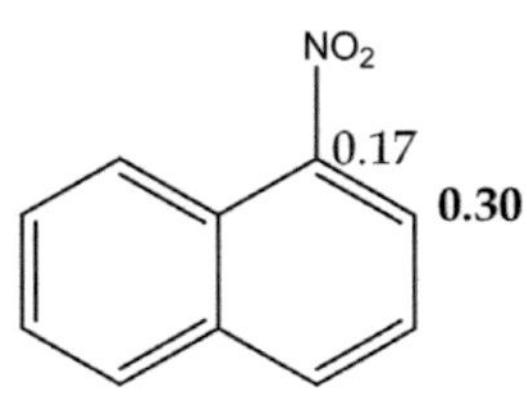

1-nitronaphtalene	
μ (a.u.)	-0.1650
η (a.u)	0.1485
ω (eV)	2.49
N (eV)	2.83

Table 21. 1-nitronaphthalene with the Danishefsky diene.

The reaction of 1-nitronaphthalene with the Danishefsky diene to obtain 3-hydroxyfenantrene has been theoretically studied using DFT methods. This reaction is a domino process that is initialized by a polar Diels–Alder reaction between the par dienophile/diene to give the formally [2 + 4] cycloadduct. The subsequent concerted elimination of nitrous acid from the primary adduct yields the precursor of the fenatrene derivative. An analysis of the global reactivity indices as well as the thermodynamic data for this domino process indicate that while the large electrophilic carácter of 1-nitronaphthalene together with the large nucleophilic character of Danishefsky diene are responsible for the participation of these reagents in a polar D-A reaction. The D-A reaction has a two-step non-intermediate mechanism characterized by the nucleophilic attack of the non-substituted methylene of the diene to the electrophilically activated C2 position of 1-nitronaphtalene. The subsequentring-closure affords the primary cycloadduct. The latter concerted elimination of nitrous acid yielded the precursor of the tricyclic aromatic final product. Spite of the large activation free energy associated with the D-A reaction and the endergonic character in the primary adduct, the irreversible extrusion of nitrous acid make feasible thermodynamically the domino reaction. (Domingo, et al, 2008)

5.5.3. Nitroquinolines

Although the global electrophilicity for the 5-nitroquinoline indicated a lightly major reactivity than the 8-nitro isomer (Figure 23). This result does not agree with the experimental data. In this respect it is possible think that the attack of the dienophile to the *para* position would be a reversible process, meanwhile the attack to the *orto* position to the nitro group evolve in form irreversible to the cycloaddition product. The major reactivity of the 8-nitro derivative could be occur due to the presence of electroelectronic factors more favourables which are produced during the nucleophilic attack of the diene, for instance a better stabilization of the negative charge in the nitro group. These effects are not considered in the reagents. (Cancian, *et al*, 2010)

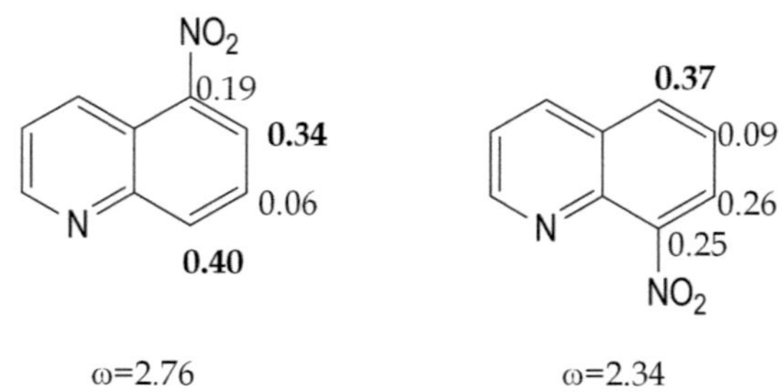

Figure 23. Global and local electrophilic indexes for 5- and 8-nitroquinolines.

Analysis of the local electrophilicities ω_k at 5-nitroquinoleine indicates that the C8 carbon, ω_{C8} = 0.40 eV, and the C6 carbon, ω_{C6} = 0.34 eV, are the electrophilic centres of the quinoleine moiety, while at 8-nitroquinoleine these centers are the C5 carbon, ω_{C5} = 0.37 eV,

http://dx.doi.org/10.5772/51656

and the C7 carbon. They correspond with the carbon that contains the nitro group and that located at the *ortho* position.

Formation of the HB between the acidic N10 hydrogen of $HMIM^+$ and the O9 oxygen of the nitro group does not only increase the electrophilicity index of the nitroquinoleine-PIL complexes, but also polarizes nitroquinoleine system. At the nitroquinoleine-PIL complexes, the most electrophilic centres are the C8 carbon, $\omega_{C5}=0.55$ eV, and the C6 carbon, $\omega_{C7}=0.54$ eV, at 5-nitroquinoleine-PIL, and at the C5 carbon, $\omega_{C5}=0.45$ eV, and the C7 carbon, $\omega_{C7}=0.40$ eV, at 8-nitroquinoleina-PIL. These results are similar to those found in 5- and 8-nitroquinoleines.

Using the 5 and 8-nitroquinoline we showed specifically the interactions models with the lowest energy between the dienophile and [HMIM][BF4] acting as reaction media.

Isoprene vs. 5-nitroquinoleine and 8-nitroquinoleine. A theoretical mechanism study.

In order to understand the catalytic role of polar ionic liquids (PILs) in P-DA reactions between isoprene and nitroquinoleines, in the absence and in the presence of [HMIM][BF4] as a model of PILs, were theoretically studied using DFT methods. These reactions are domino processes that comprises two consecutive reactions:

1. a P-DA reaction between isoprene, acting as diene, and nitroquinoleines acting as dienophiles, to yield the formal [4+2] cycloadducts (CA) and;

2. a concomitant nitroso acid extrusion at these intermediates to yield the final products (Figure 24).

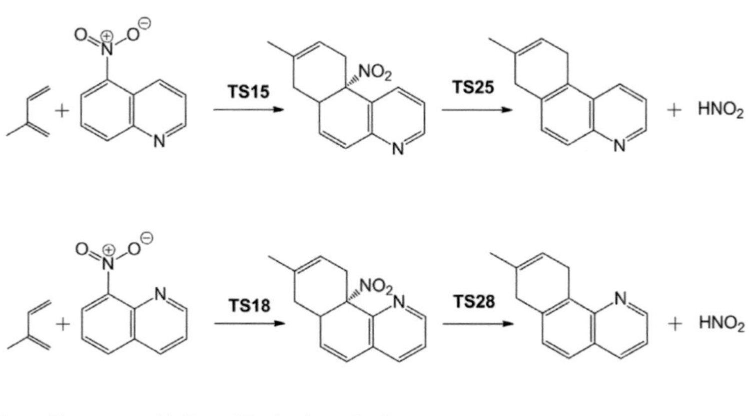

Figure 24. Reaction of isoprene with 5- and 8-nitroisoquinoline

Four reactive channels for the initial attack of isoprene on these nitroquinoleines are feasible: two pairs of stereoisomeric channels, the *endo* and the *exo* ones, and two pairs of regioisomeric channels, the *meta* and the *para* ones. Since both *endo* and *exo* channels yield the same final products after extrusion of nitroso acid, and as P-DA reactions involving isoprene present low regioselectivity, only the channels associated with the *endo/para* approach mode of isoprene, respect to the electron-withdrawing nitro group of nitroquinoleines were considered.

An analysis of the gas-phase potential energy surfaces (PES) associated with these P-DA reactions indicates that the cycloadditions take place through a one-step mechanism via high asynchronous transition states (TS). Therefore, in both cases one TS, TS15 and TS18, and the formal [4+2] CAs, were located and characterized. The second reaction of these domino processes also takes place via a one-step mechanism via a high asynchronous TSs. Thus, one TS, TS25 and TS28, and the corresponding final products were located and characterized. Total and relative gas-phase energies are given in Table 11.

	E	ΔE
Isoprene	-195.3055	
5-nitroquinoleine	-606.4260	
TS15	-801. 6914	25.2
Cycloadduct	-801.7496	-11.4
TS25	-801.7129	11.6
Product	-596.0775	
HNO_2	-205.6952	
Product + HNO_2	-801.7727	-25.9
8-nitroquinoleine	-606.4195	
TS18	-801.6819	27.0
Cycloadduct	-801.7499	-15.6
TS28	-801.7136	7.2
Product	-596.0798	
Product + HNO_2	-801.7750	-31.4

Table 22. Gas phase total energies (*E*, in au) and relative energies (*ΔE*, in kcal/mol) of the stationary points involved in the domino reaction between isoprene and nitroquinoleines (5- and 8- nitrosubstituted).

In gas phase, the activation energies associated with the nucleophilic attack of the C1 carbon of isoprene on the C6 carbon of nitroquinoleines via TS15 or TS18 present high values, 25.2 and 27.0 kcal/mol; formation of the formal [4+2] CAs are exothermic by -11.4 and -15.6 kcal/mol, respectively. Although these P-DA reactions are thermodynamically favorable, these high activation energies associated with these processes prevent the cycloaddition reactions.

The [4+2] CAs suffer a nitroso acid extrusion regenerating the aromatic system present in quinoleine. The activation energies associated with the nitroso acid extrusion via TS25 and TS28 are 23.0 and 28.8 kcal/mol; formation of the tricyclic compounds plus nitroso acid is exothermic by -14.5 and -15.8 kcal/mol, respectively. Taking into account the favorable reaction entropies associated with the extrusion processes, we can consider these reactions thermodynamically irreversible. Since TS25 and TS28 are located below TS15 and TS18, the P-DA reactions between isoprene 1 and nitroquinoleines 2 or 3 via TS15 and

TS18 become the rate-determining steps of these domino processes. The high activation energy associated with TS15 and TS18 are in agreement with the drastic reaction conditions demanded for the reactions to take place.

These activation energies are higher than those associated with the P-DA reaction and nitroso acid extrusion associated with the domino reaction between nitronaphtalene and Danishefsky's diene, 16.5 and 23.9 kcal/mol, respectively (Figure 25).

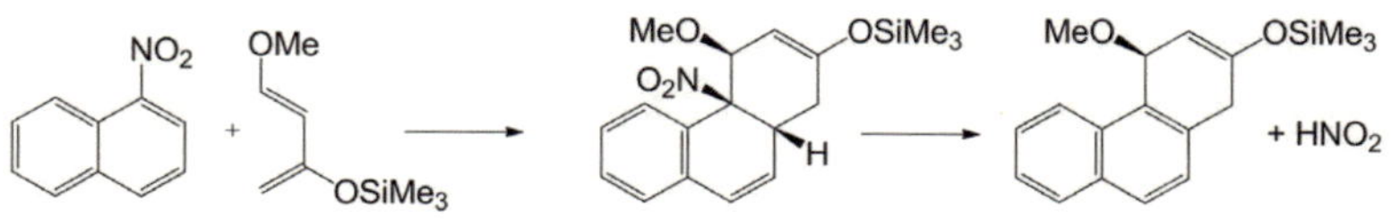

Figure 25. 1-nitronaphtalene and Danishefsky's diene

However, the lower activation energy associated with the P-DA reaction between nitronaphtalene and Danishefsky's diene, and the endothermic character of the formation of the formal [4+2] CA turn the nitroso acid extrusion into the rate-determining step in this domino reaction.

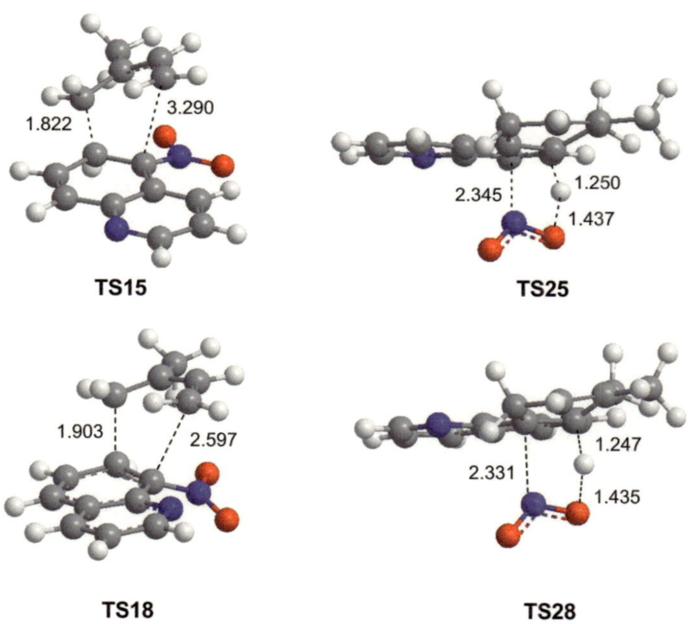

Figure 26. TS of the reaction between 5- and 8-nitroquinoline and isoprene

The gas-phase geometries of TS15, TS18, TS25 and TS28 are given in Figure 13. The lengths of the C1-C5 and C2-C8 forming bond at TSs associated with the P-DA reactions between isoprene and nitroquinoleines are 1.822 and 3.290 Å at TS15 and 1.903 and 2.597 Å at TS18, respectively. These values suggest asynchronous bond-formation processes in which the C-C bond formation between the most nucleophilic center of isoprene, the C5 carbon, and one of the most electrophilic centers of nitroquinoleines, the C1 carbon, is more advanced than the C-C bond formation between the C2 and C8 carbons. The short C1-C5 distance indicates that the cycloaddition processes are very advanced, in clear agreement with the high activation energies associated with TS15 and TS18.

At TSs associated with the nitroso acid extrusion the length of the C2-N3 breaking bond is 2.345 Å at TS25 and 2.331 Å at TS28, while the lengths of the C1-H1 breaking- and O4-H1' forming-bonds are 1.250 and 1.437 Å at TS25, and 1.247 and 1.435 Å at TS28, respectively. These values suggest asynchronous processes in which the C2-N3 breaking bond is more advanced than the H1' proton transfer process to the O4 oxygen.

The polar nature of these D-A reactions was evaluated analyzing the charge transfer (CT) at TS15 and TS18. The natural charges at these TSs were shared between the isoprene and the nitroquinoleine frameworks. At TS15 and TS18, the CT that flows from isoprene to nitroquinoleines is 0.29 eV and 0.26 eV, respectively. These values point at the zwitterionic character of these TSs. They are lower than that obtained at the TS associated with the nucleophilic attack of Danishefsky's diene on nitronaphtalene, 0.39 eV, as a consequence of the stronger nucleophilic character of Danishefsky's diene than that of isoprene.

Reaction mechanism pattern in presecence of the [HMIM][BF4] PIL.

The effects of the [HMIM][BF_4] PIL on the domino reactions between isoprene and nitroquinoleines were evaluated considering two computational models. In *Model I*, the implicit effects of the PIL were considered by forming a hydrogen bond (BH) between the acidic H10 hydrogen of HMIM and the O9 oxygen of the nitro group of nitroquinoleines (Figure 27).

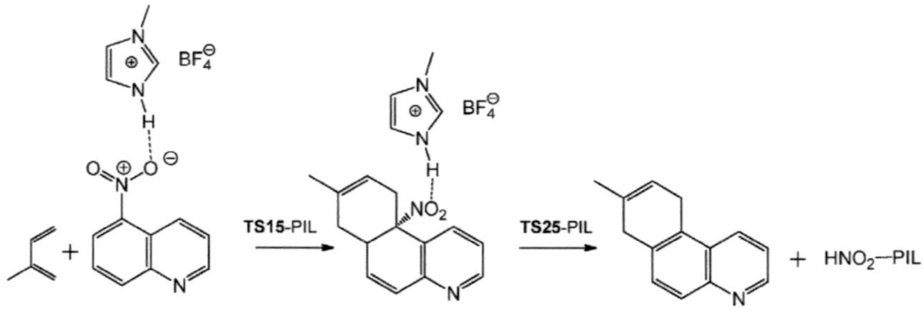

Figure 27. Reaction between 5-nitroquinoline and isoprene in presence of PIL`s

While in *Model II*, the solvent effect of the PIL is completed including electrostatic interactions modeled by the polarizable continuum model (PCM) of Tomasi's group. For the PCM calculations, 1-heptanol was considered as solvent since it has a dielectric constant closer to [HMIM][BF4]; ε= 11.3. The energy results are given in Table 12.

http://dx.doi.org/10.5772/51656

	gas phase		***1-heptanol***	
	E	**ΔE**	**E**	**ΔE**
5-nitroquinoleine-PIL	-1297.0092		-1297.0411	
TS15-PIL	-1492.2827	20.1	-1492.3173	19.4
cycloadduct-PIL	-1492.3293	-9.2	-1492.3636	-9.7
TS25-PIL	-1492.3020	8.0	-1492.3395	5.5
HNO_2 -PIL	-896.2688		-896.30298	
Product + HNO_2 -PIL	-1492.3463	-19.8	-1492.3857	-23.5
8-nitroquinoleine-PIL	-1297.0040		-1297.0391	
TS18-PIL	-1492.2736	22.5	-1492.3117	21.6
cycloadduct-PIL	-1492.3341	-15.4	-1492.3673	-13.3
TS2-PIL	-1492.3021	4.6	-1492.3397	4.1
Product + HNO_2 -PIL	-1492.3486	-24.5	1492.3873	-25.8

Table 23. Total energies (*E*, in au) and relative energies (*ΔE*, in kcal/mol), in gas phase and in 1-heptanol, of the stationary points involved in the domino reaction between isoprene and nitroquinoleines in [HMIM][BF4].

In Figure 27, the stationary points involved in the domino reactions between isoprene and [HMIM][BF_4]:nitroquinoleines complexes are given. Formation of the HB between HMIM and an oxygen atom of the nitro group decreases the activation energies associated with these P-DA reactions significantly. Now, TS1-PIL and TS2-PIL are located 20.1 and 22.5 kcal/mol above the separated reagents. In spite of this behaviour, the exothermic character of the cycloadditions, -9.2 and -15.4 kcal/mol, remains unmodified. The large acceleration found in the presence of the PIL ionic pair can be understand as an increase of the polar character of the reactions as a consequence of the increased electrophilic character of the dienophile-PIL complexes, which favors the CT process.

The second reactions of these domino processes are also slightly catalyzed by the presence of [HMIM][BF_4] PIL, since it remains hydrogen-bonded at the intermediate cycloadducts-PIL. Now, the activation energies associated with the extrusion of HNO_2-PIL are 17.2 and 20.0 kcal/mol. These reactions are exothermic by -10.6 and -9.1 kcal/mol.

Inclusion of the solvent effects by means of the PCM calculations in *Model II* stabilizes all species between 21.0 and 25.0 kcal/mol, as a consequence of the charged [HMIM][BF_4] PIL. TS15 and TS18 are slightly more stabilized than the reagents due to their zwitterionic character. As a consequence, the activation energies associated with the P-DA reactions decrease by 1.7 and 0.9 kcal/mol relative to the gas-phase calculations. Consequently, a comparison between the gas-phase relative energies associated with the domino reactions between isoprene and nitroquinoleines with those obtained in *Model I* and *II* for the domino reactions in PILs indicate that the HB formation at nitroquinoleines 2 and 3 is the main factor responsible for the acceleration of these domino reactions in PILs.

The gas-phase geometries of TS15-PIL, TS18-PIL, TS25-PIL and TS28-PIL are given in Figure 15. The lengths of the C1-C5 and C2-C8 forming bonds at the TSs associated with the P-DA reactions between isoprene and the dienophile-PIL complexes are 1.879 and 3.338 Å, and 1.835 and 3.141 Å, respectively. These values suggest a two-stage one-step mechanism channel an asynchronous TS. At TS15-PIL and TS18-PIL, the distances between the nitro O9 oxygen and the acidic H10 hydrogen of HMIM, 1.769 and 1.776 Å, suggest a strong HB interaction.

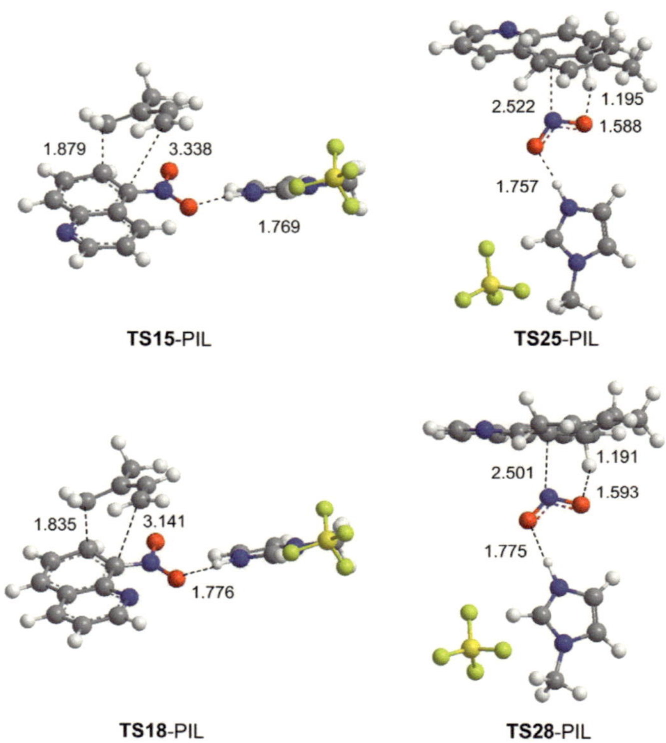

Figure 28. TS of the reaction between 5-nitroquinoline and isoprene in presence of PIL`s

At TSs associated with the extrusion of the 6-PIL complex, the length of the C2-N3 breaking bond is 2.522 Å at TS25-PIL and 2.501 Å at TS28-PIL, while the lengths of the C1-H1' breaking- and O4-H1' forming-bonds are 1.195 and 1.588 Å at TS25-PIL, and 1.191 and 1.593 Å at TS28-PIL, respectively. These values also suggest a highly asynchronous process in which the C2-N3 breaking bond is very advanced with respect to the H1' proton transfer process.

6. Conclusions

It was possible to demonstrate again the influence of the solvent in these particular type of D-A reactions. A series of aromatic carbocyclic and hetrocyclic substituted by electron with-

http://dx.doi.org/10.5772/51656

drawing groups can act as dienophiles in polar cycloaddition reactions besides different dienes in the presence of PIL´s. However, D-A reactions proceed at an appreciable rate only when either the diene or the dienophile are activated by an electron donating or electron withdrawing group, normally characterized by the presence of a heteroatom that can therefore efficiently interact with the solvent. IL´s, with their peculiar properties such as high polarizability/dipolarity, good hydrogen bond donor ability, were straight away considered to have the potential to influence the outcome of these D-A reactions, accelerating them.

In general, the products of the reactions development in PIL´s are similar to those in molecular solvents. However, the presence of PIL´s improved the reaction rate probably due to the hydrogen bonding interactions between the neoteric solvent and the dienophile. Only in a few cases we can note differences in the product distribution. For the reactions in which are possible to observe a competition between normal and hetero D-A process, the PIL´s favor the normal pathway because they improve the electrophilicity of the dienophiles.

The DFT analysis of the global properties of the interacting pair diene/dienophile illustrates the normal electron demand character of these D-A reactions. It is possible to show that the local indexes provide useful clues about the regiodirector effects, particularly of the nitro group. The presence of a solvent (molecular or neoteric) as the reaction media does not impart a prominent influence on the relative reactive sites. In few cases among those studied we can note that the relative reactive sites are affected by the solvent and the basis set (e.g. methyl 5-nitrofuran-2-carboxylate).

The site-, regio-, and stereochemistry of some of these D-A reactions has been investigated by the density functional theory, including solvent effects. Generally, these cycloadditions proceed by a concerted but asynchronous reaction mechanism. The *endo* stereochemistry is in most case preferred.

In general, the normal D-A reaction mechanism is a domino process that is initialized by the polar reaction between the diene and the dienophile to give the primary cicloadduct. These D-A reactions have a two-step non-intermediate mechanism characterized by the nucleophilic attack on the non-substituted methylene of the diene to the electrophilically activated position of the dienophile. The subsequent ring-closure affords the primary cicloadduct. This behavior makes the reaction to be regioselective. The latter concerted elimination of the nitrous acid from the primary cicloadduct yields the precursor of the final aromatic product. Spite of the large activation free energy associated with the D-A reaction and the endergonic character of formation of the primary cicloadduct, the irreversible extrusion of the nitrous acid make feasible thermodynamically the domino reaction.

DFT calculations of the electrophilicity and nucleophilicity indexes in general agree with the experimental results and they are a good reactivity and regioselectivity predictors in this type of polar cycloaddition reactions

The presence of a PIL's in the reaction media improves significantly the electrophilic character of the dienophile. However, the differences between the experimental results using PIL´s or molecular solvents are not so bigger how the calculated electrophilic values indicated.

Author details

Pedro M. E. Mancini[1*], Carla M. Ormachea[1], Claudia D. Della Rosa[1], María N. Kneeteman[1,2] and Luis R. Domingo[3]

*Address all correspondence to: pmancini@fiq.unl.edu.ar

1 Área de Química Orgánica-Departamento de Química-Facultad de Ingeniería Química-Universidad Nacional del Litoral (UNL), Santa Fe, Argentina

2 Consejo Nacional de Investigaciones Científicas y Técnicas (CONICET), de la República Argentina

3 Departamento de Química Orgánica, Facultad de Química, Universidad de Valencia, España

References

[1] Arnó, M., & Domingo, L. R. (2002). Density functional theory study of the mechanism of the proline-catalyzed intermolecular aldol reaction. *Theoretical Chemistry Accounts*, 108(4), 232-239.

[2] Biolatto, B., Kneeteman, M., Paredes, E., & Mancini, P. M. E. (2001). Reactions of 1-tosyl-3-substitutedindoles with conjugated dienes under thermal and/or hyperbaric conditions. *Journal of Organic Chemistry*, 66, 3906-3912, 0022-3263.

[3] Brasca, R., Kneeteman, M. N., Mancini, P. M. E., & Fabian, W. M. F. (2011). Comprehensive DFT Study on Site-, Regio-, and Stereoselectivity of Diels-Alder Reactions Leading to 5 -Hydroxybenzofurans. *Eur. J. Org. Chem*, 721-729.

[4] Brasca, R., Kneeteman, M. N., Mancini, P. M. E., & Fabian, W. M. F. Diels-Alder reactions for the rational design of benzo[b]thiophenes. DFT-based guidelines for synthetic chemists. *Journal of Molecular Structure*, THEOCHEM, 1010, 158.

[5] Brasca, R., Della Rosa, C., Kneeteman, M., & Mancini, P. (2011). Five-membered aromatic heterocycles in polar cycloaddition reactions: theoretical studies as a complement of the experimental researches. *Letters in Organic Chemistry*, 8(2), 82-87.

[6] Brasca, R., Della Rosa, C., Kneeteman, M., & Mancini, P. M. E. (2011). Five-Membered aromatic heterocycles in diels-alder cycloaddition reactions: Theoretical Studies as a Complement of the Experimental Researches. *Letters in Organic Chemistry*, 8(2), 82-87.

[7] Brasca, R., Kneeteman, M. N., Mancini, P. M. E., & Fabian, W. M. F. (2009). Theoretical explanation of the regioselectivity of polar cycloaddition reactions between furan derivatives and Danishefsky's diene. *Journal of Molecular Structure: THEOCHEM*, 911(1-3), 124-131.

[8] Brasca, R., Kneeteman, M. N., Mancini, P. M. E., & Fabian, W. M. F. (2011). Comprehensive DFT study on site-, regio-, and stereoselectivity of diels-alder reactions leading to 5-hydroxybenzofurans. *European Journal of Organic Chemistry*, 4, 721-729.

[9] Cancian, S., Kneeteman, M., & Mancini, P.M.E. (2010). Nitroquinolines as dienophiles in Polar Diels-Alder reactions. Influence of molecular solvent and ionic liquids. *14th International Electronic Conference on Synthetic Organic Chemistry (ECSOC-14)*, 0095-2338.

[10] Carruthers, W. (1990). *Cycloaddition Reactions in Organic Synthesis*, Pergamon Press, Oxford, UK.

[11] Corey, E.J. (2002). Catalytic enantioselective Diels-Alder reactions: methods, mechanistic fundamentals, pathways, and applications. *Angew.Chem. Int. Ed.*, 41(10), 1650-67.

[12] Della Rosa, C., Ormachea, C., Kneeteman, M. N., Adam, C., & Mancini, P. M. E. (2011). Diels-Alder reactions of N-tosylpirroles developed in protic ionic liquids. Theoretical studies using DFT methods. *Tetrahedron Lett.*, 52, 6754-6757.

[13] Della Rosa, C., Sanchez, J.P., Kneeteman, M.N., & Mancini, P.M.E. (2011). 2-Nitrobenzofuran and 3-Nitrobenzofurans as dienophiles in Polar Diels-Alder Reactions. A Simple Dibenzofurans Synthesis. Theoretical studies using DFT Methods. *Tetrahedron Lett.*, 52, 2316-2319.

[14] Della Rosa, C., Kneeteman, M., & Mancini, P. M. E. (2007). Behavior of selenophenes substituted with the electron withdrawing groups in polar Diels-Alder reactions. *Tetrahedron Letters*, 48, 7075-7078, 0040-4039.

[15] Della Rosa, C., Kneeteman, M., & Mancini, P. M. E. (2007). Comparison of the reactivity between 2- and 3-nitropyrroles in cycloaddition reactions. A simple indole synthesis. *Tetrahedron Letters*, 48, 1435-1438, 0040-4039.

[16] Della Rosa, C. D., Sanchez, J. P., Kneeteman, M. N., & Mancini, P. M. E. (2011). Diels-Alder reactions of nitrobenzofurans: A simple dibenzofuran Synthesis. Theoretical studies using DFT methods. *Tetrahedron Letters*, 52(18), 2316-2319.

[17] Della Rosa, C., Kneeteman, M., & Mancini, P. M. E. (2005). Nitrofurans as dienophiles in Diels-Alder reactions. *Tetrahedron Letters*, 46, 8711-8714, 0040-4039.

[18] Della Rosa, C., Paredes, E., Kneeteman, M., & Mancini, P. M. E. (2004). Behavior of thiophenes substituted with electron-withdrawing groups in cycloaddition reactions. *Letters in Organic Chemistry*, 1, 148-150, 1687-6865.

[19] Della Rosa, C., Sanchez, J. P., Kneeteman, M., & Mancini, P. M. E. (2010). Nitrobenzofuran as dienophile in Polar Diels-Alder reactions. A simple Dibenzofuran Synthesis. *Journal of Chemistry and Chemical Engineering*, 4(11), Serial 36, 54-59.

[20] Fringelli, F., & Tatichi, A. (2002). *The Diels-Alder Reaction*, J. Wiley & Sons, Chichester, UK.

[21] Domingo, L. R. (2002). A density functional theory study for the Diels-Alder reaction between N-acyl-1-aza-1,3-butadienes and vinylamines. Lewis acid catalyst and solvent effects. *Tetrahedron*, 58(19), 3765-3774.

[22] Domingo, L. R., Arnó, M., Contreras, R., & Pérez, P. (2002). Density functional theory study for the cycloaddition of 1,3-butadienes with dimethyl acetylenedicarboxylate. Polar stepwise vs concerted mechanisms. *Journal of Physical Chemistry A*, 106(6), 952-961.

[23] Domingo, L. R., & José, Aurell M. (2002). Density functional theory study of the cycloaddition reaction of furan derivatives with masked o-benzoquinones. Does the furan act as a dienophile in the cycloaddition reaction. *Journal of Organic Chemistry*, 67(3), 959-965.

[24] Domingo, L. R., Chamorro, E., & Pérez, P. (2008). An understanding of the electrophilic/nucleophilic behavior of electro-deficient 2,3-disubstituted 1,3-butadienes in polar Diels-Alder reactions. A density functional theory study. *Journal of Physical Chemistry A*, 112(17), 4046-4053.

[25] Hitchcock, P. B., Mohammed, T. J., Seddon, K. R., Zora, J. A., Hussey, C. L., & Ward, E. H. (1986). 1-methyl-3-ethylimidazolium hexachlorouranate(IV) and 1-methyl-3-ethylimidazolium tetrachlorodioxo-uranate(VI): Synthesis, structure, and electrochemistry in a room temperature ionic liquid. *Inorganic Chimica Acta*, 113, L25-L26, 19-191, 0020-1693.

[26] Jaramillo, P., Domingo, L. R., Chamorro, E., & Pérez, P. (2008). A further exploration of a nucleophilicity index based on the gas-phase ionization potentials. *Journal of Molecular Structure: THEOCHEM*, 865(1-3), 68-72.

[27] Mancini, P. M. E., Fortunato, G., Bravo, M. V., & Adam, A. (2012). Ionic Liquids:Binary Mixtures with Selected Molecular Solvents. Characterization of its Molecular-Microscopic Properties. Reactivity. Chapter 13, 335-362, *in Green Solvents Book 2: Properties and Applications in Chemistry, Eds. Ali Mohammad and M.P. Inamuddin*, 978-94-007-2890-5, 978-94-007-2891-2-eBook-, Springer, United Kingdom.

[28] Mancini, P. M. E., Kneeteman, M., Della Rosa, C., Bravo, V., & Adam, C. (2011). Ionic Liquids in Polar Diels-Alder Reactions using Carbocycles and Heterocycles as Dienophiles. *Ionic Liquid/Book 1. Ed. Scott Hardy*, INTECH Open Access, cap. 13, 311-344, 978-953-308-66-6.

[29] Paredes, E., Brasca, R., Kneeteman, M., & Mancini, P. M. E. (2007). A novel application of the Diels-Alder reaction: nitronaphthalene as normal electron demand dienophiles. *Tetrahedron*, 63, 3790-3799, 0040-4020.

[30] Parr, R. G., Szentpály, L. V., & Liu, S. (1999). Electrophilicity index. *Journal of the American Chemical Society*, 121(9), 1922-1924.

http://dx.doi.org/10.5772/51656

[31] Parr, R. G., & Yang, W. (1984). Density functional approach to the frontier-electron theory of chemical reactivity. *Journal of the American Chemical Society*, 106(14), 4049-4050.

[32] Welton, T. (1999). Room-Temperature Ionic Liquids. Solvents for Synthesis and Catalysis. *Chemical Review*, 99(2071), 0010-8545.

Chapter 17

Ionic Liquids as Doping Agents in Microwave Assisted Reactions

Marcos A. P. Martins, Jefferson Trindade Filho,
Guilherme S. Caleffi, Lilian Buriol and
Clarissa P. Frizzo

Additional information is available at the end of the chapter

http://dx.doi.org/10.5772/51659

1. Introduction

The use of microwave (MW) irradiation as a tool for organic synthesis has been a fast growth area [1-8]. Several examples have shown that the application of MW irradiation reduces the reaction time, increases the product yield and sometimes results in a different product distribution compared to conventional thermal heating method [1-6,9-20]. The rate acceleration observed in organic reactions using MW irradiation is due to material-wave interactions leading to thermal and nonthermal effects. The thermal effects result from a more efficient energy transfer to the reaction mixture, which is known as dielectric heating. This process relies on the ability of a substance (solvent or reactant) to absorb MW and convert them into heat. The reaction mixture is heated from the inside since the MW energy is transferred directly to the molecules (solvent, reactants, and catalysts). This process is known as 'volumetric core heating' and results in a temperature gradient that is reversed compared to the one resulting from conventional thermal heating [1,9-14]. Nonthermal effects result in differences in product distributions, yields, and reaction times. They may result from the orientation effects of polar species in the electromagnetic field that makes a new reaction path with lower activation energy [9-14, 21-23]. It has been suggested [24] that MW activation could originate from hot spots generated by dielectric relaxation on a molecular scale. Currently, thermal and nonthermal effects are being extensively studied mainly to verify the existence or not of nonthermal effects [21-23].

Several studies have reported the application of ionic liquids (ILs) in different areas and, in particular, their use in organic reactions [25-33]. ILs are generally defined as liquid electrolytes composed entirely of ions. Occasionally, a melting point criterion has been proposed to distinguish between molten salts and ILs (mp < 100 °C). However, both molten salts and ILs are better described as liquid compounds that display ionic-covalent crystalline structures [34-35]. Suitably selected, many combinations of cations and anions allow the design of ILs that meets all the requirements for the chemical reaction under study; based on this, they are also known as 'designer solvents' [36]. Properties such as solubility, density, refractive index, and viscosity can be adjusted to suit requirements simply by making changes to the structure of the anion, the cation, or both [37-43].

The junction of the use of MW irradiation with the use of ILs provides a method of high interest in organic synthesis. ILs interact very efficiently with MW irradiation through the ionic conduction mechanism [7-8] and are rapidly heated at rates easily exceeding 10 °C per second [44-50]. Despite few reports on the exact measurement of their dielectric properties and loss tangent values, the experimentally attained heating rates of ILs applying MW irradiation attest to their extremely high MW absorptivity [46,51]. This ability allows that small amounts of ILs can be employed as additives in order to increase the dielectric constant of nonpolar solvents characterizing them as doping agents [52-57]. In particular, ILs can be used as support in the synthesis of organic compounds which are carried out using MW irradiation and less polar solvents. Research groups have used ILs as doping agents for MW heating of otherwise nonpolar solvents such as hexane, toluene, tetrahydrofuran, and dioxane [52-57]. Thus, in view of the good relation between MW and IL, the following topics will be discussed in this chapter: (*i*) behavior of the solvents under MW environment with emphasis in the heating effects of adding a small quantity of ILs in solvents with different loss tangent, such as *N,N*-dimethyl formamide (DMF), acetonitrile (ACN), hexane (HEX), toluene (TOL), tetrahydrofurane (THF); (*ii*) ILs as doping agents in MW assisted reactions especially in *N*-alkylation reaction of pyrazole with alkyl halides (Figure 1).

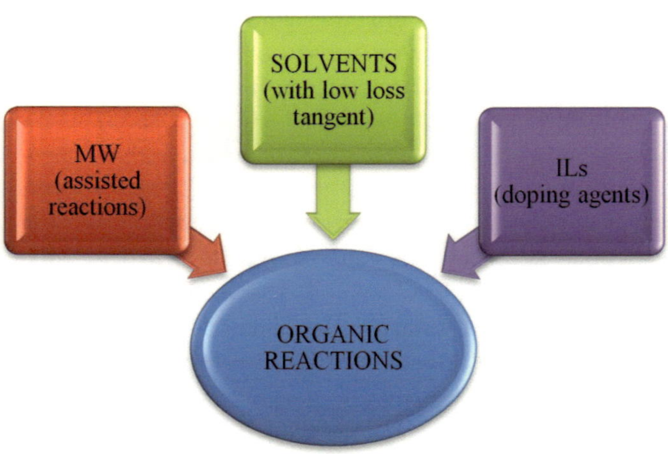

Figure 1. Ionic Liquids as doping agents in microwave assisted reactions.

http://dx.doi.org/10.5772/51659

2. Behavior of the Solvents Under Microwave Environment

Several organic solvents are used in various types of organic reactions under MW irradiation. The particular ability of the solvents to convert electromagnetic energy into thermal energy is directly related to their dielectric properties. The magnitude of the heating efficiency, in the specific temperature and frequency, is determined by the so-called 'loss tangent' (tan δ), whose formula is represented by Eq. 1 [52].

$$\tan\delta = \frac{\varepsilon''}{\varepsilon'} \quad (1)$$

In Eq. 1, ε'' is the dielectric loss and ε' is the dielectric constant. A reaction medium with a high tan δ at the standard operating frequency of a MW synthesis reactor (2.45 GHz) is required for good absorption and, consequently, for efficient heating. Solvents used for MW synthesis can be classified as high with tan δ > 0.5, medium with tan δ 0.1 – 0.5, and low MW absorbing with tan δ < 0.1 (Table 1) [14,58]. In general, the reactions which used solvents with a high tan δ have a good absorption of MW irradiation and, accordingly, an efficient heating [8,59,60]. Solvents such as DMSO and DMF are essential to reactions performed in MW. While these are great solvents for performing the reaction, the subsequent workup procedure is difficult to remove them due to their high boiling point and miscibility with the product [53]. Thus, in certain situations, it is convenient to use solvents which are less polar such as THF, TOL and HEX [14,58,59-60]. However, it is necessary to use a heating agent for the reactions carried out in solvents with low absorption in the MW irradiation. ILs, for instance, can be added to the reaction medium to increase the absorbance level of the MW irradiation [51,59]. Therefore, the use of ILs appears as a support to increase the temperature of the reactions carried out in a MW transparent solvents [53].

Solvent	tan δ	Solvent	tan δ
Ethylene glycol	1.350	1,2-Dichloroethane	0.127
Ethanol	0.941	Water	0.123
Dimethyl sulphoxide	0.825	Chloroform	0.091
Methanol	0.659	Acetonitrile (ACN)	0.062
1,2-Dichlorobenzene	0.280	Tetrahydrofurane (THF)	0.047
Methylpyrrolidone	0.275	Dichloromethane	0.042
Acetic acid	0.174	Toluene (TOL)	0.040
N,N-Dimethylformamide (DMF)	0.161	Hexane (HEX)	0.020

Table 1. Loss tangent of several solvents [59-60].

3. Heating Effects of Adding a Small Quantity of Ionic Liquid in Solvents

Systematic studies on temperature profiles and the thermal stability of IL under MW irradiation conditions were studied [52]. In these studies it was found that even the addition of a small amount of an IL resulted in dramatic changes in the heating profiles due to changes in the overall dielectric properties of the reaction medium.

Leadbeater and Torenius [53] studied the heating and contamination effects of several ILs in less polar solvents, such as HEX, TOL, THF and dioxane (DIO) (Figure 2) under MW irradiation. These authors have shown that all solvents used can be heated way above their boiling point in sealed vessels using a small quantity of an IL, thereby allowing them to be used as media for MW assisted chemistry. Table 2 shows the temperatures reached for pure solvents and for doped solvents with ILs using 200 W of power under MW irradiation.

The effects of varying the quantity of IL used to the solvent heating were investigated. The authors found that the best condition used was 2 mL of solvent and 0.2 mmol of IL, resulting in rapid heating. In these studies the contamination, if any, of the parent solvent with the IL or any decomposition products formed as they are heated were also studied [53].

Results showed that both [BMIM][PF_6] and [BMIM][BF_4] proved to be useful in MW heating of solvents, with [BMIM][PF_6] being more effective (Table 3). There was no contamination of the solvent when using [BMIM][PF_6] with any of the solvents screened or when [BMIM][BF_4] was used with HEX. There was contamination due to the decomposition of [BMIM][BF_4] when used with TOL or DIO; the extent was much less in the case of the latter. The [BMIM] [BF_4] was slightly soluble in THF thus in this case the only source of contamination at the end of the heating experiments was a trace of the parent IL rather than any decomposition. To the experiments, 100 W of the power was used [53].

Leadbeater *et al.* [54] also investigated the decomposition of some ILs and found out that when the IL was heated above 200 °C, decomposition occurred to give an alkyl halide and alkyl imidazole as shown in Scheme 1. Halide ion (X^-) acts as a nucleophile in attaching the cation with the subsequent elimination of alkyl-X. This decomposition was verified for elevated temperatures, which was not totally unexpected.

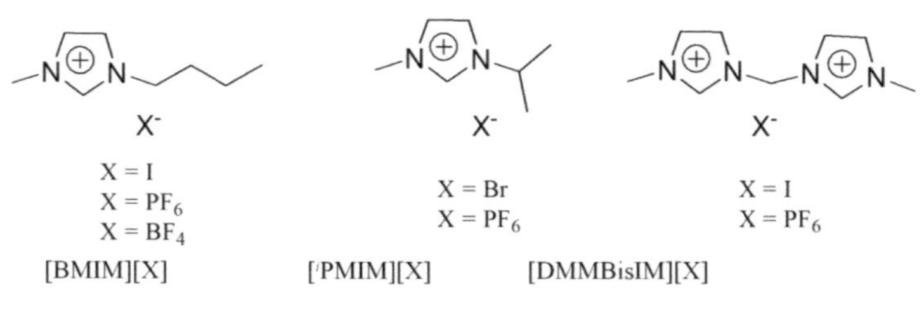

Figure 2. Ionic liquids used as doping agent.

http://dx.doi.org/10.5772/51659

Solvent	IL Added	T(attained) (°C)	Time (taken) (s)	T (without IL) (°C)	Solvent Boiling Point (°C)
HEX	[BMIM][I]	217	10	46	69
	[^{i}PMIM][Br]	228	15		
TOL	[BMIM][I]	195	150	109	111
	[^{i}PMIM][Br]	234	130		
THF	[BMIM][I]	268	70	112	66
	[^{i}PMIM][Br]	242	60		
DIO	[BMIM][I]	264	90	101	101
	[^{i}PMIM][Br]	246	90		

Table 2. The Microwave Irradiation Effects of Adding a Small Quantity of ILs in Less Polar Solvents [53].

Solvent	IL Added	T. attained (°C)	Time Taken (s)	Level of Contamination
HEX	[BMIM][PF_6]	279	20	None
	[^{i}PMIM][PF_6]	90	300	None
	[BMIM][BF_4]	192	60	None
TOL	[BMIM][PF_6]	280	60	None
	[^{i}PMIM][PF_6]	79	120	None
	[BMIM][BF_4]	165	90	Contaminated
THF	[BMIM][PF_6]	231	60	None
	[BMIM][BF_4]	95	50	contaminated[b]
DIO	[BMIM][PF_6]	149	100	None
	[BMIM][BF_4]	184	120	Contaminated

[b][BMIM][BF_4] is slightly soluble in THF and so cannot totally be removed; thus contamination is due to [BMIM][BF_4] rather than decomposition.

Table 3. Microwave irradiation effects in the presence of a small quantity of ILs in less polar solvents [53].

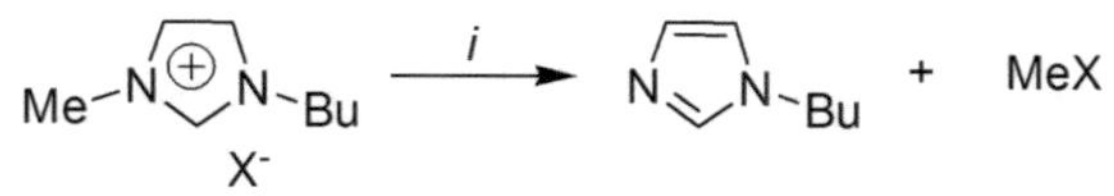

X = Cl, Br, I
i: MW

Scheme 1.

Hoffmann *et al.* [61] showed that ILs of the 1,3-dialkylimidazolium-type revealed (Figure 3) great potential for the application of MW for organic synthesis. These authors verified that the increase of MW power resulted in a drastic decrease in heating time.

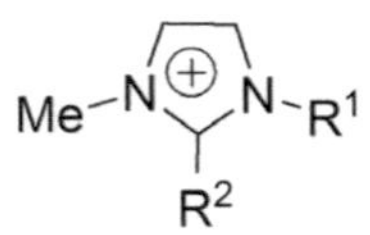

Cation	Anion
[HMIM]	[Tf_2N]
[HEIM]	[Tf_2N]
[EMIM]	[Tf_2N]
[HBIM]	[Tf_2N]
[HBIM]	[BF_4]
[HBIM]	[PF_6]

Figure 3. Ionic liquids used in the study [61].

A supplementary investigation covered the heating behavior of ILs as doping agent when mixed with solvents less polar in MW irradiation such as TOL and cyclohexane (100 mL of solvent in 1 mL, 3 mL and 5 mL of IL) (Table 4). Therefore, the authors concluded that small amounts of ILs are necessary to significantly reduce the heating time of TOL or cyclohexane under MW conditions. An increase in the MW power generates a reduction in heating time (Table 4). Also in this case, heating time approaches a limiting value even with an increase of the MW power. This was also true for the addition of ILs to both non-polar solvents (TOL and cyclohexane).

IL	**Power (W)**	**$Ht_{(35\text{-}105°C/s)}$ (TOL : IL (mL))**		
		100 : 1	100 : 3	100 : 5
[HMIM][Tf_2N]	300	318	90	70
[HMIM][Tf_2N]	400	167	66	53
[HMIM][Tf_2N]	500	112	54	39
[HBIM][PF_6]	300	548	168	143
[HBIM][PF_6]	400	319	126	92
[HBIM][PF_6]	500	229	88	86

Table 4. Heating times (Ht) of toluene/ionic liquid-mixtures.

Following the direction of these studies, we also performed some experiments using ILs as doping agents with several solvents under MW irradiation. The objective of this study was to check if the data of our MW equipment are in accordance with the data already published. Thus, we performed investigations of power profiles in different solvents with distinguished loss tangent values as DMF, ACN, THF, TOL and HEX in the presence of small quantities of [BMIM][BF_4] as doping agent. The solvents doped were submitted under MW

http://dx.doi.org/10.5772/51659

irradiation in an attempt to reach a temperature of 150°C (temperature which may be used in organic reactions) [62]. For this, we used various concentrations of IL in different solvents, as shown in Table 5. After reaching the desired temperature, the doped solvents were irradiated for 5 min and we verified that lower concentrations of IL required higher power for all solvents tested (Table 5). During the 5 min of MW irradiation the power remained substantially constant. Solvents with the low loss tangent such as HEX and TOL achieved only 99 and 108 °C, respectively, even though 300 W of power was applied.

Entry	Solvent	$[BMIM][BF_4]$ ($mmol.mL^{-1}$)	Power (W)
1	DMF	0.057	20.402
2	DMF	0.107	18.679
3	DMF	0.196	16.887
4	DMF	0.397	15.895
5	ACN	0.048	38.967
6	ACN	0.104	31.229
7	ACN	0.205	30.478
8	ACN	0.407	29.402
9	THF	0.063	240.079
10	THF	0.103	185.834
11	THF	0.218	112.582
12	THF	0.401	69.415
13	TOL	0.045	-[b]
14	TOL	0.102	119.429
15	TOL	0.197	67.805
16	TOL	0.403	47.317
17	HEX	0.049	-[c]
18	HEX	0.105	130.718
19	HEX	0.210	68.858
20	HEX	0.398	60.301

[a]In a sealed vessel, under simultaneous cooling, 150 °C for 5 min, temperature was measured with fiber-optic probe. [b]Achieved 108 °C, 300 W, 20 min. [c]Achieved 99 °C, 300 W, 20 min.

Table 5. Power dependence of IL concentration in some solvents[a].

Figure 4 illustrates the dependence between the concentrations of $[BMIM][BF_4]$ in the solvents and the power irradiated by MW equipment. At low concentrations of $[BMIM][BF_4]$ (~ 0.05 $mmol.mL^{-1}$) a significant increase in the power is required to maintain the temperature of 150 °C. Another point is that, to maintain the temperature of 150 °C, solvents such as

DMF and ACN did not require substantial variation of power as that found to HEX, TOL and THF when the concentration of IL ranged from ~ 0.05 mmol.mL^{-1} to ~ 0.4 mmol.mL^{-1}. These data corroborate previous studies reported [53,61] and highlight the efficiency of [BMIM][BF_4] as doping agent of poorly MW absorbing solvents.

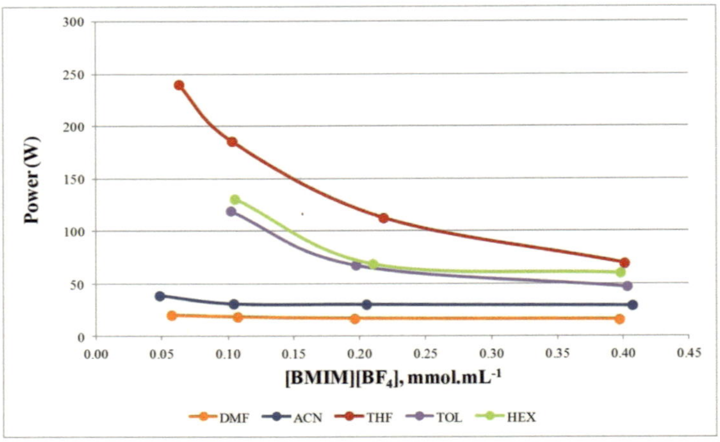

Figure 4. Power profiles of solvents with different concentrations of [BMIM][BF_4].

4. Ionic Liquids as Doping Agents (ILDA) in Microwave Assisted Reactions

The efficient use of ILs as a doping agent in reaction under MW irradiation was firstly introduced by Ley *et al.* [55]. The authors described the synthesis of thioamides from the secondary or tertiary amides (Scheme 2) and nitriles from primary amides (Scheme 3) in presence of thiophosphorylated amine resin using small quantity of IL [EMIM][PF_6] (120 mg) in TOL (2.5 mL).

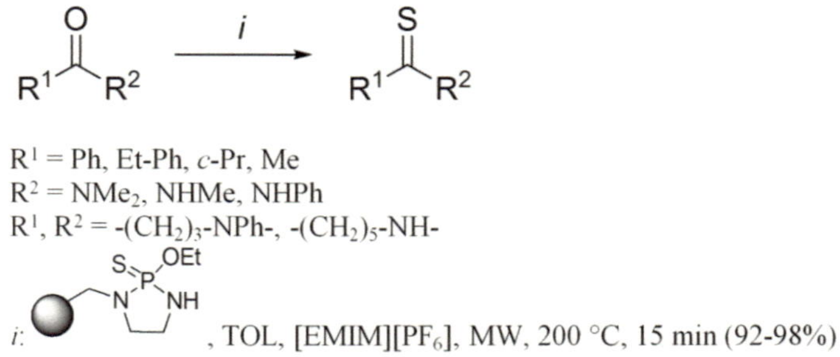

Scheme 2.

http://dx.doi.org/10.5772/51659

Protocols used the reactants thiophosphorylated amine resin and secondary or tertiary amides in a molar ratio of 1:3-20, respectively, to obtain thioamides, and used the reactants thiophosphorylated amine resin and primary amides in a molar ratio of 1:3.5, respectively to furnish nitriles. Reactions were carried out under both MW irradiation at 200 °C for 15 min to obtain the thioamides in 92-98% (Scheme 2) and nitriles in 95 - < 99% yields (Scheme 3). Acetonitrile was also investigated as an alternative MW absorbent and proved to be effective, in spite of being less efficient than the IL.

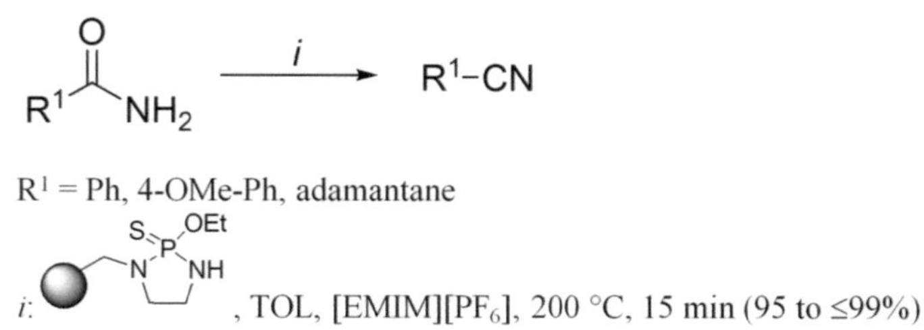

Scheme 3.

Eycken *et al.* [56] initially investigated the intramolecular hetero-Diels-Alder reaction in a series of 2(1*H*)-pyrazinones to obtain the chloro-bicycles and dione-bicycles, as showed in Scheme 4. In their initial experiments they used pyrazinone (R = Bu, n = 2) as a model substrate involving DCE as solvent to obtain the chloro-bicycles (R = Bu, n = 2). Using a preselected maximum temperature of 190 °C (300 W maximum power), neat DCE could be heated to ca. 170 °C within 10 min under sealed vessel conditions. Prolonged time heating is needed to reach higher temperatures. In an effort to promote the enhance of the maximum attainable reaction temperature, the solvent (DCE) was doped with different amounts of [BMIM][PF_6]. Adding 0.035 mmol of IL to the neat solvent (2 mL of DCE), the preselected temperature of 190 °C could be reached in 3 min upon MW heating. These results clearly demonstrated that even small amounts of IL were able to change the dielectric properties of a less polar solvent. These changes are sufficiently significant to heat more rapidly the reaction medium and to reach higher reaction temperatures. Increasing the amount of IL to 0.075 mmol led to a more rapid heating of the reaction mixture, as expected. When 0.150 mmol concentration was used, it provided a profile that allowed heating the DCE doped with IL to 190 °C in 1 min. To minimize the risk of potential contaminations or side reactions caused by the IL, all the following cycloaddition studies were carried out using this set of conditions (0.150 mmol IL for 2 mL of DCE) in 100 mg of 2(1*H*)-pyrazinones to obtain the chloro-bicycles. After, the hydrolysis reaction was carried out to obtain the dione-bicycles with yields of 57-77%.

The same authors [56] reported the synthesis of the chloro-pyridine and pyridonefrom the cycloaddition reaction of 2(1*H*)-pyrazinone with dimethylacetylenedicarboxylate (DMAD) under the MW/IL conditions (Scheme 5). The reactants 2(1*H*)-pyrazinone and DMAD were used in a molar ratio of 1:1. The reaction conditions used were the same reported previously, 190 °C, DCE/[BMIM][PF_6] (0.150 mmol IL for 2 mL of DCE) in 5 min to furnish yields of

82% of chloro-pyridine and 2% of pyridine. Another cycloaddition reaction used hetero-dienes with ethene, leading to the bicyclic cycloadducts was investigated by these authors. However, using IL as a doping agent in the DCE was not successful because this reaction was not suitable for MW irradiation.

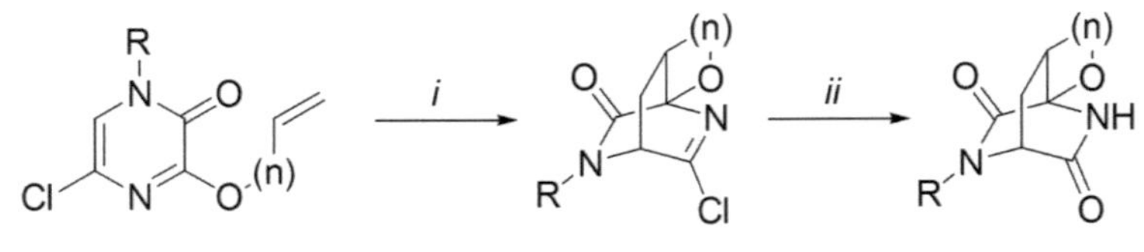

R = Bn, Ph; n = 2, 3
i : DCE, [BMIM][PF_6], MW, 190 °C, 8-15 min
ii : H_2O, MW, 190 °C, 5 min (57-77%)

Scheme 4.

Leadbeater and Torenius [53] described the Diels-Alder reaction from equimolar amounts of 2,3-dimethylbutadieneand methyl acrylate to furnish the [4 + 2] adduct cyclohex-3-ene using a mixture of TOL (2 mL) and [^{i}PrMIM][PF_6] (55 mg) under MW irradiation (Scheme 6). The mixture was irradiated at 200 °C for 5 min and led to the cyclohex-3-ene in 80% yield. The power used during the reaction performed under MW irradiation was 100 W. In a control experiment, the reaction was repeated in the absence of [^{i}PrMIM][PF_6], and it was found that after the same time (5 min at 100 W power) there was no product formed.

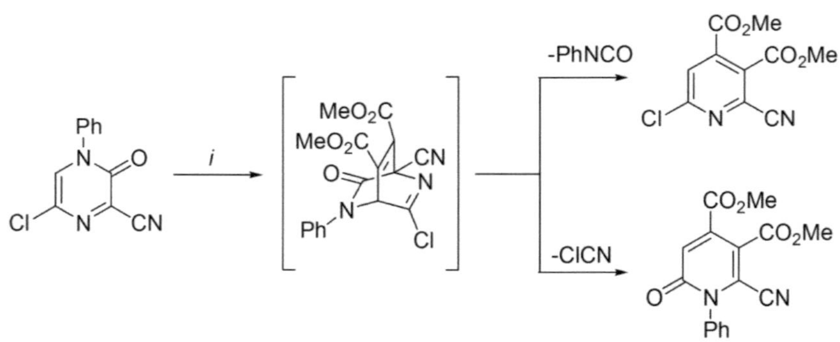

i : DMAD, DCE, [BMIM][PF_6], MW, 190 °C, 5 min (2-82%)

Scheme 5.

The same authors studied [53] the reaction of Michael addition from equimolar amounts of imidazole and methyl acrylate to furnish the methyl 3-(imidazol-1-yl) propionate (Scheme 7). The mixture of TOL (2 mL) and [^{i}PrMIM][PF_6] (55 mg) was irradiated for 2 min (200 °C, 100 W) and led to the methyl 3-(imidazol-1-yl)propionatein 75% yield. The reaction was repeated firstly in the absence of IL and TOL and secondly in the absence of TOL; in both cases after the same time and power (2 min at 100 W) there was no product formed.

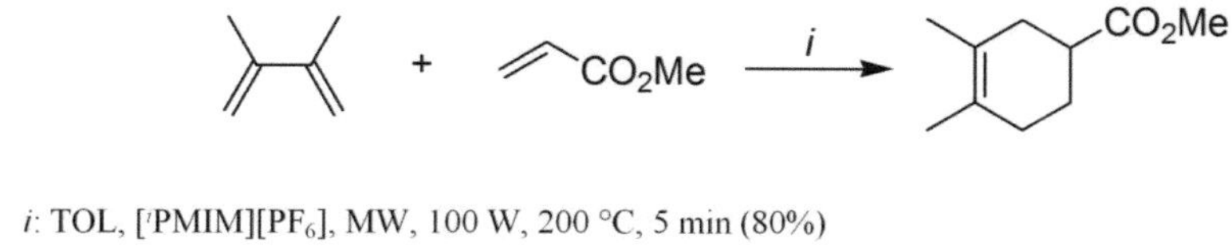

Scheme 6.

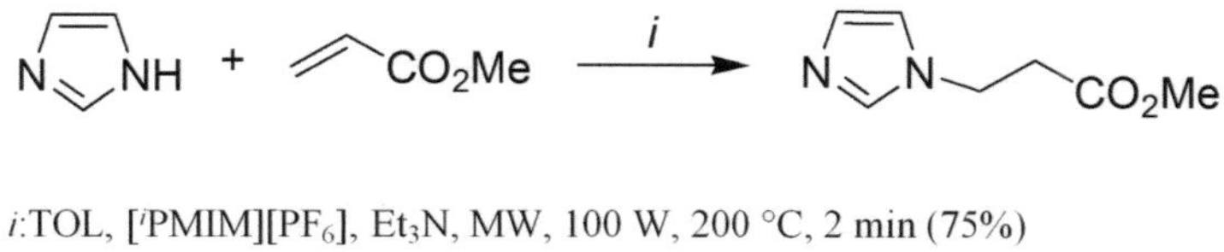

Scheme 7.

Garbacia *et al.* [57] described the ring-closing metathesis reactions (RCM) using diene substrates to furnish rings of five-, six-, or seven membered carbo- or heterocycles under MW irradiation (Scheme 8). The mixture of dienes (X = NTs, m,n = 1) and 0.5 mol% Grubbs' catalyst in the presence of DCM/[BMIM][PF_6] (0.04 M of IL) was irradiated in MW for 15 s, furnishing the desired product in > 98% yields. When neat DCM was used after the same time period only 57% conversion was observed. The authors mentioned that this was not a surprise since the reaction temperature during the full irradiation event (0-15 s) was significantly lower for the neat solvent. On the other hand, it was not possible to use the cationic ruthenium allenylidene catalyst in conjunction with an IL-doped solvent. With both [BMIM][PF_6] and [BMIM][BF_4] (0.04 M in DCM), conversions were below 30%, presumably due to catalyst deactivation.

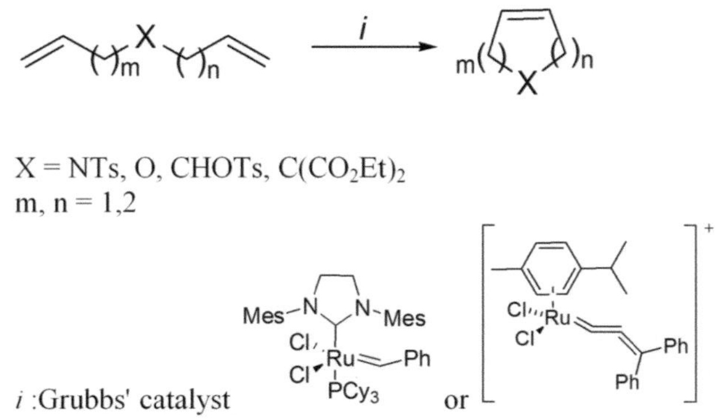

Scheme 8.

Leadbeater *et al.* [54] also reported the conversion of alcohols to alkyl halides using IL. Initially, they screened a range of reaction conditions mediated by MW irradiation using 100 W of power. Focusing on 1-octanol, they varied the MW irradiation time (0.5-10 min), the ILs ([PMIM][I], [^{i}PMIM][Br], [BMIM][Cl]) and the acid (PTSA, H_2SO_4). The reaction was performed from equimolar amounts of alcohol, IL and acid. The authors also investigated the effects of the addition of TOL as co-solvent (2 mL). When these reactions were carried out with neat IL, they reached 200 °C in a few seconds (≤ 15 s). On the other hand, using TOL as co-solvent it took a little longer to heat up but still reached 200 °C within a matter of 30–40 s (Table 6). Results showed that PTSA was more efficient than H_2SO_4 in he reactions involving the iodo, bromo and chloro ILs. Reaction times were in an increasing order: iodo < bromo < chloro substitutions with 0.5, 3 and 10 min, respectively. Most of the reactions using neat ILs presented higher product yields. The use of 2 mL of TOL as a co-solvent decreased the yield of the product formed.

IL/(Nucleofile)	Time (min)	Acid	Product	Yield[a] (%) Without co-solvent	Yield[a] (%) With co-solvent
[PMIM][I]	0.5	PTSA	$CH_3(CH_2)_7$-I	81	56
[PMIM][I]	1	PTSA	$CH_3(CH_2)_7$-I	53	38
[PMIM][I]	0.5	H_2SO_4	$CH_3(CH_2)_7$-I	3	55
[PMIM][I]	1	H_2SO_4	$CH_3(CH_2)_7$-I	38	15
[^{i}PMIM][Br]	0.5	PTSA	$CH_3(CH_2)_7$-Br	68	42
[^{i}PMIM][Br]	3	PTSA	$CH_3(CH_2)_7$-Br	95	32
[^{i}PMIM][Br]	0.5	H_2SO_4	$CH_3(CH_2)_7$-Br	73	59
[^{i}PMIM][Br]	1	H_2SO_4	$CH_3(CH_2)_7$-Br	42	40
[BMIM][Cl]	3	PTSA	$CH_3(CH_2)_7$-Cl	32	0
[BMIM][Cl]	3	H_2SO_4	$CH_3(CH_2)_7$-Cl	49	8
[BMIM][Cl]	10	PTSA	$CH_3(CH_2)_7$-Cl	42	35

[a]Yield of isolated product.

Table 6. Reaction conditions of 1-octanol with IL/(Nucleofile) [54].

Having found suitable conditions, the reaction was performed to a range of different alcohols. Further optimization of the reaction showed that the best reaction conditions for obtaining the 1-octanol were when IL was used in reaction medium. On the other hand, some dihalogenate 1,8-octanediol have furnished the best results when the co-solvent method was used as showed in Table 7. When using geraniol, not unexpectedly, geranyl iodide could not be isolated, but bromide and chloride could be obtained (Table 7). When using benzyl alcohol, it was possible to obtain the iodide in moderate yield (46%), the bromide in good yield

(68%) but only the chloride in low yield (17%). The authors believe that the co-solvent method is better because the organic product is more soluble in the organic solvent than in the IL and that once formed it moves to the organic layer and is protected from decomposition which can occur in the higher-temperature, acid IL environment.

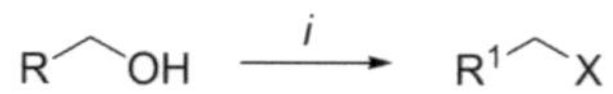

R = HO(CH$_2$)$_7$, CH=C(Me)-(CH$_2$)$_2$-CH=CMe$_2$, Ph
R^1 = X-(CH$_2$)$_7$, CH=C(Me)-(CH$_2$)$_2$-CH=CMe$_2$, Ph
i : TOL, ILs ([PMIM][I], [^{i}PMIM][Br],[BMIM][Cl]) Et$_3$N, MW, 100 W, 200 °C, 1-10 min (17-86%)

Alcohol	IL	Product	Time (min)	Yield[a] (%)
1,8-Octanediol[b]	[PMIM][I]	I-(CH$_2$)$_8$-I	3	53
1,8-Octanediol[b]	[^{i}PMIM][Br]	Br-(CH$_2$)$_8$-Br	3	86
1,8-Octanediol[b]	[BMIM][Cl]	Cl-(CH$_2$)$_8$-Cl	10	50
Geraniol	[PMIM][I]	I-CH=C(Me)-(CH$_2$)$_2$-CH=CMe$_2$	0.5	Dec[c]
Geraniol	[^{i}PMIM][Br]	Br-CH=C(Me)-(CH$_2$)$_2$-CH=CMe$_2$	3	47
Geraniol	[BMIM][Cl]	Cl-CH=C(Me)-(CH$_2$)$_2$-CH=CMe$_2$	10	30
Benzyl alcohol	[PMIM][I]	PhCH$_2$-I	1	46 (72)[d]
Benzyl alcohol	[^{i}PMIM][Br]	PhCH$_2$-Br	3	68
Benzyl alcohol	[BMIM][Cl]	PhCH$_2$-Cl	10	17

[a]Yield of isolated product. [b]0.5 mmol alcohol. [c]Dec = decomposition observed. [d]3 min.

Table 7. Conversion of alcohols to alkyl halides using co-solvent method [54].

Silva *et al.* [63] used the MW irradiation technique in the Diels–Alder reaction of tetrakis(pentafluorophenyl)porphyrin with pentacene and naphthacene. One of the synthetic methods used for the synthesis of these compounds was the use of IL-doped under MW irradiation. In order to increase the product yields, the authors used NMP and DCB as solvent systems with higher loss tangents, doped with an [BMIM][PF$_6$]. Unfortunately, none of these experiments gave better results.

5. Ionic Liquids as Doping Agents in Microwave Assisted *N*-Alkylation Reactions

Reactions of *N*-alkylation of pyrazoles using IL as doping agent under MW irradiation have been little explored. Leadbeater and Torenius [53] studied the reaction of alkylation of pyrazoles used 1*H*-pyrazole and alkyl halides to furnish 1-alkylpyrazoles under MW irradiation (Scheme 9). The authors found that to this reaction the product was not obtained using 2 mL of TOL and 55 mg of [^{i}PrMIM][PF_6], which were reaction conditions previously established for other reactions (Diels-Alder and Michael addition). Although the authors did not manage to characterize the reaction products, they affirm that *"it is clear to see that all the IL is destroyed since the biphasic starting mixture (solvent and IL) becomes a monophasic mixture after just a few seconds of MW irradiation. This shows the limitations of our protocol; it not being possible to undertake reactions which use or generate nucleophiles such as halide ions"*.

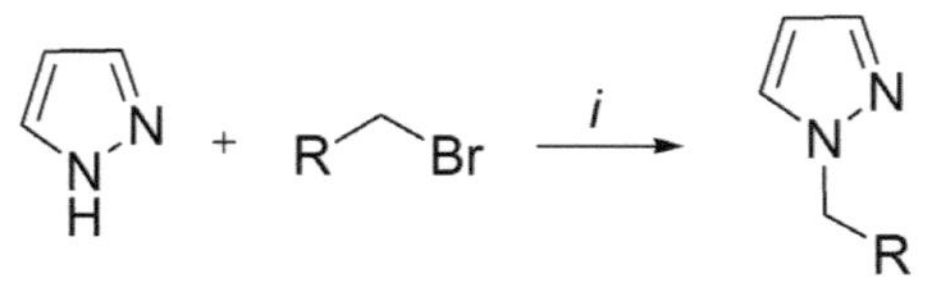

R = aryl
i : TOL, [^{i}PMIM][PF_6], Et_3N, MW, 100 W, 200 °C

Scheme 9.

Taking into account the results found by Leadbeater and Torenius [53], Kresmsner *et al.* [51] described the use of passive heating elements (PHEs) in *N*-alkylation of pyrazoles using NH-pyrazole and 1-(2-bromoethyl) benzene to obtain 1-phenethyl-1H-pyrazole. PHEs are materials which allow the compounds with low absorption of MW irradiation or poorly absorbing solvents such as HEX, carbon tetrachloride, THF, DIO, or TOL to be effectively heated to temperatures far above their boiling points (200-250 °C) under sealed vessel MW conditions. Thus, the authors used cylinders of sintered silicon carbide (SiC), PHE, which are chemically inert and strongly MW absorbing materials in the reactions of alkylation of pyrazoles.

Based on the studies mentioned above, we decided to explore the doping capacity of IL under MW irradiation in the *N*-alkylation of pyrazoles. This is a fundamental reaction of broad synthetic utility that often requires basic catalysis and thermodynamic reaction conditions. In addition, *N*-alkylation reaction of this heterocycle is a synthetic approach useful in the preparation of building blocks for the synthesis of important active compounds like pharmaceuticals [64] and agrochemicals [65]. In this way, it is clear the importance to develop a new methodology regarding this reaction. Our research group has previously reported the *N*-alkylation of pyrazoles using IL as solvent in oil bath [31]. Thus, we focused the use of efficient MW irradiation to perform the *N*-alkylation of pyrazoles in less polar solvents. Since these molecular solvents poorly absorb MW irradiation due to their lower loss tangent, the use of IL as doping agents becomes essential to achieve high temperatures. A symmetrical pyrazole and two reactive alkyl halides were chosen to perform these tests. [BMIM][BF_4] was selected due to its successful results in our previous work of *N*-alkylation in oil bath [31]. The amount of IL employed was ~ 0.1 $mmol.mL^{-1}$, which represents the minimum quantity required to achieve 150 °C in the four solvents under

http://dx.doi.org/10.5772/51659

study – HEX, TOL, THF and DIO (Figure 3) [66]. We also selected a base, KOH, to investigate the influence of basic catalysis on this reaction [31]. Initially, the reaction between butyl bromine and 3,5-dimethylpyrazole was performed in absence of basic catalysis. Based on data presented in Table 8, we could see that the reaction in HEX achieved the highest conversion followed by TOL, THF and DIO. In a basic medium, the conversion was increased only for TOL and DIO. The maintenance of moderate conversions could be explained by the low solubility of KOH in the solvents employed (Table 8).

Thus, we decided to investigate if a change in the alkylant agent reactivity could lead to higher conversions. Since iodine is a better leaving group than bromine, ethyl iodine was chosen to react with 3,5-dimethylpyrazole. Higher conversions were achieved for all tested solvents when compared with the results mentioned previously (Table 9). These results suggest that the nature of the leaving group would have greater influence than the basic catalysis on the product conversion.

Contrary to the results of Leadbeater and Torenius [53], we chose substrates for the reaction that showed moderate to good conversions. Thus, the IL is shown as an alternative to passive heating elements PHE [13].

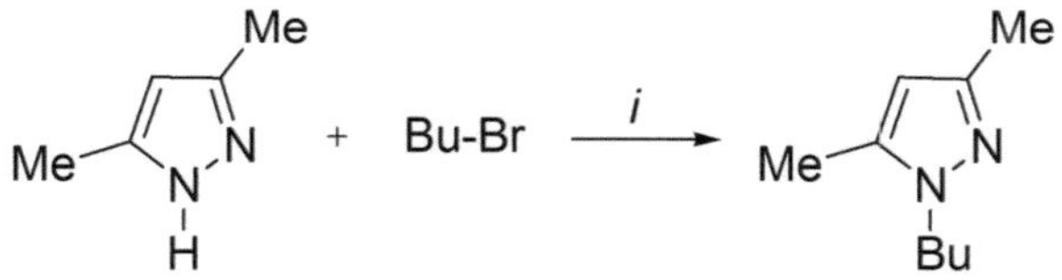

i = [BMIM][BF_4], MW, 150 °C, 30 min

Entry	Solvent[a]	Base	[BMIM][BF_4] (mmol.mL^{-1})	Conversion (%)[b]
1	HEX	-	0.117	59
2	HEX	KOH	0.123	50
3	TOL	-	0.123	17
4	TOL	KOH	0.127	37
5	THF	-	0.113	17
6	THF	KOH	0.124	17
7	DIO	-	0.108	9
8	DIO	KOH	0.118	41

[a]3mL. [b]Determined by ^1H NMR.

Table 8. Conversion of 1*H*-pyrazole in 1-butylpyrazolein low polar solvents in presence of [BMIM][BF_4].

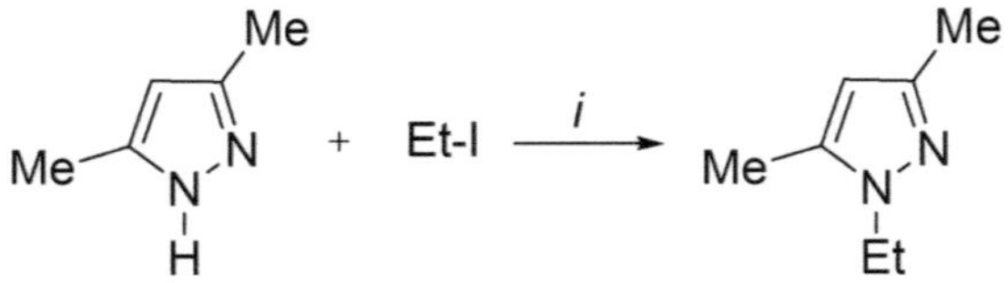

i = [BMIM][BF_4], MW, 150 °C, 30 min

Entry	Solvent[a]	[BMIM][BF_4] (mmol.mL^{-1})	Conversion (%)[b]
1	HEX	0.123	75
2	TOL	0.114	43
3	THF	0.120	71
4	DIO	0.118	43

[a]3mL.[b]Determined by ^{1}H NMR.

Table 9. Conversion of 1*H*-pyrazole in 1-ethylpyrazolein low polar solvents in presence of [BMIM][BF_4].

6. Conclusions

After analysis of the literature and results previously obtained by us about ILs as doping agents under MW irradiation, it is possible to conclude that: (*i*) the use of a small amount of IL in less polar solvents such as THF, TOL, and HEX promotes efficient heating under MW irradiation in sealed vessels; (*ii*) solvents with low tan δ when doped with small amounts of ILs are generally ideal reaction media as they allow a very rapid heating by MW irradiation in sealed vessels; (*iii*) an important limitation in the use of ILs as a doping agent is the chance of IL decomposition at temperatures higher than its thermal stability.

The examples of ILs as doping agents reviewed in this chapter showed that their applications are little explored and they have the potential to become an area of greater interest in the organic synthesis.

7. List of Abbreviations

ACN	Acetonitrile
[BMIM][BF4]	1-Butyl-3-methylimidazolium tetrafluoroborate
[BMIM][Br]	1-Butyl-3-methylimidazolium bromide

http://dx.doi.org/10.5772/51659

[BMIM][I]	1-Butyl-3-methylimidazolium iodide
[BMIM][PF6]	1-Butyl-3-methylimidazolium hexafluorophosphate
DCE	1,2-Dichloroethane
DCM	Dichloromethane
DIO	Dioxane
DMF	N,N-Dimethylformamide
DMAD	Dimethylacetalenedicarboxylate
[DMMBisIM][I]	(Bis(1-methylimidazol-3-yl))methane iodide
[DMMBisIM][PF6]	(Bis(1-methylimidazol-3-yl))methane hexafluorophosphate
[EMIM][Tf2N]	1-Ethyl-3-methylimidazolium bis(trifluoromethylsulfonyl)amide
Grubbs' catalyst	Grubbs' Catalysts (are a series of transition metal carbene complexes used as catalysts for olefin metathesis)
[HBIM][BF4]	1-Buthylimidazolium tetrafluoroborate
[HBIM][PF6]	1-Buthylimidazolium hexafluorophosphate
[HBIM][Tf2N]	1-Buthylimidazolium bis(trifluoromethylsulfonyl)amide
HEX	Hexane
[HEIM][Tf2N]	1-Ethylimidazolium bis(trifluoromethylsulfonyl)amide
[HMIM][Tf2N]	1-Methylimidazolium bis(trifluoromethylsulfonyl)amide
ILDA	Ionic Liquids as Doping Agents
[^{i}PMIM][Br]	1-*iso*-Propyl-3-methylimidazolium bromide
[^{i}PMIM][PF6]	1-*iso*-Propyl-3-methylimidazolium hexafluorophosphate
[PMIM][I]	1-Propyl-3-methylimidazolium iodide
PTSA	*p*-Toluenesulfonicacid
PHEs	Passive Heating Elements
RCM	Ring-Closing Metathesis
TOL	Toluene
THF	Tetrahydrofurane

Acknowledgements

The authors are grateful to Conselho Nacional de Desenvolvimento Científico e Tecnológico (CNPq Procs. No. 578426/2008-0; 471519/2009-0), Fundação de Amparo à Pesquisa do Estado do Rio Grande do Sul (FAPERGS/CNPq-PRONEX Edital No. 008/2009, Proc. No. 10/0037-8) and Coordenação de Aperfeiçoamento de Pessoal de Nível Superior (CAPES/PROEX) for financial support. The fellowships from CNPq (M.A.P.M., J.T.F.), and CAPES (G.S.C., C.P.F., L.B.) are also acknowledged.

Author details

Marcos A. P. Martins*, Jefferson Trindade Filho, Guilherme S. Caleffi, Lilian Buriol and Clarissa P. Frizzo

*Address all correspondence to: mmartins@base.ufsm.br

Department of Chemistry, NUQUIMHE, Federal University of Santa Maria, Brazil

References

[1] Lindström, P., Tierney, J., Wathey, B., & Westman, J. (2001). Microwave assisted organic synthesis. A review. *Tetrahedron*, 57(45), 9225-9283.

[2] Perreux, L., & Loupy, A. (2001). A tentative rationalization of microwave effects in organic synthesis according to the reaction medium, and mechanistic considerations. *Tetrahedron*, 57(45), 9199-9223.

[3] Deshayes, S., Liagre, M., Loupy, A., Luche-L, J., & Petit, A. (1999). Microwave activation in phase transfer catalysis. *Tetrahedron*, 55(36), 10851-10870.

[4] Strauss, C. R. (1999). Invited Review. A Combinatorial Approach to the Development of Environmentally Benign Organic Chemical Preparations. *Australian Journal of Chemistry*, 52(2), 83-96.

[5] Galema, S. A. (1997). Microwave chemistry. *Chemical Society Reviews*, 26(3), 233-238.

[6] Bacsa, B., Horváti, K., Bõsze, S., Andreae, F., & Kappe, C. O. (2008). Solid-Phase Synthesis of Difficult Peptide Sequences at Elevated Temperatures: A Critical Comparison of Microwave and Conventional Heating Technologies. *The Journal of Organic Chemistry*, 73(19), 7532-7542.

[7] Gabriel, C., Gabriel, S., Grant, E. H., Halstead, B. S., & Mingos, D. M. P. (1998). Dielectric parameters relevant to microwave dielectric heating. *Chemical Society Reviews*, 27(3), 213-224.

[8] Mingos, D. M. P., & Baghurst, D. R. (1991). Tilden Lecture. Applications of microwave dielectric heating effects to synthetic problems in chemistry. *Chemical Society Reviews*, 20(1), 1-47.

[9] Polshettiwar, V., & Varma, R. (2008). Microwave-Assisted Organic Synthesis and Transformations using Benign Reaction Media. *Accounts of Chemical Research*, 41(5), 629-639.

[10] Varma, R. S. (1991). Solvent-free organic syntheses. using supported reagents and microwave irradiation. *Green Chemistry*, 1(1), 43-55.

[11] Loupy, A. (2004). Solvent-free microwave organic synthesis as an efficient procedure for green chemistry. *C. R. Chim.*, 7103-112.

[12] Varma, R. S., & Polshettiwar, V. (2008). Aqueous microwave chemistry: a clean and green synthetic tool for rapid drug discovery. *Chemical Society Reviews*, 37(8), 1546-1557.

[13] Varma, R. S. (1999). Solvent-free synthesis of heterocyclic compounds using microwaves. *Journal of Heterocyclic Chemistry*, 36(6), 1565-1571.

[14] Kappe, C. O. (2004). *Controlled Microwave Heating in Modern Organic Synthesis* (Angewandte Chemie International Edition), 43(46), 6250-6284.

[15] Vargas, P. S., Rosa, F. A., Buriol, L., Rotta, M., Moreira, D. N., Frizzo, Bonacorso. H. G., Zanatta, N., & Martins, M. A. P. (2012). Efficient microwave-assisted synthesis of 1-aryl-4-dimethylamino methyleno-pyrrolidine-2, 3, 5-triones. *Tetrahedron Letters*, 53(25), 3131-3134.

[16] Buriol, L., Frizzo, Moreira. D. N., Prola, L. D. T., Marzari, M. R. B., München, T. S., Zanatta, N., Bonacorso, H. G., & Martins, M. A. P. (2011). An E-factor minimized solvent-free protocol for the preparation of 4,5-dihydro-5-(trifluoromethyl)-1H-pyrazoles. *Monatshefte für Chemie*, 142(5), 515-520.

[17] Buriol, L., Frizzo, C. P., Prola, L. D. T., Moreira, D. N., Marzari, M. R. B., Scapin, E., Zanatta, N., Bonacorso, H. G., & Martins, M. A. P. (2011). Synergic Effects of Ionic Liquid and Microwave Irradiation in Promoting Trifluoromethylpyrazole Synthesis. *Catalysis Letters*, 141(8), 1130-1135.

[18] Buriol, L., Frizzo, C. P., Marzari, M. R. B., Moreira, D. N., Prola, L. D. T., Zanatta, N., Bonacorso, H. G., & Martins, M. A. P. (2010). Pyrazole synthesis under microwave irradiation and solvent-free conditions. *Journal of the Brazilian Chemical Society*, 21(6), 1037-1044.

[19] Martins, M. A. P., Beck, P., Moreira, D. N., Buriol, L., Frizzo, C. P., Zanatta, N., & Bonacorso, H. G. (2010). Straightforward microwave-assisted synthesis of 1-carboxymethyl-5-trifluoromethyl-5-hydroxy-4,5-dihydro-1H-pyrazoles under solvent-free conditions. *Journal of Heterocyclic Chemistry*, 47(2), 301-308.

[20] Martins, M. A. P., Frizzo, Moreira. D. N., Buriol, L., & Machado, P. (2009). Solvent-Free Heterocyclic Synthesis. *Chemical Reviews*, 109(9), 4140-4182.

[21] Katritzky, A. R., & Singh, S. K. (2003). Microwave-assisted heterocyclic synthesis. AR-KIVOC (13), 68-86.

[22] Hosseini, M., Stiasni, N., Barbieri, V., & Kappe, C. O. (2007). Microwave-Assisted Asymmetric Organocatalysis. A Probe for Nonthermal Microwave Effects and the Concept of Simultaneous Cooling. *The Journal of Organic Chemistry*, 72(4), 1417-1424.

[23] Herrero, M. A., Kremsner, J. M., & Kappe, C. O. (2008). Nonthermal Microwave Effects Revisited: On the Importance of Internal Temperature Monitoring and Agitation in Microwave Chemistry. *The Journal of Organic Chemistry*, 73(1), 36-47.

[24] Laurent, R., Laporterie, A., Dubac, J., Lefeuvre, S., & Audhuy, M. (1992). Specific activation by microwaves: myth or reality? *The Journal of Organic Chemistry*, 57(26), 7099-7102.

[25] Moreira, D. N., Frizzo, C. P., Longhi, K., Soares, A. B., Marzari, M. R. B., Buriol, L., Brondani, S., Zanatta, N., Bonacorso, H. G., & Martins, M. A. P. (2011). Ionic liquid and Lewis acid combination in the synthesis of novel (E)-1-(benzylideneamino)-3-cyano-6-(trifluoromethyl)-1H-2-pyridones. *Monatshefte für Chemie*, 142(12), 1265-1270.

[26] Guarda, E. A., Marzari, M. R. B., Frizzo, Guarda. P. M., Zanatta, N., Bonacorso, H. G., & Martins, M. A. P. (2012). Enol ethers and acetals: Acylation with dichloroacetyl, acetyl and benzoyl chloride in ionic liquid medium. *Tetrahedron Letters*, 53(2), 170-172.

[27] Moreira, D. N., Longhi, K., Frizzo, C. P., Bonacorso, H. G., Zanatta, N., & Martins, M. A. P. (2010). Ionic liquid promoted cyclocondensation reactions to the formation of isoxazoles, pyrazoles and pyrimidines. *Catalysis Communications*, 11(5), 476-479.

[28] Martins, M. A. P., Guarda, E. A., Frizzo, C. P., Moreira, D. N., Marzari, M. R. B., Zanatta, N., & Bonacorso, H. G. (2009). Ionic Liquids Promoted the C-Acylation of Acetals in Solvent-free Conditions. *Catalysis Letters*, 130(1-2), 93-99.

[29] Frizzo, C. P., Marzari, M. R. B., Buriol, L., Moreira, D. N., Rosa, F. A., Vargas, P. S., Zanatta, N., Bonacorso, H. G., & Martins, M. A. P. (2009). Ionic liquid effects on the reaction of beta-enaminones and tert-butylhydrazine and applications for the synthesis of pyrazoles. *Catalysis Communications*, 10(15), 1967-1970.

[30] Moreira, D. N., Longhi, K., Frizzo, Bonacorso. H. G., Zanatta, N., & Martins, M. A. P. (2009). Ionic liquid promoted cyclocondensation reactions to the formation of isoxazoles, pyrazoles and pyrimidines. *Catalysis Communications*, 11(5), 476-479.

[31] Frizzo, C. P., Moreira, D. N., Guarda, E. A., Fiss, G. F., Marzari, M. R. B., Zanatta, N., Bonacorso, H. G., & Martins, M. A. P. (2009). Ionic liquid as catalyst in the synthesis of N-alkyl trifluoromethylpyrazoles. *Catalysis Communications*, 10(8), 1153-1156.

[32] Moreira, D. N., Frizzo, C. P., Longhi, K., Zanatta, N., Bonacorso, H. G., & Martins, M. A. P. (2008). An efficient synthesis of 1-cyanoacetyl-5-halomethyl-4,5-dihydro-1H-pyrazoles in ionic liquid. *Monatshefte für Chemie*, 139(9), 1049-1054.

[33] Martins, M. A. P., Frizzo, C. P., Moreira, D. N., Zanatta, N., & Bonacorso, H. G. (2008). Ionic Liquids in Heterocyclic Synthesis. *Chemical Reviews*, 108(6), 2015-2050.

[34] Wasserscheid, P., & Keim, W. (2000). *Ionic Liquids- New "Solutions" for Transition Metal Catalysis* (Angewandte Chemie International Edition), 39(21), 3772-3789.

[35] Seddon, K. R. (1987). *In Molten Salt Chemistry; Mamantov G, Marassi R, Eds.*, Reidel Publishing Co., Dordrecht, The Netherlands.

[36] Fremantle, M. (1998). Designer solvents- Ionic liquids may boost clean technology development. *Chemical & Engineering News*, 7632-37.

[37] Wasserscheid, P., & Welton, T. (2002). *Ionic Liquids in Synthesis*, Wiley-VCH Verlag, Stuttgart, Germany.

[38] Gordon, C. M., Holbrey, J. D., Kennedy, A. R., & Seddon, K. R. (1998). Ionic liquid crystals: hexafluorophosphate salts. *Journal of Materials Chemistry*, 8(12), 2627-2636.

[39] Seddon, K. R., Stark, A., & Torres, M. J. (2000). Influence of chloride, water, and organic solvents on the physical properties of ionic liquids. *Pure and Applied Chemistry*, 72(12), 2275-2287.

[40] Rogers, R. D., & Seddon, K. R. (2005). *Ionic Liquids III A: Fundamentals, Progress, Challenges, and Opportunities Properties and Structure*, American Chemical Society, Washington.

[41] Wilkes, J. S. (2004). Properties of ionic liquid solvents for catalysis. *Journal of Molecular Catalysis A: Chemical*, 214(1), 11-17.

[42] Holbrey, J. D., & Seddon, K. R. (1999). Ionic Liquids. *Clean Technologies and Environmental Policy*, 1(4), 223-236.

[43] Hardacre, C. (2005). Application of exafs to molten salts and ionic liquid technology. *Annual Review of Materials Research*, 35, 29-49.

[44] Horikoshi, S., Hamamura, T., Kajitani, M., Yoshizawa-Fujitaand, M., & Serpone, N. (2008). Green Chemistry with a Novel 5.8-GHz Microwave Apparatus. Prompt One-Pot Solvent-Free Synthesis of a Major Ionic Liquid: The 1-Butyl-3-methylimidazolium Tetrafluoroborate System. *Organic Process Research & Development*, 12(6), 1089-1093.

[45] Dimitrakis, G., Villar-Garcia, I. J., Lester, E., Licence, P., & Kingman, S. (2008). Dielectric spectroscopy: a technique for the determination of water coordination within ionic liquids. *Physical Chemistry Chemical Physics*, 10(20), 2947-2951.

[46] Damm, M., & Kappe, C. O. (2009). Parallel microwave chemistry in silicon carbide reactor platforms: an in-depth investigation into heating characteristics. *Molecular Diversity*, 13(4), 529-543.

[47] Martinez-Palou, R. (2009). Microwave-assisted synthesis using ionic liquids. *Molecular Diversity*, 14(1), 3-25.

[48] Leadbeater, N. E., & Torenius, H. M. (2006). *In Microwaves in Organic Synthesis, ed. A. Loupy* (2nd edition), Wiley-VCH, Weinheim, 327-361.

[49] Habermann, J., Ponzi, S., & Ley, S. V. (2005). Organic Chemistry in Ionic Liquids Using Non-Thermal Energy-Transfer Processes. *Mini-Reviews in Organic Chemistry*, 2(2), 125-137.

[50] Leadbeater, N. E., Toreniusand, H. M., & Tye, H. (2004). Microwave-Promoted Organic Synthesis Using Ionic Liquids: A Mini Review. *Combinatorial Chemistry & High Throughput Screening*, 7(5), 511-528.

[51] Kremsner, J. M., & Kappe, C. O. (2006). Silicon Carbide Passive Heating Elements in Microwave-Assisted Organic Synthesis. *The Journal of Organic Chemistry*, 71(12), 4651-4658.

[52] Kappe, C. O., Dallinger, D., & Murphree, S. S. (2009). *Pratical Microwave Synthesis for Organic Chemists*, Germany, Wiley-VCH.

[53] Leadbeater, N. E., & Torenius, H. M. (2002). A Study of the Ionic Liquid Mediated Microwave Heating of Organic Solvents. *The Journal of Organic Chemistry*, 67(9), 3145-3148.

[54] Leadbeater, N. E., Torenius, H. M., & Tye, H. (2003). Ionic liquids as reagents and solvents in conjunction with microwave heating: rapid synthesis of alkyl halides from alcohols and nitriles from aryl halides. *Tetrahedron*, 59(13), 2253-2258.

[55] Ley, S. V., Leach, A. G., & Storer, R. I. (2001). A polymer-Supported Thionating Reagent. *Journal of the Chemical Society, Perkin Transactions*, 1(4), 358-361.

[56] Eycken, E. V., der Appukkuttan, P., De Borggraeve, W., Dehaen, W., Dallinger, D., & Kappe, C. O. (2002). High-Speed Microwave-Promoted Hetero-Diels-Alder Reactions of 2(1H)-Pyrazinones in Ionic Liquid Doped Solvents. *The Journal of Organic Chemistry*, 67(22), 7904-7907.

[57] Garbacia, S., Desai, B., Lavastre, O., & Kappe, C. O. (2003). Microwave-Assisted Ring-Closing Metathesis Revisited. On the Question of the Nonthermal Microwave Effect. *The Journal of Organic Chemistry*, 68(23), 9136-9139.

[58] Obermayer, D., & Kappe, C. O. (2010). On the importance of simultaneous infrared/fiber-optic temperature monitoring in the microwave-assisted synthesis of ionic liquids. *Organic & Biomolecular Chemistry*, 8(1), 114-121.

[59] Kappe, C. O. (2008). Microwave dielectric heating in synthetic organic chemistry. *Chemical Society Reviews*, 37(6), 1127-1139.

[60] Loupy, A. (2006). *Microwaves in Organic Synthesis*, Wiley-VCH, Weinheim, 2nd edn, and references therein.

[61] Hoffmann, J., Nuchter, M., Ondruschka, B., & Wasserscheid, P. (2003). Ionic liquids and their heating behaviour during microwave irradiation- a state of the art report and challenge to assessment. *Green Chemistry*, 5(3), 296-299.

[62] The experiments were performed in a Discover CEM MW using the mode of operation: with simultaneous cooling and temperature sensor fiber optics. The power of

the equipment was established at 200 W (or 300 W when necessary). A microwave vessel (10 mL) equipped with a standard cap (vessel commercially furnished by Discover CEM) was filled with solvent (3 mL) and [BMIM][BF_4] (quantities indicated in Table 5). After the vessel was sealed, the sample was irradiated for 5 min at 150 °C, which was plotted in Synergies Version 3.5.9 software and a maximum level of internal vessel pressure of 250 psi. The irradiation powers are indicated in Table 5. The solvents doped were subsequently cooled to 50 °C by compressed air.

[63] Silva, A. M. G., Tomé, A. C., Neves, M. G. P. M. S., Cavaleiro, J. A. S., & Kappe, C. O. (2005). Porphyrins in Diels-Alder reactions. Improvements on the synthesis of barrelene-fused chlorins using microwave irradiation. *Tetrahedron Letters*, 46(28), 4723-4726.

[64] Nebel, K., Brunner-G, H., & Pissiotas, G. (1996). *Int. Appl. Pat.*, WO 96/01254.

[65] Matos, I., Pérez-Mayora, E., Soriano, E., Zukal, A., Martín-Aranda, R. M., López-Peinado, A. J., Fonseca, I., & Cejka, J. (2010). *Chemical Engineering Journal*, 161(3), 377-383.

[66] The experiments were performed in a Discover CEM MW using the mode of operation: with temperature sensor fiber optics; without simultaneous cooling. The power of the equipment was established at 200 W. A microwave vessel (10 mL) equipped with a standard cap (vessel commercially furnished by Discover CEM) was filled with 1 mmol of 3,5-dimethylpyrazole and 1.2 mmol of 1-bromobutaneor iodoethane besides the addition of ~ 0.1 mmol.mL of [BMIM][BF_4] (quantities indicated on Tables 8 and 9) and 3 mL of solvent (Tables 8 and 9). KOH was also added in equimolar amount to 3,5-dimethylpyrazole (1 mmol) at experiments indicated in Table 8. The vessel was sealed. The sample was irradiated for 30 min at 150 °C under high stirring and a maximum level of internal vessel pressure of 250 psi. The solvent of the resultant mixture was evaporated under reduced pressure. After this step, the conversion was determinate by ^1H NMR. The ^1H NMR spectra were recorded on a Bruker DPX 400 (^1H at 400.13 MHz) and in $CDCl_3$/TMS solutions at 298 K. The spectroscopy data for compounds 1-butyl-3,5-dimethyl-1*H*-pyrazole and 1-ethyl-3,5-dimethyl-1*H*-pyrazole are present in the references: [31] and Potapov A. S., Khlebnikov A. I., Ogorodnikov V. D. (2006). Synthesis of 1-Ethylpyrazole-4-carbaldehydes,1,1′-Methylenebis(3,5-dimethylpyrazole-4-carbaldehyde), and Schiff Bases Derived There from. *Russian Journal of Organic Chemistry*, 42(4), 550-554, respectively.

Chapter 18

Multicomponent Reactions in Ionic Liquids

Ahmed Al Otaibi and Adam McCluskey

Additional information is available at the end of the chapter

http://dx.doi.org/10.5772/51937

1. Introduction

In our group, we place a premium on the rapid access to a wide rage of diverse small molecules. Our current focus spans the inhibition of dynamin GTPase, protein phosphatases 1 and 2A and the development of anti-cancer lead compounds.[1]-[10] While rapid access is paramount, we also strive to develop high levels of diversity in an environmentally friendly manner. That is, we are keen to apply the tenants of green chemistry at all stages of our drug development programs. To satisfy this need we have developed a particular interest in multicomponent reactions in benign solvents.

A multicomponent reaction (MCR) can be simply classified as a reaction in which three or more components are combined together in a single reaction vessel to produce a final product or products displaying features of all inputs and thus offers greater possibilities for molecular diversity per step with a minimum of synthetic time and effort. Products from such MCRs result in high atom and step economy.

The first reported MCR was Strecker's synthesis of racemic amino acids in the 1850's.[11] Strecker's amino acid synthesis combined an aldehyde, hydrogen cyanide and ammonia in a one pot procedure leading to a range of amino acids. With over 150 years history and development, MCRs have recently seen a resurgence, in part due to the ease of access to a wide range of diverse, highly functionalized molecules, in particular the synthesis of small heterocyclic rings of medicinal chemistry importance. A discourse on the physical properties of ionic liquids is out with the scope of this work. The chemistry, reactions and properties of ionic liquids has been addressed in a number of excellent review articles in this area.[12]-[20] In this chapter we describe the current state of play associated with MCRs in ionic liquids, with a focus on 3- and 4- component MCRs (3CRs and 4CRs respectively).[21]

For ease of discussion the application of MCRs in ionic liquids is broken down into the type of product generated: heterocyclic rings containing various numbers of heteroatoms and a classification of the reaction as either a 3CR or a 4CR.

2. Three component MCRs (3CRs)

2.1. Synthesis of acyclic products

MCRs are not only applicable to the synthesis of heterocyclic systems, but represent a very facile entry point to a range of acyclic compounds such as the amido substituted naphthols shown in Scheme 1. The treatment of β-naphthol with a wide range of aldehydes (aliphatic and aromatic), substituted amides in the presence of conventional ionic liquids (Ils), such as those based on the *N*-methyl, *N*-sulfonic acid imidazolonium [MSIM] cation, afforded rapid access to 1-amidoalkyl-2-naphthols and 1-amidoaryl-2-naphthols in good to excellent yields. In this area Zolfigol *et al.* have had particular success with the functionalised ILs such as [MSIM][Cl], [DSIM][Cl] and [MSIM][$AlCl_4$], under solvent-free conditions (Scheme 1).[22]

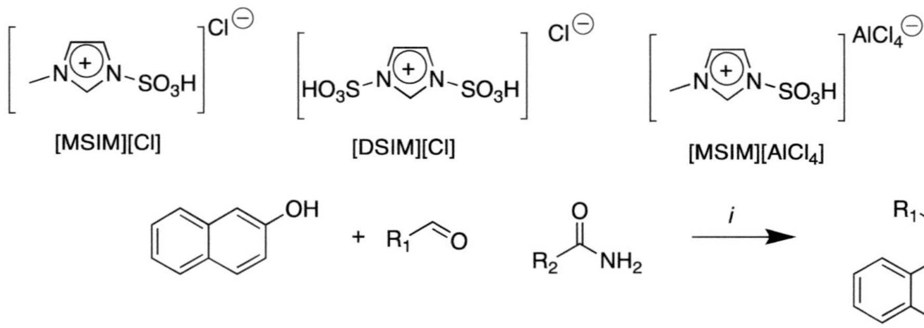

Scheme 1. Synthesis of 1-amidoalkyl-2-naphthols: (i) [MSIM][Cl] or [DSIM][Cl] or [MSIM][$AlCl_4$], 120°C, ~40 min.

Hajipour *et al.* and Hervai *et al.* effected the same transformations, and extended the methodology to allow the use of urea as the amide source using a range of Brønstead acid based Ils (BAILs).[23],[24] Hajipor *et al.*'s approach used *N*-(4-sulfonic acid)butyltriethyl ammonium hydrogen sulfate [TEBSA][HSO_4], while Hervai *et al.* applied two BAILs: 3-methyl-1-(4-sulfonic acid)-butylimidazolium hydrogen sulfate [MIM-$(CH_2)_4SO_3H$][HSO_4]) and *N*-(4-sulfonic acid)butylpyridinium hydrogen sulfate [Py-$(CH_2)_4SO_3H$][HSO_4] to effect the same transformations. [TEBSA][HSO_4] has been used previously as an efficient and reusable catalyst for nitration of aromatic compounds and esterification of various alcohols by different acids.[25]-[27] The acidic nature of BAILs has been exploited as catalysts for many other significant organic reactions, which proceed with excellent yields and selectivities and demonstrate the great potential of these ILs in catalytic technologies for chemical production.[23]

Kotadia *et al.* and Zhang independently reported the synthesis of similar 1-amidoalkylnaphthols using solid supported ionic liquids (SSILs).[28] Kotadia used a benzimidazolium based

http://dx.doi.org/10.5772/51937

ionic liquid immobilized on silica based solid support, while Zhang conducted the reaction in the presence of polyethylene glycol (PEG)-based dicationic acidic ionic liquid as a catalyst under solvent-free conditions.[29],[30] Supported reagents offer the advantages of simple and safe catalyst recycling.[28] All MCR-IL based approaches to 1-amidoalkylnaphthols where highly substituent tolerant with excellent yields observed ith both electron donating and electron withdrawing aldehydes.

Yadav and Rai reported a three component MCR approach to β-nitrocarbonitriles and β-nitrothiocyanates in [BMIM][OH] or [BMIM][BF_4]. The reaction proceeds via a Henry reaction to yield the β-nitrostyrenes follwed by Michael addition of trimethylsilyl cyanide (TMSCN) or ammonium thiocyanate to yield the β-nitrocarbonitriles and β-nitrothiocyanates in modest overall yields of 53-58% (Scheme 2).[31],[32]

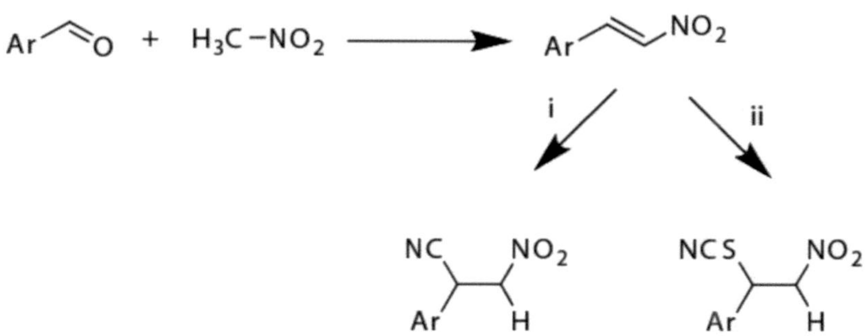

Scheme 2. Synthesis of β-nitrocarbonitriles and β–nitrothiocyanates: (i) [BMIM][OH] or [BMIM][BF_4], TMSCN, CH_3CN, 85-90°C, 6-9h; (ii) [BMIM][OH] or [BMIM][BF_4], NH_4SCN, CH_3CN, 85-90°C, 6-9h.

Zaho *et al.* has reported the combined Barbier / Friedel-Crafts alkylation of unsusbtituted benzaldehydes with allylbromide and phenols to yield 4-(2-hydroxyphenyl)-4-phenylbut-1-enes promoted by BuPyCl/SnCl•$2H_2O$ and their subsequent application to the synthesis of 4-(substituted phenyl)chromans (Scheme 3).[33]

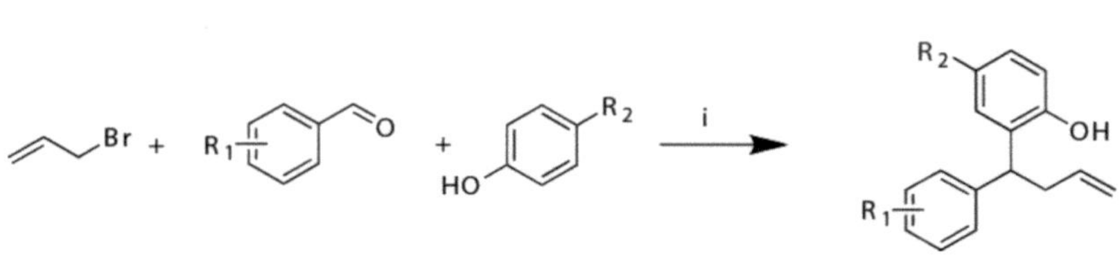

Scheme 3. Synthesis of 4-(2-hydroxyphenyl)-4-substituted phenylbut-1-enes: (i) BuPyCl/SnCl•$2H_2O$.

The use of an enolisable ketone facilitated the synthesis of a family of β-amido ketones (Scheme 4). The reaction of an enolizable ketone, aryl aldehyde and acetonitrile or benzonitrile in the presence of TMSCl using a Brønsted-acidic ionic liquid 3-methyl-1-(4-sulfonic acid) butylimidazolium hydrogen sulfate [MIM-$(CH_2)_4SO_3H$][HSO_4] as catalyst gave a family of β-amido ketones in good yield.[34]

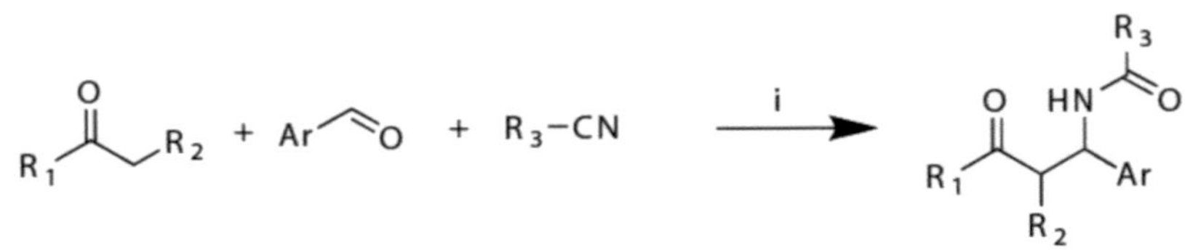

Scheme 4. Synthesis of β-amidoketones: (i) $[MIM-(CH_2)_4SO_3H][HSO_4]$, TMSCl, 80ºC.

This enolisable ketone approach also allowed Fang *et al.* to conduct a three-component Mannich-type reaction (Scheme 5) with aromatic aldehydes, aromatic amines, and ketones catalyzed by a novel functionalized ionic liquid, 3-(*N,N*-dimethyldodecylammonium)propanesulfonic acid hydrogen sulfate ($[DDPA][HSO_4]$) at room temperature to give various β-aminocarbonyl compounds in good yields.[35] [DDPA] $[HSO_4]$ was recycled and after six cycles, no loss in catalytic activity was reported. Gong *et al.* conducted the same reaction using cyclohexanone and [BMIM][OH] as the IL catalyst to afford the β-aminocarbonyl compounds in excellent yields.[36]

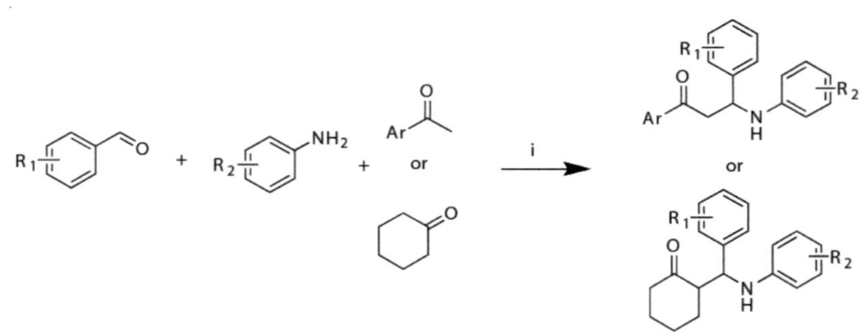

Scheme 5. Mannich-type approach to β-amidoketones: (i) $[DDPA][HSO_4]$ or [BMIM][OH].

Liu *et al.* also explored the IL mediated Mannich reaction, but utilised a series task-specific ionic liquids in developing an asymmetric of β-aminoketones from isovaleraldehyde, methyl ketones, and aromatic amines in excellent yields (ca. 90%) and %ee's (~95%).[37] L-proline was used as the chiral catalyst (Scheme 6). The reaction with $[DEMIm][BF_4]$, $[DEEIm][BF_4]$, $[BEIm][BF_4]$, $[MEIm][BF_4]$ and $[PEIm][BF_4]$ typically gave the desired product with an ee >95%. While the chemical yield dropped marginally on re-use from 96 to 85% over four cycles of IL use, the %ee remained constant.

http://dx.doi.org/10.5772/51937

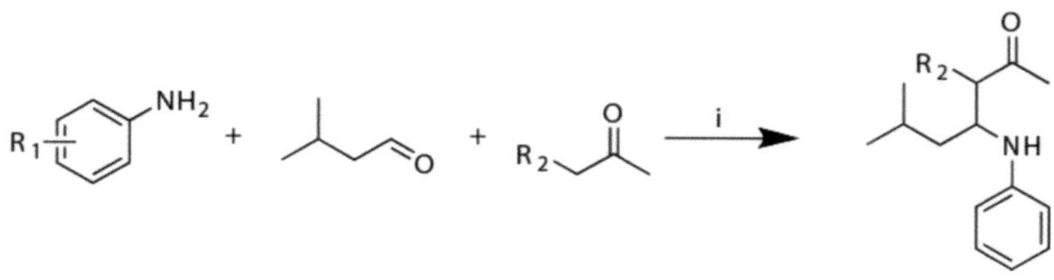

Scheme 6. Catalysed asymmetric Mannich reaction: (i) task specific ionic liquid chosen from [DEMIm][BF_4], [DEEIm][BF_4], [BEIm][BF_4], [MEIm][BF_4] and [PEIm][BF_4] / L-proline.

In a similar reaction sequence, Akbari and Heydari, replaced the activated ketone with trimethyl phosphite, in the presence of [MIM-$(CH_2)_4SO_3H$][CF_3SO_3], to affect rapid access to α-aminophosphonates (Scheme 7).[38] The reaction proceded via protonation of the carbonyl moiety, imine formation and attack at the protonated imine by trimethylphosphite. The reaction was highly tolerant of substituents on the carbonyl containing compound with pyridyl, cinnamyl, etc., affording excellent yields of the corresponding α-aminophosphonates. The ionic liquid could be recycled with no observable loss of efficacy after six cycles.

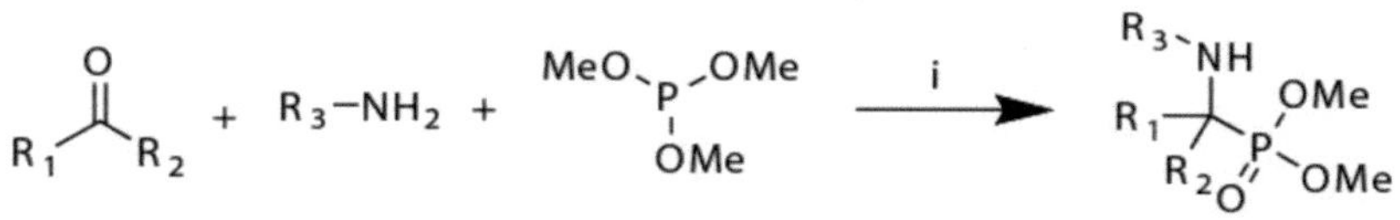

Scheme 7. Synthesis of α–aminophosphonates: (i) [MIM-$(CH_2)_4SO_3H$][CF_3SO_3] (10 mol%), H_2O.

Reddy *et al.* also reported the synthesis of α-aminophosphonates via a three-component reaction of 5-amino-2,2-difluoro-1,3-benzodioxole, aromatic aldehydes, and diethylphosphite catalysed by silica-supported boron trifluoride (BF_3.SiO_2) in [BMIM][HCl] at room temperature (Scheme 8).[39] Yields were good to excellent and reaction times were typically 5 min versus 3h using conventional solvents.

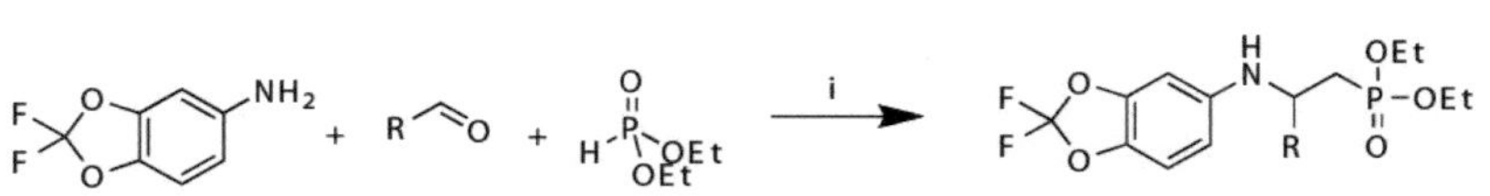

Scheme 8. Synthesis of α-aminophosphonate catalysed: (i) BF_3.SiO_2 / [BMIM][HCl].

O-Protected cyanohydrins are versatile synthetic intermediates in organic synthesis for the preparation of a wide variety of organic compounds such as α-hydroxyacids, α-hydroxy ketones, α-amino acids, and β-amino alcohols.[40]-[43] Shen and Ji developed a mild synthesis of these key intermediates via the condensation of an aldehyde, trimethylsilyl cyanide (TMSCN), and Ac_2O in [BMIM][BF_4] (Scheme 9).[44] In addition, the recovered ionic liquid could be reused for subsequent runs without the loss of activity.

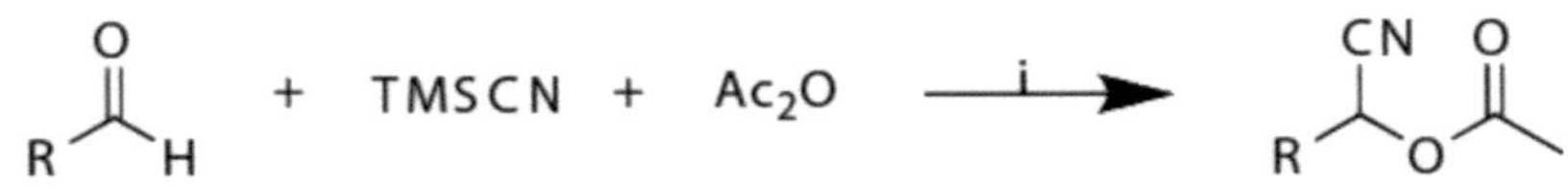

Scheme 9. One-pot synthesis of O-acetyl cyanohydrin: (i) [BMIM][BF_4].

2.1.1. 3CRs yielding heterocycles with one ring nitrogen

Arguably the major utility of MCRs is in the synthesis of highly decorated heterocyclic compounds. In our group we are interested in the synthesis of heterocyclic scaffolds that can be used in medicinal chemistry programs to instill the correct level of biological activity. Davoodnia *et al.* reported an efficient procedure for preparation of 2,4,6-triarylpyridines by treatment of acetophenones, aryl aldehydes, and NH_4OAc in the presence of [MIM-$(CH_2)_4SO_3H$][HSO_4] (Scheme 10).[45] Aromatic aldehydes with both electron donating and electron withdrawing substituents were well tolerated.

Scheme 10. Preparation of 2,4,6-triarylpyridines: (i) [MIM-$(CH_2)_4SO_3H$][HSO_4], 120ºC.

In a related study, Heravi and Fakhr, reported a high yielding ultrasonic promoted synthesis of 2-amino-6-(arylthio)-4-arylpyridine-3,5-dicarbonitrile derivatives (Scheme 11), by the reaction of aryl aldehydes, thiols and malononitrile catalyzed by $ZrOCl_2.8H_2O/NaNH_2$ in [BMIM][BF_4] at room temperature.[46] Access to the same type of pentasubsituted pyridines was also possible using [BMIM][OH] as described by Ranu (Scheme 11).[47]

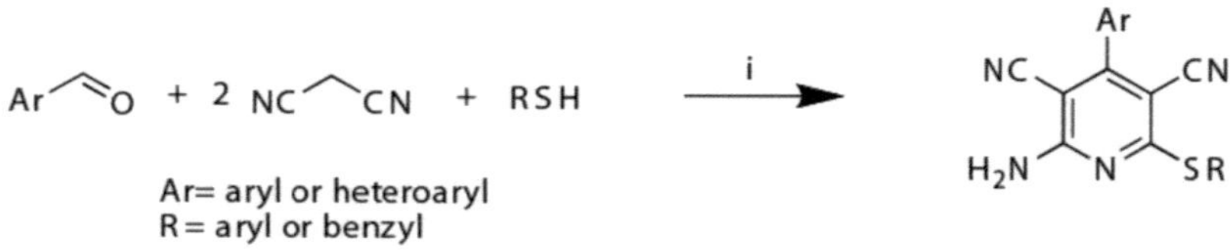

Scheme 11. Synthesis of penta substituted pyridines: (i) [BMIM][OH] / EtOH, rt; or $ZrOCl_2.8H_2O/NaNH_2$, [BMIM][BF_4], ultrasound.

The related 2,4,6-triaryl-1,4-dihydropyridines were generated in a Aldol-Michael-addition reaction cascade involving an aromatic aldehyde, acetophenone and NH_4OAc in [BMIM][BF_4] (Scheme 12).[48] The resulting 2,4,6-triaryl-1,4-dihydropyridines were then examined as potential catalysts for the the Diels-Alder reaction of *p*-quinone and cyclopentadine, and maleic anhydride and cyclopentadiene.

http://dx.doi.org/10.5772/51937

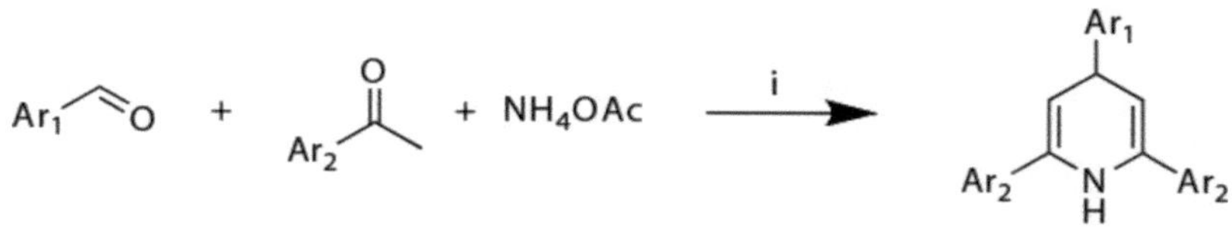

Scheme 12. Synthesis of 2,4,6,-triaryl-1,4-dihydropyridines: (i) [BMIM][BF_4], 80°C.

Wu used essentially the same reaction cascade described above, replacing acetophenone with acetoacetates which yielded, from [Bpy][BF_4], a series of 2,6-dimethyl-4-aryl-1,4-dihydropyridine-3,5-dicarboxylate esters (Scheme 13).[49] Compared with classical Hantzsch reaction conditions towards this type of product, this IL mediated reaction had the advantage of excellent yields, short reaction time, and easy workup.

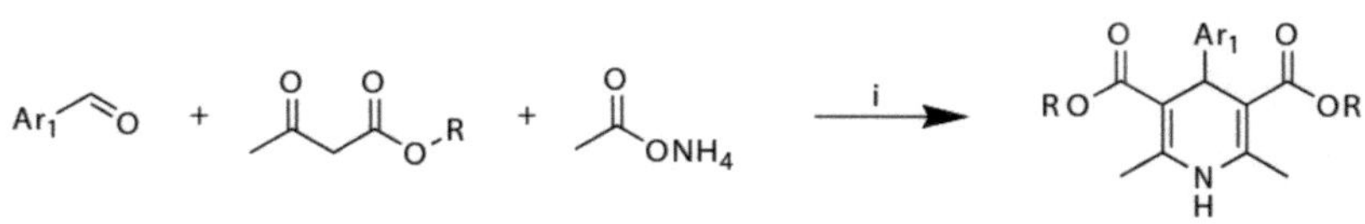

Scheme 13. Synthesis of 1,4-dihydropyridines: (i) [BPy][BF_4], 100-110°C.

Quinolin-4(1*H*)ones constitute an important class of heterocyclic compounds because of their important pharmaceutical properties, such as anti-viral,[50],[51] anti-platelet,[52] and anti-tumor effects.[53] These compounds have been exploited as precursors for anti-cancer and anti-malarial agents.[54],[55] Yadav *et al.* decribed an efficient two step synthesis of quinolin-4(1*H*)ones, 5*H*-thiazolo[3,2-*a*]pyrimidine, and 4*H*-pyrimido[2,1-*b*]benzothiazoles at room temperature.[56] The initial reaction in reaction was conducted arylamine with Meldrum's acid) and trimethylorthoformate in [BMIM][Br] at 40ºC giving the corresponding arylaminomethylene-1,3-dioxane-4,6-diones. Cyclization occured in [BMIM][BF_4] / OTf at 80ºC under nitrogen to the quiniolin-4(1*H*)ones in excellent yields (Scheme 14).[56]

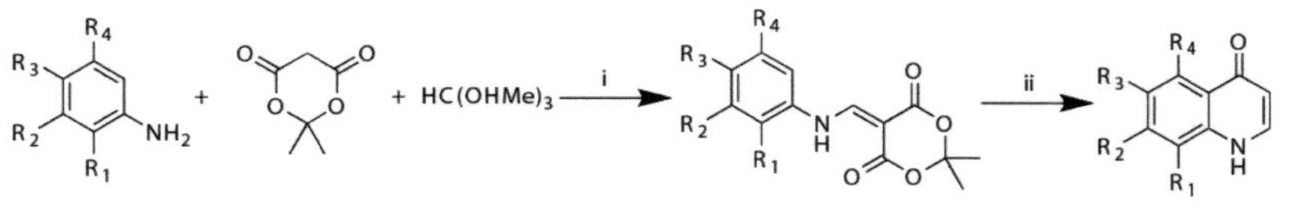

Scheme 14. Synthesis of 4(1*H*)-quinolones: (i) [BMIM][Br], 40°C, N_2; (ii) [BMIM][BF_4]/OTf, 80°C.

Wang *et al.* reported a novel reaction of 2-(1-substitutedpiperidin-4-ylidene)malononitrile, benzaldehyde, and malononitrile or cyanoacetate in the synthesis of highly substituted isoquinoline derivatives (Scheme 15).[57] The three-component reaction of benzaldehyde, malononitrile, and ethyl 4-(dicyanomethylene)piperidine-1-carboxylate was reacted in [BMIM] [BF_4] at 50ºC, delivering ethyl 6-amino-5,7,7-tricyano-3,4,7,8-tetrahydro-8-arylisoquinoline-2(1*H*)-carboxylate derivatives being obtained in excellent yields. The highest yields were obtained with [BMIM][BF_4].

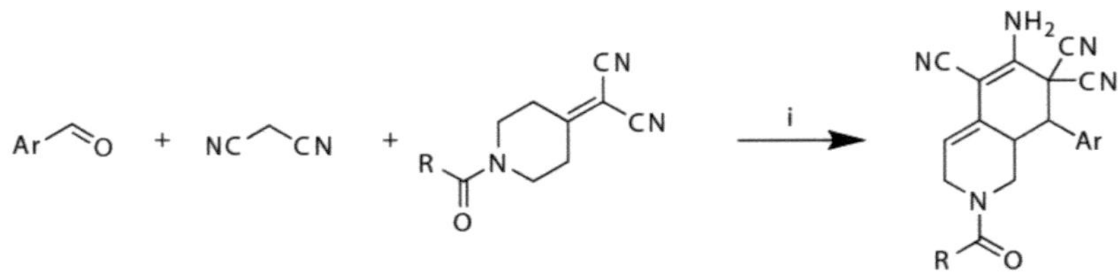

Scheme 15. Synthesis of ethyl 6-amino-5,7,7-tricyano-3,4,7,8-tetrahydro-8-arylisoquinoline-2(1*H*)-carboxylate derivatives: (i) [BMIM][BF_4], 50°C.

There are many methods for the synthesis of acridine compounds containing 1,4-dihydropyridine moieties from aldehydes, dimedone, and anilines or ammonium acetates via heating in organic solvents, or catalysis by triethyl(benzyl)ammonium chloride (TEBAC) in water, or under microwave irradiation.[58] In a much more efficient approach, Li *et al.* utilised the three component MCR in [BMIM][BF_4] at room temperature of an aromatic aldehyde, 2-hydroxy-1,4-naphthoquinone and naphthalen-2-amine giving rise to a series of 14-aryl-1,6,7,14-tetrahydrodibenzo[*a,i*]-acridine-1,6-dione derivatives (Scheme 16).[59]

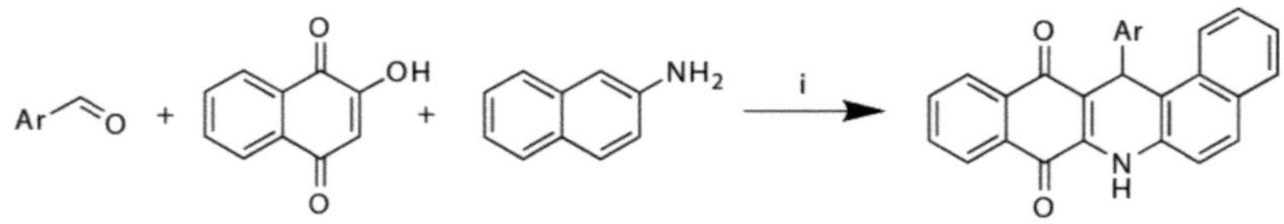

Scheme 16. Synthesis of 14-aryl-1,6,7,14-tetrahydrodibenzo[*a,i*]acridine-1,6-diones: (i) [BMIM][BF_4], rt.

Shi reported an efficient and green synthetic route to 3,3'-benzylidenebis(4-hydroxy-6-methylpyridin-2(1*H*)-ones) via condensation, addition and ammonolysis of an aldehyde, aniline and 6-methyl-4-hydroxypyran-2-one (Scheme 17).[60] Different solvents including [BMIM][Br], [BMIM][BF_4] and [BMIM][PF_6] were examined, with [BMIM][Br] giving the most favourable outcome (high yield and ease of product isolation).

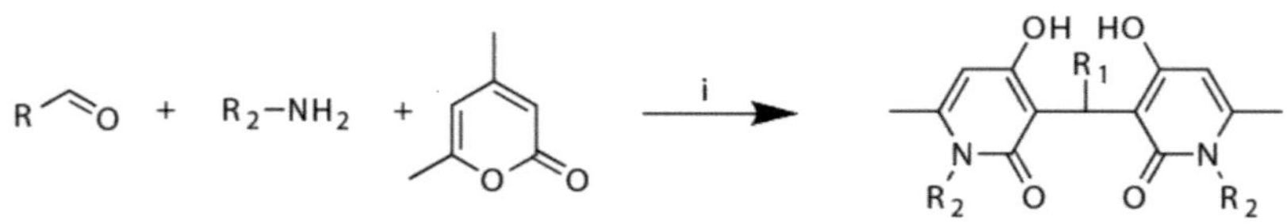

Scheme 17. Synthesis of 3,3'-benzylidenebis(4-hydroxy-6-methylpyridin-2(1*H*)-ones): [BMIM][Br], 95°C.

http://dx.doi.org/10.5772/51937

2.1.2. 3CRs yielding heterocycles with two ring nitrogens

Pyrimidine derivatives are important biologically active heterocyclic compounds which possess antimalarial.[61] Gholap's synthesis of 3,4-dihydropyrimidin-2-(1*H*)-ones from aromatic or aliphatic aldehydes with ethyl acetoacetate and urea (or thiourea), was promoted by ultrasound in [BMIM][BF_4] at room temperature affording the target compounds in excellent yields and short reaction times (Scheme 18).[62]

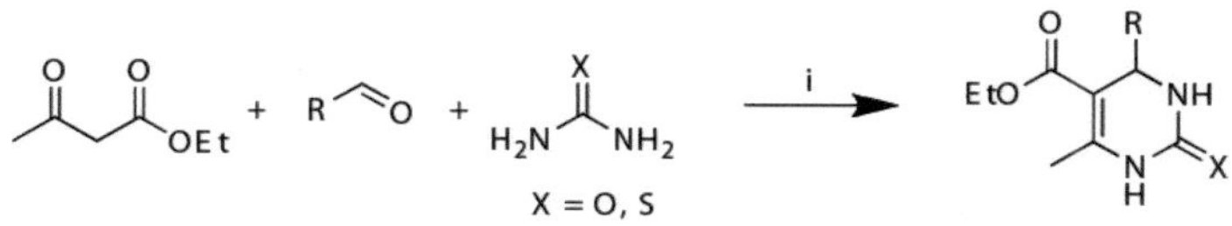

Scheme 18. Synthesis 3,4-dihydropyrimidin-2-(1*H*)-ones: (i) [BMIM][BF_4], 30°C, ultrasound.

Similar pyrimidine analogues were accessed by Gui *et al.* through the use of acidic ionic liquids such as [MIM-CH_2COOH][HSO_4], [MIM-CH_2COOH][H_2PO_4], [MIM-$(CH_2)_2$COOH][HSO_4] and [MIM-$(CH_2)_2$COOH][H_2PO_4]) which successfully promoted the Biginelli coupling of an aldehyde, 1,3-dicarbonyl compound, and urea giving easy access to 3,4-dihydropyrimidin-2(1*H*)-ones.[63] Peng and Deng, used [BMIM][BF_4] and [BMIM][PF_6] as catalysts for the same Biginelli condensation reaction at 100ºC (Scheme 19).[64]

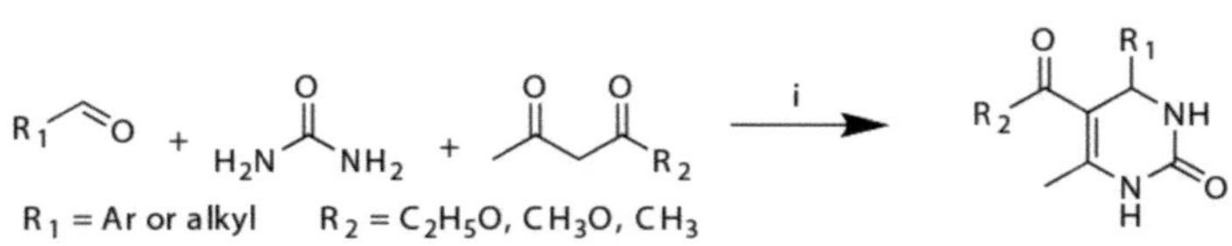

Scheme 19. The synthesis of 3,4-dihydropyrimidin-2(1*H*)-ones: (i) IL chosen from [MIMCH_2COOH][HSO_4] or [MIMCH_2COOH][H_2PO_4] or [MIM$(CH_2)_2$COOH][HSO_4] or [MIM$(CH_2)_2$COOH][H_2PO_4], 75°C.

Brønsted acidic ionic liquids have designed to replace solid acids and traditional mineral liquid acids like sulfuric acid and hydrochloric acid in chemical procedures.[65],[66] Using 3-carboxypyridinium hydrogensulfate [HCPy][HSO_4], 1,3-dicarbonyl compounds, aromatic aldehydes and urea or thiourea, Hajipour and Seddighi, successfully removed the traditional acid requirement in the synthesis of 3,4-dihydropyrimidin-2(1*H*)-ones (Scheme 20).[67]

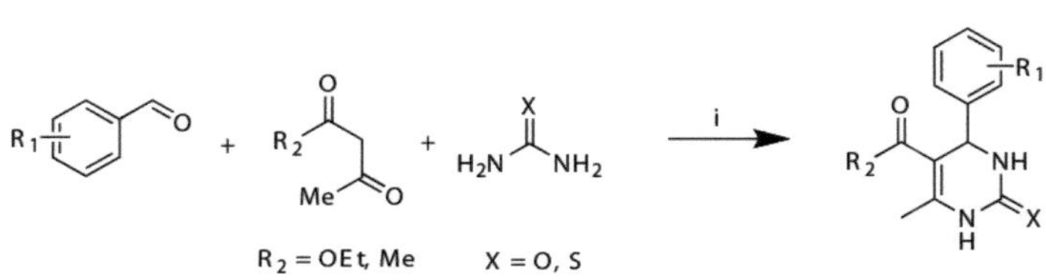

Scheme 20. Synthesis of 3,4-dihydropyrimidin-2-(1*H*)-ones: (i) [HCPy][HSO_4], 120°C.

Fang *et al.*'s dicationic acidic IL catalytic approach, in what amounted to a modified Biginelli approach, resulted in the synthesis of 3,4-dihydropyrimidin-2(1*H*)-one and 3,4-dihydropyrimidin-2(1*H*)-thione derivatives (Scheme 21), in good yields.[68] The products could be separated simply from the catalyst–water system, and the catalysts could be reused at least six times without noticeably reducing catalytic activity.

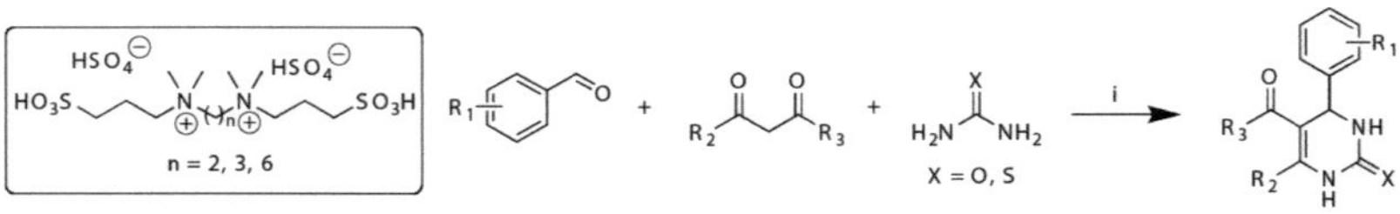

Scheme 21. Synthesis of 3,4-dihydropyrimidin-2(1*H*)-(thi)ones: (i) Dicationic acidic IL (shown in box), 90ºC.

Rather than use an acidic IL approach to dihydropyrimidinones (above) Mirzai and Valizadeh developed a microwave assisted Biginelli route using the weakly Lewis basic nitrite based ionic liquid, IL-ONO (Scheme 22).[69] These nitrite based Ils have also been used to carry out nitrosations of aromatic compounds in aqueous media.[70] Valizadeh, have reported the nitrozation of aromatic compounds using the same nitrite ionic liquid in aqueous media.[70]

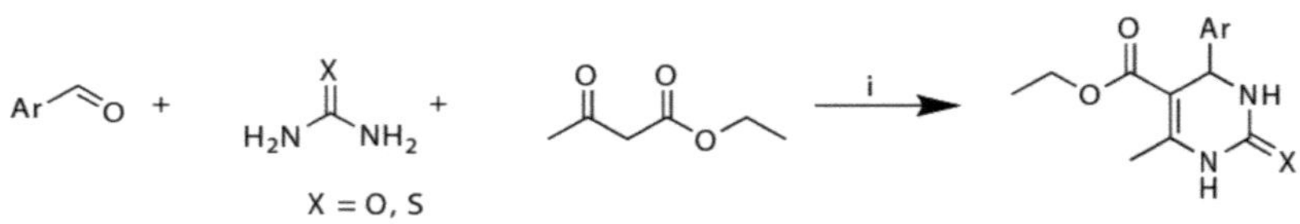

Scheme 22. Synthesis of dihydropyrimidinones: (i) IL-ONO, MW, 80ºC.

Trisubstituted imidazoles can be rapidly accessed from a one-pot condensation of 1,2-diketone or α-hydroxyketone, aldehyde, and NH_4OAc in 1,1,3,3-*N,N,N,N*-tetramethylguanidinium trifluoroacetate (TMGT) at 100ºC (Scheme 23).[71] The synthesis of trisubstituted imidazoles in TMGT as promoter and solvent for the synthesis of trisubstituted imidazoles not only represented a dramatic improvement (15-40 min, 81-94%) over conventional thermal heating but the reaction times were comparable to the recently reported microwave irradiation (20 min in HOAc, 180–200ºC). There are related routes in a four-component approach (see section 2.2.3).

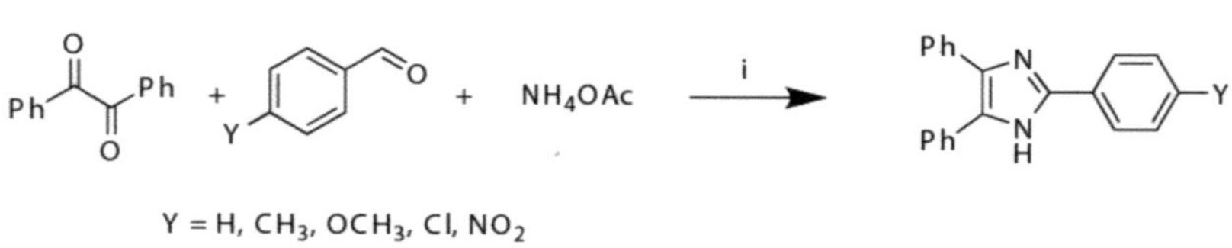

Scheme 23. Synthesis of trisubstituted imidazoles: (i) TMGT, 100ºC.

http://dx.doi.org/10.5772/51937

Khurana and Kumar have reported a simple and convenient synthesis of octahydroquinazolinone and biscoumarin derivatives (Scheme 24).[72] Despite prolonged heating (10 h) at 100ºC, the reaction of benzaldehyde, dimedone and urea in [BMIM][Br] gave only 30% of the expected product, 4,6,7,8-tetrahydro-7,7-dimethyl-4-phenyl-1*H*,3*H*-quinazoline-2,5-dione. Addition of TMSCl saw a reduction in reaction duration to 2.5 h wand a 92% isolated yield. Of the Ils examined [BMIM][Br] and [BMIM][BF_4] gave higher yields than [BMIM][Cl] and [BMIM][PF_6].

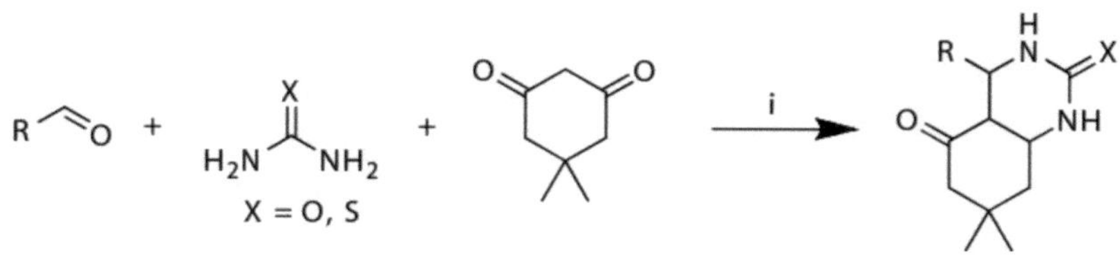

Scheme 24. Synthesis of octahydroquinazolinones: (i) [BMIM][Br], TMSCl, 100°C.

Omprakash *et al.* used catalytic [BMIM][BF_4] and ultrasonics to obtain excellent yields of quinazolin-4(3*H*)ones from anthranilic acid, primary aromatic amine and carboxylic acids (Scheme 25). Of the anilines examined, only 4-nitroaniline required elevated temperature (50℃), but this reaction was complete after 20 minutes.[73]

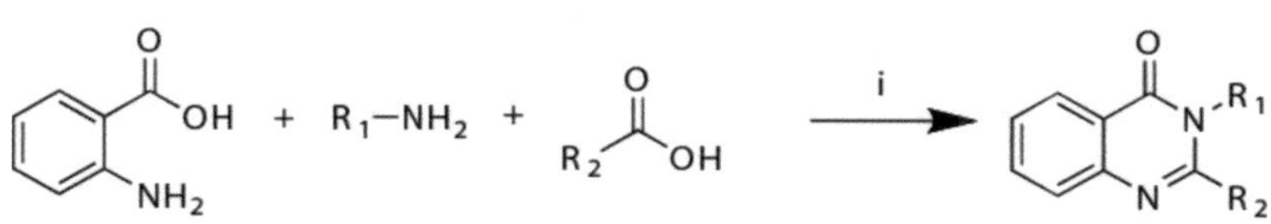

Scheme 25. Synthesis of 4(3*H*)-quinazolinones: (i) [BMIM][BF_4], ultrasound.

The related quinazoline nucleus has been prepared by Dabiri *et al.* from 2-aminobenzophenone derivatives, aldehydes and ammonium acetate in the presence of an protic ionic liquid, 1-methylimidazolium triflouroacetate, [HMIM][TFA] (Scheme 26).[74]

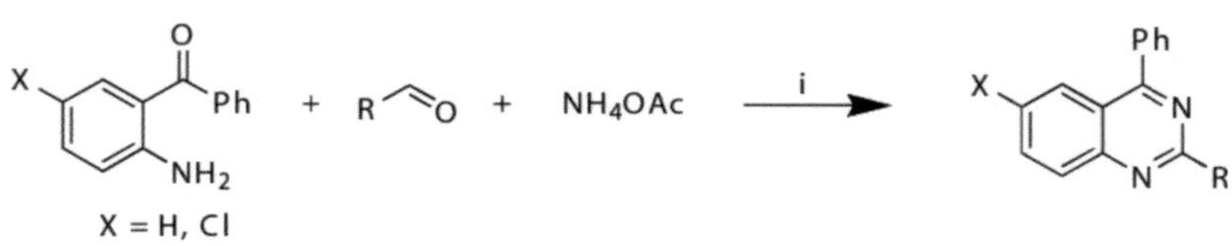

Scheme 26. Synthesis of quinazoline derivatives: (i) [HMIM][TFA], 80°C.

An analogous 6, 5-ring system, the 3-aminoimidazo[1,2-*a*]pyridine, was accessed via condensation of an aldehyde, 2-aminoazine and trimethylsilylcyanide, as an isocyanide equivalent under by simple heating in [BMIM][Br] in high yields with rather short reaction times (1-2 h) (Scheme 27).[75] Shaabani took a slighly different route to 3-arylsubstituted 3-aminoimidazo[1,2-*a*]pyridines using an isocyanide rather than an isocyanide equivalent. The use of substituted aminopyridines also allowed for the introduction of an additional C6 substituent.[76]

Scheme 27. Synthesis of 3-aminoimidazol[1,2-*a*]pyridines: (i) [BMIM][Br], 80ºC

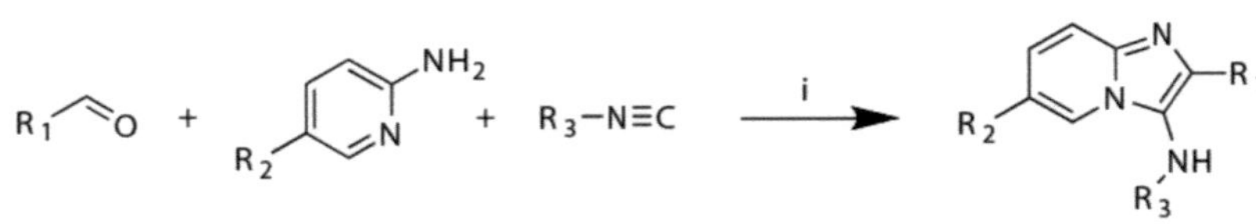

Scheme 28. Synthesis of 3-aminoimidazo[1,2-*a*]pyridines: (i) [BMIM][Br], 80ºC.

Using the Brønsted acidic ionic liquid triethylammonium hydrogen sulfate [TEBSA][HSO_4] Hajipour *et al*. synthesied pyrimidinone derivatives from a range of aromatic aldehydes, cyclopentanone, and urea or thiourea (Scheme 28).[77]

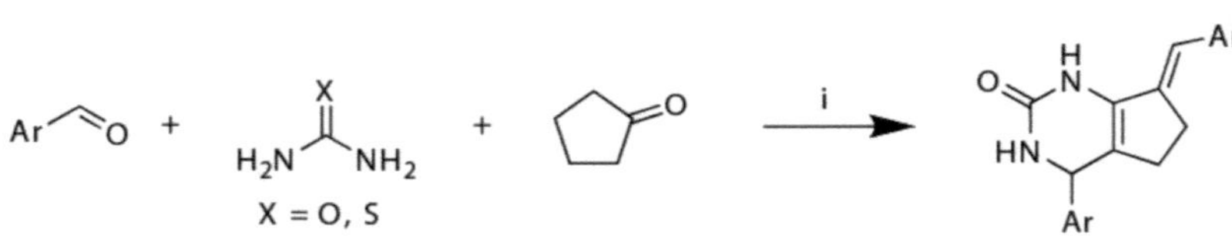

Scheme 29. Synthesis of pyrimidinone derivatives: (i) [TEBSA][HSO_4], 100ºC.

The reaction of phthalhydrazide, aromatic aldehydes, and malononitrile using controlled under microwave irradiation in the presence of [BMIM][OH] at an ambient temperature of 45ºC allowed facile access to 1*H*-pyrazolo[1,2-*b*]phthalazine-5,10-dione (Scheme 29).[78]

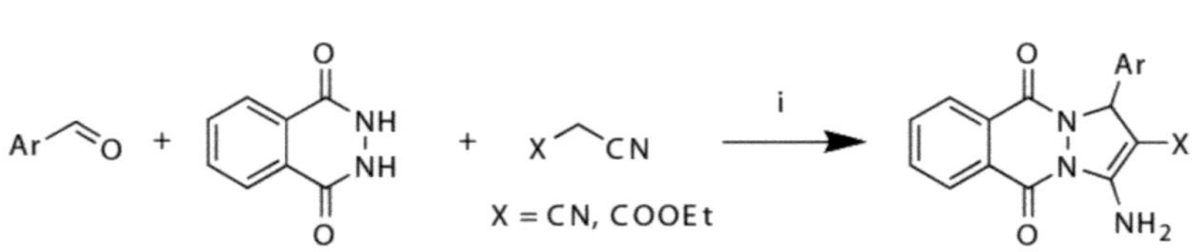

Scheme 30. Synthesis 1*H*-pyrazolo[1,2-*b*]phthalazine-5, 10-dione derivatives: (i) [BMIM][OH], microwaves, 45ºC.

Mixed solvent systems comprising [BMIM][BF_4], water and ethanol were used in the synthesis of 2*H*-indazolo[2,1-*b*]phthalazine-triones by condensation of phthalhydrazide, aromatic aldehydes, and cyclic 1,3-dicarbonyl compounds. Interestingly, the reaction required the addition of a catalytic quantity of sulfuric acid to effect the desired transformation (Scheme 30).[79]

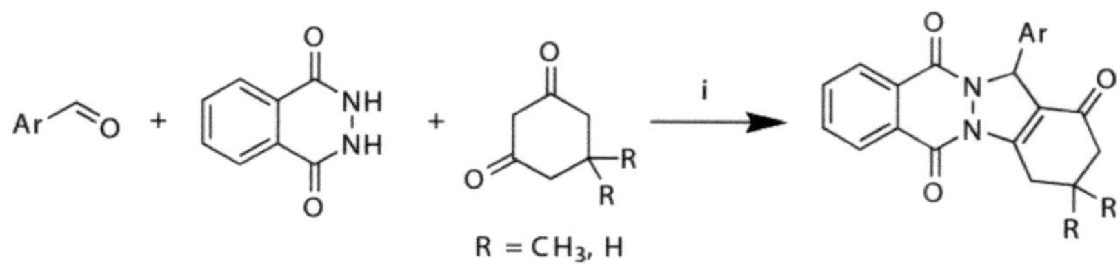

Scheme 31. Synthesis of a series of 2*H*-indazolo[2,1-*b*]phthalazinetriones: (i) [BMIM][BF_4] / H_2O-EtOH / H_2SO_4.

2.1.3. 3CRs yielding heterocycles with three ring nitrogen atoms

While not strictly speaking the synthesis of a new ring system by MCR in ILs, Wang has exploited the MCR approach in an elegant synthesis of *N*-(α-alkoxyalkyl)benzotriazoles (Scheme 31) via the condensation of benzotriazole with various aldehydes and alcohols catalysed by acidic ionic liquid [HMIM][HSO_4] at room temperature.[80] The yield was up to 99%. Wang's approach was effective when triethoxymethane was utilized instead of alcohols. Moreover, the [HMIM][HSO_4] was recyclable with no loss in catalytic activity.

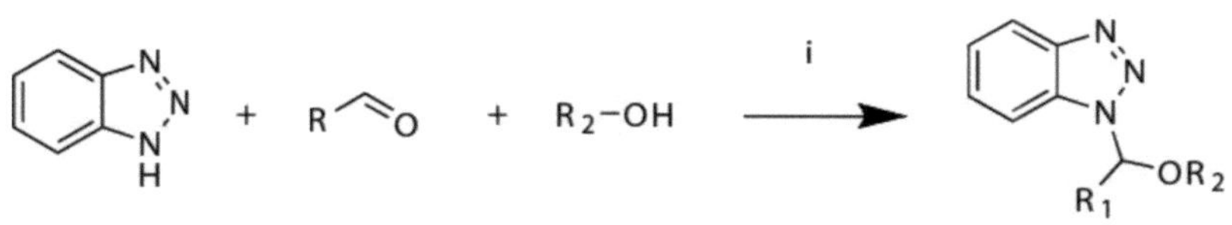

Scheme 32. *N*-(α-alkoxyalkyl)benzotriazoles: (i) [HMIM][HSO_4], rt.

Pyrazolo[3,4-b]pyridines possess a wide range of biological activities such as psychotropic and cytotoxic, and are thus a very important scaffold in medicinal chemistry.[81],[82]

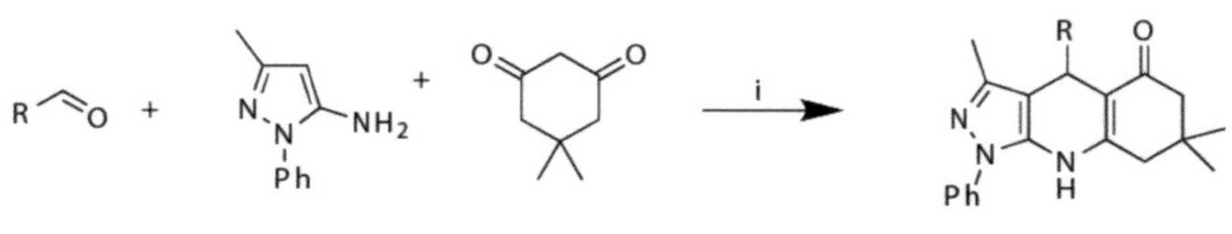

Scheme 33. Preparation of pyrazolo[3,4-*b*]pyridinone and pyrazolo[3,4-*b*]quinolinone derivatives: (i) [BMIM][BF_4], 80ºC.

Judicious choice of the 1,3-dicarbonyl source allows the synthesis of either pyrazolo[3,4-*b*]pyridinone derivatives or pyrazolo[3,4-*b*]quinolinone (Scheme 32). The combination of an aldehyde, 5-amino-3-methyl-1-phenylpyrazole, and Meldrum's acid in [BMIM][BF_4] affords the pyrazolo[3,4-*b*]pyridinone, while the use of dimedone affords the pyrazolo[3,4-*b*]quinolinones.[83]

Zhang showed that the 1,3-dicarbonyl source was not limited to Meldrum's acid or dimedone derivatives with the introduction of 1*H*-indene-1,3(2*H*)-dione for a mild synthesis of indeno[2,1-*e*]pyrazolo[3,4-*b*]pyridine-5(1*H*)-one derivatives in excellent yields (Scheme 33).[84]

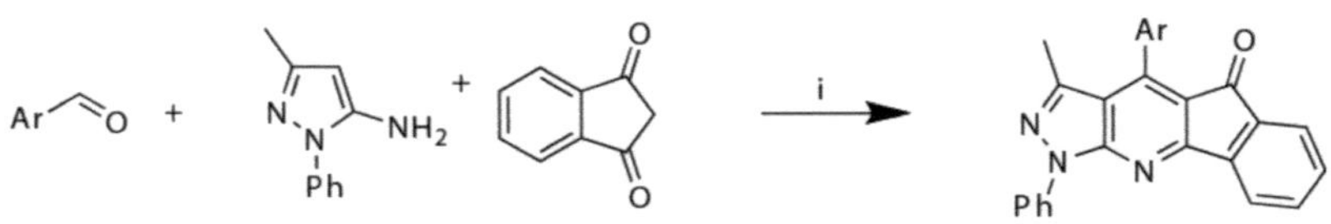

Scheme 34. Synthesise indeno[2, 1-*e*]pyrazolo[3, 4-*b*]pyridine-5(1*H*)-one: (i) [BMIM][Br], 95°C.

Zhang *et al.* reported the reaction of aldehydes, 5-amino-3-methyl-1-phenylpyrazole and malononitrile or ethyl cyanoacetate in [BMIM][BF_4] as a green route to pyrazolo[3,4-*b*]pyridines (Scheme 34).[84]

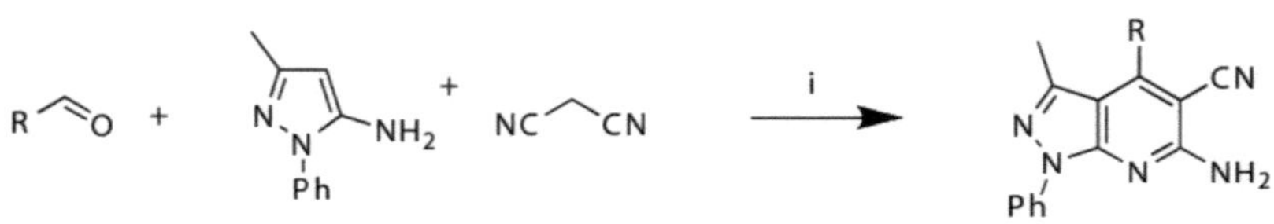

Scheme 35. Preparation of pyrazolo[3,4-*b*]pyridine derivatives: (i) [BMIM][BF_4], 80°C.

Yao *et al.* synthesised 4-aryl-3,4-dihydro-1*H*-pyrimido[1,2-*a*]benzimidazol-2-one via the reaction of aryl aldehyde, 1,3-dicarbonyl compounds and 1*H*-benzo[*d*]imidazol-2-amine in [BMIM][BF_4] (Scheme 35).[85] The reaction accomplished in [BMIM][BF_4] exhibited higher yield (75%) than other counterparts.

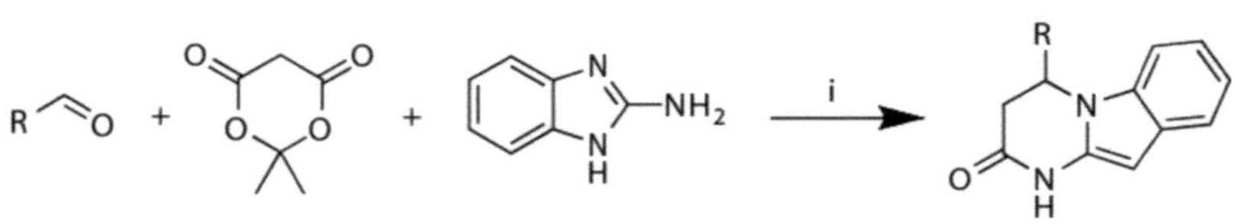

Scheme 36. Synthesis of 4-aryl-3,4-dihydro-1*H*-pyrimido[1,2-*a*]benzimidazol-2-one: (i) [BMIM][BF_4], 90°C.

Wang has a particular interest in the development novel methods for the preparation of various biologically important heterocyclic compounds by using ionic liquids. This group uses ILs as both novel reaction media and reaction promoters.[86],[87] Shi *et al.* also have similar interest and this led to the synthesis indeno[2,1:5,6]pyrido[2,3-*d*]pyrimidine and pyrimido[4,5-*b*]quinoline derivatives from aromatic aldehydes, 6-amino-3-substituted-1-methyl-

http://dx.doi.org/10.5772/51937

pyrimidine-2,4(1*H*,3*H*)-diones and dimidone analogues derivatives in ionic liquid without any catalyst (Scheme 36).[88]

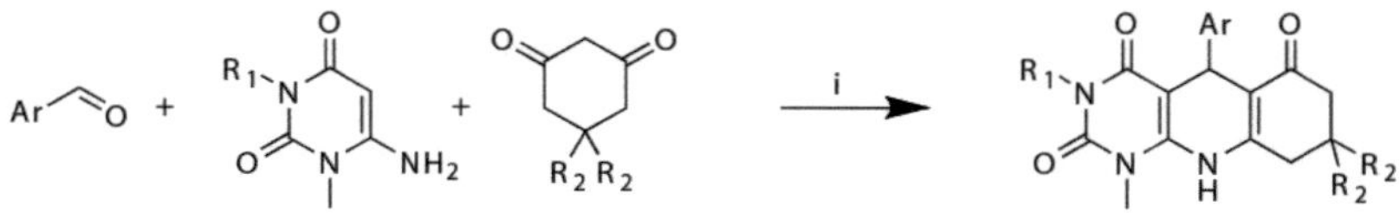

Scheme 37. Synthesis of 5-aryl-7,8,9,10-tetrahydropyrimido[4,5-*b*]quinoline-2,4,6-trione: [BMIM][Br], 95°C.

Pyrimidoquinolinedione derivatives are known to possess antitumor, anticancer, antihypertensive, antibacterial activity and are Kaposi's sarcoma-associated herpesvirus and topoisomerase inhibitors.[89],[90] Entry to this highly biologically active scaffolds can be obtained by the reaction of aldehydes, 1-naphthylamine and barbituric acid in [BMIM][BF_4] gave 7-aryl-11,12-dihydrobenzo[*h*]pyrimido-[4,5-*b*]quinoline-8,10(7*H*,9*H*)-diones (Scheme 37). While the reaction proceeded in traditional organic solvent, yield enhancements and shorter reaction tiomes were evident with the use of [BMIM][BF_4].[91]

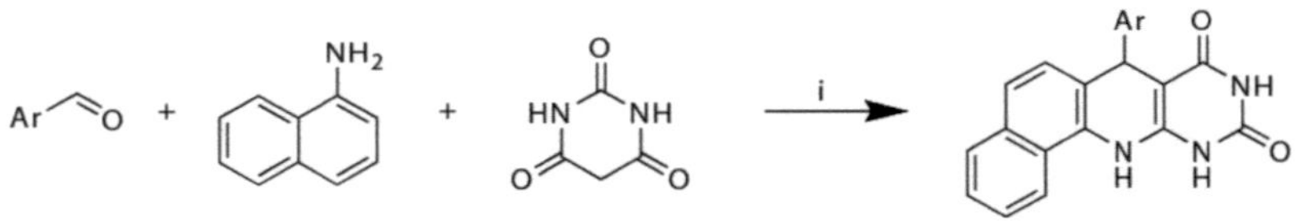

Scheme 38. The synthetic of 7-aryl-11,12-dihydrobenzo[*h*]pyrimido-[4,5-*b*]quinoline-8,10(7*H*,9*H*)-diones: (i) [BMIM][BF_4], 90°C.

2.1.4. 3CRs yielding heterocycles with >three ring nitrogens

The purine bioisosteres, [1,2,4]triazolo[1,5-*a*]pyrimidine, have been reported to possess antitumour activity.[92] Using both [BMIM][BF_4] and [Bpy][BF_4] as reaction solvents, Li synthesised 5-(trifluoromethyl)-4,7-dihydro-7-aryl-[1,2,4]-triazolo[1,5-*a*]pyrimidine derivatives from aldehydes, 3-amino-1,2,4-triazole and ethyl 4,4,4-trifluoro-3-oxobutanoate or 4,4,4-trifluoro-1-phenylbutane-1,3-dione yielding 4-, 5- and 7- substituted derivatives (Scheme 38). The 5- and 7- positions are known to be important for retention of antitumour activity.[93]

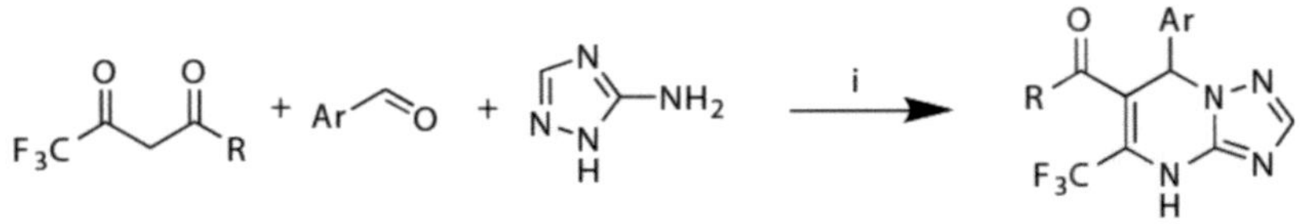

Scheme 39. Synthesis of 5-(trifluoromethyl)-4,7-dihydro-7-aryl-[1,2,4]triazolo[1,5-*a*]pyrimidine derivatives: (i) BMIM][BF_4] or [byp][BF_4].

2.1.5. 3CRs yielding heterocycles with one ring oxygen

The synthesis of heterocyclic compounds with oxygen in the ring is slightly more complex, only due to the reduced numbers of suitable oxygen nucleophiles to affect the final ring-closing step. The 4*H*-pyran core is found in a wide range of natural products and it has thus attraced a considerable degree of attention.[94]-[96] The high reactivity of 4*H*-pyran derivatives has led to their use as synthons in the synthesis of more complex species. Access to highly substituted 4*H*-pyrans is easily accomplished by the 1,1,3,3-tetramethylguanidine catalysed addition of aromatic aldehydes, malononitriles, and β-dicarbonyl in [BMIM][BF_4] (Scheme 39).[97]

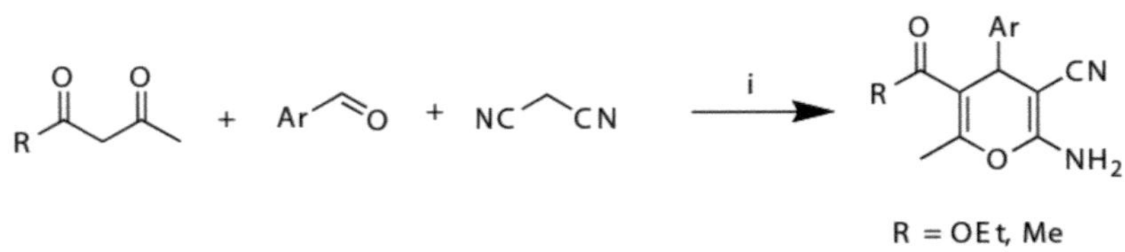

Scheme 40. Synthesis of 4*H*-pyran derivatives: (i) TMG, [BMIM][BF_4], 80°C.

With dimedone as the 1,3-dicarbonyl source the corresponding 5-oxo-5,6,7,8-tetrahydro-4*H*-benzo[*b*]pyran derivatives were accessed in [BMIM][BF_4], [HMIM][BF_4], [OMIM][BF_4], [OMIM][PF_6] and [DMIM][PF_6]. In this instance no additional catalyst was required and the reactions were complete in 2-6 h with yields ranging from 52% to 98%.[98] Fang *et al*. reported a subtle variation leading to the syntheiss of more highly substituted 5-oxo-5,6,7,8-tetrahydro-4*H*-benzo[*b*]pyrans by condensation of aromatic aldehyde, malononitrile (or ethyl cyanoacetate), and dimedone (or 1,3-cyclo-hexanedione) in water catalyzed by acidic ionic liquids such as [TEBSA][HSO_4], [TBPSA][HSO_4], [EDPSA][HSO_4] (Scheme 40).[99] The reactions gave the products in good yields between 86 to 94%.

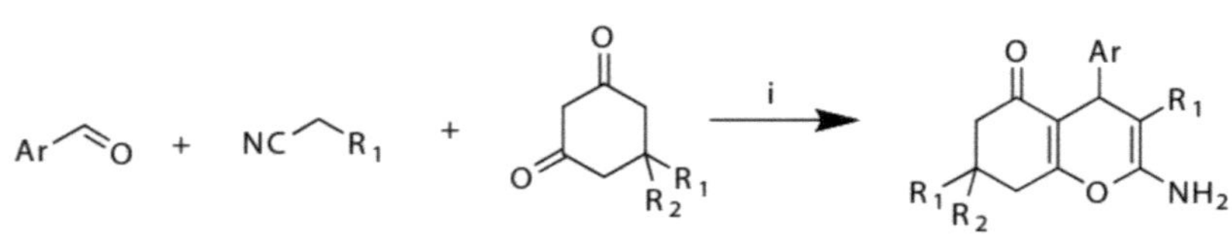

Scheme 41. Synthesis of 5-oxo-5,6,7,8-tetrahydro-4*H*-benzo[*b*]pyrans: (i) Ionic liquid chosen from [TEBSA][HSO_4], [TBPSA][HSO_4] and [EDPSA][HSO_4].

Interest in oxygen containing heterocycles is not limited to those with biological actiity. A number of analogues, such as the 2-amino-2-chromenes are natural products that have found utility in cosmetics and pigments. They also have a role as biodegradable agrochemicals.[100]-[102] Traditional approaches to this scaffold required the reaction of aldehydes, active methylene containing compounds and activated phenols. Stoichiometric quantities of organic base (piperidine) in volatile organic solvents are also required.[103],[104] By replacing the organic solvent with [BMIM][OH] the reaction proceeded with aromatic aldehydes, malononitrile with α- or β-naphthol in the absence of additional catalyst (Scheme 41).[105] After five reuses of the [BMIM][OH] the isolated product yield had dropped from 91% to 85%, which may be due to [BMIM][OH] degredation.

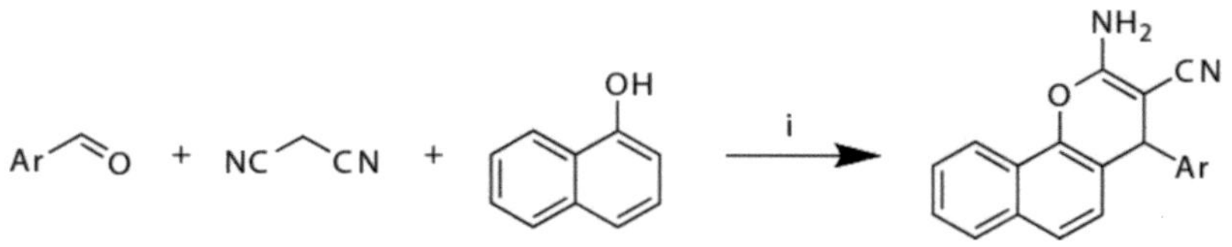

Scheme 42. Synthesis of 2-amino-2-chromenes: (i) [BMIM][OH], H_2O, reflux.

The basic 4*H*-pyran scaffold can be increased in complexity by modification of the basic building blocks described above, e.g. in Schemes 39 and 41. Both Khurana and Magoo, and Zakeri reported the synthesis of a series of 12-aryl-8,9,10,12-tetrahydrobenzo[*a*]xanthen-11-ones by the reaction of β-naphthol, aromatic aldehydes, and dimedone derivatives (Scheme 42).[106],[107] Smooth conversion was accomplished through the use of catalytic *p*-TSA in [BMIM][BF_4] at 80ºC for 35-45 min. The [BMIM][BF_4] could be recyled without a reduction in product yield.

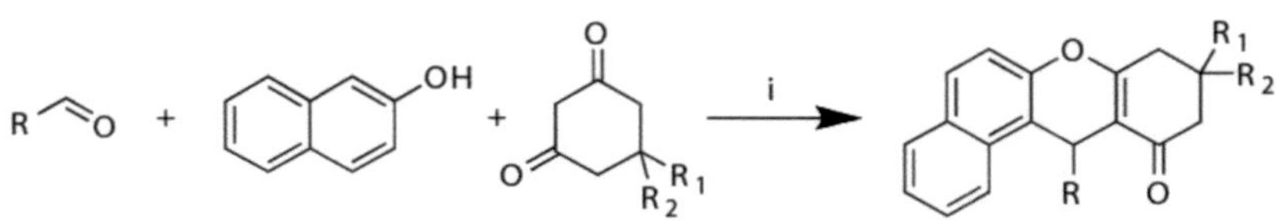

Scheme 43. Synthesis of a series of 12-aryl-8,9,10,12-tetrahydrobenzo[*a*]xanthen-11-ones: (i) [BMIM][BF_4], *p*-TSA or BAIL, 120ºC.

Zheng and Li accessed the tetrahydrobenzo[*b*]pyran and pyrano[*c*]chrome scaffoled via a series of novel Lewis basic task-specific ionic liquids. These novel Ils were used to catalyse the addition of aromatic aldehydes, dimedone and malononitrile can also be used as catalysts in multicomponent reaction accession during the mixture of tetrahydrobenzo[*b*]pyran and pyrano[*c*]chromene derivatives (Scheme 43).[108]

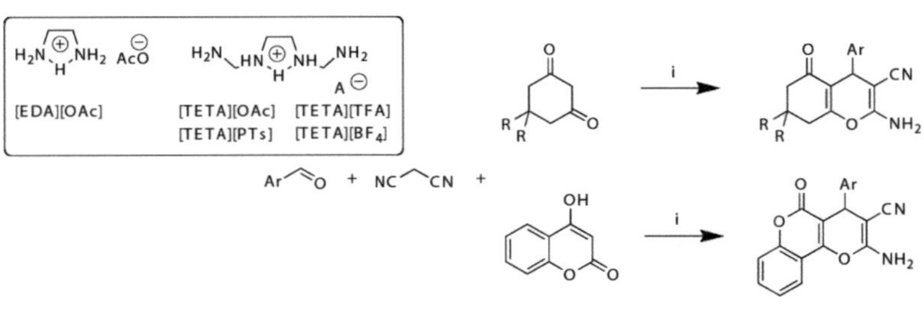

Scheme 44. Synthesis of tetrahydrobenzo[*b*]pyran and pyrano[c]chromene derivatives: (i) [TETA][TFA] (5%), H_2O-EtOH (1:1), reflux.

2.1.6. 3CRs yielding heterocycles with two ring oxygen atoms

Dihydropyrano[3,2-c]chromene-3-carbonitriles are important heterocycles with a wide range of biological properties.[109]-[111] A number of 2-amino-4Hpyrans are reported to be useful photoactive materials.[112] The task specific ionic liquid, hydroxyethanolammonium acetate [HEAA], was used to initiate a domino cascade of 4-hydroxycoumarin, aldehydes, and malononitrile at room temperature ultimately yielding 2-amino-5-oxo-4,5-dihydropyrano[3,2-*c*]chromene-3-carbonitrile derivatives (Scheme 44).[113]

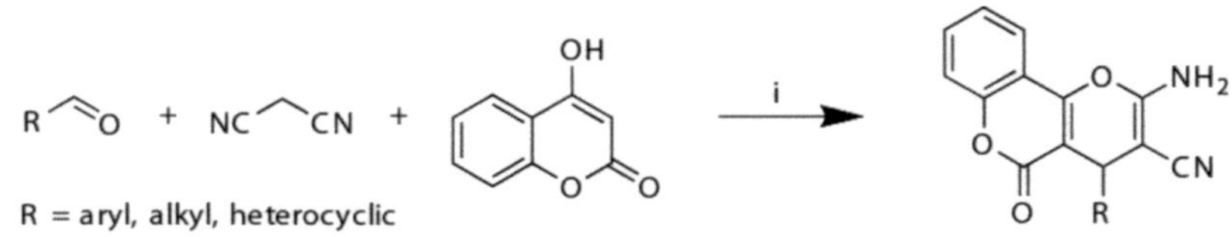

Scheme 45. Synthesis of 2-amino-5-oxo-4,5-dihydropyrano[3,2-c]chromene-3-carbonitrile derivatives: (i) [HEAA], pulverise, rt.

Gong *et al.* reported facile method for the synthesis of 4*H*-pyrans in the presence of basic ionic liquid [BMIM][OH] as catalyst in aqueous medium (Scheme 45).[114] The synthesis of 2-amino-4-aryl-5-oxo-4*H*,5H-pyrano-[3,2-*c*]chromence-3-carbonitrile was achieved by the three-component condensation of an aromatic aldehyde, malononitrile with 4-hydroxycoumarin in the presence of 10 mol% [BMIM][OH] at 100ºC.

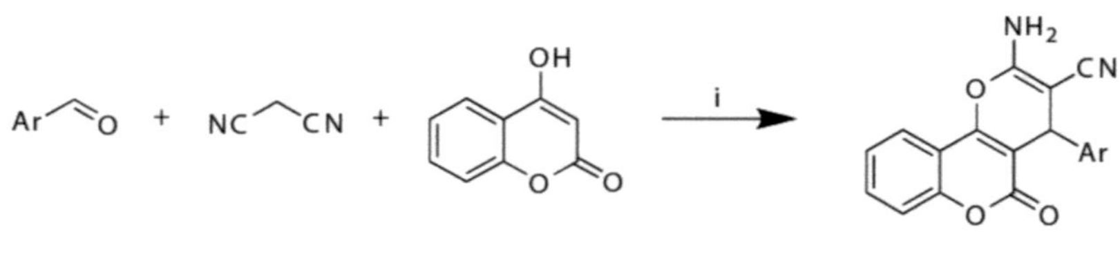

Scheme 46. Synthesis of 2-amino-4-aryl-3-cyano-5-oxo-4*H*,5*H*-pyrano[3,2-c]chromenes: (i) [BMIM][OH], H_2O, 100ºC.

The 3,4-dihydro-2*H*-furo[3,2-*c*]coumarin core is present in a number of natural products ranging including Novobiocin and warfarin. Earlier methods to this core include the oxidative cyclisation of the Michael adduct from the reaction of cyclic 1,3-diketones and chalcones using a phase transfer catalyst; from 1,3-dicarbonyl compounds and (*E*)-β-bromo-β-nitrostyrenes in the presence of tert-butylammonium bromide (TBAB) (20 mol %);[115]-[117] and the manganese acetate promoted radical cyclization of 4-hydroxycoumarin and 2-hydroxy-1,4-naphthoquinone with electron-rich alkenes.[118] Rajesh *et al.* reported a much simpler and greener approach for regio- and diastereo- selective synthesis of furocoumarins (Scheme 46).[119] These reactions proceeded chemo-, regio- and stereoselectively and furnished compounds in good to excellent yields (81-92%).

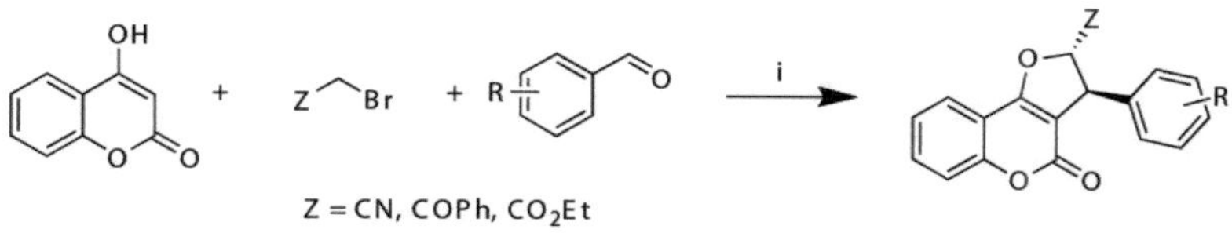

Scheme 47. Synthesis of furo[3,2-c]coumarins: (i) [BMIM][OH],pyridine, 80-90ºC.

Coumarin derivatives have received considerable attention because they possess several types of pharmacological properties, such as antibacterial, anticancer.[120] The coumarins have attracted the attention of a number of research groups interested not only in their biological activity, but also in developing more activity, but also in debeloping more environmentally friendly approaches to their synthesis. Gong *et al.* reported the condensation of 4-hydroxycoumarin, aldehydes, and Meldrum's acids or malononitrile or α-cyanocinnamonitriles in the presence of [BMIM][OH].[114],[121] While Chen *et al.* used 4-hydroxycoumarin, benzaldehyde and 1,3-dicarbonyl by use 1,3-dimethyl-2-oxoimidazolidine-1,3-diium cation [DMDBSI][2HSO$_4$] were employed as the model reactions in the presence of different catalysts (Scheme 47).[122] Both approaches offered ease of access and considerable improvements over the traditional approaches to this class of compounds.[123]-[127]

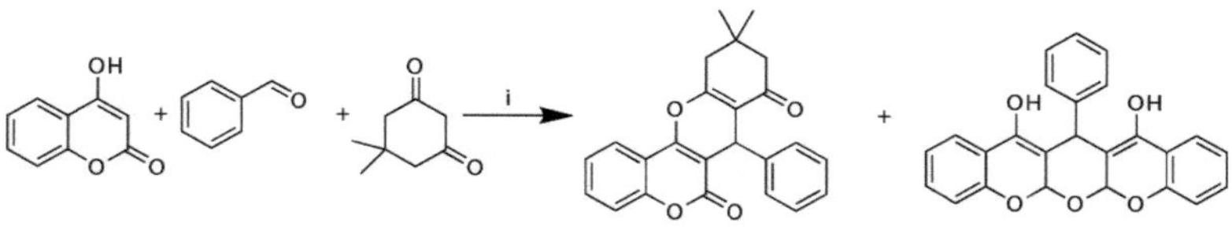

Scheme 48. Synthesis of 4-hydroxycoumarin: (i) [DMDBSI][2H$_2$SO$_4$], H$_2$O, reflux.

2.1.7. 3CRs yielding heterocycles with one ring sulfur atom

Thiophenes, dihydrothiophenes and tetrahydrothiophenes are known important constituents of a range of pharmacologically active compounds.[128]-[130] While these compounds are of significant interest to medicinal and synthetic chemists, the synthetic routes to highly functionalised sulfur heterocycles are not well developed. Notwithstanding this, Zhang *et al.* have reported the synthesis of thiopyrans from aldehydes, malononitrile and cyanothioacetamide in an ionic liquid [BMIM][BF$_4$] as a recyclable solvent and promoter without the need of a catalyst (Scheme 48).[131]

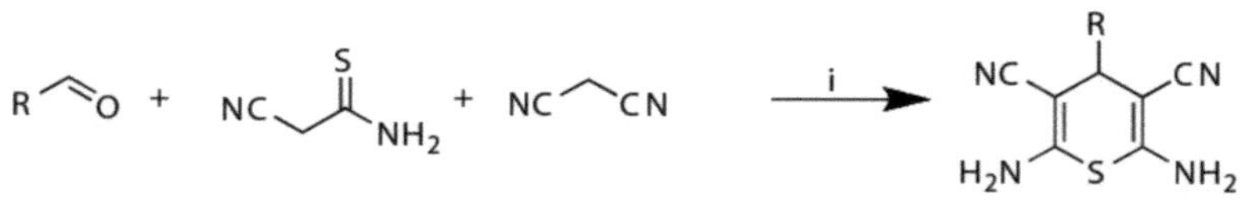

Scheme 49. Synthesis of thiopyrans: (i) [BMIM][BF$_4$], 80ºC.

Given that most of the synthetic procedures towards sulfur heterocycles suffer from some drawbacks such as low yields, long reaction times, the requirement for harsh reaction conditions, it is not surprising that a number of groups have risen to the challenge and examined the use of ionic liquids as a potential method for enhancing the reaction outcomes whilst increasing the efficiency of the synthesis.[132]-[134] Kumar *et al.* have developed a series of novel amino acid derived functional ionic liquids that facilitated the synthesis of dihydrothiophene and tarcine derivatives in good yield under mild conditions from 2-arylidenemalononitrile, 1,3-thiazolidinedione, aliphatic or aromatic amines were added with ionic liquid [Bz-His(n-propyl)$_2$-OMe][Br] and water (Scheme 49). While the products were shown as single diastereoisomers, no details of the level of diastereoselectivity were provided.[135]

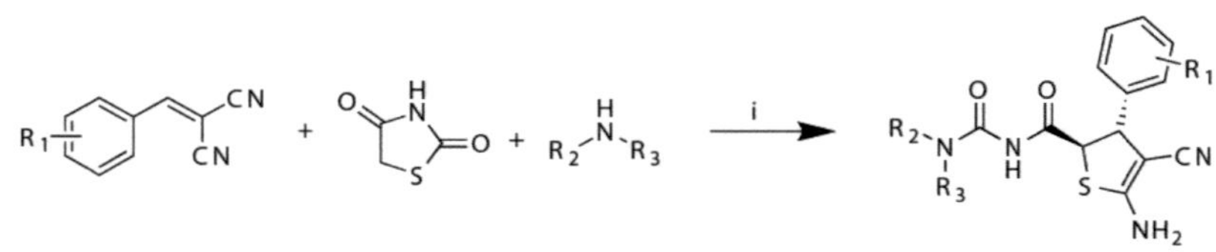

Scheme 50. Synthesis of dihydrothiophenes: (i) [Bz-His(n-propyl)$_2$-OMe][Br], 70°C.

2.1.8. 3CRs yielding heterocycles with ring oxygen and nitrogen atoms

The synthesis of simple heterocycles with a single type of heteroatom is important, but a considerable number of biologically active compounds have different types of heteroatom within a single structure. The benzo[*b*][1,4]oxazin scaffold as a privileged structure for the generation of drug-like libraries in drug-discovery programs has been amply demonstrated. Benzo[*b*][1,4]oxazin derivatives have been used as the basic framework for substances of interest in numerous therapeutic areas, such as anti-Candina albicans agents,[136] antifungals, [137] and kinase inhibitors.[138] Ebrahim *et al.* used [BMIM][Br] as both the solvent and reaction promotor for the room temperature three-component condensation of 2-aminophenole, an aldehyde and isocyanide to prepare benzo[*b*][1,4]oxazines (Scheme 50).[139]

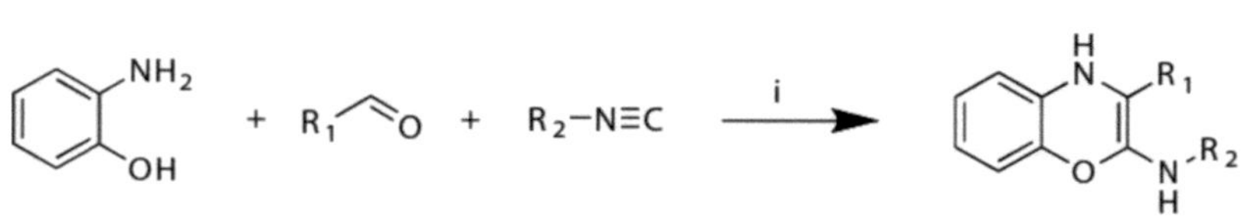

Scheme 51. Synthesis of benzo[*b*][1,4]oxazines: (i) [BMIM][Br], rt.

Asri *et al.* described use of ionic liquids as complementary new media for multicomponent reactions leading to the 2,6-diazabicyclo[2.2.2]octane core (Scheme 51).[140] Interestingly both hydrophobic and hydrophillic Ils [BMIM][BF$_4$] and [BMIM][NTf$_2$] gave acceptable yields, as the original synthesis required the use of toluene and a significant excess of molecular sives to remove water and drive the reaction forward. The original synthesis of these

2,6-diazabicyclo[2.2.2]octanes required 6g of molecular seives per 200 mg reagent. Thus the use of Ils represents a significant greening of this synthesis.[141]

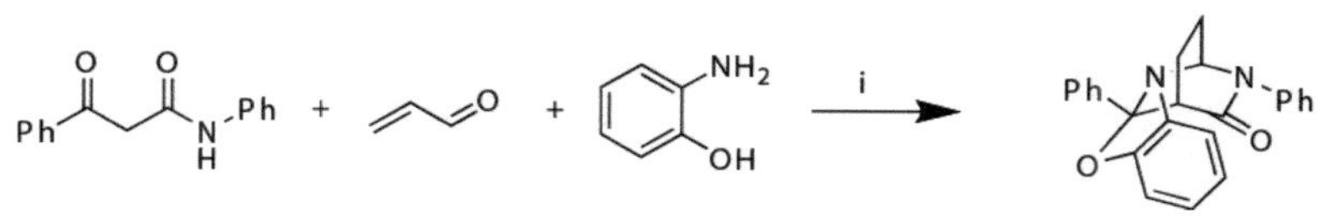

Scheme 52. Synthesis 2,6-diazabicyclo[2.2.2]octane core: (i) [BMIM][BF_4] or [BMIM][NTf_2], 110°C, 4 Å sieves.

The synthesis of 6-amino-4-aryl-5-cyano-3-methyl-1-phenyl-1,4-dihydropyrano[2,3-*c*]pyrazoles was first reported by Otto in 1974.[142] These molecules have since been shown to possess interesting biological activity.[143] Balaskar *et al.* simplified the synthesis of this important class of compounds in a triethylammonium acetate [TEAA] ionic liquid catalyzed reaction of aromatic aldehydes, malononitrile and 3-methyl-1-phenyl-2-pyrazolin-5-one at room temperature (Scheme 52). TEAA plays dual role as reaction media and catalyst. These reactions are rapid, complete in 25 min, and typically high yielding (>90%).[144]

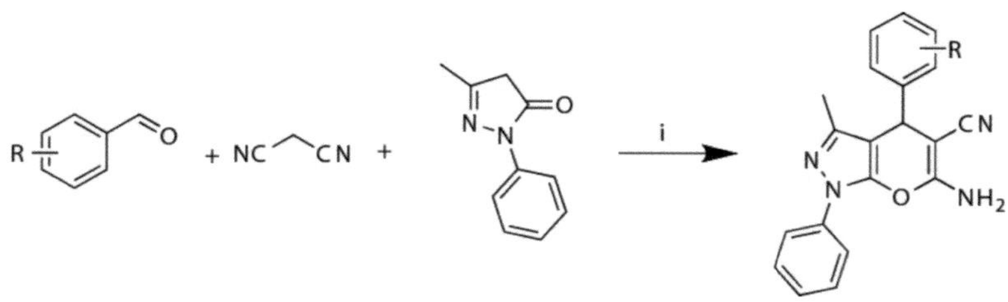

Scheme 53. Synthesis of 6-amino-4-aryl-5-cyano-3-methyl-1-phenyl-1,4 dihydropyrano[2,3-c]pyrazoles: (i) [TEAA], rt.

The furopyridine core is another privileged scaffold in medicinal chemistry.[145],[146] Shi *et al.* rectified what they perceived as an oversight in this area with their synthesis of the furo[3,4-*b*]pyridine motifs by reaction of an aldehyde, 5-amino-3-methyl-1-phenylpyrazole and tetronic acid. They explored the use of [BMIM][Br], [BMIM][BF_4], [PMIM][Br], water, glacial acetic acid, acetone, and ethanol as potential solvents for the synthesis of furo[3,4-*e*]pyrazolo[3,4-*b*]pyridine-5(7*H*)-one derivatives (Scheme 53).[147] Across the range of aromatic aldehydes examined, the ILS [BMIM][Br], [BMIM][BF_4], and [PMIM][Br] consistently gave the highest product yields and the shortest reaction times.

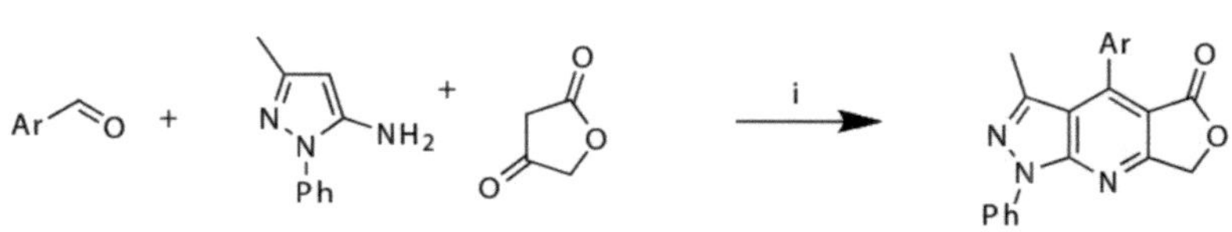

Scheme 54. Synthesise the furo[3,4-e]pyrazolo[3,4-*b*]pyridine-5(7*H*)-one derivatives: (i) ILs, 95°C.

Moghaddam reported the synthesis of novel spiro[chromeno[2,3-*d*]pyrimidine-5,3'-indoline]tetraone derivatives by the combination of isatin, barbituric acid, and cyclohexane-1,3-dione derivatives in the presence of alum ($KAl(SO_4)_2 \bullet 12H_2O$) as a catalyst for 15 min and [BMIM][PF_6] (Scheme 54).[148] The unique structural array and the highly pronounced pharmacological activity displayed by the class of spirooxindole compounds have made them attractive synthetic targets.[149]

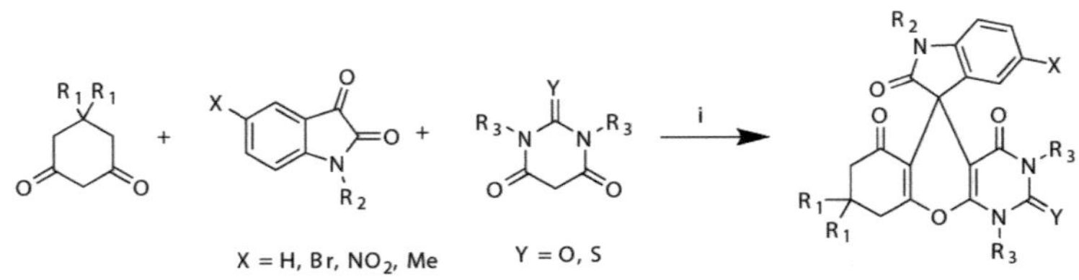

Scheme 55. Synthesis of spiro[chromeno[2,3-*d*]pyrimidine-5,3'-indoline]tetraone derivatives: (i) [BMIM][PF_6], alum, 100°C.

The synthesis of furopyrimidines and 2-aminofurans have received little attention with only a few procedures reported. Among these, the furo[2,3-*d*]pyrimidines have been shown to possess sedative, antihistamine, diuretic, muscle relaxant, and antiulcer properties.[150]-[156] The condensation of an aldehyde, *N,N*-dimethylbarbituric acid and alkyl or aryl isocyanide in [BMIM][Br] gave furo[2,3-*d*]pyrimidine-2,4(1*H*,3*H*)-diones in high yields at room temperature within 20 minutes (Scheme 55).[157]

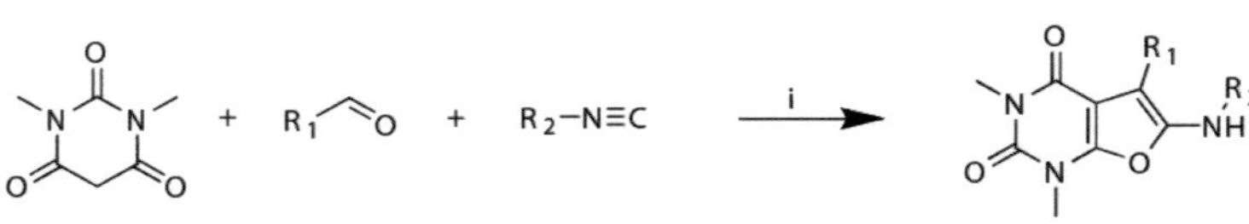

Scheme 56. Synthesis of furo[2,3-*d*]pyrimidine-2,4(1*H*,3*H*)-diones: (i) [BMIM][Br], rt, 15-20 min.

The potential antitumour pharmacophore, benzopyrano[2,3-*d*]pyrimidine,[158] was accessed by Gupta *et al.* by the condensation of the salicyladehyde, malononitrile, and dimethylamine at room temperature in [BMIM][BF_4] at room temperature (Scheme 56). Howver, this approach was limited to the use of dimethlamine, with the diethylamine resulted in no reaction.[159]

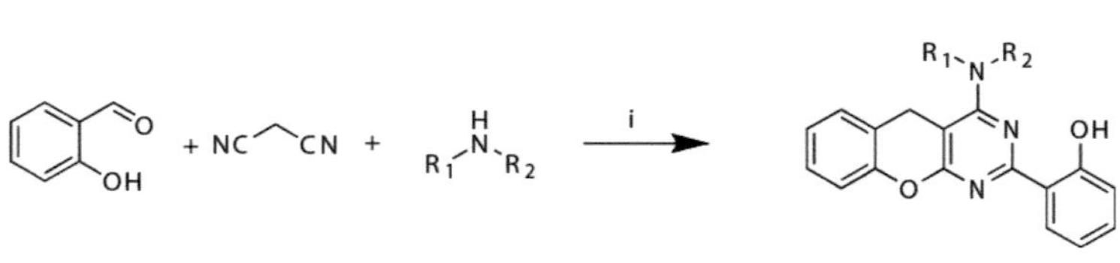

Scheme 57. Synthesis of benzopyranopyrimidines: (i) [BMIM][BF_4], rt.

2.1.9. 3CRs yielding heterocycles with ring sulfur and nitrogen atoms

4-Thiazolidinones have been exploited as potential bactericidal, antifungal, anticonvulsant, anti-HIV, and antituberculotic agents.[160],[161] While there have been multiple synthestic approaches, there is still considerable scope to develop a more environmentally friendy and efficient approach to this scaffold.[162],[163] Lingampalle *et al.* have developed a rapid entry to 4-thiazolidinones via the *N*-methylpyridinium tosylate [NMP][Ts] cyclocondensation of amines, aromatic ketones, and mercaptoacetic acid.[164],[165] The reaction proceeds via imine formation, followed by rapid cyclocondensation at 120 ºC (Scheme 57).

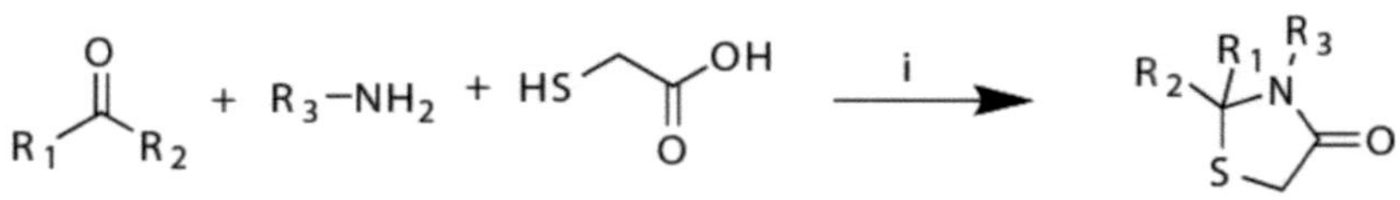

Scheme 58. Synthesis of 4-thiazolidinones: (i) [NMP][Ts], 120ºC, 3h.

The (1*H*)-quinolones, thiazolo[3,2-*a*]pyrimidines and pyrimido[2,1-*b*]benzothiazoles display considerable bioactivity and are important lead compounds in the development of anti-viral, anti-platelet, anti-cancer and anti-malarial agents.[166]-[168] While there are many reported synthesis of pyrimido[2,1-*b*]benzothiazoles, arguably Yadav *et al.* 's 1-methoxyethyl-3-methylimidazolium trifluoroacetate [MOEMim][TFA] mediated MCR is the most direct and efficient reported thus far (Scheme 58).[169]

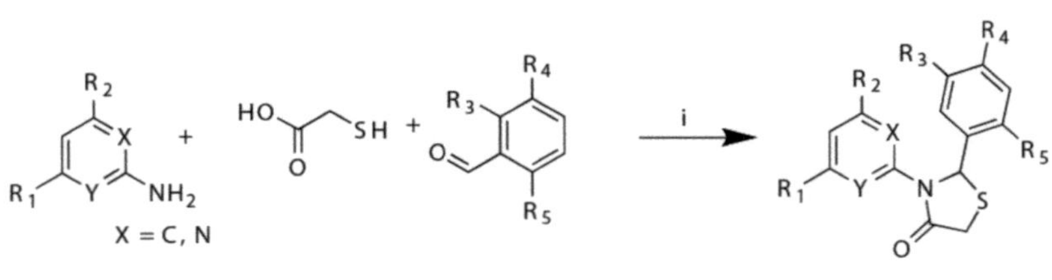

Scheme 59. Synthesis of 2,3-diaryl/2-aryl-3-heteroaryl-1,3-thiazolidin-4-ones: (i) [MOEMim][TFA].

2.1.10. 3CRs yielding heterocycles with ring oxygen, sulfur and nitrogen atoms

Application of a tandem Knoevenagel, Michael and ring transformation reactions involving 3-arylrhodanines, aromatic aldehydes and a mercaptoacetyl transfer agent, 2-methyl-2-phenyl-1,3-oxathiolan-5-one, in a chiral ionic liquid L-prolinium sulfate [Pro_2SO_4], gave 6-mercaptopyranothiazoles with diastereoselectivities of 88-95%ee (Scheme 59).[170] The reactions were conducted at room temperature 25-30 h, followed by isolation to yield a single diastereomer in 76-90% yields.

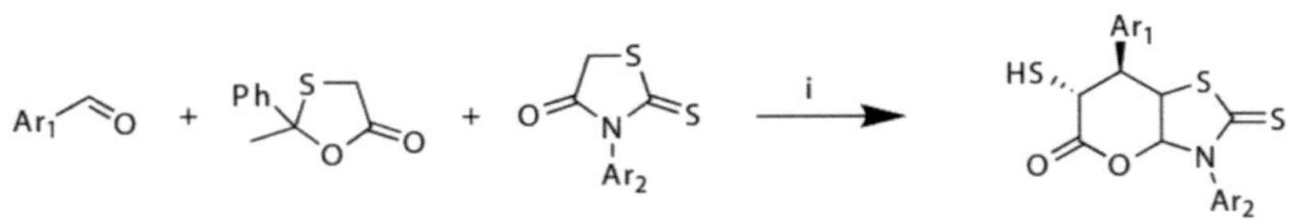

Scheme 60. Synthesis of 2-methyl-2-phenyl-1,3-oxathiolan-5-one: (i) [Pro_2SO_4], rt.

The biologically important 4*H*-benzo[*b*]pyrans can be smoothly accessed as shown in Scheme 60 in excellent yields (77-95%) after stirring for 30-50 min at ethanol reflux. In this instance [BMIM][OH] was used as a catalyst.[121]

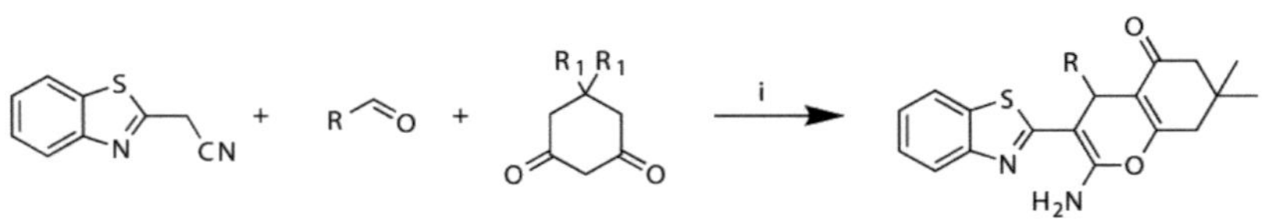

Scheme 61. Benzo[*b*]pyrans: (i) [BMIM][OH], EtOH, reflux.

2.2. 4CRs yielding heterocycles

2.2.1. 4CRs yielding heterocycles with one nitrogen in the ring

Increasing the number of components in MCRs from three to four offers the potential to increase substituent diversity, atom and step economy. This increased structural complexity allows for a facile access to highly decorated scaffolds, but interestingly in the four component IL mediated MCR, this has been limited to the synthesis of heterocyclic compounds. For example, rapid access to both alkyl and aromatic substituted 1,4-dihydropyridine derivatives can be accomplished via the reaction of an aldehyde, a 1,3-dicarbonyl compound, Meldrum's acid and ammonium acetate as the nitrogen source in [BMIM][BF_4] (Scheme 61).[172],[173]

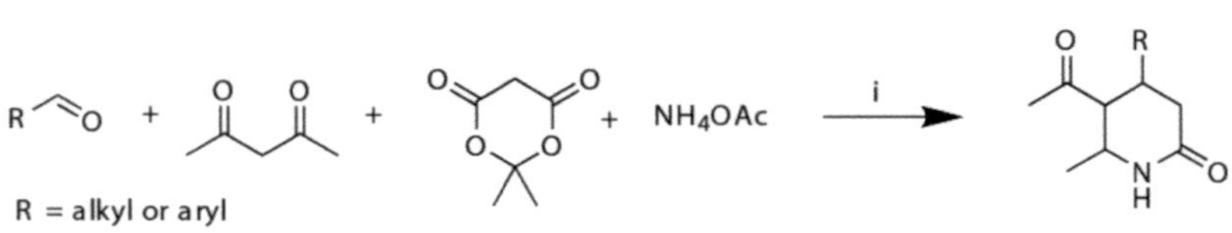

Scheme 62. Synthesis of 1, 4-dihydropyridine derivatives: (i) [BMIM][BF_4].

2.2.2. 4CRs yielding heterocycles with ring oxygen and nitrogen atoms

Kanakarajuwe *et al.* have exploited the four-component MCR for the synthesis of novel antibacterial chromeno[2,3-*d*]pyrimidin-8-amines in [BMIM][BF_4] (Scheme 62). Simple stirring of a mixture of α-naphthol, malononitrile, aryl aldehydes and NH_4Cl in [BMIM][BF_4] and with trace triethylamine (TEA) in DMF allowed direct isolation of the desired analogue, bypassing the more traditional route which involved isolation of corresponding iminochromenes.[174]

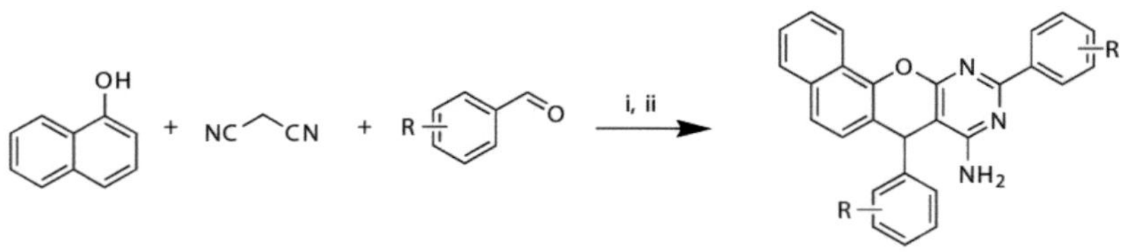

Scheme 63. General synthetic route of chromeno[2,3-*d*]pyrimidine-8-amine derivatives: (i) [BMIM][BF_4], TEA/DMF; (ii) NH_4Cl, 100°C.

2.2.3. 4CRs yielding heterocycles with two ring nitrogens

The biologiocal roles of substituted imidazoles are well documented and numerous biologically active analoges have been reported.[175] Shaterian *et al.* have developed both a three and four component MCR route from benzil (or benzoin), substituted benzaldehydes and ammonium acetate (3-component), or with the addition of phenyl hydrazine (4-component) for the synthesis of 2,4,5-trisubstitutedimidazoles and 1,2,4,5-tetrasubstituted imidazoles respectively.[176] *N*-Methyl-2-pyrrolidonium hydrogen sulfate [NMP][HSO_4] at 100ºC was found to be superior to all previous reports which used a wide variety of catalys to conduct the same transformations. Recyling of the [NMP][HSO_4] saw a gradual diminution of the product yield from 98% to 82% over seven cycles. While this team examined the four component route (addition of phenylhydazine) giving 1,2,4,5-substituted imidazoles in high yields using Brønsted acidic ionic liquid, [NMP][HSO_4] (Scheme 63).

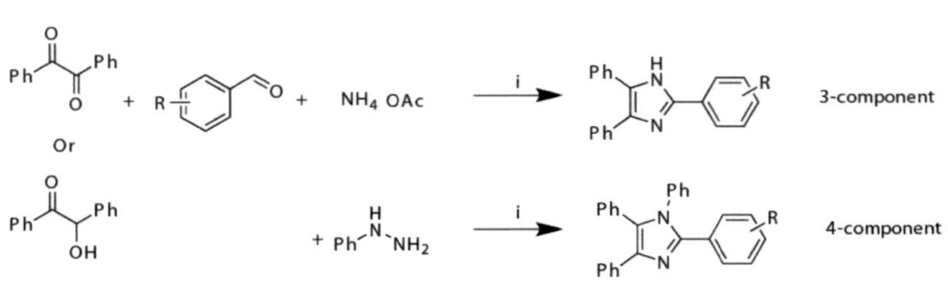

Scheme 64. Synthesis of 2,4,5- and 1,2,4,5-substituted imidazoles by a three or four component MCR: (i) [NMP][HSO_4].

Pyrano[2,3-*c*]pyrazoles represent an important scaffold in medicinal chemistry with multiple synthetic approaches developed. These approaches include synthesis in water, ethanol reflux, microwave assisted and solvent free aporaches. Each approach comes replete with its own set of advantages and disadvantages from excess solvent requirements, long reaction times and poor yields.[177]-[180] Khurana *et al.* synthesis of 4*H*-pyrano[2,3-*c*]pyrazoles avoids most of these disadvantages by providing for a high yielding (typically >85%), short duration cyclocondensation of hydrazine monohydrate or phenyl hydrazine, ethyl acetoacetate, aldehydes, and malononitrile in [BMIM][BF_4] with catalytic quantities of L-proline (10 mol%) (Scheme 64).[181]

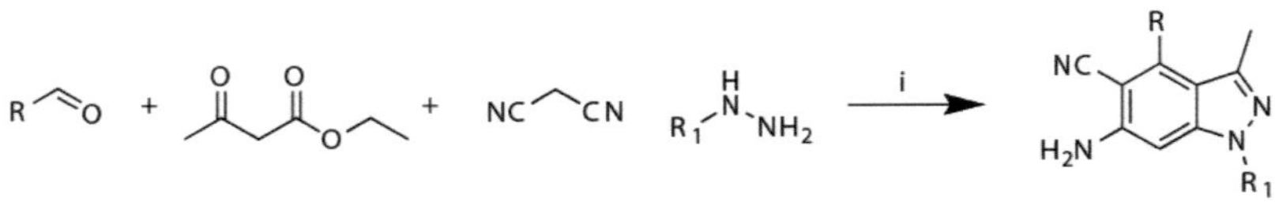

Scheme 65. Synthesis of pyrano[2,3-c]pyrazoles: (i) [BMIM][BF_4], L-proline (10 mol%), 50ºC.

Xiao *et al.* have described a novel, efficient, and green procedure for the synthesis of 3-(5-amino-3-methyl-1*H*-pyrazol-4-yl)-3-arylpropanoic acid derivatives through the four-component reaction in [BMIM][BF_4] (Scheme 65).[182] Reactions were rapid (5 min) and the product isolated by pouring onto water and recrystallisation from EtOH / H_2O to afford pure product.

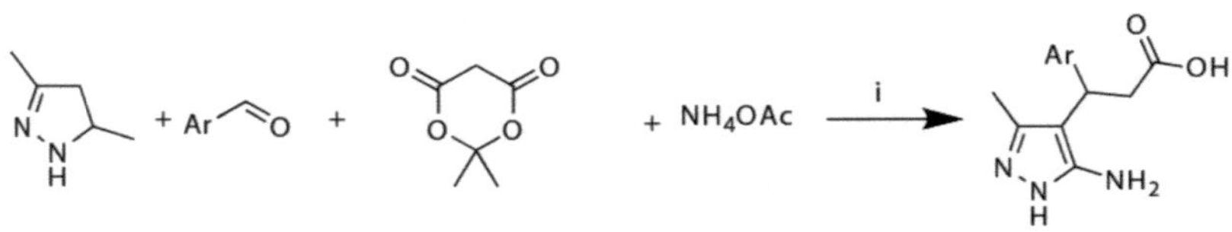

Scheme 66. Synthesis of 3-(5-amino-3-methyl-1*H*-pyrazol-4-yl)-3-arylpropanoic acid derivatives: (i) [BMIM][BF_4].

2.2.4. 4CRs yielding heterocycles with >three ring nitrogens

Ghahremanzadeh *et al.* reported the green synthesis of 1*H*-indolo[2,3-*b*]pyrazolo[4,3-*e*]pyridines from indolin-2-one, 3-oxo-3-phenylpropanenitrile, phenylhydrazine and benzaldehyde under a variety of conditions (Scheme 63).[183] The best reaction outcome was observed with the use of *p*-TSA in [BMIM][Br] at 140ºC (Scheme 66).

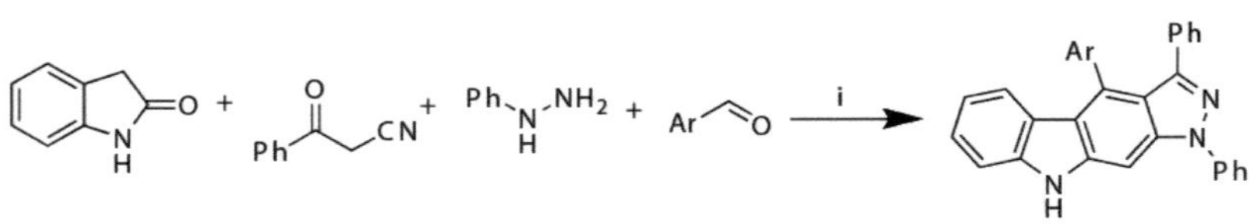

Scheme 67. Synthesis of 1*H*-indolo[2,3-*b*]pyrazolo[4,3-e]pyridines: (i) [BMIM][Br], p-TSA, 140ºC.

Building on their earlier report on the synthesis of spiro[indolinepyrazolo[4',3':5,6]pyrido[2,3-*d*]pyrimidine]triones from barbituric acid, phenylhydrazine, 3-oxo-3-phenylpropanenitrile and isatin, Ghahremanzadeh *et al.* noted that the use of mixture of alum

http://dx.doi.org/10.5772/51937

($KAl(SO_4)_2{\cdot}_{12}H_2O$) and [BMIM][$PF_6$] was a green approach to the same class of compounds (Scheme 67).[184],[185]

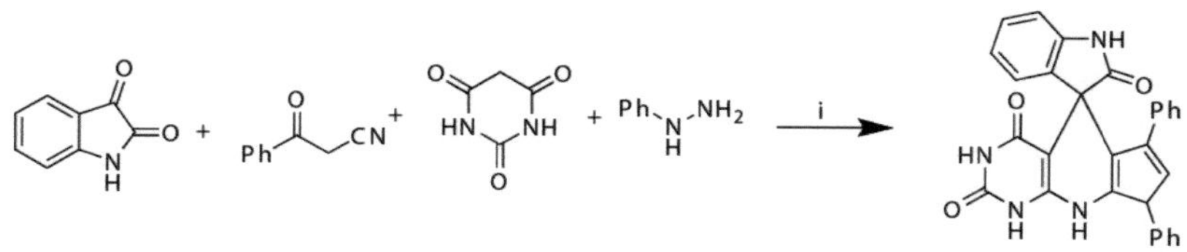

Scheme 68. Synthesis of spiro[indolinepyrazolo[4',3':5,6]pyrido[2,3-*d*]pyrimidine]triones: (i) [BMIM][PF_6], alum.

3. Conclusions

In this brief review we have demonstrated the considerable utility of room temperature ionic liquids in multicomponent reactions. Almost universally, the addition of an ionic liquid increases the speed of reaction and reaction yields. In many cases the ionic liquid was used as both the solvent and the reaction promotor. It was possible to add catalytic quantities of ionic liquids in conventional solvent and still achieve a much greener reaction outcome.

While the linear variant of the four-component MCR in ionic liquids is currently poorly described, there is little doubt that room temperature ionic liquids will aid in the synthesis of such species. Overall the IL-MCR approach is an extremely useful one, especially for the rapid entry to highly functionalised heterocyclic molecules of potentials use in medicianl chemistry.

Author details

Ahmed Al Otaibi and Adam McCluskey

Chemistry, School of Environmental & Life Sciences, The University of Newcastle, University Drive, Callaghan NSW, Australia

References

[1] Hill, T.A.; Odell, L.R.; Quan, A.; Ferguson, G.; Robinson, P.J.; McCluskey, A. Long Chain Amines and Long Chain Ammonium Salts as Novel Inhibitors of Dynamin GTPase Activity. Bioorganic & Medicinal Chemistry Letters 2004;14 3275-3278.

[2] Hill, T.; Odell, L. R.; Edwards, J. K.; Graham, M. E.; McGeachie, A. B.; Rusak, J.; Quan, A.; Abagyan, R.; Scott, J. L.; Robinson, P. J.; McCluskey, A. Small Molecule In-

hibitors of Dynamin I GTPase Activity: Development of Dimeric Tyrphostins. Journal of Medicinal Chemistry 2005;48 7781-7788.

[3] Hill, T. A.; Stewart, S. G.; Gordon, C. P.; Ackland, S. P.; Gilbert, J.; Sauer, B.; Sakoff, J. A.; McCluskey, A. Norcantharidin Analogues: Synthesis, Anticancer Activity and Protein Phosphatase 1 and 2A Inhibition. Chemistry Medicinal Chemistry 2008;3 1878-1892.

[4] Tarleton, M.; Robertson, M. J.; Gilbert, J.; Sakoff, J. A.; McCluskey, A. Library Synthesis and Cytotoxicity of A Family of 2-Phenylacrylonitriles and Discovery of An Estrogen Dependent Breast Cancer Lead Compound. Medicinal Chemistry Communications 2011;2 31-37.

[5] Odell, L. R.; Howan, D.; Gordon, C. P.; Robertson, M. J.; Chau, N.; Mariana, A.; Whiting, A. E.; Abagyan, R.; Daniel, J. A.; Gorgani, N. N.; Robinson, P. J.; McCluskey, A. The Pthaladyns: GTP Competitive Inhibitors of Dynamin I and II GTPase Derived from Virtual Screening. Journal of Medicinal Chemistry 2010;53 5267-5280.

[6] Hill, T. A.; Stewart, S. G.; Sauer, B.; Gilbert, J.; Ackland, S. P.; Sakoff, J. A.; McCluskey, A. Heterocyclic Substituted Cantharidin and Norcantharidin Analogues—Synthesis, Protein Phosphatase (1 and 2A) Inhibition, and Anti-cancer Activity. Bioorganic & Medicinal Chemistry Letters 2007;17 3392-3397.

[7] Hill, T. A.; Stewart, S. G.; Ackland, S. P.; Gilbert, J.; Sauer, B.; Sakoff, J. A.; McCluskey, A. Norcantharimides, Synthesis and Anticancer Activity: Synthesis of New Norcantharidin Analogues and their Anticancer Evaluation. Bioorganic & Medicinal Chemistry. 2007;15 6126-6134.

[8] Quan, A.; McGeachie, A. B.; Keating, D. J.; van Dam, E. M.; Rusak, J.; Chau, N.; Malladi, C. S.; Chen, C.; McCluskey, A.; Cousin, M. A.; Robinson, P. J. Myristyl Trimethyl Ammonium Bromide and Octadecyl Trimethyl Ammonium Bromide Are Surface-Active Small Molecule Dynamin Inhibitors that Block Endocytosis Mediated by Dynamin I or Dynamin II. Molecular Pharmacology 2007;72 1425-1439.

[9] Stewart, S. G.; Hill, T. A.; Gilbert, J.; Ackland, S. P.; Sakoff, J. A.; McCluskey, A. Synthesis and Biological Evaluation of Norcantharidin Analogues: Towards PP1 Selectivity. Bioorganic & Medicinal Chemistry 2007;15 7301-7310.

[10] Hill, T. A.; Gordon, C. P.; McGeachie, A. B.; Venn-Brown, B.; Odell, L. R.; Chau, N.; Quan, A.; Mariana, A.; Sakoff, J. A.; Chircop, M.; Robinson, P. J.; McCluskey, A. Inhibition of Dynamin Mediated Endocytosis by the Dynoles Synthesis and Functional Activity of a Family of Indoles. Journal of Medicinal Chemistry 2009;52 3762-3773.

[11] Strecker, A. Ueber die Künstliche Bildung der Mitchsaüre und einen Neuen, dem Glycocoll Homologen Körper. Liebigs Annalen der Chemie 1850;75 27.

[12] Hallett, J. P.; Welton, T. Room-Temperature Ionic Liquids: Solvents for Synthesis and Catalysis. 2. Chemical Reviews 2011;111 3508-3576.

[13] Dupont, J.; de Souza, R.; Suarez, P. A. Z. Ionic Liquid (Molten Salt) Phase Organometallic Catalysis. Chemical Reviews 2002;102 3667-3692.

[14] Fisher, T.; Sethi, A.; Welton, T.; Woolf, J. Diels-Alder Reactions in Room-temperature Ionic Liquids. Tetrahedron Letters 1999;40 793-796.

[15] Earle, M. J.; McCormac, P. B.; Seddon, K. R. Regioselective alkylation in ionic liquids. Chemical Communications 1998; 2245-2246.

[16] Boon, J. A.; Levisky, J. A.; Pflug, J. L.; Wilkes, J. S. Friedel-Crafts reactions in ambient-temperature molten salts. Journal of Organic Chemistry 1986;51 480-483.

[17] Ellis, B.; Keim, W.; Wasserscheid, P. Linear dimerisation of but-1-ene in biphasic mode using buffered chloroaluminate ionic liquid solvents. Chemical Communications 1999; 337-340.

[18] Earle, M. J.; McCormac, P. B.; Seddon, K. R. Diels–Alder Reactions in Ionic Liquids. A Safe Recyclable Alternative to Lithium Perchlorate–diethyl Ether Mixtures. Green Chemistry 1999;1 23-25.

[19] Corey, E. J.; Zhang, F. Y. Highly Enantioselective Michael Reactions Catalyzed by a Chiral Quaternary Ammonium Salt. Illustration by Asymmetric Syntheses of (S)-Ornithine and Chiral 2-Cyclohexenones. Organic Letters 2000;2 1097-1100.

[20] Schoefer, S. H.; Kaftzik, N.; Kragl, U.; Wasserscheid, P. Enzyme Catalysis in Ionic Liquids: Lipase Catalysed Kinetic Resolution of 1-Phenylethanol with Improved Enantioselectivity. Chemical Communications 2001; 425-426.

[21] Bertozzi, F.; Gustafsson, M.; Olsson, R. A Novel Metal Iodide Promoted Three-Component Synthesis of Substituted Pyrrolidines. Organic Letters 2002;4 3147-3150.

[22] Zolfigol, M. A.; Khazaei, A.; Moosavi-Zare, A. R.; Zare, A.; Khakyzadeh, V. Rapid Synthesis of 1-amidoalkyl-2-naphthols over Sulfonic Acid Functionalized Imidazolium Salts. Applied Catalysis A: General 2011;400 70-81.

[23] Hajipour, A. R.; Rajaei, A.; Ruoho, A. E. A Mild and Efficient Method for Preparation of Azides from Alcohols Using Acidic Ionic Liquid [H-NMP]HSO_4. Tetrahedron Letters 2009;50 708-711.

[24] Heravi, M. M.; Tavakoli-Hoseini, N.; Bamoharram, F. F. Brønsted Acidic Ionic Liquids as Efficient Catalysts for the Synthesis of Amidoalkyl Naphthols. Synthetic Communications 2011;41 298-306.

[25] Fang, D.; Shi, Q. R.; Cheng, J.; Gong, K.; Liu, Z. L. Regioselective mononitration of aromatic compounds using Brønsted acidic ionic liquids as recoverable catalysts. Applied Catalysis A: General 2008;345 158-163.

[26] Gui, J.; Cong, X.; Liu, D.; Zhang, X.; Hu, Z.; Sun, Z. Novel Brønsted Acidic Ionic Liquid as Efficient and Reusable Catalyst System for Esterification. Catalysis Communications 2004;5 473-477.

[27] Hajipour, A. R.; Ghayeb, Y.; Sheikhan, N.; Ruoho, A. E. Brønsted Acidic Ionic Liquid as An Efficient and Reusable Catalyst for one-pot Synthesis of 1-amidoalkyl 2-Naphthols Under Solvent-free Conditions. Tetrahedron Letters 2009;50 5649-5651.

[28] Zhang, Q.; Luo, J.; Wei, Y. A Silica Gel Supported Dual Acidic Ionic Liquid: An Efficient and Recyclable Heterogeneous Catalyst for the One-pot Synthesis of Amidoalkyl Naphthols. Green Chemistry 2010;12 2246-2254.

[29] Kotadia, D. A.; Soni, S. S. Silica Gel Supported–SO_3H Functionalised Benzimidazolium Based Ionic Liquid as A Mild and Effective Catalyst for Rapid Synthesis of 1-Amidoalkyl Naphthols. Journal of Molecular Catalysis A: Chemical 2012;353-354 44-49.

[30] Luo, J.; Zhang, Q. A One-Pot Multicomponent Reaction for Synthesis of 1-Amidoalkyl-2-Naphthols Catalyzed by PEG-based Dicationic Acidic Ionic Liquids Under Solvent-free conditions. Monatshefte für Chemie 2011;142 923-930.

[31] Yadav, L. D. S.; Rai, A. The first Ionic Liquid-promoted Three-component Coupling Strategy for An Expeditious Synthesis of β-nitrocarbonitriles/thiocyanates. Tetrahedron Letters 2009;50 640-643.

[32] Yan, S.; Gao, Y.; Xing, R.; Shen, Y.; Liu, Y.; Wu, P.; Wu, H. An Efficient Synthesis of (*E*)-nitroalkenes Catalyzed by Recoverable Diamino-Functionalized Mesostructured Polymers. Tetrahedron 2008;64 6294-6299.

[33] Zhao, X. L.; Liu, L.; Chen, Y. J.; Wang, D. Three-Component Barbier Allylation-Friedel-Crafts Alkylation Promoted by BuPyCl/$SnCl_2$•$2H_2O$: Application to The Synthesis of 4-(Substituted Phenyl)Chromans. Chinese Journal of Chemistry 2007;25 1312-1322.

[34] Davoodnia, A.; Heravi, M. M.; Rezaei-Daghigh, L.; Tavakoli-Hoseini, N. A Modified and Green Procedure for the Synthesis of β-Amido Ketones Using a Brønsted-Acidic Ionic Liquid as Novel and Reusable Catalyst. Chinese Journal of Chemistry 2010;28 429-433.

[35] Fang, D.; Gong, K.; Zhang, D. Z.; Liu, Z. L. One-pot, Three-component Mannich-type Reaction Catalyzed by Functionalized Ionic Liquid. Monatshefte für Chemie 2009;140 1325-1329.

[36] Gong, K.; Fang, D.; Wang, H.L.; and; Liu, Z.L. Basic Functionalized Ionic Liquid Catalyzed One-pot Mannich-type Reaction: Three Component Synthesis of ▪-Amino Carbonyl Compounds. Monatshefte fur Chemie 2007;138 1195-1198.

[37] Liu, B.; Xu, D.; Dong, J.; Yang, H.; Zhao, D.; Luo, S.; Xu, Z. Highly Efficient AILs/L-Proline Synergistic Catalyzed Three-Component Asymmetric Mannich Reaction. Synthetic Communications 2007;37 3003-3010.

[38] Akbari, J.; Heydari, A. A Sulfonic Acid Functionalized Iionic Liquid as A homogeneous and recyclable catalyst for the one-pot synthesis of α-Aminophosphonates. Tetrahedron Letters 2009;50 4236-4238.

[39] Reddy, M. V.; Dindulkar, S. D.; Jeong, Y. T. BF_3SiO_2-Catalyzed One-pot Synthesis of α-Aminophosphonates in Ionic Liquid and Neat Conditions. Tetrahedron Letters 2011;52 4764-4767.

[40] Gregory, R. J. H. Cyanohydrins In Nature and the Laboratory: Biology, Preparations, and Synthetic Applications. Chemical Reviews 1999;99 3649-3682.

[41] Brunel, J. M.; Holmes, I. P. Chemically Catalyzed Asymmetric Cyanohydrin Syntheses. Angewandte Chemie International Edition 2004;43 2752-2778.

[42] North, M. Synthesis and Applications of Non-racemic Cyanohydrins. Tetrahedron Asymmetry 2003;14 147-176.

[43] Chen, F. X.; Feng, X. M. Synthesis of Racemic Tertiary Cyanohydrins. Synthesis Letters 2005; 892-899.

[44] Shen, Z. L.; Ji, S. J. Ionic Liquid [bmim]BF4 as An Efficient and Recyclable Reaction Medium for the Synthesis of *O*-Acetyl Cyanohydrin via One-Pot Condensation of Aldehyde, TMSCN, and Ac2O. Synthetic Communications 2009;39 808-818.

[45] Davoodnia, A.; Bakavoli, M.; Moloudi, R.; Tavakoli-Hoseini, N.; Khashi, M. Highly Efficient, One-pot, Solvent-free Synthesis of 2,4,6-Triarylpyridines using A Brønsted-Acidic Ionic Liquid
as Reusable Catalyst. Monatshefte für Chemie 2010;141 867-870.

[46] Heravi, M. R. P.; Fakhr, F. Ultrasound-Promoted Synthesis of 2-Amino-6-(Arylthio)-4-Arylpyridine-3,5-Dicarbonitriles using ZrOCl2 8H2O/NaNH2 as the Catalyst in the Ionic Liquid [bmim]BF4 at Room Temperature. Tetrahedron Letters 2011;5 6779-6782.

[47] Ranu, B. C.; Jana, R.; Sowmiah, S. An Improved Procedure for the Three-Component Synthesis of Highly Substituted Pyridines Using Ionic Liquid. Journal of Organic Chemistry 2007;72 3152-3154.

[48] Wu, H.; Wan, Y.; Lu, L. L.; Shen, Y.; Ye, L.; Zhang, F. R. Catalyst-Free One-Pot Synthesis
of 2,4,6-Triaryl-1,4-dihydropyridines in Ionic Liquid and Their Catalyzed Activity on Two Simple Diels–Alder Reactions. Synthetic Communications 2008;38 666-673.

[49] Wu, X. Y. Facile and Green Synthesis of 1,4-Dihydropyridine Derivatives in n-Butyl Pyridinium Tetrafluoroborate. Synthetic Communications. 2012;[42] 454-459.

[50] Llinas-Brunet, M.; Bailey, M. D.; Ghiro, E.; Gorys, V.; Halmos, T.; Poirier, M.; Rancourt, J.; Goureare, N. A Systematic Approach to the Optimization of Substrate-Based Inhibitors of the Hepatitis C Virus NS3 Protease: Discovery of Potent and Specific Tripeptide Inhibitors. Journal of Medicinal Chemistry 2004;47 6584-6594.

[51] Frutos, R. P.; Haddad, N.; Houpis, I. N.; Johnson, M.; Smith-Keenan, L. L.; Fuchs, V.; Yee, N. K.; Farina, V.; Faucher, A. M.; Brochu, C.; Hache, B.; Duceepe, J. -S.; Beaulieu, P. Synthesis 2006; 256-XX.

[52] Huang, L. J.; Hsich, M. C.; Teng, C. M.; Lee, K. H.; Kno, S. C. Synthesis and Antiplatelet Activity of Phenyl Quinolones. Bioorganic & Medicinal Chemistry 1998;6 1657-1662.

[53] Gasparotto, V.; Castagliuolo, I.; Chiarelotto, G.; Pezzi, V.; Montanaro, D.; Brun, P.; Palu, G.; Viola, G.; Ferlin, M. G. Synthesis and Biological Activity of 7-Phenyl-6,9-dihydro-3*H*-pyrrolo[3,2-*f*]quinolin-9-ones: A New Class of Antimitotic Agents Devoid of Aromatase Activity. Journal of Medicinal Chemistry 2006;49 1910-1915.

[54] Ruchelman, A. L.; Singh, S. K.; Ray, A.; Wu, X. H.; Yang, J.M.; Li, T.K.; Liu, A.; Liu, L. F.; LaVoie, E. J. 5*H*-Dibenzo[*c,h*]1,6-naphthyridin-6-ones: Novel Topoisomerase I-Targeting Anticancer Agents With Potent Cytotoxic Activity. Bioorganic & Medicinal Chemistry 2003;11 2061-2073.

[55] Theeraladanon, C.; Arisawa, M.; Nishidi, A.; Nakagawa, M. A Novel Synthesis of Substituted Quinolines Using Ring-closing Metathesis (RCM): its Application to the Synthesis of Key Intermediates for Anti-malarial Agents. Tetrahedron 2004;60 3017-3035.

[56] Yadav, A. K.; Sharma, G. R.; Dhakad, P.; Yadav, T. A Novel Ionic Liquid Mediated Synthesis of 4(1*H*)-Quinolones, 5*H*-thiazolo[3,2-*a*]pyrimidin-5-one and 4*H*-pyrimido[2,1-*b*]benzothiazol-4-ones. Tetrahedron Letters 2012;53 859-862.

[57] Wang, X. S.; Wu, J. R.; Li, Q.; Zhang, M. M. A Novel and Green Method for the Synthesis of Highly Substituted Isoquinoline Derivatives in Ionic Liquid. Journal of Heterocyclic Chemistry 2009;46 1355-1363.

[58] Martin, N.; Quinteiro, M.; Seoane, C.; Soto, J. L.; Mora, A.; Suárez, M.; Ochoa, E.; Morales, A.; Del Bosque, J. R. Synthesis and Conformational Study of Acridine Derivatives Related to 1,4-dihydropyridines. Journal of Heterocyclic Chemistry 1995;51 235-238.

[59] Li, Y.; Xu, X.; Shi, D.; Ji, S. One-pot Synthesis of 14-Aryl-1,6,7,14-tetrahydrodibenzo-[*a,i*]acridine-1,6-dione in Ionic Liquids. Chinese Journal of Chemistry 2009;27 1510-1514.

[60] Shi, D.; Ni, S.; Yang, F.; Ji, S. An Efficient and Green Synthesis of 3,3'-Benzylidenebis(4-hydroxy-6-methylpyridin-2(1*H*)-one) Derivatives through Multi-Component Reaction in Ionic Liquid. Journal of Heterocyclic Chemistry 2008;45 1275-1280.

[61] Agarwal, A.; Srinivas, K.; Puri, S. K.; Chauhan, P. M. S. Synthesis of 2,4,6-trisubstituted pyrimidines as antimalarial agents. Bioorganic & Medicinal Chemistry. 2005;13 4645-4650.

[62] Gholap, A. R.; Venkatesan, K.; Daniel, T.; Lahoti, R. J.; Srinivasan, K. V. Ionic liquid Promoted Novel and Efficient Onepot Synthesis of 3,4-dihydropyrimidin-2-(1*H*)-ones at Ambient Temperature Under Ultrasound Irradiation. Green Chemistry 2004;6 147-150.

[63] Gui, J.; Liu, D.; Wang, C.; Lu, F.; Lian, J.; Jiang, H.; Sun, Z. TITLES Synthetic Communications. 2009;[39] 3436-3443.

[64] Peng, J.; Deng, Y. Ionic liquid Catalyzed Biginelli Reaction Under Solvent-free Conditions. Tetrahedron Letters 2001;42 5917-19.

[65] Zolfigol, M. A.; Khazaei, A.; Moosavi-Zare, A. R.; Zare, A. 3-Methyl-1-Sulfonic Acid Imidazolium Chloride as A New, Efficient and Recyclable Catalyst and Solvent for the Preparation of *N*-sulfonyl Imines at Room Temperature. Journal of the Iranian Chemical Society 2010;7 646-651.

[66] Cole, A. C.; Jensen, J. L.; Ntai, I.; Tran, K. L. T.; Weaver, K. J.; Forbes, D. C.; Davis, J.; H. J. Novel Brønsted Acidic Ionic Liquids and Their Use as Dual Solvent–Catalysts. Journal of the American Chemical Society 2002;124 5962-5963.

[67] Hajipour, A. R.; Seddighi, M. Pyridinium-Based Brønsted Acidic Ionic Liquid as a Highly Efficient Catalyst for One-Pot Synthesis of Dihydropyrimidinones. Synthetic Communications 2012; 42 227-235.

[68] Fang, D.; Zhang, D. Z.; Liu, Z. L. One-Pot Three-Component Biginelli-Type Reaction Catalyzed by Ionic Liquids in Aqueous Media. Monatshefte fur Chemie 2010;141 419-423.

[69] Mirzai, M.; Valizadeh, H. Microwave-Promoted Synthesis of 3,4-Dihydropyrimidin-2(1*H*)-(thio)ones Using IL-ONO as Recyclable Base Catalyst Under Solvent-Free Conditions. Synthetic Communications 2012;42 1268-1277.

[70] Valizadeh, H.; Heravi, M. M.; Amiri, M. Unexpected Synthesis of N-Methylbenzo[*d*]isoxazolium Hydroxides Under Microwave Irradiation Conditions Molecular Diversity JOURNAL 2010;14 575-579.

[71] Shaabani, A.; Rahmati, A. 1,1,3,3-*N,N,N,N*-Tetramethylguanidinium Trifluoroacetate Ionic Liquid–Promoted Efficient One-Pot Synthesis of Trisubstituted Imidazoles. Synthetic Communications 2006;36 65-70.

[72] Khurana, J. M.; Kumar, S. Ionic liquid: An Efficient and Recyclable Medium for the Synthesis of Octahydroquinazolinone and Biscoumarin Derivatives. Monatshefte fur Chemie 2010;141 561-564.

[73] Pawar, O. B.; Chavan, F. R.; Sakate, S. S.; Shinde, N. D. Ultrasound Promoted and Ionic Liquid Catalyzed Cyclocon densation Reaction for the Synthesis of 4(3*H*)-Quinazolinones. Chinese Journal of Chemistry 2010;28 69-71.

[74] Dabiri, M.; Salehi, P.; Bahramnejad, M. Ecofriendly and Efficient One-Pot Procedure for The Synthesis of Quinazoline Derivatives Catalyzed by an Acidic Ionic Liquid Under Aerobic Oxidation Conditions. Synthetic Communications 2010;40 3214-3225.

[75] Shaabani, A.; Maleki, A. Ionic Liquid Promoted One-Pot Three-Component Reaction: Synthesis of Annulated Imidazo[1,2-*a*]azines Using Trimethylsilylcyanide. Monatshefte fur Chemie 2007;138 51-56.

[76] Shaabani, A.; Soleimani, E.; Maleki, A. Ionic liquid Promoted One-Pot Synthesis of 3-Aminoimidazo[1,2-*a*]Pyridines. Tetrahedron Letters 2006;47 3031-3034.

[77] Hajipour, A. R.; Ghayeb, Y.; Sheikhan, N.; Ruoho, A. E. Brønsted Acidic Ionic Liquid as an Efficient and Reusable Catalyst for One-Pot, Three-Component Synthesis of Pyrimidinone Derivatives via Biginelli-Type Reaction Under Solvent-Free Conditions. Synthetic Communications 2011;41 2226-2233.

[78] Raghuvanshi, D. S.; Singh, K. N. A Highly Efficient Green Synthesis of 1*H*-pyrazolo[1,2-*b*]Phthalazine-5,10-dione Derivatives and their Photophysical Studies. Tetrahedron Letters 2011;52 5702-5705.

[79] Khurana, J. M.; Magoo, D. Efficient One-Pot Syntheses of 2H-Indazolo[2,1-*b*]Phthalazine-triones by Catalytic H2SO4 in Water–ethanol or Ionic Liquid. Tetrahedron Letters 2009;50 7300-7303.

[80] Wang, H. X.; JI, S. J.; GU, D. G. Synthesis of N-(α-Alkoxyalkyl)benzotriazoles Catalyzed by Acidic Ionic Liquid at Room Temperature. Chinese Journal of Chemistry 2007;25 1041-1043.

[81] Ehlert, F. J.; Ragan, P.; Chen, A.; Roeske, W. R.; Yamamura, H. I. Modulation of Benzodiazepine Receptor binding: Insight into Pharmacological Efficacy. European Journal of Pharmacology 1982;78 249-253.

[82] Sanghvi, Y. S.; Larson, S. B.; Willis, R. C.; Robins, R. K.; Revankar, G. R. Synthesis and Biological Evaluation of Certain C-4 Substituted Pyrazolo[3,4-*b*]Pyridine Nucleosides. Journal of Medicinal Chemistry 1989;32 945-951.

[83] Zhang, X.; Li, D.; Fan, X.; Wang, X.; Li, X.; Qu, G.; Wang, J. Ionic Liquid-Promoted mMlti-component Reaction: Novel
and Efficient Preparation of Pyrazolo[3,4-*b*]Pyridinone, Pyrazolo[3,4-*b*]Quinolinone and their Hybrids with Pyrimidine Nucleoside. Molecular Diversity 2010;14 159-167.

[84] Zhang, X. Y.; Li, X. Y.; Fan, X. S.; Wang, X.; Wang, J. J.; Qu, G. R. A Novel Synthesis of Pyrazolo[3,4-*b*]Pyridine Derivatives through Multi-component Reaction in Ionic Liquids. Chinese Chemical Letters 2008;19 153-156.

[85] Yao, C.; Lei, S.; Wang, C.; Li, T.; Yu, C.; Wang, X.; Tua, S. Three-Component Synthesis of 4-Aryl-1*H*-Pyrimido[1,2-*a*] Benzimidazole Derivatives in Ionic Liquid. Journal of Heterocyclic Chemistry 2010;26 47-32.

[86] Zhang, X.; Fan, X.; Niu, H.; Wang, J. An Ionic Liquid as A Recyclable Medium for the Green Preparation of ▪ ▪ ▪ -bis (substituted benzylidene)Cycloalkanones Catalyzed by FeCl3 6H2O. Green Chemistry 2003;5 267-269.

[87] Fan, X.; Hu, X.; Zhang, X.; Wang, J. Ionic Liquid Promoted Knoevenagel and Michael Reactions. Australian Journal of Chemistry 2004;57 1067-1071.

[88] Shi, D. Q.; Ni, S. N.; Yang, F.; Shi, J. W.; Dou, G. L.; Li, X. Y.; Wang, X. S.; Ji, S. J. An Efficient Synthesis of Pyrimido[4,5-*b*]Quinoline and Indeno[2',1':5,6]pyrido[2,3-*d*]Pyr-

imidine Derivatives via Multicomponent Reactions in Ionic Liquid. Journal of Heterocyclic Chemistry 2008;45 963-702.

[89] Kimachi, T.; Yoneda, F.; Sasaki, T. New Synthesis of 5-Amino-5-Deazaflavin Derivatives by Direct Coupling of 5-Deazaflavins and Amines. Journal of Heterocyclic Chemistry 1992;29 763-765.

[90] Bond, A.; Reichert, Z.; Stivers, J. T. Novel and Specific Inhibitors of a Poxvirus Type I Topoisomerase. Molecular Pharmacology 2006;69 547-557.

[91] Guo, H. Y.; Yu, Y. One-pot Synthesis of 7-Aryl-11,12-Dihydrobenzo[*h*]Pyrimido-[4,5-*b*]Quinoline-8,10(7*H*,9*H*)-diones via Three-Component Reaction in Ionic Liquid. Chinese Chemical Letters 2010;21 1435-1438.

[92] Zhang, N.; Ayral-Kaloustian, S.; Nguyen, T.; Afragola, J.; Hernandez, R.; Lucas, J. Synthesis and SAR of [1,2,4]Triazolo[1,5-*a*]pyrimidines, a Class of Anticancer Agents with a Unique Mechanism of Tubulin Inhibition. Journal of Medicinal Chemistry 2007; 50 319-327.

[93] Li, T.; Yao, C.; Lei, S.; Yu, C.; Tu, S. A Facile One-Pot Three-Component Synthesis of 5-(Trifluoromethyl)-4,7-dihydro-[1,2,4]-Triazolo[1,5-*a*]Pyrimidine Derivatives in Ionic Liquid. Chinese Journal of Chemistry 2011;29 2427-2432.

[94] Kuthan, J. New Developments in the Chemistry of Pyrans. Advances in Heterocyclic Chemistry 1995;62, 19-135.

[95] Hatakeyama, S.; Ochi, N.; Numata, H.; Takano, S. A New Route to Substituted 3-methoxycarbonyldihydropyrans: Enantioselective Synthesis of (–)-Methyl Elenolate. Chemical Communications 1988; 1202-1204.

[96] Cingolant, G. M.; Pigini, M. Research in the Field of Antiviral Compounds. Mannich Bases of 3-Hydroxycoumarin. Journal of Medicinal Chemistry 1969;12 531-532.

[97] Peng, Y.; Song, G.; Huang, F. Tetramethylguanidine-[bmim][BF4].
An Efficient and Recyclable Catalytic System for One-Pot Synthesis of 4H-Pyrans. Monatshefte fur Chemie 2005;[136] 727-731.

[98] Jiang, Z. Q.; JI, S. J.; Lu, J.; Yang, J. M. A Mild and Efficient Synthesis of 5-Oxo-5,6,7,8-tetrahydro-4*H*-benzo[*b*]pyran Derivatives in Room Temperature Ionic Liquids. Chinese Journal of Chemistry 2005;23 1085-1089.

[99] Fang, D.; Zhang, H. B.; Liu, Z. L. Synthesis of 4H-Benzopyrans Catalyzed by Acyclic Acidic Ionic Liquids in Aqueous Media. Journal of Heterocyclic Chemistry 2010;47 63-67.

[100] Hafez, E. A. A.; Elnagdi, M. H. Nitriles in Heterocyclic Synthesis: Novel Synthesis of Benzo[*c*]coumarin and of Benzo[*c*]pyrano[3,2-*c*]quinoline Derivatives. Heterocycles 1987;26 903-907.

[101] Kidwai, M.; Saxena, S. Aqua mediated synthesis of substituted 2-amino-4*H*-chromenes and in vitro study as antibacterial agents. Bioorganic & Medicinal Chemistry Letters 2005;15 4295-4298.

[102] Shestopalov, A. M.; Niazimbetova, Z. I.; Evans, D. H. Synthesis of 2-Amino-4-aryl-3-cyano-6-methyl-5-ethoxycarbonyl-4*H*-pyrans. Heterocycles 1999;51 1101-1107.

[103] Bloxham, J.; Dell, C. P.; Smith, C. W. Preparation of Some New Benzylidenemalononitriles by an SNAr Reaction: Application to Naphtho[1,2-*b*]pyran Synthesis. Heterocycles 1994;38 399-408.

[104] Zhuang, Q. Y.; Rong, L. C.; Shi, D. Q. Synthesis and Crystal Structure of Substituted Naphthopyran. Chinese Journal of Organic Chemistry 2003;23 671-673.

[105] Gong, K.; Wang, H. L.; Fang, D.; Liu, Z. L. Basic Ionic Liquid as Catalyst for the Rapid and Green Synthesis of Substituted 2-amino-2-chromenes in Aqueous Media. Catalysis Communications 2008;9 650-653.

[106] Khurana, J. M.; Magoo, D. pTSA-Catalyzed One-Oot synthesis of 12-aryl-8,9,10,12-Tetrahydrobenzo[*a*]xanthen-11-Ones in Ionic Liquid and Neat Conditions. Tetrahedron Letters 2009;50 4777-4780.

[107] Zakeri, M.; Heravi, M. M.; Saeedi, M.; Karimi, N.; Oskooie, H. A.; Tavakoli-Hoseini, N. One-pot Green Procedure for Synthesis of Tetrahydrobenzo[a]- xanthene-11-One Catalyzed by Brønsted Ionic Liquids under Solvent-free Conditions. Chinese Journal of Chemistry 2011;29 1441-1445.

[108] Zheng, J.; Li, Y. Basic ionic liquid-catalyzed multicomponent synthesis of tetrahydrobenzo[b]pyrans and pyrano[c]chromenes. Mendeleev Communications 2011;21 280-281.

[109] Burgard, A.; Lang, H.J.; Gerlach, U. Asymmetric synthesis of 4-amino-3,4-dihydro-2,2-dimethyl-2*H*-1-benzopyrans. Tetrahedron 1999;55 7555-7562.

[110] Evans, J. M.; Fake, C. S.; Hamilton, T. C.; Poyser, R. H.; Showell, G. A. Synthesis and Antihypertensive Activity of 6,7-disubstituted trans-4-amino-3,4-dihydro-2,2-dimethyl-2*H*-1-benzopyran-3-ols. Journal of Medicinal Chemistry 1984;27 1127-1131.

[111] Evans, J. M.; Fake, C. S.; Hamilton, T. C.; Poyser, R. H.; Watts, E. A. Synthesis and antihypertensive activity of substituted trans-4-amino-3,4-dihydro-2,2-dimethyl-2H-1-benzopyran-3-ols. Journal of Medicinal Chemistry 1983;26 1582-1589.

[112] D. Arnesto, D.; Horspool, W. M.; Martin, N.; Ramos, A.; Seaone, C. Synthesis of Cyclobutenes by the Novel Photochemical Ring Contraction of 4-substituted 2-amino-3,5-dicyano-6-phenyl-4*H*-pyrans. The Journal of Organic Chemistry 1989;54 3069-3072.

[113] Shaterian, H. R.; Honarmand, M. Task-Specific Ionic Liquid as the Recyclable Catalyst for the Rapid and Green Synthesis of Dihydropyrano[3,2-*c*]chromene Derivatives. Synthetic Communications 2011;41 3573-3581.

http://dx.doi.org/10.5772/51937

[114] Gong, K.; Wang, H. L.; Luo, J.; Liu, Z. L. One-Pot Synthesis of Polyfunctionalized Pyrans Catalyzed by 1145 Basic Ionic Liquid in Aqueous Media. Journal of Heterocyclic Chemistry 2009;46 1145-1150.

[115] Xie, J.W.; Li, P.; Wang, T.; Zhou, F.T. Efficient and Mild Synthesis of Functionalized 2,3-dihydrofuran Derivatives via Domino Reaction in Water. Tetrahedron Letters 2011;52 2379-2382.

[116] Fan, L.P.; Li, P.; Li, X.S.; Xu, D.C.; Ge, M.M.; Zhu, W.D.; Xie, J.W. Facile Domino Access to Chiral Mono-, Bi-, and Tricyclic 2,3-Dihydrofurans. The Journal of Organic Chemistry 2010;75 8716-8719.

[117] Rueping, M.; Parra, A.; Uria, U.; Besselièvre, F.; Merino, E. Catalytic Asymmetric Domino Michael Addition–Alkylation Reaction: Enantioselective Synthesis of Dihydrofurans. Organic Letters 2010;12 5680-5683.

[118] Yılmaz, M.; Yakut, M.; Pekel, A. T. Synthesis of 2,3-Dihydro-4*H*-furo[3,2-*c*]chromen-4-ones and 2,3-Dihydronaphtho[2,3-*b*]furan-4,9-diones by the Radical Cyclizations of Hydroxyenones with Electron-Rich Alkenes using Manganese(III) Acetate. Synthetic Communications 2008;38 914-927.

[119] Rajesh, S. M.; Perumal, S.; Menéndez, J. C.; Pandian, S.; Murugesan, R. Facile Ionic Liquid-mediated, Three-Component Sequential Reactions for the Green, Regio- and Diastereo-selective Synthesis of Furocoumarins. Tetrahedron 2012;68 5631-5636.

[120] Manolov, I.; Maichle-Moessmer, C.; Danchev, N. D. Synthesis, structure, toxicological and pharmacological investigations of 4-hydroxycoumarin derivatives. European Journal of Medicinal Chemistry. 2006;[41] 882-890.

[121] Gong, K.; Wang, H. L.; Luo, J.; Liu, Z. L. One-Pot Synthesis of Polyfunctionalized Pyrans Catalyzed by 1145 Basic Ionic Liquid in Aqueous Media. Journal of Heterocyclic Chemistry. 2009;[46] 1145-1150.

[122] Chen, Z.; Zhu, Q.; Su, W. A Novel Sulfonic Acid Functionalized Ionic Liquid Catalyzed Multicomponent Synthesis of 10,11-dihydrochromeno[4,3-*b*]chromene-6,8(7*H*, 9*H*)-dione Derivatives in Water. Tetrahedron Letters 2011;52 2601-2604.

[123] Bloxham, J.; Dell, C. P.; Smith, C. W. Preparation of Some New Benzylidenemalononitriles by an SNAr Reaction: Application to Naphtho[1,2-*b*]pyran Synthesis. Heterocycles 1994;38 399-408.

[124] Wang, X. S.; Shi, D. Q.; Yu, H. Z. Synthesis of 2-Aminochromene Derivatives Catalyzed by KF/Al2O3. Synthetic Communications 2004;34 509-514.

[125] Jin, T. J.; Xiao, J. C.; Wang, S. J.; Li, T. S. An Efficient and Convenient Approach to the Synthesis of Benzopyrans by a Three-Component Coupling of One-Pot Reaction. Synthesis Letters 2003;13 2001-2004.

[126] Wang, X. S.; Shi, D. Q.; Tu, S. J. A Convenient Synthesis of 5-Oxo-5,6,7,8-tetrahydro-4*H*-benzo[*b*]pyran Derivatives Catalyzed by KF-Alumina. Synthetic Communications 2003;33 119-126.

[127] Jin, T. S.; Wang, A. Q.; Wang, X. A Clean One-pot Synthesis of Tetrahydrobenzo[*b*]pyran Derivatives Catalyzed by Hexadecyltrimethyl Ammonium Bromide in Aqueous Media. Synthesis Letters 2004;5 871-873.

[128] M. J. de Groot, M. J.; Alex, A. A.; Jones, B. C. Development of a Combined Protein and Pharmacophore Model for Cytochrome P450 2C9. Journal of Medicinal Chemistry 2002;45 1983-1993.

[129] M. Lee, M.; D. Hesek, D.; and S. Mobashery, S. A Practical Synthesis of Nitrocefin. The Journal of Organic Chemistry 2005;70 367-369.

[130] Flynn, B. L.; Flynn, G. P.; Hamel, E.; Jung, M. K. The Synthesis and Tubulin Binding Activity of Thiophene-Based Analogues of Combretastatin A-4. Bioorganic & Medicinal Chemistry Letters 2001;11 2341-2343.

[131] Zhang, X.; Li, X.; Fan, X.; Wang, X.; Li, D.; Qu, G.; Wang, J. Ionic liquid promoted preparation of 4*H*-thiopyran
and pyrimidine nucleoside-thiopyran hybrids through one-pot multi-component reaction of thioamide. Molecular Diversity 2009;13 5761-5765.

[132] Shvekhgeimer, M.G. A. Dihydrothiophenes. Synthesis and properties (review). Chemistry of Heterocyclic Compounds 1998;34 1101-1122.

[133] McIntosh, J. M.; Goodbrand, H. B.; Masse, G. M. Dihydrothiophenes. II. Preparation and properties of some alkylated 2,5-dihydrothiophenes. Journal of Organic Chemistry 1974;39 202-206.

[134] Leusen, A. M.; Berg, K. J. Formation and Reactions of 2,3-dimethylene-2,3-dihydrothiophene. Tetrahedron Letters 1988;29 2689-2692.

[135] Kumar, A.; Gupta, G.; Srivastava, S. Functional Ionic Liquid Mediated Synthesis (FILMS) of dihydrothiophenes and Tacrine Derivatives. Green Chemistry 2011;13 2459-24963.

[136] Fringuelli, R.; Pietrella, D.; Schiaffella, F.; Guarrac, I. A.; Perito, S.; Bistoni, F.; Vecchiarelli, A. Anti-Candida albicans Properties of Novel Benzoxazine Analogues. Bioorganic & Medicinal Chemistry 2002;10 1681-1686.

[137] Macchiarulo, A.; Costantino, G.; Fringuelli, F.; Vecchiarelli, A.; Schiaffella, F.; Fringuelli, R. 1,4-Benzothiazine and 1,4-Benzoxazine Imidazole Derivatives with Antifungal Activity: A Docking Study. Bioorganic & Medicinal Chemistry 2002;10 3415-3423.

[138] Adams, N. D.; Darcy, M. G.; Dhanak, D.; Duffy, K. J.; Fitch, D. M.; Knight, S. D.; Newlander, K. A.; Shaw, A. N. WO Int. Patent 2006113432, 2006.

[139] Soleimani, E.; Khodaei, M. M.; Koshvandi, A. T. K. Three-Component, One-Pot Synthesis of Benzo[*b*][1,4]oxazines in Ionic Liquid 1-Butyl-3-methylimidazolium Bromide. Synthetic Communications 2012;42 1367-1371.

[140] Asri, Z. E.; Nisson, Y. G.; Guillen, F. D. R.; Basle, O.; Isambert, N.; Duque, M. D. M. S.; Ladeira, S.; Rodriguez, J.; Constantieux, T.; Plaquevent, J. C. Multicomponent Reactions in Ionic Liquids: Convenient and Ecocompatible Access to the 2,6-DABCO Core. Green Chemistry 2011;13 2549-2552.

[141] F. Liéby-Muller, F.; Constantieux, T.; Rodriguez, J. Multicomponent Domino Reaction from β-Ketoamides: Highly Efficient Access to Original Polyfunctionalized 2,6-Diazabicyclo[2.2.2]octane Cores. Journal of the American Chemical Society 2005;127 17176-17177.

[142] Otto, H.H. Synthesis of some 4*H*-pyrano[2,3-*c*]pyrazoles. Archiv der Pharmazie 1974; [307] 444-447.

[143] Lehmann, F.; Holm, M.; Laufer, S. Three-Component Combinatorial Synthesis of Novel Dihydropyrano[2,3-*c*]pyrazoles. Journal of Combinatorial Chemistry 2008;10 364-367.

[144] Balaskar, R. S.; Gavade, S. N.; Mane, M. S.; Shingate, B. B.; Shingare, M. S.; Mane, D. V. Greener Approach Towards the Facile Synthesis of 1,4-dihydropyrano[2,3-*c*]pyrazol-5-yl Cyanide Derivatives at Room Temperature. Chinese Chemical Letters 2010;21 1175-1179.

[145] New, J. S.; Christopher, W. L.; Yevich, J. P.; Butler, R.; Schlemmer, R. F., Jr; VanderMaelen, C. P.; Cipollina, J. A. The Thieno[3,2-*c*]pyridine and Furo[3,2-*c*]pyridine Rings: New Pharmacophores with Potential Antipsychotic Activity. Journal of Medicinal Chemistryy 1989;32 1147-1156.

[146] Bukoski, R. D.; Bo, J.; Xue, H.; Bian, K. Antiproliferative and Endothelium-Dependent Vasodilator Properties of 1,3-Dihydro-3-*p*-chlorophenyl-7-hydroxy-6-methylfuro[(3,4*c*]pyridine hydrochloride (Cicletanine). Journal of Pharmacology and Experimental Therapeutics 1993;265 30-35.

[147] Shi, D. Q.; Yang, F.; Ni, S. N. A Facile Synthesis of Furo[3,4-*e*]pyrazolo[3,4-*b*]pyridine-5(7*H*)one Derivatives via Three-Component Reaction in Ionic Liquid without Any Catalyst. Journal of Heterocyclic. Chemistry 2009;46 469-476.

[148] Moghaddam, M. M.; Bazgir, A.; Mehdi, A. M.; Ghahremanzadeh, R. Alum (KAl(SO4)2•12H2O) Catalyzed Multicomponent Transformation: Simple, Efficient, and Green Route to Synthesis of Functionalized Spiro[chromeno[2,3-*d*]pyrimidine-5,3'-indoline]tetraones in Ionic Liquid Media. Chinese Journal of Chemistry 2012;30 709-714.

[149] Alper, P. B.; Meyers, C.; Lerchner, A.; Siegel, D. R.; Carreira, E. M. Facile, Novel Methodology for the Synthesis of Spiro[pyrrolidin-3,3′-oxindoles]: Catalyzed Ring

Expansion Reactions of Cyclopropanes by Aldimines. Angewandte Chemie International Edition 1999;38 3186-3189.

[150] Figueroa-Villar, J. D.; Carneiro, C. L.; Cruz, E. R. Synthesis of 6-Phenylaminofuro[2,3-*d*]pyrimidine-2,4(1*H*,3*H*)diones from Barbiturylbenzylidenes and Isonitriles. Heterocycles 1992;34 891-894.

[151] Kobayashi, K.; Tanaka, H.; Tanaka, K.; Yoneda, K.; Morikawa, O.; Konishi, H. One-Step Synthesis of Furo[2,3-*d*]Pyrimidine-2,4(1*H*,3*H*)-Diones Using the CAN-Mediated Furan Ring Formation. Synthetic Communications 2000;30 4277-4291.

[152] Vilsmaier, E.; Baumheier, R.; Lemmert, M. Trapping of Diacceptor-Substituted Methylenecyclopropanes with Isocyanides A Further Application of the Principle of the Two-fold Nucleophilic Substitution at a Cyclopropane. Synthesis 1990; 995-998.

[153] Kawahara, N.; Nakajima T.; Itoh T.; Ogura, H. Simple Syntheses of Pyrrolo- and Furopyrimidine Derivatives. Heterocycles 1984;22 2217-2220.

[154] Qian, C. Y.; Nishino, H.; Kurosawa, K.; Korp, J. D. Manganese(II) Acetate-Mediated Double 2-hydroperoxyalkylations of Barbituric Acid and its Derivatives. Journal of Organic Chemistry. 1993;58 4448-4451.

[155] Nair, V.; Vinod, A. U.; Abhilash, N.; Menon, R. S.; Santhi, V.; Varma, R. L.; Viji, S.; Mathewa, S.; Srinivasb, R. Multicomponent reactions involving zwitterionic intermediates for the construction of heterocyclic systems: one pot synthesis of aminofurans and iminolactones. Tetrahedron 2003;59 10279-10286.

[156] Yadav, J. S.; Subba, Reddy, B. V.; Shubashree, S.; Sadashiv, K.; Naidu, J. J. Ionic Liquids-Promoted Multi-Component Reaction: Green Approach for Highly Substituted 2-Aminofuran Derivatives. Synthesis 2004;2376-2380.

[157] Shaabani, A.; Soleimani, E.; Darvishi, M. Ionic Liquid Promoted One-Pot Synthesis of Furo[2,3-*d*]pyrimidine-2,4(1*H*,3*H*)diones. Monatshefte fur Chemi 2007;138 43-46.

[158] Hadfield, J. A.; Pavlidis, V. H.; Perry, P. J.; McGown, A. T. Synthesis and anticancer activities of 4-oxobenzopyrano[2,3-*d*]pyrimidines. Anti-Cancer Drugs 1999;10 591-595.

[159] Gupta, A. K.; Kumari, K.; Singh, N.; Singh, D.; Raghuvanshi; Singh, K. N. An Eco-safe Approach to Benzopyranopyrimidines and 4*H*-chromenes in Ionic Liquid at Room Temperature. Tetrahedron Letters. 2012;53 650-653.

[160] Rawal, R. K.; Tripathi, R.; Katti, S. B.; Pannecouque, C.; Clercq, E. D. Design, synthesis, and evaluation of 2-aryl-3-heteroaryl-1,3-thiazolidin-4-ones as anti-HIV agents. Bioorganic & Medicinal Chemistry 2007;15 1725-1731.

[161] Srivastava, T.; Gaikwad, A. K.; Haq, W.; Sinha, S.; Katti, S. B. Synthesis and biological evaluation of 4-thiazolidinone derivatives as potential antimycobacterial agents (NA-1265FP). Arkivoc 2005;2 120-130.

[162] Veeresena, G.; Vie, N.; James, T. D.; Duane, D. M. Efficient Microwave Enhanced Synthesis of 4-Thiazolidinones. Synthesis Letters 2004;13 2357-2358.

[163] Fraga-Dubreuil, J.; Bazureau, J. P. Efficient combination of task-specific ionic liquid and microwave dielectric heating applied to one-pot three component synthesis of a small library of 4-thiazolidinones. Tetrahedron 2003;59 6121-6130.

[164] Lingampalle, D.; Jawale, D.; Waghmare, R.; Mane, R. Ionic Liquid–Mediated, One-Pot Synthesis for 4-Thiazolidinones. Synthetic Communications 2010;40 2397-2401.

[165] Kroutil, J.; Budesınsky, M. Preparation of diamino pseudodisaccharide derivatives from 1,6-anhydro-β-d-hexopyranoses via aziridine-ring cleavage. Carbohydrate Research 2007;342 147-153.

[166] Llinas-Brunet, M.; Bailey, M. D.; Ghiro, E.; Gorys, V.; Halmos, T.; Poirier, M.; Rancourt, J.; Goudreau, N. A Systematic Approach to the Optimization of Substrate-Based Inhibitors of the Hepatitis C Virus NS3 Protease: Discovery of Potent and Specific Tripeptide Inhibitors. Journal of Medicinal Chemistry 2004;47 6584.

[167] Huang, L. J.; Hsich, M. C.; Teng, C. M.; Lee, K. H.; Kno, S. C. Synthesis and antiplatelet activity of phenyl quinolones. Bioorganic & Medicinal Chemistry 1998;6 1657-1662.

[168] Theeraladanon, C.; Arisawa, M.; Nishidi, A.; Nakagawa, M. A novel synthesis of substituted quinolines using ring-closing metathesis (RCM): its application to the synthesis of key intermediates for anti-malarial agents. Tetrahedron 2004;60 3017-3035.

[169] Yadav, A. K.; Kumar, M.; Yadav, T.; Jain, R. An Ionic Liquid Mediated One-Pot Synthesis of Substituted Thiazolidinones and Benzimidazoles. Tetrahedron Letters 2009;50 5031-5034.

[170] Yadav, L. D. S.; Yadav, B. S.; Rai, V. K. Multicomponent Reactions in Chiral Ionic Liquids: A Stereocontrolled Route to Mercaptopyranothiazoles. Journal of Heterocyclic. Chemistry 2008;45 1315-1319.

[171] Wen, L. R.; Xie, H. Y.; Li, M. A Basic Ionic Liquid Catalyzed Reaction of Benzothiazole, Aldehydes, and 5,5-Dimethyl-1,3-cyclohexanedione: Efficient Synthesis of Tetrahydrobenzo[*b*]pyrans. Journal of Heterocyclic Chemistry 2009;46 954-959.

[172] Zhang, X. Y.; Li, Y. Z.; Fan, X. S.; Qu, G. R.; Hu, X. Y.; Wang, J. Multicomponent Reaction in Ionic Liquid: A Novel and Green Synthesis of 1, 4-Dihydropyridine Derivatives. Chinese Chemical Letters 2006;17 150-152.

[173] Tu, S.; Zhu, X.; Zhang, J.; Xu, J.; Zhang, Y.; Wang, Q.; Jia, R.; Jiang, B.; Zhang, J.; Yao, C. New potential biologically active compounds: Design and an efficient synthesis of *N*-substituted 4-aryl-4,6,7,8-tetrahydroquinoline-2,5(1*H*,3*H*)diones under microwave irradiation. Bioorganic & Medicinal Chemistry Letters 2006;16 2925-2928.

[174] Kanakaraju, S.; Prasanna, B.; Basavoju, S.; Chandramouli, G. V. P. Ionic Liquid Catalyzed One-Pot Multi-Component Synthesis, Characterization

and Antibacterial Activity of Novel Chromeno[2,3-*d*]pyrimidin-8-amine derivatives. Journal of Molecular Structure 2012;1017 60-64.

[175] R. Breslow, R. Biomimetic Chemistry and Artificial Enzymes: Catalysis by Design. Accounts of Chemical Research 1995;28 146-153.

[176] Shaterian, H. R.; Ranjbar, M. An Environmental Friendly Approach for The Synthesis of Highly Substituted Imidazoles Using Brønsted Acidic Ionic Liquid, N-methyl-2-pyrrolidonium hydrogen sulfate, as reusable catalyst. Journal of Molecular Liquids 2011;160 40-49.

[177] Lehmann, F.; Holm, M.; Laufer, S. T. Three-Component Combinatorial Synthesis of Novel Dihydropyrano[2,3-*c*]pyrazoles. Journal of Combinatorial Chemistry 2008;10 364-367.

[178] Zhou, J. F.; Tu, S. J.; Zhu, H. Q.; Zhi, S. J. A Facile One-Pot Synthesis Of Pyrano[2,3-*c*]Pyrazole Derivatives Under Microwave Irradiation. Synthetic Communications 2002;32 3363.

[179] Guo, S. B.; Wang, S. X.; Li, J. T. D,L-Proline-Catalyzed One-Pot Synthesis of Pyrans and Pyrano[2,3-*c*]pyrazole Derivatives by a Grinding Method under Solvent-Free Conditions. Synthetic Communications 2007;37 2111-2120.

[180] Ren, Z.; Cao, W.; Tong, W.; Jin, Z. Solvent-Free, One-Pot Synthesis of Pyrano[2,3-*c*]pyrazole Derivatives in the Presence of KF $2H_2O$ by Grinding. Synthetic Communications 2005;35 2509-2513.

[181] Khurana, J. M.; Nand, B.; Kumar, S. Rapid Synthesis of Polyfunctionalized Pyrano[2,3-*c*]pyrazoles via Multicomponent Condensation in Room- Temperature Ionic Liquids. Synthetic Communications 2011;41 41, 405-410.

[182] Xiao, Z.; Lei, M.; Hu, L. An Unexpected Multi-Component reaction to synthesis of 3-(5-amino-3-methyl-1*H*-pyrazol-4-yl)-3-arylpropanoic acids in ionic liquid. Tetrahedron Letters 2011;52 7099-7102.

[183] Ghahremanzadeh, R.; Ahadi, S.; Bazgir, A. A one-pot, four-component synthesis of α-carboline derivatives. Tetrahedron Letters 2009;50 7379-7381.

[184] Ghahremanzadeh, R.; Sayyafi, M.; Ahadi, S.; Bazgir, A. Novel One-Pot, Three-Component Synthesis of Spiro[Indoline-pyrazolo[4′,3′:5,6]pyrido[2,3-*d*]pyrimidine]trione Library. Journal of Combinatorial Chemistry 2009;11 393-396.

[185] Ghahremanzadeh, R.; Moghaddam, M. M.; Ayoob Bazgir, A.; Akhondi, M. M. An Efficient Four-component Synthesis of Spiro[indolinepyrazolo[4',3':5,6]pyrido[2,3-*d*]pyrimidine]triones. Chinese Journal of Chemistry 2012;30 321-326.

Chapter 19

Safer and Greener Catalysts – Design of High Performance, Biodegradable and Low Toxicity Ionic Liquids

Rohitkumar G. Gore and Nicholas Gathergood

Additional information is available at the end of the chapter

http://dx.doi.org/10.5772/45605

1. Introduction

Molten salts which are ionic (i.e. a mixture of cation and anion) in nature and have a melting point below 100 °C are termed as Ionic liquids (ILs). [1] Preferably salts which are liquid at room temperature are called room temperature ionic liquids (RTILs). ILs have received great attention in the last couple of decades due to their unique properties such as low vapour pressure, high thermal stability, recyclability, non-flammability, and control over the product distribution. [2,3] Due to the control over fugative emission; ILs can be a replacement for volatile organic compounds (VOCs) which are commonly used as solvents in organic processes. Since the first ionic liquid was reported, [4] there has been a large number of articles been published with different types of cations and anions. One can easily design 10^{18} possible structures of ILs by varying cations and anions. This makes them "designer" molecules. [5-9] These designed combinations have already been found useful in different fields of chemistry, such as organic chemistry, [10-14] electrochemistry, [15-19] analytical chemistry, [20-24] and biochemistry. [15,25]

There are five major classes of cations in ILs e.g. ammonium, pyridinium, imidazolium, phosphonium and sulfonium (Figure 1).

Along with these, there are a large number of commonly used anions such as halides (chloride, bromide, iodide), bis(trifluoromethanesulfonimide) (NTf_2), tetrafluoroborate (BF_4), hexafluorophosphate (PF_6), octyl sulfate ($OctOSO_3$), acetate (OAc) and dicynamide ($N(CN)_2$) to name a few. Change in the anionic component can drastically affect physical properties of an ionic liquid such as hydrophilicity, viscosity and melting point.

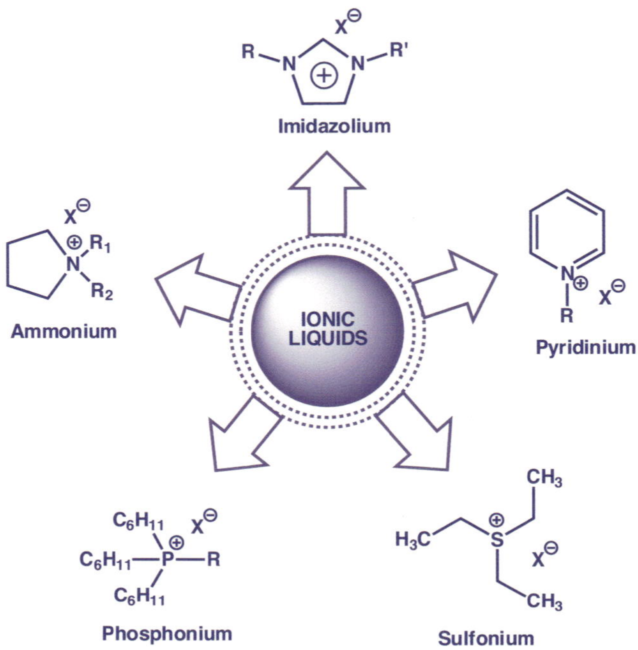

Figure 1. Major types of cations in ILs

2. Applications of ionic liquids in organic synthesis

Ionic liquids have been widely exploited in numerous organic reactions due to the versatility in the physical properties such as ease of product separation, [26,27] enhancement in rate of reaction, [28-30] catalyst immobilization, [31-33] and recyclability. [34-36] Modifications in cations and/or anions have facilitated their use in organic reactions while playing a role of reagent, solvent or catalyst. This can be reflected in a huge number of publications. Hence we are discussing, in our opinion only interesting representative examples here in this chapter.

In this chapter the aim is to demonstrate the versatility of ionic liquids in organic synthesis. We are also going to discuss the environmental fate of ionic liquids by addressing the importance of toxicity, eco(toxicity), biodegradation and green chemistry metrics. By exploring these parameters one can design and synthesise safer and greener catalyst/solvent.

2.1. Heck reaction

The palladium catalysed C-C bond forming reaction between aryl halide or vinyl halide (or triflate) and activated alkene in presence of base is known as the Heck reaction. [37] This reaction is named after Prof. Richard F. Heck, for which he was awarded Nobel Prize in Chemistry 2010, *"for palladium-catalyzed cross couplings in organic synthesis"* jointly with Prof. Ei-ichi Negishi and Prof. Akira Suzuki. This reaction is also known as Mizoroki-Heck reaction, as Tsutomu Mizoroki was the first to report this reaction. [38] A large variety of organic and inorganic bases can be used in this reaction. Phosphine ligands have been used to stabilize the catalytic system in molecular solvents. Although the reaction conditions are mild, the major drawback is that it is difficult to recycle the palladium catalyst in traditional solvents. Kaufmann and co-workers (1996) were first to demonstrate the use of tetraalkylammonium salts as an effective solvent in Heck reaction. [39] Since then a large number of publications have shown that different class of ILs can be used as solvent, catalyst or as a ligand in the Heck reaction. [40] L. Wang and co-workers have reported the Heck reaction of aryl halide and styrene (Scheme 1) in an ethanolamine-functionalized quaternary ammonium bromide which act as base, ligand and solvent. [41]

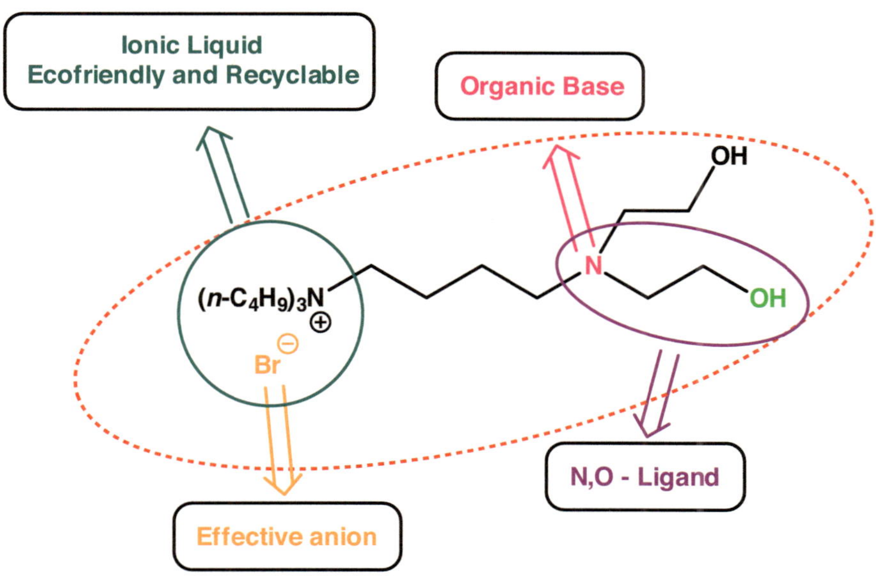

Figure 2. The functions of DHEABTBAB IL in the Heck reaction

The task specific ionic liquid i.e. 4-Di(hydroxyethyl)aminobutyl tributylammonium bromide (DHEABTBAB) (Figure 2) and palladium acetate served as an excellent catalytic system for the cross-coupling of a variety of olefins and aryl halides to give good to excellent results. The Heck reactions of styrene and iodobenzene/bromobenzene have shown excellent conversions and yield (>99%), whereas reaction of styrene and chlorobenzene have given only 66% yield. The reactions of activated and deactivated bromobenzenes and styrene/acrylates generated good to excellent yields (82 to 99%). This catalytic system was also successfully recycled and reused up to 6 times without significant loss of activity. Transmission electron microscopy (TEM) image of Pd-nanoparticles formed in (DHEABTBAB) showed even distribution due to ethanolamine moiety of the IL, which can either coordinate to the palladium or point away from the surface of the nanoparticle.

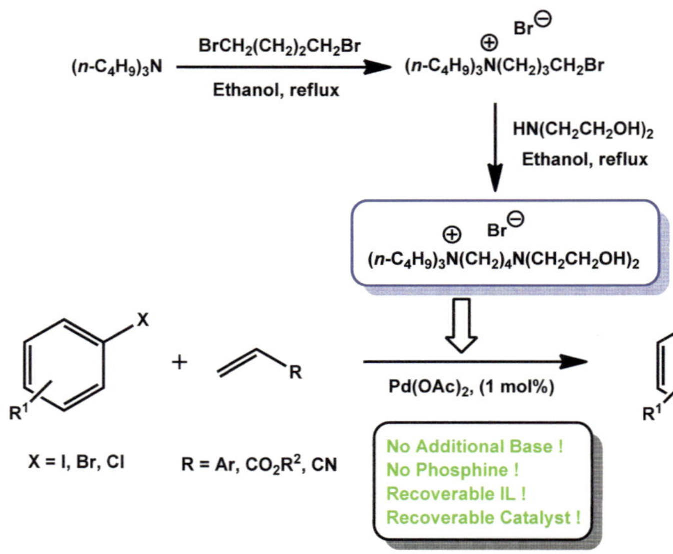

Scheme 1. Synthesis of ionic liquid and its application in the Heck reaction

In an attempt to eliminate the use of phosphine ligands, Xaio and co-workers demonstrated the *in situ* formation of a *N*-heterocyclic carbene complex with palladium when 1,3-dialkylimidazolium ILs were used as a solvent under basic conditions to generate carbene ligand. [42] They had successfully isolated the palladium carbene complex, by deprotonation of imidazolium-based ionic liquids in presence of base to form the catalytic precursor. Such participation of *N*-heterocyclic carbene as a ligand was predicted by Seddon. [43]

http://dx.doi.org/10.5772/45605

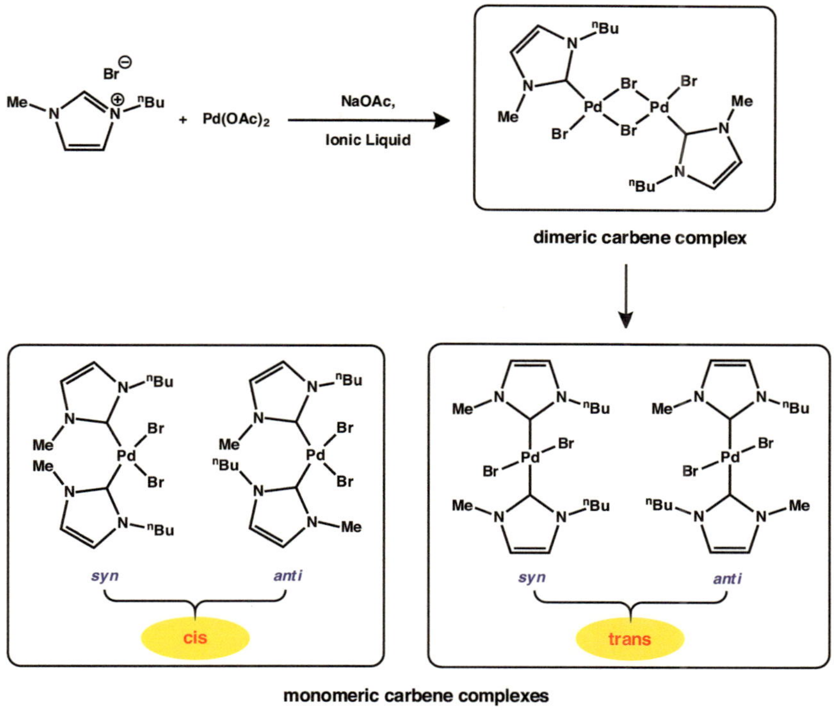

Scheme 2. Stepwise formation of *N*-hetrocyclic carbene complex of palladium

[BMIM] based ionic liquid and palladium acetate in presence of base such as sodium acetate first formed dimeric carbene complex, which eventually gave monomeric carbene complexes (Scheme 2). Existence of all four isomers of monomeric carbene complex was confirmed by ^{1}H-NMR. The Heck reaction of aryl halides with acrylates/styrene in ionic liquid under the reaction conditions have performed better than the isolated *trans* isomer of *N*-hetrocyclic carbene complex in ionic liquid. This might be due to the presence of other active palladium species formed in situ. Shrinivasan and co-workers have further supported such formation of Pd-carbene complex in [BMIM] based IL and accelerated the reaction under ultrasonic irradiation even at room temperature. [44]

Attempts and further efforts into increasing the recyclability of palladium catalyst and to reduce the use of solvent resulted in exploration of solid supported ionic liquids for use in the Heck reaction. [45,46] B. Han and co-workers have reported copolymerized ionic liquid supported palladium nanoparticles as an effective catalyst for the Heck reaction under solvent-free conditions. [47] The 1-aminoethyl-3-vinylimidazolium bromide i.e. [VAIM][Br] ionic liquid was grafted on cross-linked polymer polydivinylbenzene (PDVB). The palladium nanoparticles were anchored onto the polymer via the amino group in the ionic liquid (Scheme 3). Formation of the catalyst was confirmed by a number of analytical techniques such as X-ray photoelectron spectroscopy, transmission electron microscopy, Fourier transform infrared spectroscopy, etc.

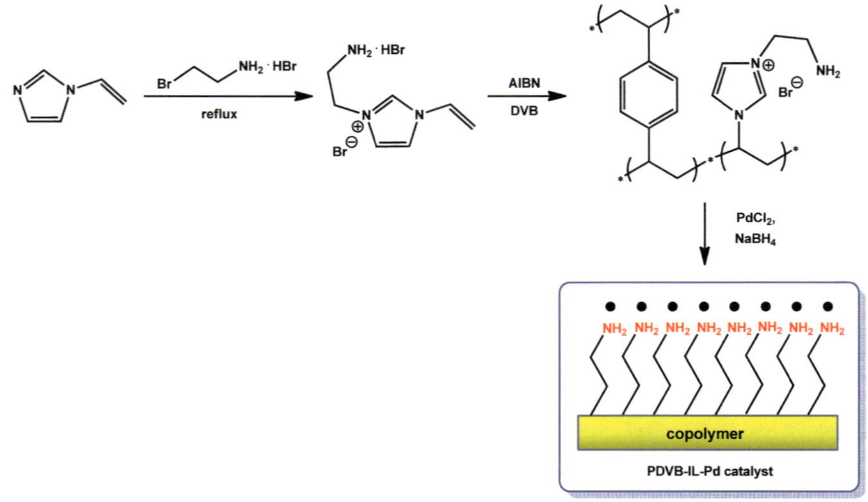

Scheme 3. Preparation of copolymerized ionic liquid supported palladium nanoparticles

The Heck reactions of a variety of iodobenzenes and acrylates have shown excellent conversions (above 93%) irrespective to the substitution on benzene ring. Triethylamine served as a good base under the reaction conditions. Due to the insoluble nature of cross-linked polymer and strong co-ordination between amino group and palladium nanoparticles, the catalyst was recovered very easily by filtration and washed with ethanol. The PDVB-IL-Pd catalyst was very active even after the 4th recycle and was confirmed by TEM image. The excellent stability of the catalyst was due to its insoluble nature in both reactants and product, and high thermal stability i.e. up to above 220 °C.

2.2. Sonogashira reaction

The palladium catalysed C-C coupling reaction of aryl halide and terminal acetylene is known as the Sonogashira reaction. Copper iodides have also been used as a co-catalyst in this reaction. [48] A stoichiometric amount of base is always used as acid (HX) scavenger. This is a widely used and efficient way to prepare substituted or unsubstituted acetylenes. Although Sonogashira coupling reactions works well under mild conditions, the drawback of this reaction is that copper catalysts used can promote side reactions, such as Glaster-type homocoupling of acetylenes. [49-53]

Ryu and co-workers have reported a palladium(II) catalysed efficient Sonogashira coupling in ionic liquid, without any copper co-catalyst. [54] The reactions with an aryl halides and alkyl/aryl acetylenes were carried out in [BMIM][PF_6] as a solvent and diisopropylamine or piperidine as a base. A number of palladium catalyst were screened in the Sonogashira reaction, where bis(triphenylphosphine)palladium(II) dichloride showed high catalytic activity in absence of copper co-catalyst. The Sonogashira reactions of aryl halides and alkyl/aryl acetylenes gave respective dialkyl/diaryl acetylenes in good yields (87-97%).

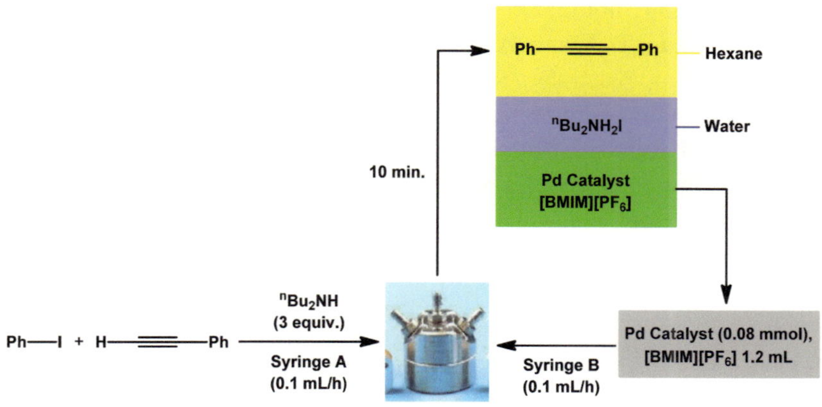

Scheme 4. Sonogashira reaction in a Microflow system

The group has successfully demonstrated the application of this reaction in a microflow reactor with IMM micromixer. Iodobenzene, phenylacetylene and base dibutylamine (syringe A) was added via one inlet to IMM's micromixer and Pd catalyst and [BMIM][PF_6] (syringe B) at the other inlet by using syringe pump (Scheme 4). After reacting in micromixer for 10 min., the product was easily isolated by Hexane/Water extraction, where Pd catalyst in ionic liquid was recycled and reused with slight loss of activity.

In an effort to develop an air stable copper free Sonogashira reaction, Wu, Liu and co-workers have reported palladium complex functionalized ionic liquid as a catalyst in Sonogashira reaction in [BMIM][PF_6] under aerobic and copper free conditions. [55]

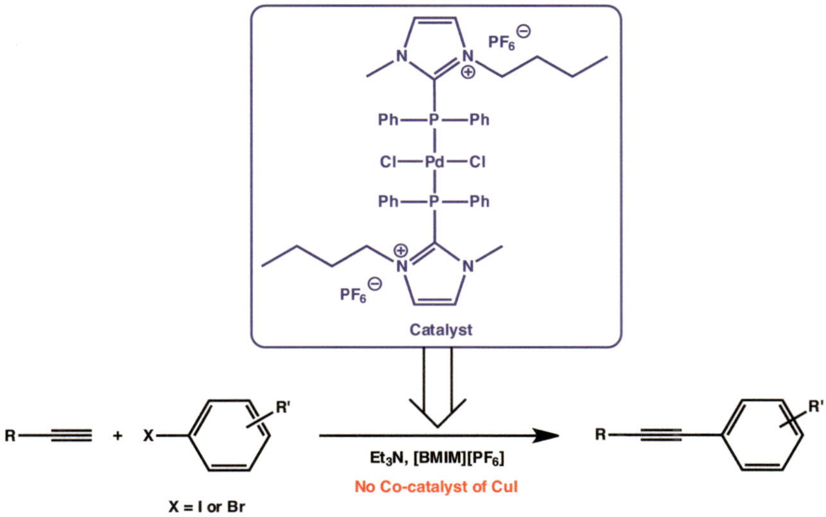

Scheme 5. The Sonogashira reactions in [BMIM][PF_6]

The functionalized ionic liquid i.e. di-(1-butyl-2-diphenylphosphino-3-methylimidazolium)-dichloridopalladium(II) hexafluorophosphate showed efficient catalytic activity and recyclability in coupling reactions (Scheme 5). A clear trend i.e. I>Br>Cl was observed in aryl halide and phenylacetylene couplings. Iodobenzene have shown excellent reactivity with variety of terminal acetylenes (90-99%) in Sonogashira coupling. The phosphine-ligated palladium complex functionalized ionic liquid was easily recycled and reused. Recyclability experiments displayed a gradual loss of activity of the catalyst in [BMIM][PF_6] after 6 recycles (100% to 68% yields), whereas rapid loss of activity in CH_3CN after 4 recycles (98% to 48% yields).

2.3. Suzuki coupling

The Suzuki reaction is a palladium catalysed coupling between aryl/vinyl boronic acid and aryl/vinyl halide in presence of base. This reaction is named after Prof. Akira Suzuki (Nobel Prize in Chemistry, 2010) and also referred to as Suzuki-Miyaura coupling. [56-58] It is one of the important C-C bond forming reaction in the synthesis of styrene and substituted biaryl compounds.

Wei and co-workers have developed a highly efficient silica supported ionic liquid with palladium incorporated anion catalyst for the Suzuki-Miyaura cross-coupling in water under reflux conditions. [59] The catalyst was prepared by an anion exchange reaction between silica-immobilized diimidazolium ionic liquid brushes with the sodium salt of Pd-EDTA (Scheme 6). This catalyst has shown great stability in air and excellent reactivity without any phosphine ligands. The Suzuki coupling of a large variety of aryl bromide and aryl iodides with phenylboric acid in water and PdEDTA-Ionic liquid brush as a catalyst gave very high yields ranging from 89% to 100%. This catalyst did not show loss of activity even after 10 recycles. Another advantage of SiO_2-BisILsOct[PdEDTA] catalyst was that it also act as a phase transfer catalyst in the reaction of water insoluble aryl halides.

Lombardo and co-workers have reported the triethylammonium ion-tagged diphenylphosphine palladium(II) complex for Suzuki-Miyaura reaction in pyrrolidinium ionic liquids under mild reaction conditions. [60] In an effort towards increasing the recyclability of palladium catalyst, triethylammonium ionic liquid supported diphenylphosphine ligand have been prepared. 1-Butyl-1-methyl-pyrrolidinium bis(trifluoromethanesulfonimide) ([bmpy][NTf_2]) prove to be the best solvent along with water in presence of potassium phosphate as a base in the coupling of *o*-bromotoluene and phenylboronic acid. A Suzuki reaction of a number of electron donating and electron withdrawing groups on aryl halides and aryl boronic acids showed good to excellent results. 2-methylphenylboronic acid has given 99% yields when coupled with 1-naphthylbromide and also with 4-bromobiphenyl. The coupling of *p*-anisylboronic acid and 4-bromobiphenyl gave 84% yield with triethylammonium ion-tagged diphenylphosphine palladium(II) complex.

http://dx.doi.org/10.5772/45605

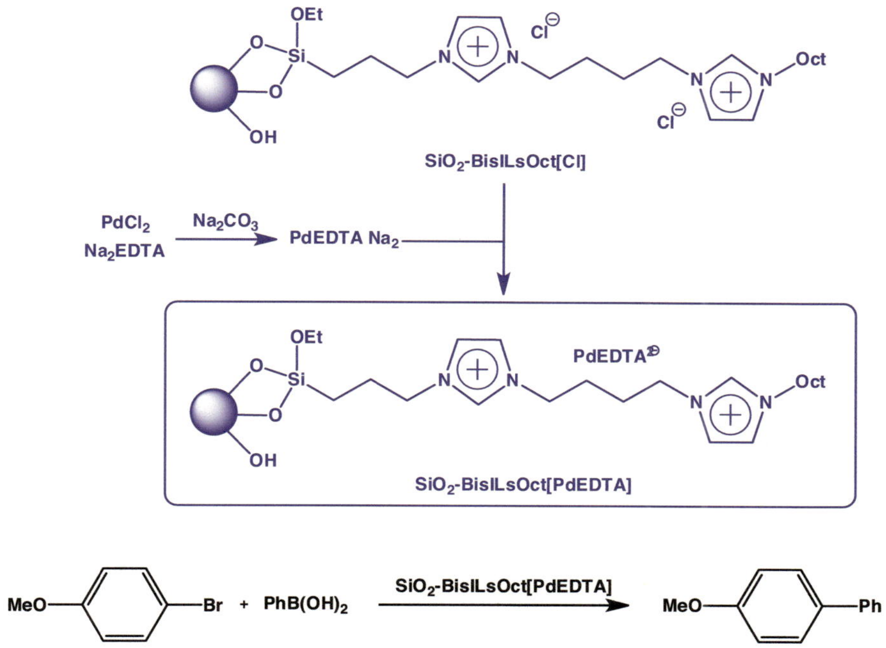

Scheme 6. Suzuki reaction with SiO_2-BisILsOct[PdEDTA] catalyst

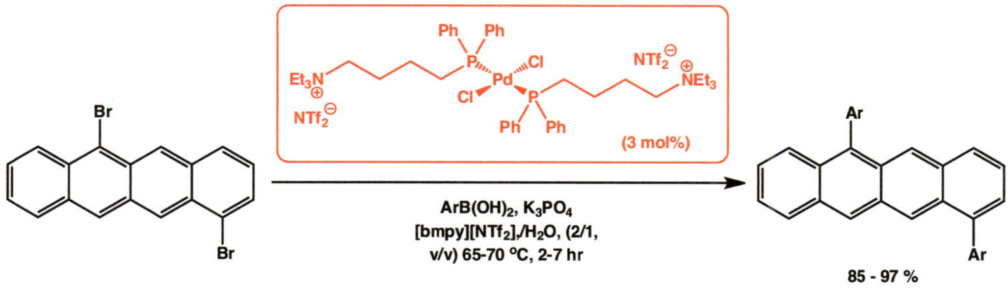

Scheme 7. Suzuki cross-coupling between 5,11-dibromotetracene with arylboronic acids

Miozzo and co-workers have demonstrated an excellent use of such ionic liquid ligated palladium complex in the challenging Suzuki cross-coupling between 5,11-dibromotetracene with arylboronic acids under mild conditions (Scheme 7). [61] These couplings have given very high yields with phenyl and substituted phenylboronic acid (93-97%) even with 2-naphthylboronic acid (95% yield).

2.4. Stille coupling

The palladium catalyzed C-C bond formation reaction between organotin reagents and sp^2-hybridised organohalides are typically classed as Stille coupling reactions. [62,63] It is an important method of alkylation/arylation of vinyl/aryl halide. Organotin reagents used in this reaction are stable and easily stored in air. But the major drawbacks of Stille reaction are the toxicity of organotin reagents and recyclability of palladium catalyst.

To increase the recyclability of palladium catalyst and solvent, Handy and Zhang have reported the use of ionic liquid as a effective media for Stille coupling. [64] Stille coupling reactions were compared between NMP and [BMIM][BF_4] as a solvent with bis(benzonitrile)palladium(II) chloride as a catalyst and in the presence of triphenylarsine and Copper(I) iodide. These reactions demonstrated the compatibility of ionic liquid in Stille reactions.

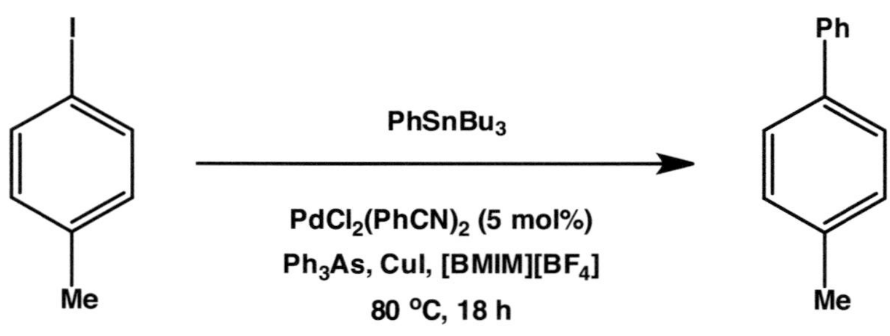

Scheme 8. Stille coupling of 4-iodotoluene and tributylphenyltin in [BMIM][BF_4]

Aryl coupling of a variety of aryl iodides and bromides and tributylphenyltin afforded the respective products with good yields (Scheme 8). The coupling of 4-iodotoluene and tributylphenyltin showed the highest conversion with 95% yield, whereas p-bromoanisole and tributylphenyltin gave only 15% yield towards the desired product. Ionic liquid and the catalyst bis(benzonitrile)palladium(II) chloride were recycled without loss of activity even after 5th cycle.

Due to the toxicity of organotin reagents and contamination of product by tin, organotin reagents have been boycotted by the pharmaceutical industry. Legoupy and co-workers reported an ionic liquid supported organotin reagent which can recycled and minimise contamination of product by organotin compound, without the use of solvent and additives. [65] Ionic liquids supported dibutylphenyltin was successfully synthesized and used as a reagent in palladium catalyzed Stille cross-coupling reactions involving brominated substrates under solvent-free conditions.

http://dx.doi.org/10.5772/45605

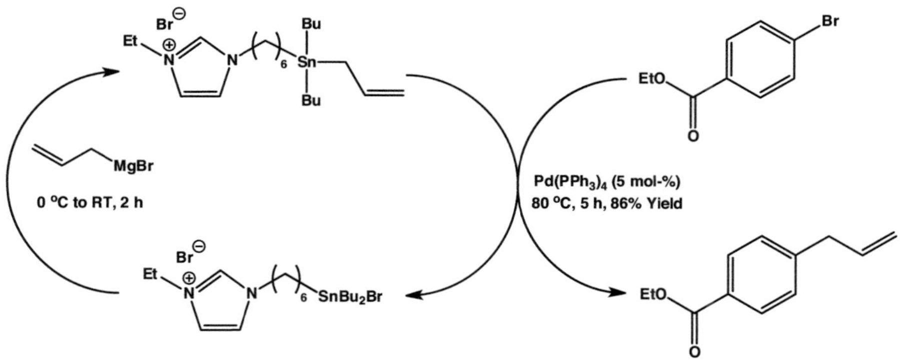

Scheme 9. Recyclable organotin reagent for Stille coupling

An effective use of different ionic liquid supported vinyl, allyl, aryl and heteroaryl organotin reagents with aryl bromides have seen use in Stille cross-coupling reaction. Such ionic liquid incorporated organotin reagent was recycled and reused 5 times with good yields and without loss of reactivity by using Grignard reaction (Scheme 9). It also helped to minimise tin contamination to less than 3 ppm.

2.5. Diels-Alder reaction

The cycloaddition reaction between the conjugated diene and dienophile/substituted alkene is known as Diels-Alder reaction. [66] Prof. Otto Paul Hermann Diels and Prof. Kurt Alder was awarded Nobel Prize in Chemistry in 1950 for "*for their discovery and development of the diene synthesis*". Diels-Alder reaction is an important tool in synthesis of huge and complex cyclic molecules such as cholesterol, reserpine, etc. Heterocyclic compounds can also be prepared with this reaction by using heteroatom (most of the times N and O) either as the diene or dienophile component. Diels-Alder reaction has immense importance due to the 100% atom economy in product formation. The reaction can be performed either by heating or by using Lewis/Brønsted acid catalysts such as $ZnCl_2$, HBF_4, $Sc(OTf)_3$ etc. in organic solvents.

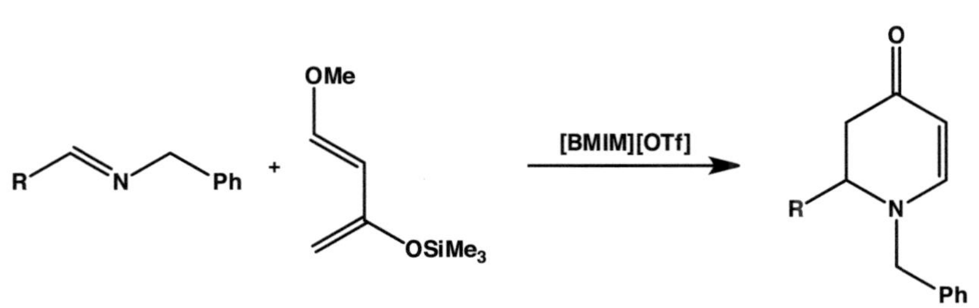

Scheme 10. Aza-Diels-Alder reaction of Danishefsky's diene with imines

Pégot and Vo-Thanh have reported aza-Diels-Alder reaction of Danishefsky's diene with imines in ionic liquids, at room temperature without any acid catalyst and organic solvents. [67] The reaction of *N*-benzylidenebenzylamine and Danishefsky's diene in [BMIM][OTf] showed high i.e. 94% conversion (91% yield) in 1 hour at room temperature (Scheme 10). Only half an equivalent amount of ionic liquid as used with respect to *N*-benzylidenebenzylamine. In the study of an effect of counter anion of [BMIM] cation, triflate (OTf) and bis(trifluoromethanesulfonimide) (NTf_2) has shown high yields i.e. 91% and 94% respectively in comparison with tetrafluoroborate (BF_4) and hexafluorophosphate (PF_6) i.e. 62% and 53% respectively. Reactions using pyridinium and ammonium cations with triflate anion gave good yields (91% and 89% respectively). These studies have shown that ionic liquids can be used as both polar solvent and as a catalyst in Aza-Diels-Alder reaction.

Zhou and co-workers reported C_2-symmetric ionic liquid-tagged bis(oxazoline) copper catalyst for Diels-Alder reaction in ionic liquid. [68] Bis(oxazoline)-copper(II) complexes have already been used as a Lewis acid catalyst in enantioselective Diels-Alder reactions. [69] In order to increase recyclability of the catalyst, the imidazolium-tagged bis(oxazoline) ligand copper catalyst was synthesized. (Scheme 11) The ionic liquid part of the ligand increased the insolubility of the copper catalyst in typical reaction solvents like diethyl ether, which makes workup procedure very simple. The product was separated from catalyst just by washing with diethyl ether.

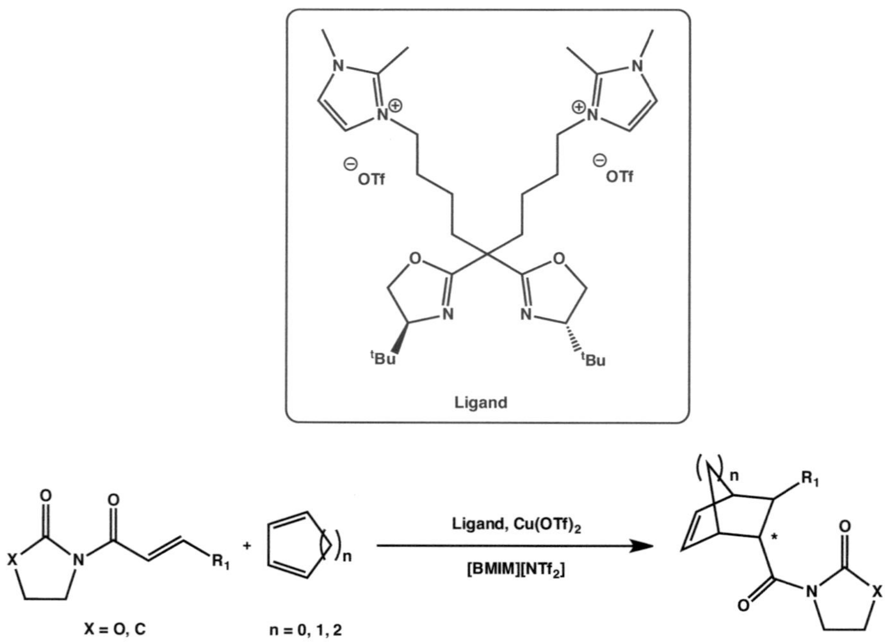

Scheme 11. Screening of the ligands in an asymmetric Diels-Alder reaction

http://dx.doi.org/10.5772/45605

The ionic liquid-tagged bis(oxazoline) copper catalyst did not show any conversion in the Diels-Alder reaction of *N*-acryloyloxazolidinone and cyclohexa-1,3-diene in DCM as a solvent. When the same reaction was carried out in [BMIM][NTf_2] has given required product with 98% conversion and 97% ee. *N*-Acryloyloxazolidinone was found to be more active than *N*-acryloylpyrrolidinone with cyclopentadine/cyclohexadiene in presence of C_2-symmetric ionic liquid-tagged (*S,S*)-*t*-Bu-box copper catalyst in [BMIM][NTf_2]. This efficient catalytic system (catalyst + IL) was recycled 20 times without loss of activity or enantioselectivity. This excellent recyclability was due to the ionic character of the ligand. The toxicity testing of the ligands synthesized were carried out on luminescent bacteria. The traditional *t*-Bu-box ligand has shown higher LC50 values (45 μg/mL) than most active ligand (11 μg/mL).

2.6. Acetalisation reactions

The acid catalysed nucleophilic addition of an alcohol to aldehyde or ketone to form respective acetal or ketal is termed as an acetalisation reaction. This is one of the important reactions in organic synthesis. As the carbonyl functionality is very reactive, it is important to protect against the attack of nucleophiles, acidic, basic or reducing agents. [70] There are several methods to protect aldehydes/ketones. Acetalisation i.e. formation of acetal has its own advantages, as it is stable to all nucleophilic and basic reagents. This reaction can be catalysed by traditional liquid acids such as HCl, H_2SO_4, etc. and also by solid acids i.e. Lewis/Brønsted acid catalysts such as $ZnCl_2$, $FeCl_3$, Zeolites, p-TSA, etc. [70-72] A water molecule is formed as a by-product in this reaction, which is important from a Green Chemistry perspective. The major drawback of this reaction is involvement of harmful liquid acids, which also involves handling hazards.

Forbes, Davis and co-workers reported Brønsted acidic ionic liquids with covalently bonded sulfonic acid functionality containing imidazolium and phosphonium cations. [73] These ILs has shown dual use as both catalyst and solvent in Fisher esterification and pinacol/benzopinacol rearrangement. Fang and co-workers further exploited such covalently bonded sulfonic acid functionality in ionic liquid and its dual use in acetalisation reaction. [74]

The Brønsted acidic ionic liquid *N,N,N*-trimethyl-*N*-propanesulfonic acid ammonium hydrogen sulfate ([TMPSA][HSO_4]) has been prepared economically and used as a catalyst and as a solvent in acetalisation reactions. A number of aldehydes and ketones were reacted with 1,2-diols and methanol to form acetal/ketal in [TMPSA][HSO_4] (Scheme 12). All reactions showed 100% selectivity with excellent conversions within 5-60 minutes. Most of the reactions gave quantitative yields, except the protection of acetophenone with methanol with 65% conversion. This Brønsted acidic ionic liquid was recycled 9 times without loss of catalytic activity and selectivity.

Du and Tian have also demonstrated the use of simple protonated 1-methylimidazolium ionic liquids as a Brønsted acid catalyst in the protection of aldehydes and ketones. [75] The IL catalyst was inexpensively prepared by protonation of 1-methylimidazole. The protection of various aldehydes and ketones with triethyl orthoformate in presence of 1-methylimidazolium tetrafluoroborate displayed very high conversions (84% - 93% yields) at room temperature (Scheme 13). The catalyst was easily recycled just by filtration and reused without any loss of activity.

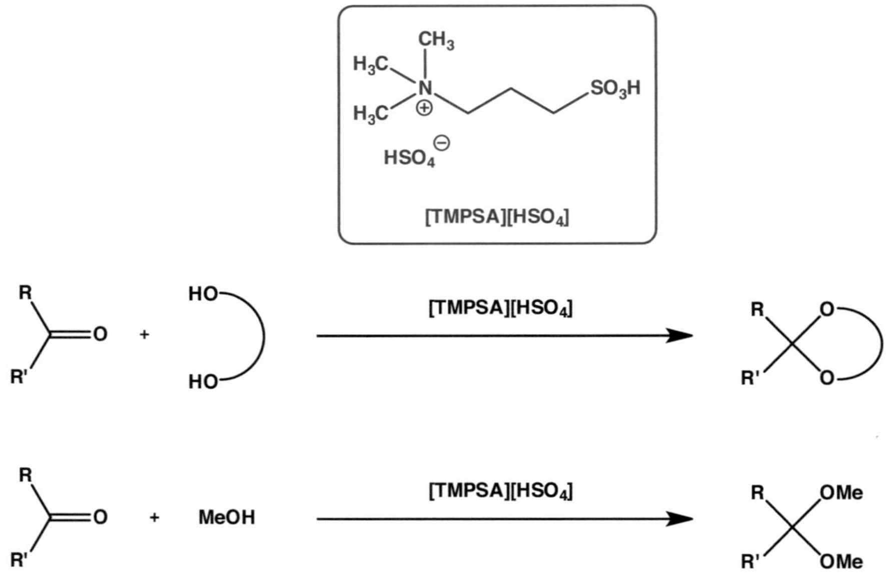

Scheme 12. Protection of aldehydes/ketones with alcohols in presence of [TMPSA][HSO_4]

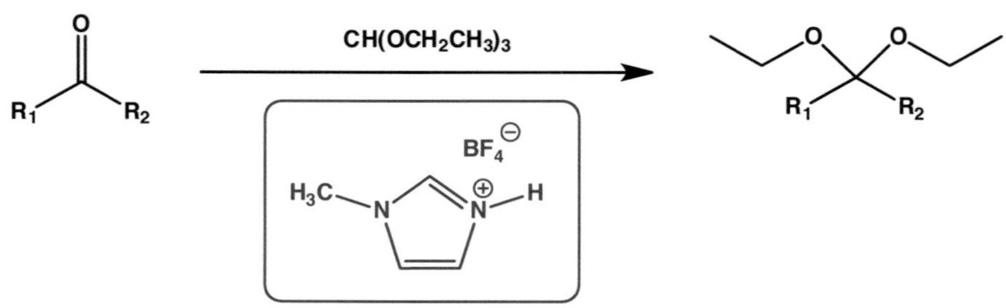

Scheme 13. Protection of aldehydes/ketones with triethyl orthoformate and IL catalyst

3. Environmental fate of ionic liquids

Due to the wide range of applications and versatility, ionic liquids are continually being used extensively in industry, [2] which has triggered an issue of waste management. Also, many of these are totally synthetic novel compounds. Hence it is important to study the environmental impact of such ionic liquids before releasing into the natural environment. Due to their low vapour pressure, ILs can reduce the possibility of air pollution. But bearing an ionic nature; ILs have a notably high solubility in water [7,76,77] (except NTf_2^- & PF_6^-) which is a viable and common means by which these ILs get released in nature. In order to check the biocompatibility of ILs, toxicity, eco-toxicity and biodegradation studies have to be carried out. ILs are usually referred to as "Green" alternatives to Volatile Organic Compounds (VOCs). Instead of the "Green" label, ILs can be categorized in the pattern of '*Traffic Signal Lights*' as discussed at the BATIL (Biodegradation And Toxicity of Ionic Liquids) meeting in DECHEMA, Frankfurt, 2009 (Fig. 1.3). [78] As we start classifying ILs in three colours (Red, Yellow and Green), we can find that most ILs are in the Red and Yellow regions, although this information was solely based on toxicity data. For an IL to be classified more accurately by a 'Traffic Signal Lights' pattern, detailed information about the toxicity, biodegradation and ease of synthesis etc. are required. Similar classification can be applied to commonly used organic solvents (Figure 3). [79]

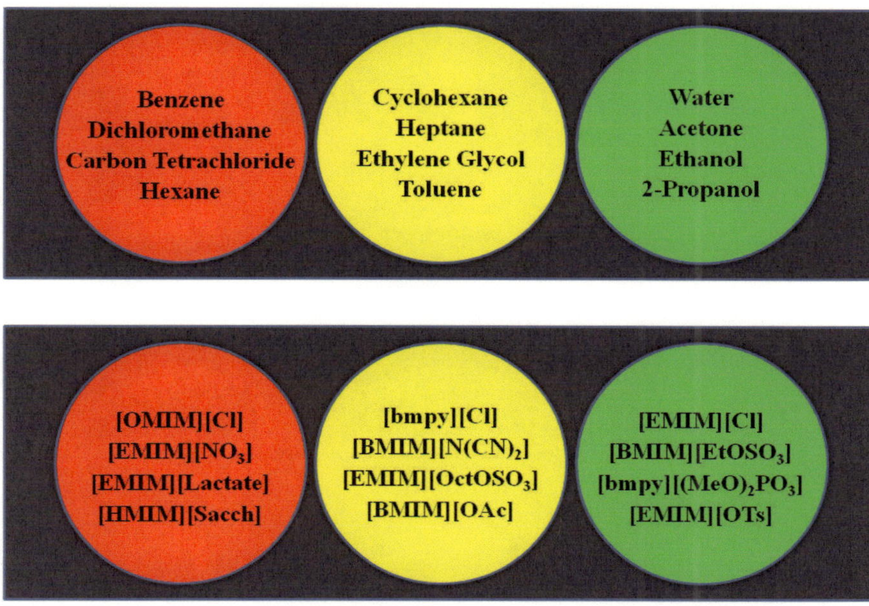

Figure 3. Recommendation for data representation of toxicity of ionic liquids and commonly used organic solvents [78,79]

3.1. Toxicity and eco(toxicity) of ionic liquids

As mentioned before, many ionic liquids are non-natural (synthetic) molecules. While a single toxicological test yields useful, albeit limited data, over the last decade, a large number of publications have demonstrated a wide variety of 'biological test systems' for toxicity testing of ionic liquids (Figure 4). [80,81] This includes fungi, bacteria, algae, enzymes, rat cell line, fish, etc. Only by assessing the toxicity of IL across a broad range of organisms can a 'true' understanding of how environmentally friendly the compound is?

Stock and co-workers reported the effect of ionic liquids on acetylcholinesterase. [82] Enzymes are a crucial part of the human nervous system. Acetylcholinesterase is known to catalyse the hydrolysis of the neurotransmitter acetylcholine, to acetate and choline. Inhibition of acetylcholinesterase results in muscular paralysis and other medically significant nervous problems. Organophosphates are a major class of acetylcholinesterase inhibitors. A range of commonly used imidazolium, pyridinium and phosphonium ionic liquids were tested in this assay. Imidazolium and pyridinium ionic liquids showed high toxicity to acetylcholinesterase at very low concentrations, whereas phosphonium ionic liquids were non-toxic within the test limits. This testing showed that toxicity of these ionic liquids lies in the cationic part and alkyl side chain and not in the anionic part.

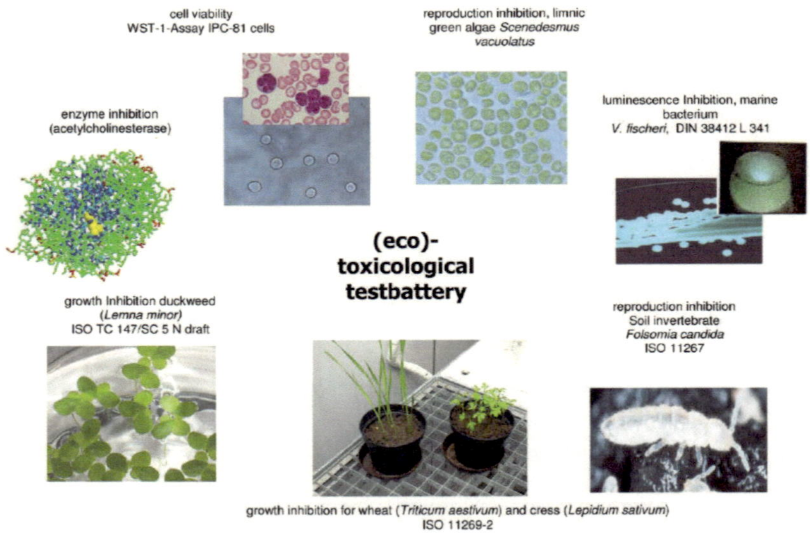

Figure 4. Toxicological test battery [80]

Another important finding of this assay was that increasing the length of alkyl side chains increase the toxicity. This can be explained as long alkyl chain increases lipophilic nature of the ionic liquids, which can then easily incorporate within the biological membrane of nerve cell synapses. [83] Similar trends between the toxicity and length of alkyl chain on luminescence inhibition of *Vibrio fischeri* and *promyelocytic leukemia rat cell line* IPC-81 were reported by Ranke and co-workers. [84] Leukemia rat cell line IPC-81 was also used to observe the cytotoxic effect of commercially available anions. [85] No significant anion effect was found under the test system.

Bernot and co-workers demonstrated that acute toxicity of certain 1-butyl-3-methyl imidazolium ionic liquids on *Daphnia Magna* were mainly due to the cationic part. [86] *Daphnia Magna* has been extensively used for ecotoxicological evaluation of chemicals in invertebrates. Ionic liquids were found to influence the reproduction of *Daphnia Magna*. 1-Butyl-3-methylimidazolium bromide was found to be most toxic in the test system (LC_{50}: 8.03 mg/L). This study demonstrated that the toxicity of ionic liquids was influenced by the cation component, which was confirmed by high LC_{50} values for sodium salts of similar anions. Yu and co-workers reported the toxicity study of 1-alkyl-3-methylimidazolium bromide ionic liquids towards the antioxidant defence system of *Daphnia Magna*. [87] Increasing the length of alkyl side chain was found again to increase toxicity. Toxicity of ionic liquids in this case was due to oxidative stress in *Daphnia Magna,* which was evaluated by measuring the activity of antioxidant defence enzymes, levels of the antioxidant glutathione and malondialdehyde i.e. peroxidation by-product of lipid. [C_{12}MIM][Br] showed very high toxicity with an LC_{50} of 0.05 mg/L under 48h incubation time. Samorì and co-workers reported the toxicity effect of oxygenated alkyl side chain imidazolium ionic liquids in *Daphnia Magna* and *Vibrio Fischeri.* [88] A direct comparison between the toxicity of 1-butyl-3-methylimidazolium tetrafluoroborate ([BMIM][BF_4]) and 1-methoxyethyl-3-methylimidazolium tetrafluoroborate and dicyanamide ([MOEMIM][BF_4] and [MOEMIM][$N(CN)_2$]) proved that incorporation of oxygen functionality helps to lower the toxicity of the ionic liquids (Figure 5). The 50% effective concentration (EC_{50}) for [BMIM][BF_4] towards the inhibition of *Daphnia Magna* and *V. Fischeri* was lower (5.18 and 300 mg/L, respectively) than for [MOEMIM][BF_4] (209-222 and 3196 mg/L, respectively) and [MOEMIM][$N(CN)_2$] (209 and 2406 mg/L, respectively).

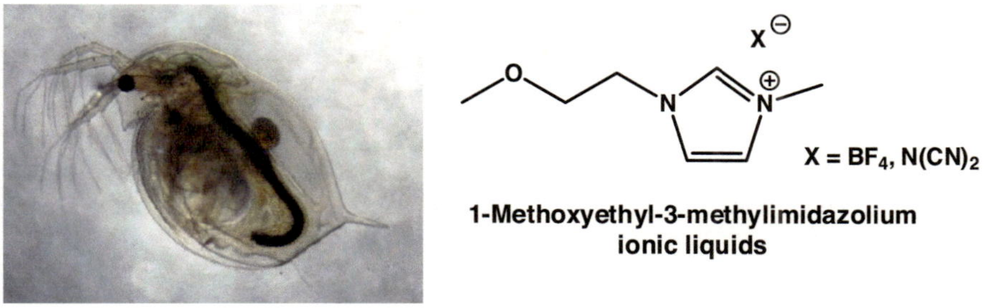

Figure 5. Schematic of 1-methoxyethyl-3-methylimidazolium ILs towards Daphnia Magna

Gathergood and co-workers further demonstrated that imidazolium based ionic liquids with an oxygen functionality i.e. ester and ether side chains, have reduced antimicrobial activity to a great extent. [89] Four Gram negative bacteria (*Pseudomonas aeruginosa, Escherichia coli, Klebsiella sp., Salmonella sp.*) and three Gram positive bacteria (*Staphylococcus aureus, Enterococcus sp., Bacillus subtilis*) were screened in the assay. A range of long ether and poly ether ester side chain imidazolium ionic liquids showed a huge reduction in the toxicity in this test system, compared with similar number of atoms in long alkyl side chains.

In order to check the toxicity effect of ionic liquids in humans, a cytotoxicity assay with human cell lines was designed. HeLa is one of the most extensively used cell lines in medicinal research. HeLa is a human tumor cell line, which is a prototype of epithelium. Due to the first contact of an organism with toxic materials, HeLa cell line has great importance. Stepnowski and co-workers reported the cytotoxic effect of imidazolium ionic liquids in Hela cell line. [90] The EC_{50} values of a range of 1-alkyl-3-methylimidazolium ionic liquids were evaluated on human epithelium HeLa cells. Ionic liquids with a decyl side chain with tetrafluoroborate as the anion component demonstrated high toxicity (EC_{50} = 0.07 mM). This was supportive of the results with other test systems. The cytotoxicity of ionic liquids were compared with the known 50% effective concentrations (EC_{50} values) of traditional organic solvents such as dichloromethane (71.43 mM), phenol (42.68 mM), xylene (52.43 mM) and ethanol (1501.43 mM). These studies revealed that the tested ionic liquids had significant toxicity against human cell line HeLa, compared with organic solvents. Lu and co-workers utilised this assay for testing the cytotoxicity of a large range of ionic liquids containing imidazolium, pyridinium, choline, triethylammonium and phosphonium cations with halide, NTf_2^-, and BF_4^- anions. [91] In general, choline and alkyl-triethylammonium ionic liquids were found to be less toxic than their imidazolium and pyridinium salt counterparts.

In an effort to evaluate the eco(toxicity) of ionic liquids, Yun and co-workers reported an assay of freshwater microalgae *Selenastrum capricornutum*. [92] The bromide salts of commonly used 1-butyl-3-methylimidazolium, 1-butyl-3-methylpyridinium, 1-butyl-1-methylpyrrolidinium, tetrabutylammonium, and tetrabutylphosphonium ILs were tested against the *S. capricornutum* and compared with traditional water miscible organic solvents such as dimethylformamide, 2-propanol and methanol. Increase in the toxicity of imidazolium and pyridinium cations were observed with an increase in incubation time, whereas the opposite trend was found in the case of tetrabutylammonium, and tetrabutylphosphonium ILs. The growth inhibition of *S. capricornutum* was higher in ionic liquids than organic solvents. A similar test system was applied to investigate the toxicological effect of anions. [93] Toxicity of various anions incorporated with 1-butyl-3-methylimidazolium cation were compared with their respective sodium and potassium salts. The anions were found to inhibit the growth of freshwater algae *S. capricornutum*. The clear trend in algae toxicity was observed as hexafluoroantimonate (SbF_6^-) > hexafluorophosphate (PF_6^-) > tetrafluoroborate (BF_4^-) > triflate ($CF_3SO_3^-$) > octyl sulphate ($OctOSO_3^-$) > halide (Br^-, Cl^-). Toxicity studies (in fish, aquatic plants/invertebrates) on anionic surfactants have shown that toxicity is dependent on a number of factors such as alkyl chain length, solubility and stability in water. [94] As the length of alkyl chain increases, toxicity increases until certain limits. Further increase in chain length can decrease the hydrophilic nature of these materials, reducing bioavaibility of compound which results in a general decrease in the toxicity. [95]

3.2. Biodegradation of ionic liquids

Ionic liquids are well known for being stable to heating and in a variety of reaction conditions. Although this is an important property in their applications, it can raise issues regarding degradation and bioaccumulation when released in nature. Accumulated data on the anti-microbial toxicity of novel ionic liquids can be used as a preliminary guideline before performing the biodegradation tests. The biological test system has its limitations, such as when reported toxicity data is only available for certain individual organisms, whereas biodegradation assays usually have a large sample group of organisms. Also, breakdown products/intermediates of ionic liquids can be toxic, which can be resistant to

http://dx.doi.org/10.5772/45605

further degradation, which leads to the issue of bioaccummulation. Hence it is important to perform biodegradation studies of ionic liquids. [96] Boethling and co-workers in their review article *"Designing Small Molecules for Biodegradability"*, gave useful and general guidelines for the design and synthesis of environmental friendly chemicals. [97] According to their observations, compounds containing unsubstituted alkyl chains, benzene rings, oxygen functionalities such as esters, aldehydes, and carboxylic acids (potential sites for enzymatic hydrolysis) greatly increase biodegradation. Whereas compounds containing halogens, branched chains, heterocycles, functional groups such as nitro, nitroso and arylamines motifs, adversely affect the biodegradation. There are several biodegradation study methods approved by the Organisation for Economic Cooperation and Development (OECD) (See Table 1).

Test No.	Name	Analytical method
OECD 301 A	DOC Die-Away	Dissolved organic carbon
OECD 301 B	CO_2 evolution	CO_2 evolution
OECD 301 C	MITI (Ministry of International Trade and Industry, Japan)	Oxygen consumption
OECD 301 D	Closed bottle	Dissolved oxygen
OECD 301 E	Modified OECD screening	Dissolved organic carbon
OECD 301 F	Manometric respirometry	Oxygen consumption
ISO 14593	CO_2 headspace test	CO_2 evolution
OECD 309	OECD 309	^{14}C labelling
ASTM 5988	ASTM 5988	CO_2 production / Biochemical oxygen demand

Table 1. Biodegradation methods in use

Data collected from all of the tests mentioned in Table 1 can be catagorised according to OECD guidelines as - (a) Ultimate biodegradation: Denotes complete degradation/utilisation of a test compound to produce carbon dioxide (CO_2), water, biomass and inorganic substances. Such biodegradation can achieved due to the mineralisation by microorganisms. This is one of the significant characteristics, before classed as a 'biocompatible' compound. (b) Readily biodegradable: These are positive results showing rapid ultimate degradation of the test compound under aerobic conditions in stringent screening tests. Both mineralisation and elimination/alteration (abiotic process such as hydrolysis, oxidation and photolysis) of the test substance can be observed. (c) Primary biodegradation: An elimination or alteration of the test sample by microorganisms, in order to lose its specific properties. [98]

Gathergood and Scammells reported the synthesis of ester and amide functionalised side chain imidazolium ionic liquids [99] according to the guidelines outlined by Dr. Boethling. [100] All of the novel alkyl ester and amide side chain methylimidazolium ionic liquids were subjected to biodegradation studies. A biodegradation study of bromide salts of these ionic liquids along with commonly used [BMIM][BF_4] and [BMIM][PF_6] ILs was carried out under the 'Closed Bottle Test' (OECD 301D) [101] although none of the tested ionic liquids passed the minimum 60% biodegradation threshold in order to be classified as 'readily biodegradable'. However, [BMIM][BF_4] and [BMIM][PF_6] did not show biodegradation in the test system. Ester functionalised side chain ionic liquids demonstrated improved biodegradation. Increasing the length of the ester side chain increased the biodegradation, for example a methyl ester derivative showed 17% biodegradation, whereas biodegradation of an octyl ester derivative was 32% after 28 days. Another important observation was that amide side chain ionic liquids showed very negligible biodegradation. The effect of different anion on the rate of biodegradation was tested under the same test system. A range of 1-butyl-3-methylimidazolium and ester functionalised 3-methyl-1-(propoxycarbonylmethyl)-imidazolium ionic liquids were prepared with different anions such as Br^-, BF_4^-, PF_6^-, NTf_2^-, $N(CN)_2^-$ and $OctOSO_3^-$. $BMIM^+$ based ionic liquids showed poor biodegradation, except with an octyl sulphate as anion. The biodegradation of 3-methyl-1-(propoxycarbonylmethyl)-imidazolium ionic liquid was increased from 19% with bromide anion to 49% with octyl sulphate anion after 28 days. Gathergood and co-workers further demonstrated that incorporation of ether and polyether linkages along with ester functionality in the side chain can increase biodegradation to a great extent. [89] The biodegradation of a large range of ether and polyether ester side chain methylimidazolium ionic liquids were studied using the CO_2 Headspace test. Octyl sulphate salts of 1-methylimidazolium ionic liquids with propoxyethoxy and butoxyethoxy esters were found to be readily biodegradable (> 60% biodegradation in 28 days).

Docherty and co-workers reported biodegradation studies of imidazolium and pyridinium based ionic liquids by OECD dissolved organic carbon Die-Away test. The test was carried out with the activated sludge microorganisms from wastewater treatment plant. [102] Denaturing gradient gel electrophoresis (DGGE) was also used to investigate the microbial community profile. The results showed that the tested pyridinium based ionic liquids has better biodegradability than the corresponding imidazolium salts. Octyl-3-methyl-pyridinium bromide was found to be readily biodegradable with complete degradation within 15-25 days of incubation, whereas hexyl-3-methyl-pyridinium bromide was degraded within 40-50 days. Such complete biodegradation was further supported by the work of Stolte and co-workers in their investigation of the primary biodegradation of a variety of

http://dx.doi.org/10.5772/45605

ionic liquids by the modified OECD 301 D test. [103] A range of 1-alkyl-3-methylimidazolium, 1-alkylpyridinium and 4-(dimethylamino)pyridinium halide salts were screened under the test system. The biodegradation products were identified by HPLC-MS analysis. 1-Octyl-3-methylimidazolium chloride gave a result of 100% primary biodegradation within 31 days. The stepwise degradation by two possible metabolic pathways was predicted based on HPLC-MS analysis. This predicted breakdown pathway was due to enzymatic oxidation of the terminal carbon.

Scammells and co-workers also reported biodegradation of ester functionalised pyridinium ionic liquids using the CO_2 Headspace test (ISO 14593). [104] Pyridinium ionic liquids with alkyl and ester side chains, along with ester derivatives of nicotinic acid with alkyl side chain were tested under aerobic conditions. The halide salts of pyridinium ionic liquids with alkyl side chain (C_4, C_{10} and C_{16}) showed poor biodegradation, whereas it was improved up to a max. of 45% after 28 days by the use of octyl sulfate anion. Switching from alkyl chain to ester side chain in pyridinium ILs increased the biodegradation dramatically. Such substitution made them readily biodegradable, independent of the chosen anionic component. Bromide, hexafluorophosphate and octyl sulfate showed a high level of biodegradability (85% to 90%), whereas NTf_2^- gave 64% biodegradation after 28 days. Ionic liquids derived from the nicotinic acid ester derivative i.e. ester group at 3-position of the pyridine ring were found to be readily biodegradable even with methyl or butyl side chain. Amide side chain derivatives showed low biodegradability, even with using octyl sulfate as the anion (30% biodegradation after 28 days). Scammells and co-workers further studied the effect of the incorporation of ester, ether and hydroxyl side chain in phosphonium based ionic liquids. [105] All phosphonium ionic liquids showed poor biodegradation independent of ester, ether and hydroxyl side chains and anions in CO_2 Headspace test. Only heptyl ester side chain ionic liquid with octyl sulfate anion showed highest 30% biodegradation after 28 days.

Most of the ionic liquids prepared are not 'readily biodegradable'; however, several structural modifications have shown a positive improvement in the biodegradability of ionic liquids. [96,106] Striving towards compounds with 'Ultimate biodegradation' is preferred over 'readily biodegradable' examples and is a major research area. Hence it is important to study the biodegradation pathways of ionic liquids along with kinetics and metabolite studies to assist the 'benign by design' approach.

3.3. Guidelines for designing 'Green' ionic liquid catalysts/solvents

From the literature, the following observations were made and are summarised in Figure 6:

- Linear alkyl chains in general can increase biodegradation compared to branched hydrocarbon chains.

- Oxygen containing functionalities, such as ester and hydroxyl groups in the side chain of imidazolium cation, not only reduces microbial toxicity but also increases rate of biodegradation. This is, however, not effective in phosphonium based ionic liquids.

- Ether substitution reduces bactericidal toxicity.

- Ester substitutions at 1 and 3 position of pyridinium cation can improve biodegradation.

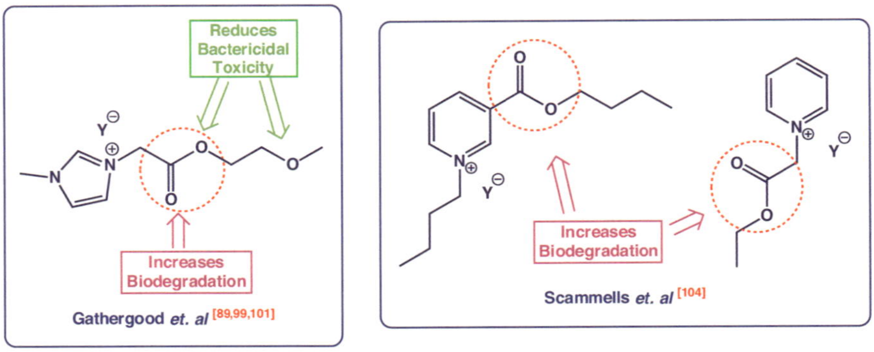

Figure 6. Guidelines for designing 'Green' ionic liquids (left, Gathergood and Scammells; right Scammells)

4. Green chemistry metrics

To achieve 'Green' synthesis of any chemical, the *'12 Principles of Green Chemistry'* given by Anastas and Warner serves as the most useful set of guidelines and gives us the important message that 'Prevention is better than cure'. [107] These principles suggest not only to consider toxicity and biodegradation, but also to measure the 'greenness' of the chemical process. To evaluate the 'greenness' of any process, a number of factors associated with the chemical process has to studied. Green Chemistry metrics can help to measure the efficiency and 'greenness' of the chemical process. There are several well established methods to determine the sustainability of a chemical process under areas such as the economical, technical and social effects of such processes. Economical methods mainly consist of profit related analysis, whereas technical methods analyses quality, productivity and related issues. Social methods concern the society and environmental aspect of the chemical process. [108] Porteous has shown that it is easy to correlate Green Chemistry metrics with the "12 Principles of Green chemistry". [109] There are several metrics available to measure efficiency, use of energy and resources, toxicity, biodegradation, safety and life cycle impact of the chemical process, which are closely related to the 12 Principles.

The Environmental (E) factor is one of the most widely used and efficient methods to measure the amount of waste generated in the process. [110] This is also known as Sheldon's E-factor.

$$\text{E factor} = \frac{\text{Total mass of waste (kg)}}{\text{Mass of product (kg)}}$$

E-factor calculations is one of the important methods to calculate the waste associated with the process, which includes all unwanted side products, reagents, solvents and energy. Water used in the process is usually excluded from the calculations. The higher the E-factor value, the higher the waste generated which has an adverse effect on environment. Economically this adds on to the profit and the cost of disposal. Many modifications on the measurement of E-factor were adopted in industry. GlaxoSmithKline (GSK) has introduced a concept of 'Mass Intensity' based on Sheldon's E-factor. [111]

$$\text{Mass Intensity} = \frac{\text{Mass all materials used (excluding water)}}{\text{Mass of product}}$$

Mass Intensity measures the amount of reagents, solvents and workup reagents used in the process. Hence it takes an account of product yields and stoichiometry of the reagents. In order to measure the synthetic efficiency of the process, the "Atom Economy" concept was found to be useful. Atom economy in chemical reactions is one of the 12 Principles of Green Chemistry, which gives indications as to the overall efficiency of a chemical reaction. [112]

$$\text{Atom Economy} = \frac{\text{Molecular weight of desired product}}{\text{Molecular weight of all products / reactants}} \text{ X } 100\%$$

Although this is an important concept, which suggests to minimise the waste. It does not consider the actual mass, yield, solvents and other reagents used in the process. Reaction mass efficiency (RME) has overcome these drawbacks. [111] Reaction mass efficiency considers actual mass of reactants and product. This is one of the commonly used metrics to evaluate the efficiency of the chemical process.

$$\text{Reaction Mass Efficiency} = \frac{\text{Mass of product}}{\text{Mass of all reactants}} \text{ X } 100\%$$

Apart from these, real time analysis is also important to analyse the chemical process. Real time analysis is again one of the "12 Principles of Green Chemistry", which enables the chemist to identify the formation of waste along the process. A number of analytical techniques such as HPLC, GC, NMR, FT-IR, and sensors etc. were already found to be useful in real time analysis. It's also important to determine the robustness of the process, which will allow preparing chemicals on a large scale. The measurement of toxicity and biodegradation are also important metrics to evaluate 'greenness' of the chemical products.

5. Case study

In order to design and synthesise 'Green' ionic liquid catalysts, guidelines prepared from toxicity and biodegradation studies can be helpful. Novel ionic liquids have been extensively prepared and used in organic synthetic applications, but only few research groups have published complementary toxicity and biodegradation data to support its environmental impact. [69,104,113]

Connon and co-workers reported an application of pyridinium salts as an effective catalyst in acetalisation reactions. [114-116] Ester group substitution at either the 3 or both 3 and 5 positions of the pyridinium ring displayed excellent catalytic activity with very low catalyst loading in the acetalisation of benzaldehyde with methanol. The remarkable feature of the catalyst was that it is not acidic in nature, but can act as a Brønsted acid in the presence of protic media. The highest catalytic activity pyridinium salts, 3,5-bis(ethoxycarbonyl)-1-(phenylmethyl) bromide, showed excellent catalytic activity in the protection of a variety of aldehydes with methanol. The catalyst was also found to be useful in diol and dithiol protections (Scheme 14). Catalytic activity of the catalyst was predicted based on anticipated nucleophilic attack of the alcohol to the pyridinium to generate the Brønsted acidic active species.

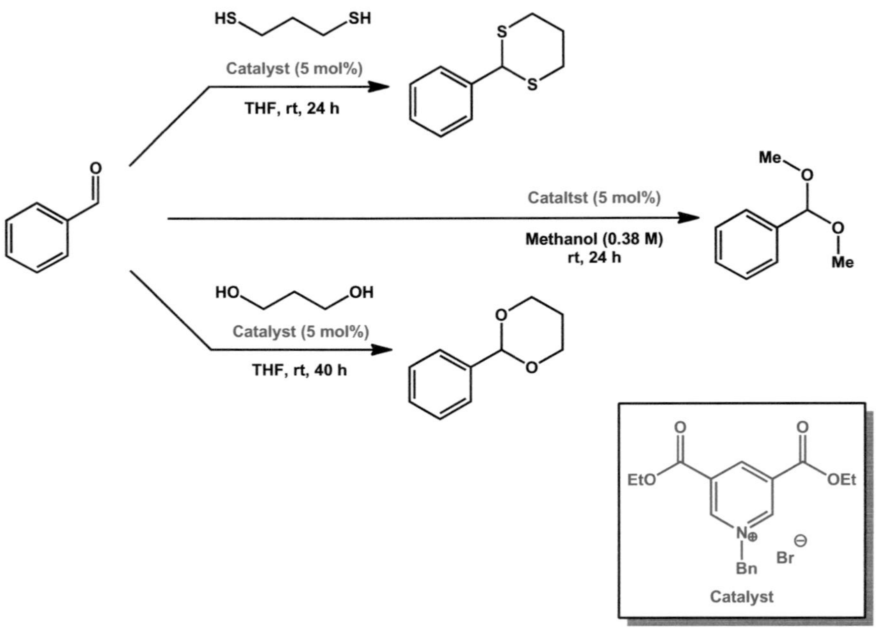

Scheme 14. Acetalisation of benzaldehyde with methanol catalysed by pyridinium IL

On the basis of previous findings, Gathergood, Connon and co-workers reported the design, synthesis and application of imidazolium ionic liquid catalysts in acetalisation reactions. [113] A range of ester and amide side chain imidazolium ionic liquids was

http://dx.doi.org/10.5772/45605

prepared. The catalytic activity of all of these imidazolium salts was evaluated in the acetalisation of benzaldehyde in methanol (Scheme 15). The absence of catalyst resulted in no formation of the corresponding acetal after 24 hours. All of the imidazolium bromide salts showed poor catalytic activity independent of ester or amide side chain (9 to 13% conversions). When switched to the NTf_2 anion, reaction conversions in amide side chain ionic liquids were increased marginally. Whereas ester side chain ionic liquids with NTf_2 anion gave 51% conversion. A tetrafluoroborate salt of a methyl ester side chain imidazolium ionic liquid gave high (85%) conversion towards required product. Hexafluorophosphate and octyl sulfate anions performed poorly in this reaction. The anion exchange from bromide to tetrafluoroborate had greatly influenced the acetalisation reaction of benzaldehyde with methanol, which gave 85% conversion.

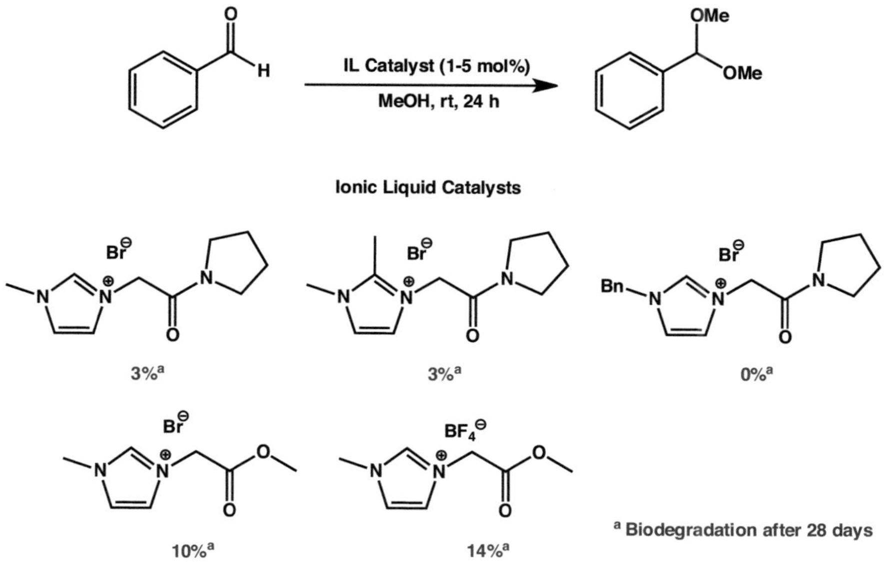

Scheme 15. Acetalisation of benzaldehyde using imidazolium IL as a catalyst. Biodegradation data for IL/Catalyst (ISO 14593) included.

The ionic liquid did not appear to have an acidic nature, but in the presence of a protic medium was proposed to generate a Brønsted acid species to catalyse the acetalisation reaction. (Figure 7) The most active catalyst with tetrafluoroborate anion was further exploited in the acetalisation reactions of a variety of aldehydes with methanol at room temperature. These reactions showed good to excellent conversions with 5-10% catalyst loading. When saturated aldehyde such as 3-phenylpropanal reacted with deuterated methanol in presence of 1 mol% catalyst, the reaction gave quantitative conversion in only 1 minute. Diol and dithiol protection of benzaldehyde showed very good results. The BF_4 catalyst promoted protection of benzaldehyde with 1,2-ethanedithiol gave 92% conversion, whereas 1,3-propanedithiol and 1,3-propanediol gave 65% and 86% conversion respectively. Recyclability of the most active BF_4 anion catalyst was performed on the 1,3-dithiolane protection of benzaldehyde. The catalyst was recycled and reused 15 times without significant loss of activity.

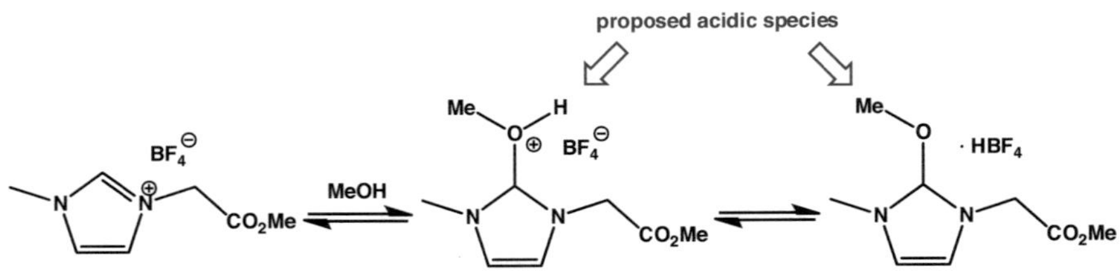

Figure 7. Proposed mode of action of catalytic aprotic imidazolium ions

Biodegradation studies of pyridinium-based ILs have shown that esters substitution at either the 1 or 3-position have a beneficial effect on degradation of the heterocyclic core, independent of the anion. [104] Also biodegradation studies in the literature have shown that only the side chain of the imidazolium ionic liquids undergo degradation, whereas imidazole core was found to persist in most of the OECD tests. [103]

Biodegradation studies of ester and amide side chain ionic liquids along with substituted imidazolium salts was also carried out by using the "CO_2 Headspace" test (ISO 14593) (Scheme 14, 15). [113, 117] All imidazolium ionic liquids prepared failed to pass the minimum 60% biodegradation threshold value in order to be classified as 'readily biodegradable'. Ester functionalised ionic liquids displayed higher biodegradation levels than amide functionalised ionic liquids in 28 days. The first generation imidazolium ionic liquid catalysts showed 10% to 14% biodegradation for methyl ester side chain ionic liquids, where as maximum 3% biodegradation was observed in amide side chain ionic liquids after 28 days.

The toxicity of all ionic liquids was tested in an environmental and medicinally significant microbial assay including 12 fungal and 8 bacterial strains. [113,118] *In vitro* antifungal activities of the compounds were evaluated on a panel of four ATCC strains (*Candida albicans* ATCC 44859, *Candida albicans* ATCC 90028, *Candida parapsilosis* ATCC 22019, *Candida krusei* ATCC 6258) and eight clinical isolates of yeasts (*Candida krusei* E28, *Candida tropicalis* 156, *Candida glabrata* 20/I, *Candida lusitaniae* 2446/I, *Trichosporon asahii* 1188) and filamentous fungi (*Aspergillus fumigatus* 231, *Absidia corymbifera* 272, *Trichophyton mentagrophytes* 445). Whereas *In vitro* antibacterial activities of the compounds were evaluated on a panel of three ATCC strains (*Staphylococcus aureus* ATCC 6538, *Escherichia coli* ATCC 8739, *Pseudomonas aeruginosa* ATCC 9027) and five clinical isolates (*Staphylococcus aureus* MRSA HK5996/08, *Staphylococcus epidermidis* HK6966/08, *Enterococcus* sp. HK14365/08, *Klebsiella pneumoniae* HK11750/08, *Klebsiella pneumoniae* ESBL HK14368/08). All ester and amide side chain ionic liquids shown in Scheme 15 were non-toxic up to the 2000 μM concentration. This is less toxic than values reported for antimicrobial QAC (IC_{95} >10 μM). [80,81]

In order to evaluate the 'greenness' of the synthesis of ionic liquid catalysts, Gathergood and Connon have also applied some important Green Chemistry metrics (Section 4), such as -

http://dx.doi.org/10.5772/45605

- Sheldon E-factor
- GSK Reaction Mass Efficiency,
- Andraos Reaction Mass Efficiency
- Atom economy
- 1 / stoichiom. factor (excess reagents)

These metrics assisted improvements in the synthetic process, by reducing amount of solvents in the work-up and purification procedure, reduce number of steps to make required compounds. For example, as the synthesis of tetrafluoroborate ionic liquids involves preparation of halide salts followed by anion exchange metathesis. In an effort to reduce the number of steps and amount of reagents and solvents in the synthesis, alkyl imidazoles were directly reacted to Meerwein's salt i.e. trimethyloxonium tetrafluoroborate to give tetrafluoroborate ionic liquids in excellent yield. (For full analysis see ref. 117) Hence Green Chemistry metrics assessment is important to achieve green synthesis.

6. Conclusion

In this chapter we have demonstrated that ionic liquids have great potential and versatility in organic synthesis, with the dualistic ability to act as a solvent and as a catalyst. Ionic liquids were found to be possible replacements over traditional volatile organic solvents. Such ionic liquid solvents were found to be useful in transition metal catalysed reactions. They not only enabled catalyst immobilisation, but also increased recyclability of expensive transition metal catalysts. Ease of product separation and their stability against a variety of reagents has proved their important characteristics. We have seen that modification in the cationic part of the ionic liquids, according to the requirements of the aforementioned reactions, enabled them to act as organocatalysts or ligands for transition metal catalysts. These ionic liquid catalysts have shown comparative catalytic activity against known organocatalysts but such materials had a distinct advantage in that the ionic liquids could be recycled, with no discernible loss of activity. In order to increase recyclability, ionic liquid catalysts/ ligands were grafted onto a solid/ polymer support. These modifications helped to separate ionic liquid catalysts from the reaction mixture.

We have also illustrated the efforts attempted by the scientific community to evaluate the 'greenness' of ionic liquids, by using toxicity and biodegradation methods. The majority of ionic liquids are non-natural molecules, hence it's important to check their biocompatibility. Toxicity studies can serve as a "first post" primary evaluation of biodegradation. A variety of test systems including fungi, bacteria, algae, enzymes, rat cell line, human cell line and fish etc. were implemented to check the toxicity of ionic liquids. Most of these test systems have shown that toxicity of the ionic liquids comes from the cationic component. Important observations from such test systems also included that (a) long alkyl chains increase the toxicity with increase in the length of hydrocarbons and (b) incorporation of oxygen functionalities (ether, ester, and hydroxyl etc.) reduces the toxicity.

A number of Organisation for Economic Cooperation and Development (OECD) tests were found to be useful in the estimation of biodegradation. These tests mainly involve calculation of CO_2 evolution and oxygen consumed. These tests showed that most of the 1,3-dialkyl imidazolium ionic liquids are non-biodegradable, in most of the test systems. Although the alkyl side chain can undergo degradation, the imidazole core can still persist during biodegradation studies. Introduction of oxygen functionality such as ether / ester, either in the side chain of the imidazolium cation or at the C1 or C3 position of the pyridinium cation, were found to increase the rate of biodegradation.

In the case study, we have demonstrated the schematic approach towards the design and synthesis of ionic liquid catalysts for acetalisation reactions and evaluation of the biocompatibility of such catalysts, by using toxicity and biodegradation methods. According to the guidelines laid by the toxicity and biodegradation testing, ionic liquid catalysts were designed and prepared. Such ester and amide side chain imidazolium catalysts were found to be useful in acetalisation and thioacetalisation reactions. These catalysts have shown low toxicity against a variety of fungal and bacterial strains, but poor biodegradability in the "CO_2 Headspace Test". Further modifications in the cationic part are underway with the goal of increasing biodegradation and catalytic activity. Although biodegradation was not improved by such modifications, the catalytic activity was modified and increased to a great extent. Green Chemistry metrics had given useful information about the 'greenness' of the synthetic route. Which allowed to modify the process by reducing the number of steps, amount of solvents, etc. This case study was a clear example of the importance, in the design of organocatalysts, of what the potential environmental impact of such compounds might be, and why it is important to understand that a truly "green" catalyst needs to possess a balance of both activity and eco-friendliness.

Acknowledgements

The authors wish to thank Enterprise Ireland (EI), the Irish Research Council for Science, Engineering and Technology (IRCSET) and the Environmental Protection Agency (EPA) in Ireland for funding green chemistry research in Nicholas Gathergood's group. The case study was funded by EPA STRIVE project 2008-ET-MS-6-S2 (Rohitkumar Gore & Nicholas Gathergood (DCU) and Lauren Myles & Stephen Connon (TCD)). We also thank Alan Coughlan for assistance with proof-reading.

Author details

Rohitkumar G. Gore and Nicholas Gathergood*

*Address all correspondence to: Nick.Gathergood@dcu.ie

http://dx.doi.org/10.5772/45605

School of Chemical Sciences and National Institute for Cellular Biotechnology, Dublin City University, Glasnevin, Dublin, Ireland

References

[1] Rogers R. D. and Seddon K. R. (eds.), Ionic Liquids: Industrial Applications for Green Chemistry, ACS Symposium Series 818, American Chemical Society, USA, 2002

[2] Plechkova N. V. and Seddon K. R. Applications of ionic liquids in the chemical industries. Chemical Society Reviews, 2008; 37, 123-150

[3] Wasserscheid P. and Stark A. (eds.), Handbook of Green Chemistry, Volume 6: Ionic Liquids, Wiley-VCH, 2010

[4] Wilkes J. S. A short history of ionic liquids - from molten salts to neoteric solvents. Green Chemistry, 2002; 4, 73-80

[5] Seddon K. R. Ionic Liquids for Clean Technology. Journal of Chemical Technology & Biotechnology, 1997; 68, 351-356

[6] Marsh K. N., Boxall J. A., Lichtenthaler R. Room temperature ionic liquids and their mixtures - a review, Fluid Phase Equilibria, 2004; 219, 93-98

[7] McFarlane J., Ridenour W. B., Luo H., Hunt R. D., Depaoli D. W., Ren R. X. Room Temperature Ionic Liquids for Separating Organics from Produced Water. Separation Science and Technology, 2005; 40, 1245-1265

[8] Sheldon R. A., Green solvents for sustainable organic synthesis: state of the art. Green Chemistry, 2005; 7, 267-278

[9] Freemantle M. Designer solvents - Ionic liquids may boost clean technology development. Chemical English News 76, (30th March), 1998, 32-37

[10] Hallett J. P., Welton T. Room-temperature ionic liquids: Solvents for synthesis and catalysis. Chemical Reviews (Washington, DC, United States), 2011; 111(5), 3508-3576

[11] Hubbard C. D., Illner P., Eldik R. Understanding chemical reaction mechanisms in ionic liquids: successes and challenges. Chemical Society Reviews, 2011; 40, 272-290

[12] Wasserscheid P., Joni J. Green organic synthesis in ionic liquids, in Wasserscheid P., Stark A. (eds.), Handbook of Green Chemistry, Volume 6: Ionic Liquids, Wiley-VCH, Weinheim, Germany 2010; 41-63

[13] Chowdhury S. M., Ram S., Scott J. L. Reactivity of ionic liquids. Tetrahedron, 2007; 63(11), 2363-2389

[14] Stark A. and Seddon K. R. Ionic Liquids, in Kirk-Othmer Encyclopaedia of Chemical Technology, 5th Edit., Ed. A. Seidel, Vol. 26 (John Wiley & Sons, Inc., Hoboken, New Jersey) 2007; 836-920

[15] Opallo M., Lesniewski A. A review on electrodes modified with ionic liquids. Journal of Electroanalytical Chemistry, 2011; 656, 2-16

[16] Shiddiky M. J. A., Torriero A. A. J. Application of ionic liquids in electrochemical sensing systems. Biosensors & Bioelectronics, 2011; 26(5), 1775-1787

[17] Liu H., Liu Y., Li J., Ionic liquids in surface electrochemistry. Physical Chemistry Chemical Physics, 2010; 12(8), 1685-1697

[18] Pitner W. R., Kirsch P., Kawata K., Shinohara H., Applications of Ionic Liquids in Electrolyte Systems, in Wasserscheid P., Stark A. (eds.), Handbook of Green Chemistry, Volume 6: Ionic Liquids, Wiley-VCH, Weinheim, Germany 2010; 191-201

[19] Buzzeo M. C., Evans R. G., Compton R. G. Non-Haloaluminate Room-Temperature Ionic Liquids in Electrochemistry-A Review. ChemPhysChem, 2004; 5, 1106-1120

[20] Poole C. F., Poole S. K. Ionic liquid stationary phases for gas chromatography. Journal of Separation Science, 2011; 34(8), 888-900

[21] Ho T. D., Canestraro A. J., Anderson J. L. Ionic liquids in solid-phase microextraction: A review. Analytica Chimica Acta, 2011; 695, 18-43

[22] Sun P., Armstrong D. W. Ionic liquids in analytical chemistry. Analytica Chimica Acta, 2010; 661(1), 1-16

[23] Pandey S. Analytical applications of room-temperature ionic liquids: A review of recent efforts. Analytica Chimica Acta, 2006; 556, 38-45

[24] Koel M. Ionic liquids in chemical analysis. Critical Reviews in Analalytical Chemistry, 2005; 35, 177-192

[25] Dominguez de Maria P., Maugeri Z. Ionic Liquids in Biotransformations: From proof-of-concept to emerging deep-eutectic-solvents. Current Opinion in Chemical Biology, 2011; 15(2), 220-225

[26] Klingshirn M. A., Rogers R. D., Shaughnessy K. H. Palladium-catalyzed hydroesterification of styrene derivatives in the presence of ionic liquids. Journal of Organometallic Chemistry, 2005; 690, 3620-3626

[27] Mizushima E., Hayashi T., Tanaka M. Palladium-catalysed carbonylation of aryl halides in ionic liquid media: high catalyst stability and significant rate-enhancement in alkoxycarbonylation. Green Chemistry, 2001; 3, 76-79

[28] Earle M. J., McCormac P. B., Seddon K. R. Diels-Alder reactions in ionic liquids. A safe recyclable alternative to lithium perchlorate–diethyl ether mixtures. Green Chemistry, 1999; 1, 23-25

[29] Vijayaraghavan R., MacFarlane D. R. Charge Transfer Polymerization in Ionic Liquids. Australian Journal of Chemistry, 2004; 57, 129-133

[30] Rosa J. N., Afonso C. A. M., Santos A. G. Ionic liquids as a recyclable reaction medium for the Baylis-Hillman reaction. Tetrahedron, 2001; 57, 4189-4193

[31] Yadav J. S., Reddy B. V. S., Baishya G., Reddy K. V., Narsaiah A. V. Conjugate addition of indoles to α,β-unsaturated ketones using $Cu(OTf)_2$ immobilized in ionic liquids. Tetrahedron, 2005; 61, 9541-9544

[32] Johansson M., Linden A. A., Baeckvall J.-E. Osmium-catalyzed dihydroxylation of alkenes by H_2O_2 in room temperature ionic liquid co-catalyzed by $VO(acac)_2$ or $MeReO_3$. Journal of Organometallic Chemistry, 2005; 690, 3614-3619

[33] Serbanovic A., Branco L. C., Nunes da Ponte M., Afonso C. A. M. Osmium catalyzed asymmetric dihydroxylation of methyl trans-cinnamate in ionic liquids, followed by supercritical CO_2 product recovery. Journal of Organometallic Chemistry, 2005; 690, 3600-3608

[34] Picquet M., Stutzmann S., Tkatchenko I., Tommasi I., Zimmermann J., Wasserscheid P. Selective palladium-catalysed dimerisation of methyl acrylate in ionic liquids: towards a continuous process. Green Chemistry, 2003; 5, 153-162

[35] Forsyth S. A., Gunaratne H. Q. N., Hardacre C., McKeown A., Rooney D. W., Seddon K. R. Utilisation of ionic liquid solvents for the synthesis of Lily-of-the-Valley fragrance {β-Lilial®; 3-(4-t-butylphenyl)-2-methylpropanal}. Journal of Molecular Catalysis A: Chemical, 2005; 231, 61-66

[36] Reetz M. T., Wiesenhoefer W., Francio G., Leitner W. Biocatalysis in ionic liquids: batchwise and continuous flow processes using supercritical carbon dioxide as the mobile phase. Chemical Communications, 2002; 992-993

[37] Heck R. F. Palladium-catalyzed reactions of organic halides with olefins. Accounts of Chemical Research, 1979; 12, 146-151

[38] Mizoroki T., Mori K., Ozaki A. Arylation of Olefin with Aryl Iodide Catalyzed by Palladium. Bulletin of the Chemical Society of Japan, 1971; 44, 581

[39] Kaufmann D. E., Nouroozian M., Henze H. Molten Salts as an Efficient Medium for Palladium Catalyzed C-C Coupling Reactions. Synlett, 1996; 11, 1091-1092

[40] Bellina F. and Chiappe C. The Heck Reaction in Ionic Liquids: Progress and Challenges. Molecules, 2010; 15, 2211-2245

[41] Wang L., Li H., Li P. Task-specific ionic liquid as base, ligand and reaction medium for the palladium-catalyzed Heck reaction. Tetrahedron, 2009; 65, 364-368

[42] Xu L., Chen W., Xiao J. Heck Reaction in Ionic Liquids and the in Situ Identification of N-Heterocyclic Carbene Complexes of Palladium. Journal of Organometallic Chemistry, 2000; 19, 1123-1127

[43] Carmichael A. J., Earle M. J., Holbrey J. D., McCormac P. B., Seddon K. R. The Heck Reaction in Ionic Liquids: A Multiphasic Catalyst System. Organic Letters, 1999; 1(7), 997-1000

[44] Deshmukh R. R., Rajagopal R., Srinivasan K. V. Ultrasound promoted C-C bond formation: Heck reaction at ambient conditions in room temperature ionic liquids. Chemical Communications, 2001; 17, 1544-1545

[45] Crudden C. M., Sateesh M., Lewis R. Mercaptopropyl-Modified Mesoporous Silica: A Remarkable Support for the Preparation of a Reusable, Heterogeneous Palladium Catalyst for Coupling Reactions. Journal of American Chemical Society, 2005; 127, 10045-10050

[46] Ma X., Zhou Y., Zhang J., Zhu A., Jiang T., Han B. Solvent-free Heck reaction catalyzed by a recyclable Pd catalyst supported on SBA-15 via an ionic liquid. Green Chemistry, 2008; 10, 59-66

[47] Liu G., Hou M., Song J., Jiang T., Fan H., Zhang Z., Han B. Immobilization of Pd nanoparticles with functional ionic liquid grafted onto cross-linked polymer for solvent-free Heck reaction. Green Chemistry, 2010; 12, 65-69

[48] Sonogashira K., Tohda Y., Hagihara N. A convenient synthesis of acetylenes: catalytic substitutions of acetylenic hydrogen with bromoalkenes, iodoarenes and bromopyridines. Tetrahedron Letters, 1975; 16, 4467-4470

[49] Glaser C. Beiträge zur Kenntniss des Acetenylbenzols, Chemische Berichte, 1869; 2, 422-424

[50] Hay A. S., Oxidative Coupling of Acetylenes and Systems Containing a Silicon-Oxygen-Vanadium Linkage. Journal of Organic Chemistry, 1962; 27, 3320-3323

[51] Rossi R., Carpita A., Begelli C. A palladium-promoted route to 3-alkyl-4-(1-alkynyl)-hexa-1,5-dyn-3-enes and/or 1,3-diynes. Tetrahedron Letters, 1985; 26, 523-526

[52] Liu Q., Burton D. J. A facile synthesis of diynes. Tetrahedron Letters, 1997; 38, 4371-4374

[53] For a review of alkyne coupling, see: Siemsen P., Livingston R. C., Diederich F. Acetylenic Coupling: A Powerful Tool in Molecular Construction. Angewandte Chemie International Edition (English), 2000, 39, 2632-2657

[54] Fukuyama T., Shinmen M., Nishitani S., Sato M., Ryu I. A Copper-Free Sonogashira Coupling Reaction in Ionic Liquids and Its Application to a Microflow System for Efficient Catalyst Recycling, Organic Letters, 2002; 4(10), 1691-1694

[55] Zhang J., Dakovic M., Popovic Z., Wu H., Liu Y. A functionalized ionic liquid containing phosphine-ligated palladium complex for the Sonogashira reactions under aerobic and CuI-free conditions. Catalysis Communications, 2012; 17, 160-163

[56] Miyaura N., Yamada K., Suzuki A. A new stereospecific cross-coupling by the palladium-catalyzed reaction of 1-alkenylboranes with 1-alkenyl or 1-alkynyl halides. Tetrahedron Letters, 1979; 20(36), 3437-3440

http://dx.doi.org/10.5772/45605

[57] Miyaura N. and Suzuki A. Stereoselective Synthesis of Arylated (E) -Alkenes by the Reaction of Alk-1-enylboranes with Aryl Halides in the Presence of Palladium Catalyst. Journal of Chemical Society, Chemical Communications, 1979; 866-867

[58] Miyura N. and Suzuki A., Palladium-Catalyzed Cross-Coupling Reactions of Organoboron Compounds. Chemical Reviews, 1995; 95, 2457-2493

[59] Wei J., Jiao J., Feng J., Lv J., Zhang X., Shi X., Chen Z. PdEDTA Held in an Ionic Liquid Brush as a Highly Efficient and Reusable Catalyst for Suzuki Reactions in Water. Journal of Organic Chemistry, 2009; 74, 6283-6286

[60] Lombardo M., Chiarucci M., Trombini C. A recyclable triethylammonium ion-tagged diphenylphosphine palladium complex for the Suzuki–Miyaura reaction in ionic liquids. Green Chemistry, 2009; 11, 574-579

[61] Papagni A., Trombini C., Lombardo M., Bergantin S., Chams A., Chiarucci M., Miozzo L., Parravicini M. Cross-Coupling of 5,11-Dibromotetracene Catalyzed by a Triethylammonium Ion Tagged Diphenylphosphine Palladium Complex in Ionic Liquids. Organometallics, 2011; 30, 4325-4329

[62] Stille J. K. Angewandte Chemie, 1986; 98, 504-519

[63] Stille J. K. The Palladium-Catalyzed Cross-Coupling Reactions of Organotin Reagents with Organic Electrophiles [New Synthetic Methods (58)]. Angewandte Chemie International Edition (English), 1986; 25, 508-524

[64] Handy S. T. and Zhang X. Organic Synthesis in Ionic Liquids: The Stille Coupling. Organic Letters, 2001; 3(2), 233-236

[65] Louaisil N., Pham P. D., Boeda F., Faye D., Castanet A.-S., Legoupy S. Ionic Liquid Supported Organotin Reagents: Green Tools for Stille Cross-Coupling Reactions with Brominated Substrates. European Journal of Organic Chemistry, 2011; 1, 143-149

[66] Diels O., Alder K. Synthesen in der hydroaromatischen Reihe. Justus Liebigs Annalen der Chemie, 1928; 460, 98-122

[67] Pégot B., Vo-Thanh G. Ionic Liquid Promoted Aza-Diels-Alder Reaction of Danishefsky's Diene with Imines. Synlett, 2005; 9, 1409-1412

[68] Zhou Z., Li Z., Hao X., Dong X., Li X., Dai L., Liu Y., Zhang J., Huang H., Li X., Wang J. Recyclable copper catalysts based on imidazolium-tagged C2-symmetric bis(oxazoline) and their application in D-A reactions in ionic liquids. Green Chemistry, 2011; 13, 2963-2971

[69] Evans D. A., Miller S. J., Lectka T., von Matt P. Chiral Bis(oxazoline)copper(II) Complexes as Lewis Acid Catalysts for the Enantioselective Diels-Alder Reaction. Journal of American Chemical Society, 1999; 121, 7559-7573

[70] Greene T. W. Protective groups in Organic Synthesis, Wiley-Interscience, New York, 1981, p. 178

[71] Bornstein J., Bedell S. F., Drummond P. E., Kosoloki C. F. Alkylations of Diphenylacetonitrile with Certain Halides by Potassium Amide in Liquid Ammonia. Dehydrocyanations of Polyphenyl Nitriles to Form Olefins. Journal of American Chemical Society, 1956; 78, 83-86

[72] McKinzie C. A., Stocker J. H. Preparation of ketals. A Reaction mechanism. Journal of Organic Chemistry, 1955; 20, 1695-1701

[73] Cole A. C., Jensen J. L., Ntai I., Tran T., Weaver K. J., Forbes D. C., Davis Jr. J. H. Novel Brønsted Acidic Ionic Liquids and Their Use as Dual Solvent-Catalysts. Journal of American Chemical Society, 2002; 124, 5962-5963

[74] Fang D., Gong K., Shi Q., Liu Z. A green procedure for the protection of carbonyls catalyzed by novel task-specific room-temperature ionic liquid. Catalysis Communications, 2007; 8, 1463-1466

[75] Du Y. and Tian F. Brønsted Acidic Ionic Liquids as Efficient and Recyclable Catalysts for Protection of Carbonyls to Acetals and Ketals Under Mild Conditions. Synthetic Communications, 2005; 35, 2703-2708

[76] Anthony J. L., Maginn E. J., Brennecke J. F. Solution thermodynamics of imidazolium-based ionic liquids and water. Journal of Physical Chemistry B, 2001, 105, 10942-10949

[77] Wong D. S. H., Chen J. P., Chang J. M., Chou C. H. Experimental study on the transport properties of fluorinated ethers Fluid. Journal of Phase Equilibria and Diffusion, 2002; 194-197, 1089-1095

[78] Wood N. and Stephens G. Accelerating the discovery of biocompatible ionic liquids. Physical Chemistry Chemical Physics, 2010; 12, 1670-1674

[79] Alfonsi K., Colberg J., Dunn P. J., Fevig T., Jennings S., Johnson T. A., Kleine H. P., Knight C., Nagy M. A., Perry D. A., Stefaniak M., Green chemistry tools to influence a medicinal chemistry and research chemistry based organisation. Green Chemistry, 2008; 10, 31-36

[80] Matzke M., Stolte S., Thiele K., Juffernholz T., Arning J., Ranke J., Welz-Biermann U., Jastorff B. The influence of anion species on the toxicity of 1-alkyl-3-methylimidazolium ionic liquids observed in an (eco)toxicological test battery. Green Chemistry, 2007; 9, 1198-1207

[81] Pham T. P. T., Cho C.-W., Yun Y.-S. Environmental fate and toxicity of ionic liquids: A review. Water Research, 2010; 44, 352-372

[82] Stock F., Hoffmann J., Ranke J., Störmann R., Ondruschka B., Jastorff B. Effects of ionic liquids on the acetylcholinesterase - a structure–activity relationship consideration. Green Chemistry, 2004; 6, 286-290

[83] Couling D. J., Bernot R. J., Docherty K. M., Dixon J. K., Maginn E. J. Assessing the factors responsible for ionic liquid toxicity to aquatic organisms via quantitative structure–property relationship modeling. Green Chemistry, 2006; 8, 82-90

[84] Ranke J., Mölter K., Stock F., Bottin-Weber U., Poczobutt J., Hoffmann J., Ondruschka B., Filser J., Jastorff B. Biological effects of imidazolium ionic liquids with varying chain lengths in acute Vibrio fischeri and WST-1 cell viability assays. Ecotoxicology and Environmental Safety, 2004; 58, 396-404

[85] Stolte S., Arning J., Bottin-Weber U., Matzke M., Stock F., Thiele K., Uerdingen M., Welz-Biermann U., Jastorff B., Ranke J. Anion effects on the cytotoxicity of ionic liquids. Green Chemistry, 2006; 8, 621-629

[86] Bernot R. J., Brueseke M. A., Evans-White M. A., Lamberti G. A. Acute and chronic toxicity of imidazolium-based ionic liquids on Daphnia magna. Environmental Toxicology and Chemistry, 2005; 21, 87-92

[87] Yu M., Wang S.-H., Luo Y.-R., Han Y.-W., Li X.-Y., Zhang B.-J., Wang J.-J. Effects of the 1-alkyl-3-methylimidazolium bromide ionic liquids on the antioxidant defense system of Daphnia magna. Ecotoxicology and Environmental Safety, 2009; 72, 1798-1804

[88] Samorì C., Pasteris A., Galletti P., Tagliavini E. Acute toxicity of oxygenated and non-oxygenated imidazolium-based ionic liquids to Daphnia magna and Vibrio fischeri, Environmental Toxicology and Chemistry, 2007; 26, 2379-2382

[89] Morrissey S., Pegot B., Coleman D., Garcia M. T., Ferguson D., Quilty B., Gathergood N. Biodegradable, non-bactericidal oxygen-functionalised imidazolium esters: A step towards 'greener' ionic liquids. Green Chemistry, 2009; 11, 475-483

[90] Stepnowski P., Składanowski A. C., Ludwiczak A., Łaczyńska E. Evaluating the cytotoxicity of ionic liquids using human cell line HeLa. Human & Experimental Toxicology, 2004; 23, 513-517

[91] Wang X., Ohlin C. A., Lu Q., Fei Z., Hu J., Dyson P. J. Cytotoxicity of ionic liquids and precursor compounds towards human cell line HeLa. Green Chemistry, 2007; 9, 1191-1197

[92] Cho C.-W., Jeon Y.-C., Pham T. P. T., Vijayaraghavan K., Yun Y.-S. The ecotoxicity of ionic liquids and traditional organic solvents on microalga Selenastrum capricornutum. Ecotoxicology and Environmental Safety, 2008; 71, 166-171

[93] Cho C.-W., Jeon Y.-C., Pham T. P. T., Yun Y.-S. Influence of anions on the toxic effects of ionic liquids to a phytoplankton Selenastrum capricornutum. Green Chemistry, 2008; 10, 67-72

[94] Könnecker G., Regelmann J., Belanger S., Gamon K., Sedlak R. Environmental properties and aquatic hazard assessment of anionic surfactants: Physico-chemical, environmental fate and ecotoxicity properties. Ecotoxicology and Environmental Safety. 2011; 74, 1445-1460

[95] Dyer S. D., Lauth J. R., Morrall S. W., Herzog R. R., Cherry D. S. Development of a Chronic Toxicity Structure–Activity Relationship for Alkyl Sulfates. Environmental Toxicology and Water Quality. 1997; 12, 295-303

[96] Coleman D. and Gathergood N. Biodegradation studies of ionic liquids. Chemical Society Reviews, 2010; 39, 600-637

[97] Boethling R. S., Sommer E., DiFiore D. Designing Small Molecules for Biodegradability. Chemical Reviews, 2007; 107, 2207-2227

[98] Introduction to the OECD guidelines for testing of chemicals, Section 3, 2003

[99] Gathergood N. and Scammells P. J. Design and Preparation of Room-Temperature Ionic Liquids Containing Biodegradable Side Chains. Australian Journal of Chemistry, 2002; 55, 557-560

[100] Boethling R. S. Designing Biodegradable Chemicals. ACS Symposium Series, 1996; 640, 156-171

[101] Garcia M. T., Gathergood N., Scammells P. J. Biodegradable ionic liquids: Part I. Concept, preliminary targets and evaluation. Green Chemistry, 2004; 6, 166-175

[102] Docherty K. M., Dixon J. K., Kulpa Jr. C. F. Biodegradability of imidazolium and pyridinium ionic liquids by an activated sludge microbial community. Biodegradation, 2007; 18, 481-493

[103] Stolte S., Abdulkarim S., Arning J., Blomeyer-Nienstedt A., Bottin-Weber U., Matzke M., Ranke J., Jastorff B., Thoeming J. Primary biodegradation of ionic liquid cations, identification of degradation products of 1-methyl-3-octylimidazolium chloride and electrochemical wastewater treatment of poorly biodegradable compounds. Green Chemistry, 2008; 10, 214-224

[104] Harjani J. R., Singer R. D., Garcia M. T., Scammells P. J. Biodegradable pyridinium ionic liquids: design, synthesis and evaluation. Green Chemistry, 2009; 11, 83-90

[105] Atefi F., Garcia M. T., Singer R. D., Scammells P. J. Phosphonium ionic liquids: design, synthesis and evaluation of biodegradability. Green Chemistry, 2009; 11, 1595-1604

[106] Stolte S., Steudte S., Igartua A., Stepnowski P. The Biodegradation of Ionic Liquids - the View from a Chemical Structure Perspective. Current Organic Chemistry, 2011; 15, 1946-1973

[107] Anastas P. T., Warner J. C. Green Chemistry: Theory and Practice, Oxford University Press, Oxford, UK, 1998

[108] Heaton C. A. An Introduction to Industrial Chemistry, 3rd edn, Blackie Academic and Professional (Springer), London, UK, 1995

[109] Porteous A. Dictionary of Environmental Science and Technology, John Wiley & Sons Ltd, Chichester, UK, 1992

[110] Sheldon R. A. Chemistry & Industry (London), 1992; 903-906

[111] Constable D. J. C., Curzons A. D., Cunningham V. L. Metrics to 'green' chemistry - which are the best? Green Chemistry, 2002; 4, 521-527

[112] Trost B. M. The atom economy-a search for synthetic efficiency. Science, 1991; 254, 1471-1477

[113] Myles L., Gore R., Gathergood N., Connon S. J. Highly recyclable, imidazolium derived ionic liquids of low antimicrobial and antifungal toxicity: A new strategy for acid catalysis. Green Chemistry, 2010; 12, 1157-1162

[114] Procuranti B. and Connon S. J. A reductase-mimicking thiourea organocatalyst incorporating a covalently bound NADH analogue: efficient 1,2-diketone reduction with in situ prosthetic group generation and recycling. Chemical Communications, 2007; 1421-1423

[115] Procuranti B. and Connon S. J. Unexpected catalysis: aprotic pyridinium ions as active and recyclable Brønsted acid catalysts in protic media. Organic Letters, 2008; 10, 4935-4938

[116] Procuranti B., Myles L., Gathergood N., Connon S. J. Pyridinium Ion Catalysis of Carbonyl Protection Reactions. Synthesis, 2009; 23, 4082-4086

[117] Rohitkumar Gore, Ph.D. thesis, Dublin City University

Chapter 20

New Generations of Ionic Liquids Applied to Enzymatic Biocatalysis

Ana P.M. Tavares, Oscar Rodríguez and
Eugénia A. Macedo

Additional information is available at the end of the chapter

1. Introduction

Ionic liquids are salts in a liquid state, combinations of cations and anions that are liquid at temperatures below 100 °C. Thus, they have been called Room-Temperature Ionic Liquids (RTILs, or just ILs) in order to differentiate them from traditional salts, which melt at much higher temperatures and receive the name of "molten salts". In contrast to conventional organic solvents, ILs usually have extremely low volatility. Indeed, vapor pressures for ILs are scarce in the literature exactly because they are extremely low (< 1 Pa) and have to be obtained at high temperatures (400-500 K) [1]. For this "negligible" vapor pressure, ILs are often said to be "green" solvents when compared to traditional, environmentally harmful volatile organic compounds (VOCs). A big goal in the use of ILs in enzymatic reactions is the replacement of VOCs by ILs. In addition, ILs have other potential advantageous properties such as reasonable thermal stability; ability to dissolve a wide range of organic, inorganic and organometallic compounds; controlled miscibility with organic solvents (which is relevant for applications in biphasic systems) among others. All these properties make them very attractive non-aqueous solvents for biocatalysis. As they have been extensively described, ILs offer new possibilities for the application of solvent engineering to enzymatic reactions. Biocatalysis with ILs as reaction medium was first showed in the beginning of 2000 [2-4]. During the last decade, ILs have fast increased their attention as reaction media for enzymes with some remarkable results [2-4]. The advantage of using ILs in enzymatic biocatalysis, as compared to VOCs, is the enhancement in the solubility of substrates or products without inactivation of the enzymes, high conversion rates and high activity and stability [5]. ILs are also being used as co-solvents in aqueous biocatalytic reactions, since ILs help to dissolve nonpolar substrates, while avoiding enzyme inactivation like water-miscible organic solvents, as DMSO or acetonitrile, often do [6].

Another mentioned characteristic of ILs is the possibility of obtaining the desired physico-chemical properties by selecting combinations of cations and anions ("tunability"), which makes them "designer solvents". For example, ILs can be produced to be water-miscible, partially miscible or totally immiscible, and can also be synthesized with different viscosities. These interesting properties make them a very important reaction media for enzyme stabilization and reaction. The use of organic solvents in bioprocess presents a number of further problems. The main concerns are the toxicity of the organic solvents to both the process operators and the environment (eco-toxicity), and also the volatile and flammable nature of these solvents, which make them a potential explosion hazard [7]. Thus, ILs have emerged as a potential replacement for organic solvents in biocatalytic processes at both laboratory and industrial scale. The negligible vapor pressure means that they emit no volatile compounds, and also introduces the additional possibility of removal of products by distillation without further contamination by the solvent. It also facilitates the recycling of ILs, decreasing operation costs. All these properties make ILs very important for the stabilization and activation of enzymes; therefore, numerous enzymatic reactions have been investigated in different types of ILs as will be shown in the next sections. Several topics about biocatalysis in ILs will be reviewed: their effect on the activity and stability of enzymes, toxicity of ILs, new generation of ILs and methods to stabilize enzymes will be discussed.

2. Enzymatic activity and stability in ionic liquids

The most important criteria for selecting an enzyme-IL system are the activity and stability of the enzyme within the reaction medium. ILs have been reported to be an appropriated medium to increase the stability and activity of enzymes, as opposed to common organic solvents [8-11]. However, depending on the enzyme nature, the IL can be or not suitable for the reaction [12].

2.1. Lipases

Lipases, and *Candida Antarctica* Lipase B (CALB) in particular, are the most studied enzymes in ILs. Most of these reactions in ILs are carried out with, no or low content of water as co-solvent. Therefore, hydrophobic ILs are used and the enzyme activity and stability is dependent on the IL. Several studies in the literature show that enzymes, majority lipases, exhibit greater stability in pure ILs than in traditional organic solvents [13, 14]. A review of Zhao 2005 [15] shows that ILs with larger cations are better for enzyme activity than smaller cations. The reasons for that are the longer hydrophobic alkyl chains in the cation presents less tendency to take away the essential water molecules from the enzyme. In fact, one of the most interest conclusions of Zhao is that hydrophobic ILs maintain lipase activity and stability better than hydrophilic ILs, as the latter will take water molecules away from enzyme structure. According to Diego et al. [16] the enzyme stabilization by water immiscible ILs (such as $[(CF_3SO_2)_2N]^-$ types) can be explained by a more compact enzyme conformation/ confinement formed from the evolution of α-helix to β-sheet secondary structure of the enzyme. On the other hand, hydrophobic ILs may decrease the stability and activity of the enzyme due to: (1) the interaction with the substrates or products, as organic solvents [17]; (2) interaction by electrostatic forces [18] and (3) removing essential water molecules from the enzyme [17]. Lau et al. [19] observed that enzyme activity in ILs was related with the conformation of enzyme; the hydrogen bonding could be the key to understanding the interactions of enzymes and ILs. Another work of Lozano et al. [20] showed that lipase and α-chymotrypsin were strongly stabilized in two ILs ([btma][NTf_2] and [emim][NTf_2]) due to the maintenance of the native structure of the enzymes, as observed by both fluorescence and circular dichroism spectroscopy.

2.2. Cellulases

ILs are also used in the pretreatment of cellulose hydrolysis by cellulase for production of biofuels and other products. However, cellulases can be inactivated in the presence of ILs, even when present at low concentrations. In order to explore these ILs abilities, it is important to find a compatible cellulose-IL system [10]. The IL must solubilize the lignocellulosic biomass and at the same time, keep the enzyme active. It was shown that pretreatment of cellulose with ILs such as [bmim][Cl], [mmim][Cl], and [HEMA] resulted in faster conversion to glucose and thermostability than hydrolysis with cellulose that was not pretreated [22, 23]. A similar behavior was found for cellulases from different sources with imidazolium-based ILs, which enhanced the enzyme and thermal stability [24]. The stability of cellulases from *Penicillium janthinellum* mutants was evaluated in 10-50% (v/v) of [bmim][Cl] and the enzymes were significantly stable in 10% (v/v) of IL [25]. Another work investigated the stability and activity of commercial cellulases in aqueous solutions of 1-ethyl-3-methylimidazolium acetate [emim][OAc]. Cellulases retained 77% of their original activity in 15% and 20% (w/v) of IL and presented an avicel (a model substrate for cellulose) conversion efficiency of 91% [26].

2.3. Oxidoreductases

Several oxidoreductases, such as laccase, peroxidase, chloroperoxidase, D-amino acid oxidase and alcohol dehydrogenases, have been reported as active enzymes in aqueous solution with ILs [27]. When compared to organic solvents, these enzymes are more active and stable in the presence of ILs [27].

Laccases and peroxidases are the most effective enzymes capable to catalyze the degradation of phenolic compounds. Phenolics such as hydroquinone, catechols, guaiacol, and 2,6-dimethoxyphenol are good substrates for these enzymes in either aqueous and non-aqueous media. Recent reports have been addressing the activity and stability of both enzymes in ILs [6, 28-31]. For example, laccase activity and stability was well maintained in the presence of several imidazolium-based ILs [31] such as [C_4mim][Cl], [emim][$MDEGSO_4$], [emim][$EtSO_4$] and [emim][$MeSO_3$] [6], but are inactivated in the presence of [C_{10}mim][Cl] [29]. Peroxidase was also described to mantain its activitiy and stability in imidazolium-based ILs for concentrations up to 25 % v/v [30].

Alcohol dehydrogenases are enzymes that catalyze the reduction of ketones. Due to the vast field of aplication of alcohol dehydrogenases, the study of this enzyme in ILs is promising. A recent work presented the effect of 10 different ILs (with either imidazolium or ammonium cations) on the enzyme stability. Improved storage stabilities and improved enzyme activities were found in the most promising, ammonium-based, AMMOENG™ 101 IL [32]. Later, the same group [33] proved the feasibility of continuous production using the previously recommended IL, combined with product separation using a membrane bioreactor (the so-called process integration). Hussain and co-workers [34] showed that the use of 10% (v/v) [bmp][NTf_2] facilitated the conversion of ketone to the chiral alcohol. Dabirmanesh et al. [21], showed the influence of different imidazolium based ILs on the structure and stability of alcohol dehydrogenase and the results exhibited that the ILs could affected the enzyme stability, but not the tertiary structure, suggesting that the enzyme was reversibly inhibited.

There are only few reports investigating the enzyme activity of D-amino acid oxidase in ILs. This enzyme catalyzes the deamination of various d-amino acids into imino acids. The activity and stability of free and immobilized d-amino acid oxidase in five imidazolium ILs were evaluated, and the most promising ILs were [bmim][BF_4] and [mmim][$MMPO_4$]. Total conversion of substrate in presence of 20% [mmim][$MMPO_4$] was obtained [35].

3. Factors affecting enzymes in ionic liquids

The section before showed the stabilization and activation of enzymes in ILs. However, it is also very important to understand the factors affecting the enzymes activity and stability in IL media. It has been reported that enzyme reactions in ILs can be affected by several factors such as the water activity, pH, excipients and impurities [36]. Several properties of ILs have also been related to the activity and stability of enzymes. The most important include: polarity, hydrogen-bonding capacity, viscosity, kosmotropicity/chaotropicity and hydrophobicity, among others. It is clear from this set of properties that the type and strength of interactions ILs can establish with enzyme molecules will certainly influence their 3D structure. Such influence may produce or not changes in enzyme activity.

A few works have related the ILs polarity with the activity of enzymes. Lozano and co-workers [37] observed that in less polar IL, lower activities of α-chymotrypsin were obtained. The same behavior was obtained for lipase, the enzyme activity increased with the increase in IL polarity during the acetylation of racemic 1-phenylethanol with vinyl acetate [38] and for the synthesis of methylglucose fatty acid esters [39].

The negative effect of hydrogen-bonding on the enzyme activity in the presence of ILs can be associated with the anions effect and their action as hydrogen-bonding acceptors for the protein (lipase) [40]. Another work suggested a similar reasoning for the effect of anions: the decrease of lipase activity in [bmim][lactate] was caused by secondary structure changes of the protein, due to hydrogen-bonding interactions between lactate anions and peptide chains [19]. However, due to the limited number of ILs and enzymes investigated, deeper studies are required for a better understanding of this interaction.

As the majority of ILs are viscous fluids, the mass transfer limitations should be considered when the reaction is rapid and the IL is relatively viscous. Many enzymatic reactions in pure ILs can be heterogeneous due to the low solubility of the enzymes in ILs. Some studies have reported that the activity of enzymes is dependent on the IL viscosity: Bose et al. [23] attributed the lower activity of cellulase to the high IL ([HEMA]) viscosity. Lozano et al. [37] indicated that the activity of α-chymotrypsin was dependent on the IL viscosity, and thus higher enzyme activities were observed in less viscous ILs. On the other hand, the work of Zhao et al. [41] suggested that IL viscosity was not directly related to the lipase activity, but mass transfer limitations. The high viscosity may reduce the reaction rate, however the IL structure was responsible for lipase stabilization. So the author concludes that IL viscosity could influence the enzymatic reaction rates, however it is not the principal factor for the enzyme stabilization. Basso et al. [42] suggested that in the reactions for amide synthesis by immobilized penicillin G amidase, the high viscosities of the ILs did not affect the initial rates. Concluding, the effect of IL viscosity can affect the reaction rate, but this behavior is not the same for all enzymatic reactions in ILs, specially when reaction rates are measured in equilibrium instead of kinetics [42].

The kosmotropicity/chaotropicity (Hofmeister series) is related with the effect of water structure (and thus, protein salting in/out). There are reports in the literature that try to correlate the ion kosmotropicity with the enzyme behavior in aqueous solutions of ILs [43-48]. The reviews by Zhao et al. [15] and by Yang [49] discuss the probable mechanisms of Hofmeister effects of ILs. Kosmotropic anions (PO_4^{3-}, CO_3^{2-}, SO_4^{2-}, …) and chaotropic cations (Cs^+, Rb^+, K^+, NH^{4+}, …) stabilize enzymes, while chaotropic anions (NO_3^-, I^-, BF_4^-, PF_6^-,...) and kosmotropic cations ($(C_4H_9)N^+$, $(C_3H_7)_4N^+$, $(C_2H_5)_4N^+$,…) destabilize it [15].

Attending to the solubility of ILs in water, they can be divided into hydrophobic (water immiscible) and hydrophilic (water miscible). Most often, water miscibility depends on the ILs anions rather than the cations [50]. The hydrophobicity in ILs is generally determined by the log *P* scale, based on the partition coefficient of ILs between 1-octanol and water [51]. The stablility of enzymes can also be related to the log *P*. Usually, enzymes are more stable in solvents with a larger log *P* (>3) [106]. Many works from literature have reported that for lipases, activity increases with the increase in the IL hydrophobicity [13, 51-55]. Nevertheless, this conclusion is in contradiction with the polarity effect mentioned at the beggining of the section (more polar ILs promote enzyme stability). In our opinion, the vast pool of ILs and enzymes, and the large differences in their chemical structure, make it very difficult to extract general trends and conclusions. Just as an example, several authors [37] have proved that hydrophobic ILs (thus, less polar) maintain better immobilized lipase activity in the pure IL (low water content). At the same time, our group has shown that laccase and peroxidase activities are best maintained in more polar (less hydrophobic, more hydrophilic) ILs, when used in aqueous solutions [6; 29]. Both statements are correct, because the reaction conditions are completely different: very low water content in the former study, and water excess in the latter. Thus, the IL affinity for water will be dramatic at low water content, but not when there is plenty of water.

4. Green aspects of ionic liquids

The interest in the development of biocatalytic processes in ILs media is desired to obtain green technologies and unconventional properties to replace organic solvents (namely VOCs). ILs appear free of many problems associated with the use of VOCs due to their non-volatility, non-flammable character and both high thermal and chemical stability. However, the use of certain ILs raises some concerns regarding environmental impact, attending to their potential toxicity and biodegradability. As the use of ILs has been increasing in different fields from biology to electrochemistry, the assessment of their environmental, health and safety impact is highly required. In recent years, environmental aspects related to ILs have been strongly addressed, stating that many ILs commonly used cannot be regarded as 'green solvents'. In general, ILs used in biocatalysis have not been designed for biocompatibility and harmless. There are some recent reports showing that the ecotoxicity of alkylmethylimidazolium cations (the most used in biocatalysis) is undesirable, and ecotoxicity increases with the length of alkyl chains in cation [56-58]. Thus, for future applications it is necessary to improve the green aspects of ILs. These improvements are currently going on. The best examples are the choline-derived cations (which are based on food grade choline chloride) or imidazolium derivatives designed for biodegradability (*e.g.*, adding ether groups in the alkyl side chains) [59], and ILs based on amino acids [60-61]. It is expected that much improved and green ILs will become available soon. Currently, three different generations of ILs can be identified, as described below.

4.1. First generation of ionic liquids

The first IL known was ethylammonium nitrate, reported in 1914 by Walden [62], but attracted little interest. The first generation of ILs with widespread utilization was mainly composed of cations like dialkylimidazolium and alkylpyridinium derivatives, and anions like chloroaluminate and other metal halides which have been described as toxic and non-biodegradable [57]. The most common anions are chloroaluminate or other metal halide anions that react with water and thus are not suitable for biotransformations. This generation of ILs was also oxygen-sensitive [63] and can only be handled under inert-gas atmosphere due to the hygroscopic nature of $AlCl_3$ [64]. In the 1980s, Wilkes et al. started the extensive research on first generation ILs [65]. However, due to these limitations, the progress in their use was limited. For this reason, research was directed towards the synthesis of air- and water-stable ILs, the second generation of ILs.

4.2. Second generation of ionic liquids

After one decade the second generation of ILs [66] appeared. The water- and oxygen-reactive anions were replaced by halides (Cl^-, Br^-, I^-) or anions such as BF_4^-, PF_6^- and $C_6H_5CO_2$, which are stable to water and air. Cations such as dialkylimidazolium or alkylpyridinium were maintained, and ammonium and phosphonium were added. These ILs present interesting properties such as lower melting points, different solubilities in classic organic solvents, viscosities, etc. Due to these properties, the second generation attracted a great interest in various fields, and research in ILs experienced an important boost from the 1990's. The first reports of biocatalysis with ILs were published in the beginning of 2000's [2, 4, 38, 67]. One of the disadvantages of these ILs is the high cost. According to Gorke et al. [66], the high costs are related to starting materials (namely fluorinated components) and purification of final product required in the preparation. The most important disadvantage of the second generation is the toxicity, which in general is similar to those of chlorinated and aromatic solvents [56]. However, this second generation of ILs attracted the attention of the wide scientific community and has been providing interesting and novel applications in differents areas. This generation of ILs is the most studied and a great number of applications in biocatalysis have been published. The activity, stability, kinetic and thermal stability of different enzymes such as oxidases, lipases or cellulases has been studied, and sinthesis of various products has been carried out.

4.3. Third generation of ionic liquids

The third generation of ILs (advanced ILs) is based on more hydrophobic and stable anions such as [$(CF_3SO_2)_2N^-$], sugars, amino or organic acids, alkylsulfates, or alkylphosphates and cations such as choline. The cations and/or anions used are biodegradable, readily available, and present lower toxicities. Besides, a new class of solvent systems, called deep eutectic solvents (DES), is more hydrophilic than the second generation, and in general is water-miscible [66]. DES are mixtures of salts (in general they are not liquids at room temperature) such as choline chloride, and uncharged hydrogen bond donors such as amines, amides, alcohols, carboxylic acids, urea, or glycerol [28]. A typical example is the choline chloride/urea mixture, which produces a DES with a melting point of 12°C at concentrations around 50% [66]

The advantages of the third generation are: lower costs (similar to organic solvents), simple to prepare, biodegradable, do not require purification, the purity of the starting materials determines the final purity and uses anions and cations with low toxicity. As this generation is recent, few works have been published [68, 69]. The transesterification of ethyl valerate with 1-butanol, showed good activity in DES, and in choline chloride: glycine the activity was similar to activity in toluene for all lipases [69]. The third generation will reach the commercial level soon [70].

5. Methods for stabilization of enzymes in ionic liquids

Stabilization of enzymes in ILs is one of the keys for the development of more efficient biocatalytical processes for industrial, environmental, or biomedical applications. As discussed in previous sections, stabilization of enzymes in ILs is one of the keys for the development of more efficient biocatalytic processes for industrial, environmental, or biomedical applications. The use of enzymes in ILs presents different advantages when compared to conventional organic solvents. On the other hand, in some cases the application of enzymes can be limited by the low solubility, activity or stability in ILs. The improvement of enzyme functionality is crucial for large-scale applications in order to be economically viable. The methods to stabilize and activate enzymes in ILs can be divided into two different strategies: the modification of enzymes and/or the modification of the solvent (ILs). The modification of enzymes includes lyophilization (to change the morphology of the solid enzyme), chemical modification (for the chemical addition of functionalities into the enzyme biomolecule) and immobilization in a suitable support. The second strategy includes the modification of the IL reaction media, such as IL coating, additives or use of microemulsions with ILs. These methodologies have been used with promissory results [5].

5.1. Modification of solvent media

In order to avoid the enzyme insolubility, some works have reported the introduction of functional groups in IL structure such as hydroxyl, ether, and amide (which present high affinity for enzymes) [19]. For enzymes that are active in pure solvents, such as lipases, the most hydrophilic ILs can remove enzyme-bound water molecules that are essential to maintain protein structure and active function. In such case, these ILs (hydrophilic) are not adequate.

Another strategy is the addition of water in IL (co-solvent), but the enzyme may present low catalytic activity due to a changed conformation in ILs [71]. Several researchers have reported enzymatic reactions, especially for oxidative enzymes, in hydrophilic ILs with a high concentration of water (in the range 5 – 50%) and promissory results have been found [6, 29-31].

Water-in-IL microemulsions, or reverse micelles, have been used as a very efficient technique for solubilizing enzymes in hydrophobic ILs. The advantage of this approach is that the enzyme is protected of the contact with the solvent by a layer of water and surfactant molecules. As an example, the use of water-in-IL microemulsions was reported by Moniruzzaman et al. [50] as a new medium for dissolving various enzymes and proteins. Additionally, several authors have reported the use of different microemulsions systems with good results for enzyme stability [72-74].

5.2. Modification of the enzyme

The most common methodology for enzyme modification is immobilization. It is well known that immobilization of enzymes presents excellent advantages for biocatalysis, namely in the recovery of the enzyme for reutilization, product separation and recovery from the reaction media, application in continuous systems, and for enzyme stabilization. Indeed, enzyme immobilization increases thermal and operational stability of the biocatalysts compared to the free enzyme.

The use of immobilized enzymes in IL media has been reported by many research groups, using different methods of immobilization and supports. The most frequently used enzyme immobilization techniques are: physical adsorption, covalent attachment, entrapment in polymeric matrixes and cross-linking of enzyme molecules. For lipases, it was found that reaction rates in ILs were comparable or higher than in organic solvents and also immobilized lipase was more active than its free form [75-79]. The same behavior was found for proteinase [80], papain [81] and for heme-containing proteins [82].

The chemical modification of enzymes with poly(ethylene glycol) (PEG) is a well-known method (the so-called PEGylation) for enzyme stabilization in denaturing environments. PEG presents both hydrophilic and hydrophobic properties, so the modified enzymes can increase their solubility in some ILs [83]. Turner et al. [84] also reported higher activity of PEGylated cellulase than free cellulase in IL solutions.

Another method for activating and stabilizing enzymes in non-aqueous media is co-lyophilization of the enzyme. Maruyama et al. [85] lyophilized lipase with poly(ethylene glycol) (PEG) to prepare PEG–lipase complexes, finding that the activity of lipase in ILs increased more than 14-fold. Wang and Mei [86] also lyophilized lipase with cyclodextrins, and the activity of lipase in ILs ([bmim][PF_6] and [bmim][BF_4]) was improved.

6. Applications of ionic liquids in biocatalysis

The use of ILs as solvents or co-solvents for reaction media of enzymes is well recognized in biocatalysis. Examples available in the open literature include: polymerizations, biosensors, production of biofuels, synthesis of sugar- and ester-derivatives, among many others.

A large number of examples of the use of ILs for the enzymatic production of esters by lipase have been published [87]. The common esters synthetized in ILs are aliphatic and aromatic esters, for applications in polymers, biodiesel, and in the perfume, flavour and pharmaceutical industries. The synthesis of a wide range of aliphatic organic esters was carried out by transesterification from vinyl esters and alcohols and catalyzed by lipase in different 1,3-dialkylimidazolium ILs [88, 89]. Aromatic esters have also been synthetized with lipase in two ILs, [bmim][PF_6] and [bmim][BF_4] [90]. The esterification of 2-substituted-propanoic acids with 1-butanol was catalyzed by lipase in ILs [bmim][PF_6] and [omim][PF_6] [91]. Yuan et al. [92] studied the enantioselective esterification of menthol with propionic anhydride using lipase in [bmim][PF_6] and [bmim][BF_4]. The resolution of (R,S)-ibuprofen by esterification with lipases in the same ILs is another interesting example [93]. The aliphatic polyester synthesis by lipase, also in [bmim][PF_6], was reported by Nara et al. [94]. Later, the enzymatic preparation of polyesters by ring-opening polymerization and by polycondensation with lipase in [bmim][Tf_2N], [bmim][PF_6] and [bmim][BF_4] was also investigated [95]. According to these authors, the use of ILs could be an advantage in the polymerization of highly polar monomers with low solubility in organic solvents.

The production of biofuels, such as biodiesel (fatty acid methyl esters) has been also investigated in ILs through the transesterification of a triglyceride with methanol. Biodiesel is a renewable and environmentally-friendly fuel. Several ILs have been utilized for biodiesel production. Most often, the synthesis of biodiesel by enzymatic reactions in ILs is based on a short-chain 1,3-dialkylimidazolium cation, such as [bmim][PF_6] or [bmim][NTf_2], and the reaction is carried out in a biphasic system with lipase and using an adequate substrate (e.g., soybean oil) [96]. For homogeneous one-phase systems, imidazolium ILs with long alkyl chains such as [C_{16}mim][NTf_2] and [C_{18}mim][NTf_2] have been used [97,98]. These long chain, lipophilic ILs create a nonaqueous system suitable for oil transesterification. Ha et al. [99] studied the biodiesel production using immobilized lipase in 23 ILs. Among the ILs tested, it was found that highest biodiesel production yield was obtained in [emim][TfO]. But it is important to highlight that several works have been published for biodiesel production by lipases [100-102].

In recent years, a significant number of publications have showed the direct electron-transfer reaction between redox proteins or enzymes and IL-based composite electrodes. Biosensors are small devices which convert the biological recognition event into an electrical signal, so it can be used for selective analysis [103]. Several composite electrodes based on ILs have been prepared. Many of them can be found in a recent review by Shiddiky and Torriero[104], such as: hemoglobin biosensor; myoglobin and cytochrome c biosensors; catalase biosensors; glucose oxidase biosensors; horseradish peroxidase biosensors.

Sugar-based compounds are widely used in pharmaceuticals, cosmetics, detergents and food. A recent review by Galonde et al. [105] shows the synthesis of glycosylated compounds in ILs.

7. Conclusion

Ionic liquids have demonstrated to be suitable solvents for enzymatic reactions. They can be beneficial regarding to activity, (enantio)selectivity and stability of enzymes. The use of enzymes in ILs opens new possibilities for non-aqueous enzymology with high efficiency in several areas. Here, it was shown that a large variety of enzymes tolerate ILs or aqueous-IL mixtures as reaction medium. Moreover, the development of green and biodegradable ILs is reinforcing enzymatic applications of ILs, as stated in this work. Indeed, it is expected to become a standard in biotransformations, thus contributing to a greener chemical and biochemical industries.

Nomenclature

tris-(2-hydroxyethyl)-methylammonium methylsulfate [HEMA]

Cations:

[mmim] = 1,3-dimethylimidazolium

[emim] = 1-ethyl-3-methylimidazolium

[bmim] or [C_4mim] = 1-butyl-3-methylimidazolium

[bmp] = butylmethylpyrrolidinium

[btma] = butyl-trimethylammonium

[omim] = 1-octyl-3-methylimidazolium

[C_{10}mim] = 1-decyl-3-methylimidazolium

[C_{16}mim] = 1-hexadecyl-3-methylimidazolium

[C_{18}mim] = 1-octadecyl-3-methyl- imidazolium

Anions

[OAc] = acetate

[$MDEGSO_4$] = 1-ethyl-3-methylimidazolium 2-(2-methoxyethoxy)

[$MeSO_3$] = methanesulfonate

[TfO] = trifluoromethanesulfonate

[BF_4] = tetrafluoroborate

[$MMPO_4$] = dimethylphosphate

[Cl] = chloride

[$EtSO_4$] = ethyl sulfate

[$(CF_3SO_2)_2N^-$] = bis(trifluoromethylsulfonyl)amide

[$MeSO_4$] = methyl sulfate

[PF_6] = hexafluorophosphate

[NTf_2] = bis(trifluoromethylsulfonyl)imide

Acknowledgements

This work was supported by project PEst-C/EQB/LA0020/2011, financed by FEDER through COMPETE - Programa Operacional Factores de Competitividade and by Fundação para a

Ciência e a Tecnologia (FCT, Portugal). A.P.M. Tavares and O. Rodríguez acknowledge the financial support (Programme Ciência 2008 and Programme Ciência 2007, respectively) from FCT.

Author details

Ana P.M. Tavares*, Oscar Rodríguez and Eugénia A. Macedo

LSRE - Laboratory of Separation and Reaction Engineering - Associate Laboratory LSRE/LCM, Faculdade de Engenharia, Universidade do Porto, Porto, Portugal

References

[1] Esperança JMSS, Canongia JNL, Tariq M, Santos LMNBF, Magee JW, Rebelo LPN. Volatility of Aprotic Ionic Liquids A Review. Journal of Chemical Engineering Data 2010;55 3–12.

[2] Cull SG, Holbrey JD, Vargas-Mora V, Seddon KR, and Lye GJ. Room-temperature ionic liquids as replacements for organic solvents in multiphase bioprocess operations. Biotechnology and Bioengineering 2000;69(2) 227-233.

[3] Erbeldinger M, Mesiano AJ, Russell AJ. Enzymatic catalysis of formation of Z-aspartame in ionic liquid - an alternative to enzymatic catalysis in organic solvents. Biotechnology Progress 2000;16 1129-1131.

[4] Lau RM, van Rantwijk F, Seddon KR, Sheldon RA. Lipase-catalyzed reactions in ionic liquids. Organic Letters 2000;2 4189-4191.

[5] Moniruzzaman M, Kamiya N, Nakashima K, Goto M. Water-inionic liquid microemulsions as a new medium for enzymatic reactions. Green Chemistry 2008;10 497–500.

[6] Tavares APM, Rodriguez O, Macedo EA. Ionic liquids as alternative co-solvents for laccase: Study of enzyme activity and stability. Biotechnology and Bioengineering 2008;101 201-207.

[7] Schmid A, Kollmer A, Mathys RG, Withot B. Development toward large-scale bacterial bioprocesses in the presence of bulk amounts of organic solvents. Extremophiles 1998;2 249–256.

[8] Datta S, Holmes B, Park JI, Chen Z, Dibble DC, Hadi M, Blanch HW, Simmons BA, Sapra R. Ionic liquid tolerant hyperthermophilic cellulases for biomass pretreatment and hydrolysis. Green Chemistry 2010;12(2) 338–345.

[9] Lozano P, Diego TD, Carrie D, Vaultier M, Iborra JL. Continuous green biocatalytic processes using ionic liquids and supercritical carbon dioxide. Chemical Communication 2002;7: 692–693.

[10] Wang Y, Radosevich M, Hayes D, Labbe N. Compatible ionic liquidcellulases system for hydrolysis of lignocellulosic biomass. Biotechnology and Bioengineering 2011;108(5) 1042–1048.

[11] Persson M, Bornscheuer UT. Increased stability of an esterase from Bacillus stearothermophilus in ionic liquids as compared to organic solvents. Journal of Molecular Catalysis B: Enzymatic 2003;22 21–27.

[12] Mantarosie L, Coman S, Parvulescu VI. Comparative behavior of various lipases in benign water and ionic liquids solvents. Journal of Molecular Catalysis A: Chemical 2008;279 223–229.

[13] Kaar JL, Jesionowski AM, Berberich JA, Moulton R, Russell AJ. Impact of ionic liquid physical properties on lipase activity and stability. Journal of the American Chemical Society 2003;125 4125–4131.

[14] Shan H, Li Z, Li M, Ren G, Fang Y. Improved activity and stability of pseudomonas capaci lipase in a novel biocompatible ionic liquid, 1-isobutyl-3-methylimidazolium hexafluorophosphate. Journal of Chemical Technology and Biotechnology 2008;83 886–891.

[15] Zhao H. Effect of ions and other compatible solutes on enzyme activity, and its implication for biocatalysis using ionic liquids. Journal of Molecular Catalysis B: Enzymatic 2005;37 16–25.

[16] De Diego T, Lozano P, Gmouh S, Vaultier M, Iborra JL. Understanding structure-stability relationships of Candida antartica Lipase B in ionic liquids. Biomacromolecules 2005;6 1457-1464.

[17] Yang Z, Russell AJ in: Koskinen AMP, Kilbanov AM (Ed) Enzymatic Reactions in Organic Media. Blackie Academic & Professional, New York; 1996. p43.

[18] Park C, Raines RT. Quantitative Analysis of the Effect of Salt Concentration on Enzymatic Catalysis. Journal of the American Chemical Society. 2001;123 11472-11479.

[19] Lau RM, Sorgedrager MJ, Carrea G, van Rantwijk F, Secundo F, Sheldon RA. Dissolution of Candida antarctica lipase B in ionic liquids: effects on structure and activity. Green Chemistry 2004;6 483–487.

[20] Lozano P, De Diego T, Gmouh S, Vaultier M, Iborra JL. Dynamic structure_/function relationships in enzyme stabilization by ionic liquids. Biocatalysis and Biotransformation 2005;23 169-176.

[21] Dabirmanesh B, Khajeh K, Ranjbar B, Ghazi F, Heydari A. Inhibition mediated stabilization effect of imidazolium based ionic liquids on alcohol dehydrogenase. Journal of Molecular Liquids 2012;170 66–71.

[22] Bose S, Armstrong DW, Petrich JW. Enzyme-catalyzed hydrolysis of cellulose in ionic liquids: A green approach toward the production of biofuels. The Journal of Physical Chemistry B 2010;114 8221–8227.

[23] Bose S, Barnes CA, Petrich JW. Enhanced stability and activity of cellulase in an ionic liquid and the effect of pretreatment on cellulose hydrolysis. Biotechnology and Bioengineering 2012;109 434-443.

[24] Ilmberger N, Meske D, Juergensen J, Schulte M, Barthen P, Rabausch U, Angelov A, Mientus M, Liebl W, Schmitz RA, Streit WR. Metagenomic cellulases highly tolerant towards the presence of ionic liquids—linking thermostability and halotolerance. Applied Microbiology and Biotechnology 2012;95 135–146.

[25] Adsul MG, Terwadkar AP, Varma, AJ, Gokhale DV. Cellulases from penicillium janthinellum mutants: Solid-state production and their stability in ionic liquids. BioResources 2009;4 1670-1681.

[26] Wang Y, Radosevich M, Hayes D, Nicole L. Compatible Ionic liquid-cellulases system for hydrolysis of lignocellulosic biomass. Biotechnology and Bioengineering 2011;108 1042-1048.

[27] Tavares APM, Rodriguez O, Raquel Cristóvão, Macedo EA. Ionic Liquids: Alternative Reactive Media for Oxidative Enzymes. In: Kokorin A (Ed) Ionic Liquids: Applications and Perspectives. Rijeka: InTech; 2011 p449-516.

[28] Domínguez de María P. Ionic Liquids in Biotransformations and Organocatalysis: Solvents and Beyond. John Wiley & Sons; 2012.

[29] Rodríguez O, Cristóvão RO, Tavares APM, Macedo EA. Effect of the alkyl chain length on enzymatic kinetic with imidazolium ionic liquids. Applied Biochemistry and Biotechnology 2011;164 524-533.

[30] Carneiro AP, Rodríguez O, Mota FL, Tavares APM, Macedo EA. Kinetic and stability study of the peroxidase inhibition in ionic liquids. Industrial & Engineering Chemistry Research 2009; 10810-10815.

[31] Domínguez A, Rodríguez O, Tavares APM, Macedo EA, Longo MA, Ma. Sanromán A. Studies of laccase from Trametes versicolor in aqueous solutions of several methylimidazolium ionic liquids. Bioresource Technology 2011;102 7494–7499.

[32] Kohlmann C, Robertz N, Leuchs S, Dogan Z, Lütz S, Bitzer K, Naamnieh S, Greiner L. Ionic liquid facilitates biocatalytic conversion of hardly water soluble ketones. Journal of Molecular Catalysis B: Enzymatic 2011;68 147–153.

[33] Kohlmann C, Leuchs S, GreinerL. Walter Leitner Continuous biocatalytic synthesis of (R)-2-octanol with integrated product separation. Green Chemistry 2011;13 1430-1436.

[34] Hussain W, Pollard DJ, Truppo M, Lye GJ. Enzymatic ketone reductions with co-factor recycling: Improved reactions with ionic liquid co-solvents. Journal of Molecular Catalysis B: Enzymatic 2008;55 19–29.

[35] Lutz-Wahl S, Trost EM, Wagner B, Manns A, Fischer L. Performance of d-amino acid oxidase in presence of ionic liquids. Journal of Biotechnology 2006;124 163–171.

[36] Yang Z, Pan W. Ionic liquids: green solvents for nonaqueous biocatalysis. Enzyme Microbiology and Technology 2005;37 19–28.

[37] Lozano P, de Diego T, Guegan J-P, Vaultier M, Iborra JL. Stabilization of α-chymotrypsin by ionic liquids in transesterification reactions. Biotechnology and Bioengineering 2001;75 563–569.

[38] Park S, Kazlauskas RJ. Improved preparation and use of room-temperature ionic liquids in lipasecatalyzed enantio- and regioselective acylations. Journal of Organic Chemistry 2001;66 8395-8401.

[39] Mutschler J, Rausis T, Bourgeois J-M, Bastian C, Zufferey D, Mohrenz. Ionic liquid-coated immobilized lipase for the synthesis of methylglucose fatty acid esters. Green Chemistry 2009; 1793–1800.

[40] Ventura SPM, Santos LDF, Saraiva JA, Coutinho JAP. Concentration effect of hydrophilic ionic liquids on the enzymatic activity of Candida antarctica lipase B. World Journal of Microbiology and Biotechnology 2012;28 2303–2310.

[41] Zhao H, Baker GA, Song Z, Olubajo O, Zanders L, Campbell SM. Effect of ionic liquid properties on lipase stabilization under microwave irradiation. Journal of Molecular Catalysis B: Enzymatic 2009;57: 149–157.

[42] Basso A, Cantone S, Linda P, Ebert C. Stability and activity of immobilised penicillin G amidase in ionic liquids at controlled. Green Chemistry 2005;7 671-676.

[43] Zhao H, Olubajo O, Song Z, SimsAL , Person TE, Lawal RA, Holley LA. Effect of kosmotropicity of ionic liquids on the enzyme stability in aqueous solutions. Bioorganic Chemistry 2006;34 15–25.

[44] Fujita K, MacFarlane DR, Forsyth M, Yoshizawa-Fujita M, Murata K, Nakamura N. Solubility and stability of cytochrome c in hydrated ionic liquids: effect of oxo acid residues and kosmotropicity. Biomacromolecules 2007;8 2080–2086.

[45] Constantinescu D, Weingartner H, Herrmann C, Protein denaturation by ionic liquids and the Hofmeister series: a case study of aqueous solutions of ribonuclease A. Angewandte Chemie International Edition 2007;6 8887–8889.

[46] Kaftzik N, Wasserscheid P, Kragl U. Use of ionic liquids to increase the yield and enzyme stability in the β-galactosidase catalyzed synthesis of N-acetyllactosamine. Organic Process Research and Development 2002;6 553–557.

[47] Lang M, Kamrat T, Nidetzky B. Influence of ionic liquid cosolvent on transgalactosylation reactions catalyzed by thermostable β-glycosylhydrolase CelB from Pyrococcus furiosus. Biotechnology and Bioengineering 2006;95 1093–1100.

[48] Zhao H, Song Z. Nuclear magnetic relaxation of water in ionic-liquid solutions: Determining the kosmotropicity of ionic liquids and its relationship with the enzyme enantioselectivity. Journal of Chemical Technology and Biotechnology 2007;82 304-312.

[49] Yang Z. Hofmeister effects: an explanation for the impact of ionic liquids on biocatalysis. Journal of Biotechnology 2009;144 12–22.

[50] Moniruzzaman M, Nakashima K, Kamiya N, Goto M. Recent advances of enzymatic reactions in ionic liquids. Biochemical Engineering Journal 2010;48 295-314.

[51] Zhao H, Baker GA, Song Z, Olubajo O, Zanders L, Campbell SM. Effect of ionic liquid properties on lipase stabilization under microwave irradiation. Journal of Molecular Catalysis B: Enzymatic 2009;57 149–157.

[52] de los Rios AP, Hernandez-Fernandez FJ, Martinez FA, Rubio M, Villora G. The effect of ionic liquid media on activity, selectivity and stability of Candida antarctica lipase B in transesterification reactions. Biocatalysis Biotransformation 2007;25 151–156.

[53] Hernandez-Fernandez FJ, de los Rios AP, Tomas-Alonso F, Gomez Dand Villora G. Stability of hydrolase enzymes in ionic liquids. The Canadian Journal of Chemical Engineering 2009;87 910–914.

[54] Nara SJ, Harjani JR and Salunkhe MM. Lipase-catalysed transesterification in ionic liquids and organic solvents: a comparative study. Tetrahedron Letter 2002;43 2979–2982.

[55] Shen Z-L, Zhou W-J, Liu Y-T, Ji S-J, Loh T-P. One-potchemo enzymatic syntheses of enantiomerically-enrichedO-acetylcyanohydrins from aldehydes in ionic liquid. Green Chemistry 2008;10 283–286.

[56] Docherty KM, Kulpa CF. Toxicity and antimicrobial activity of imidazolium and pyridinium ionic liquids. Green Chemistry 2005;7 185-189.

[57] Wells AS; Coombe VT. On the freshwater ecotoxicity and biodegradation properties of some common ionic liquids. Organic Process Research and Development 2006;10 794-798.

[58] Stolte S, Arning J, Bottin-Weber U, Matzke M, Stock F, Thiele K, Uerdingen M, Welz-Biermann U, Jastorff B, Ranke J. Anion effects on the cytotoxicity of ionic liquids. Green Chemistry 2006;8 621-629.

[59] Gathergood N, Scammels PJ, Garcia MT. Biodegradable ionic liquids. Part III. The first readily biodegradable ionic liquids. Green Chemistry 2006;8 156-160.

[60] Fukumotu K, Yoshizawa M, Ohno H. Room Temperature Ionic Liquids from 20 Natural Amino Acids. Journal of the American Chemical Society 2005;127 2398-2399.

[61] Tao G, He L, Liu W, Xu L, Xiong W, Wang T, Kou Y. Preparation, characterization and application of amino acid-based green ionic liquids. Green Chemistry 2006;8 639-646.

[62] Walden P. Molecular weights and electrical conductivity of several fused salts. Bulletin of the Imperial Academy of Sciences (St. Petersburg) 1914;1800 405-422.

[63] Endres F, Abedinw SZ. Air and water stable ionic liquids in physical chemistry. Physical Chemistry Chemical Physics 2006,8 2101–2116.

[64] Moustafa EM, Abedin SZ, Shkurankov A, Zschippang E, Saad AY, Bund A, Endres F. Electrodeposition of Al in 1-Butyl-1-methylpyrrolidinium Bis(trifluoromethylsulfonyl) amide and 1-Ethyl-3-methylimidazolium Bis(trifluoromethylsulfonyl)amide Ionic Liquids: In Situ STM and EQCM Studies. The Journal of Physical Chemistry B 2007;111 4693-4704.

[65] Wilkes JS, Levisky JA, Wilson RA, Hussey CL. Dialkylimidazolium chloroaluminate melts: a new class of room-temperature ionic liquids for electrochemistry, spectroscopy and synthesis. Inorganic Chemistry 1982;21 1263-1264.

[66] Gorke J, Srienc F, Kazlauskas RJ. Toward advanced ionic liquids. Polar, enzyme-friendly solvents for biocatalysis. Biotechnology Bioprocess Engineering 2010;15 40-53.

[67] Itoh T, Akasaki E, Kudo K, Shirakami S. Lipase-catalyzed enantioselective acylation in the ionic liquid solvent system: reaction of enzyme anchored to the solvent. Chemical Letter 2001;30 262-263.

[68] Lindberg D, de la Fuente Revenga M, Widersten M. Deep eutectic solvents (DESs) are viable cosolvents for enzyme-catalyzed expoxide hydrolysis. Journal of Biotechnology 2010;147 169-171.

[69] Gorke J, Srienc F, Kazlauskas RJ. Hydrolasecatalyzed biotransformations in deep euthetic solvents. Chemical Communication 2008 1235-1237.

[70] Domínguez PM, Maugeri Z. Ionic liquids in biotransformations: from proof-of-concept to emerging deep-eutectic-solvents. Current Opinion in Chemical Biology 2011;15 220–225.

[71] Eckstein M, Sesing M, Kragl U, Adlercreutz P. At low water activity α-chymotrypsin is more active in an ionic liquid than in non-ionic organic solvents. Biotechnology Letters 2002;24 867–872.

[72] Zhou G-P, Zhang Y, Huang X-R, Shi C-H, Liu W-F, Li Y-Z. Catalytic activities of fungal oxidases in hydrophobic ionic liquid 1-butyl-3- methylimidazolium hexafluorophosphate-based microemulsion. Colloids and Surfaces B: Biointerfaces 2008;66 146–149.

[73] Zhang Y, Huang X, Li Y. Negative effect of [bmim][PF6] on the catalytic activity of alcohol dehydrogenase: mechanism and prevention. Journal of Chemical Technology and Biotechnology 2008;83 1230–1235.

[74] Pavlidis IV, Gournis D, Papadopoulos GK, Stamatis H. Lipases in water-in-ionic liquid microemulsions. Structural and activity studies. Journal of Molecular Catalysis B: Enzymatic 2009;60 50–56.

[75] Toral AR, de los Rios AP, Hernandez FJ, Janssen MHA, Schoevaart R, van Rantwijk F. Cross-linked Candida antarctica lipase B is active in denaturing ionic liquids. Enzyme Microbiology and Technology 2007;40 1095–1099.

[76] Shah S, Solanki K, Gupta MN. Enhancement of lipase activity in non-aqueous media upon immobilization on multi-walled carbon nanotubes. Chemistry Central Journal 2007 1-30.

[77] Lee SH, Dang DT, Ha SH, Chang WJ, Koo YM. Lipase-catalyzed synthesis of fatty acid sugar ester using extremely supersaturated sugar solution in ionic liquids. Biotechnology and Bioengineering 2008;99 1-8.

[78] Schofer SH, Kaftzik N, Wasserscheid P, Kragl U. Enzyme catalysis in ionic liquids: lipase catalysed kinetic resolution of 1-phenylethanol with improved enantioselectivity. Chemical Communications 2001 425-426.

[79] Zhao H, Jones CL, Cowins JV. Lipase dissolution and stabilization in ether-functio nalized ionic liquids Green Chemistry 2009;11 1128-1138.

[80] Eker B, Asuri P, Murugesan S, Linhardt RJ, Dordick JS. Enzyme carbon nanotube conjugates in room-temperature ionic liquids. Applied Biochemistry and Biotechnology 2007;143 153–163.

[81] Bian W, Yan B, Shi N, Qiu F, Lou, LL, Qi B, Liu S. Room temperature ionic liquid (RTIL)-decorated mesoporous silica SBA-15 for papain immobilization: RTIL increased the amount and activity of immobilized enzyme. Materials Science and Engineering 2012;32 364-368.

[82] Du P, Liu S, Wu P, Cai C. Preparation and characterization of room temperature ionic liquid/single-walled carbon nanotube nanocomposites and their application to the direct electrochemistry of heme-containing proteins/enzymes. Electrochimica Acta 2007;52 6534–6547.

[83] Nakashima K, Okada J, Maruyama T, Kamiya N,Goto M. Activation of lipase in ionic liquids by modification with comb-shaped poly(ethylene glycol). Science and Technology of Advanced Materials 2006;7 692–698.

[84] Turner MB, Spear SK, Huddleston JG, Holbrey JD, Rogers RD. Ionic liquid salt-induced inactivation and unfolding of cellulase from Trichoderma reesei. Green Chemistry 2003;5 443–447.

[85] Maruyama T, Nagasawa S, Goto M. Poly(ethylene glycol)-lipase complex that is catalytically active for alcoholysis reactions in ionic liquids. Biotechnol Letters 2002;24 1341–1345.

[86] Wang Y, Mei L. Lyophilization of lipase with cyclodextrins for efficient catalysis in ionic liquids. Journal of Bioscience and Bioengineering 2007;103 345-349.

[87] Hernández-Fernández FJ, Rios AP, Lozano-Blanco LJ, Godiınez C. Biocatalytic ester synthesis in ionic liquid media. Journal of Chemical Technology and Biotechnology 2010;85 1423–1435.

[88] Rios AP, Hernandez-Fernandez FJ, Tomas-Alonso F, Gomez D and Villora G. Synthesis of esters in ionic liquids. The effect of vinyl esters and alcohols. Process Biochemistry 2008;43 892–895.

[89] Rios AP, Hernandez-Fernandez FJ, Tomas-Alonso F, Gomez D and Villora G. Synthesis of flavour esters using free Candida antarctica lipase B in ionic liquids. Flavour and Fragrance Journal 2008;23 319–322.

[90] De Diego T, Lozano P, Abada MA, Steffenskya K, Vaultier M and Iborra JL. On the nature of ionic liquids and their effects on lipases that catalyze ester synthesis. Journal of Biotechnology 2009;140 234–241.

[91] Ulbert O, Frater T, Belafi-Bako K, Gubicza L. Enhanced enantioselectivity of Candida rugosa lipase in ionic liquids as compared to organic solvents. Journal of Molecular Catalysis B: Enzymatic 2004;31 39–45.

[92] Yuan Y, Bai S, Sun Y. Comparison of lipase-catalysed enantioselective esterification of (9)-menthol in ionic liquids and organic solvents. Food Chemistry 2006;97 324–330.

[93] Contesini FJ, de Oliveira Carvalho P. Esterification of (RS)- Ibuprofen by native and commercial lipases in a two-phase system containing ionic liquids. Tetrahedron Asymmetry 2006;17 2069–2073.

[94] Nara SJ, Harjani JR, Salunkhe MM, Mane AT, Wadgaonkar PP. Lipase-catalysed polyester synthesis in 1-butyl-3- methylimidazolium hexafluorophosphate ionic liquid. Tetrahedron Letter 2003;44 1371–1373.

[95] Marcilla R, de Geus M, Mecerreyes D, Duxbury CJ, Koning CE, Heise A. Enzymatic polyester synthesis in ionic liquids. European Polymer Journal 2006;42 1215–1221.

[96] Gamba M, Lapis AAM, Dupont J. Supported ionic liquid enzymatic catalysis for the production of biodiesel. Advanced Synthesis and Catalysis 2008;350 160–4.

[97] De Diego T, Manjón A, Lozano P, Vaultier M, Iborra JL, Vaultier M. An efficient activity ionic liquid–enzyme system for biodiesel production. Green Chemistry 2011;13 444–51.

[98] Lozano P, Bernal JM, Piamtongkam R, Fetzer D, Vaultier M. One-phase ionic liquid reaction medium for biocatalytic production of biodiesel. ChemSusChem 2010;3 1359–63.

[99] Ha SH, Lan MN, Lee SH, Hwang SM, Koo Y-M. Lipase-catalyzed biodiesel production from soybean oil in ionic liquids. Enzyme and Microbial Technology 2007;41 480–483.

[100] De Diego T, Arturo M, Pedro L, Iborra JL. A recyclable enzymatic biodiesel production process in ionic liquids. Bioresource Technology 02 6336-6339

[101] Ruzich NI, Bassi AS. Investigation of enzymatic biodiesel production using ionic liquid as a co-solvent. Canadian Journal of Chemical Engineering 2010;88 277-282.

[102] Lai J-Q, Hu Z-L, Wang P-W, Yang Z. Enzymatic production of microalgal biodiesel in ionic liquid [BMIm][PF6]. Fuel 2012;95 329-333.

[103] Wang J. Real-Time Electrochemical Monitoring: Toward Green Analytical Chemistry. Accounts of Chemical Research 2002;35 811–816.

[104] Shiddiky MJA, Torriero AAJ. Application of ionic liquids in electrochemical sensing systems. Biosensors and Bioelectronics 2011;26 1775–1787.

[105] Galonde N, Nott K, Debuigne A, Deleu M, Jerome C, Paquot M, Wathelet J-P. Use of ionic liquids for biocatalytic synthesis of sugar derivatives. Journal of Chemical Technology and Biotechnology 2012;87 451–471.

[106] Laane C, Boeren S, Vos K, Veeger C. Rules for optimization of biocatalysis in organic solvents. Biotechnology and Bioengineering 1987;30 81-87.

Chapter 21

Pharmaceutical Salts: Solids to Liquids by Using Ionic Liquid Design

Clarissa P. Frizzo, Izabelle M. Gindri, Aniele Z. Tier, Lilian Buriol, Dayse N. Moreira and Marcos A. P. Martins

Additional information is available at the end of the chapter

http://dx.doi.org/10.5772/51655

1. Introduction

Ionic liquids (ILs) have attracted increasing interest lately in several areas such as chemistry, physics, engineering, material science, molecular biochemistry, energy and fuels, among others. Scientific literature has been daily invaded by papers that show a variety of new ionic liquids and new applications. Furthermore, the range of ILs used has been broadened, and there has been a significant increase in the scope of both physical and chemical IL properties [1, 2]. ILs are defined as liquid organic salts composed entirely of ions, and a melting point criterion has been proposed to distinguish between molten salts and ionic liquids (mp< 100 °C) [3].

When ILs based on 1-alkyl-3-methylimidazolium salts were first reported in 1982 by Wilkes et al. as tetrachloroaluminates, they were called ILs of first generation [4]. Replacement of this moisture-sensitive anion by the tetrafluoroborate ion and other anions led, in 1992, to air- and water-stable ILs, called then second generation, [5] which have found increasing applications such as reaction media for various kinds of organic reactions. At the onset of the new millennium, the concept of task-specific ILs, called third generation was introduced by Davis [6] (Figure 1). These compounds are defined as ILs in which the anion, cation, or both covalently incorporate a functional group (designed to endow them with particular properties, such as physical, chemical or in terms of reactivity) as a part of the ion structure [6,7]. Simultaneously, Rogers et al. [8] proposed that the ILs can be grouped into three generations in according to their properties and applications (Figure 1). The use of these compounds as solvents characterized them as ILs of first generation due to a unique and accessible physical property set characterized by low or no volatility, thermal stability, or large liquid ranges. Second generation of ILs has potential application such as energetic materials, lubricants and scavenger materials. In these cases, ILs provide a platform where the properties of both cation and anion can be independently modified, permitting the design of new functional materials, while retaining the desired features of an IL. Third generation of ILs has been described as the one where the biological activity is a primary IL property. Thus, ILs are seen as active pharmacological ingredient (API).

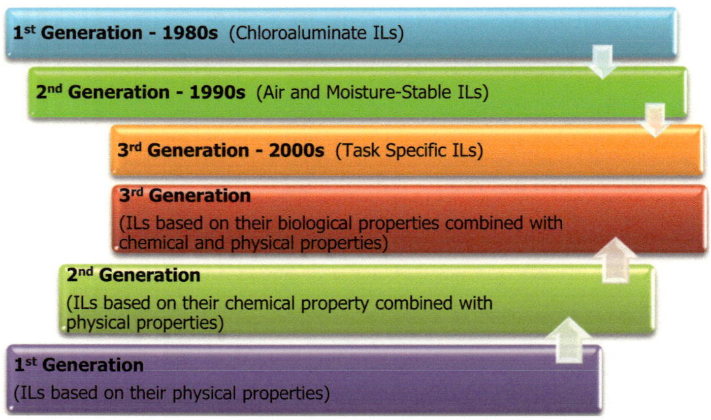

Figure 1. Historical evolution of ILs: chronological and useful development.

Potential pharmaceutical applications of ILs have been showed initially by studies of their toxicity and antimicrobial activity [9,10]. Currently, applications expanded to the use of pharmacologically active ions to develop novel ILs (3rd generation), the use in the formation of microemulsion droplets to transport and release of drugs, and as stabilizing agents of actives, additives and polymers in pharmaceutics [11]. One of the most import pharmaceutical applications is the use of pharmacologically active ions to develop novel liquid salts, since more than 50% of the drugs in the market today are sold as organic salts [12]. Hough and Rogers consider in their review [13] that the conversion of a drug into a salt is a crucial step in the drug development and can have a huge impact on its properties, including solubility, dissolution rate, hygroscopicity, stability, impurity profile and particle characteristics. Thus, the authors believe that an IL approach seems more than appropriate in the design of APIs, where a delicate balance exists between the exact chemical functionality needed for the desired effect in the absence of adverse side effects and the physical properties required for manufacturing, stability, solubility, transport and bioavailability [14,15]. Historically, the pharmaceutical industry depends mostly on crystalline APIs. However, many formulations fail during testing because of issues as, for example, delivery mechanisms such as dissolution, transport, and bioavailability or poor control over polymorphism which can dramatically change properties such as solubility [16,17,18]. In the context of APIs, the counter ions could be selected to synergistically enhance the desired effects or to neutralize unwanted side effects of the active entity. They could also be chosen to pharmacologically act independently [12,19] or to improve the pharmacokinetics properties [20] (Figure 2). Over the past few years, there have been three

http://dx.doi.org/10.5772/51655

reviews published in which ILs from APIs occupied a central theme [12,13,20]. In these reviews, the approach of ILs from APIs is discussed from different points of view: i) historical approach of ILs (from solvents to ILs from APIs), ii) focus on the use of ILs from APIs to solve the toxicity of ILs and polymorphism and iii) a good review where all these topics are shortly discussed. However, there is no concern with the issue of the synthesis and physical and chemical characterization of new salts like ILs. Thus, considering the lack of complete and deep survey about all questions (advantages and disadvantages) in the literature and in continuation of our research on ILs [21], we propose this chapter to show the application of IL approach to obtain liquid pharmaceutical salt. This denotes that the material to be covered here includes only papers where pharmaceutical activities (pharmacokinetic and pharmacological) are present at the cation or anion and there is focus on the obtainment of ILs from APIs (Table 1). Here, we consider biologically active the ILs whose components interact with any biological system, and pharmaceutically active the ILs that present any pharmacokinetics and/or pharmacological activities (Table 3). Thus, it was necessary to mention that papers describing ILs with only one of the biologically active components or with no pharmaceutically active were excluded. Thus, the ions alkyimidazolium, phosphoniun, derivatives of non-nutritive sugars, and N-trifluoromethanesulfonate were not included in the scope of this chapter. Another scope limitation of this chapter is about the use of mechanical or thermal methods to the liquefaction of a salt from APIs. This means that pharmacologically active salts such as the procainamide and verapamil hydrochloride that pass from a crystalline state to amorphous state through changes of conditions such as temperature and pressure were excluded [22,23].

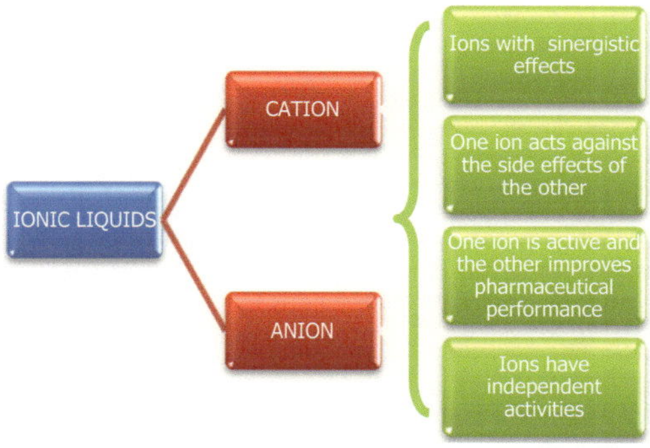

Figure 2. Cation and Anion combination in ILs from APIs and their activities.

Hence, in this chapter we will present the main problems of pharmaceutical industry in relation to solid salt APIs and how the proprieties as well as the limitations of the ILs affect the salification (i.e., salt formation) of APIs. The synthesis, the characterization of physical and chemical properties as well as the pharmaceutical performance of new ILs will be discussed. For this propose, we emphasize that the ILs selected to this chapter (Table 1) present in their structure at least one pharmacologically active entity, and the other (cation or anion) was introduced with the objective to increase the activity, reduce side effects or improve pharmaceutical performance by changing physical or chemical properties (Figure 2).

Compound	Name	Ref.
1	3-hydroxy-1-octyloxymethylpyridinium acesulfamate ([1-(OctOMe)-3-OH-Py][Ace])	[47]
2	3-hydroxy-1-octyloxymethylpyridinium Saccharinate ([1-(OctOMe)-3-OH-Py][Sac])	[47]
3	Benzalkonium Acesulfamate [BA][Ace]	[47]
4	Benzalkonium Saccharinate [BA][Sac]	[47]
5	Benzalkonium Salicylate [BA][Sal]	[37]
6	Benzethonium Acetylsalicylate [BE][Asp]	[37]
7	Benzethonium Aaccharinate[Ben][Sac]	[32]
8	Benzethonium Salicylate [BE][Sal]	[37]
9	Cetylpyridinium Acetylsalicylate [CetPy][Asp]	[37]
10	Cetylpyridinium Salicylate [CetPy][Sal]	[37]
11	Cetylpyridinium Ampicillin [C_{16}pyr][Amp]	[35]
12	Choline Ampicillin [Col][Amp]	[35]
13	Choline Phenytoin[Col][Phe]	[32]
14, 15, 16	Choline-derivative Acesulfamate [Col][Ace]	[46]
17	Didecyldimethylammonium Acesulfamate ([DDA][Ace])	[47]
18	Didecyldimethylammonium Ibuprofenate [DDA][Ibu]	[8]
19	Didecyl-dimethyl-ammonium Saccharinate ([DDA][Sac]	[47]
20	Hexadecylpyridinium Acesulfamate ([Hex][Ace])	[47]
21	Hexadecylpyridinium Aaccharinate ([Hex][Sac])	[47]
22	Hexetidinium Salicylate [Hext][Sal]	[37]
23	Lidocainium Acetylsalicylate [LID][Asp]	[37]
24	Lidocainium Docusate [Lid][Doc]	[8]
25	Lidocainium Salicylate [LID][Sal]	[37]
26	Mepenzolate Acesulfamate[Mep][Ace]	[32]
27	Mepenzolate Saccharinate[Mep][Sac]	[32]
28	Procainium Salicylate [Proc][Sal]	[37]
29	Procainium Amidesalicylate [PA][Sal]	[37]
30	Propantheline Acesulfamate[Pro][Ace]	[32]
31	Propantheline Cyclamate[Pro][Cyc]	[32]
32	Propantheline *p*-toluenesulfonate[Pro][pTO]	[32]
33	Propantheline Saccharinate[Pro][Sac]	[32]
34	Pyridostigmine Saccharinate[Pyr][Sac]	[32]
35	Ranitidine Docusate [Ran]Doc]	[8]

http://dx.doi.org/10.5772/51655

Compound	Name	Ref.
36	Tetrabutylphosphonium Salicylate [P(BU)$_4$][Sal]	[37]
37	Tetraethylammonium Ampicillin[TEA][Amp]	[35]
38	Tramadolium Acetyl-salicylate [Tram][Asp]	[37]
39	Tramadolium Salicylate [Tram][Sal]	[37]
40	Trihexyltetradecylphosphonium Ampicillin [P$_{6,6,6,14}$][Amp]	[35]

Table 1. ILs from active pharmaceutical ingredients found in this chapter.

2. Fundamentals

In this section we will deal with some important points of ILs approach salification of APIs and the strategies used in the search and characterization of new ILs from APIs. The synthetic procedure as well as physical and chemical properties (main thermal properties) of ILs will also be discussed.

2.1. Salification of APIs

The physical form of a drug substance is of great importance since it directly affects the manner in which the material is formulated and presented to the consumer, as well as influence more fundamental characteristics such as solubility and dissolution rate, which, in turn, impact on bioavailability [8, 17, 18].

The drugs converted into salts were found to be more stable and water soluble in comparison to free bases or acid which qualifies them as the preferred forms to use as therapeutic agents [19, 24-26]. Such salts may offer advantages over the corresponding free drug in terms of physical properties such as melting point (thermal stability), crystallinity, hygroscopicity, dissolution rate, or solubility (bioavailability). From a pharmaceutical viewpoint the melting enthalpy, melting temperature and solubility are of particular importance, both because of their routine measurement and their influence on processing and bioavailability [19, 26]. Taking these advantages into account, the pharmaceutical industry relies predominantly on solid, primarily crystalline forms for the delivery of APIs, mainly for reasons of purity, thermal stability, manufacturability, and ease of handling [8, 17, 18]. However, solid forms of APIs often suffer from polymorphic conversion, low solubility, and a variety of factors which affect bioavailability associated with the final solid form [17, 18, 27]. Many phase II trials of new APIs end in failure due to their efficacy, often related to bioavailability and thus solubility [16-18]. These factors motivate the screening for novel solid forms, including salts, polymorphs, pseudopolymorphs (or solvates), and co-crystals (Figure 3).

Liquid drug formulations from salification are rarely found and are usually based on eutectic mixtures [24], however, salt drug generation can result in a liquid salt, known as ionic liquid. The main advantage of ILs in most of the cases is that their salt properties are retained in a wide liquid range. This fundamental property of ILs is because ions are generally organic with low symmetry and diffuse charge. In addition, ILs have properties such as negligible vapor pressure resulting in reduced inhalatory exposure, absence of flammability, and their high variability concerning organic chemical structure in order to optimize technological features like solvation properties. This tunable solubility with several organic compounds, viscosity, conductivity, as well as thermal and electrochemical stability is ideal in terms of technical applicability [3]. The control of the properties of an IL is based on the manipulation of the interactions between the ions. The suppression of these interactions reduces lattice energies and the extreme suppression of these interactions leads to glass formation upon cooling, polymorphism, multiple phase transitions, and ion dissociation [28-31]. An understanding of the physical and chemical properties of ILs allows the proper selection of a specific IL for a given application. Thus, for example, by choosing ionic components capable of solubilizing specific solutes, one can control the critical solubility to crystallization processes [29].

Stoimenovski et al. [12] cited in their work that studies have been suggest that ILs do not dissolve as independent ions but keep a nanostructured organization in aqueous media. This fact, constitute other important advantage of the obtainment of an API as IL. Drugs that are highly ionic have difficulty crossing the membrane in order to reach their site of action. Ion-pair formation enhances the transport of various ionic drugs through the skin and across the absorbing membrane [12]. Therefore, highly ion-associated pharmaceutically active ILs would be highly beneficial forms of the original pharmaceutical active salts, as they could cross the membrane more rapidly [12]. Therefore, targeted alterations of a final drug form based on the various property sets obtainable through an IL approach may help to enhance efficacy, while retaining, improving or even introducing a second activity. The potential of this approach as a drug phase is to date poorly exploited [13, 13, 20].

Clearly, a salt that does not exhibit a crystalline phase will not present polymorphism, but there are further advantages to be realized and exploited in the delivery of the API [32]. In particular, a non-crystalline salt in a liquid or glassy phase will probably exhibit the enhanced solubility exhibited by amorphous phases [17, 18, 32]. This concept bears further discussion as it is potentially one of the most important advantages of the formulation of APIs as ionic liquid phase [32].

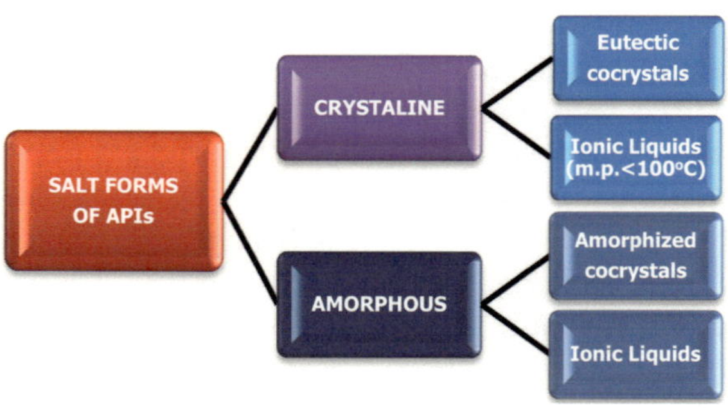

Figure 3. Salt forms of Active Pharmaceutical Ingredients (APIs).

http://dx.doi.org/10.5772/51655

2.2. Synthesis of ILs from APIs

The selection of pairs of ions to form ILs is carried out with candidate ions that have low symmetry and charge diffuse, traits that also characterize several typical APIs. Even the nitrogen-containing heterocycles, commonly used in ILs today, are frequently found in APIs or API precursors [8,33]. The process generally is a simple way to modify the properties of a drug with ionizable functional groups to overcome undesirable features of the parent drug [12].

Care must be taken when choosing appropriate IL-forming ion pairs. Many of the important APIs are not permanent ions, but rather are protonated or deprotonated to form the commonly used salts; thus suitable pKa differences need to be considered [8,34]. MacFarlane and Seddon [8,1] have recently proposed that protic ILs can only be considered ILs if the pKa difference is such that more than 99% of the salt exists in ionized form. For an API this distinction may not be needed, since having a balance between ionized and neutral forms may have advantages. There may be a significant advantage of drugs with low degree of ionisation over the fully ionised ones due to their ability to cross membranes more efficiently. An example of a partially ionized pharmaceutically active IL is 1-methylhexylammonium salicylate [12]. Salicylic acid, an analgesic with a pKa value of 2.98, was reacted with 1-methylhexylamine, a nasal decongestant with a pKa value of 10.5, to produce a liquid at room temperature with a glass transition at −40°C and a ΔpKa of 7.52.

Most of the syntheses found in the papers selected to this chapter consist of metathesis reactions. The cation and anion in their available salt forms were separately dissolved in a solvent (e.g., water, methanol, ethanol, acetone) allowed to stir with heating to ca. 90 °C (if necessary) or at room temperature. Some alternative methods to methatesis reaction to specific ILs were also described. Ferraz et al. [35] used a method to change the anion using ion exchange resin described by Ohno et al. [36]. Ferraz et al. [35] employed Amberlite resin (in the OH form) in order to exchange halides (bromide or chloride) to the hydroxide form and then this basic solution was neutralized by the addition of an adequate acid solution. The acid–base reaction yielded the desired IL. The organic cations were selected from salts which were first transformed into hydroxides by the use of an ionic exchange column (Amberlite IRA-400 OH) in methanol. Next, the β-lactam antibiotic previously dissolved in a moderately basic ammonia solution was used to neutralize the selected cations. In this case, pure ILs were obtained after eliminating the excess ammonia and/or β-lactam antibiotic by evaporation and crystallization, respectively. Bica et al. [37, 38] also showed an alternative synthesis in solvent-free conditions. The compounds **25** and **28** were also prepared by melting a stoichiometric mixture of base and salicylic acid at ~100 °C to obtain a liquid. Similarly, **22** was directly synthesized by the reaction of hexetidine with salicylic acid. This solvent-free preparation is clearly advantageous compared to conventional metathesis, since solvents and stoichiometric NaCl waste are prevented. Furthermore, ILs are obtained in high purity without halide, metal, or solvent impurities, as necessary for pharmaceutical applications. The isolation of the product occurred considering that usually the inorganic salt precipitates. Thus, in most of cases, the product was extracted by filtration of inorganic salt. The solvent was removed with a rotary evaporator. The resulting product was placed on a high vacuum line to remove any residual solvent. In some cases, when inorganic salt is partially soluble, ILs had to undergo a process of extraction typically with chloroform or dichloromethane. Following that step, the organic phase was then washed with water to remove any inorganic salt (e.g., NaCl, which was monitored by a silver nitrate test), and solvent was removed with a rotary evaporator. The resulting product was placed on a high vacuum line to remove any residual solvent. In some cases, an extra purification was described, mainly to remove excess of halides.

2.3. ILs characterization

When searching for an IL from APIs, one has to care about its physical state and properties because it may not be an IL, but a crystalline solid (Figure 3). A significant number of drugs currently on the market are formulated as amorphous materials, and the most common means of preparing the amorphous phase is by quenching: rapid cooling from the melt; rapid precipitation from solution (for example on addition of an antisolvent); spray drying; flash evaporation; lyophilisation, in other words, methods that allow the disordered glassy phase to be "frozen in" before nucleation and growth that would lead to the appearance of crystals [32, 17, 18]. Dean et al. [32] have showed a schematic comparison of the accessible phases and the relative free energy of each phase. The authors highlight that in cases where a stable crystalline form exists at ambient temperature, ΔGf^{o} for the crystalline phase is lower than that for the amorphous material formed by quenching. In this manner, the quenching results in a form that is not the most thermodynamically stable phase at that temperature. The authors also emphasize that in this situation, any event that initiates nucleation may lead to growth of a crystalline phase at the expense of the amorphous glass. Such nucleation initiators may include heating (yielding a plastic phase that then crystallizes) or localized shock, such as that applied during size reduction (grinding) or formulation. In some cases, even the desired increased solubility leads immediately to the crystallization of a stable crystalline phase. In relation to the search by ILs, the authors emphasized that in the preparation of a salt that has an amorphous phase as its most thermodynamically stable form (in the temperature range of interest), a less preferred, but an effective form would be one with fusion temperature below the temperature of interest [32].

In the light of the extensive literature on crystalline drug forms, currently supported by crystal engineering [39,40] by using supramolecular synthons [41] concept and considering the abovementioned data, their benefits are clear to develop strategies for increasing solubility of drug compounds [42] and to chose co-crystal formers [43] for poorly soluble drugs [44]. Therefore, combining an understanding of the effect of robust supramolecular synthons on the process of crystallization, with concepts involving IL, Dean et al. [32] proposed the development of an "anti-crystal engineering" approach to the synthesis of ILs from APIs. This would comprise identifying and intentionally avoiding the pairing of cations and anions that might yield common supramolecular synthons, with the goal of decreasing the likelihood of crystallization of salt. This "anti-crystal engineering" approach is postulated as a means of narrowing the search for ILs from API to cation and anion pairs that have a higher likelihood of not crystallizing. Dean et al. [32] illustrated this approach by preparing and analyzing a series of API salts; some of which crystallized readily, while others were characterized as ILs and remained in an amorphous glass or liquid form in spite of vigorous attempts to bring about crystallization. To achieve the combination, the authors studied the possibility of cations and anions form supramolecular synthons mainly from interactions of hydrogen that usually considered imparting a tendency toward crystallization that should be avoided in the quest for ILs from APIs. As a result, they observed that all cation/anion combinations bearing both hydrogen bond donor and acceptor groups yield crystalline salts (**13**, **26**, **27** and **31**) [32]. For example, **13** (Table 1 and 2) which might be predicted to form the greatest number of strong, directional hydrogen bonds between different ions yields the salt with the highest melting point. Of greater interest, however, are those salts that are crystalline solids and even do not (at first examination of individual ions) exhibit the capability to form hydrogen bonded synthons (**33** and **34**), but have some energy stabilization resulting in crystallization. On the other hand, some salts were observed as salt without a crystalline phase (or with a sub-ambient melting point) indicating that they are in the most thermodynamically stable phase as a liquid or glass at ambient temperature. Thus, these salts (**7**, **30** and **32**) are considered ILs by the authors [32]. In another work in search by ILs from APIs, Bica et al. [38] found that generating oligomeric ions from tautomer's protons has a tremendous influence on physical properties that allow the expansion of the liquid ranges of some salts. They proposed that for this type of liquid salt formulation, the term ionic liquids might be controversial. They suppose (say) this based on initial experiments and further investigation concerning ionicity and simple eutectic behavior are currently ongoing in their laboratories. The authors define oligomeric ions as those that enable liquefaction of solid ILs (or other salts) by simply changing the stoichiometry or complexity of the ions. They highlight that the strategy does not need to employ the parent of the anion or cation in use and that this can be particularly useful for pharmaceutically active salts or ILs. According to the authors, another advantage of the oligomeric ions strategy is that the design of pharmaceutical IL may benefit since it is permitted to modify the physical properties of a given salt form once obtained [38].

Therefore, considering the "anti-crystal strategy" proposed by Dean et al. [32] and the "oligomeric ions" proposed by Bica et al. [38] in the search for ILs from API, the possibility of formation of solid crystals (eutectics), amorphous, or liquid phase is a fundamental question (Figure 3). Thus, one cannot imagine the search for new ILs from API without performing a complete calorimetric characterization of what an IL may be. Currently, unfortunately, in most of the cases, the ions are selected by facility from synthesis or purification routes rather than rational choice or screening.

The calorimetric data reported in papers collected to this chapter showed that most of the salts synthesized can be classified as ILs, while the remaining were crystalline salts. Those who were characterized as IL can be shared in three general types of behavior. The first group of ILs exhibits just melting points below 100 °C, allowing their classification as ILs (**1**, **2**, **4**, **9**, **10**, **14**, **16** and **21**). The second type of behavior is characterized by formation of an amorphous glass. These ILs have no melting, but only glass-transition (**6-8**, **17**, **18**, **23-25**, **28-30**, **32-35** and **38**). The third group of ILs is characterized by compounds that have melting points and glass transition (**2**, **5**, **11**, **12**, **15**, **19**, **20**, **34**, **36** and **37**)). Compounds **13**, **22**, **26**, **27**, **31**, **33** and **39** were found to have high melting not fitting the definition of ILs (Table 2). Other physical and chemical properties also should be evaluated to characterize an organic salt as an IL. Considering the scope of this chapter, some of the properties were found in a few papers (Table 2). These properties were density, solubility and thermal stability (Table 2). Viscosity is an important physical property to characterize an IL. The viscosity of ionic liquids is essentially determined by their tendency to form hydrogen bonding and by the strength of their van der Waals interactions. The structure of the cation strongly influences the viscosity of the IL. However, this property was not reported in any of the selected papers. Density was reported only for two ILs (**14, 15**). These ILs were denser than water thus the density of comparable ILs decreases as the bulkiness of the organic cation increases. It is evident that the density of such compounds increases with increasing molecular weight of the anion which confirms the results shown by Fredlake et al [45]. Density reported to ILs from APIs in the selected papers to this chapter is characteristic of ILs[3]. At room temperature, the ILs from API reported in this chapter were grouped into miscible in water and other polar organic solvents (hydrophilic) and partially miscible or immiscible in water and hexane (hydrophobic). Water miscible ILs were choline derivatives, ampilicin, saccharinate and water imiscible were choline, ammoniun and pyridiniun derivatives. IL choline derivatives were present in both groups on dependence of anion. Solubility data of other IL derivatives of other cations and anions were not reported by the authors. The thermal stability of ILs is limited by the strength of their heteroatom-carbon and their heteroatom-hydrogen bonds, respectively [3]. The ILs **9, 14, 15** and **23** showed a lower thermal stability (115.14-126.5 °C) while **1-6, 8, 10-12, 17-22, 24, 25, 28, 29** and **35-40** have shown more stable (154.22-307.94 °C).

Compound	Thermal Data (°C) and Density at 25°C (g.mL-1)	Ref.
1	m.p = 79-81[d], TGA = 267	[47]
2	m.p = 95-98[d], TGA = 301	[47]
3	m.p = 90, Tg = -36[a], TGA = 187/249/394[c]	[47]
4	m.p = 74, TGA = 204	[47]
5	m.p = 96.02, Tg = 51.01, TGA = 171.95[b]	[37]
6	Tg =2.84, TGA = 154.22[b]	[37]
7	Tg = -4	[32]
8	Tg = -13.72, TGA = 167.76[b]	[37]
9	m.p = 61.31, TGA = 115.14[b]	[37]
10	m.p = 73.97, TGA = 205.61[b]	[37]
11	m.p = 86.0, Tg = -19.64, TGA = 269.39	[35]
12	m.p = 58.0, Tg = -20.12, TGA = 221.29	[35]
13	m.p = 215 – 217	[32]

http://dx.doi.org/10.5772/51655

Compound	Thermal Data (°C) and Density at 25°C (g.mL-1)	Ref.
14	m.p. = 31-36 and82-84 TGA = 126.5, density = 1.103 - 1.277	[46]
15	m.p. = 85-86,Tg = -49, TGA = 122, density = 1.041-1.270	[46]
16	m.p.= 87-88	[46]
17	Tg = -53, TGA = 232/426[c]	[47]
18	Tg = -73, TGA = 168[b]	[8]
19	m.p = 16, Tg = -33, TGA = 214	[47]
20	m.p = 57, Tg = -11, TGA = 267/494[c]	[47]
21	m.p = 66, TGA = 253/412[c]	[47]
22	m.p = 106.81, TGA = 182.14[b]	[37]
23	Tg = -13.97, TGA = 120.71[b]	[37]
24	Tg = -29, TGA = 222 [b]	[8]
25	Tg = 19.78, TGA = 158.46[b]	[37]
26	m.p = 135 – 137, Tg = 34	[32]
27	m.p = 187 – 189, Tg = 53	[32]
28	Tg = 13.87, TGA = 187.33[b]	[37]
29	Tg = 19.87, TGA = 159.21[b]	[37]
30	Tg = -20	[32]
31	m.p = 133 – 137, Tg = 20	[32]
32	Tg = 7	[32]
33	m.p = 133 – 135, Tg = 18	[32]
34	m.p = 94 – 96, Tg = 4	[32]
35	Tg = -12, TGA = 249	[8]
36	m.p = 57.32, Tg = -56.47, TGA = 307.94[b]	[37]
37	m.p = 79.0, Tg = -18.64, TGA = 214.75	[35]
38	Tg = 13.78, TGA = 169.64[b]	[37]
39	m.p = 176.17, TGA = 177.16[b]	[37]
40	TGA = 297.65	[35]

[a]Solid-solid transition.[b]$T_{onset\ 5\%}$.[c]Multiple decomposition steps. [d]Visual melting point range *via* hot-plate apparatus.

Table 2. Physical and Chemical Properties of some ILs from APIs.

3. Pharmaceutical activity assessment

In view of the objective of our chapter, it is important to evaluate the pharmaceutical profile of ILs from API. Pharmaceutical profile includes changes on the pharmacokinetics and/or pharmacological behavior of the salts when they turn into liquids by changing their cation or/anion.

No specific pharmacokinetic property was evaluated. However important observation reporting by Hough et al [8] and related in a review by Stoimenovski et al [12] is that strongly hydrophilic ionic actives often possess insufficient ability to penetrate biological membranes. Combining such an active ion with another of a more lipophilic character may offer a solution to this problem. For example, lidocaine docusate [Lid][Doc], an IL form of the local surface anaesthetic lidocaine, combines the relatively hydrophobic lidocaine cation with a hydrophobic anion, docusate (an emollient), to produce a hydrophobic IL, which exhibits reduced or controlled water solubility (Figure 2) and thus should exhibit extended residence time on the skin [8].

From the selected papers to this chapter, only three of them bring evaluation of pharmaceutical properties of new API (IL). Pharmacological activities were evaluated in four papers. Pharmacological activity evaluated were antinociception [8], suppression of PC12 neuritic outgrowth [8], antibactericidal [46, 47] and antifungal [46, 47] activities and skin irritation [47]. Antinociception activity [8] was assessed using a modification of the tail-withdrawal procedure. Two antinociceptive models were used: warm water tail-withdrawal from 49 °C water in intact mice, and warm water tail-withdrawal from a 47 °C bath, following tail injury. In this test it was observed that [Lid][Doc] produced a longer duration of antinociceptive effect than lidocaine hydrochloride [Lid][HCl]. The authors suggest that the high hydrophobicity for [Lid][Doc] in relation to [Lid][HCl] account for the increased duration of [Lid][Doc] over [Lid][HCl] observed *in vivo* models. Also, this may constitute a slow-release mechanism unique to any hydrophobic IL [8].

In the same paper [8], the evaluation of the suppression of PC12 neuritic outgrowth by [Lid][Doc] and [Lid][HCl] was evaluated. The suppression of PC12 neuritic outgrowth is related to the local anesthetic effects. These anesthetics suppress nerve growth factor (NGF) mediated neuronal differentiation in rat pheochromocytoma (PC12) cells. This was used as a bioassay for detecting that killed PC12 cells treated with [Lid][HCl] was higher than [Lid][Doc]. Authors suggest that the PC12-NGF data indicate potential differences between [Lid][Doc] and [Lid][HCl] at the cellular level and showed a mechanism of action entirely different for [Lid][Doc] than that for [Lid][HCl]. Docusate may enhance membrane permeability which may suggest at least one mechanism associated with the apparent increase in [Lid][Docl] efficacy *in vivo*. However, while an increase in permeability may enhance transdermal transport and account for the longer duration and greater efficacy of [Lid][Doc]*in vivo*, the longer duration of [Lid][Doc] on the mouse tail-withdrawal indicates an alternative mechanism.

http://dx.doi.org/10.5772/51655

Antimicrobial, antifungical and antibactericidal activities were evaluated to ILs **14** and **15** [46], **3**, **4**, **17** and **19** [47]. Results were expressed in terms of mean minimum inhibitory concentration (MIC) and minimum bactericidal concentration (MBC). The efficacies of **14** and **15** were low (but still high enough to be effective) in comparison with that of the widely used benzalkonium chloride. No MIC and MBC values could be established for the [Ace] salts **16**, because of their hydrophobic characters. Salts **3**, **4**, **17** and **19** were also evaluated; benzalkonium and didecylmethylamonnium chloride which inherently exhibit anti-microbial, anti-bacterial and anti-fungal activities were used as standard for comparison. ILs activities are similar to those of commercially available, although the ILs were not found to be limited to a specific class of bacteria or fungi. Skin irritation of salts **17** and **19** [47] was also determined. Each IL was tested on 3 male New Zealand albino rabbits, where the fur was previously removed from the back of the rabbit. Half a milliliter of the ILs (100%, pure) was distributed on two 6 cm^3 sites of the same animal. The application site was then covered with a porous gauze dressing and secured in place with tape. After a 4h exposure, the dressing was removed and the application site was gently washed with water. Observations were then conducted at 1, 24, 48, and 72 h, where the test sites were evaluated for erythema and edema using a prescribed scale. The skin irritation of these ILs is defined as category 4 (the highest) by standard organization for economic co-operation and development (OECD) grading.

Finally, the acute oral toxicities of salts **17** and **19** were determined. The toxicity was tested according to the method of acute toxic class. Wistar rats male and female were used for each IL tested. Results indicated the acute toxicity range for both ILs was between 300–2000 mg/kg in male and female rats. Thus, these ILs would be classified as category 4 (harmful) toxins according to standard organization for economic co-operation and development (OECD) grading. Table 3 depicted the cation and anion covered in this review. Structures and pharmacological activities of each of them are also showed.

Structure	Name	Activity
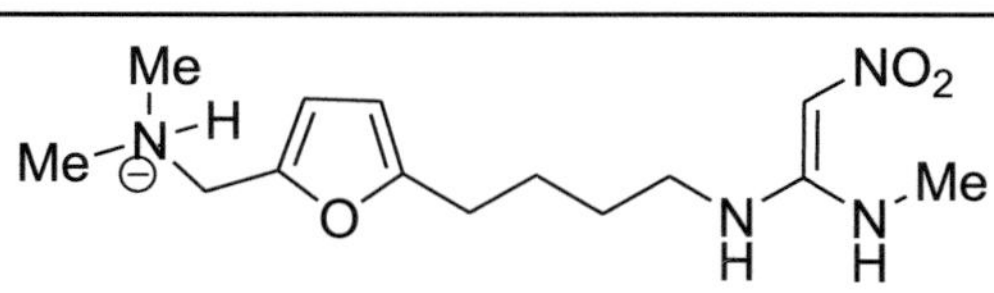	Ranitidine [Ran]	Histamine H2- receptor antagonist
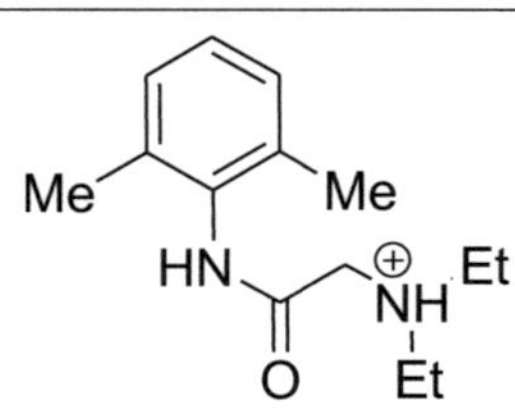	Lidocainum [Lid]	local anesthetic,

Structure	Name	Activity
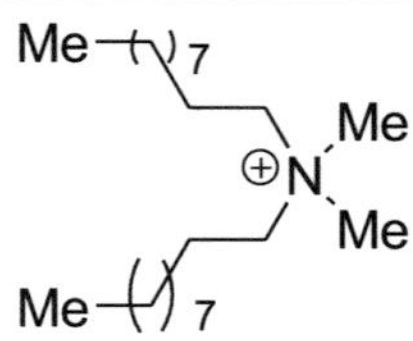	Didecylmethylammonium [DDA]	antibacterial
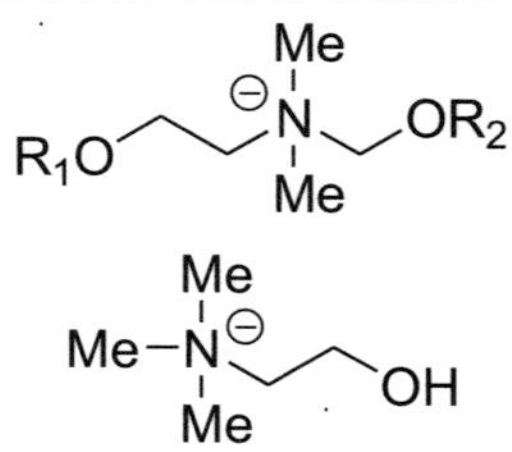 R¹=H, R²=from Et to $C_{14}H_{29}$ and $C_{12}H_{23}$(4) R¹=Ac, R²=from Et to $C_{14}H_{29}$ and $C_{12}H_{23}$(5) R¹ =C_9H_{19},R² = from Etto $C_{12}H_{25}$ and $C_{12}H_{23}$(6)[a]	Cholinederivatives[Col]	acetylcholine precursor
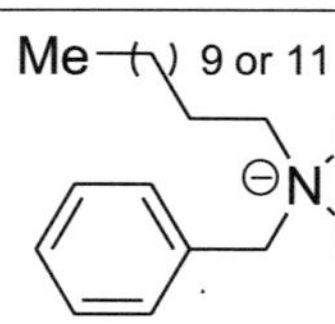	Benzalkonium [BA]	antibacterial
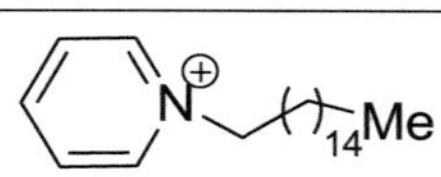	Hexadecylpyridinium [Hex]	antibacterial
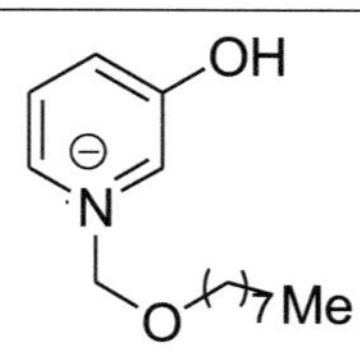	3-Hydroxy-1-octyloxymethylpyridinium [1-(OctOMe)-3-OH-Py]	antimicrobial
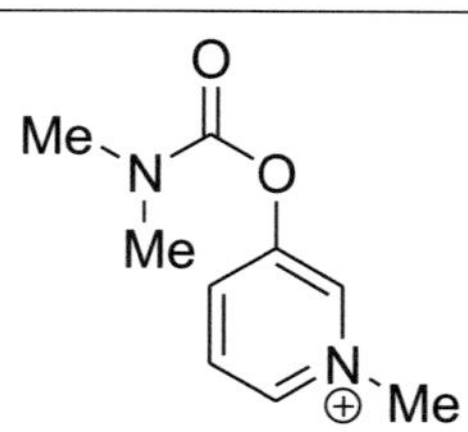	Pyridostigmine [Pyr]	reversible acethylcholinesteterase inhibitor
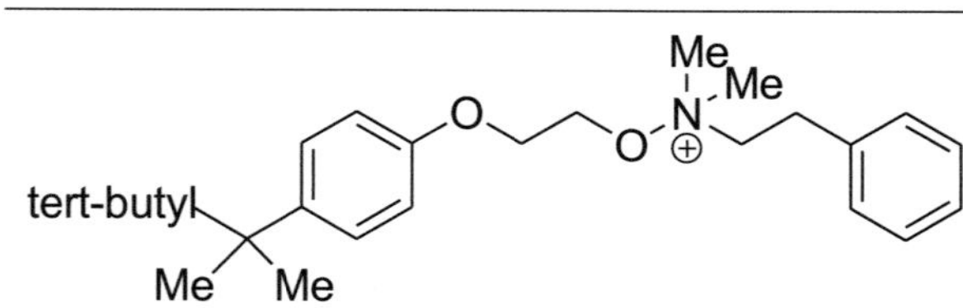	Benzethonium[Ben]	antibacterial

Structure	Name	Activity
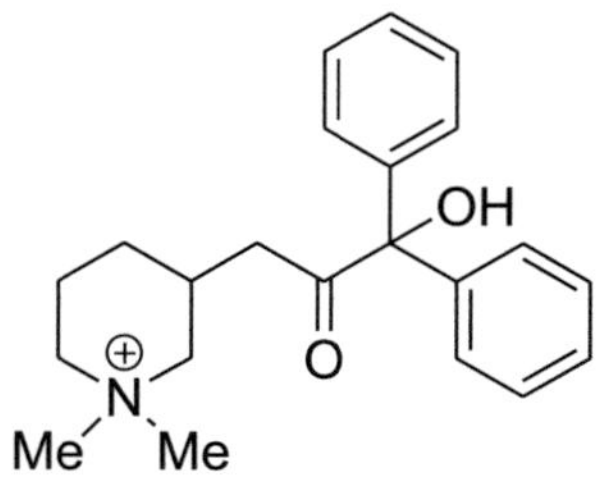	Mepenzolate[Mep]	skeletalmusclerelaxant
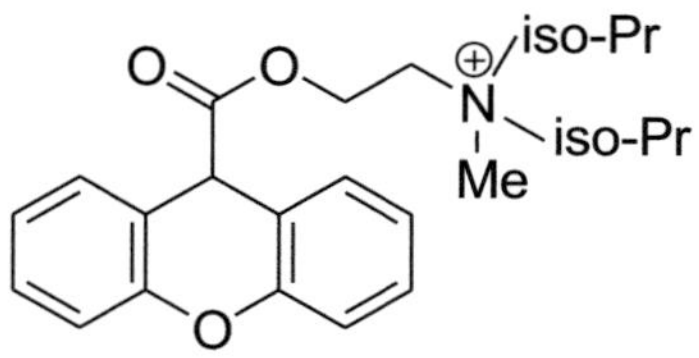	Propantheline[Pro]	antimuscarinic
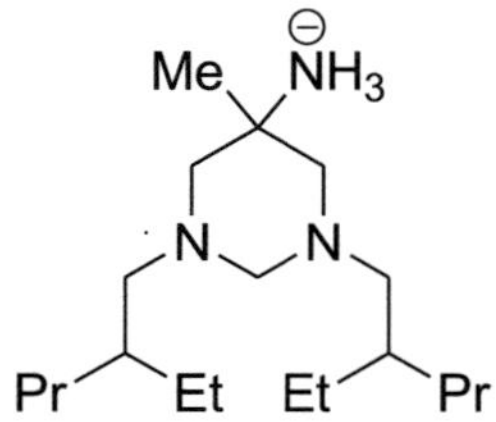	Hexetidinium[Hext]	antibacterial
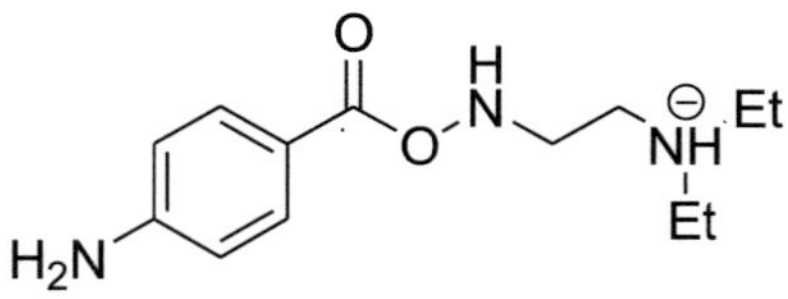	Procainiunamide[PA]	antiarrhythmic
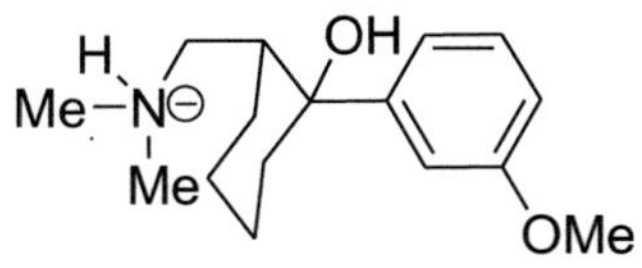	Tramadolium[Tram]	analgesic
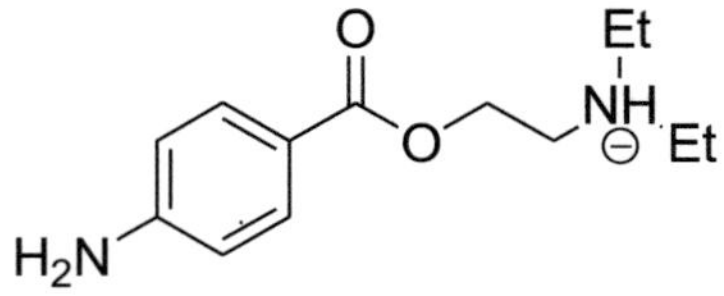	Procainiun [Proc]	localanesthesic
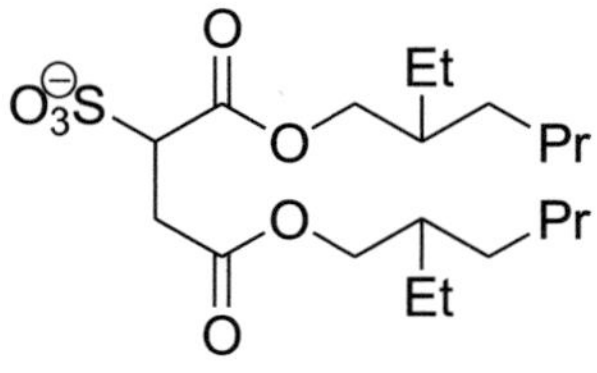	Docusate [Doc]	emolient

Structure	Name	Activity
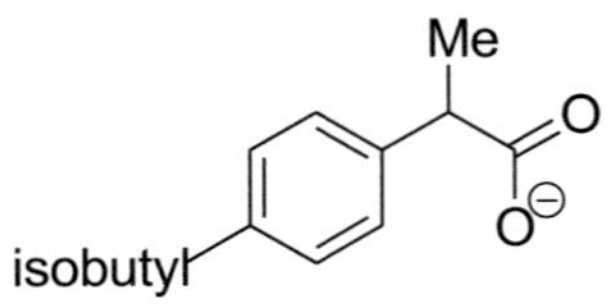	Ibuprofenate [Ibu]	anti-inflammatory
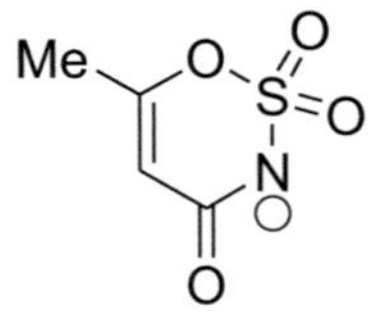	Acesulfamate [Ace]	noncaloric sugar
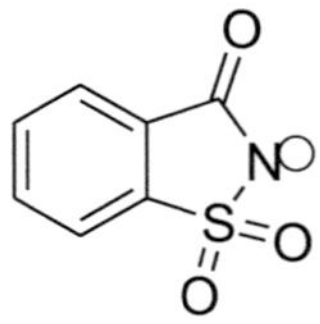	Saccharinate[Sac]	noncaloric sugar
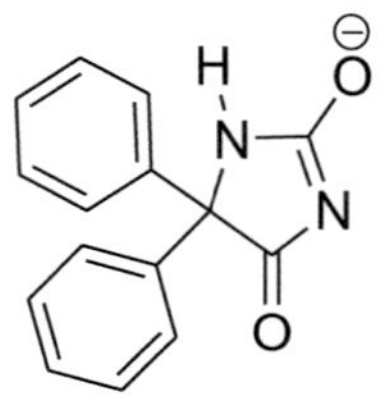	Phenytoin[Phe]	antiepileptic
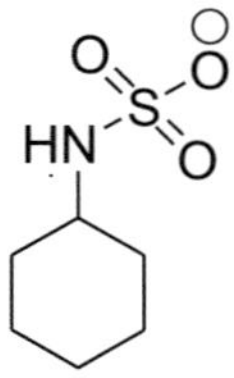	Cyclamate [Cyc]	noncaloric sugar
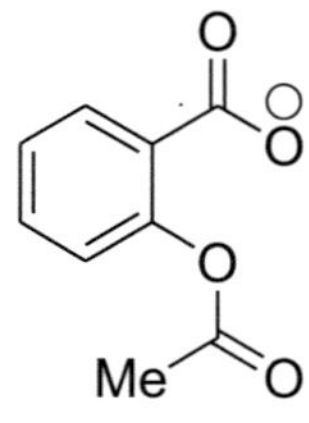 and 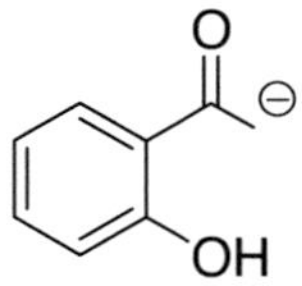	Acetylsalicylate [Asp] salicylate [Sal]	anti-inflammatory, analgesic, anti-pyretic

Structure	Name	Activity
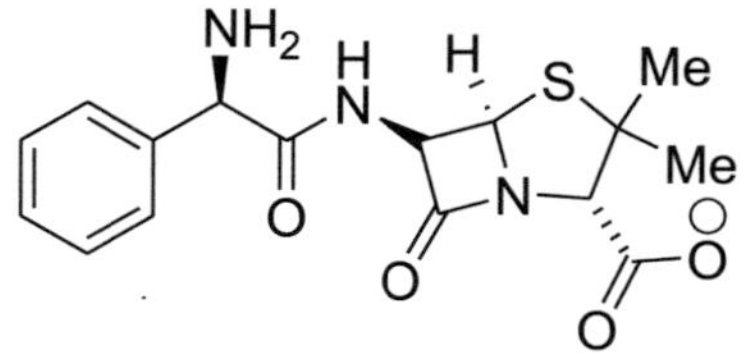	Ampicillin [Amp]	antibacterial

[a]Number **4**, **5** and **6** are of the ILs formed from these cations (See Table 1).

Table 3. Structure and activity of ions found in this chapter.

4. Conclusion

After having examined the literature in ILs from APIs it is possible to conclude that: *(i)* this is a research area developed by few groups; *(ii)* a complete and elaborated work including synthesis, physical and chemical properties studies and pharmacological activity estimation is necessary to produce significant results in this area (some groups have already performed more elaborated works); *(iii)* evaluation of physical and chemical properties, main thermal behavior is fundamental to develop new liquid APIs from IL approach; *(iv)* modification in the physical state of an API can result in modification or modulation of pharmaceutical properties of drugs. For example, co-formation of two separate solid actives in a solid dosage form significantly differ from a dual functional IL formulation. The ions in an IL dissolve in the body fluids exactly the same way—since one ion cannot dissolve without the other; this is not true to separate solid forms administered at the same time since each may dissolve at quite different rates. In addition, increase in solubility and bioavailability can enable a new formulation and/or a new doseage. Consequently, pharmacokinetic and pharmacological profiles studies necessarily lead to a potential patent protection for each of new forms of drugs [12].

In this chapter, we hope to have given a clear idea of the use of IL approach in the obtainment of liquid or amorphous API. We would like to conclude with an optimistic view for the future expansion of the development of new drug profiles. This positive view comes from the certainty that the results reported here are the beginning of a great advance in this promising field in the near future.

List of Abbreviations

API – Active Pharmacological Ingredient

IL –Ionic Liquid

MBC - Minimum Bactericidal Concentration

MIC - Minimum Inhibitory Concentration

NGF - Nerve Growth Factor

OECD - Organization Economic Co-operation and Development

PC12 - Pheochromocytoma Cells

$\Delta Gf°$– Free Energy of Fusion

Acknowledgements

The authors are greatful to Conselho Nacional de Desenvolvimento Científico e Tecnológico (CNPq/Universal Proc. No. 578426/2008-0; 471519/2009-0), Fundação de Amparo à Pesquisa do Estado do Rio Grande do Sul (FAPERGS/CNPq-PRONEX Edital No. 008/2009, Proc. No. 10/0037-8), and Coordenação de Aperfeiçoamento de Pessoal de Nível Superior (CAPES/PROEX) for financial support. The fellowships from CNPq (M.A.P.M., D.N.M., A.Z.T.) and CAPES (C.P.F., L.B., I.M.G.) are also acknowledged.

Author details

Clarissa P. Frizzo*, Izabelle M. Gindri, Aniele Z. Tier, Lilian Buriol, Dayse N. Moreira and Marcos A. P. Martins

*Address all correspondence to: clarissa.frizzo@yahoo.com.br

Department of Chemistry, NUQUIMHE, Federal University of Santa Maria, Santa Maria, RS, Brazil

References

[1] MacFarlane D R., Seddon K R.Ionic Liquids- Progress on the Fundamental Issues.Australian Journal of Chemistry (2007). http://www.publish.csiro.au/paper/CH06478.htmaccessed 20 june 2012)., 60(1), 3-5.

[2] Torimoto, T, Tsuda, T, Okazaki, K, & Kuwabata, S. New Frontiers in Materials Science Opened by Ionic Liquids. Advanced Materials (2010). http://onlinelibrary.wiley.com/doi/10.1002/adma.200902184/abstractaccessed 20 june 2012)., 22(11), 1196-1221.

[3] Wilkes, J. S, Wasserscheid, P, & Welton, T. Ionic Liquids in Synthesis, Weinheim: Wiley-VCH; (2007). http://onlinelibrary.wiley.com/doi/10.1002/9783527621194.ch1/summary(accessed june 2012).

[4] Wilkes, J S, Levisky, J A, Wilson, R A, & Hussey, C L. Dialkylimidazoliumchloroaluminate melts: a new class of room-temperature ionic liquids for electrochemistry, spectroscopy and synthesis. Inorganic Chemistry1982;http://pubs.acs.org/doi/abs/10.1021/ic00133a078(accessed june (2012). , 21(3), 1263-1264.

[5] Wilkes, J S, & Zaworotko, M J. Air and water stable 1-ethyl-3-methylimidazolium based ionic liquids. Journal of the Chemical Society, Chemical Communications1992; (13): 965-967. http://pubs.rsc.org/en/content/articlelanding/1992/c3/c39920000965(accessed june (2012).

[6] Davis, J. H Jr. Task-Specific Ionic Liquids. Chemistry Letters 2004;33(9), 1072-1077. http://www.csj.jp/journals/chem-lett/cl-cont/cl2004-9.html(accessed june (2012).

[7] Visser, A E, Swatloski, R P, Reichert, W M, Mayton, R, Sheff, S, Wierzbicki, A, & Davis, J. H Jr, Rogers R D. Task-specific ionic liquids for the extraction of metal ions from aqueous solutions. Chemical Communications(2001).

[8] Hough, W L, Smiglak, M, Rodríguez, H, Swatloski, R P, Spear, S K, Daly, D T, Pernak, J, Grisel, J E, Carliss, R D, Soutullo, M D, & Davis, J. H Jr, Rogers R D. The third evolution of ionic liquids: active pharmaceutical ingredients. New Journal of Chemistry (2007). http://pubs.rsc.org/en/content/articlelanding/2007/nj/b706677paccessed 20 june 2012)., 31(8), 1429-1436.

[9] Carter, E B, Culver, S L, Fox, P A, Goode, R D, Ntai, I, Tickell, M D, Traylor, R K, Hoffman, N W, & Davis, J. H Jr. Sweet succes: ionic liquids derived from non-nutritive sweeteners. Chemical Communications (2004). http://pubs.rsc.org/en/content/articlelanding/2004/cc/b313068aaccessed 20 june 2012).

[10] Pernak, J, Stefaniak, F, & Weglewski, J. PhosphoniumAcesulfamate Based Ionic Liquids. European Journal of Organic Chemistry (2005). http://onlinelibrary.wiley.com/doi/10.1002/ejoc.200400658/abstractaccessed 20 june 2012).

[11] Moniruzzamana, M, Kamiya, N, & Goto, M. Ionic liquid based microemulsion with pharmaceutically accepted components: Formulation and potential applications. Journal of colloid and Interface Science (2010). http://www.sciencedirect.com/science/article/pii/S0021979710009306(accessed june 2012).

[12] Stoimenovski, J. MacFarlane D R, Bica K, Rogers R D. Crystalline vs. Ionic Liquid Salt Forms of Active Pharmaceutical Ingredients: A Position Paper. Pharmaceutical Research (2010). http://www.springerlink.com/content/4g14306n81x7t8n0/accessed 20 june 2012)., 27(4), 521-526.

[13] Hough, W L, & Rogers, R D. Ionic Liquids Then and Now: From Solvents to Materials to Active Pharmaceutical Ingredients. Bulletin of the Chemical Society of Japan

(2007). https://www.jstage.jst.go.jp/article/bcsj/80/12/80_A7019/_article(accessed june 2012)., 80(12), 2262-2269.

[14] Bennett, B, & Cole, G. Pharmaceutical Production- An Engineering Guide. United Kingdom: Institution of Chemical Engineers; (2003). http://www.knovel.com/web/portal/browse/display?_EXT_KNOVEL_DISPLAY_bookid=1113(accessed june 2012).

[15] Byrn, S. R, Pfeiffer, R. R, & Stephenson, G. Grant WJD., GleasonWB. Solid-State Chemistry of Drugs. West Lafayette: SSCI; (1999).

[16] Schuster, D, Laggner, C, & Langer, T. Why Drugs Fail- A Study on Side Effects in New Chemical Entities. Current Pharmaceutical Design (2005). http://www.benthamdirect.org/pages/b_viewarticle.php?articleID=3137721(accessed june 2012)., 11(27), 3545-3559.

[17] Datta, S, & Grant, D. J W. Crystal structures of drugs: advances in determination, prediction and engineering. Nature Reviews Drug Discovery (2004). http://www.nature.com/nrd/journal/3n1/abs/nrd1280.html(accessed june 2012).

[18] Yu, L, Reutzel, S M, & Stephenson, G A. Physical characterization of polymorphic drugs: an integrated characterization strategy. Pharmaceutical Science & Technology Today (1998). http://www.sciencedirect.com/science/article/pii/S1461534798000315(accessed june 2012)., 1(3), 118-127.

[19] Kumar, V, & Malhotra, S. V. IonicLiquids as PharmaceuticalSalts: A Historical Perspective. In: Malhotra SV. (ed.) Ionic Liquid Applications: Pharmaceuticals, Therapeutics, and Biotechnology. Washington: American Chemical Society; (2010). http://pubs.acs.org/978-0-84122-547-3(accessed june 2012).

[20] Ferraz, R, Branco, L C, Prudêncio, C, Noronha, J P, & Petrovski, Z. IonicLiquids as Active PharmaceuticalIngredients. ChemMedChem (2011). http://www.ncbi.nlm.nih.gov/pubmed/21557480(accessed june 2012)., 6(6), 975-985.

[21] Martins, M, Frizzo, P, Moreira, C P, Zanatta, D N, & Bonacorso, N. H G. Ionic Liquids in Heterocyclic Synthesis. Chemical Reviews.(2008). http://pubs.acs.org/doi/abs/10.1021/cr078399y(accessed june 2012)., 108(6), 2015-2050.

[22] Wojnarowska, Z, Roland, C M, Swiety-pospiech, A, Grzybowska, K, & Paluch, M. Anomalous Electrical Conductivity Behavior at Elevated Pressure in the Protic Ionic Liquid Procainamide Hydrochloride. Physical Review Letters (2012). http://prl.aps.org/abstract/PRL/108i1/e015701accessed 20 june 2012).

[23] Wojnarowska, Z, Paluch, M, Grzybowski, A, Adrjanowicz, K, Grzybowska, K, Kaminski, K, Wlodarczyk, P, & Pionteck, J. Study of molecular dynamics of pharmaceutically important protic ionic liquid-verapamil hydrochloride. The Journal of Chemical Physics (2009). http://jcp.aip.org/resource/1/jcpsa6/131i10/104505s1accessed 20 june 2012).

[24] Stahl, P. H, & Wermuth, C. G. Handbook of Pharmaceutical Salts; Properties, Selection, and Use. Zurich: Verlag Helvetica ChimicaActa; (2008).

[25] Paulekuhn, G S, Dressman, J B, & Saal, C. Trends in Active Pharmaceutical Ingredient Salt Selection based on Analysis of the Orange Book Database. Journal of Medicinal Chemistry (2007). http://pubs.acs.org/doi/abs/10.1021/jm701032y(accessed june 2012)., 50(26), 6665-6672.

[26] Brodin, A, Nyqvist-mayer, A, Broberg, F, Wadsten, T, & Forslund, B. Phase diagram and aqueous solubility of the lidocaine-prilocaine binary system. Journal of Pharmaceutical Sciences (1984). http://onlinelibrary.wiley.com/doi/10.1002/jps.2600730413/abstract(accessed june 2012)., 73(4), 481-484.

[27] Karpinski, P H. Polymorphism of Active Pharmaceutical Ingredients. Chemical Engineering & Technology (2006). http://onlinelibrary.wiley.com/doi/10.1002/ceat.200500397/abstract(accessed june 2012)., 29(2), 233-237.

[28] Lu, J, & Rohani, S. Polymorphism and Crystallization of Active Pharmaceutical Ingredients (APIs) Current Medicinal Chemistry (2009). http://www.benthamdirect.org/pages/content.php?CMC/2009/00000016/00000007/0006C.SGM(accessed june 2012)., 16, 884-905.

[29] Reichert, W M, Holbrey, J D, Vigour, K B, Morgan, T D, Broker, G A, & Rogers, R D. Approaches to Crystallization from Ionic Liquids: Complex Solvents-Complex Results, or, a Strategy for Controlled Formation of New Supramolecular Architectures? ChemInform (2007). http://onlinelibrary.wiley.com/doi/10.1002/chin.200712247/abstract?deniedAccessCustomisedMessage=&userIsAuthenticated=true(accessed june 2012)., 38(12), 4767-4779.

[30] Golding, J, & Forsyth, S. MacFarlane D R, Forsyth M, Deacon G B. Methanesulfonate and *p*-toluenesulfonate salts of the N-methyl-N-alkylpyrrolidinium and quaternary ammonium cations: novel low cost ionic liquids. Green Chemistry (2002). http://pubs.rsc.org/en/content/articlelanding/2002/gc/b201063a(accessed june 2012)., 4(3), 223-229.

[31] Abdul-sada, A K, Elaiwi, A E, & Greenway, A M. European Journal of Mass Spectrometry (1997). http://www.impublications.com/content/abstract?code=E03_0245(accessed june 2012)., 3(3), 245-247.

[32] Dean, M P, Turanjanin, J, & Yoshizawa-fujita, M. MacFarlane D R, Scott J L. Exploring an Anti-Crystal Engineering Approach to the Preparation of Pharmaceutically Active Ionic Liquids. Crystal Growth & Design (2009). http://pubs.acs.org/doi/abs/10.1021/cg8009496accessed 20 june 2012)., 9(2), 1137-1145.

[33] Higasio, Y S, & Shoji, T. Heterocyclic compounds such as pyrroles, pyridines, pyrollidins, piperdines, indoles, imidazol and pyrazins. Applied Catalysis A (2001). http://www.sciencedirect.com/science/article/pii/S0926860X01008158(accessed june 2012).

[34] Belieres, J-P, & Angell, C A. Protic Ionic Liquids: Preparation, Characterization, and Proton Free Energy Level Representation. The Journal of Physical Chemistry B (2007). http://pubs.acs.org/doi/abs/10.1021/jp067589u(accessed june 2012)., 111(18), 4926-4937.

[35] Ferraz, R, Branco, L C, Marrucho, I M, Araújo, J, Rebelo, M, Ponte, L P N, Prudêncio, M N, Noronha, C, & Petrovski, J P. Z. Developmentof novel ionicliquidsbasedonampicillin. Medicinal Chemistry Communication (2012). http://pubs.rsc.org/en/content/articlelanding/2012/md/c2md00269haccessed 20 june 2012)., 3(4), 494-497.

[36] Fukumoto, K, Yoshizawa, M, & Ohno, H. Room Temperature Ionic Liquids from 20 natural Amino Acids. Journal of the Americal Chemical Society (2005). http://pubs.acs.org/doi/abs/10.1021/ja043451iaccessed 20 june 2012)., 127(8), 2398-2399.

[37] Bica, K, Rijksen, C, Nieuwenhuyzen, M, & Rogers, R D. In search of pure liquid salt forms of aspirin: ionic liquid approaches with acetylsalicylic acid and salicylic acid. Physical Chemistry Chemical Physics (2010). http://pubs.rsc.org/en/content/articlelanding/2010/cp/b923855gaccessed 20 june 2012)., 12(8), 2011-2017.

[38] Bica, K, & Rogers, R D. Confused ionic liquid ions-a ''liquification'' and dosage strategy for pharmaceutically active salts. Chemical Communications (2010). http://pubs.rsc.org/en/content/articlelanding/2010/cc/b925147b(accessed june 2012)., 46(8), 1215-1217.

[39] Thalladi, V R, Goud, B S, Hoy, V J, Allen, F H, Howard, J, & Desiraju, K. G R. Supramolecularsynthons in crystal engineering. Structure simplification, synthon robustness and supramolecularretrosynthesis. Chemical Communications (1996). http://pubs.rsc.org/en/content/articlelanding/1996/cc/cc9960000401(accessed june 2012).

[40] Desiraju, G R. Crystal Engineering: A Holistic View. AngewandteChemie International Edition (2007). http://onlinelibrary.wiley.com/doi/10.1002/anie.200700534/abstract(accessed june 2012)., 46(44), 8342-8356.

[41] Desiraju, G R. SupramolecularSynthons in Crystal Engineering- A New Organic Synthesis. AngewandteChemie International Edition in English (1995). http://onlinelibrary.wiley.com/doi/10.1002/anie.199523111/abstract(accessed june 2012)., 34(21), 2311-2327.

[42] Blagden, N, Matas, M, Gavan, P T, & York, P. Crystal engineering of active pharmaceutical ingredients to improve solubility and dissolution rates. Advanced Drug Delivery Reviews (2007). http://www.sciencedirect.com/science/article/pii/S0169409X07000828(accessed june 2012)., 59(7), 617-630.

[43] Trask, A V, & Jones, W. Crystal Engineering of Organic Cocrystals by the Solid-State Grinding Approach. Topics in Current Chemistry (2005). http://www.springerlink.com/content/j54wclkmeug3y254/(accessed june 2012)., 254, 41-70.

[44] Vishweshwar, P, Mcmahon, J A, Bis, J A, & Zaworotko, M J. Pharmaceutical co-crystals. Journal of Pharmaceutical Sciences (2006). http://onlinelibrary.wiley.com/doi/10.1002/jps.20578/abstract(accessed june 2012)., 95(3), 499-516.

[45] Fredlake, C P, Crosthwaite, J M, Hert, D G, Aki, S, & Brennecke, V K. J F. Thermophysical Properties of Imidazolium-Based Ionic Liquids. Journal of Chemical & Engineering Data (2004). http://pubs.acs.org/doi/abs/10.1021/je034261aaccessed 20 june 2012)., 49(4), 954-964.

[46] Pernak, J, Syguda, A, Mirska, I, Pernak, A, Nawrot, J, Pradzynska, A, Griffin, S T, & Rogers, R D. Choline-Derivative-Based Ionic Liquids. Chemistry A European Journal (2007). http://onlinelibrary.wiley.com/doi/10.1002/chem.200700285/abstractaccessed 20 june 2012)., 13(24), 6817-6827.

[47] Hough-troutman, W L, Smiglak, M, Griffin, S, Reichert, W M, Mirska, I, Jodynis-liebert, J, Adamska, T, Nawrot, J, Stasiewicz, M, Rogers, R D, & Pernak, J. Ionic liquids with dual biological function: sweet and anti-microbial, hydrophobic quaternary ammonium-based salts. New Journal of Chemistry (2009). http://pubs.rsc.org/en/content/articlelanding/2009/nj/b813213paccessed 20 june 2012)., 33(1), 26-33.

[48] Viau, L, Tourne-peteilh, C, Devoisselle, J-M, & Vioux, A. Ionogels as drug delivery system: one-step sol-gel synthesis using imidazoliumibuprofenate ionic liquid. Chemical Communications. (2010). http://pubs.rsc.org/en/content/articlelanding/2010/cc/b913879jaccessed 20 june 2012)., 46(2), 228-230.

Chapter 22

Increase in Thermal Stability of Proteins by Aprotic Ionic Liquids

Hidetaka Noritomi

Additional information is available at the end of the chapter

http://dx.doi.org/10.5772/51231

1. Introduction

Proteins are biomolecules of great importance in the biochemical processes such as the medical, pharmaceutical, and food fields, since they exhibit their outstanding biological activities under mild condition. However, most of proteins dissolved in an aqueous solution are immediately denatured and inactivated at high temperatures due to the disruption of weak interactions including ionic interactions, hydrogen bonds, and hydrophobic interactions, which are prime determinants of protein tertiary structures [5-3]. In particular, protein aggregation easily occurs upon the exposure of the hydrophobic parts of proteins, which are usually located in the inside of native proteins, and this phenomenon becomes the major problem because of the fast irreversible inactivation. Thermal denaturation of proteins is a serious problem not only in the separation and storage of proteins but also in the processes of biotransformation, biosensing, drug production, and food manufacturing. Several strategies have so far been proposed in order to prevent thermal denaturation of proteins [4-14]. They include chemical modification, immobilization, genetic modification, and addition of stabilizing agents. The addition of stabilizing agents is one of the most convenient methods for minimizing thermal denaturation, compared to other methods. It has been reported that inorganic salts, polyols, sugars, amino acids, amino acid derivatives, chaotropic reagents, and water-miscible organic solvents are available for improving protein stability. However, these additives do not sufficiently prevent irreversible protein aggregation or some of them are no longer stable at high temperatures.

Ionic solvent that is liquid at room temperature has attracted increasing attention as a green solvent for the chemical processes because of the lack of vapor pressure, the thermal stability, and the high polarity [15, 16]. Chemical and physical properties of ionic liquids can be changed by the appropriate modification of organic cations and anions, which are constituents of ionic liquids. It has recently been reported that protic ionic liquids such as alkylammonium salts keep the stability of proteins in an aqueous solution at high temperatures [17, 18], and amyloid fibrils of proteins are dissolved in protic ionic liquids and are refolded by dilution with an aqueous solution [19]. On the other hand, biotransformation in ionic liquids has increasingly been studied [20-2]. Aprotic ionic liquids such as immidazolium salts have mainly been employed as reaction media, since the high activity of enzymes is exhibited as usual. We have found that the activity of protease is highly maintained not only in water-immiscible aprotic ionic liquids but also in water-miscible aprotic ionic liquids [23, 24].

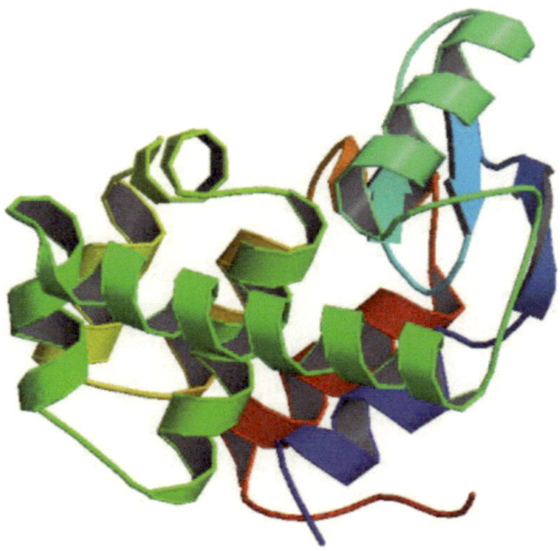

Figure 1. Structure of hen egg white lysozyme.

Ionic liquid	Structure	m.p.	Water miscibility
[emim][Tf]	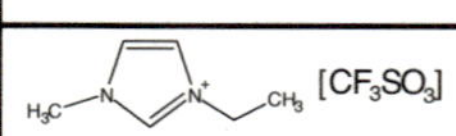	-9	Miscible
[emim][BF_4]	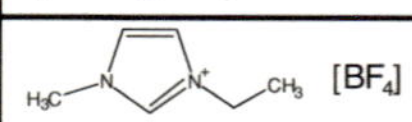	14.6	Miscible
[emim][Cl]	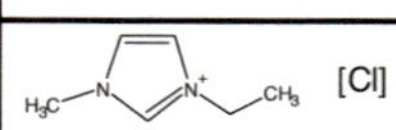	78	Miscible

Figure 2. Structures of ionic liquids used in the present work.

http://dx.doi.org/10.5772/51231

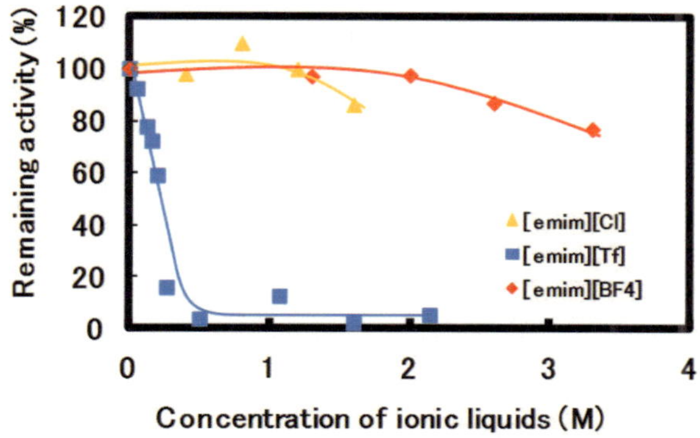

Figure 3. Effect of concentration of ionic liquids on remaining activity of lysozyme during incubation at 25 ℃. The aqueous solution of 100 μM lysozyme with requisite concentration of ionic liquids was incubated in a water bath thermostated at 25 ℃ for 30 min.

However, despite such potential capability of aprotic ionic liquids, there have not been any works on the thermal stability of proteins in aqueous solutions containing water-miscible aprotic ionic liquids.

In the chapter, the effect of water-miscible aprotic ionic liquids consisting of 1-ethyl-3-methylimidazolium cations and several kinds of anions on thermal stability of proteins in aqueous solutions is mainly discussed [25].

2. Dependence of the remaining activity of lysozyme on the concentration of ionic liquids after the incubation at 25 °C

As a model protein, hen egg white lysozyme has been employed as shown in Fig. 1, since it has been well investigated regarding its structure, properties, and functions [26]. Lysozyme is a compact protein of 129 amino acids which folds into a compact globular structure. The molecular weight of lysozyme is 14,300, and the structure of lysozyme includes α-helices, β sheets, random coils, β turns, and disulfide bonds, which are typical structures of proteins. Lysozyme attacks peptidoglycans in the cell walls of Gram-positive bacteria, and catalyzes hydrolysis of 1,4-beta-linkages between N-acetylmuramic acid and N-acetyl-D-glucosamine residues in a peptidoglycan. Accordingly, lysozyme has been used as an anti-inflammatory agent, a preservative, a freshness-keeping agent, an antibacterial agent, a disinfectant, and so on.

Room temperature ionic liquids of alkyl imidazolium cations are widely used, and are commercially available. Figure 2 shows structures and properties of ionic liquids introduced in this chapter. 1-Ethyl-3-methylimidazolium trifluoromethanesulfonate, 1-ethyl-3-methylimidazolium tetrafluoroborate, and 1-ethyl-3-methylimidazolium chloride are abbreviated to [emim][Tf], [emim][BF_4], and [emim][Cl], respectively. Their properties such as melting point alter by switching from one anion to another.

It has been well known that ions and other compatible solutes affect enzyme activity [22]. Figure 3 shows the plot of the remaining activity of lysozyme against the concentration of ionic liquids after the incubation at 25 ℃ for 30 min. The remaining activity in the presence of [emim][Cl] or [emim][BF_4] was independent on the concentration of ionic liquids till 1.2 M [emim][Cl] or 2.0 M [emim][BF_4] and gradually dropped, while it in the presence of [emim][Tf] dramatically decreased with an increase in the concentration of [emim][Tf]. These results indicate that [emim][Tf] tends to strongly function as a denaturant, compared with [emim][Cl] and [emim][BF_4]. Electrolytes promote or inhibit the stability of proteins according to the kind of electrolytes [27-28].

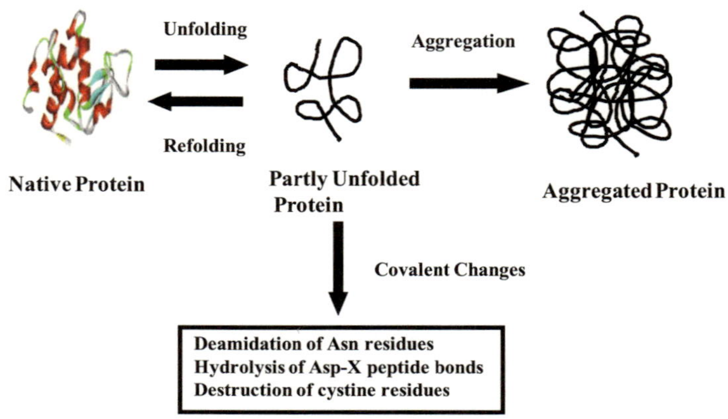

Figure 4. Schematic illustration of thermal denaturation of proteins.

3. Thermal stabilization of lysozyme by aprotic ionic liquids

3.1. Thermal inactivation of lysozyme

When proteins dissolved in an aqueous solution are placed at high temperatures, most of proteins are immediately unfolded due to the disruption of weak interactions including ionic effects, hydrogen bonds, and hydrophobic interactions, which are prime determinants of protein tertiary structures. In addition, the intermolecular aggregation among unfolded proteins and the chemical deterioration reactions in unfolded proteins proceed as shown in Fig. 4 [2, 29, 30]. In particular, protein aggregation easily occurs upon the exposure of the hydrophobic surfaces of a protein, and this phenomenon becomes the major problem because of the fast irreversible inactivation. On the other hand, when a heated solution of denatured proteins without protein aggregation is slowly cooled back to its normal biological temperature, the reverse process, which is renaturation with restoration of protein function, tends to occur. Accordingly, if stabilizing agents can sufficiently prevent irreversible aggregation of unfolded proteins, it is expected that unfolded proteins are refolded by cooling treatment, and the high remaining activity is obtained.

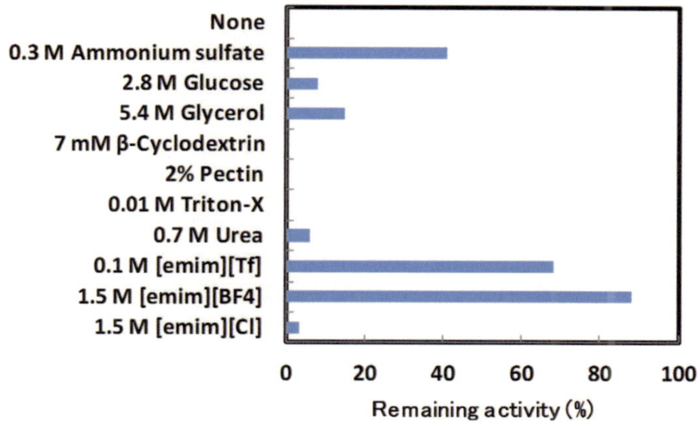

Figure 5. Remaining activities of lysozyme in the presence of various kinds of additives after heat treatment at 90 ℃ for 30 min.

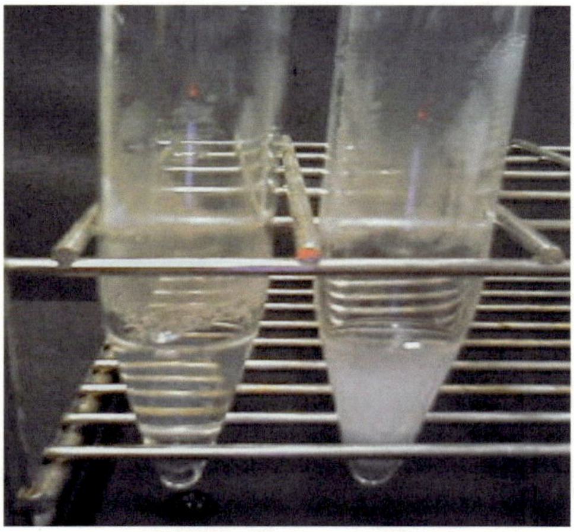

(a) (b)

Figure 6. Photographs of lysozyme solutions after heat treatment at 90 ℃ for 30 min: (a) lysozyme solution with 1.5 M [emim][BF_4], (b) lysozyme solution without [emim][BF_4].

Figure 5 represents the remaining activities of lysozyme in the presence of various kinds of additives after heat treatment at 90 ℃ for 30 min as an accelerated test. Lysozyme without additives lost its activity perfectly after heat treatment. Native lysozyme solution immediately became turbid due to the aggregation of thermally-denatured proteins, as soon as heat treatment was carried out, as shown in Fig. 6(b). It has been reported that the precipitation due to protein aggregation at high temperatures is observed above 10 μM lysozyme [31]. As lysozyme concentration in the present work was 100 μM (1.4 mg/mL) which was ten times higher than that, the formation of protein aggregation was dramatically accelerated. Ammonium sulfate, which was an inorganic salt, glucose and glycerol, which were polyols, and urea, which was a chemical denaturant, inhibited the formation of protein aggregation, and exhibited thermal stabilization to some extent. β-Cyclodextrin, which was an inclusion compound, pectin, which was a thickener, and Triton-X, which was a nonionic surfactant, could not maintain the stability of lysozyme at high temperatures, although they were widely used as a stabilizer. On the other hand, [emim][BF_4] and [emim][Tf] showed high remaining activities. The lysozyme solution in the presence of ionic liquids was transparent after heat treatment, as seen in Fig. 6(a). When lysozyme solution in the presence of protic ionic liquids such as alkylammonium formates is heated at 90 ℃, protein aggregation is prevented, and any cloudy appearance is absent [18]. The hydrophobic core of lysozyme unfolded by heat interacts with the cation of ionic liquids, and cation adsorption results in acquisition of a net positive charge preventing aggregation via electrostatic repulsion [17].

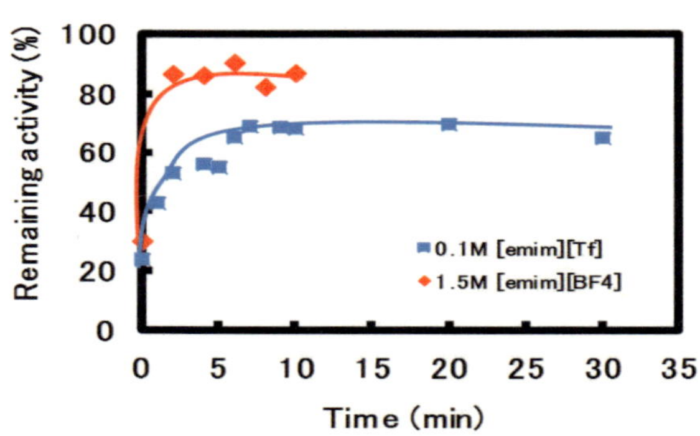

Figure 7. Time dependence of remaining activity of lysozyme with ionic liquids under cooling at 25 ℃ after heat treatment at 90 ℃ for 30 min. After heat treatment, the aqueous solution of 100 μM lysozyme with 0.1 M [emim][Tf] or 1.5 M [emim][BF_4] was incubated in a water bath thermostated at 25 ℃.

3.2. Refolding of lysozyme by ionic liquids

When the formation of protein aggregation is inhibited at high temperatures by ionic liquids, and thermally-denatured proteins are individually dispersed in an aqueous solution, it is probably that denatured proteins are gradually refolded under cooling conditions. Figure 7 shows the time course of remaining activity in the presence of ionic liquids under cooling treatment at 25 ℃ after the heat treatment at 90 ℃ for 30 min. The remaining activities of lysozyme in the presence of 1.5 M [emim][BF_4] and 0.1 M [emim][Tf] exhibited 30 and 24%, respectively, just after heat treatment. The remaining activity of lysozyme with 1.5 M

[emim][BF_4] or 0.1 M [emim][Tf] rapidly increased with incubation time at 25 ℃, and reached a plateau around 2 and 7 min, respectively. In sufficiently low concentration of proteins, where protein aggregation is not formed, when the hydrophobic core of proteins is exposed, but the disulfide bonds keep intact, denatured proteins gradually tend to refold to their native structures on cooling after thermal denaturation [32-36]. The refolding of thermally-denatured proteins is enhanced in the presence of protic ionic liquids such as alkylammonium nitrate and alkylammonium formates [32, 21]. Moreover, N'-alkyl and N'-(ω-hydroxyalkyl) N-methylimidazolium chlorides refold denatured proteins such as hen egg white lysozyme and the single-chain antibody fragment ScFvOx [37].

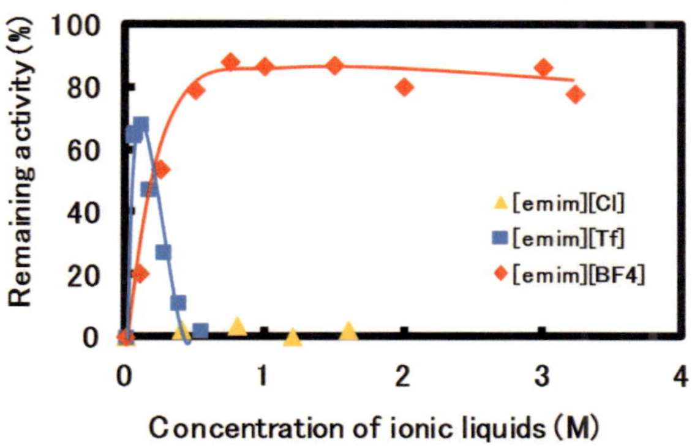

Figure 8. Effect of concentration of ionic liquids on remaining activity of lysozyme after heat treatment at 90 ℃ for 30 min. The aqueous solution of 100 μM lysozyme with requisite concentration of ionic liquids was incubated in a silicone oil bath thermostated at 90 ℃ for 30 min.

3.3. Dependence of the remaining activity of lysozyme on the concentration of ionic liquids via heat treatment

The stability of proteins depends upon the kind and concentration of electrolytes [27, 28]. Figure 8 shows the plot of the remaining activity of lysozyme against the concentration of ionic liquids after the heat treatment at 90 ℃ for 30 min. The remaining activity was strongly dependent on the concentration of [emim][BF_4] or [emim][Tf], while the effect of concentration of [emim][Cl] was not observed. The remaining activity in the presence of [emim][BF_4] increased with an increase in the concentration of [emim][BF_4] and reached a plateau around 0.8 M. The remaining activity in the presence of [emim][Tf] dramatically increased with increasing the concentration of [emim][Tf], the maximal remaining activity was obtained at 0.1 M [emim][Tf], and then decreased steeply. As seen in Fig. 3, the remaining activity decreased at 25 ℃ with an increase in the concentration of [emim][Tf]. Chemical denaturants, such as urea and guanidine hydrochloride, can promote dissolution of inclusion bodies, which are protein aggregation formed by prokaryotic expression systems [38]. Similarly, [emim][Tf] inhibits the formation of protein aggregation at low [emim][Tf] concentrations, but it mainly denatures proteins at higher [emim][Tf] concentrations. Moreover, it has been reported that after heat treatment the remaining activity of lysozyme increases with an increase in the concentration of ethylammonium formate and 2-methoxyethylammonium formate, while the remaining activity increases at low concentration of propylammonium formate, but at higher concentrations of propylammonium formate the protein spontaneously denatures [18]. Thus, the dependence of concentration of ionic liquids on the remaining activity of proteins changes by switching from one ionic liquid to another.

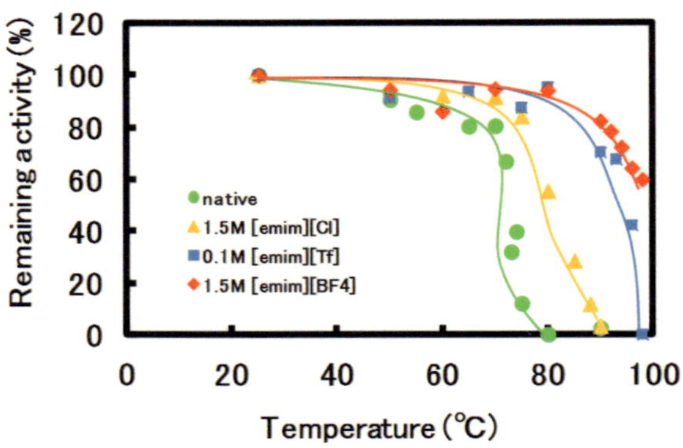

Figure 9. Thermal denaturation curves of lysozyme with or without ionic liquids. The aqueous solution of 100 μM lysozyme with or without ionic liquids was incubated in a silicone oil bath thermostated at requisite temperature for 30 min.

3.4. Dependence of the remaining activity of lysozyme on the temperature of heat treatment

The thermal inactivation of proteins more rapidly proceeds by higher temperatures. Figure 9 shows the relationship between the temperature of heat treatment and the remaining activity of lysozyme in aqueous solutions containing water-miscible ionic liquids after the heat treatment for 30 min. As seen in the figure, the remaining activity of lysozyme without ionic liquids gradually decreased with an increase in temperature below 70 ℃, accompanied with the formation of precipitation due to protein aggregation, drastically dropped in the range from 70 to 80 ℃, and was then lost at temperatures of 80 ℃ or higher. The transition temperature was exhibited around 75 ℃, similar to the case measured by differential scanning calorimetry [17]. On the other hand, the remaining activity of lysozyme with 1.5 M [emim][Cl] gradually decreased with an increase in temperature below 75 ℃, and drastically dropped in the range from 75 to 90 ℃. The remaining activity of lysozyme with 1.5 M [emim][BF_4] was highly maintained below 80 ℃, gradually decreased with temperature, and the remaining activity depicted 60% at 98 ℃. Similarly, the remaining activity of lysozyme with 0.1 M [emim][Tf] was highly retained below 80 ℃, gradually decreased with temperature below 92 ℃, drastically dropped in the range from 92 to 98 ℃, and was then lost at 98 ℃. These results

http://dx.doi.org/10.5772/51231

indicate that the addition of aprotic ionic liquids to an aqueous solution of lysozyme effectively improves the thermal stability of lysozyme at high temperatures.

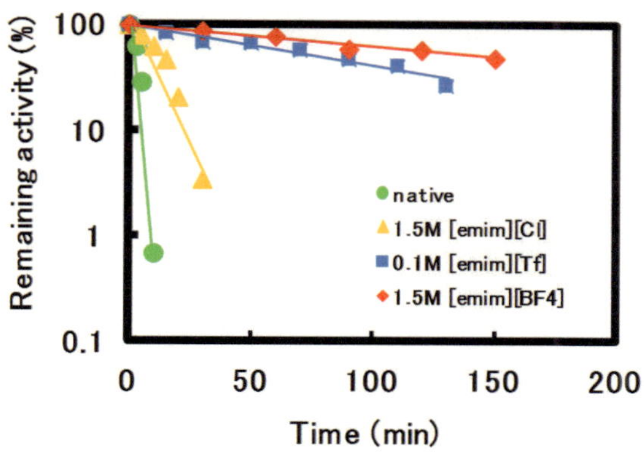

Figure 10. Time dependence of remaining activity with or without ionic liquids after heat treatment at 90 ℃. The aqueous solution of 100 μM lysozyme with or without ionic liquids was incubated in a silicone oil bath thermostated at 90 ℃.

Ionic liquid	Rate constant (min^{-1})	Half line (min)
none	0.43	1.6
1.5 M [emim][Cl]	0.065	22
0.1 M [emim][Tf]	0.0081	86
1.5 M [emim][BF_4]	0.0049	141

Table 1. Rate constants and half lives of inactivation of lysozyme at 90 ℃.

3.5. Time course of remaining activity of lysozyme via heat treatment with or without ionic liquids

Heating time directly enhances the thermal inactivation of proteins. Figure 10 shows time course of remaining activity of lysozyme with or without ionic liquids through the heat treatment at 90 ℃. The remaining activity of lysozyme without ionic liquids dramatically decreased with an increase in time, accompanied with the formation of protein aggregation, and was almost lost at 10 min. It has been reported that the remaining activity in the thermal denaturation process accompanied with the formation of protein aggregation follows first-order kinetics [31]. As seen in the figure, the relationship of the remaining activity of proteins in the absence of ionic liquids with the time of heat treatment could be correlated by first-order kinetics. On the other hand, 1.5 M [emim][BF_4] or 0.1 M [emim][Tf] effectively prevented the thermal inactivation of lysozyme. In the presence of ionic liquids the turbidity of solutions due to protein aggregation was not observed during heat treatment. This indicates that the thermal inactivation mainly results from the covalent change as shown in Fig. 4. The plots of remaining activity versus heat treatment time on thermal inactivation of lysozyme in the presence of ionic liquids followed first-order kinetics on linearity. It has been reported that the thermal inactivation of lysozyme obeyed first-order kinetics when it irreversibly proceeded by the covalent change without the formation of protein aggregation [39]. Table 1 represents rate constants and half lives of inactivation of lysozyme with or without ionic liquids calculated from the fitting curves in Fig. 10. The half lives with 1.5 M [emim][BF_4], 0.1 M [emim][Tf], and 1.5 M [emim][Cl] were 88, 54, or 6.9 times longer than that without ionic liquids, respectively.

4. Conclusion

In this chapter the effect of water-miscible aprotic ionic liquids on thermal stability of lysozyme has been described. Aprotic ionic liquids could sufficiently prevent thermally-denatured proteins from aggregating. The activity of lysozyme in the presence of aprotic ionic liquids was kept to some extent, just after heat treatment at high temperatures. Moreover, thermally-denatured lysozyme was effectively refolded by cooling. Consequently, the high remaining activity of lysozyme was obtained. These results indicate that aprotic ionic liquids act not only as an inhibitor of protein aggregation, but also as a protective agent in the native structure of protein and an accelerator in the refolding of thermally-denatured proteins. The remaining activity of lysozyme markedly depended upon the kind of anions and the concentration of ionic liquids. Especially, [emim][Tf] exhibited the thermostabilization effect of proteins at low concentrations, but mainly worked as a denaturant of proteins at high concentrations. The effect of [emim][BF_4] and [emim][Tf] upon thermal stabilization at high temperatures was much superior to that of [emim][Cl]. As chemical and physical properties of ionic liquids can be changed by the appropriate modification of organic cations and anions, which are constituents of ionic liquids, it is expected that the ionic liquid, which is more suitable for the thermal stabilization of proteins, is prepared by tailoring the constituents of ionic liquids.

Author details

Hidetaka Noritomi[1]

1 Tokyo Metropolitan University Japan

References

[1] Ballesteros, A., Plou, F. J., Iborra, J. L., & Halling, P.J. (1998). *Stability and Stabilization of Biocatalysts*, Elsevier, Amsterdam.

[2] Volkin, D.B., & Klibanov, A.M. (1989). Minimizing protein inactivation. *Protein Function: Practical Approach*, 1-24, IRL Press, Oxford.

[3] Klibanov, A.M. (1983). Stabilization of enzymes against thermal inactivation. *Adv. Appl. Microbiol.*, 29, 1-28.

[4] Illanes, A. (1999). Stability of biocatalysts. *Electro. J. Biotechnol.*, 2, 1-9.

[5] Noritomi, H., Kai, R., Iwai, D., Tanaka, H., Kamiya, R., Tanaka, M., Muneki, K., & Kato, S. (2011b). Increase in thermal stability of proteins adsorbed on biomass charcoal powder prepared from plant biomass wastes. *J. Biomedical, Sci. Eng.*, 4, 692-698.

[6] Gerlsma, S.Y. (1968). Reversible denaturation of ribonuclease in aqueous solutions is influenced by polyhydric alcohols and some other additives. *J. Biological Chem.*, 243, 957-961.

[7] Kaushik, J.K., & Bhat, R. (1998). Thermal stability of proteins in aqueous polyol solutions: role of the surface tension of water in the stabilizing effect of polyols. *J. Phys. Chem. B*, 102, 7058-7066.

[8] Back, J. F., Oakenfull, D., & Smith, M. B. (1979). Increased thermal stability of proteins in the presence of sugars and polyols. *Biochemistry*, 18, 5191-5196.

[9] Lee, J. C., & Timasheff, S.N. (1981). The stabilization of proteins by sucrose. *J. Biological Chem.*, 256, 7193-7201.

[10] Santoro, M.M., Liu, Y., Khan, S.M.A., Hou, L.X., & Bolen, D.W. (1992). Increased thermal stability of proteins in the presence of naturally occurring osmolytes. *Biochemistry*, 31, 5278-5283.

[11] Yancey, P.H., Clark, M.E., Hand, S. C., Bowlus, R.D., & Somero, G.N. (1982). Living with water stress: evolution of osmolyte systems. *Science,*, 217, 1214-1222.

[12] Arakawa, T., Bhat, R., & Timasheff, S.N. (1990). Why preferential hydration does not always stabilize the native structure of globular proteins. *Biochemistry*, 29, 1924-1931.

[13] Ikegaya, K. (2005). Kinetic analysis about the effects of neutral salts on the thermal stability of yeast alcohol dehydrogenase. *J. Biochemistry,*, 137, 349.

[14] Cioci, F., & Lavecchia, R. (1998). Thermostabilization of proteins by water-miscible additives. *Chem. Biochem. Eng. Q.*, 12, 191-199.

[15] Welton, T. (1999). Room-temperature ionic liquids. Solvents for synthesis and calalysis. *Chem. Rev.*, 99, 2071-2083.

[16] Greaves, T.L., & Drummond, C.J. (2008). Protic ionic liquids: properties and applications. *Chem. Rev.*, 108, 206-237.

[17] Summers, C.A., & Fowers, I. I. R. A. (2000). Protein renaturation by the liquid organic salt ethylammonium nitrate. *Protein Sci.*, 9, 2001-2008.

[18] Mann, J.P., Mc Cluskey, A., & Atkin, R. (2009). Activity and thermal stability of lysozyme in alkylammonium formate ionic liquids- influence of cation modification. *Green Chem.*, 11, 785-792.

[19] Byrne, N., & Angell, C. A. (2009). Formation and dissolution of hen egg white lysozyme amyloid fibrils in protic liquids. *Chem Comm.*, 1046-1048.

[20] Moniruzzaman, M., Nakashima, K., Kamiya, N., & Goto, M. (2010). Recent advances of enzymatic reactions in ionic liquids. *Biochem. Eng. J.*, 48, 295-314.

[21] Yang, Z., & Pan, W. (2005). Ionic liquids: Green solvents for nonaqueous biocatalysis. *Enz. Microbial Technol.*, 37, 19-28.

[22] Zhao, H. (2011a). Effect of ions and other compatible solutes on enzyme activity, and its implication for biocatalysis using ionic liquids. *J. Mol. Catal. B: Enzymatic*, 37, 16-25.

[23] Noritomi, H., Nishida, S., & Kato, S. (2007). Protease-catalyzed esterification of amino acid in water-miscible ionic liquid. *Biotechnol Lett.*, 29, 1509-1512.

[24] Noritomi, H., Suzuki, K., Kikuta, M., & Kato, S. (2009). Catalytic activity of α-chymotrypsin in enzymatic peptide synthesis in ionic liquids. *Biochem. Eng. J.*, 47, 27-30.

[25] Noritomi, H., Minamisawa, K., Kamiya, R., & Kato, S. (2011b). Thermal stability of proteins in the presence of aprotic ionic liquids. *J. Biomedical Sci. Eng.*, 4, 94-99.

[26] Jollès, P. (1996). Lysozymes:. *Model Enzymes in Biochemistry and Biology.*, Birkhäuser Verlag, Basel.

[27] Von Hippel, P.H., & Schleich, T. (1969). The effects of neutral salts on the structure and conformational stability of macromolecules in solution. *Structure and Stability of Biological Macromolecules*, 417-574, Marcel-Dekker, New York.

[28] Nostro, P.L., & Ninham, B.W. (2012). Hofmeister phenomena: an update on ion specificity in biology. *Chem. Rev.*, 112, 2286-2322.

[29] Lumry, R., & Eyring, H. (1954). Conformation changes of proteins. *J. Phys. Chem.*, 58, 110-120.

[30] Zale, S.E., & Klibanov, A.M. (1983). On the role of reversible denaturation (unfolding) in the irreversible thermal inactivation of enzymes. *Biotechnol. Bioeng*, 25, 2221-2230.

[31] Nohara, D., Mizutani, A., & Sakai, T. (1999). Kinetic study on thermal denaturation of hen egg-white lysozyme involving precipitation. *J. Biosci. Bioeng.*, 87, 199-205.

[32] Ibara-Molero, B., & Sanchez-Ruiz, J.M. (1997). Are there equilibrium intermediate states in the urea-induced unfolding of hen egg-white lysozyme? *Biochemistry,*, 36, 9616-9624.

[33] Griko, Y.V., Freire, E., Privalov, G., Dael, H.V., & Privalov, P.L. (1995). The unfolding thermodynamics of c-type lysozyme- a calorimetric study of the heat denaturation of equine lysozyme. *J. Mol. Biol.*, 252, 447-459.

[34] Privalov, P.L., & Khechinashvili, N.N. (1974). A thermodynamic approach to the problem of stabilization of globular protein structure. *J. Mol. Biol.*, 86, 665-684.

[35] Khechinashvili, N.N., Privalov, P.L., & Tiktopulo, E.I. (1973). Calorimetric investigation of lysozyme thermal denaturation. FEBS Lett., , 30, 57-60.

[36] Anfinsen, C.B. (1973). Principles that govern the folding of protein chains. *Sci.*, 181, 223-230.

[37] Lange, C., Patil, G., & Rudolph, R. (2005). Ionic liquids as refolding additivesN′-alkyl and N′-(ω-hydroxyalkyl) N-methylimidazolium chlorides. *Protein Sci.*, 14, 2693-2701.

[38] Rudolph, R., & Lilie, H. (1996). In vitro folding of inclusion body proteins. *FASEB J.*, 10, 49-56.

[39] Ahern, T.J., & Klibanov, A.M. (1985). The mechanism of irreversible enzyme inactivation at 100 ℃. *Science*, 228, 1280-1284.

Section 4

Materials and Processing

Chapter 23

Use of Ionic Liquid Under Vacuum Conditions

Susumu Kuwabata, Tsukasa Torimoto,
Akihito Imanishi and Tetsuya Tsuda

Additional information is available at the end of the chapter

http://dx.doi.org/10.5772/ 52597

1. Introduction

Ionic liquid (IL) is a kind of salt that can stay as a liquid phase even at room temperature. However, researchers sometimes prefer to call it room temperature ionic liquid (RTIL) to distinguish between liquid salt at around room temperature and that at high temperature. Fig. 1 shows structural formulas of well-known ILs with their names and abbreviations. The liquid possesses several attracting characteristics like high ionic conductivity, wide electrochemical windows, and negligible vapor pressure [1-5]. All features imply that IL has high stability and inertness, which have become very useful to utilize IL as electrolytes for Li-ion secondary batteries and PEM fuel cells, reaction solvents for organic synthesis and nanoparticle preparation, and lubricants usable in cosmic space [4].

The negligible vapor pressure of most ILs at room-temperature means that such the ILs can be put in a vacuum chamber without any vaporization. This fact invented a new technological concept because there are several instruments that require vacuum conditions for sample analyses and material manufacturing. Those instruments are basically designed for dealing with solid samples because it is quite common sense that the vacuum conditions should provide dry atmosphere. In other words, conventional procedures with such the instruments cannot be applied to any wet sample, although researchers frequently meet the cases where they would like to deal with wet samples and liquid itself in vacuum equipments. Possibility to introduce ILs to the vacuum instruments could innovatively change the techniques requiring vacuum conditions. Then, some researchers including our research group started to put ILs in vacuum chambers of several instruments.

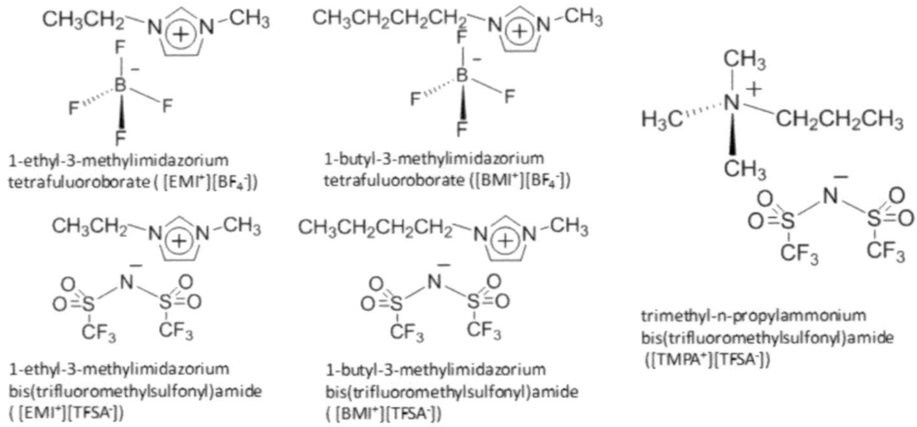

Figure 1. Structure formulas of typical ionic liquids with names and their abbreviations.

In this chapter, new techniques developed by putting ILs in the vacuum chambers of instruments will be introduced. Actually there are some other liquids having very low vapor pressure like silicon grease, which also can be introduced into the vacuum chamber. However, such oily medias cannot work as solvent for chemical and physical reaction because of their extremely high viscosity. On the other hand, ILs, which work well as solvent and electrolyte, make the dry vacuum conditions become wonderful wet world as will be introduced in this chapter.

2. XPS analysis

XPS is an instrument for analyzing the composition of solid materials and chemical state of each element. To detect generated photoelectrons with high sensitivity, vacuum condition is required. This condition is also effective for avoiding contamination of the sample surfaces. Although many researchers would like to put liquid into this instrument, vacuum condition in its sample chamber does not allow it. Nevertheless, some papers have reported the aggressive attempt to analyze the liquid surface by XPS with intricately designed sample stages [6-9]. In contrast, IL that is not vaporized under ultra-vacuum conditions can easily be put in the XPS chamber without any specific technique of modification.

$[EMI^+][C_2H_5SO_3^-]$ was the first IL that has been subjected to XPS analyses. It was found that IL emitted stable photoelectron flux, giving XPS spectra with high resolution [10-13]. Then, the obtained spectra enable a peak separation by the fitting calculations. Figure 2(a) shows the C1s spectrum of $[EMI^+][C_2H_5SO_3^-]$ together with the fitting curves based on C elements in five locations in the IL [14]. The XPS analysis of IL also allows detection of species dissolved in the IL, enabling in situ analysis of chemical reactions. An example is shown in Fig. 2(b) which http://dx.doi.org/10.5772/ 52597

were XPS spectra of $Pd(OAc)_2(PPh_3)$ dissolved in $[EMI^+][C_2H_5SO_3{}^-]$ [14]. Successive measurements showed that the intensity of spectra due to Pd(II) decreased, while the spectra due to Pd^o increased, showing decomposition of Pd complex known as the Heck catalyst.

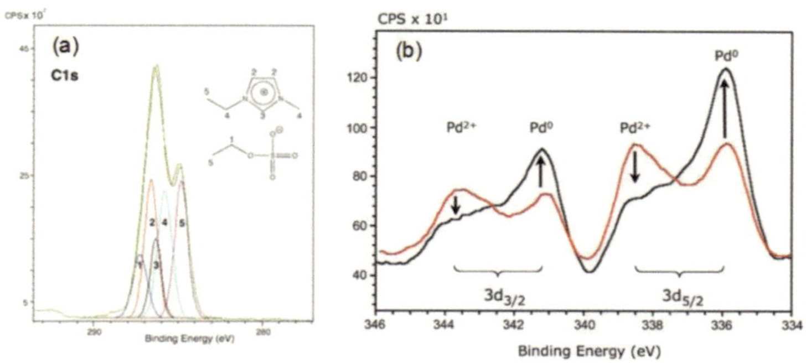

Figure 2. High resolution XPS spectra of $[EMI^+][C_2H_5OSO_3{}^-]$ detailing the C1s photoemission (a) and those of $Pd(Oac)_2$ in $[EMI^+][C_2H_5OSO_3{}^-]$. The red line shows data recorded at the start of the XPs experiment and the black line presents data recorded 6 h later.

3. MALDI Mass Spectroscopy

The matrix-assisted laser desorption/ionization mass spectrometry (MALDI-MS) is one of the groundbreaking analysis techniques that vaporize samples by laser irradiation with assistance of an appropriate matrix, as schematically illustrated in Fig. 3(a). This way allows to vaporize very large molecules like proteins that could not been analyzed without decomposition of the sample by the conventional mass spectrometry. The Matrix selection is quite essential, and an ideal matrix is a material possessing sufficient absorption coefficient for the laser beam, low vapor pressure, ability to dissolve or co-crystallize with sample, and ability to promote ionization of sample without its significant decomposition. The liquid matrices having low vapor pressure like glycerol and 3-nitrobenzyl alcohol have been utilized in the early research but their inherent volatility still causes some problems, such as decrease in their amounts with time. Another problem is that these liquids possess no UV absorbability, requiring addition of another photosensitive component.

Since IL possesses both no volatility and UV absorbability, it seems to be an ideal solution matrix for MALDI. However, usual ILs such as $[BMI^+][BF_4{}^-]$ and $[BMI^+][PF_6{}^-]$ were unfortunately unable to ionize samples dissolved in them [15]. Then, new ionic liquid family for the liquid matrixes were synthesized using solid acidic compounds of α-cyano-4-hydroxycinnamic acid (CHCA), sinapinic acid (SA), and 2,5-dihydroxybenzoic acid (DHB), which are widely used as solid matrixes for MALDI-MS [15,16]. Then, some of them were found to keep liquid state at room temperature and work as liquid matrices for detection of polymer and some biomolecules by MALDI-MS [15-18].

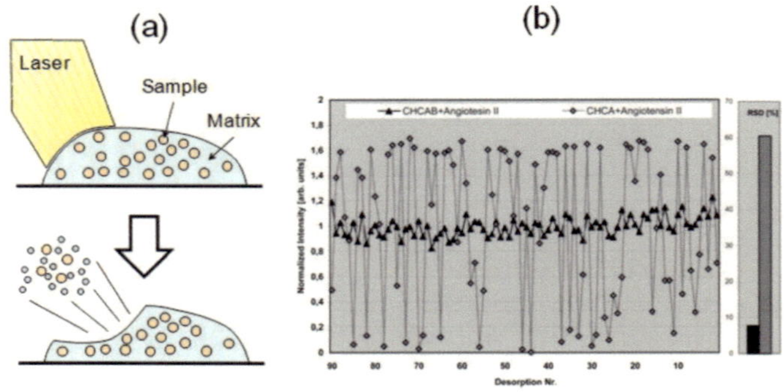

Figure 3. a) Schematic illustration of MALDI. (b) $[M + H]^+$ ion intensities from 90 positions on a human angiotensin II preparation with ionic liquid matrix CHCAB (black triangles) and with traditional CHCA matrix (grey squares). The relative standard deviations (RSD) of the data series are given as bar graphs. Black bars indicate RSD values found using ionic liquid matrixes, and gray bars indicate RSD values of the data series yielded by the respective traditional MALDI matrixes.

Fig.3(b) shows change in signal intensities of $[M + H]^+$ obtained at 90 different positions on a spot of sample-matrix mixture. As expected, the IL matrix of α-cyano-4-hydroxycinnamic acid butylamine (CHCAB) gave much narrow data dispersion than that obtained for the solid matrix of CHCA, indicating evidently usefulness of IL matrix for improvement of reproducibility. Another feature of the IL matrix is higher ability to suppress decomposition of sample than the conventional solid matrix. Use of CHCA-based guanidium salt and its analogous salts as IL matrixes enabled detection of oligosaccharides, which exhibit poor ionization efficiencies and tend to get thermal fragmentation through the loss of SO_3 groups, with suppression of loss of SO_3 [17,18].

4. Use of ILs for SEM observations

The first attempt was to observe ILs with a scanning electron microscope (SEM) [19]. The fact that ILs possess ionic conductivity but they do not possess electric conductivity gave an anticipation of charging of a IL drop during SEM observation. As a matter of fact, nonvolatile silicon oil, which can also be put in a vacuum chamber without vaporization, exhibited a white image with lots of noise because of charging behavior (Fig. 4(a)). Surprisingly, however, IL droplet gave a dark contrast images without any noise (Figure 4(b) -(d)), indicating that ILs are not charged by electron beam irradiation. Pulse radiolysis studies on ILs have revealed that electrons injected in ILs with high accelerated voltage are stabilized in condensed ions, allowing electrons to move in the liquid [20]. Consequently, ILs behave like electrically conducting materials for SEM observations. Based on this fact, several attempts have been done for SEM observations using ILs. The simplest way must be putting conductivity in place of metal or carbon deposition to insulating materials to observe them with a SEM. However, if neat IL was put onto the surface of the insulating material, existence of the liquid pools interfered observations of surface details of the abrasive paper, as shown in Fig. 5 (b), as compared with the case of the Au-deposited sample (Fig. 5 (a)). This troublesome can be resolved by dilution of IL with volatile solvent like alcohol. The abrasive paper was soaked in 2 mol dm^{-3} $[BMI^+][TFSA^-]$ / ethanol solution in a couple of seconds and was taken out of the solution. Leaving it in air for several ten seconds allowed vaporization of ethanol, resulting in stay of a thin IL layer on the sample. In fact, its SEM image as shown in Fig. 5c was quite similar to that obtained for the Au-coated abrasive paper [21].

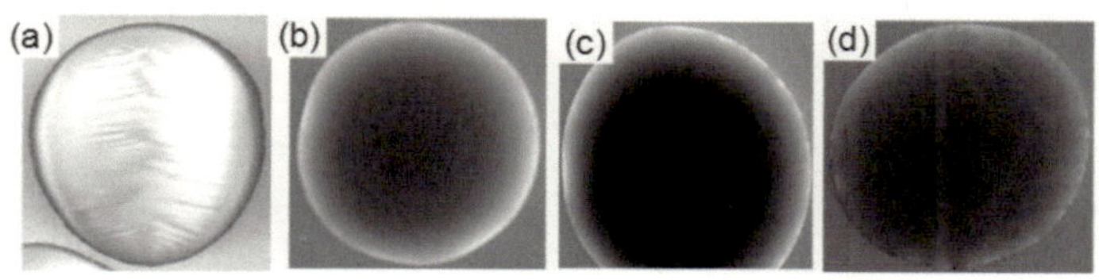

Figure 4. SEM images of droplets of silicon oil (a), $[BMI^+][BF_4^-]$ (b), $[EMI^+][BF_4]$ (c), and $[EMI^+][TFSA^-]$

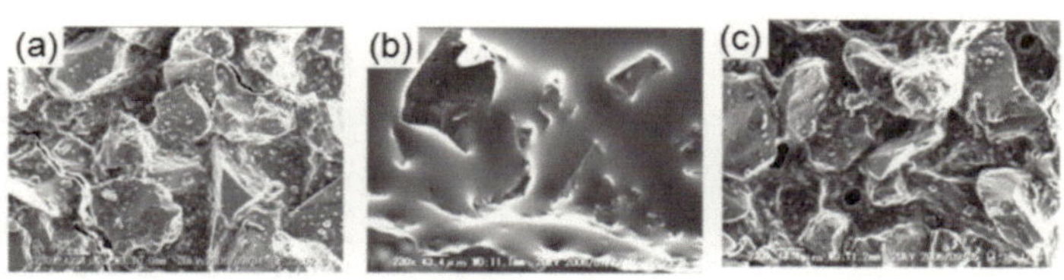

Figure 5. SEM images of surfaces of abrasive paper coated with gold (a), neat $[BMI^+][TFSA^-]$ (b), and /ethanol solution (c).

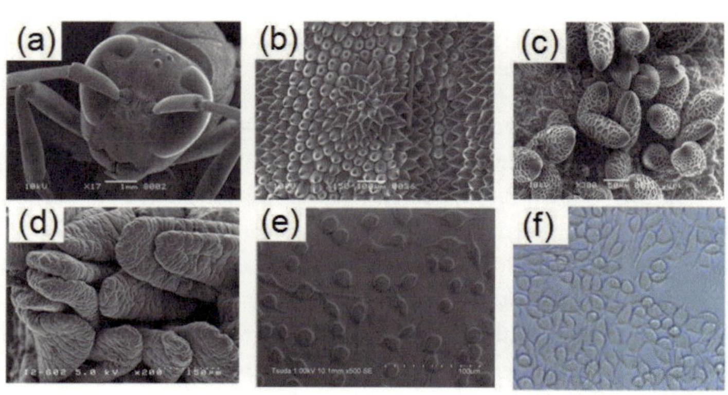

Figure 6. SEM images of insect, flower, tissue, and cell pretreated with IL; a) head of a yellow jacket, b) stamen of asteraceae, c) pollens of lily, d) villi of mouse small intestine, and mouse-derived fibroblast L929 cells (e). A picture (f) is optical microscope image of L929 cells for comparison.

Application of IL as an electric conducting material to an insulating sample gives another advantages, as compared with metal or carbon deposition. The liquid can keep the sample wet conditions even in vacuum chamber. This possibility has in particular a positive effect on observation of biological specimens [22-25]. Some examples are shown in Fig. 6 [22]. Since biological specimens have complex surface structures, metal or carbon deposition cannot perfectly deposit conducting films on the surfaces having dimples and indented places. However, liquid can reach anywhere on the complex surfaces, resulting in complete suppression of the charging behavior. Also replacement of water contained in the biological specimens with IL keeps the sample wet condition. As a result, SEM image of IL-treated fibrous blast cells, as shown in Fig. 6(e) was quite similar to shapes of cells containing water, which were observed by an optical microscope (Fig. 6(f)).

The IL treatment became now significant for biological and medical studies by electron microscope observations. An example is shown in Fig. 7 [23,24]. The metastasis is one of the most serious problems during cancer treatment. Such action has never been seen for the normal cells which prefer to adhere to each other during and after the cell fissions. This means that the normal cells must have some specific means to keep connections with neighboring cells. It is already known that something works well to make cell-to-cell junction, but direct observation of the something by an electron microscope has not yet been succeeded.

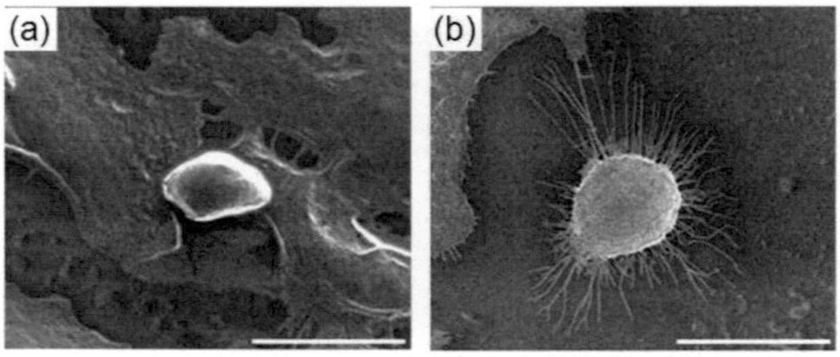

Figure 7. SEM images of A549 cells with (a) and without (b) pre-TGF-β1 treatment for 18 h. The IL treatment was conducted for the both samples before SEM observation.

The cells shown in Fig. 7 are human lung epithelial cells (A549). Before SEM observation of the cells, fixation treatment and metal or carbon coating are required to keep the cell morphology under vacuum conditions and to put electrical conductivity. Sometimes, however, such the treatments deform the delicate moieties of the cells. The A549 cells, which were subjected to dehydration and Pt-sputtering gave their SEM images showing relatively smooth surface with short microvilli. On the other hand, if IL was applied to the cells in place of the Pt-sputtering to put electrical conductivity to the cell, the SEM image shown in Fig. 7(b) was obtained. It indicates evidently long microvilli around the cell and ruffled cell surface. It is noteworthy that some of long microvilli made connections between separated cells, forming cellular bridges. The transforming growth factor (TFG)-β1 is a representative epthelial-mesenchymal transition (EMT), which is a key event in cancer metastasis. In fact, it is well known that the cells pre-treated with (TFG)-β1 lose their polarity and cell-to-cell contact. Fig. 7(a) shows A549 cells, which were subjected to the (TFG)-β1 treatment before SEM observation using IL. It is apparent by comparison of Fig. 7(a) with (b) that the A549 treated with (TFG)-β1 completely lost microvilli, implying that EMT-inducing TGF- β1 on the regulation of filopodia formation in mitotic cells. As a result the cells, which are free from other cells, enable their metastases.

5. Observation of electrochemical reactions

ILs work as a favorable electrolyte for several kinds of electrochemical reactions, implying that such the electrochemical reactions can be induced in a vacuum chamber, allowing observation of the reactions by an electron microscope [26,27]. As the first attempt, SEM observation of electrochemical Ag deposition has been conducted. Since the electrochemical reaction proceeds in IL as an electrolyte, IL may disturb the observation. However, the reaction occurring at several μm from the IL surface can be observed because the accelerated electron beam of SEM can penetrate such a thing IL layer.

In situ SEM observation of silver deposition was made by applying electrode potential to the working electrode, while observing the electrode surface from the top. The polarization potential chosen was -0.22 V, which was a little more negative than the onset potential of silver deposition (-0.15 V), and -1.14 V vs. Ag/Ag^+ where the reaction rate is determined by diffusion of Ag^+. It is well known that silver deposition with nucleus growth is dominant when overpotential is small, whereas aciculate deposits becomes dominant at the potentials where diffusion determines the reaction rate. Such the natural rule was well represented by the SEM images taken with the reaction time, as shown in Fig. 8 [26].

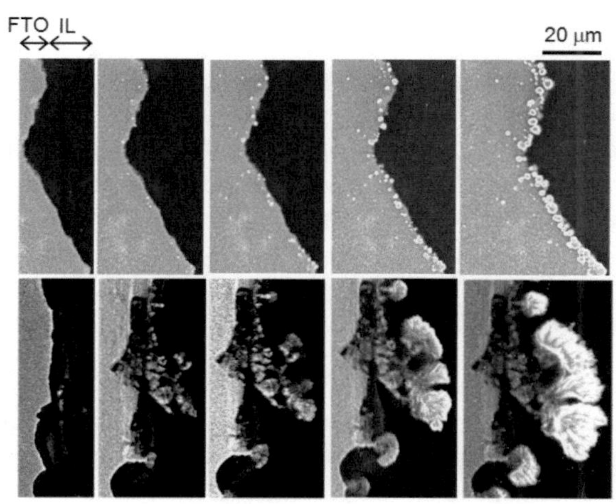

Figure 8. SEM images of gradual deposition of silver particles polarized at 0.22 V (upper) and -1.14 V (lower) vs. Ag/Ag^+ for 0, 15, 30, 60 min, and180 min.

6. EDX analysis

Energy dispersive X-ray spectrometry (EDX) has become powerful for elemental analysis at small parts when it is combined with electron microscopes. Our previous studies revealed that EDX analysis is effective for detecting changes in components caused by electrochemical reactions in ILs. As a concrete application, we utilized this technique to reveal reaction mechanism of the electrochemical actuator.

The actuator device was prepared using a film of poly(vinylidene fluoride-co-hexafluoropropylene) (PVdF-HFP) containing IL. This composite film was sandwitched between two thin Au or Pt layers that were deposited by a metal sputtering. When $[EMI^+][TFSA^-]$ was used to prepare the PVdF-HFP-IL composite, the resulting actuator bent toward the positive side (Fig. 10 (a)). The same tendency was observed for other polymer-IL composite actuators. Based on such the results, some researchers explained that bending is caused by difference in size between cation and anion of IL, the former is a little larger than the latter. Therefore, larger cations and smaller anions are attracted to the negative and the positive metal layer, respectively, resulting in expansion of the former side more than the latter.

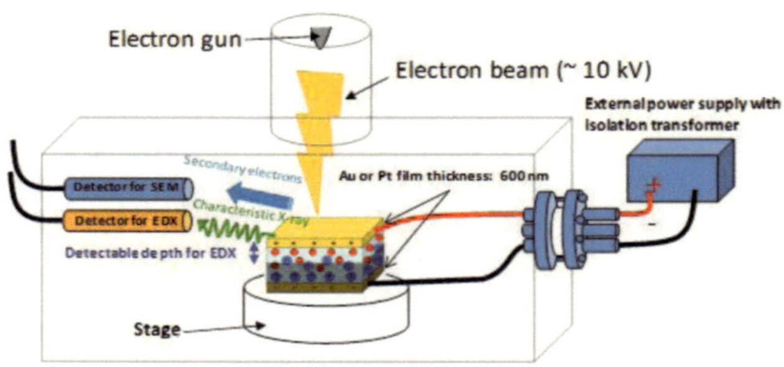

Figure 9. Schematic illustration of in situ SEM system for an actuator to investigate component changes by its reaction.

However, unexpected result was obtained when the fluorohydrogenate IL, $[EMI^+][(FH)_{2.3}F^-]$ was used for the composite preparation. The actuator fabricated using this composite bent toward its negative side, as shown in Fig. 10 (b) although size of $[(FH)_{2.3}F^-]$ is smaller than $[TFSA^-]$, requiring another reaction mechanism that can explain reasonably the bending toward positive and negative sides. Then, we attempted in situ EDX analysis using the specifically modified SEM instrument, as shown in Fig. 9.

The prepared actuator was put on the sample stage of the SEM and DC voltage of + 3.5 V was applied to the Pt layers from a power supply outside of the SEM chamber. The one side of the actuator was observed by the SEM and change in amount of ions at vicinity of the Pt layer was detected by EDX while changing polarity of the DC supply. Typical change in the EDX spectra were shown in Fig. 10 (c) and (d). In case of the $[EMI^+][TFSA^-]$ composite, the peak intensities for the anion components such as O, F, and S did not show significant change at all after changing polarity, while the intensity of carbon, which is mainly contained in the cation, markedly increased when the polarity was changed from plus to minus. On the contrary, in case of the $[EMI^+][(FH)_{2.3}F^-]$ composite, the F intensity decreased by changing voltage from +3.5 to -3.5 V, while almost no change was observed for the carbon intensity. Such the results indicated that cations and anions moved dominantly by changing the voltage polarity. Those behavior may be comprehensible because the transport numbers of cations and anions are larger for $[EMI^+][TFSA^-]$ and $[EMI^+][(FH)_{2.3}F^-]$, respectively. Based on those results and information, schematic illustrations shown in Fig. 10 (e) and (f) can be depicted to explain the actuator's bending. If only cations move in the polymer-IL composite, population of ions existing in the vicinity of the positive side decreases and that in the vicinity of the negative side increases, resulting in bending toward the positive side. It is, therefore, explainable that the bending toward the negative side in case of the IL, anion of which has larger transport number than cation, as shown in Fig. 10 (f) [28].

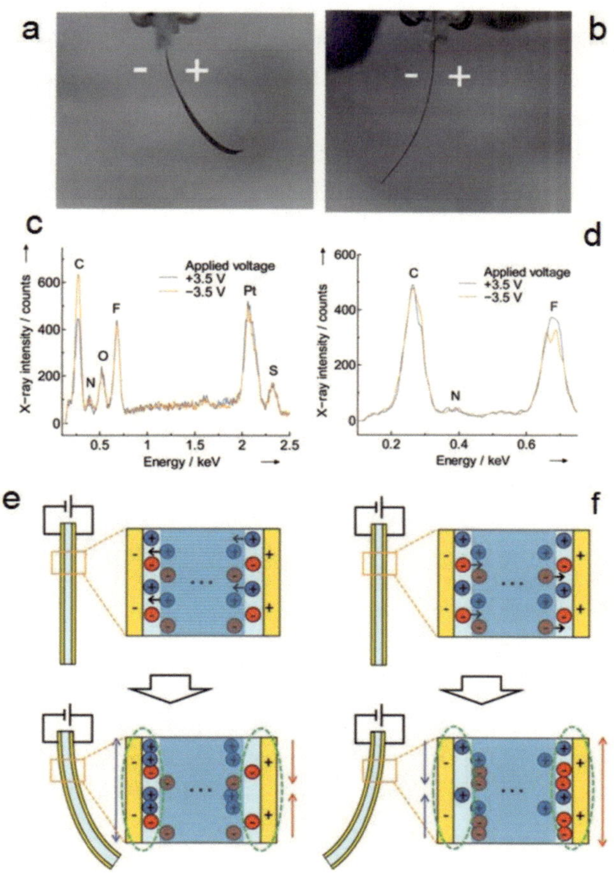

Figure 10. Motion of the electrochemical actuators prepared using ILs of $[EMI^+][TFSA^-]$ (a) and $[EMI^+][(FH)_{2.3}F^-]$ (b), their in situ EDX results (c, d), and plausible reaction mechanisms (e, f).

7. Preparation of metal nanoparticles by plasma deposition method

Ag, Cu, and Al nanoparticles have been successfully synthesized by plasma deposition method [29-31]. This approach called glow discharge electrolysis, is based on historical articles reported about 100 years ago. Schematic illustration of the system for the plasma deposition is illustrated in Fig. 11 (a). The plasma generation does not need vacuum conditions, but appropriate gas of low pressure is required to generate a stable plasma. However, the use of ILs is also essential in this case because the presence of vapor of a volatile liquid in the gas phase would inhibit the plasma generation. A typical plasma experiment is shown in Fig. 11 (b) [31]. The reaction media was [EMI^+][$TFSA^-$] with 62 mmol L^{-1}Cu(I). A dark brown-layer that appeared at the interphase between the IL and plasma phase was growing with plasma irradiation time, indicating that Cu(I) was reduced to Cu nanoparticles with an average size of ~11 nm. However, the surface was covered with a copper oxide layer. If the metal nanoparticles are yielded under vacuum or inert gas condition, this is a common issue for which it is very difficult to collect metallic state nanoparticles, especially base metal nanoparticles that are oxidized readily under atmospheric condition. One of solution methodologies about this will be introduced in a later section. Very recently, the plasma deposition method was adopted for preparation of Au [32,33] and Pt [33] nanoparticles. The relationship between deposition conditions and the characteristics of the prepared nanoparticles was studied in detail. One of interesting findings is that nanoparticles were prepared even if the cathode was placed in the IL phase and not the gas phase.

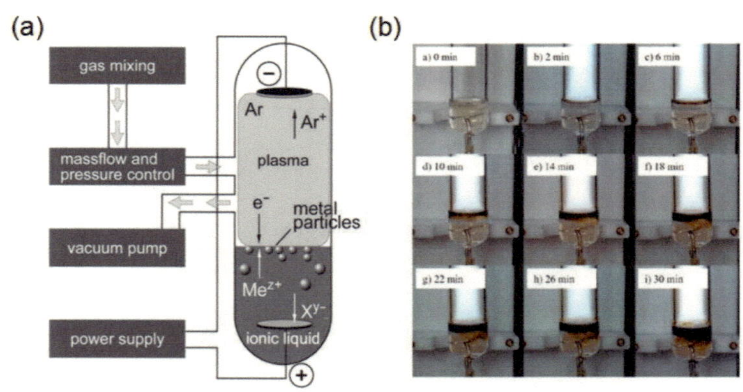

Figure 11. Schematic illustration of the experimental setup for plasma electrochemical reduction of metal ions dissolved in ILs (a) and photographs of the plasma electrochemical reduction experiment of Cu(I) dissolved in a IL at different reduction times (b).

8. Nanoparticle synthesis by sputtering

It is well known that metal vapor-deposition is a method to prepare ultrapure metal nanoparticles or films on solid substrates. Several research groups have developed the sputter deposition of metal nanoparticle onto pure ILs, as schematically illustrated in Fig. 12 (a), in order to prepare ultrapure metal nanoparticles [32-38]. The simple sputter deposition of Au onto ILs resulted in a solution containing highly dispersed Au nanoparticles whose size was dependent on the IL used [32]. In situ TEM observations of Au-deposited ILs revealed that highly-dispersed nanoparticles with no aggregation were seen in ILs, as shown in Fig. 12 (b), (c). Sputter deposition onto [EMI^+][BF_4 $^-$] gave spherical Au nanoparticles having an average diameter of 5.5 nm, while much smaller nanoparticles with size of 1.9 nm were obtained for the experiment using [Me_3PrN^+][$TFSA^-$]. Although sputtered species were assumed not to considerably suffer gas-phase collisions in the space between Au foil and IL solution because of low gas pressure, their injection into IL solution could make high concentration enough to coalesce with each other. Their coalescence would proceed until Au nanoparticles were stabilized by the adsorption of IL ions, the degree being dependent on IL. Since it is well-known that $TFSA^-$ anion makes a coordination bond with metal ions, it is expected that strong adsorption of $TFSA^-$ suppresses the growth and/or coalescence of particles.

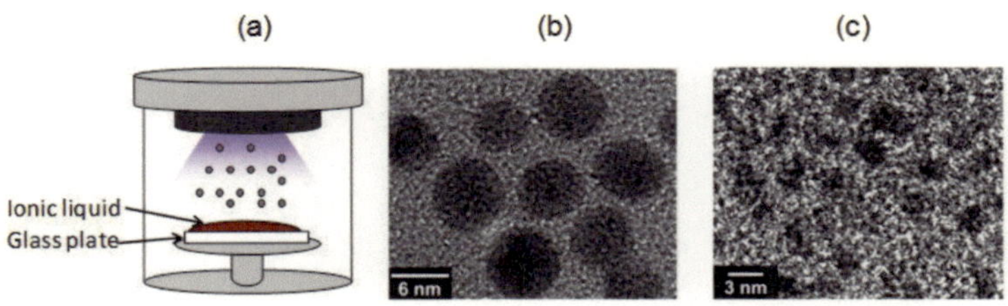

Figure 12. Schematic illustration of the metal nanoparticle formation by sputtering (a) and TEM image of Au nanoparticles sputter-deposited in $EtMeImBF_4$.

This method has achieved the preparation of various pure metal nanoparticles, such as Au [32-36], Ag [37,38], Pt [39,40] and so forth, possessing particle sizes less than 10 nm in diameter without any specific stabilizing agent. A small-angle X-ray scatting study revealed the initial formation mechanism of the gold nanoparticles during the sputtering process onto several [1,3-dialkylimidazolium][BF_4] [36]. The proposed formation mechanism is divided into two phases where it was concluded that both surface tension and viscosity of the IL are important factors for the Au nanoparticle growth and its stabilization.

The simultaneous sputter deposition of different pure metals on IL is a facile synthetic method to prepare bimetallic alloy nanoparticles [37]. The use of metal targets composed of radially-arranged Au and Ag foils allowed simultaneous sputter deposition of Au and Ag onto [BMI^+][PF_6 $^-$] in argon atmosphere at ca. 20 Pa. The color of [BMI^+][PF_6 $^-$] subjected to the sputter deposition was varied (Fig. 11), depending on a fraction of gold foils on targets. Each absorption spectrum of the resulting solution exhibited a single peak or shoulder assigned to the surface plasmon resonance (SPR) band of the metal particles and the peak wavelength increased almost linearly with the Au fraction. These results indicated that the simultaneous injection of sputtered species of Au and Ag into an IL caused coalescence with each other in the solution, resulting in the formation of AuAg alloy nanoparticles, as shown in Fig. 13. The chemical composition and optical properties of the alloy nanoparticles are easily controlled just by varying the area ratio of the individual pure metal foils in a sputtering target.

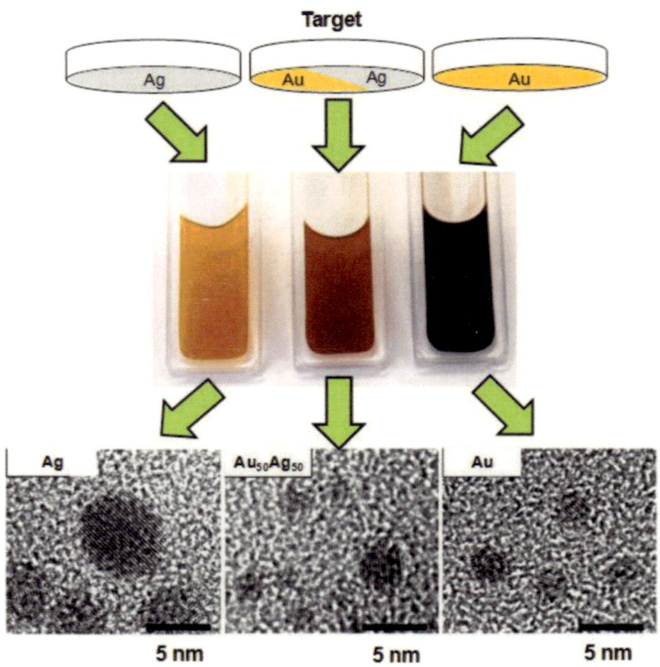

Figure 13. Photographs of [BuMeIm][PF_6] ILs after sputtering experiments at Au-Ag targets having different surface area ratios and TEM images of the resulting nanoparticles obtained at each Au-Ag target.

Furthermore, it was discovered that hollow nanoparticles can be synthesized by some modification of the sputtering method [39]. This fact was found when we attempted to produce indium metal nanoparticles by sputtering of In onto [BMI^+][BF_4^-]. SEM observation and some analyses of the obtained nanoparticles revealed that they had In/In_2O_3 core/shell configuration, as shown in Fig. 14(a). Since the melting point of In is 156.6 °C, we attempted to heat the resulting In/In_2O_3-dispersed IL at 250 °C, giving the In_2O_3 hollow nanoparticles, a TEM image of which is shown in Fig. 14(b). The plausible reaction mechanisms for synthesis of In/In_2O_3 and In_2O_3 hollow nanoparticles are schematically illustrated in Fig. 12(c) and (d), respectively. The oxidation of the In nanoparticle surfaces and the melted In by heating might be caused by oxygen that was dissolved in the IL when the IL was taken out from the sputtering instrument.

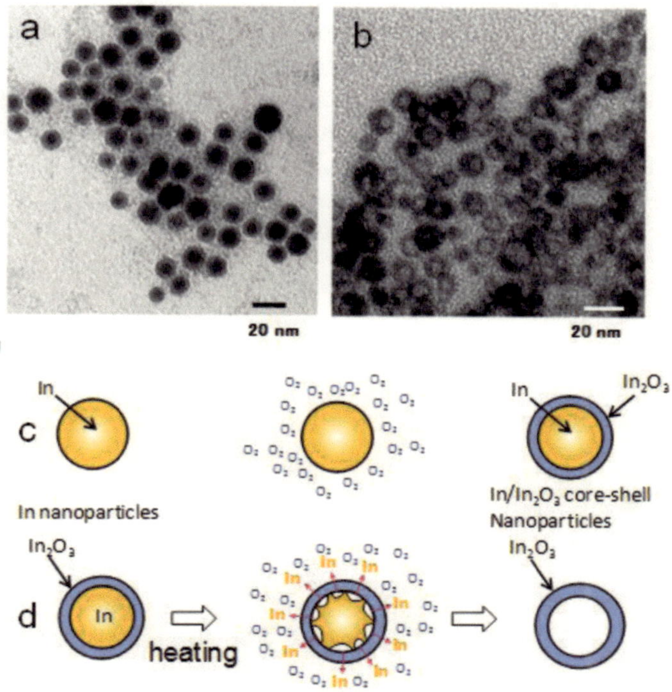

Figure 14. Nanoparticles synthesized by sputtering of In onto [EMI⁺][BF_4^-] (a), those after heating at 250 °C, and the plausible reaction mechanisms for the In sputtering (c) and the heating (d) procedures.

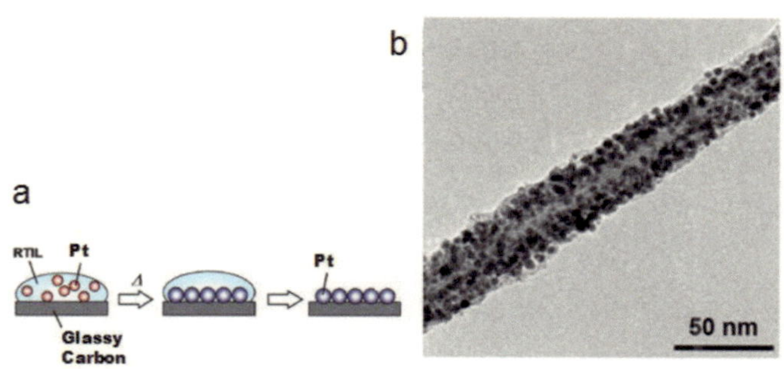

Figure 15. A way to immobilize Pt nanoparticles on surface of carbon substance (a) and Pt nanoparticles-immobilized carbon nanotube (b) prepared by this way.

The produced metal nanoparticles are stably dispersed in ILs for long time without any specific stabilizing agent. However, the nanoparticles can be immobilized onto carbon substances by putting the nanoparticles-dispersed IL on a carbon substrate followed by heating and then removal of IL by washing with acetonitrile (Fig. 15(a)) [34]. When the resulting Pt nanoparticles-immobilized carbon substance was used as an electrode, it exhibited high electrocatalytic activities toward O_2 reduction, indicating that the immobilized Pt nanoparticles kept their catalytic acitvities [40,41]. Similar method was found to be useful to immobilize Pt nanoparticles onto surfaces of carbon nanotubes, as shown in Fig. 15 (b) [42]. In this case, Pt-dispersed IL and carbon nanotubes are vigorously mixed and the resulting mixture was heated, followed by washing with acetonitrile. The Pt nanoparticles on the carbon nanotubes also exhibited high electrocatalytic activities, as shown in Fig. 14.

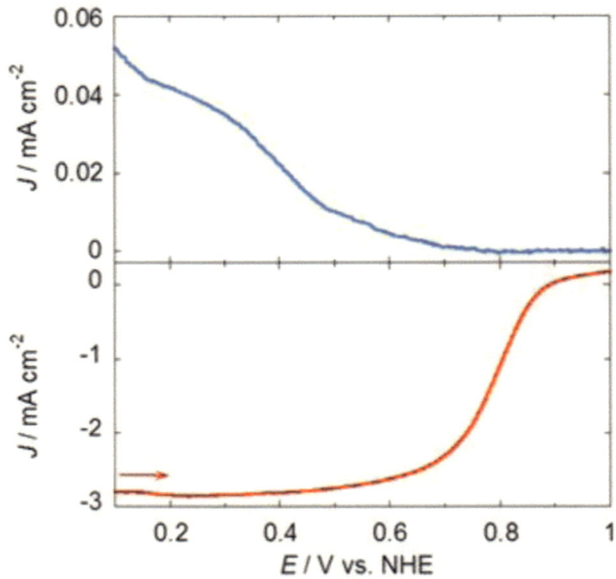

Figure 16. Hydrodynamic voltammograms for O_2 reduction at a Pt–SWCNT modified rotating ring disk electrode in O_2-saturated 0.1 M $HClO_4$ aqueous solution at 298 K. The electrodes were (top) Pt ring and (bottom) Pt–SWCNT modified GC disk. The IL used for Pt nanoparticle preparation was [Me_3PrN^+][$TFSA^-$]. The potential for the ring electrode was 1.20 V. The scan rate was 10 mV s^{-1}; the rotation rate was 1200 rpm.

9. Nanoparticle preparation by quantum Beam

Irradiation of electron beam to IL is another way to synthesize metal particles. This fact was found first when we observed [BMI^+][$TFSA^-$] containing 0.1 mol dm^{-3} $NaAuCl_4$ [43]. As shown in Fig. 17(a)-(c), many bright lines appeared in the IL droplet on a FTO with observation time. When the bright line was observed with higher magnification, a SEM image of Fig. 17 (d) was obtained, showing Several experiments and analyses have revealed that Au particles are produced by reduction of Au^{3+} ions by electron beam, and that particle size depends on the experimental conditions [44].

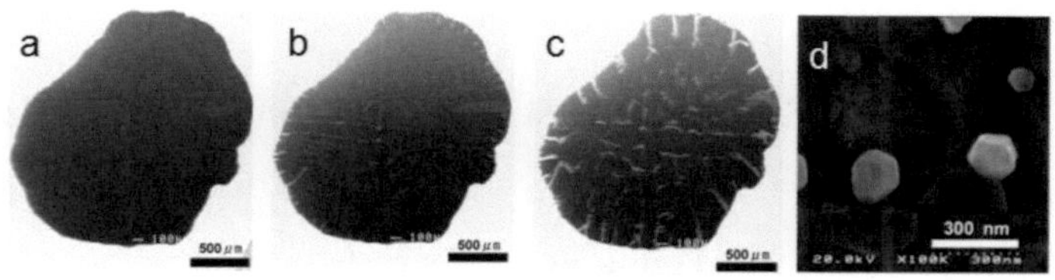

Figure 17. SEM images of the ionic liquid irradiated with an electron beam for 0 s (a), 90 s (b), and 300 s (c), and generated Au particles (d).

Accelerated electron beam and γ-ray generated by industrial plants, which are usually utilized for sterilizing medical kits, are also available for reduction of metal ions to metal nanoparticles [45]. In this case, since irradiation is made by conveying IL containing metal ions, which is sealed in a sample vial, through the generator, mass production of metal nanoparticles is possible. This technique was able to synthesize several kinds of metal nanoparticles, such as Au, Ag, Cu, Ni, Pd, Pt, Mg, Fe, Zn, Al, Sn, and FePt alloy [46].

10. Conclusion Remarks

Chemists prefer wet conditions than dry conditions because wet conditions are more desired for inducing chemical reactions. Since wet conditions are also required for all living things, investigation of biomaterials under wet conditions are much better than that under dry conditions. Unfortunately, however, many instruments for precise analyses and those for material production with micro or nano scales require vacuum conditions because the air intererferes precise proving and controlling. So far, vacuum and wet are contradictory words because there is no liquid which can stand in vacuum without vaporization. IL is the first liquid that can set the relationship between vacuum and wet on the right footing. There are more instruments requiring vacuum conditions than those introduced in this article. We would like to apply IL to more instruments to make many scientists know that IL is the key material to solve the mysteries of our wet world.

Author details

Susumu Kuwabata[1,2], Tsukasa Torimoto[2,3], Akihito Imanishi[2,4] and Tetsuya Tsuda[1,4]

*Address all correspondence to: kuwabata@chem.eng.osaka-u.ac.jp

1 Department of Applied Chemistry, Graduate School of Engineering, Osaka University, Japan

2 Japan Science and Technology Agency, CREST, Japan

3 Department of Crystalline Material Sciences, Graduate School of Engineering, Nagoya University, Japan

4 Department of Chemistry, Graduate School of Engineering Science, Osaka University, Japan

Frontier Research Base for Global Young Researchers, Graduate School of Engineering, Japan

References

[1] Ohno, H. (2005). Electrochemical Aspects of Ionic Liquids. *Wiley-Interscience, New Jersey*.

[2] Wasserscheid, P., & Welton, T. (2007). Ionic Liquids in Synthesis. *Wiley-VCH, Weinheim, Germany*.

[3] Koel, M. (2009). Ionic Liquids in Chemical Analysis. *CRC Press, Boca Raton*.

[4] Torimoto, T., Tsuda, T., Okazaki, K. I., & Kuwabata, S. (2010). New Frontiers in Materials Science Opened by Ionic Liquids. *Adv. Mater*, 22(5), 1196-1221.

[5] Kuwabata, S., Tsuda, T., & Torimoto, T. (2010). Room-Temperature Ionic Liquid. A New Medium for Material Production and Analyses under Vacuum Conditions. *J. Phys. Chem. Lett.*, 1(21), 3177-3188.

[6] Oliveira, F. C. C., Rossi, L. M., Jardim, R. F., & Rubim, J. C. (2009). Magnetic Fluids Based on γ-Fe_2O_3 and $CoFe_2O_4$ Nanoparticles Dispersed in Ionic Liquids. *J. Phys. Chem. C*, 113(20), 8566-8572.

[7] Swatloski, R. P., Spear, S. K., Holbrey, JD, & Rogers, R. D. (2002). Dissolution of cellose with ionic liquids. *J. Am. Chem. Soc.*, 124(18), 4974-4975.

[8] Zhu, S., Wu, Y., Chen, Q., Yu, Z., Wang, C., Jin, S., Ding, Y., & Wu, G. (2006). Dissolution of cellulose with ionic liquids and its application: a mini-review. *Green Chem*, 8(4), 325-327.

[9] Feng, L., & Chen, Z. (2008). Research progress on dissolution and functional modification of cellulose in ionic liquids. *J. Mol. Liq*, 142(1-3), 1-5.

[10] Smith, E. F., Rutten, F. J. M., Villar-Garcia, I. J., Briggs, D., & Licence, P. (2006). Ionic Liquids in Vacuo: Analysis of Liquid Surfaces Using Ultra-High-Vacuum Techniques. *Langmuir*, 22(22), 9386-9392.

[11] Hofft, O., Bahr, S., Himmerlich, M., Krischok, S., Schaefer, J. A., & Kempter, V. (2006). Electronic Structure of the Surface of the Ionic Liquid [EMIM][Tf_2N] Studied by Metastable Impact Electron Spectroscopy (MIES), UPS, and XPS. *Langmuir*, 22(17), 7120-7123.

[12] Caporali, S., Bardi, U., & Lavacchi, A. (2006). X-ray Photoelectron Spectroscopy and Low Energy Ion Scattering Studies on 1-Butyl-3-methyl-imidazolium Bis(Trifluoromethane) Sulfonamide. *J. Electron Spectrosc. Relat. Phenom*, 151(1), 4-8.

[13] Lovelock, K. R. J., Kolbec, C., Cremer, K. T., Paape, N., Schulz, P. S., Wasserscheid, P., Maier, F., & Steinruck, H. P. (2009). Influence of Different Substituents on the Surface Composition of Ionic Liquids Studied Using ARXPS. *J. Phys. Chem. B*, 113(9), 2854-2864.

[14] Smith, E. F., Villar, Garcia. I. J., Briggs, D., & Licence, P. (2005). Ionic Liquids in Vacuo; Solution-Phase X-ray Photoelectron Spectroscopy. *Chem. Commun*, 5633-5635.

[15] Armstrong, D. W., Zhang, L. K., He, L., & Gross, M. L. (2001). Ionic Liquids as Matrixes for Matrix-Assisted Laser Desorption/Ionization Mass Spectrometry. *Anal. Chem*, 73(15), 3679-86.

[16] Mank, M., Stahl, B., & Boehm, G. (2004). Dihydroxybenzoic Acid Butylamine and Other Ionic Liquid Matrixes for Enhanced MALDI-MS Analysis of Biomolecules. *Anal. Chem*, 76(10), 2938-2950.

[17] Laremore, T. N., Zhang, F., & Linhardt, R. J. (2007). Ionic Liquid Matrix for Direc UV-MALDI-TOF-MS Analysis of Dermatan Sulfate and Chondroitin Sulfate Oligosaccharides. *Anal. Chem.*, 79(4), 1604-1610.

[18] Fukuyama, Y., Nakaya, S., Yamazaki, Y., & Tanaka, K. (2008). Ionic Liquid Matrixes Optimized for MALDI-MS of Sulfated/Sialylated/Neutral Oligosaccharides and Glycopeptides. *Anal. Chem*, 80(6), 2171-2179.

[19] Kuwabata, S., Kongkanand, A., Oyamatsu, D., & Torimoto, T. (2006). Observation of Ionic Liquid by Scanning Electron Micrsocope. *Chem. Lett*, 35(6), 600-601.

[20] Wishart, J. F., & Neta, P. (2003). Spectrum and Reactivity of the Solvated Electron in the Ionic Liquid Methyltributylammonium Bis(trifluoromethylsulfonyl)imide. *J. Phys. Chem. B.*, 107(30), 7261-7267.

[21] Arimoto, S., Sugimura, M., Kageyama, H., Torimoto, T., & Kuwabata, S. (2008). Development of New Techniques for Scanning Electron Microscope Observation Using Ionic Liquid. *Electrochim. Acta*, 53(21), 6228-6234.

[22] Tsuda, T., Nemoto, N., Kawakami, K., Mochizuki, E., Kishida, S., Tajiri, T., Kushibiki, T., & Kuwabata, S. (2011). SEM Observation of Wet Biological Specimens Pretreated with Room-Temperature Ionic Liquid. *ChemBioChem*, 12(17), 2547-2550.

[23] Ishigaki, Y., Nakamura, Y., Takehara, T., Nemoto, N., Kurihara, T., Koga, H., Nakagawa, H., Takegami, T., Tomosugi, N., Miyazawa, S., & Kuwabata, S. (2011). Ionic liquid enables simple and rapid sample preparation of human culturing cells for scanning electron microscope analysis. *Microsc. Res. Tech*, 74(5), 415-420.

[24] Ishigaki, Y., Nakamura, Y., Takehara, T., Shimasaki, T., Tatsuno, T., Takano, F., Ueda, Y., Motoo, Y., Takegami, T., Nakagawa, H., Kuwabata, S., Nemoto, N., Tomosugi, N.,

& Miyazawa, S. (2011). Scanning electron microscopy with an ionic liquid reveals the loss of mitotic protrusions of cells during the epithelial-mesenchymal transition. *Microsc. Res. Tech*, 74(11), 1024-1031.

[25] Yanaga, K., Maekawa, N., Shimomura, N., Ishigaki, Y., Nakamura, Y., Takegami, T., Tomosugi, N., Miyazawa, S., & Kuwabata, S. (2012). Use of ionic liquid in fungal taxonomic study of ultrastructure of basidiospore ornamentation. *Mycolog. Prog.*, 11(1), 343-347.

[26] Arimoto, S., Kageyama, H., Torimoto, T., & Kuwabata, S. (2008). Development of In Situ Scanning Electron Microscope System for Real Time Observation of Metal Deposition from Ionic Liquid. *Electrochem. Commun*, 10(12), 1901-1904.

[27] Arimoto, S., Oyamatsu, D., Torimoto, T., & Kuwabata, S. (2008). Development of in situ electrochemical scanning electron microscopy with ionic liquids as electrolytes. *Chem Phys Chem*, 9(5), 763-767.

[28] Tsuda, T., Baba, M., Sato, T., Sakao, R., Matsumoto, K., Hagiwara, R., & Kuwabata, S. (2011). Nonvolatile IL-based artificial muscle: Actuation mechanism identified by in situ EDX analysis. *Chem. Europ. J.*, 17(40), 11122-11126.

[29] Meiss, S. A., Rohnke, M., Kienle, L., Abedin, S. E., Endres, F., & Janek, J. (2007). Employing Plasmas as Gaseous Electrodes at the Free Surface of Ionic Liquids: Deposition of Nanocrystalline Silver Particles. *ChemPhysChem*, 8(1), 50-3.

[30] Abedin, S. Z. E., Pölleth, M., Meiss, S. A., Janek, J., & Endres, F. (2007). Ionic Liquids as Green Electrolytes for the Electrodeposition Nanomaterials. *Green Chem.*, 9(6), 549-553.

[31] Brettholle, M., Höfft, O., Klarhöfer, L., Mathes, S., Friedrichs, M., Abedin, S. Z. E., Krischok, S., Janek, J., & Endres, F. (2010). Plasma electrochemistry in ionic liquids: deposition of copper nanoparticles. *Phys. Chem. Chem. Phys.*, 12(8), 1750-1755.

[32] Torimoto, T., Okazaki, K., Kiyama, T., Hirahara, K., Tanaka, N., & Kuwabata, S. (2006). Sputter Deposition onto Ionic Liquids: Simple and Clean Synthesis of Highly Dispersed Ultrafine Metal Nanoparticles. *Appl. Phys. Lett*, 89(24), 243117/1-243117/3.

[33] Khatri, O. P., Adachi, K., Murase, K., Okazaki, K., Torimoto, T., Tanaka, N., Kuwabata, S., & Sugimura, H. (2008). Self-Assembly of Ionic Liquid (BMI-PF_6)-Stabilized Gold Nanoparticles on a Silicon Surface: Chemical and Structural Aspects. *Langmuir*, 24(15), 7785-7792.

[34] Okazaki, K., Kiyama, T., Suzuki, T., Kuwabata, S., & Torimoto, T. (2009). Thermally Induced Self-assembly of Gold Nanoparticles Sputterdeposited in Ionic Liquids on Highly Ordered Pyrolytic Graphite Surfaces. *Chem. Lett.*, 38(4), 330-331.

[35] Kameyama, T., Ohno, Y., Kurimoto, T., Okazaki, K., Uematsu, T., Kuwabata, S., & Torimoto, T. (2010). Size Control and Immobilization of Gold Nanoparticles Stabilized in an Ionic Liquid on Glass Substrates for Plasmonic Applications. *Phys. Chem. Chem. Phys.*, 12(8), 804-811.

[36] Hatakeyama, Y., Okamoto, M., Torimoto, T., Kuwabata, S., & Nishikawa, K. (2009). Small-Angle X-ray Scattering Study of Au Nanoparticles Dispersed in the Ionic Liquids 1-Alkyl-3-methylimidazolium Tetrafluoroborate. *J. Phys. Chem. C*, 113(10), 3917-3922.

[37] Okazaki, K., Kiyama, T., Hirahara, K., Tanaka, N., Kuwabata, S., & Torimoto, T. (2008). Single-Step Synthesis of Gold-Silver Alloy Nanoparticles in Ionic Liquids by a Sputter Deposition Technique. *Chem. Commun*, 691-693.

[38] Suzuki, T., Okazaki, K., Kiyama, T., Kuwabata, S., & Torimoto, T. (2009). A Facile Synthesis of AuAg Alloy Nanoparticles Using a Chemical Reaction Induced by Sputter Deposition of Metal onto Ionic Liquids. *Electrochemistry*, 77(8), 636-638.

[39] Suzuki, T., Okazaki-I, K., Suzuki, S., Shibayama, T., Kuwabata, S., & Torimoto, T. (2010). Nanosize-controlled syntheses of indium metal particles and hollow indium oxide particles via the sputter deposition technique in ionic liquids. *Chem. Mater.*, 22(18), 5209-5215.

[40] Tsuda, T., Kurihara, T., Hoshino, Y., Kiyama, T., Okazaki-I, K., Torimoto, T., & Kuwabata, S. (2009). Electrocatalytic activity of platinum nanoparticles synthesized by room-temperature ionic liquid-sputtering method. *Electrochemistry*, 77(8), 693-695.

[41] Tsuda, T., Yoshii, K., Torimoto, T., & Kuwabata, S. (2010). J. Power Sources , 195, 5980-5985.

[42] Yoshii, K., Tsuda, T., Arimura, T., Imanishi, A., Torimoto, T., & Kuwabata, S. (2012). Platinum nanoparticle immobilization onto carbon nanotubes using Pt-sputtered room-temperature ionic liquid. *RSC Adv*, 10.1039/C2RA21243A.

[43] Imanishi, A., Tamura, M., & Kuwabata, S. (2009). Formation of Au nanoparticles in an ionic liquid by electron beam irradiation. *Chem. Commun*, 1775-1777.

[44] Imanishi, A., Gonsui, S., Tsuda, T., Kuwabata, S., & Fukui-I, K. (2011). Size and shape of Au nanoparticles formed in ionic liquids by electron beam irradiation. *Phys. Chem. Chem. Phys.*, 13(33), 14823-14830.

[45] Tsuda, T., Seino, S., & Kuwabata, S. (2009). Gold nanoparticles prepared with a room-temperature ionic liquid-radiation irradiation method. *Chem. Comm*, 6792-6794.

[46] Tsuda, T., Sakamoto, T., Nishimura, Y., Seino, S., Imanishi, A., & Kuwabata, S. (2012). Various metal nanoparticles produced by accelerated electron beam irradiation of room-temperature ionic liquid. *Chem. Commun.*, 925-927.

Chapter 24

Plasma Process on Ionic Liquid Substrate for Morphology Controlled Nanoparticles

Toshiro Kaneko, Shohei Takahashi and
Rikizo Hatakeyama

Additional information is available at the end of the chapter

http://dx.doi.org/10.5772/52095

1. Introduction

The interaction of discharge plasmas with liquids [1-5] is one of the active topics in the realm of recent plasma science and science technology. It has pioneered new channels relating to nano material creation based on their distinct properties such as ultra-high density, high reactivity, high process rate, and so on. Especially, the boundary between the plasmas and the liquids, which activates physical processes and chemical reactions, has attracted much attention as a novel field in the nano-bio material creation. For example, the nanoparticle synthesis using the plasma-liquid interfaces [6-10] is especially advantageous in that a reducing agent is the plasma itself, and then, toxic stabilizers and reducing agents are unnecessary and the synthesis is continuous during the plasma irradiation. In these methods, although it has been reported that the metal salt is reduced by an electron or an active hydrogen, the precise control of the synthesis in terms of the synthesis rate, morphology (size, shape, structure, and so on) control remain unclear because the inevitable high voltage discharge in the atmospheric pressure and the consequential dynamic behavior of the gas-liquid interface prevent us from analyzing the precise properties of the plasmas in the interfacial region.

In this sense, for the purpose of the generation of the static and stable plasma contacting with the liquid, we adopt ionic liquids [11,12] which have the interesting characteristics such as their composition consisting of only positive and negative ions, i.e., no neutral solvent, extremely low vapor pressure, high heat capacity, and nonflammability. These characteristics enable us to introduce the ionic liquids to the vacuum system and the discharge plasma. Therefore, the ionic liquids are the most suitable liquid for the formation of nano-composite materials using the discharge plasmas in contact with the liquids [13-21].

On the other hand, recently, highly-ordered periodic structures of metal nanoparticles have attracted much attention due to their high catalytic activity, unique photosensitive reactivity, bio sensitivity, and so on [22-26]. One possibility is use of nano-carbons such as carbon nanotubes or graphenes as template for synthesis of the nanoparticles [27-32]. However, the structures of the nanoparticles are decided by the chemical properties of the nano-carbons and are difficult to be freely controlled by the external parameters. To realize the easy and flexible control of the periodic structure of the nanoparticles, we adopt a novel plasma technique combined with introduction of ionic liquids under strong magnetic fields up to several tesla (T), whose concept is schematically shown in Figure 1 [33].

Since the plasma generated under the strong magnetic field keeps its structure due to confinement along the magnetic field lines, the plasma structure can be transcribed to the liquid surface, resulting in the synthesis of the structured nanoparticles at the gas-liquid interface when the plasma reduces the metal chlorides in the liquid. This method could contribute to supplying a considerable amount of spatially-periodic nanoparticles available for the development of unique optoelectronic devices [34].

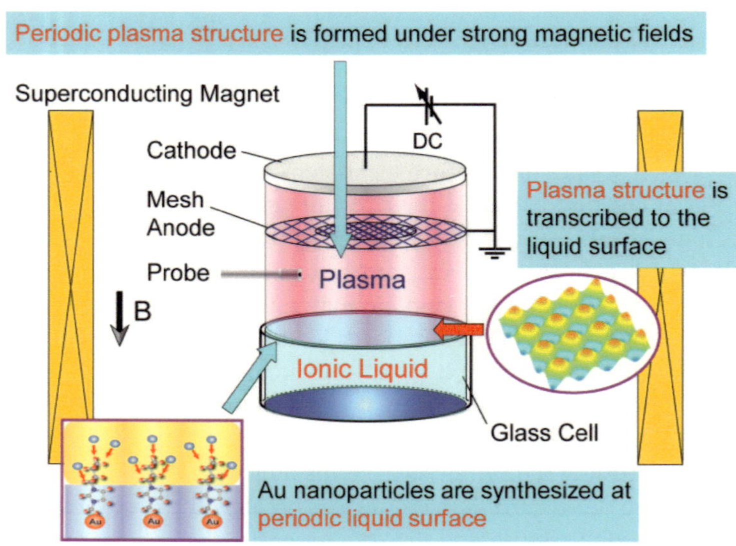

Figure 1. Concept of synthesis of periodic metal nanoparticles using discharge plasmas in contact with ionic liquid substrate.

2. Gas-liquid interfacial plasma process using ionic liquid substrate

Figure 2 shows schematic model of the gas-liquid interfacial plasmas for Au nanoparticle synthesis. An electrode which is made of a platinum (Pt) plate is located inside the glass cell, and a popular ionic liquid (1-buthl-3-methyl-imidazolium tetrafluoroborate: $[C_8H_{15}N_2]^+[BF_4]^-$) is in-

http://dx.doi.org/10.5772/52095

troduced on the Pt electrode as cathode electrodes for the purpose of investigating the effects of the ionic liquid on the discharge. On the other hand, a grounded anode electrode which is made of the SUS plate is set in a gas phase (plasma) region at a distance of 60 mm from the cathode electrodes. This discharge configuration, in which the ionic liquid cathode electrode is in the glass cell, is defined as "ion irradiation mode", because the positive ions in the plasma are accelerated by the electric field formed on the ionic liquid as shown in Figure 2(a).

In order to examine the effects of the ion irradiation to the ionic liquid on discharge-related phenomena, the cathode electrode is switched to the SUS plate located at the top of the gas plasma region, which is defined as "electron irradiation mode" and the anode electrode consisting of the ionic liquid in the glass cell is grounded as shown in Figure 2(b). Removal of the water dissolved in the ionic liquid is performed under the vacuum condition for 2 hours after introducing the ionic liquid into the glass chamber. A negative direct current (DC) voltage is supplied to the cathode electrode, where typical discharge voltage V_D, discharge current I_D, plasma irradiation time t_{pi} are V_D = 500 ~ 1500 V, I_D = 1 ~ 5 mA, t_{pi} = 1 ~ 40 min, respectively. The argon gas is adopted as a discharge medium, and the gas pressure P_{gas} is varied from 20 Pa to 200 Pa approximately. A Langmuir probe is inserted at the position of z = 0 ~ 60 mm to measure parameters of the plasma in contact with the ionic liquid substrate (z = 0: surface of the ionic liquid substrate in the glass cell).

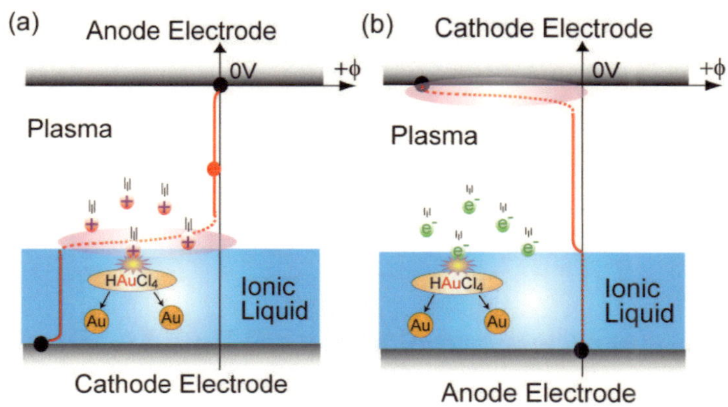

Figure 2. Schematic model of nanoparticle synthesis in the gas-liquid interface on the ionic liquid substrate. (a) ion irradiation mode, (b) electron irradiation mode.

Using this ion or electron irradiation, gold nanoparticles are synthesized in the ionic liquid by the reduction of Au chloride such as $HAuCl_4$ dissolved in the ionic liquid. Figure 3 shows transmission electron microscopy (TEM) images of the Au nanoparticles synthesized in (a) the ion irradiation mode and (b) the electron irradiation mode for P_{gas} = 60 Pa, I_D = 1 mA, and t_{pi} = 40 min. In both cases, the Au nanoparticles can be formed, however, it is found that, in ion irradiation mode, the average diameter of the Au nanoparticles is smaller and the particle number is larger than that in electron irradiation mode. The reduction reaction of the Au ions is believed to be caused by electrons injected from the plasma in electron irradiation mode, while in ion irradiation mode, the reduction may be caused by the hydrogen radical H*, which is generated by the dissociation of the ionic liquid. Based on this mechanism, the hydrogen radical is considered to be more effective for the reduction of Au ions than electrons, and efficient Au nanoparticle synthesis is realized using ion irradiation.

Since Au nanoparticles with diameter less than 100 nm are known to exhibit localized surface plasmon resonance, visible absorption spectra are obtained for a quantitative observation of the Au nanoparticle concentration. Figure 3(c) shows visible absorption spectra of the Au nanoparticles synthesized by an Ar plasma in ion and electron irradiation mode. The absorption peak appears around 550 nm, corresponding to the Au plasmon resonance, and the absorption-peak intensity in ion irradiation mode is obviously larger than that in electron irradiation mode. Ar ions with high energy can penetrate deep into the ionic liquid, promoting the generation of hydrogen radicals. The increased concentration of hydrogen radicals may reduce Au ions more effectively in ion irradiation mode than electron irradiation mode. The rate of Au nanoparticle synthesis could be controlled by the irradiation energy of inert gas ions such as Ar.

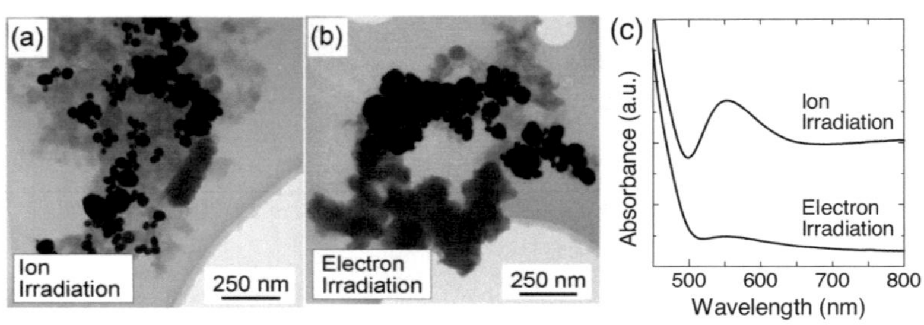

Figure 3. TEM images of Au nanoparticles synthesized in (a) the ion irradiation mode and (b) the electron irradiation mode. (c) UV-Vis absorption spectra of Au nanoparticles. P_{gas} = 60 Pa, I_D = 1 mA, t_{pi} = 40 min.

3. Control of nanoparticle morphology by gas-liquid interfacial plasmas

3.1. Periodic nanoparticle structure formed by periodic plasma

Figure 4(a) shows the schematic of an experimental setup for nanoparticle structure formation using the gas-liquid interfacial discharge plasma under strong magnetic fields, which has a glass cell with 15 mm inner diameter and 10 mm depth in a cylindrical glass chamber with 75 mm diameter and 200 mm length. A DC voltage V_D is supplied to an upper cathode electrode composed of a stainless steel (SUS) plate and a SUS mesh grid (10 meshes/inch) is used as an anode electrode to promote a spatial diffusion of the plasma. Typical discharge current is I_D=3 mA. Nitrogen gas is adopted as a discharge gas, and the gas pressure P_{gas} is varied from 20 to 100 Pa.

The new kind of the ionic liquid (N.N.N.-Trimethyl-N-propyl-ammonium Bis (trifluoro methane sulfonyl) imide) put in the glass cell is placed on a peltier element which is located at a distance of 50 mm from the anode electrode. Since this ionic liquid does not become supercooled state, we can make the ionic liquid solid state by cooling the ionic liquid using the peltier element located under the glass cell. When the strong magnetic fields are applied along the machine axis, the generated plasma is strongly magnetized, and then, the periodic plasma structure formed by the mesh anode is maintained just above the ionic liquid as shown in Figure 4(b).

The Au nanoparticles are synthesized in the ionic liquid by the plasma reduction of $HAuCl_4$. The ionic liquid can be cooled by the peltier element and becomes the solid state as mentioned above, with keeping the structure of Au nanoparticles synthesized by the plasma irradiation at the liquid interface.

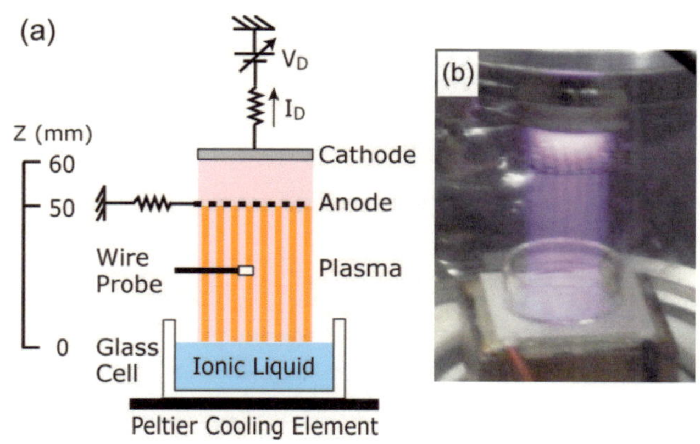

Figure 4. a) Schematic of the experimental setup and (b) photo of the synthesized periodic plasma structure for B = 1 T, I_D = 3 mA, P_{gas} = 20 Pa.

Figure 5 shows the radial profiles of the electron density n_e of the plasma as functions of (a) magnetic fields B and (b) gas pressure P_{gas}. When the strong mantic fields is applied (closed circles), the electron density has the depression periodically in the radial direction. The electron density in the high density region is about 10^8 cm^{-3}, and that in the depression region becomes one order smaller. The interval of the depression is about 2.5 mm, which is corresponding to the distance between the SUS wire of the mesh anode electrode. Therefore, these density depressions are caused by shielding of the plasma under the wire of the mesh anode, where the generated plasma cannot pass through toward the ionic liquid substrate. In the absence of the magnetic field, on the other hand, the radial profile of the electron density is almost flat even using the mesh anode, because the plasma diffuses in the radial direction and becomes uniform.

When the gas pressure is changed from P_{gas}=20 Pa to 40 Pa, the radial density profile drastically changes as shown in Figure 5(b). Since the collision between the ions and neutral particles becomes frequent with an increase in the gas pressure, the formed periodic structure of the electron density collapses and becomes relatively flat.

Based on these results, it is found that the structure of the plasma is sensitive to the magnetic field and gas pressure, which are necessary to be carefully adjusted to obtain the desired plasma structure.

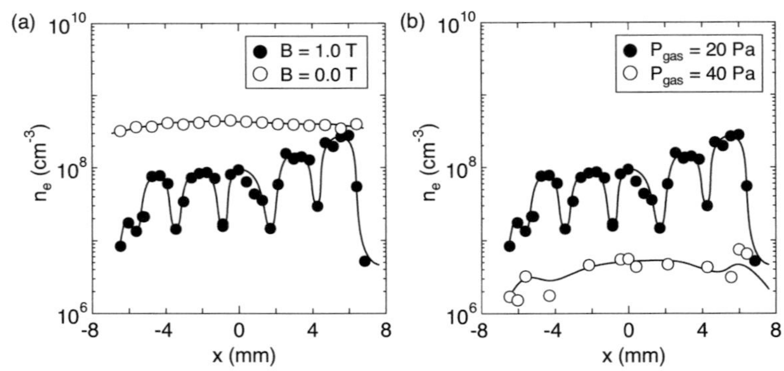

Figure 5. Radial profiles of the electron density of the plasmas (a) as a function of magnetic fields B for P_{gas} = 20 Pa and (b) as a function of gas pressure P_{gas} for B = 1 T. I_D = 3 mA.

Using this periodic plasma structure, we attempt to synthesize the Au nanoparticles with periodic morphology. Figure 6 shows photos of the temporal evolution of the periodic Au nanoparticle structure which is formed at the gas-liquid interface in accordance with the periodic plasma structure under the condition of the strong magnetic field of B = 1 T. The density of the periodic structured Au nanoparticles increases with plasma irradiation time t_{pi}, and the obvious structure is formed typically within t_{pi} = 5 min. The interval of the nanoparticle structure is about 2.5 mm, which is same as the distance between the wires of the mesh anode. As mentioned above, the electron density decreases under the shadow region of the mesh anode. Therefore, the nanoparticles are considered to be synthesized in the electron density depression region. This synthesis mechanism is discussed later.

In order to analyze the properties of the periodic structured Au nanoparticles, the ionic liquid is cooled using the peltier element located under the glass cell and is changed to the solid phase. The solid phase ionic liquid containing the Au nanoparticles can be extracted from the vacuum chamber with keeping its structure. Figure 7 shows the picture of the periodic structure of the Au nanoparticles which are synthesized on the ionic liquid substrate at room temperature, and (a) kept at room temperature and (b) cooled under T_{sub} ~ 0 °C after the synthesis. The ionic liquid becomes solid state under T_{sub} = 0 °C and the Au nanoparticles are fixed in the solid state of the ionic liquid.

However, when the plasma is irradiated for longer than 5 min, the Au nanoparticles diffuse and the structure is broadened as shown in Figure 7(b). In order to suppress the diffusion of the Au nanoparticles, the temperature of the ionic liquid is reduced during the plasma process using the peltier element, resulting in the increase in viscosity of the ionic liquid. Although the diffusion of the Au nanoparticles is suppressed, the Au nanoparticle synthesis rate becomes low. Therefore, it is necessary to precisely control the temperature of the ionic liquid for the fine periodic Au nanoparticle structure.

http://dx.doi.org/10.5772/52095

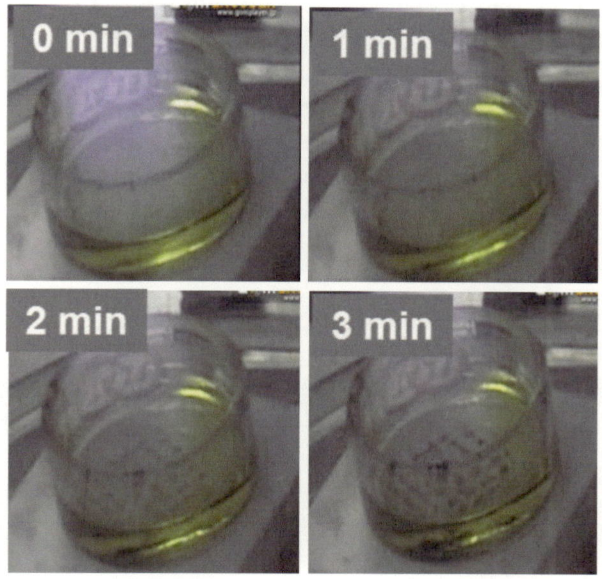

Figure 6. Temporal evolution of the periodic nanoparticle structure formed by controlled gas-liquid interfacial plasmas. $P_{gas} = 20$ Pa, $I_D = 3$ mA, $B = 1$ T.

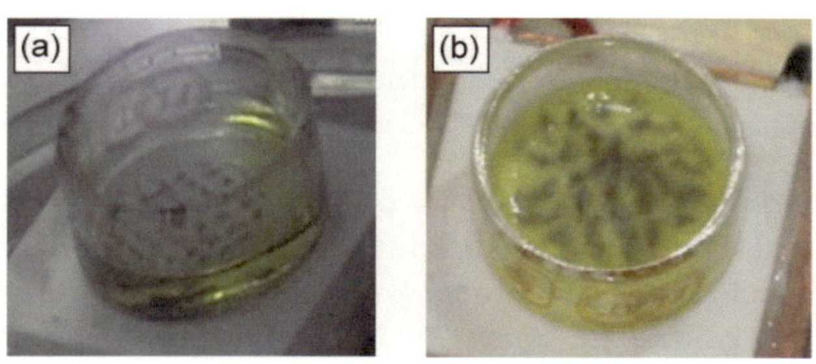

Figure 7. Pictures of the periodic nanoparticle structures which are formed at room temperature and (a) kept at room temperature and (b) cooled under $T_{sub} = 0$ °C. $P_{gas} = 20$ Pa, $I_D = 3$ mA, $B = 1$ T.

3.2. Ring-shaped nanoparticle structure formed by structure-controlled plasma

As the next step, we attempt to form more finely periodic structures of the Au nanoparticles based on the self-organizing behavior of turbulent plasmas generated by the nonlinear development of plasma fluctuations. For this purpose, a ring electrode is inserted in the plasma column as shown in Figure 8, and a positive DC bias voltage V_{RG} is applied to the electrode. It is found that the high frequency fluctuation (100 kHz – 1 MHz) is excited by the positive bias voltage, however the self-organized plasma structure is not observed at present. Therefore, in this experiment, the ring electrode is used for the plasma structure control without bias voltage.

Using this configuration, the Au nanoparticles are synthesized by reducing $HAuCl_4$. It is found that the ring shaped Au nanoparticle structure is formed corresponding to the shape of the inserted ring electrode as shown in Figure 8(b). This result means that the Au nanoparticles are synthesized at the region without plasma irradiation due to the shielding by the ring electrode. Although the Au nanoparticles are usually synthesized by the reduction effect of the electrons in the plasma [14-16], the electron is absent in the shadow region of the ring electrode in this experiment. Therefore, the reducing agent is not electrons in this case.

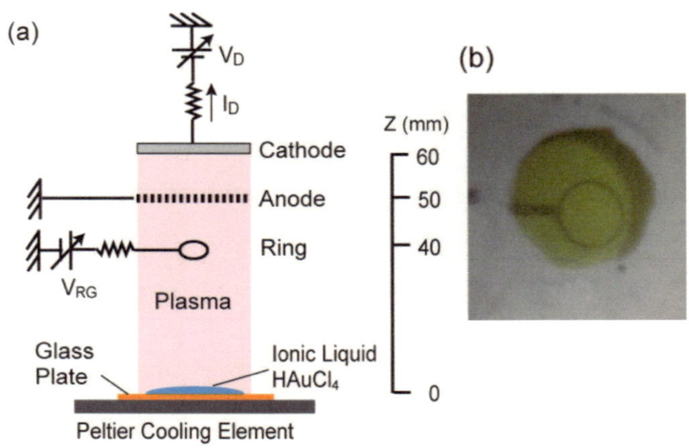

Figure 8. a) Experimental apparatus using ring electrode for nanoparticle structure control and (b) the synthesized ring shaped nanoparticle structure $B = 1$ T, $I_D = 2$ mA, $P_{gas} = 20$ Pa.

To explain the phenomena, we use the model of the Au nanoparticle synthesis as shown in Figure 9(a). As mentioned above, the charged particles such as the electrons and the positive ions cannot reach to the shadow region of the ring electrode, namely, only neutral radicals can arrive at the shadow region. The ionic liquid used in this experiment is described in Figure 9(b), which has C-H bond in cation (positive ion) and C-F bond in anion (negative ion). When the radicals in the plasma are irradiated to the ionic liquid, the C-H bond of the ionic liquid is considered to be dissociated, and the generated hydrogen radicals reduce the Au ions, resulting in the synthesis of Au nanoparticles in the shadow region of the ring electrode. On the other hand, in the plasma irradiation region, relatively high-energy ions are irradiated to the ionic liquid, and the C-F bond whose dissociation energy (D=5.07 eV) is larger than that of the C-H bond (D=4.29 eV), can be dissociated by the high-energy ions. Therefore, the Au nanoparticles are destroyed by the oxidation effect of the fluorine radicals which come from the dissociation of the ionic liquid by the collision with the high energy charged particle.

To understand the mechanism of this ring-shaped nanoparticle structure formation, the ring electrode is changed to disk electrode as shown in Figure 10, where the bias voltage of the disk electrode V_{DK} is floated in this experiment. In this case, since the shadow region of the disk electrode is wide and clear compare with the ring electrode, it is possible to clarify the species in the plasma, which are necessary for the synthesis of the Au nanoparticles. To change the plasma characteristics irradiated to the ionic liquid, we change the discharge current I_D.

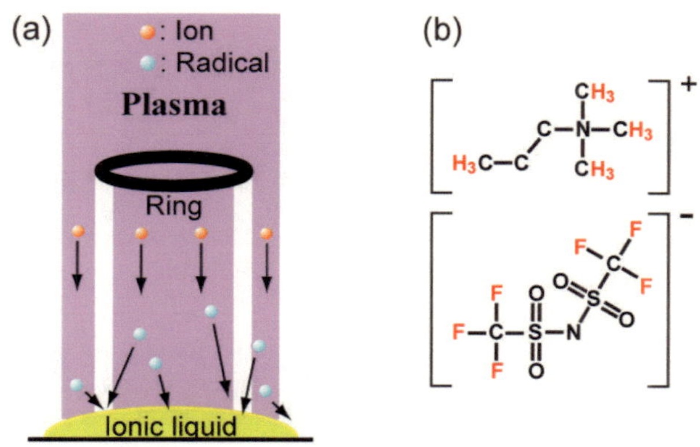

Figure 9. a) Model of the synthesis mechanism of the ring shaped Au nanoparticles and (b) chemical formula of the ionic liquid used in this experiment.

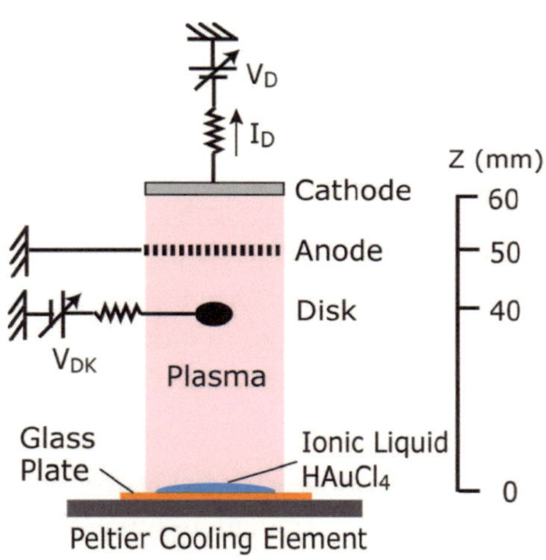

Figure 10. Experimental apparatus using disk electrode for understanding the mechanism of nanoparticle structure formation.

Figure 11 shows snapshots of the temporal evolution of the Au nanoparticle synthesis for (a) I_D= 1 mA and (b) I_D=3 mA. In the case for I_D=1 mA, the Au nanoparticles are synthesized from the outside of the ionic liquid in the plasma irradiation region, and are absent in the shadow region of the disk electrode for the plasma irradiation time t_{pi}=5 min. In the case of I_D=3 mA, on the other hand, the Au nanoparticles are synthesized not only from the outside of the ionic liquid but also the edge region of the disk electrode for t_{pi}=40 sec. Furthermore, the Au nanoparticles are not synthesized in the plasma irradiation region.

Figure 11. Photos of the Au nanoparticle structure as a function of plasma irradiation time t_{pi} for (a) I_D = 1 mA and (b) I_D = 3 mA. B = 1 T, P_{gas} = 20 Pa.

Since the electron density for I_D=1 mA is lower than that for I_D=3 mA, the sheath electric field formed above the ionic liquid is small for I_D=1 mA. Therefore, the ion irradiation energy to the ionic liquid is small, and as a result, the ion cannot dissociate the C-F bond of the ionic liquid. However, the C-H bond which has relatively low dissociation energy (4.29 eV) is dissociated by the low-energy ions or neutral nitrogen radicals. The dissociated hydrogen radicals can reduce the Au ions, resulting in the formation of the Au nanoparticles in the plasma irradiation region for I_D=1 mA.

For I_D=3 mA, on the other hand, the sheath electric field becomes large in the plasma irradiation region, and then, the high-energy ion irradiation can dissociate the C-F bond of the ionic liquid. Therefore, the Au ions are difficult to be reduced by the oxidation effect of the fluorine radical in the plasma irradiation region. However, in the shadow region of the disk electrode, the C-H bond of the ionic liquid is dissociated by the nitrogen radical of the plasma, and the dissociated hydrogen radical can synthesize the Au nanoparticles. In the center of the shadow region of the disk electrode, the Au nanoparticles are not synthesized because the nitrogen radicals cannot reach to the center of the shadow region.

Based on these results, it is found that the Au nanoparticles are synthesized by the reduction effect via the neutral radical irradiation, and are destroyed by the oxidation effect via the high-energy ion irradiation in the plasma irradiation region.

3.3. Size control of ring-shaped nanoparticle structure under inhomogeneous conversing magnetic fields

For the purpose of size control of the ring-shaped nanoparticle structure, we use the plasma - ionic liquid interface under the inhomogeneous converging magnetic fields (B) to shrink the size of the nanoparticle structure [33].

Figure 12 shows the schematic of a new experimental setup for the discharge plasma in contact with the ionic liquid containing a gold chloride ($HAuCl_4$), where the distance between a ring electrode and the ionic liquid on the glass plate is defined as z_r. The plasma is generated between a plate cathode electrode and a mesh anode electrode, and is irradiated to the ionic liquid. Here, a mirror ratio R_m is defined as the ratio of the magnetic field strength at the ring electrode position (B_{RG}) to that at the ionic liquid substrate position (B_{IL}).

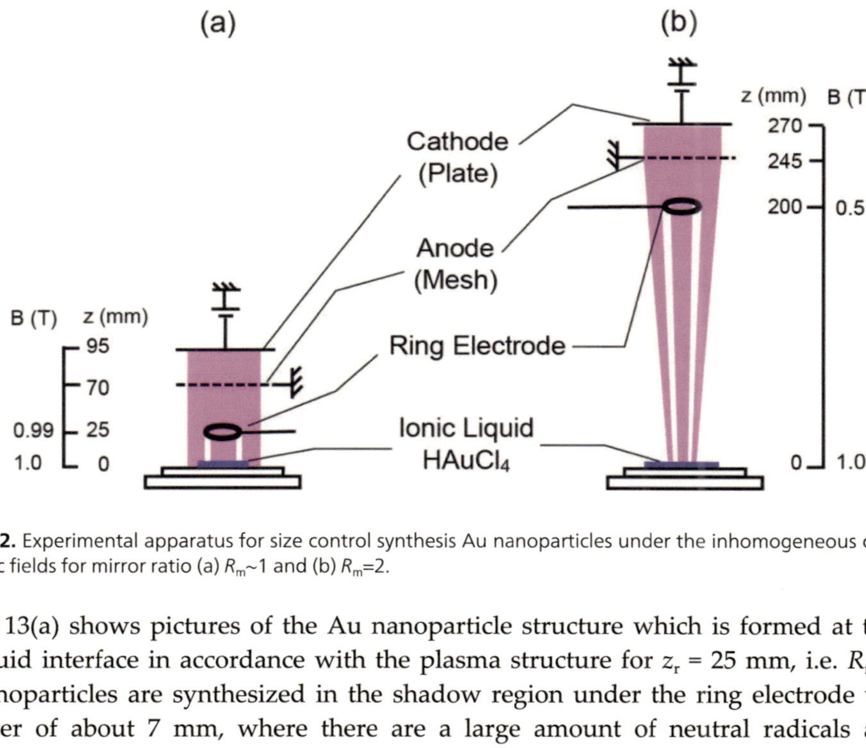

Figure 12. Experimental apparatus for size control synthesis Au nanoparticles under the inhomogeneous converging magnetic fields for mirror ratio (a) R_m~1 and (b) R_m=2.

Figure 13(a) shows pictures of the Au nanoparticle structure which is formed at the plasma-liquid interface in accordance with the plasma structure for z_r = 25 mm, i.e. R_m~1. The Au nanoparticles are synthesized in the shadow region under the ring electrode with the diameter of about 7 mm, where there are a large amount of neutral radicals and few charged ions. The Au nanoparticles are synthesized by the reduction effect of the radical in the plasma, while the charged ions inhibit the synthesis of the Au nanoparticles by the oxidation effect.

It is possible to control the Au nanoparticle structure by using shrinkage of the plasma diameter under the converging magnetic field. Figures 13(b) and 13(c) show pictures of the Au nanoparticle structure for z_r = 200 mm (R_m=2) and 550 mm (R_m=10), respectively.

The diameters of the Au nanoparticle structures are observed to be 5 mm and 2.2 mm for R_m=2 and R_m=10, respectively. Since the diameter of the plasma is proportional to the square root of the mirror ratio, the diameter of the ring shaped nanoparticles is also changed corresponding to the plasma diameter. Therefore, the diameter of the ring-shaped Au nanoparticle structure for z_r = 200 mm and 550 mm should be $1/\sqrt{2}$ and$1/\sqrt{10}$, respectively. We can easily control the size of the nanoparticle structure by simply change the magnetic field configuration, which is useful for the future nanoparticle nano-electronics devices.

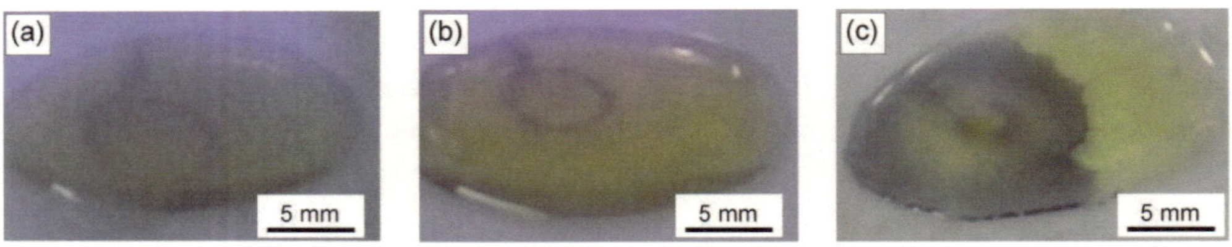

Figure 13. Picture of nanoparticle structures for (a) R_m = 1, (b) R_m = 2, and (c) R_m = 10. I_D=2 mA, P_{gas}=20 Pa, B_{IL}=1 T.

4. Conclusion

A direct current (DC) discharge plasma has stably been generated just above the ionic liquid by applying the DC voltage to an electrode immersed in the ionic liquid. The precise potential structure and the resultant plasma ion or electron irradiation to the ionic liquid are controlled. This ion irradiation is found to be effective for the synthesis of gold nanoparticles in comparison with the conventional electron irradiation system, and the control of the plasma-ion irradiation to the ionic liquid has the possibility of application to the synthesis of the various kinds of size- and yield-controlled nanoparticles.

Furthermore, the periodic and ring shaped Au nanoparticle structures are formed, which correspond to the shape of the strongly-magnetized plasmas generated using the mesh anode electrode or the ring/disk electrode inserted in the plasma.

It is very interesting that the structure of the Au nanoparticles depends on the discharge current, namely, the Au nanoparticles are synthesized from the periphery and absence in the shadow region of the disk electrode for small discharge current, while the Au nanoparticles are synthesized at the boundary of the disk electrode for relatively large discharge current. These phenomena are well explained by the reduction and oxidation effects of the radicals which are generated by the plasma irradiation to the ionic liquid and resultant dissociation of the ionic liquid.

Finally, the size of the Au nanoparticle structure can be controlled by using shrinkage of the plasma diameter under the converging magnetic field, which enables us to freely form the micro- or nano-sized nanoparticle structures.

http://dx.doi.org/10.5772/52095

Acknowledgements

The authors thank Prof. K. Tohji, K. Motomiya, T. Miyazaki, and H. Ishida for their technical assistance. We express our gratitude to Dr. K. Baba, Dr. Q. Chen, T. Harada, and T. Okuno for their collaboration. This work was supported by a Grant-in-Aid for Scientific Research from the Ministry of Education, Culture, Sports, Science and Technology, Japan.

Author details

Toshiro Kaneko, Shohei Takahashi and Rikizo Hatakeyama

Department of Electronic Engineering, Tohoku University, Sendai, Japan

References

[1] Gubkin J. Electrolytische Metallabscheidung an der freien Oberfläche einer Salzlösung. Annalen der Physik 1887; 268(9), 114-115.

[2] Kanzaki Y., Hirabe M., Matsumoto O. Glow Discharge Electrolysis of Aqueous Sulfuric Acid Solution in Various Atmosphere. Journal of The Electrochemical Society 1986; 133(11), 2267-2270.

[3] Baba K., Kaneko T., Hatakeyama R. Ion Irradiation Effects on Ionic Liquids Interfaced with RF Discharge Plasmas. Applied Physics Letters 2007; 90(20), 201501-1-3.

[4] Kaneko T., Baba K., Hatakeyama R. Static Gas-Liquid Interfacial Direct Current Discharge Plasmas Using Ionic Liquid Cathode. Journal of Applied Physics 2009; 105(10), 103306-1-5.

[5] Bruggeman P., Leys C. Non-Thermal Plasmas in and in Contact with Liquids. Journal of Physics D: Applied Physics 2009; 42(5), 053001-1-28.

[6] Koo I. G., Lee M. S., Shim J. H., Ahn J. H., Lee W. M. Platinum Nanoparticles Prepared by a Plasma-Chemical Reduction Method. Journal of Materials Chemistry 2005; 15(38), 4125-4128.

[7] Hieda J., Saito N., Takai O. Exotic Shapes of Gold Nanoparticles Synthesized Using Plasma in Aqueous Solution. Journal of Vacuum Science and Technology A 2008; 26(4), 854-856.

[8] Richmonds C., Sankaran R. M. Plasma-Liquid Electrochemistry: Rapid Synthesis of Colloidal Metal Nanoparticles by Microplasma Reduction of Aqueous Cations. Applied Physics Letters 2008; 93(13), 131501-1-3.

[9] Sato S., Mori K., Ariyada O., Atsushi H., Yonezawa T. Synthesis of Nanoparticles of Silver and Platinum by Microwave-Induced Plasma in Liquid. Surface and Coatings Technology 2011; 206(5), 955-958.

[10] Chen L., Iwamoto C., Omurzak E., Takebe S., Okudera H., Yoshiasa A., Sulaimankulova S., Mashimo, T. Synthesis of Zirconium Carbide (ZrC) Nanoparticles Covered with Graphitic "Windows" by Pulsed Plasma in Liquid. RSC Advances 2011; 1(6), 1083–1088.

[11] Seddon K. R. Ionic Liquids: A Taste of the Future. Nature Materials 2003; 2(6), 363-365.

[12] Rogers R. D., Seddon K. R. Ionic Liquids - Solvents of the Future?. Science 2003; 302(5646), 792-793.

[13] Baba K., Kaneko T., Hatakeyama R. Efficient Synthesis of Gold Nanoparticles Using Ion Irradiation in Gas-Liquid Interfacial Plasmas. Applied Physics Express 2009; 2(3), 035006-1-3.

[14] Kaneko T., Baba K., Harada T., Hatakeyama R. Novel Gas-Liquid Interfacial Plasmas for Synthesis of Metal Nanoparticles. Plasma Processes and Polymers 2009; 6(11), 713-718.

[15] Kaneko T., Chen Q., Harada T., Hatakeyama R. Structural and Reactive Kinetics in Gas-Liquid Interfacial Plasmas. Plasma Sources Science and Technology 2011; 20(3), 034014-1-8.

[16] Meiss S. A., Rohnke M., Kienle L., Zein El Abedin S., Endres F., Janek J. Employing Plasmas as Gaseous Electrodes at the Free Surface of Ionic Liquids: Deposition of Nanocrystalline Silver Particles. ChemPhysChem 2007; 8(1), 50-53.

[17] Torimoto T., Okazaki K., Kiyama T., Hirahara K., Tanaka N., Kuwabata S. Sputter Deposition onto Ionic Liquids: Simple and Clean Synthesis of Highly Dispersed Ultrafine Metal Nanoparticles. Applied Physics Letters 2006; 89(24), 243117-1-3.

[18] Xie Y. B., Liu C. J. Stability of Ionic Liquids under the Influence of Glow Discharge Plasmas. Plasma Processes and Polymers 2008; 5(3), 239-245.

[19] Kuwabata S., Tsuda T., Torimoto T. Room-Temperature Ionic Liquid. A New Medium for Material Production and Analyses under Vacuum Conditions. The Journal of Physical Chemistry Letters 2010; 1(21), 3177–3188.

[20] Wei Z., Liu C. J. Synthesis of Monodisperse Gold Nanoparticles in Ionic Liquid by Applying Room Temperature Plasma. Materials Letters 2011; 65(2), 353–355.

[21] Kareem T. A., Kaliani A. A. I-V Characteristics and the Synthesis of ZnS Nanoparticles by Glow Discharge at the Metal–Ionic Liquid Interface. Journal of Plasma Physics 2012; 78(2), 189-197.

http://dx.doi.org/10.5772/52095

[22] Shaw C. P., Fernig D. G., Levy R. Gold Nanoparticles as Advanced Building Blocks for Nanoscale Self-Assembled Systems. Journal of Materials Chemistry 2011; 21(33), 12181-12187.

[23] Huang L., Tu C. C., Lin L. Y. Colloidal Quantum Dot Photodetectors Enhanced by Self-Assembled Plasmonic Nanoparticles. Applied Physics Letters 2011; 98(11), 113110-1-3.

[24] Maye M. M., Nykypanchuk D., van der Lelie D., Gang O. DNA-Regulated Micro- and Nanoparticle Assembly. Small 2007; 3(10), 1678-1682.

[25] Chen Q., Kaneko T., Hatakeyama R. Rapid Synthesis of Water-Soluble Gold Nanoparticles with Control of Size and Assembly Using Gas-Liquid Interfacial Discharge Plasma. Chemical Physics Letters 2012; 521, 113–117.

[26] Yu J., Rance G. A., Khlobystov A. N. Electrostatic Interactions for Directed Assembly of Nanostructured Materials: Composites of Titanium Dioxide Nanotubes with Gold Nanoparticles. Journal of Materials Chemistry 2009; 19(47), 8928-8935.

[27] Kaneko T., R. Hatakeyama R. Creation of Nanoparticle-Nanotube Conjugates for Life-Science Application Using Gas-Liquid Interfacial Plasmas. Japanese Journal of Applied Physics 2012; in press.

[28] Baba K., Kaneko T., Hatakeyama R., Motomiya K., Tohji K. Synthesis of Monodispersed Nanoparticles Functionalizing Carbon Nanotubes in Plasma-Ionic Liquid Interfacial Fields. Chemical Communications 2010; 46(2), 255-257.

[29] Georgakilas V., Gournis D., Tzitzios V., Pasquato L., Guldie D. M., Prato M. Decorating Carbon Nanotubes with Metal or Semiconductor Nanoparticles. Journal of Materials Chemistry 2007; 17(26), 2679–2694.

[30] Wildgoose G. G., Banks C. E., Compton R. G. Metal Nanoparticles and Related Materials Supported on Carbon Nanotubes: Methods and Applications. Small 2006; 2(2), 182-193.

[31] Ye X., Lin Y., Wang C., Engelhard M. H., Wang Y., Wai C. M. Supercritical Fluid Synthesis and Characterization of Catalytic Metal Nanoparticles on Carbon Nanotubes. Journal of Materials Chemistry 2004; 14(5), 908-913.

[32] Han L., Wu W., Kirk F. L., Luo J., Maye M. M., Kariuki N. N., Lin Y., Wang C., Zhong C.-J. A Direct Route toward Assembly of Nanoparticle–Carbon Nanotube Composite Materials. Langmuir 2004; 20(14), 6019-6025.

[33] Kaneko T., Takahashi S., Hatakeyama R. Control of Nanoparticle Synthesis Using Physical and Chemical Dynamics of Gas-Liquid Interfacial Non-Equilibrium Plasmas. Plasma Physics and Controlled Fusion 2012; in press.

[34] Krenn J. R. Nanoparticle Waveguides: Watching Energy Transfer. Nature Materials 2003; 2(4), 210-211.

Chapter 25

Ionic-Liquid-Assisted Synthesis of Hierarchical Ceramic Nanomaterials as Nanofillers for Electromagnetic-Absorbing Coatings

Elaheh Kowsari

Additional information is available at the end of the chapter

http://dx.doi.org/10.5772/51653

1. Introduction

The use of electronic devices is experiencing an exponential growth in all the fields of human life, and most of them (personal computers, communication, medical and analytic devices, a lot of domestic appliances) work in the microwave frequency range. This growth gives rise to an increase of Electro Magnetic Interference (EMI) so that it is mandatory to develop systems to protect electronic devices from external interferences.

Electromagnetic (EM) wave absorption materials have attracted much attention owing to the expanded EM interference problems. The EM absorbers are now requested to have not only strong absorption characteristics and wide absorption frequency, but also light weight and antioxidation. The electric permittivity (ε) and magnetic permeability (μ) are parameters related to the dielectric and magnetic properties of a material, and directly associated to their absorbing characteristics [1-5].

The relative permittivity and permeability are represented by Equations 1 and 2, respectively; the values of these parameters are calculated from the experimental values of the transmission and refection coeffcients of the material.

$$\varepsilon_r = \varepsilon' + i\varepsilon'' \quad (1)$$

$$\mu_r = \mu' + i\mu'' \quad (2)$$

INTECH
open science | open minds

In these equations, the primed and double-primed symbols denote real and imaginary components. When the material is lossy, the permittivity and permeability of are complex and some of the incident electromagnetic energy is dissipated [6-7].

In the case of a magnetic material, losses are produced by changes in the alignment and rotation of the magnetization spin [1,8, 9].

The traditional absorbers such as ferrite have strong absorption characteristics, but the thickness required is too large. Therefore, many nanostructures have recently been studied for attenuation of EM wave. These materials are involved with carbon nanotubes, iron and zinc oxide, etc [10-14]. EM wave absorption capability depends on the nature, shape, and size of an absorber.

The dramatic effect that shape anisotropy has on the electronic, optical, and catalytic properties of noble ceramic nanostructures makes the development of morphology-controlled synthesis strategies a main step toward the design of future nanodevices [15-19].

In this regard, the last years have been very prolific in the design of new procedures dealing with the synthesis of noble metal anisotropic structures (Au, Ag, Pt, Pd) such as nanowires, [20-22] nanoplates, [23-26] nanocubes, [27,28] and nanorods [29] with well-controlled size. Recently, some examples concerning the synthesis of branched nanostructures (monopod,bipod, tripod, tetrapod,multipod),[30-35] star polyhedral crystals [36], nanoflowers [37], and ringlike nanostructures[38] have also been reported.

Hierarchical nano-/micro-structures with specific morphology have fascinated scientists all over the world because of their sophisticated architectures which are expected to provide some unique and exciting properties. To date, many recent efforts have been devoted to the synthesis of inorganic materials with hierarchical shapes, including metal [39,40], metal oxide [41], sulfide [42], hydrate [43], and other minerals [44,45].

Recently, ionic liquids have aroused increasing interest because of their unique properties and the potential applications. Ionic liquid can act as a new reaction medium for reactants and morphology templates for the products at the same time, which enables the synthesis of hierarchical nano-/micro-structures with novel or improved properties [46]. Novel nanostructures can be produced by selecting suitable ionic liquids reaction systems. Various nano- or microstructured materials, such as Bi_2S_3 nanostructures [47, 48],Bi_2Se_3 nanosheets [49] and hollow TiO_2 microspheres [50] have been synthesized in ionic liquids.

In this chapter, an attempt has been made to develop hierarchical ceramic nanomaterials with a wide range of morphologies and sizes with improved reflection losses (RL). The duration of experimental processes parameters like amount of IL, pH and temperature have been extensively optimized to obtain morphologies and sizes.

1.1. Development of Innovative Synthesis Methods of Ceramic Nano-materials Using Ionic Liquid

Synthesis of nanomaterials is an increasing active area [51,52]. This interest arises from not only their unusual chemical and physical properties but also their potential application in many fields, which have stimulated the search for new synthetic methods for these materials. Size, morphology and dimensionality can strongly affect the properties of nanostructured materials. Recently, nanostructured metallic and semiconducting materials with various structures and morphologies have received muchattention due to their novel applications, intriguing properties, and quantum size effects [53]. Especially, a three-dimensional (3D) integrated platform of nanostructured materials is highly desirable for applications in advanced nanoelectronic, optoelectronic, solar cells, sensor, etc., [54–56].

Novel nanostructures can be produced by selecting suitable ILs reaction systems. Various nano- or microstructured materials, have been synthesized in ILs

Cupric oxide (CuO) with leaf-like, chrysanthemum-like and rod shapes have been synthesized by microwave-assisted approach using an ionic liquid 1-n-butyl-3-methyl imidazolium tetrafluoroborate ([BMIM]BF4) by Xu and coworkers [57]. By controlling the concentration of [BMIM]BF_4 and reaction time, shape transformation of CuO nanostructures could be achieved in a short period of time. The morphologies of the samples were shown in Figure. 1. Leaf-like CuO nanosheets with uniform shape and size were obtained on a large scale (Figure. 1(a)).

Figure 1. FESEM images of sample A (a–b), EDS results of sample A (c), and TEM image of sample A (d). The inset showed the SAED pattern taken from d.(Reproduced from X Xu, M Zhang, J Feng, M Zhang. Shape-controlled synthesis of single-crystalline cupric oxide by microwave heating using an ionic liquid. Mat Lett. 2008; 62:2787–2790, Copyright (2008), with permeation from Elsevier).

MoS_2 microspheres were successfully synthesized via a facile hydrothermal route assisted by an ionic liquid [BMIM][BF4] by Ma and coworkers [58]. SEM images showed that the MoS_2 microspheres had uniform sizes with mean diameter about 2.1 μm. The MoS_2 microspheres had rough surfaces and were constructed with sheetlike structures. Ionic liquid played a crucial role as a templating reagent in the formation of MoS_2 microspheres. A possible formation mechanism of MoS_2 microspheres was preliminarily presented. The size and morphology of the samples were examined by SEM. Figure. 2A shows that the as-synthesized MoS_2 products display a uniform spherical morphology with mean diameter of 2.1 μm.

Figure 2. SEM images of MoS2 microspheres prepared by IL-assisted hydrothermal process. (A) and (B) as-synthesized samples and (C) after annealing at 800 °C for 2 h. [Reproduced from Ma L, Chen WX, Li H, Zheng YF, Xu ZD. Ionic Liquid-Assisted Hydrothermal Synthesis of MoS2 Microspheres. Mat Lett. 2008; 62: 797–799, Copyright (2008), with permeation from Elsevier)].

Wurtzite CdSe nanoparticles-assembled microspheres with macropores have been successfully synthesized through a modified hydrothermal method with $Cd(NO_3)_2$ and Na_2SeO_3 as precursors and hydrazine hydrate as a reductant in the presence of 1-n-butyl-3-methylimidazolium bromide ([Bmim]Br) by Liu and coworkers [59]. The results indicated that the CdSe microspheres have an average size of about 3 μm and were assembled by CdSe nanoparticles with size ranging from 20 to 40 nm. It was found that the pH and [Bmim]Br have influence on the morphologies of the products. Figure. 3 shows Wurtzite CdSe nanoparticles-assembled microspheres.

ZnO/SnO_2 nanostructured have been successfully synthesized by a hydrothermal method in the presence of the chiral ionic liquid (CIL) ditetrabutylammonium tartrate, [TBA]2[L-Tar] by Kowsari and coworkers [60].The results revealed that using different ratios of

Zn^{2+}/Sn^{4+} affects the phase and morphology of the ZnO/SnO_2 nanocomposite materials. Figure. 4 shows ZnO/SnO_2 nanostructures.

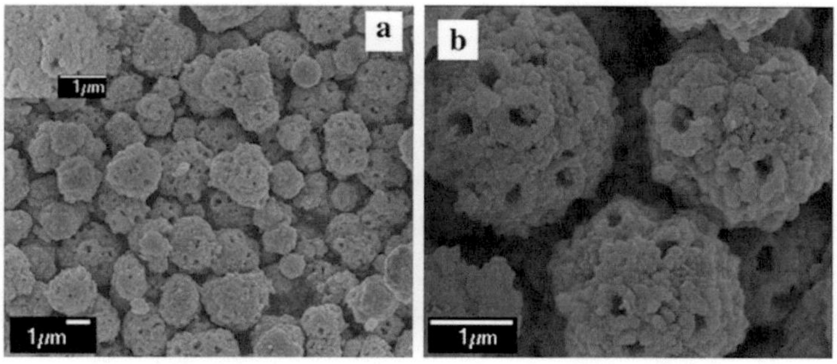

Figure 3. (a) Low magnification SEM image of the CdSe microspheres, the inset shows the structure on the surface of the microspheres; (b) Magnified SEM image of three individual CdSe microspheres. [Reproduced from Liu X, Peng P, Ma J, Zheng W. Preparation of novel CdSe microstructure by modified hydrothermal method. Mat Lett. 2009; 63: 673–675. ,Copyright (2009), with permeation from Elsevier)].

Figure 4. SEM images of products obtained at different Zn^{2+}/Sn^{4+} molar ratios: (a) 2:1, (b) 1:1, (c) 4:1, (d) 1:2, (e) 0:1, and (f) 1:0; the reaction time was kept constant at 24 h and the reaction temperature was 170 °C, CIL=0.05 g. [Reproduced from Kowsari E, Ghezelbash MR. Ionic liquid-assisted, Facile Synthesis of ZnO/SnO_2 Nanocomposites, and Investigation of Their Photocatalytic Activity. Mat Lett. 2012; 68: 17–20,Copyright (2009), with permeation from Elsevier)].

A hydrothermal method has been employed to prepare cactus-like zincoxysulfide ZnO_xS_{1-x} nanostructures with the assistance of a dicationic task-specific ionic liquid (TSIL), [mim]{$(CH)_2$}$_3$[imm]$(SCN)_2$ by kowsari and coworkers[61]. To the best of our knowledge, this is the first time that this TSIL with the SCN anion has been used in place of conventional reagents as a source of S to prepare a ZnO_xS_{1-x} nanostructure. The effect of the TSIL concentration on the morphology of the products shows in Figure. 5.

Figure 5. SEM images of products obtained with different amounts of FIL: (a) 0.22 g (S-1), (b) 0.1 g (S-2); the reaction time was constant at 24 h and Zn2+/OH−=1:20; the reaction temperature was 170 °C. [Reproduced from Kowsari E, Ghezelbash MR. Synthesis of Cactus-Like Zincoxysulfide (ZnO_xS_{1-x}) Nanostructures Assisted by a Task-Specific Ionic Liquid and Their Photocatalytic Activities. Mat Lett.2011; 65: 3371–3373.Copyright (2011), with permeation from Elsevier)].

2. Influence of ionic liquids on the growth of nanofillers for electromagnetic-absorbing coatings

2.1. The fabrication of $BaCO_3$ nanostructures as nanofillers for electromagnetic-absorbing coatings

An economical and efficient ionic liquid-assisted chemical method was demonstrated for the first time for the fabrication of $BaCO_3$ nanostructures. The shape of these $BaCO_3$ nanostructures could be readily controlled by changing the chemical conditions and the amount of the chiral ionic liquid (CIL), ditetrabutylammonium tartrate, $[TBA]_2$[L-Tar] by Kowsari and coworkers [62]. The CIL is a reagent and templating agent for the fabrication of $BaCO_3$ nanostructures. It was demonstrated that $[TBA]_2$[L-Tar] served as a modifier in the reactionsystem.Figure 6 shows typical SEM images of $BaCO_3$ nanostructures synthesized with 0.05 g CIL at 170 °C for 24 h. From the SEM image Figure 6, it is clear that the typical $BaCO_3$ dendritic nanostructures assembled were hyacinth-like with rod-like nanostructures with lengths of up to several micrometers.

http://dx.doi.org/10.5772/51653

Figure 6. Typical scanning electron micrographs of $BaCO_3$ nanostructures synthesized with 0.05 g CIL, NaOH = 0.04 M (2 ml) at 170 °C for 24 h, the $Ba(NO_3)_2$ = 0.53 g. [Reproduced from Kowsari E, Karimzadeh AH. Using a Chiral Ionic Liquid for Morphological Evolution of $BaCO_3$ and its Radar Absorbing Properties as a Dendritic Nanofiller. Mat Lett. 2012;74, 33-36, Copyright (2012), with permeation from Elsevier)].

With increase in the amount of CIL to 0.1 g, flower-like $BaCO_3$ structures composed of dendritic petals were produced, which appeared as a result of oriented attachment and self-assembly, as exhibited in Fig 7.

Figure 7. Typical scanning electron micrographs of $BaCO_3$ nanostructures synthesized with 0.1g CIL, NaOH = 0.04 M (2 ml) at 170 °C for 24 h, the $Ba(NO_3)_2$ = 0.53 g. [Reproduced from Kowsari E, Karimzadeh AH. Using a Chiral Ionic Liquid for Morphological Evolution of $BaCO_3$ and its Radar Absorbing Properties as a Dendritic Nanofiller. Mat Lett.2012;74, 33-36, Copyright (2012), with permeation from Elsevier)].

The effects of the synthetic parameters, such as the concentration of NaOH, reaction temperature, reaction time and [Ba^{2+}], on the morphologies of the resulting products were investigated. Typical SEM images of the products with different morphologies brought about by varying the amount of NaOH are presented in Figure. 8.

Too much NaOH was found to be harmful to the formation of flower-like structures and resulted in atrophic flowers, which are analogs of a sphere as shown in Figure. 8. NaOH affected the reaction kinetics through tuning the dissolution-deposition equation of $Ba(OH)_2$, because soluble $Ba(NO_3)_2$ will first react with NaOH to form $Ba(OH)_2$ precipitate (Ksp) 1.09 × 10-15). The trend for $Ba(OH)_2$ to release Ba2+ decreased if the concentration of OH- is relatively high, leading to the thermodynamic change of nucleation and growth velocities, which is beneficial for the formation of spherical structures.

Figure 9 shows XRD patterns of the products obtained under different reaction conditions. It is clear that all of the peaks can be readily indexed to the pure orthorhombic phase of BaCO3 (JCPDS card no. 05-0378). The sharp diffraction peaks of the sample indicate that well-crystallized BaCO3 crystals can be easily obtained under the current synthetic conditions. On comparing the XRD patterns of the four products, It was shown that the relative intensity of the peaks varied slightly.

Figure 8. Typical scanning electron micrographs of $BaCO_3$ nanostructures synthesized with (a) 2.5 ml, CIL = 0.05 g at 170 °C for 24 h, $Ba(NO_3)_2$ = 0.53 g.] [Reproduced from Kowsari E, Karimzadeh AH. Using a Chiral Ionic Liquid for Morphological Evolution of $BaCO_3$ and its Radar Absorbing Properties as a Dendritic Nanofiller. Mat Lett.2012;74, 33-36, Copyright (2012), with permeation from Elsevier)].

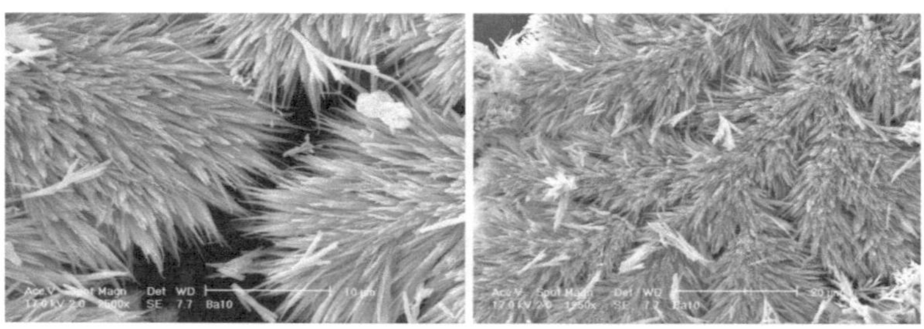

Figure 9. Typical scanning electron micrographs of $BaCO_3$ nanostructures synthesized with CIL = 0.05 g at 170 °C for 24 h, $Ba(NO_3)_2$ = 0.53 g. [Reproduced from Kowsari E, Karimzadeh AH. Using a Chiral Ionic Liquid for Morphological Evolution of $BaCO_3$ and its Radar Absorbing Properties as a Dendritic Nanofiller. Mat Lett.2012;74, 33-36, Copyright (2012), with permeation from Elsevier)].

The value of the minimum reflection loss for the $BaCO_3$ dendritic nanostructures composite was -40 dB at 10.2 GHz (for a thickness of 4.0 mm) [Figure. 10]. According to the results shown above, $BaCO_3$ dendritic nanostructures showed very strong absorption of microwave compared with other samples. It is noteworthy that the $BaCO_3$ dendritic nanostructures have special geometrical morphology. Such isotropic crystal symmetry can form isotropic quasi antennas and some continuous networks in the composites. It is possible for the electromagnetic waves to penetrate the nanocomposites formed by the numerous antenna-like semiconducting $BaCO_3$ dendritic nanostructures and for the energy to be induced into a dissipative current; then the current will be consumed in the continuous networks, which leads to the energy attenuation (19-20). More importantly, the interfacial electric polarization should be considered. However, further experimental and theoretical work is needed to clarify this mechanism.

http://dx.doi.org/10.5772/51653

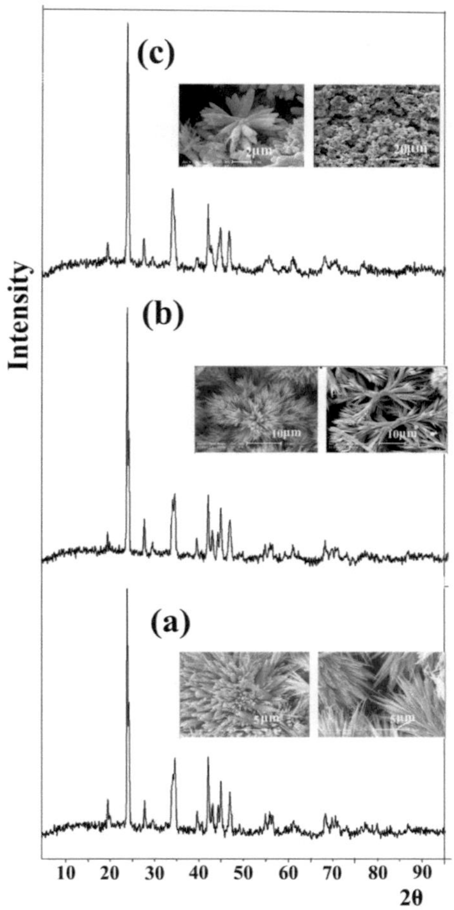

Figure 10. X-ray diffraction patterns of the $BaCO_3$ nanostructures synthesized with (a) 5 ml, (b) 2.5 ml, (NaOH (0.04 M) [CIL = 1.25 g /L at 170 °C for 24 h, $Ba(NO_3)_2$ = 13.25 g/L.], (c) 2.5 g/L CIL. [NaOH = 0.04 M (2 ml) at 170 °C for 24 h, the $Ba(NO_3)_2$ = 13.25 g/L]; [Reproduced from Kowsari E, Karimzadeh AH. Using a Chiral Ionic Liquid for Morphological Evolution of $BaCO_3$ and its Radar Absorbing Properties as a Dendritic Nanofiller. Mat Lett.2012;74, 33-36, Copyright (2012), with permeation from Elsevier)].

2.2. Fern-like, fish skeleton-like, bunched cubic, and butterfly-like BaO nanostructures as nanofillers for Radar-absorbing coatings

Fern-like, fish skeleton-like, bunched cubic, and butterfly-like BaO nanostructures have been synthesized by a hydrothermal method at 170 °C from $Ba(NO_3)_2$ •3H2O and NaOH in the presence of ammonium persulfate $(NH_4)_2S_2O_8$ (APS) by Kowsari and coworkers[63]. The effect of the chiral ionic liquid (CIL) ditetrabutylammonium tartrate, $[TBA]_2[L\text{-}Tar]$, on the morphologies of the products has been investigated. It was demonstrated that $[TBA]_2[L\text{-}Tar]$ served as a modifier in the reaction system. Furthermore, the BaO nanostructures have been used as fillers in high-performance microwave-absorbing coatings.

By controlling the concentrations of CIL, Ba^{2+}, and APS, shape transformations of the BaO could be achieved. When the amount of CIL was zero, rotor-like BaO was formed (Figure 11).

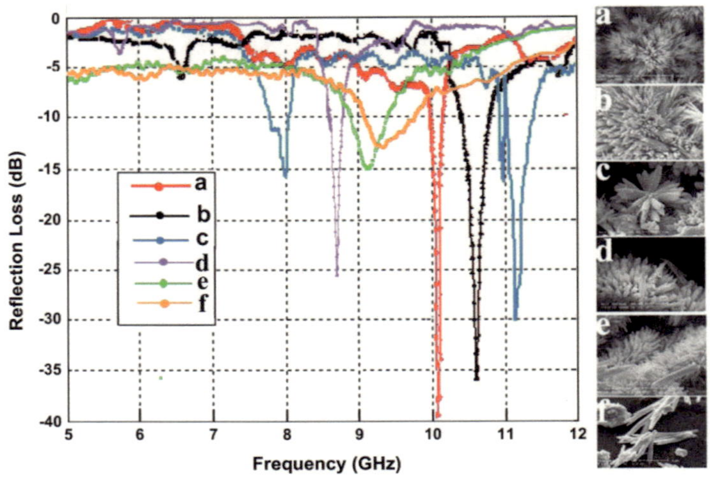

Figure 11. Reflection loss (dB) as a function of frequency (GHz) of $BaCO_3$ nanostructures synthesized with: (a) NaOH = 2.5 mL (0.04 M), CIL = 1.25 g /L for 24 h $Ba(NO_3)_2$ = 13.25 g/L; (b) NaOH = 5 mL (0.04 M) and CIL = 1.25 g /L for 24 h, $Ba(NO_3)_2$ = 13.25 g/L; (c) NaOH = 2 mL (0.04 M) and CIL = 2.5 g/L, for 24 h, amount of $Ba(NO_3)_2$ = 13.25 g/L; (d) NaOH = 2 mL (0.04 M), CIL = 1.25 g /L for 24 h, $Ba(NO_3)_2$ = 6.52 g/L; (e) NaOH = 2 mL (0.04 M), CIL = 1.25 g /L for 24 h, $Ba(NO_3)_2$ = 13.25 g/L; (f) NaOH- = 2 mL (0.04 M), CIL = 1.25 g /L for 48 h, $Ba(NO_3)_2$ = 13.25 g/L, (at 170 °C and with filler = 0.025 g in coating). [Reproduced from Kowsari E, Karimzadeh AH. Using a Chiral Ionic Liquid for Morphological Evolution of $BaCO_3$ and its Radar Absorbing Properties as a Dendritic Nanofiller. Mat Lett.2012;74, 33-36, Copyright (2012), with permeation from Elsevier)].

Figure 12. SEM images of products obtained with : 0.1 g CIL, keeping the reaction time fixed at 24 h; $Ba(NO_3)_2$ = 0.2 g, APS = 0.27 g, reaction temperature 170 °C. [Reproduced from Kowsari E, Karimzadeh AH. Fabrication of Fern-Like, Fish Skeleton-Like, and Butterfly-Like BaO Nanostructures as Nanofillers for Radar-Absorbing Nanocomposites. Mat Lett. 2012; 74:33–36, Copyright (2012), with permeation from Elsevier].

When the amount of CIL was increased gradually, while maintaining the same reaction time, rotor-like BaO changed to bunched cubic BaO, as shown in Figure 14. The cubes have a porous surface structure.

Figure 13. SEM images of products obtained with 0.1 g, CL= 0.15 g keeping the reaction time fixed at 24 h; $Ba(NO_3)_2$ = 0.2 g, APS = 0.27 g, reaction temperature 170 °C. [Reproduced from Kowsari E, Karimzadeh AH. Fabrication of Fern-Like, Fish Skeleton-Like, and Butterfly-Like BaO Nanostructures as Nanofillers for Radar-Absorbing Nanocomposites. Mat Lett.2012; 74:33–36, Copyright (2012), with permeation from Elsevier].

Figure 14. SEM images of products obtained with $Ba(NO_3)_2$ = 0.8 g, CIL = 0; APS = 0.27 g; (the reaction time was kept constant at 24 h and the reaction temperature was 170 °C). [Reproduced from Kowsari E, Karimzadeh AH. Fabrication of Fern-Like, Fish Skeleton-Like, and Butterfly-Like BaO Nanostructures as Nanofillers for Radar-Absorbing Nanocomposites. Mat Lett.2012; 74:33–36, Copyright (2012), with permeation from Elsevier].

The CIL clearly plays a key role in tailoring the form of the resultant BaO nanostructures. It is thought that hydrogen bonds formed between the hydrogen atom at the position-2 (connected to oxygen) of the CIL cation and the oxygen atoms of O–Ba crystal cores may act as effective bridges in connecting the produced BaO nuclei and CIL cations, playing a crucial role in the directional growth of the 2D nanocrystals.

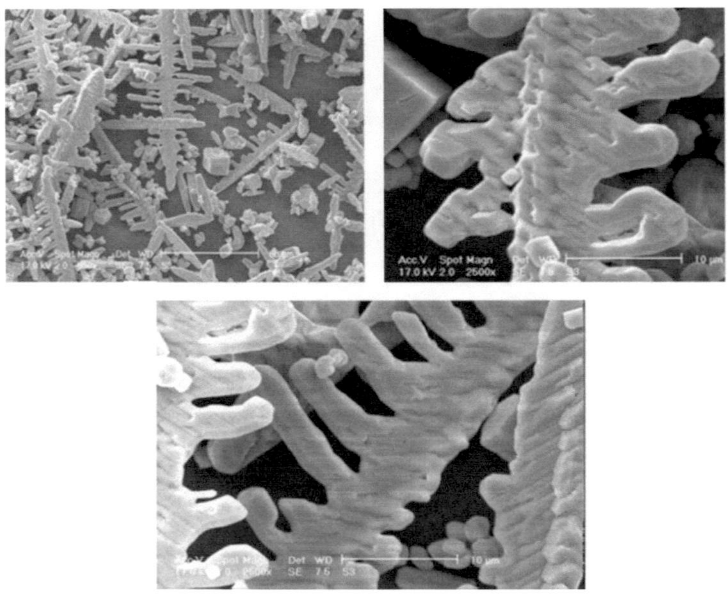

Figure 15. SEM images of products obtained with: $Ba(NO_3)_2$ = 0.4 g, CIL = 0, APS = 0.27 g; (the reaction time was kept constant at 24 h and the reaction temperature was 170 C). [Reproduced from Kowsari E, Karimzadeh AH. Fabrication of Fern-Like, Fish Skeleton-Like, and Butterfly-Like BaO Nanostructures as Nanofillers for Radar-Absorbing Nanocomposites. Mat Lett.2012; 74:33–36, Copyright (2012), with permeation from Elsevier].

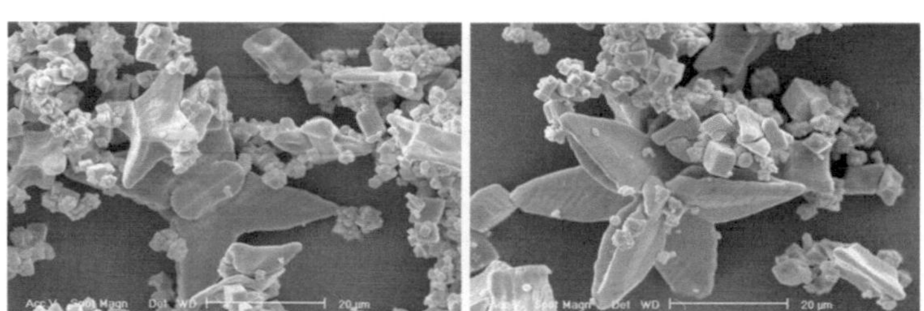

Figure 16. SEM images of products obtained with $Ba(NO_3)_2$ = 0.2 g, CIL = 0, APS = 0.27 g; (the reaction time was kept constant at 24 h and the reaction temperature was 170 C). [Reproduced from Kowsari E, Karimzadeh AH. Fabrication of Fern-Like, Fish Skeleton-Like, and Butterfly-Like BaO Nanostructures as Nanofillers for Radar-Absorbing Nanocomposites. Mat Lett.2012; 74:33–36, Copyright (2012), with permeation from Elsevier].

We also investigated the effect of the amount of $Ba(NO_3)_2$ on the morphology of the products. The respective products obtained from this series of experiments are depicted in Figure 13, 14, 15.

The value of the maximum reflection loss for the composite with BaO cubic nanostructures was measured as ◎20 dB at 10.5 GHz for a thickness of 4.0 mm. According to the results shown in Figure 4, the BaO cubic nanostructures showed very strong absorption of microwaves compared with the other samples. It may be noted that the BaO cubic nanostructures had a special geometrical morphology.

2.3. Morphology evolution of the ZnO/Zn(OH)2 nanofillers using ionic liquids

An efficient ionic-liquid-assisted chemical method for the fabrication of $ZnO/Zn(OH)_2$ nanoplates is demonstrated by kowsari and coworkers [64]. The shape of the resulting $ZnO/Zn(OH)_2$ nanostructures could be readily controlled by changing the chemical conditions and the amount of $[Cn(mim)]^+$ $H2PO_{4-}$ as a task-specific ionic liquid (TSIL). The TSIL thus serves as a reagent and templating agent for the fabrication of $ZnO/Zn(OH)_2$ nanoplates. Furthermore, a possible growth mechanism of the $ZnO/Zn(OH)_2$ nanostructures is proposed. The effects of different morphologies of the nanofillers on electromagnetic properties have been investigated.

By controlling the concentration of the TSIL and the reaction time, shape transformations of the $ZnO/Zn(OH)_2$ nanostructures could be achieved. Our results show that the quantity of the TSIL affected the morphology of the $Zn(OH)_2$ and caused changes in the interplanar separation and the arrangement of the nanoplates. Salient SEM images are compared in Fig. 1. In this figure, it can be seen that as the quantity of the TSIL was increased from 0.1 g to 0.2 g, the separation between the nanoplates decreased and the number of nanoplates increased.

The TSIL clearly played a key role in tailoring the form of the resultant $ZnO/Zn(OH)_2$ nanocomposites.To analyze the effect of the reaction time on morphology, the reaction was carried out for 12, 24, and 48 h, while keeping the amounts of Zn2+ and TSIL constant. Fig. 2 shows the morphologies of the products. It can be seen that with increasing reaction time, the separation between the nanoplates increased and the number of nanoplates decreased

Figure 17. SEM images of products obtained with different time of reaction: (a) 12 h, (b) 48 h, $Zn(NO_3)_2$ = 0.8 g, OH = 0.27 g, reaction temperature 170 °C, TSIL =0.1 g.

Fig. 18 clearly shows the process of TSIL-assisted hydrothermal growth of ZnO and Zn(OH)2 crystals. The amount of the growth unit $[Zn(OH)_2(H_2PO_4)2]^{2-}$ is greatly increased, and further ZnO and $Zn(OH)_2$ nuclei directly conglomerate to form a two-dimensional nanosheet structure in order to lower the surface potential. Thereafter, the nanosheets self-assemble to produce a cabbage-like structure.

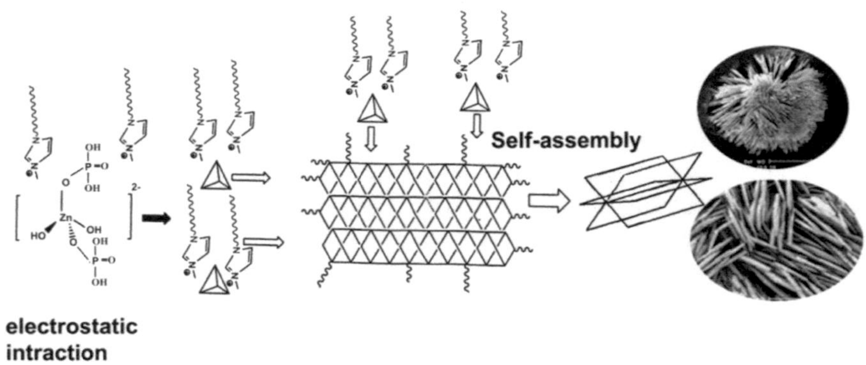

Figure 18. Schematic illustration of the processes of $Zn(OH)_2/ZnO$ nanostructure assembly by IL-assisted hydrothermal growth.

The TSIL clearly played a key role in tailoring the form of the resultant $ZnO/Zn(OH)_2$ nanocomposites. In a general manner, the produced coatings showed broadband absorber behavior, which may be attributed to the dielectric properties and the particular shape of the $Zn(OH)_2/ZnO$ nanofillers. When radar impinges on a radar -absorbing coating, the incident radiation is not totally absorbed immediately. The radar attenuation by the radar-absorbing coating is significantly complex and different attenuation mechanisms can occur. Equal amounts of three types of $Zn(OH)_2/ZnO$ nanostructures with different morphologies and different separations between their nanoplates, as synthesized under different conditions, were used as fillers in coatings.

The value of the maximum reflection loss for the composite with $Zn(OH)_{2/}ZnO$ nanostructures was measured ~20 dB at 8 GHz for a thickness of 4.0 mm. The samples incorporating $Zn(OH)_2/ZnO$ nanostructures of a morphology with a shorter inter-plate separation displayed higher absorptions. The absorption was also shifted toward higher frequencies. Moreover, a higher bandwidth was observed. It may be noted that the $ZnO/Zn(OH)_2$ nanostructures had a special geometrical morphology, namely that of a multi-layer microwave absorber. More importantly, the interfacial electric polarization should be considered. It is well known that the permittivity mainly originates from electronic polarization, ion polarization, and intrinsic electric-dipole polarization, on which the crystal structure, size, and shape of nanomaterials may have important influences. On the basis of the electron microscopy characterization results, we can conclude that the radar-absorbing properties are associated with the crystal structure, the crystallization, and the degree of aggregation of the nanocrystal building blocks. This also explains why the $ZnO/Zn(OH)_2$ nanostructures studied here show different radar-absorption properties. However, further experimental and theoretical work is needed to clarify the mechanism.

http://dx.doi.org/10.5772/51653

3. Conclusion

Over the past years, there has been an increasing interest in developing hierarchically structural materials on a nanometer scale due to their novel or enhanced properties. By controlling the condition of reaction, shape transformations of the resulting hierarchical ceramic nanomaterials could be achieved. In this chapter, the ceramic nanomaterials have been synthesized in a single reaction system by simply adjusting the reaction conditions and used as fillers in high-performance microwave-absorbing coatings, with epoxy resin as the polymer matrix. The task specific ionic liquids thus serves as a reagent or templating agent for the fabrication of ceramic oxides. The effects of different morphologies of the nanofillers on electromagnetic properties have been investigated. The maximum reflection attenuation was measured as −20 dB in the frequency range 5–12 GHz.

Author details

Elaheh Kowsari*

Address all correspondence to: kowsarie@aut.ac.ir

Department of Chemistry, Amirkabir University of Technology, Tehran, Iran

References

[1] Balanis, C. A. (1989). *Advanced Engineering Electromagnetics*, New York, John Wiley and Sons.

[2] Lee, S. M. (1991). *International Encyclopedia of Composites*, New York, VCH Publishers.

[3] Clark, D. E., Diane, C. F., Stephen, J. O., & Richards, S. (1995). *Microwaves: Theory and Application in Materials Processing III*, Westerville, The American Ceramic Society.

[4] Hippel, A. (1954). *Dielectric Materials and Applications*, London, Artech House.

[5] Yusoffa, A. N., Abdullah, M. H., Ahmad, S. H., Jusoh, S. F., Mansor, A. A., & Hamid, S. A. A. (2002). Electromagnetic and absorption properties of some microwave absorbers. *J.Appl Phy*, 92(2), 876-882.

[6] Orfanidis, S. J. (2008). *Electromagnetic Waves and Antennas*, Available from, http://www.ece.rutgers.edu/~orfanidi/ewa/, accessed Oct.

[7] Sucher, M. ., & Fox, J. (1963). *Handbook of the microwave measurements* (3 ed.), New York, John Wiley and Sons.

[8] Thostenson, E. T., & Chou, T. W. (1999). Microwave processing: fundamentals and applications. *Composites: Part A: Applied Science and Manufacturing*, 30(9), 1055-1071.

[9] Jarem, L. M., Johnson, J. B. ., & Scott, W. (1995). Measuring the permittivity and permeability of sample at Ka Band using a partially flled waveguide. *IEEE Transactions on Microwave and Techniques*, 43(12), 2654-2667.

[10] Watts, P. C. P., Hsu, P. C. P., Barnes, W. K., & Chambers, A. (2003). B High Permittivity from Defective Multiwalled Carbon Nanotubes in the X-Band. *Adv Mater.*, 15(7-8), 600-603.

[11] Wadhawan, A., Garrett, D., & Perez, J. M. (2003). Nanoparticle-assisted microwave absorption by single-wall carbon nanotubes. *Appl Phys Lett.*, 83(13), 2683-2685.

[12] Deng, L. J., & Han, M. (2007). Microwave Absorbing Performances of Multiwalled Carbon Nanotube Composites with Negative Permeability. *Appl Phys Lett.*, 91(2), 023119-023121.

[13] Liu, J. R., Itoh, M., Terada, M., Horikawa, T., & Machida, K. I. (2007). Enhanced Electromagnetic Wave Absorption Properties of Fe Nanowires in Gigaherz Range. *Appl Phys Lett*, 91(9), 093101-093103.

[14] Chen, Y. J., Cao, M. S., Wang, T. H., & Wan, Q. (2004). Microwave absorption properties of the ZnO nanowire-polyester composites. *Appl. Phys. Lett.*, 84(17), 3367-3370.

[15] Tang, Z., & Kotov, N. A. (2005). One-dimensional assemblies of nanoparticles: Preparation, properties, and promise. *AdV Mater.*, 17(8), 951-962.

[16] Murphy, C., Sau, T. K., Gole, A. M., Orendorff, C. J., Gao, J., Gou, L., Hunyadi, S., & Li, T. (2005). Anisotropic Metal Nanoparticles: Synthesis, Assembly, and Optical Applications. *J. Phys. Chem. B*, 109(29), 13857-13870.

[17] Xia, Y., & Halas, N. (2005). Shape-Controlled Synthesis and Surface Plasmonic Properties of Metallic Nanostructures. *J MRS Bull.*, 30(5), 338-348.

[18] Liz-Marza´n, L. M. (2006). Tailoring Surface Plasmons through the Morphology and Assembly of Metal Nanoparticles. *Langmuir*, 22(1), 32-41.

[19] Link, S., & El -Sayed, M. A. (1999). Spectral Properties and Relaxation Dynamics of Surface Plasmon Electronic Oscillations in Gold and Silver Nanodots and Nanorods. *J. Phys. Chem. B*, 103(40), 8410-8426.

[20] Vasilev, K., Zhu, T., Wilms, M., Gillies, G., Lieberwirth, I., Mittler, S., Knoll, W., & Kreiter, M. I. (2005). One-step Synthesis of Gold Nanowires in Aqueous Solution. *Langmuir*, 21(26), 12399-12403.

[21] Wiley, B., Sun, Y., Mayers, B., & Xia, Y. (2005). Shape-Controlled Synthesis of Metal Nanostructures: The Case of Silver. *Chem Eur J*, 11(2), 454-463.

[22] Giersig, M., Pastoriza-Santos, I., & Liz-Marza´n, L. M. (2004). Evidence of an Aggregative Mechanism During the Formation of Silver Nanowires in N,N-dimethylformamide. *J Mater Chem*, 14, 607-610.

[23] Umar, A. A., & Oyama, M. (2006). Formation of Gold Nanoplates on Indium Tin Oxide Surface: Two-Dimensional Crystal Growth from Gold Nanoseed Particles in the Presence of Poly(vinylpyrrolidone). *Cryst Growth Des.*, 6(4), 818-821.

[24] Shankar, S. S., Rai, A., Ankamwar, B., Singh, A., Ahmad, A., & Sastry, M. (2004). Biological Synthesis of Triangular Gold Nanoprisms. *Nat Mater*, 3(7), 482-488.

[25] Xiong, Y., Mc Lellan, J. M., Chen, J., Yin, Y., Li, Z., & Xia, Y. (2005). Kinetically Controlled Synthesis of Triangular and Hexagonal Nanoplates of Palladium and Their SPR/SERS Properties. *J Am Chem Soc.*, 127(48), 17118-17127.

[26] Yang, J., Lu, L., Wang, H., Shi, W., & Zhang, H. (2006). Glycyl Glycine Templating Synthesis of Single-Crystal Silver Nanoplates. *Cryst Growth Des.*, 6(9), 2155-2158.

[27] Sun, Y., & Xia, Y. (2002). Shape-Controlled Synthesis of Gold and Silver Nanoparticles. *Science*, 298(5601), 2176-2179.

[28] Yu, D., & Yam, V. W. (2004). Controlled Synthesis of Monodisperse Silver Nanocubes in Water. *J. Am Chem Soc.*, 126(41), 13200-13201.

[29] Gou, L., & Murphy, C. J. (2005). Fine-Tuning the Shape of Gold Nanorods. *Chem. Mater.*, 17(14), 3668-3672.

[30] Chen, S., Wang, Z. L., Ballato, J., Foulger, S. H., & Carrol, D. L. (2003). Monopod, Bipod, Tripod, and Tetrapod Gold Nanocrystals. *J Am ChemSoc.*, 125(52), 16186-16187.

[31] Sau, T. K., & Murphy, C. J. (2004). Room Temperature, High-Yield Synthesis of Multiple Shapes of Gold Nanoparticles in Aqueous Solution. *J Am Chem Soc*, 126(28), 8648-8649.

[32] Yamamoto, M., Kashiwagi, Y., Sakata, T., Mori, H., & Nakamoto, M. (2005). Synthesis and Morphology of Star-Shaped Gold Nanoplates Protected by Poly(N-vinyl-2-pyrrolidone). *Chem Mater.*, 17(22), 5391.

[33] Kuo, C., & Huang, M. H. (2005). Synthesis of Branched Gold Nanocrystals by a Seeding Growth Approach. *Langmuir*, 21(5), 2012-2016.

[34] Hao, E., Bailey, R. C., Schatz, G. C., Hupp, J. T., & Li, S. (2004). Synthesis and Optical Properties of "Branched" Gold Nanocrystals. *Nano Lett.*, 4(2), 327-330.

[35] Bakr, O. M., Wunsch, B. H., & Stellacci, F. (2006). High-Yield Synthesis of Multi-Branched Urchin-Like Gold Nanoparticles. *Chem Mater.*, 18(14), 3297-3301.

[36] Burt, J. L., Elechiguerra, J. L., Reyes-Gasga, J., Montejano-Carrizales, J. M., & Yacaman, M. J. (2005). Beyond Archimedean Solids: Star Polyhedral Gold Nanocrystals. *J Cryst Growth*, 285(4), 681-691.

[37] Wang, T., Hu, X., & Dong, S. (2006). Surfactantless Synthesis of Multiple Shapes of Gold Nanostructures and Their Shape-Dependent SERS Spectroscopy. *J Phys Chem. B*, 110(34), 16930-16936.

[38] Ma, H., Huang, S., Feng, X., Zhang, X., Tian, F., Yong, F., Pan, W., Wang, Y., & Chen, S. (2006). Electrochemical Synthesis and Fabrication of Gold Nanostructures Based on Poly(N-vinylpyrrolidone). *Chem Phys Chem;*, 7(2), 333-335.

[39] Shen, G. Z., Bando, Y., & Golberg, D. (2007). Self-Assembled Hierarchical Single-Crystalline β-SiC Nanoarchitectures. *Cryst Growth Des.*, 7(1), 35-38.

[40] Teng, X. W., & Yang, H. (2005). Synthesis of platinum multipods: an induced anisotropic growth. *Nano Lett.*, 5(5), 885-891.

[41] Liu, B., & Zeng, H. C. (2004). Fabrication of ZnO "Dandelions" via a Modified Kirkendall Process. *J Am Chem Soc.*, 126(51), 16744-16746.

[42] Xie, S. H., & Zhao, D. Y. (2002). A Simple Route for the Synthesis of Multi-Armed CdS Nanorod-Based Materials. *Adv Mater.*, 14(21), 1537-1540.

[43] Zhang, Z., Sun, H., Shao, X., Li, D., Yu, H., & Han, M. (2005). Three-Dimensionally Oriented Aggregation of a Few Hundred Nanoparticles into Monocrystalline Architectures. *Adv Mater.*, 17(1), 42-47.

[44] Kowsari, E., & Faraghi, G. (2010). Synthesis by an ionic liquid assisted method and optical properties of nanoflower Y_2O_3. *J Mater Res Bull*, 45, 939-945.

[45] Murray, C. B., Kagan, C. R., & Bawendi, M. G. (1995). Self-Organization of CdSe Nanocrystallites into Three-Dimensional Quantum Dot Superlattices. *Science*, 270(5240), 1335-1338.

[46] Andreas, T. (2004). CuCl Nanoplatelets from an Ionic Liquid-Crystal Precursor. *Angew Chem Int. Ed. Engl.*, 43(40), 5380-5382.

[47] Jiang, Y., & Zhu, Y. J. (2005). Microwave-Assisted Synthesis of Sulfide M2S3 (M = Bi, Sb) Nanorods Using an Ionic Liquid. *J Phys Chem., B*, 109(10), 4361-4364.

[48] Jiang, J., Yu, S. H., Yao, W. T., Ge, H., & Zhang, G. Z. (2005). Morphogenesis and Crystallization of Bi_2S_3 Nanostructures by an Ionic Liquid-Assisted Templating Route: Synthesis, Formation Mechanism, and Properties. *Chem. Mater.*, 17(24), 6094-6100.

[49] Jiang, Y., Zhu, Y. J., & Cheng, G. F. (2006). Synthesis of Bi_2Se_3 Nanosheets by Microwave Heating Using an Ionic Liquid Cryst Growth Des. 9, 2174-2176.

[50] Nskashima, T., & Kimizuka, N. (2003). Interfacial Synthesis of Hollow TiO_2 Microspheres in Ionic Liquids. *J Am Chem Soc.*, 125(21), 6386-6387.

[51] Rai, P., Jo, J. N., Lee, I. H., & Yu, Y. T. (2010). Fabrication of 3D rotorlike ZnO nanostructure from 1D ZnO nanorods and their morphology dependent photoluminescence property. *J Solid State Sci*, 12, 1703-1710.

[52] Patzke, G. R., Krumeich, F., & Nesper, R. (2002). Oxidic Nanotubes and Nanorods-Anisotropic Modules for a Future Nanotechnology Angew. *Chem., Int. Ed. Engl.*, 41(14), 2446-2461.

[53] Cui, Y., & Lieber, C. M. (2001). Functional Nanoscale Electronic Devices Assembled Using Silicon Nanowire Building Blocks. *Science*, 291(5505), 851-853.

[54] Chen, A. C., Peng, X. P., Koczkur, K., & Miller, B. (2004). Super-hydrophobic tin oxide nanoflowers. *Chem Commun.*, 1964-1965.

[55] Gur, I., Fromer, N. A., Geier, M. L., & Alivisatos, A. P. (2005). Air-stable all-inorganic nanocrystal solar cells processed from solution. *Science*, 310(5747), 462-465.

[56] Antonietti, M., Kuang, D. B., Smarsly, B., & Yong, Z. (2004). Ionic liquids for the convenient synthesis of functional nanoparticles and other inorganic nanostructures. *Angew ChemInt Ed*, 43(38), 4988-4992.

[57] Xu, X., Zhang, M., Feng, J., & Zhang, M. (2008). Shape-controlled Synthesis of Single-Crystalline Cupric Oxide by Microwave Heating Using an Ionic Liquid. *Mat Lett.*, 62(17-18), 2728-2790.

[58] Ma, L., Chen, W. X., Li, H., Zheng, Y. F., & Xu, Z. D. (2008). Ionic Liquid-Assisted Hydrothermal Synthesis of MoS_2 Microspheres. *Mat Lett.*, 62(6-7), 797-799.

[59] Liu, X., Peng, P., Ma, J., & Zheng, W. (2009). Preparation of Novel CdSe Microstructure by Modified Hydrothermal Method. *Mat Lett.*, 63(8), 673-675.

[60] Kowsari, E., & Ghezelbash, M. R. (2012). Ionic Liquid-Assisted, Facile Synthesis of ZnO/SnO_2 Nanocomposites, and Investigation of Their Photocatalytic Activity. *Mat Lett.*, 68, 17-20.

[61] Kowsari, E., & Ghezelbash, Mohammad Reza. (2011). Synthesis of cactus-like zincoxysulfide (ZnO_xS_{1-x}) nanostructures assisted by a task-specific ionic liquid and their photocatalytic activities. *Mat Lett.*, 65-3371.

[62] Kowsari, E., & Karimzadeh, A. H. (2012). Using a Chiral Ionic Liquid for Morphological Evolution of $BaCO_3$ and its Radar Absorbing Properties as a Dendritic Nanofiller. *Mat Lett.*, 74, 33-36.

[63] Kowsari, E., & Karimzadeh, A. H. (2012). Fabrication of Fern-Like, Fish Skeleton-Like, and Butterfly-Like BaO Nanostructures as Nanofillers for Radar-Absorbing Nanocomposites. *Mat Lett.*, 74, 33-36.

Chapter 26

Ionic Liquids as Components in Fluorescent Functional Materials

Jun-ichi Kadokawa

Additional information is available at the end of the chapter

http://dx.doi.org/10.5772/51160

1. Introduction

In recent years, photo functional materials have been increasing much attention because of their attractive characteristics such as good specificity, excellent sensitivity, and easy handling [1]. Fluorescent-emitting materials are one of the most practically used photo functional materials in the many application fields such as color sensors and probes in biological science, key elements in color devises and displays, organic light-emitting diodes, and organic field-effect transistors [2]. Furthermore, a variety of polymers bearing covalently linked fluorescent dye moieties, exampling polymethacrylate, polyacrylamide, and conjugated polymer, have been synthesized to provide novel polymeric fluorescent materials [3-5]. To develop new fluorescent functional materials, the author has noted ionic liquids (ILs) as material components. ILs are low-melting-point molten salts, defined as which form liquids at room temperature or even at temperatures lower than a boiling point of water. The property is owing to that the liquid state is thermodynamically favorable due to the large size and conformational flexibility of the ions, in which these behaviors lead to small lattice enthalpies and large entropy changes that favor the liquid state [6]. In the past more than a decade, ILs have attracted much attention due to their specific characteristics such as a negligible vapor pressure, excellent thermal stabilities, and controllable physical and chemical properties [7]. Beyond these traditional properties of ILs, recently, interests and applications on ILs have been extended to the researches related to functional materials as designer substrates with controllable physical and chemical properties or even specific functions [8], so-called 'task-specific ILs' [9,10]. As one of the unique and specific properties of ILs, it has been reported that imidazolium-type ILs exhibit excitation-wavelength-dependent fluorescent behavior due to the presence of energetically different associated species [11-14]. For example, 1-butyl-3-methylimidazolium chloride (BMIMCl) typically exhibits emissions maxima at around 450-600 nm depending on the excitation wavelengths (Figure 1). The imidazolium-type ILs which form such different species have potential as components to contribute to developing new fluorescent functional materials.

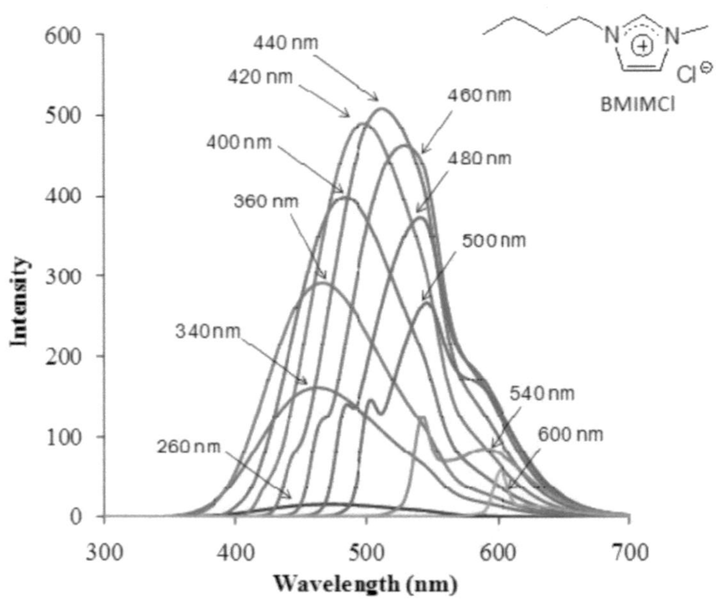

Figure 1. Fluorescence spectra of a liquid BMIMCl by excitation at 260-600 nm.

In this chapter, the author describes the use of the imidazolium-type ILs as components to prepare new fluorescent functional materials. A first topic deals with the appearance of fluorescent resonance-energy-transfer (FRET) in solutions of fluorescent dyes with BMIMCl. As a second topic, on the basis of this unique FRET system, the preparation and FRET functions of polymeric IL (PIL) films carrying fluorescent dye moieties are disclosed. Furthermore, a third topic deals with the preparation of ion gel materials from BMIMCl which exhibit the FRET function and other unique fluorescent properties.

2. Fluorescent Properties in Solutions of Rhodamine 6G with Ionic Liquid

2.1. FRET

Besides exhibiting emission by excitation at a characteristic wavelength of each fluorescent dye, the fluorescent materials in practical applications are often required to exhibit fluorescent emissions by excitation at different wavelength areas. For the purpose to develop such dye materials, the author has noted the FRET technique [15], which has been used in designed fluorescent materials to obtain a large shift of the excitation wavelength from that the dyes natively show. FRET is an interaction between the electronic excited states of two fluorescent substrates, a donor and an acceptor, in which excitation energy is transferred from the former to the latter without emission of a photon (Figure 2) [16]. By means of FRET, new high performance biosensors, fluorescence imaging, and quantification systems of selective interaction have been developed for targets of biological molecules, such as proteins and lipids [17,18].

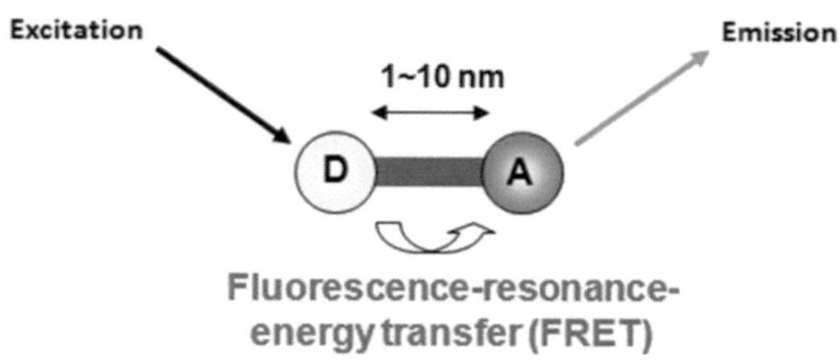

Figure 2. Image of FRET from a donor to an acceptor.

2.2. FRET in Systems of Rhodamine 6G with BMIMCl

To develop the basic technique for the preparation of new functional fluorescent materials, the author found a unique FRET system in a solution of rhodamine 6G with BMIMCl, where the former and latter acted as an acceptor and a donor, respectively [19]. Rhodamine 6G is a representative red fluorescent dye and exhibits emission maxima at ca. 540-610 nm by excitation at around 520 nm [20]. When the fluorescence spectra of the solution of rhodamine 6G with BMIMCl (2.5 mmol/L) were measured by excitation at 260-600 nm, emissions at ca. 608 nm due to the dye were observed in all the spectra, whereas fluorescence peaks due to BMIMCl were not detected (Figure 3). From these results, the occurrence of FRET from BMIMCl to rhodamine 6G in the solution was supposed. Indeed, all the fluorescence spectra of a sole BMIMCl liquid excited at various wavelengths were overlapped with an absorption peak of rhodamine 6G at 545 nm. The occurrence of FRET in the solution of rhodamine 6G with BMIMCl was confirmed further using the Stern-Volmer relation [21].

On the other hand, the fluorescence spectra of a solution of another fluorescent dye, pyrene, with BMIMCl by excitation at 260-600 nm showed the emissions due to BMIMCl. This was owing to no occurrence of FRET in the solution because an absorption of pyrene was not overlapped with the emissions of BMIMCl. Moreover, when the fluorescence spectra of a solution of a dye with no fluorescent emission, that is, Congo red, were measured by excitation at various wavelengths, the emissions due to BMIMCl were not observed. This result was explained by the energy transfer from BMIMCl to Congo red because an absorption of the dye overlapped with the emissions of BMIMCl.

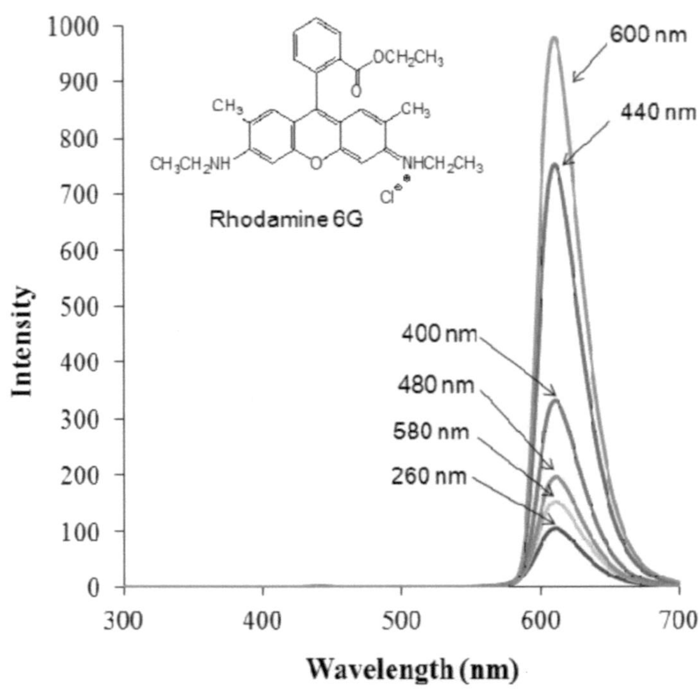

Figure 3. Fluorescence spectra of rhodamine 6G/BMIMCl solution by excitation at 260-600 nm.

3. Tunable Multicolor Emissions of Polymeric Ionic Liquid Films Carrying Fluorescent Dye Moieties

3.1. Polymeric Ionic Liquids

Polymeric ionic liquids (PILs) are defined as the polymers obtained by polymerization of ILs having polymerizable groups (polymerizable ILs) [22,23]. Thus, 'PILs' are termed just the polymeric forms of ILs, but they are not necessary to show liquid forms at room temperature or even at some ambient temperatures. The polymeric ILs, therefore, are often called 'polymerized ILs' too. The major advantages for providing the PILs are to be enhanced stability, and improved processability and feasibility in application as practical materials. Polymerizable ILs as a source of the PILs can be available by incorporating the polymerizable groups at both anionic and cationic sites (Figure 4). In the former case, polymerizable anions are ionically exchanged from some anions of general ILs (Figure 4(a)), giving the polymerizable ILs. In the latter case, vinyl, meth(acryloyl), and vinylbenzyl groups have typically been appeared as the polymerizable group (Figure 4(b)). Because 1-vinylimidazole is a commercially available, the vinylimidazolium-type polymerizable ILs are prepared by quaternization of 1-vinylimidazole with a variety of alkyl halides. The reaction of vinylbenzyl halides or haloalkyl (meth)acrylates with 1-alkylimidazoles gives the corresponding imidazolium-type polymerizable ILs having the vinylbenzyl or (meth)acryloyl polymerizable group (Figure 5). Furthermore, when vinylbenzyl halides or haloalkyl (meth)acrylates are reacted with 1-vinylimidazole, the polymerizable ILs having two polymerizable groups are produced. Because these polymerizable ILs can be converted into insoluble and stable PILs with the cross-linked structure by the radical polymerization (Figure 6), they have a highly potential as the source of the components in the practical materials.

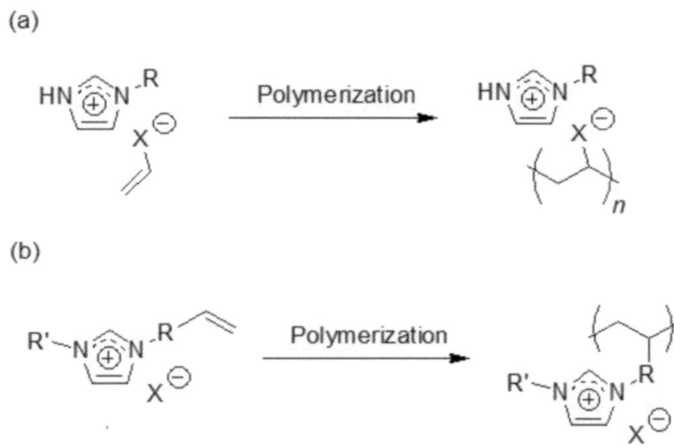

Figure 4. Polymerization of polymerizable ILs having a polymerizable group at anionic site (a) and cationic site (b).

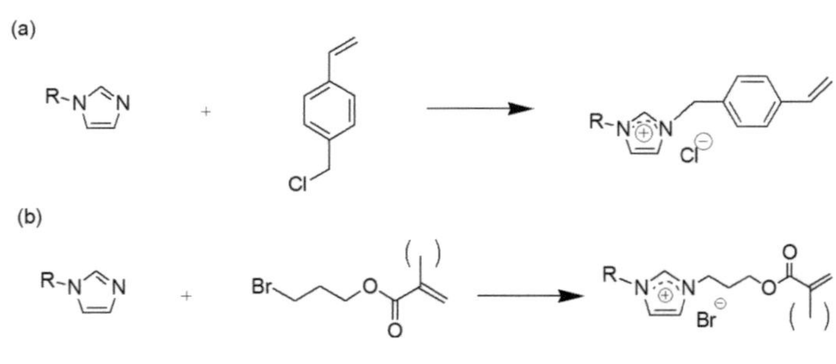

Figure 5. Typical synthetic schemes for polymerizable ILs having vinylbenzyl (a) and (meth)acrylate (b) groups.

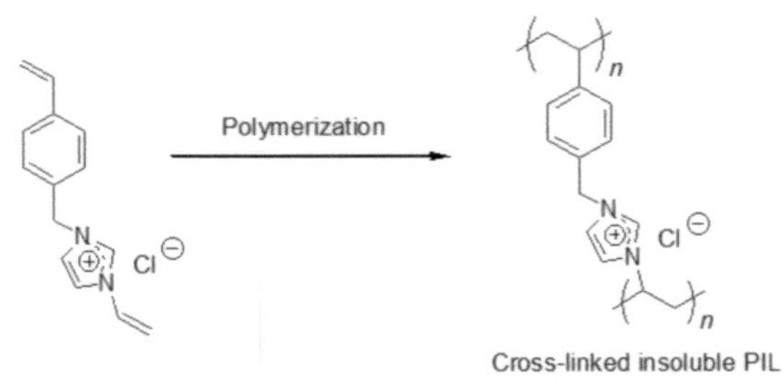

Figure 6. Polymerization of a polymerizable IL having two polymerizable groups to produce a cross-linked insoluble PIL.

3.2. Preparation of Transparent Polymeric Ionic Liquid Films

To incorporate the aforementioned unique FRET function into a film material, the preparation of a transparent PIL film was attempted by radical polymerization of the appropriate polymerizable ILs [24]. For this purpose, the two imidazolium-type polymerizable ILs, 1-methyl-3-(4-vinylbenzyl)imidazolium chloride (1) and 1-(3-methacryloyloxypropyl-3-vinylimidazolium bromide (2) were employed to obtain a cross-linked PIL (Figure 7). For the preparation of the film form of PIL, a solution of 1 and 2 (10:1), and AIBN as a radical initiator (1mol% for 1+2) was sandwiched between two glass plates. Then, the system was heated at 65 °C for 30 min and subsequently at 75 °C for 2 h to occur the copolymerization. The resulting material had the film form with transparent property.

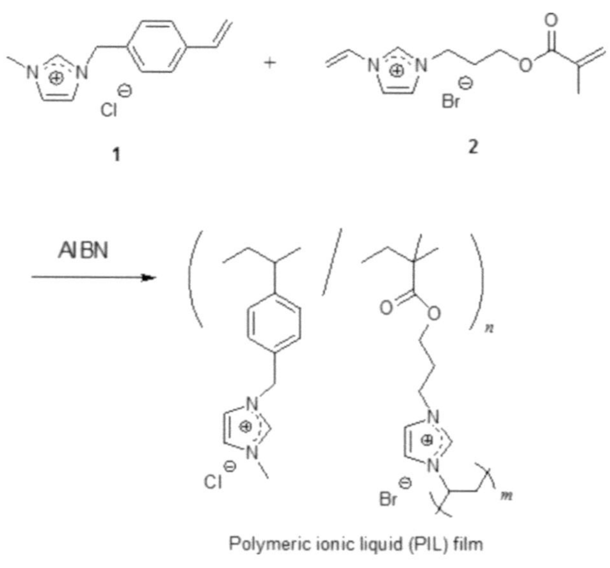

Figure 7. Radical copolymerization of 1 with 2 by AIBN to give PIL film.

The UV-vis spectrum of the film showed small absorptions at 280-550 nm, which were probably related to the fluorescent emissions of the imidazolium-type ILs, besides large absorptions at the wavelengths below 280 nm. The fluorescence spectra of the film showed excitation-wavelength-dependent fluorescent emission maxima at around 430-470 nm by excitation at 260-400 nm (Figure 8). Indeed, the film exhibited blue emission by UV light irradiation at 365 nm (Figure 9). The fluorescent behavior of the film was similar as that of the general imidazolium-type IL such as BMIMCl.

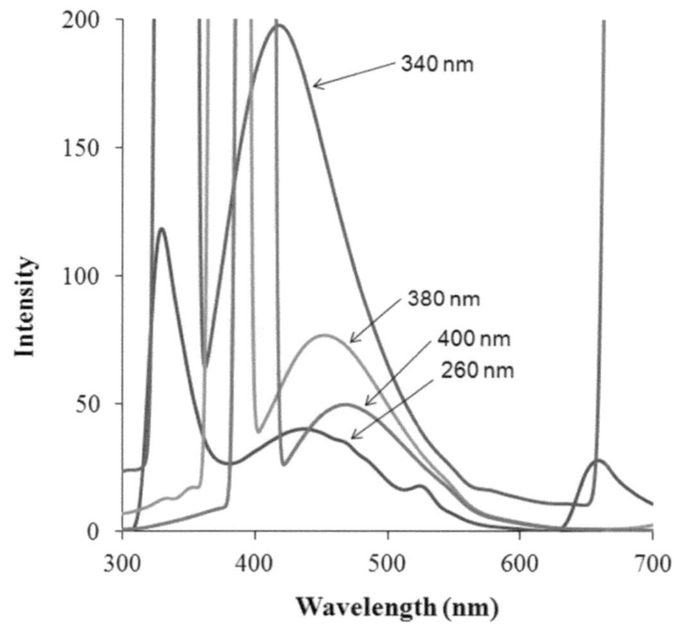

Figure 8. Fluorescence spectra of PIL film by excitation at 260-400 nm.

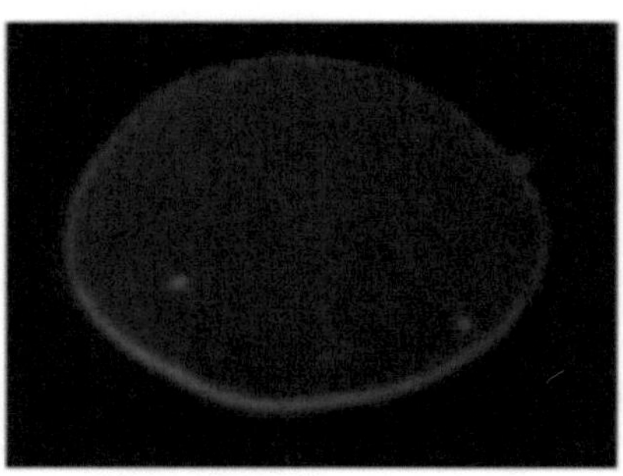

Figure 9. Photograph of PIL film under UV light irradiation at 365 nm.

3.3. Preparation and Multicolor Emissions of Fluorescent Polymeric Ionic Liquid Films

On the basis of the principle of three primary colors, the PIL films which exhibit multicolor emissions depending on combinations of the primary colors have considerably been designed [24,25]. For this purpose, three fluorescent dyes, rhodamine (red emission), 7-(diethylamino)coumarin-3-carboxylic acid (DEAC, green emission), and pyranine (blue emission) were selected, and thus, polymerizable rhodamine, DEAC, and pyranine derivatives (3, 4, and 5) having a methacrylate group were synthesized. Then, radical copolymerization of 1, 2, with 3, 4, or 5 was conducted by a similar procedure as aforementioned for PIL film to produce the PIL films 6, 7, and 8 carrying respective dye moieties (Figure 10).

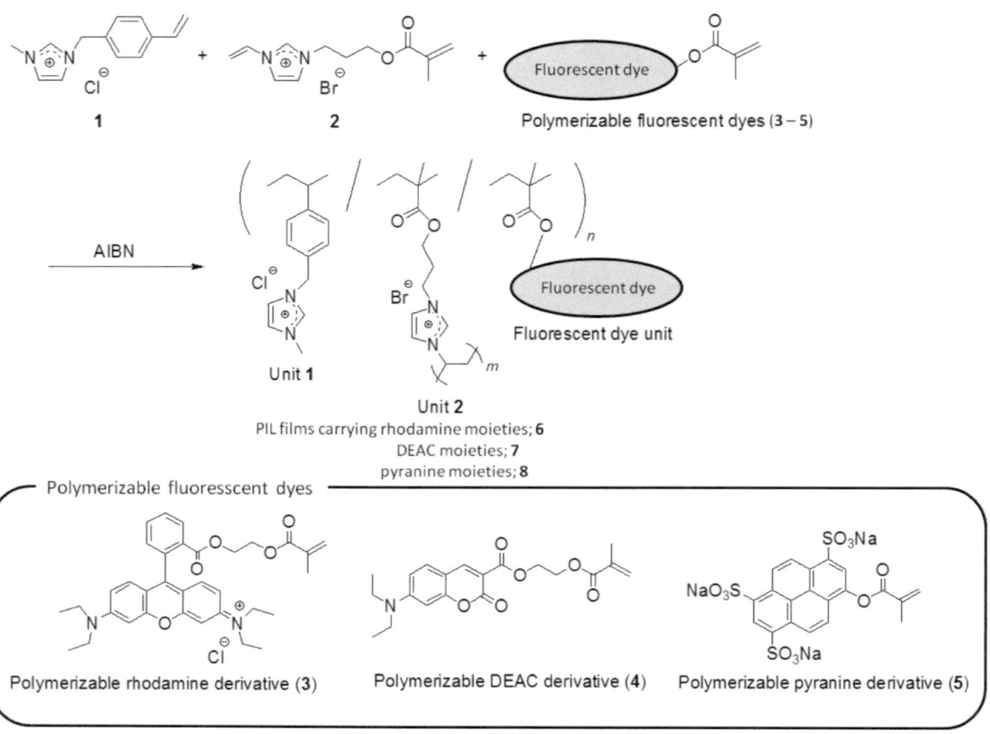

Figure 10. Radical copolymerization of 1, 2, with 3, 4, or 5 to give PIL films carrying primary color fluorescent dye moieties.

When the fluorescence spectra of the film 6 were measured by excitation at 260-400 nm, emissions at ca. 620 nm due to the rhodamine group in addition to scattering peaks of excitation lights were observed in all the spectra (Figure 11(a)). On the other hand, fluorescent emissions at around 430-470 nm due to the units 1 and 2 did not appear. These results suggested the occurrence of FRET from the units 1 and 2 to the rhodamine group in the film. Indeed, all the emissions of the PIL film composed of the units 1 and 2 (without fluorescent dye moieties; hereafter, this film is named the basic PIL film) excited at various wavelengths were partially overlapped with an absorption peak of the film 6 at wavelength areas of around 450-600 nm.

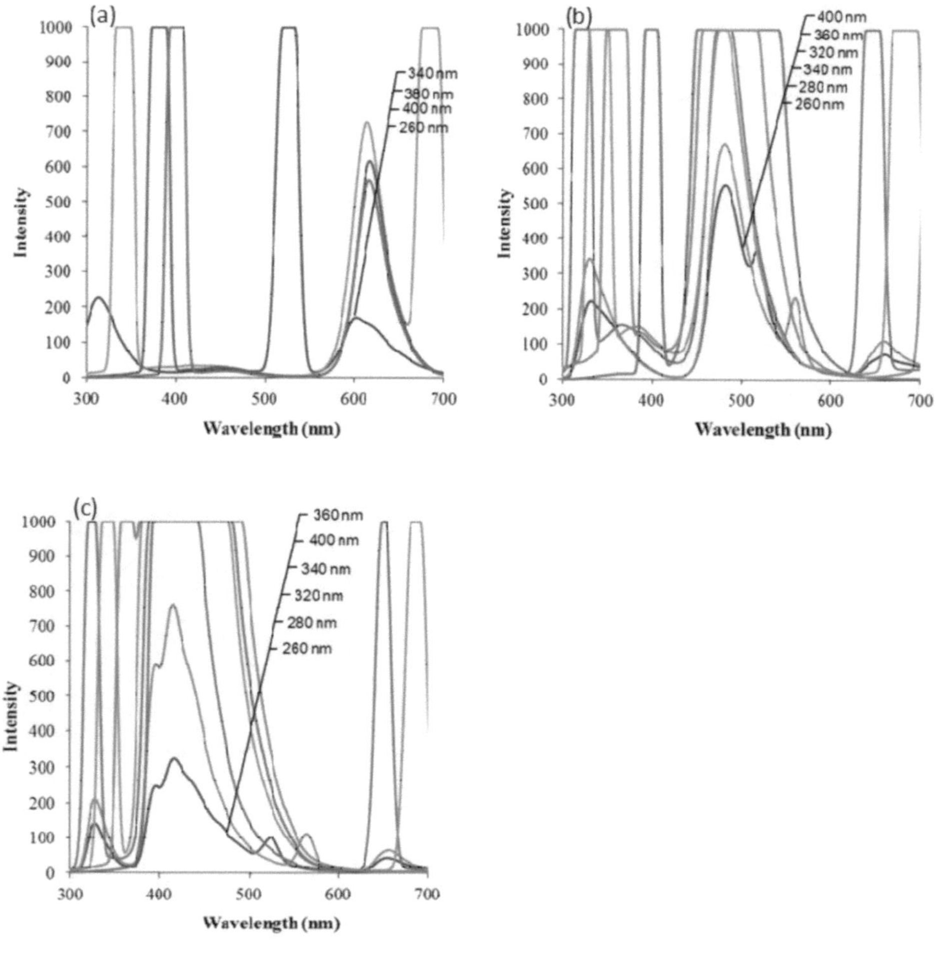

Figure 11. Fluorescence spectra of PIL films 6, 7, and 8 ((a)-(c), respectively) by excitation at 260-400 nm.

When the fluorescence spectra of the film 7 were also measured by excitation at 260-400 nm, emissions at ca. 470 nm due to the DEAC group were observed (Figure 11(b)). Furthermore, all the emissions of the basic PIL film excited at various wavelengths were totally or even partially overlapped with absorptions of the film 7. Taking the UV-vis spectrum of the film 7 into consideration, it was also supposed that the DEAC moieties in 7 emitted by excitation at around the wavelengths areas where the absorptions of 7 appeared. Therefore, the above results suggested that the emissions due to the DEAC group in 7 excited at wide wavelength areas were owing to either direct excitation of the DEAC group or FRET from the units 1 and 2 to the DEAC group in the film.

Similarly, emissions due to pyranine moieties were observed at ca. 420 nm in the fluorescence spectra of the film 8 excited at 260-400 nm (Figure 11(c)). The fluorescent emissions of the basic PIL film excited at shorter wavelength areas, i.e., 260-360 nm were partially overlapped with absorptions of the film 8 at around 300-400 nm. On the other hand, the emissions of the basic PIL film by excitation at longer wavelength area such as 380 and 400 nm appeared at wavelengths longer than ca. 400 nm, which were not mostly overlapped with the absorptions of the film 8. Moreover, the pyranine moieties in 8 emitted by excitation at around the wavelength areas where the absorptions of 8 appeared. Therefore, it was supposed that the emissions of the pyranine group in the film 8 by excitation at shorter wavelength area were owing to either direct excitation of the pyranine group or FRET from the units 1 and 2 to the pyranine group, whereas those excited at longer wavelength areas were probably caused by only direct excitation of the pyranine group in the film.

Actually, the film 6, 7, and 8 showed the red, green, and blue emissions, respectively, by the UV-vis light irradiations at 365 nm (Figure 12).

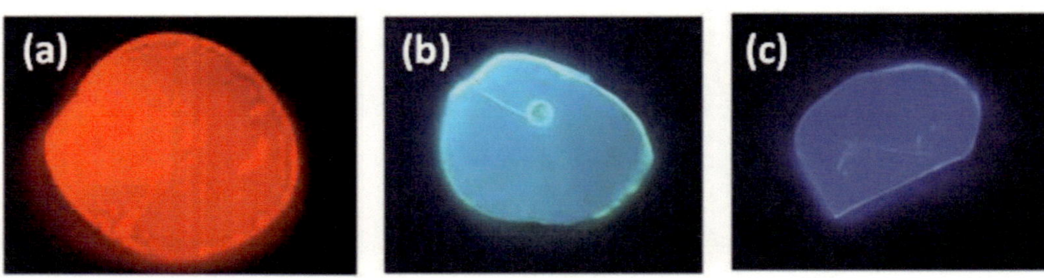

Figure 12. Photographs of PIL films 6, 7, and 8 ((a)-(c), respectively) under UV light irradiations at 365 nm.

By means of possible combinations among the rhodamine, DEAC, and pyranine dyes, which emitted the three primary colors, the PIL films exhibiting tunable color emissions were prepared. Three combinations of polymerizable dyes, that is, 3 and 4, 3 and 5, and 4 and 5, were copolymerized with 1 and 2 by AIBN according to the same experimental manner as that for the basic PIL film (Figure 13). The fluorescence spectra of the resulting films showed two kinds of emissions due to the incorporated dye moieties by excitation at 260-400 nm. These data suggested that the respective dye groups in the PIL films were individually emitted by direct excitation or FRET. The PIL film carrying three dye moieties was similarly prepared by copolymerization of 1, 2, with the three polymerizable dyes. The fluorescence spectra of the resulting film also showed three kinds of emissions excited at 260-400 nm. Thus, the resulting films exhibited yellow, magenta, cyan, and white fluorescent emissions, respectively, by UV light irradiation at 365 nm (Figure 13). These results indicated that the PIL films carrying proper fluorescent dye moieties emitted tunable multicolors by excitation at a sole wavelength.

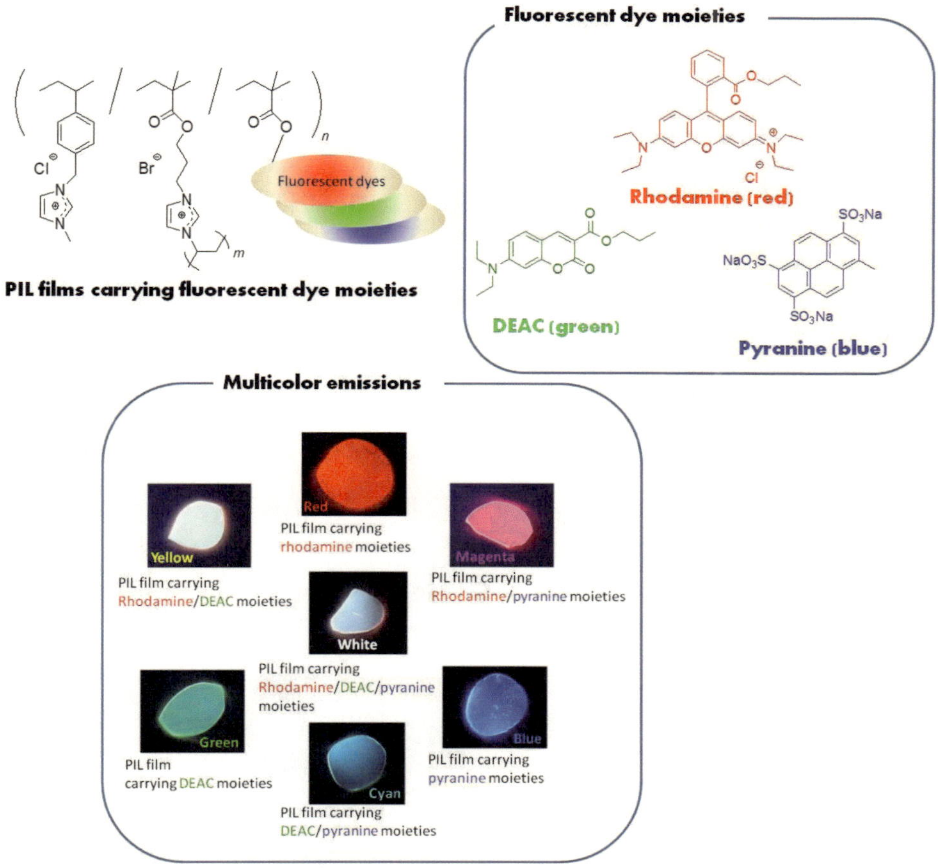

Figure 13. Multicolor emissions of PIL films carrying various combinations of fluorescent dye moieties.

4. Preparation of Photo Functional Ion Gels of Polysaccharides with an ionic liquid

4.1. Ion Gels of Polysaccharides with Ionic Liquids

Because ILs have been found to be used as good solvents for natural polysaccharides such as cellulose [26-29], and accordingly, can be considered to have a specific affinity for polysaccharides, efficient methods to produce new polysaccharide-based materials compatibilized with ILs have the potential to lead to the practical use of natural polysaccharides as the promising biomass resources [30,31]. On the basis of these viewpoints, the author has reported the facile preparation of gel materials of abundant polysaccharides such as cellulose, starch, and chitin, which include ILs as disperse media in the polysaccharide network matrixes, so-called ion gels [32-35]. Besides such abundant polysaccharides, many kinds of natural polysaccharides from various sources have been known [36]. For example, some polysaccharides such as guar gum and xanthan gum are used as hydrocolloid polysaccharides for a stabilizer, a viscous agent, and a structure provider in food industries [37]. Guar gum is a galactomannan extracted from the seed of the leguminous shrub *Cyamopsis tetragonoloba* and consists of a (1→ 4)-linked β-D-mannopyranose main-chain with a branched α-D-galactopyranose unit at 6 position (Figure 14). Xanthan gum produced by *Xanthomonas campestris* has a cellulose-type main-chain (β-(1→ 4)-glucan) with trisaccharide side chains attached to alternate glucose units in the main-chain (Figure 14). The author has reported that functional ion gels of hydrocolloid polysaccharides, e.g., guar gum and xanthan gum, with BMIMCl were obtained when the corresponding solutions of the polysaccharides in BMIMCl in appropriate concentrations were left standing at room temperature [38-42]. These ion gels have been applied to providing functional materials by means of the specific fluorescent behaviors of ILs.

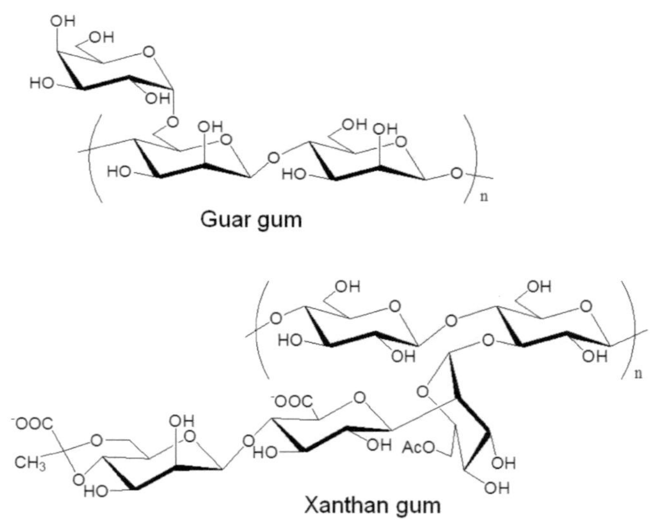

Figure 14. Structures of guar and xanthan gums.

4.2. FRET Function of Ion Gel of Guar Gum with an Ionic Liquid

For the preparation of gel materials exhibiting the aforementioned unique FRET function, the gelling system of BMIMCl using guar gum was employed. When the fluorescence spectra of the guar gum/BMIMCl ion gel was measured by excitation at 260-600 nm, the similar excitation-wavelength-dependent fluorescence behavior as that of a sole BMIMCl was appeared (Figure 15). Accordingly, the guar gum/BMIMCl ion gel containing rhodamine 6G (1.5 mmol/L) was prepared from the mixture of rhodamine 6G and guar gum with BMIMCl. The fluorescence spectra of the resulting ion gel exhibited emissions due to rhodamine 6G by excitation at 260-600 nm, whereas no emissions due to BMIMCl were observed (Figure 16). These results indicated the occurrence of FRET from BMIMCl to rhodamine 6G in the ion gel. Indeed, the gel showed the red emissions by photo irradiation at various wavelengths (Figure 17).

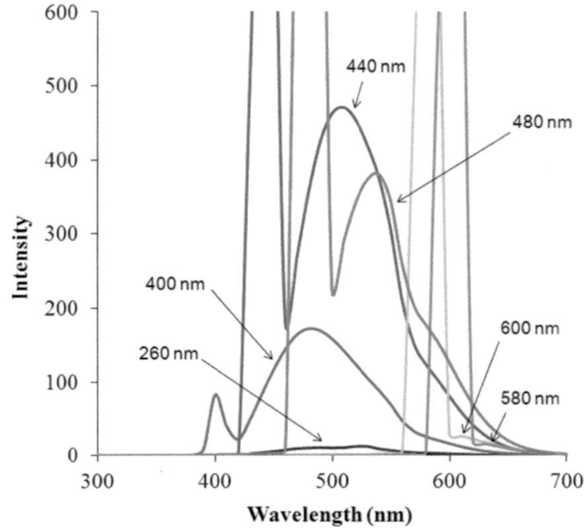

Figure 15. Fluorescence spectra of guar gum/BMIMCl ion gel by excitation at 260-600 nm.

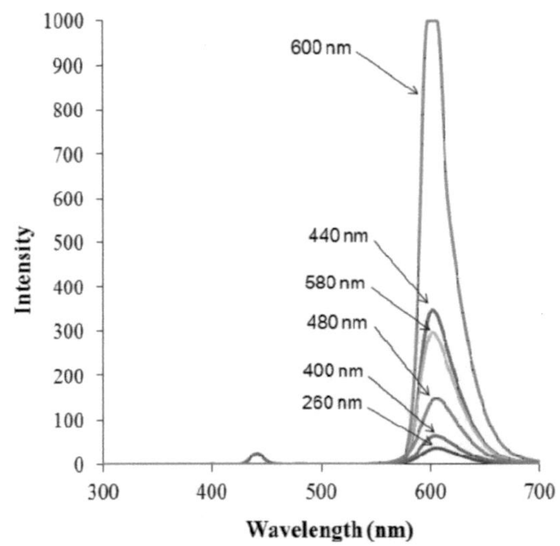

Figure 16. Fluorescence spectra of guar gum/rhodamine 6G/BMIMCl ion gel by excitation at 260-600 nm.

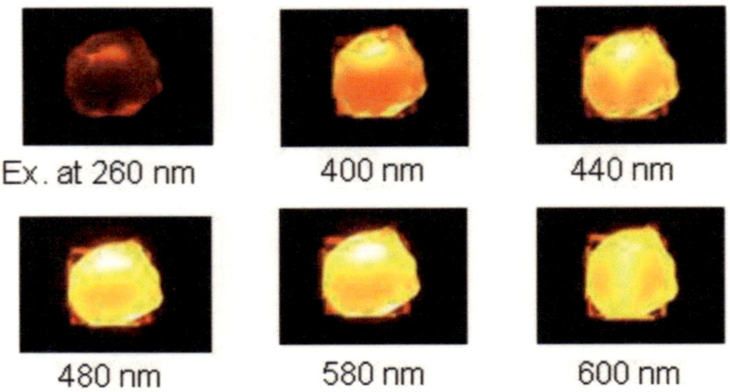

Figure 17. Photographs of guar gum/rodamine 6G/BMIMCl ion gel by excitation at 260-600 nm.

4.3. Fluorescent Behaviors of Ion Gel of Xanthan Gum with an Ionic Liquid

The author has been interested in the association state of BMIMCl in the xanthan gum/BMIMCl ion gels because nano-ordered association of 1-butyl-3-methylimidazolium-type ionic liquids in the liquid state was suggested in previous report [43]. The UV-vis spectra of the ion gels were measured to evaluate the association states of BMIMCl [41]. A liquid BMIMCl showed significant absorptions at wavelengths below 250 nm besides very small absorptions at 250-450 nm (Figure 18(a)). However, the strong absorptions in a wide range from 200 to 450 nm were observed in the UV-vis spectra of the ion gels with different contents (10 and 30% (w/w), Figure 18(b) and (c)). Such strong absorption was not observed in the UV-vis spectrum of guar gum/BMIMCl ion gel. These results suggested the presence of the different association state of BMIMCl in the xanthan gum/BIMICl ion gel from that in the liquid and the guar gum/BIMICl ion gel. The presence of the specific association state of BMIMCl in the xanthan gum/BMIMCl ion gel was also confirmed by the ^{1}H NMR analysis.

On the basis of the above findings, the fluorescent behaviors of the xanthan gum/BMIMCl ion gels were investigated. Figure 19 shows the fluorescence spectra of the ion gels in the different xanthan gum contents (10, 20, 40, and 60% (w/w)) by excitation at 360-480 nm. Emission maxima were obviously shifted to the longer wavelength areas with increasing xanthan gum contents. Such red-shift was probably due to the presence of the specific association states of BMIMCl depending to the xanthan gum contents in the gels. Actually, the colors of the gels were changed from yellow to red-brown with increasing the xanthan gum contents (Figure 20). These results suggest that the present xanthan gum/BMIMCl ion gels can be applied to the new fluorescent gel materials in the future.

http://dx.doi.org/10.5772/51160

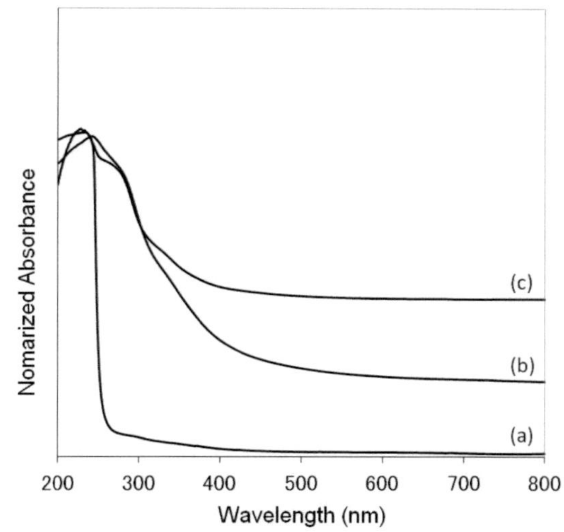

Figure 18. UV-vis spectra of a liquid BMIMCl (a), and xanthan gum/BMIMCl ion gels in 10 and 30% (w/w) contents ((b) and (c), respectively.

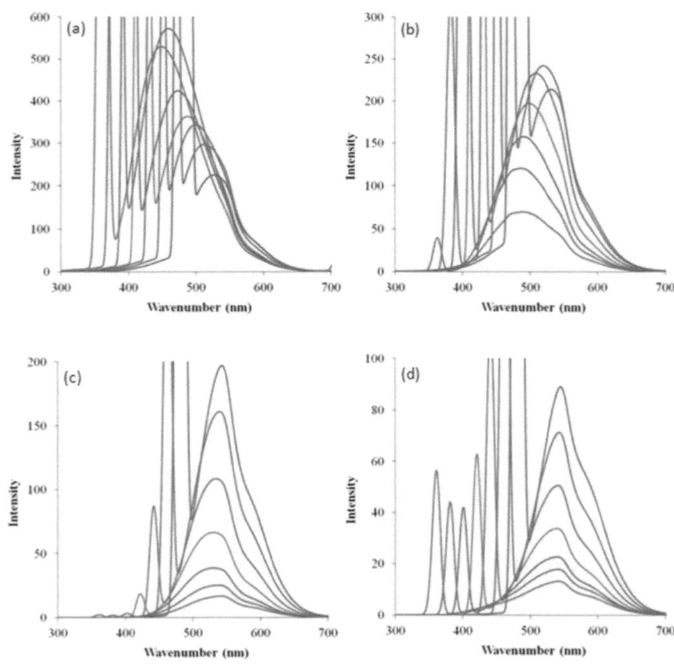

Figure 19. Fluorescence spectra of xanthan gum/BMIMCl ion gels in 10, 20, 40, and 60% (w/w) contents by excitation at 360-480 nm ((a) – (d), respectively).

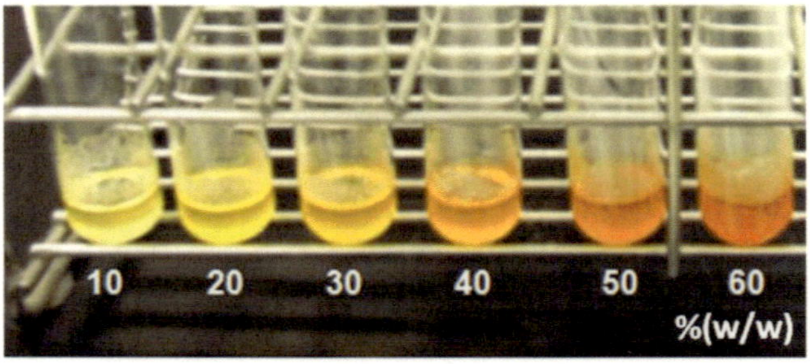

Figure 20. Photographs of xanthan gum/BMIMCl ion gels in 10-60% (w/w) contents.

5. Conclusion

This chapter overviewed the preparation of new fluorescent materials composed of the ILs as components which exhibited specific and unique photo functions. The unique FRET system using rhodamine 6G and the imidazolium-type IL, BMIMCl, was successfully appeared. The radical copolymerization of two PILs, which had one and two polymerizable groups, respectively, was carried out with AIBN as an initiator to give the transparent polymeric ionic liquid (PIL) film. The fluorescence spectra of the film exhibited excitation-wavelength-dependence fluorescent emission maxima at around 430 – 470 nm by excitation at 260 – 400 nm. On the basis of the above results, the PIL films carrying fluorescent dye moieties were prepared by radical copolymerization of polymerizable ionic liquids with appropriate polymerizable fluorescent dye derivatives. The films carrying rhodamine, 7-(diethylamino)coumarin-3-carboxylic acid (DEAC), and pyranine moieties exhibited the three primary color emissions, i.e., red, green, and blue, respectively, by excitation at wide wavelength areas. By incorporating possible combinations of the dye moieties in the PIL backbones, furthermore, the PIL films, which emitted tunable multicolors, were successfully obtained.

For the preparation of materials exhibiting the unique fluorescent behaviors, the gelling system of BMIMCl using guar gum of a natural polysaccharide containing rhodamine 6G was employed. The fluorescence spectra of the resulting ion gel showed the emissions due to rhodamine 6G by excitation at 260 – 600 nm, whereas no emissions due to BMIMCl were observed, indicating the occurrence of FRET from BMIMCl to rhodamine 6G in the gel. The fluorescent behaviors of xanthan gum/BMIMCl ion gels were also investigated. The gels exhibited the xnathan gum content-dependent emission changes, probably owing to the presence of specific association states of BMIMCl in the gels.

The specific fluorescent functions of the materials described in this chapter are realized by the unique photo properties of the ILs. The present materials have the potential for the practical applications in the various fields in the future.

http://dx.doi.org/10.5772/51160

Acknowledgements

The author is indebted to the co-workers, whose names are found in references from his papers, for their enthusiasistic collaborations.

Author details

Jun-ichi Kadokawa*

Address all correspondence to: kadokawa@eng.kagoshima-u.ac.jp

Graduate School of Science and Engineering, Kagoshima University, Japan

References

[1] Simmons, J., & Potter, K. S. (2000). Optical Materials. Waltham: Academic Press

[2] Kim, E., & Park, S. B. (2009). Chemistry as a Prism: A Review of Light-emitting Materials Having Tunable Emission Wavelengths. *Chemistry- A European Journal*, 4, 1646-1658.

[3] Obata, M., Morita, M., Nakase, K., Mitsuo, K., Asai, K., Hirohara, S., & Yano, S. (2007). Synthesis and Photophysical Properties of Rhodamine B Dye-bearing Poly(isobutyl Methacrylate-co-2, 2, 2 trifluoroethyl Methacrylate) as a Temperature-sensing Polymer Film. *Journal of Polymer Science, Part A: Polymer Chemistry*, 45, 2876, 2885.

[4] Shiraishi, Y., Miyamoto, R., & Hirai, T. (2008). Rhodamine-conjugated Acrylamide Polymers Exhibiting Selective Fluorescence Enhancement at Specific Temperature Ranges. *Journal of Photochemistry and Photobiology A: Chemistry*, 200, 432-437.

[5] Zhu, M., Zhou, C., Zhao, Y., Li, Y., Liu, H., & Li, Y. (2009). Synthesis of a Fluorescent Polymer Bearing Covalently Linked Thienylene Moieties and Rhodamine for Efficient Sensing. *Macromolecular Rapid Communications*, 30, 1339-1344.

[6] Welton, T. (1999). Room-temperature Ionic Liquids. Solvents for Synthesis and Catalysis. *Chemical Reviews*, 99, 2071-2083.

[7] Plechkova, N. V., & Seddon, K. R. (2008). Applications of Ionic liquids in the Chemical Industry. *Chemical Society Reviews*, 37, 123-150.

[8] Giernoth, R. (2010). Task-specific Ionic Liquids. Angewandte Chemie International Edition , 49, 2834-2839.

[9] Davis, J. H. (2004). Task-specific Ionic Liquids. *Chemistry Letters*, 33, 1072-1077.

[10] Lee, S. G. (2006). Functionalized Imidazolium Salts for Task-specific Ionic Liquids and Their Applications. *Chemical Communications*, 1049-1063.

[11] Paul, A., Mandal, P. K., & Samanta, A. (2005). How Transparent are The Imidazolium Ionic Liquids? A Case Study with 1 -Methyl-3-butylimidazolium Hexafluorophosphate, [bmim][PF_6]. *Chemical Physics Letters*, 402, 375-379.

[12] Paul, A., Mandal, P. K., & Samanta, A. (2005). On The Optical Properties of The Imidazolium Ionic Liquids. *The Journal of Physical Chemistry B*, 109, 9148-9153.

[13] Paul, A., & Samanta, A. (2006). Optical Absorption and Fluorescence Studies on Imidazolium Ionic Liquids Comprising The Bis(trifluoromethanesulphonyl)imide anion. *Journal of Chemical Sciences*, 118, 335-340.

[14] Mandal, P. K., Paul, A., & Samanta, A. (2006). Excitation Wavelength Dependent Fluorescence Behavior of The Room Temperature Ionic Liquids and Dissolved Dipolar Solutes. *Journal of Photochemistry and Photobiology A: Chemistry*, 182, 113-120.

[15] Miyawaki, A., Llopis, J., Heim, R., Mc Caffery, J. M., Adams, J. A., Ikura, M., & Tsien, R. Y. (1997). Fluorescent Indicators for Ca^{2+} Based on Green Fluorescent Proteins and Calmodulin. *Nature*, 388, 882-887.

[16] Lakowicz, J. R. (1999). Principle of Fluorescence Spectroscopy. , 2nd Edition. New York Plenum

[17] Kikuchi, K. (2010). Design, Synthesis, and Biological Application of Fluorescent Sensor Molecules for Cellular Imaging. In: Endo I, Nagamune T. (eds.) Nano/Micro Biotechnology- Advances in Biochemical Engineering-Biotechnology. Berlin Springer , 119, 63-78.

[18] Kikuchi, K. (2010). Design, Synthesis and Biological Application of Chemical Probes for Bio-imaging. *Chemcal Society Reviews*, 39, 2048-2053.

[19] Izawa, H., Wakizono, S., & Kadokawa, J. (2010). Fluorescence Resonance-energy-transfer in Systems of Rhodamine 6G with Ionic Liquid Showing Emissions by Excitation at Wide Wavelength Areas. *Chemical Communications*, 46, 6359-6363.

[20] Beija, M., Afonso, C. A. M., & Martinho, J. M. G. (2009). Synthesis and Applications of Rhodamine Derivatives as Fluorescent probes. *Chemical Society Reviews*, 38, 2410-2433.

[21] Bhattar, S. L., Kolekar, G. B., & Patil, S. R. (2008). Fluorescence Resonance Energy Transfer between Perylene and Riboflavin in Micellar Solution and Analytical Application on Determination of Vitamin B-2. *Journal of Luminescence*, 128, 306-310.

[22] Green, O., Grubjesic, S., Lee, S., & Firestone, M. A. (2009). The Design of Polymeric Ionic Liquids for The Preparation of Functional Materials. *Journal of Macromolecular Science: Part C, Polymer Reviews*, 49, 339-360.

[23] Mecerreyes, D. (2011). Polymeric Ionic Liquids: Broadening The Properties and Applications of Polyelectrolytes. *Progress in Polymer Science*, 36, 1629-1648.

[24] Wakizono, S., Yamamoto, K., & Kadokawa, J. (2011). FRET Function of Polymeric Ionic Liquid Film Containing Rhodamine Moieties for Exhibiting Emissions by Excitation at Wide Wavelength Areas. *Journal of Photochemistry and Photobiology A: Chemistry*, 222, 283-287.

[25] Wakizono, S., Yamamoto, K., & Kadokawa, J. (2012). Tunable Multicolor Emissions of Polymeric Ionic Liquid Films Carrying Proper Fluorescent Dye Moieties. Journal of Materials Chemistry , 22, 10619-10624.

[26] Seoud, O. A. E., Koschella, A., Fidale, L. C., Dorn, S., & Heinze, T. (2007). Applications of Ionic Liquids in Carbohydrate Chemistry: A Windows of Opportunities. *Biomacromolecules*, 8, 2629-2647.

[27] Liebert, T., & Heinze, T. (2008). Interaction of Ionic Liquids with Polysaccharides 5. Solvents and Reaction Media for The Modification of Cellulose. *BioResources*, 3, 576-601.

[28] Feng, L., & Chen, Z. I. (2008). Research Progress on Dissolution and Functional Modification of Cellulose in Ionic Liquids. *Journal of Molecular Liquids*, 142, 1-5.

[29] Pinkert, A., Marsh, K. N., Pang, S., & Staiger, M. P. (2009). Ionic Liquids and Their Interaction with Cellulose. *Chemical Reviews*, 109, 6712-6728.

[30] Kadokawa, J. (2011). Preparation of Polysaccharide-based Materials Compatibilized with Ionic Liquids. In: Kokorin A. (ed.) Ionic Liquids, Application and Perspectives. Rijeka InTech , 95-114.

[31] Kadokawa, J. (2012). Preparation of Functional Ion Gels of Polysaccharides with Ionic Liquids. In: Mun J, Sim H. (eds.) Handbook of Ionic Liquids: Properties, Applications and Hazards. Hauppauge Nova Science Publishers , 455-466.

[32] Kadokawa, J., Murakami, M., & Kaneko, Y. (2008). A Facile Preparation of Gel Materials from a Solution of Cellulose in Ionic Liquid. *Carbohydrate Research*, 343, 769-772.

[33] Kadokawa, J., Murakami, M., Takegawa, A., & Kaneko, Y. (2009). Preparation of Cellulose-starch Composite Gel and Fibrous Material from a Mixture of The Polysaccharides in Ionic Liquid. *Carbohydrate Polymers*, 75, 180-183.

[34] Prasad, K., Murakami, M., Kaneko, Y., Takada, A., Nakamura, Y., & Kadokawa, J. Weak. (2009). Gel of Chitin with Ionic Liquid, 1Allyl-3-methylimidazolium Bromide. *International Journal of Biological Macromolecules*, 45, 221-225.

[35] Takegawa, A., Murakami, M., Kaneko, Y., & Kadokawa, J. (2010). Preparation of Chitin/cellulose Composite Gels and Films with Ionic Liquids. *Carbohydrate Polymers*, 79, 85-90.

[36] Schuerch, C. (1986). Polysaccharides. In: Mark HF, Bilkales N, Overberger CG. (eds.) Encyclopedia of Polymer Science and Engineering, 2nd Edition. New York John Wiley & Sons , 13, 87-162.

[37] Stephen, A. M., Philips, G. O., & Williams, P. A. (1995). Food Polysaccharides and Their Applications. London, Taylor & Francis.

[38] Prasad, K., Kaneko, Y., & Kadokawa, J. (2009). Novel Gelling Systems of κ-, ι- and λ-Carrageenans and Their Composite Gels with Cellulose Using Ionic Liquid. *Macromolecular Bioscience*, 9, 376-382.

[39] Prasad, K., Izawa, H., Kaneko, Y., & Kadokawa, J. (2009). Preparation of Temperature-induced Shapeable Film Material from Guar Gum-based Gel with an Ionic Liquid. *Journal of Materials Chemistry*, 19, 4088-4090.

[40] Izawa, H., Kaneko, Y., & Kadokawa, J. (2009). Unique Gel of Xanthan Gum with Ionic Liquid and Its Conversion into High Performance Hydrogel. *Journal of Materials Chemistry*, 19, 6969-6972.

[41] Izawa, H., & Kadokawa, J. (2010). Preparation and Characterizations of Functional Ionic Liquid-gel and Hydrogel of Xanthan Gum. *Journal of Materials Chemistry*, 20, 5235-5241.

[42] Mine, S., Prasad, K., Izawa, H., Sonoda, K., & Kadokawa, J. (2010). Preparation of Guar Gum-based Functional Materials Using Ionic Liquid. *Journal of Materials Chemistry*, 20, 9220-9225.

[43] Romero, C., Moore, H. J., Lee, T. R., & Baldelli, S. (2007). Orientation of 1-Butyl-3-methylimidazolium Based Ionic Liquids at a Hydrophobic Quartz Interface Using Sum Frequency Generation Spectroscopy. *The Journal of Physical Chemistry C*, 111, 240-247.

Chapter 27

Preparation, Physicochemical Properties and Battery Applications of a Novel Poly(Ionic Liquid)

Takaya Sato, Takashi Morinaga and Takeo Ishizuka

Additional information is available at the end of the chapter

http://dx.doi.org/10.5772/51162

1. Introduction

Ionic liquids (ILs) are generally defined as a salt with a melting point lower than 100 °C whose properties include non-volatility, non-flammability, and a relatively high ionic conductivity [1]. Recently, therefore, interest has increased in the possible use of this type of liquid as an electrolyte in energy storage devices, for example, a lithium rechargeable battery [2–9] and an electric double layer capacitor [10–14].

At the same time, to further the development of large, thin, prismatic electrochemical devices for which there is a high market demand, solid electrolytes are generally preferred over liquid electrolytes from the view point of ease of production and long device lifetime. In the last decade, many researchers and battery companies have been developing a "solid state polymer electrolyte" or a "gel polymer battery" with a film-like shape. More recently, growing attention has been paid to poly(ionic liquids) as a class of polymeric materials that are highly non-flammable. Examples of new polyelectrolytes poly(ILs) have been produced from polymerizable ionic liquid monomers by several polymerization processes. Research into the application of poly(ILs) for polymer electrolytes were under intense study by Ono et al around at 2005 [15-17]. The ensuing intensive studies on poly(ILs) in the last five years significantly expanded the research scope of this new type of polymer, and some valuable review [18-19] and feature articles have been published [20].

Amid such developments, we also developed a polymerizable ionic liquid, *N*, *N*-diethyl- *N*-(2-methacryloylethyl)-*N*-methylammonium bis(trifluoromethylsulfonyl)imide (DEMM-TFSI, whose molecular structure is shown in Fig.1). From many previous studies, it was known that aliphatic quaternary ammonium based ionic liquids had an obviously higher cathodic stability than the aromatic type ionic liquids. Therefore, polymer materials made from these polymerizable ILs could have high durability in use as an electrolyte in various energy devices. However, because there have been only a few reports on this topic, we describe in this chapter the preparation, polymerization, physicochemical properties, and the application of quaternary ammonium type polymerizable ionic liquids and related polymers.

On the other hand, the use of ILs as the solvent in free radical polymerization media instead of a conventional organic solvent markedly affects the rate and degree of polymerization. It has been reported that poly(methylmethacrylate) (PMMA) reactions are much more rapid in an ionic liquid than in a nonpolar solvents, and that a PMMA prepared in an ionic liquid has a molecular weight approximately five times higher than in benzene and toluene [21]. The ionic liquid monomer, DEMM-TFSI, also gives the ultra high molecular weight poly(DEMM-TFSI) by bulk polymerization with a conventional 2-2'-azoisobutyronitrile (AIBN) initiator [22]. However, control of the molecular weight was difficult due to the strong enhancement of the propagation rate coefficients (k_p) in ILs. Although molecular weight control in the polymerization of polymerizable ILs is more successful using an atom transfer radical polymerization (ATRP) [20, 23, 24], there are few investigation of length control by other polymerization methods, such as conventional chain transfer radical (CTR) polymerization, adding chain transfer agents, or reversible addition fragmentation chain transfer (RAFT) polymerization. In this chapter, we describe the molecular weight-controlled polymerization of DEMM-TFSI by CTR polymerization and RAFT polymerization. We also detail the physicochemical properties of the resulting molecular weight-controlled ionic liquid polymer, poly(DEMM-TFSI), especially the thermal properties (glass transition temperature; T_g), ionic conductivity and the self-diffusion coefficient (D) of poly-cation and anion in solution using pulsed-gradient spin echo NMR (PGSE-NMR) spectroscopy.

For electrochemical device applications, a good method to produce a flexible polymer electrolyte membrane with high conductivity and non-flammability is to use poly(ILs) as a host polymer for the gelation of ILs [22, 25]. Poly(ILs) have a higher affinity for ionic liquids than that of conventional polymers such as PMMA, poly(ethylene oxide) (PEO) and poly(vinylidenefluoride) (PVdF). In fact, the ionic liquid gel composite material including poly(ILs) showed higher ionic conductivity than that of the materials including conventional polymer such as a PMMA. A high degree of compatibility with the ionic liquid results in a high ionic conductivity and good physical properties even when only a small amount of polymer material is added. This allows the IL to be completely encapsulated, thereby avoiding liquid leakage accidents. Work in this area will be also described in this chapter.

2. Preparation and physicochemical properties of DEMM-TFSI

We developed a new type of polymerizable IL, *N*, *N*-diethyl- *N*- (2-methoxyethyl)- *N*-methylammonium bis(trifluoromethylsulfonyl)imide (DEMM-TFSI in Fig. 1), with electrochemical stability in a wide range of potential. The preparation route of the DEMM-TFSI, was given elsewhere [22]. In brief, 2-(diethylamino)ethylmethacrylate was treated with 1.2 equiv. of methyliodide in tetrahydrofuran at 0 °C and stirred overnight, giving *N*,*N*-diethyl-*N*-(2-

methacryloylethyl)-*N*-methylammonium iodide as a precipitate, which was filtered off and recrystallized in tetrahydrofuran-ethanol solvent. The recrystallized product was treated with exactly 1.0 equiv. of lithium bis(trifluoromethylsulfonyl)imide in deionized water for 5 h. After the reaction, the mixture was separated into two phases, the bottom phase being *N,N*-diethyl-*N*-(2-methacryloylethyl)-*N*- methylammonium bis(trifluoromethylsulfonyl)imide.

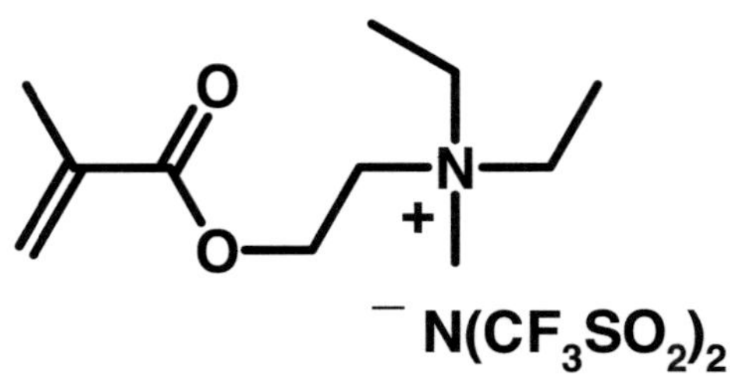

Figure 1. Molecular structure of DEMM-TFSI

The ionic conductivity is 0.6 mScm^{-1} at 25 °C measured by a conductivity meter (HM-30R, DKK-TOA Corporation).

Fig.2 depicts the differential scanning calorimetry (DSC) and thermal gravimetric analysis (TGA) curves of the DEMM-TFSI. This polymerizable IL did not exhibit a clear melting or freezing temperature in a typical DSC measurement; however, the glass transition temperature T_g was observed at −68 °C. The decomposition temperature corresponding to a 10% weight loss according to TGA measurements occurred at 329 °C. DEMM-TFSI exists as an ionic liquid over a very wide temperature range, approximately 400 °C.

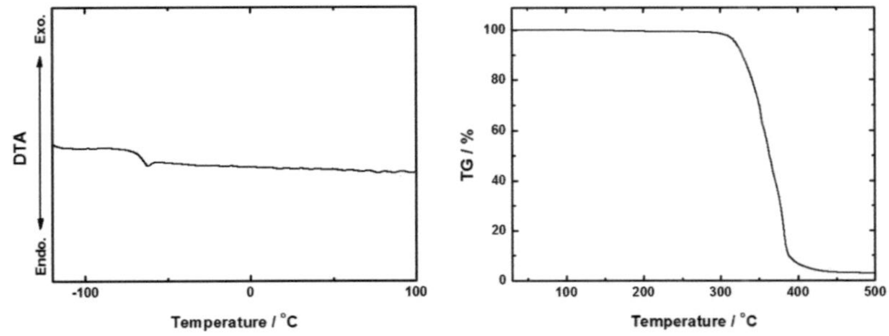

Figure 2. (a) Differential scanning calorimetry (DSC) and (b) thermal gravimetric analysis (TGA) results for the ionic liquid monomer, DEMM-TFSI at a heating rate of 5 °C min^{-1}.

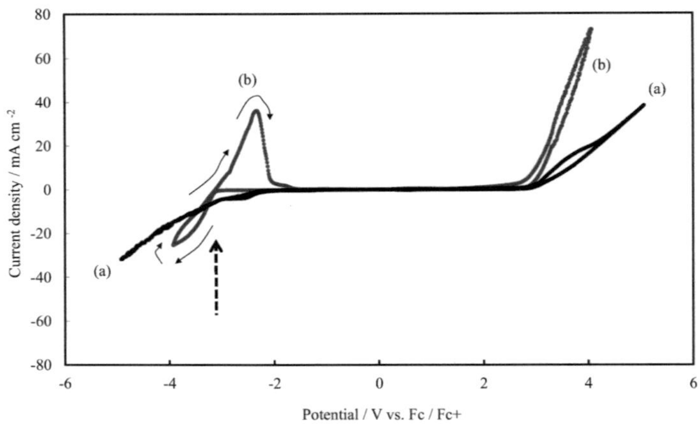

Figure 3. Cyclic voltammogram of an ionic liquid monomer, an organic electrolyte at 25 °C. Scan rate: $10mVs^{-1}$; platinum working and counter electrodes; Ag/Ag^{+} reference electrode. The potential value (V) was referenced to the ferrocene (Fc)/ferrocenium (Fc^{+}) redox couple in each salt. (a) 0.1 mol kg^{-1} of DEMM-TFSI in propylene carbonate (PC) solution, (b) 0.1 mol kg^{-1} of LiTFSI in PC solution. The concentration is defined in terms of molality = (mol solute/kg solvent). The dashed line arrow indicates the potential of the Li/Li^{+} couple.

Fig. 3 shows both the limiting reduction potentials ($E_{red.}$) on platinum of the ionic liquid monomer as measured by cyclic voltammetry at room temperature, and a voltammogram of an ordinary organic electrolyte LiTFSI in PC. The sharp peaks around −3.0V are probably due to the deposition and dissolution of Li metal, since a color change of the working electrode by the metal deposition appears with a current loop at that potential. The $E_{red.}$ and the limiting oxidation potentials ($E_{oxd.}$) were defined as the potential where the limiting current density reached $1mAcm^{-2}$. The $E_{red.}$ of the DEMM-TFSI ionic liquid monomer was positioned about 0.7V positive against the Li/Li^{+}. We confidently expected this result, because the monomer molecule had an easily reduced double bond.

3. Preparation of poly(DEMM-TFSI)

3.1. Radical polymerization of ionic liquid monomer; DEMM-TFSI

The poly(IL), poly(DEMM-TFSI), was synthesized by the bulk polymerization method. First, the monomer was dissolved in acetonitrile, and the solution was treated with activated carbon; the resultant acetonitrile solution was evaporated and the purified monomer was dried in vacuum at 25 °C. The ionic liquid monomer and 2,2'-Azoisobutyronitrile (AIBN), at a ratio of 1.0 mol% to the amount of methacryl groups present in the monomer, were mixed until they became homogeneous. The mixture was degassed in vacuum at 50 °C, and kept standing at 70 °C for 15 h. After polymerization, the product polymer was dissolved in acetonitrile and precipitated into ethanol and water, before a final drying in vacuum at 70 °C. The preparation scheme is shown in Fig. 4.

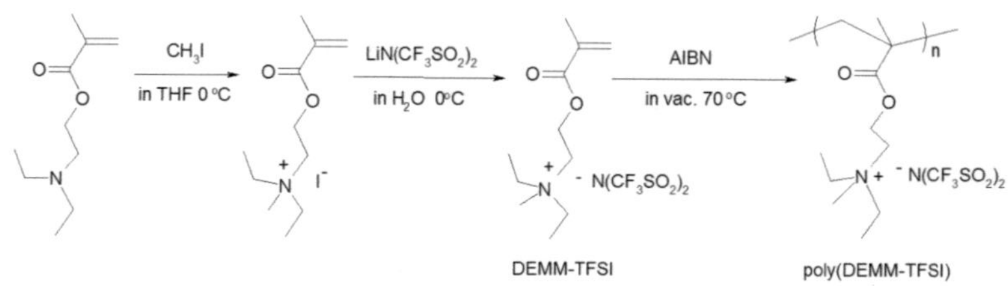

Figure 4. Synthesis of ionic liquid monomer (DEMM-TFSI) and poly(DEMM-TFSI).

It has been reported that poly(methyl methacrylate) (PMMA) reactions are much more rapid in an ionic liquid than in a nonpolar solvent such as benzene, and that a PMMA prepared in an ionic liquid has a molecular weight approximately five times higher than in benzene [21]. The ionic liquid monomer DEMM-TFSI also gives the ultra high molecular weight poly(DEMM-TFSI) by bulk polymerization with an AIBN initiator. Fig. 5 illustrates the GPC trace of the poly(DEMM-TFSI) product polymerized using a mol ratio of [monomer]/[AIBN] = 1:0.01. The resulting polymer was a rubbery solid, with a weight average molecular weight (M_w) of 1,084,000 and a polydispersity index by GPC analysis of 2.95 [PDI = (M_w)/number-average molecular weight (M_n)].

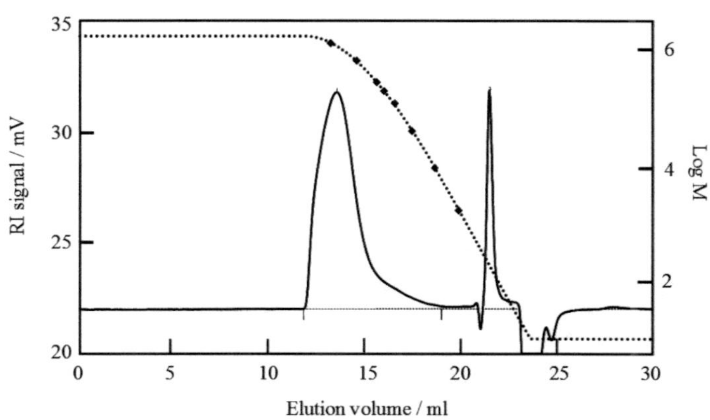

Figure 5. GPC trace of poly(DEMM-TFSI). M_n = 368,000; M_w = 1,084,000; M_w / M_n = 2.95. GPC analyses were performed at 40 °C, with a Shodex GPC-101 equipped with two, series-connected OHpak SB-806M HQ columns with a solution of 0.5 M acetic acid and 0.2 M sodium nitrate in acetonitrile and water (1/1 v/v) as the eluent. The weight- and number-average molecular weight were estimated on the basis of the calibration curve established with standard poly(ethylene oxide)s with the Shodex 480-II data station.

3.2. Molecular weight control of poly(DEMM-TFSI)

It is well known when the radical polymerizations are conducted in an ionic liquid, a significant incerase of the k_p / k_t ratio is normally observed in comparison to those carried out in other polar solvents [18, 26]. Not only the rates of polymerization, but also the molecular weights of the polymers produced were considerably higher for several kinds of monomers. Also, the polymerizable ionic liquids, in our case DEMM-TFSI, give considerably higher molecular weight polymers in bulk radical polymerizations. The difficulty in molecular weight control might become disadvantageous in material development. In this section, we show how we succeeded in the molecular weight control of the radical polymerization process using DEMM-TFSI by simply adding a chain transfer reagent (CTA).

Fig. 6 illustrates the relationships between the number average molecular weights and the concentration of the charge transfer reagent, 3-mercapto-1-hexanol in bulk radical polymerization media using AIBN as a initiator at 50 °C. We were able to synthesize poly (DEMM-TFSI) over M_n range from 5000 to 50,000, indicating that 3-mercapto-1-hexanol is an effective CTA under these conditions. The polydispersity index (PDI = M_w / M_n) was approximately 1.8 and 2.7 for polymer with a M_n of ten thousand or less and several tens of thousands or more, respectively; these PDI values are not very different from those found with a conventional free radical polymerization.

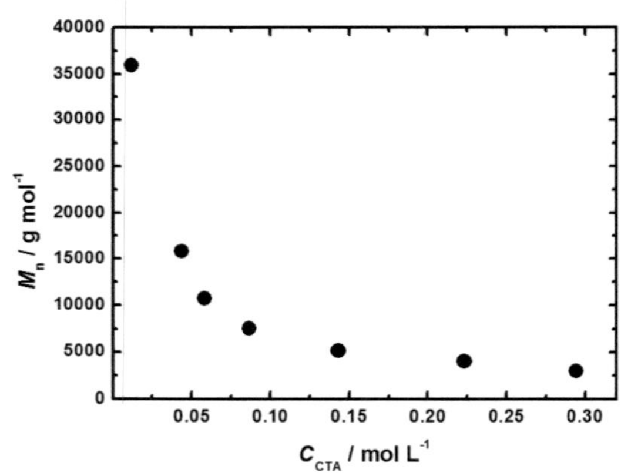

Figure 6. The relationship between the number average molecular weight of polymerized ionic liquid, DEMM-TFSI and the concentration of the charge transfer reagent 3-mercapto-1-hexanol, using 2,2'-azobis(isobutyronitrile) (AIBN) as initiator at 50 °C. Molar ratio: [DEMM-TFSI] / [AIBN] = 1 / 0.015. Number-average molecular weights were as determined by GPC.

Using the controlled/living radical polymerization techniques that have shown remarkable progress recently is perhaps the best choice for preparation of the poly(ILs) with controlled molecular weight and a narrow PDI. Two such possibilities are atom transfer radical polymerization (ATRP) and reversible addition-fragmentation chain transfer (RAFT) polymerization [27]. We have already achieved exceptionally dense grafting of well-defined pol(DEMM-TFSI) on a solid surface, known as a concentrated polymer brush prepared by surface initiated ATRP. Such a polymer brush surface showed unique properties, including high modulus, super lubrication, and unique size exclusion that were quite different and even unpredictable from those of the previously studied semidilute polymer brushes [24]. The polymerization of DEMM-TFSI was well controlled and exhibited living characteristics when a Cu(I)Cl and Cu(II)Cl_2 mixture and 2,2′-bipyridine complex was used as the catalyst system and ethyl-2-bromoisobutyrate as initiator. However, if such synthesized poly(ILs) are used for the electrochemical devices, some residue of the metallic catalyst may ruin the reliability of the devices. An apparent disadvantage of applying ATRP for the synthesis of poly(ILs) is the unavoidable complexation of polymers with the catalytic copper ions. However, RAFT polymerization is free of this problem, as no metal source is involved.

Compared to ATRP, there are not many examples of research into RAFT polymerization. Because papers concerning the RAFT polymerization of polymerizable quaternary ammonium type ionic liquids are very few in number, we investigated the kinetics of the RAFT polymerization of DEMM-TFSI. Figure 7(a) shows the variation in ln($[M]_0$ / $[M]_t$) versus polymerization time for the polymerization of DEMM-TFSI in acetonitrile at 50 °C with AIBN in the presence of 2-cyano-2-propyl benzodithioate (CPBT). The reaction was conducted at a ratio of $[\text{DEMM-TFSI}]_0$ / $[\text{CPBT}]_0$= 200 / 1, the concentration of DEMM-TFSI and AIBN being 75 wt.% and 0.5 wt.% in acetonitrile, respectively. Almost full conversion was reached after 5 hours and an almost linear first-order kinetic plot is seen until almost 100% conversion. Nevertheless, a linear increase in the number-average molecular weight determined by GPC spectroscopy with conversion is observed, indicating a constant number of propagating chains throughout the polymerization (Figure 7(b)). The PDI value is consistently small (M_w / M_n = 1.45-1.24) from the first stage of the polymerization to the end. This is an indication that the polymer chain end is capped with the fragments of CPBT, as expected according to the general mechanism of the RAFT process. These data indicated the molecular weight controlled synthesis of poly(DEMM-TFSI) with a narrow PDI can be successfully performed by RAFT polymerization of DEMM-TFSI.

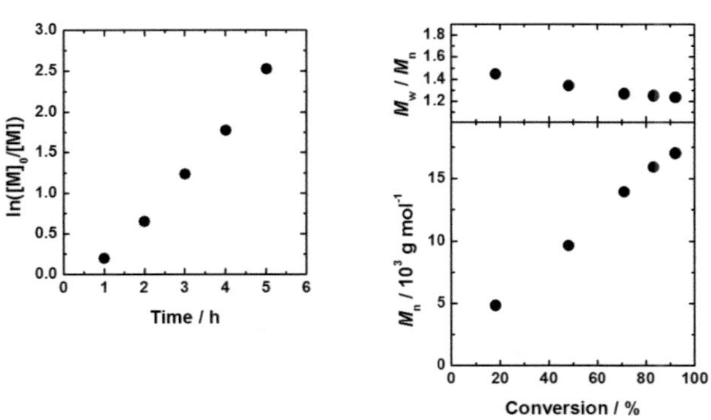

Figure 7. (a) Plot of ln($[M]_0$/$[M]_t$) versus time and (b) evolution of number average molecular weight (M_n) and polydispersity index (M_w/M_n) for the RAFT polymerization of DEMM-TFSI with 2,2′-azobis(isobutyronitrile) (AIBN) in the presence of 2-cyano-2-propyl benzodithioate (CPBT) in acetonitrile at 50 °C: $[\text{DEMM-TFSI}]_0$ / $[\text{CPBT}]_0$ / $[\text{AIBN}]_0$ = 200 / 1 / 3.91. The number average molecular weights and conversion were determined by GPC and ^{1}H-NMR.

4. Physicochemical properties of poly(DEMM-TFSI)

In Fig. 8 are shown the T_g values of poly(DEMM-TFSI) with various M_n values and relatively small polydispersity index values ($M_w / M_n < 1.2$) measured by differential scanning calorimetry (DSC) analysis during heating at 10 deg/min. versus the number average molecular weight (M_n). The polymer was prepared by atom transfer radical polymerization (ATRP) using a complex catalyst consisting of copper chloride and 2,2′-bipyridine in acetonitrile [24]. ATRP represents one of the branches of living radical polymerization (LRP). The M_n value was estimated as an absolute value, assuming a 100% initiation, from the monomer-to-initiator molar ratio and the conversion determined by ^{1}H-NMR using a JEOL JEM-ECX400 spectrometer. The M_w / M_n value was determined by poly(ethyleneglycol)-calibrated gel permeation chromatography (GPC) using a Shodex GPC-101 high-speed liquid chromatography system.

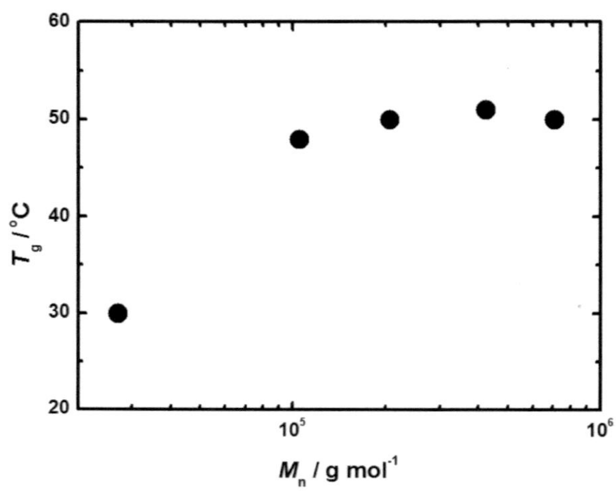

Figure 8. Glass transition temperatures, measured calorimetrically during heating at 10 deg. / min, versus number average molecular weight.

The glass transition temperature is related to the start of the segmental motion of polymers. In general, polymers with more free volume have a lower T_g value and that temperature is essentially independent of molecular weight because the free volume ratio per segment of polymer materials is the same even if the length of a polymer main chain changes. However, it is known that a lower molecular weight polymer usually has lower T_g value. Because the chain ends have high mobility and more free volume than the middle part of polymer, the free volume increases with the numbers of terminals, and T_g decreases as a result.

A poly-cation type material such as poly(DEMM-TFSI) has bulky tetra-alkyl ammonium functional groups in each unit and has a relatively low T_g value due to its large free volume due to the effect of the electrostatic repulsions of cationic groups. A polymer with a molecular weight that exceeds 100,000 has roughly a T_g of 50 °C while, on the other hand, a 20 degrees lower value of T_g of 30 °C was found for a polymer with a M_n about 10,000, presumably due to the chain ends effect. The relationships between M_n and T_g of this poly(IL) had a similar trend to that of polystyrene [28, 29]. We realized that the T_g rose greatly on polymerization only up to about 50 of degree of polymerization, though the T_g of the monomer had shown an extremely low value of -68 °C.

It was thought that a excellent solid polymer electrolyte with high physical strength and high ionic conductivity could be developed by the polymerization of ionic liquid monomers, so much attention has been directed toward developing poly(ILs) in order to avoid some disadvantages of liquid electrolytes, such as leakage and flammability, in energy device applications, including the lithium ion rechargeable battery, the electric double layer capacitor and dye-sensitized solar cell. However, the ionic conductivity of the polymerized ionic liquids was generally considerably lowered due to both the considerable elevation of the glass transition temperature and a reduced number of mobile ions, as one of ions is fixed to the polymer chain and so cannot move.

Fig. 9 shows the temperature dependence of the ionic conductivity for a series of poly(DEMM-TFSI)s with different M_n values. In the case of our polymer, the ionic conductivity of poly(DEMM-TFSI) was four digits or more lower than that of an ionic liquid such as DEME-TFSI (molecular structure is indicated in Fig. 11.) in the room temperature region. Unfortunately, this level of conductivity is not practicable for devices. However, an interesting point is that neither the value of the ionic conductivity nor the temperature dependency of the conductivity do not depend on the molecular weight of the polymer, and exhibit an almost constant value. Thus, polymers with M_n of 27,000, 250,000 and 710,000 have T_g values of 40, 50 and 50 °C, respectively. The ionic conductivity of these polymers continuously changes without an inflection point in the area before and after the T_g. There seems to be no correlation between the ionic conductivity value and the glass transition temperature. It is widely accepted that ion conduction in an amorphous polymer matrix should occur above the glass transition temperature as it is coupled with a segmental motion of polymer chain [30]. The matrix polymer solvated the mobile ions and created a liquid-like environment around the ions. However, in the poly(DEMM-TFSI) matrix, it seems the anion moves independently of the polymer chain which stops its segmental motion below the glass transition temperature. The mobility of the anion in this poly(IL) seems to be different from that of the solvated ions in the polymer without any dissociable ionic substituent, such as polyethylene oxide (PEO) that contains a lithium salt. Ionic conduction does not appear in a PEO matrix below the glass transition temperature. In the poly(DEMM-TFSI), an alternative anion conduction mechanism that does not involve polymer motion might exist. There are only a few previous studies related to the ionic conduction mechanism in poly-cation and poly-anion systems below T_g. We are planning to combine DSC and ionic conductivity measurements using a dielectric relaxation spectroscopy measurement in order to improve our understanding of this interesting phenomenon.

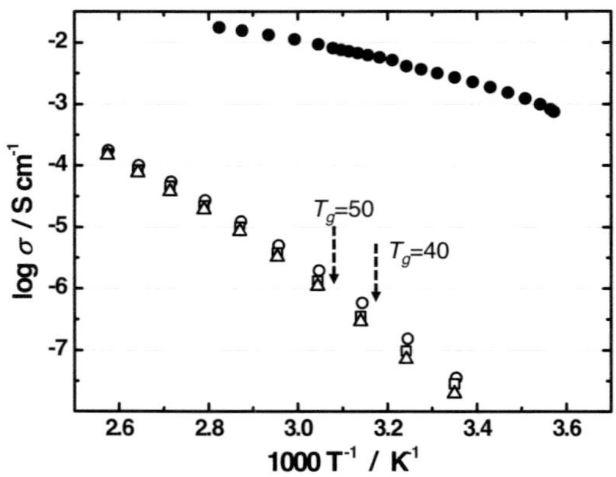

Figure 9. Temperature dependence of the ionic conductivity for poly(DEMM-TFSI)s with different M_n values and an ionic liquid by the complex impedance method [24]. Open circle, Poly(DEMM-TFSI) with M_n=27,000; open triangle, M_n=250,000; open square, M_n=710,000; filled circle, ionic liquid (DEME-TFSI). Ionic conductivity measurements were made by an AC-impedance method using a multi-frequency LCR meter (Agilent Technology E4980A Precision LCR). The sample was loaded at 100 °C between two polished stainless-steel discs acting as ion-blocking electrodes with ceramic-sheet spacer with a 100 μm thickness. The measurement cell was set in a thermostat oven chamber and collected at various frequencies ranging from 20 Hz to 2 MHz and various temperatures ranging from −5 to 125 °C. The conductivity of the ionic liquid was measured at various temperatures by a conductivity meter (HM-30R, DKK-TOA Corporation).

Hayamizu et al. have succeeded in measuring the self diffusion coefficients (D) values of the solvent molecule, the cation, the anion, and the polymer molecule in a PEO-type gel electrolyte system by the method of pulsed gradient spin-echo multinuclear NMR (PGSE-NMR) [31, 32, 33]. We also carried out PGSE-NMR measurements in order to evaluate the D value of the poly-cation and counter anion (TFSI) in a relatively high viscosity poly(DEMM-TFSI) solution in acetonitrile. The polymer concentration of the solution used for the measurement is 20 percent by weight, a value that far exceeded the overlap concentration, C*, where the polycation chains can contact and entangle with each other. So, the movement of the polymer chain was suppressed in this concentration region, hence it can be said that the environment was in a gel-like state.

Fig. 10 represents the NMR echo signal attenuation of a poly(DEMM-TFSI) solution in acetonitrile at 25 °C probed by ^{1}H and ^{19}F nuclear signals attributed to the methyl proton on the ammonium cation and acetonitrile and the TFSI anion. The attenuation due to free diffusion in the Stejskal and Tanner sequence using half-sine-shaped gradient pulses is given by

$$E = S / S_0 = \exp(-\gamma^2 g^2 \delta^2 D(4\Delta - \delta) / \pi^2) \quad (1)$$

where γ is the gyromagnetic ratio, S is the amplitude of the echo signal and S_0 is the amplitude where g = 0, g is the amplitude of the gradient pulse, δ is the duration of the gradient pulse, and Δ is the interval between the gradient pulses. Thus, D could be determined from the slope of a plot of ln E against varying g. In the present experiments, the maximum g value was 13.5 T/m, the Δ was set to 10 ms, and the δ values varied in the range of 0.1 to 3 ms. The self-diffusion coefficient D of the poly-cation, anion, and solvent molecules are 1.6×10^{-11}, 4.5×10^{-10} and 2.8×10^{-9} m^2/s, respectively. The D value of the anion molecule was about 1/6 of that of the solvent molecule. However, one might think that the anion has the same degree of mobility as the solvent molecule because it has a several times larger molecular volume than the solvent molecule. That is to say, the anion can move like a solvent molecule in a very high concentration solution of poly-cation without suffering a strong restraint due to electrostatic effects. The anion seems to be able to escape from the restraint of the poly-cation even in a very high concentration poly-cation solution.

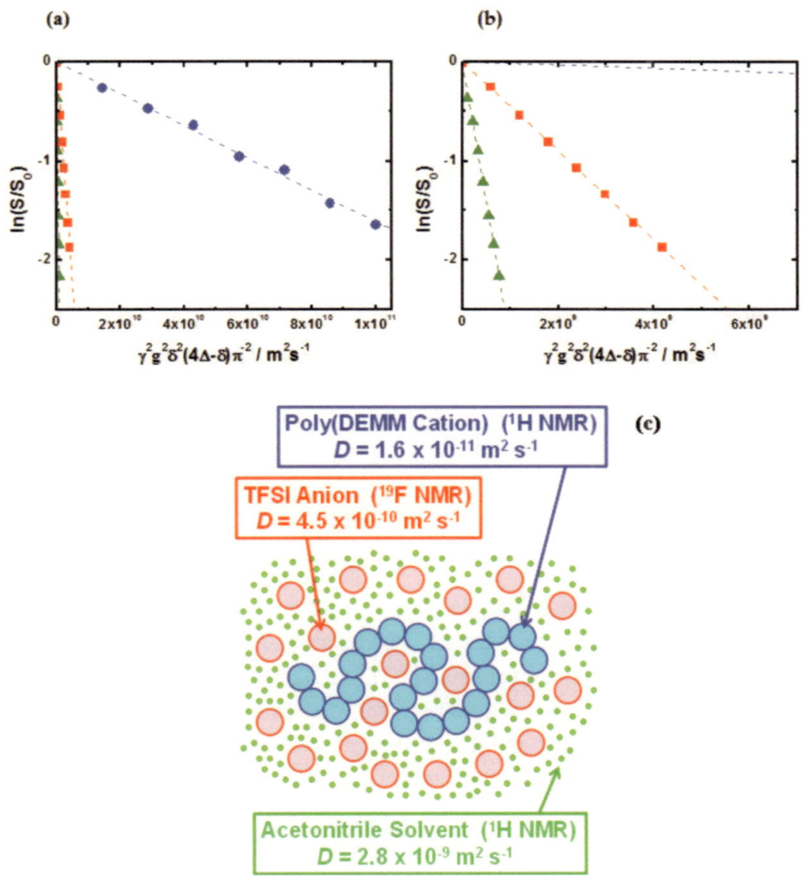

Figure 10. (a) PGSE attenuation plots for the single N-CH_3 of [Poly(DEMM-TFSI)] and the CH_3 signal of acetonitrile probed by the ^{1}H nucleus and the TFSI anion probed by ^{19}F at 25 °C obtained by varying δ at different g values for Δ = 10 ms. (b) Maginified plots area. (c) Self-diffusion coefficient values in high concentration poly(DEMM-TFSI) in acetonitorile solution.

5. High rate performance of a lithium polymer battery using an ionic liquid polymer composite

Some investigators have attempted to develop a non-flammable polymer electrolyte system; we have developed a polymer-gel electrolyte system consisting of a lithium salt in an ionic liquid and poly(ILs), which we have called a *LILP* composite system (Li-salt + ionic liquids + poly(ILs)). Generally, a binary Li-IL, specifically, a lithium salt dissolved in an ionic liquid having the same anion are used for *LILP* system. Several *LILP* systems with conductivities over 10^{-3} S cm^{-1} at room temperature have been developed [34, 25, 17]. However, since the binary Li-IL have a considerably high viscosity, the cell containing such liquids has a poor charge / discharge performance at a relatively large current, namely a lower power density, compared to conventional cells using the flammable organic solvent. And, there have been few reports of the performance of Li ion cells incorporating a *LILP* system. We are aiming to develop a truly safe Li ion polymer cell with a good charge and discharge performance at a large current, we discovered that the choices of $LiMn_2O_4$ and $Li_4Ti_5O_{12}$ [36] as, respectively, the cathode and anode active material, produced a faster charge / discharge reaction than conventional $LiCoO_2$ and graphite systems, and we combined these active materials and poly(DEMM-TFSI), which is very compatible with ionic liquids, as a novel *LILP* system. Our novel Li polymer cell has the following structure: negative electrode: $Li_4Ti_5O_{12}$/*LILP*-including ultrahigh molecular weight ionic liquid polymer / positive electrode: $LiMn_2O_4$. In this section, we will discuss the performance of this cell.

Poly (DEMM-TFSI) has the capability to dissolve a lithium salt independent of the presence of a liquid electrolyte. In a polarizing microscope analysis of Poly(DEMM-TFSI) which contained 1M concentration of dissolved lithium salt $LiClO_4$, we did not observe the birefringence indicating the existence of crystals. Thus, there must have been a complete dissociation of the lithium salt in the polymer matrix as the polymer itself has a non-crystalline nature. DSC measurements indicate that prepared poly(DEMM-TFSI) has a glass transition temperature (T_g) of approximately 50 °C. Thus, a poly(DEMM-TFSI) / lithium salt composite could potentially serve as an all-polymer electrolyte at temperatures over 50 °C.

Moreover, poly(DEMM-TFSI) dissolves in a variety of quaternary ammonium ionic liquids to make a gel. For example, the ionic liquid, *N,N*-diethyl-*N*-(2-methoxyethyl)-*N*-methylammonium bis(trifluoro- methylsulfonyl) imide (DEME-TFSI, molecular structure is indicated in Fig. 11.) containing only 5% of the ultrahigh molecular weight poly(DEMM-TFSI) lost its liquid characteristics and became a gel. It seems that the strong cohesiveness and loss of liquidity appear because of the entanglement effect of long polymer chains. When we adjusted the solution to a suitable viscosity by adding a supplementary solvent, in our case a propylene carbonate (PC) and vinylene carbonate (VC) mixture, it filled the pore space in the electrode and the separator.

The ionic liquid/poly(ILs) composite from which the solvent is removed by vacuum evaporation at a relatively high temperature has no components that leak out of the electrode. However, the high-polarity *LILP* matrix interacted with the supplementary solvent (PC + VC) and probably obstructed perfect evaporation of added solvents. The weight percentage

http://dx.doi.org/10.5772/51162

of remained solvent in the composite was 1.7 wt%. We reported that the vinylene carbonate (10 wt%) in the electrolyte Li-DEME-TFSI, composed of an ionic liquid, DEME-TFSI and Li-TFSI was effective as solid electrolyte interface (SEI) forming additives on the carbon materials such as graphite used as an active material in the anode of a lithium ion cell [9]. Holzapfel and co-workers reported that 2% VC to ionic liquid, EMI-TFSI, contributed to the SEI formation, although the effect was not perfect [37]. Also in this case, there will be a possibility that the remained VC in composite contributes to the SEI formation. However, we realized that the effective SEI to prevent capacitance deterioration with charge/discharge cycles was not formed by the too small amounts (1.7 wt%) of carbonate solvents in *LILP* composite from the cycling behavior of the cell. A highly reliable polymer battery is not easily obtained when the polymerization is carried out in the battery bag. In such an "in-situ radical polymerization", some initiator and unreacted monomer may remain in the polymer matrix. However, our process makes possible the preparation of an electrolyte with few impurities through the use of the purified polymer combined with the binary Li-IL.

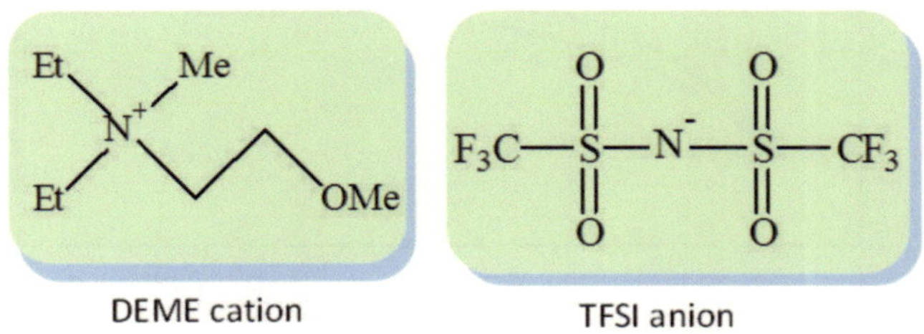

Figure 11. Molecular structure of ionic liquid, DEME-TFSI

Fig. 12 shows both the limiting reduction ($E_{red.}$) and oxidation ($E_{oxd.}$) potentials of the poly(DEMM-TFSI) in PC as measured by cyclic voltammetry at room temperature. The E_{red}. of the poly(DEMM-TFSI) can definitely be seen at around 2.0–2.5V positive relative to Li/Li +. However, its presence is not clear, and the current density is small because, we carried out the measurement in rather dilute conditions to avoid turbulence caused by an increased viscosity in the more concentrated polymer solution. In addition, the $E_{red.}$ of the DEME based ionic liquids was merely somewhat positive against the Li/Li+ [9]. Thus, we realized the need either to select an electrode that would avoid cathodic decomposition during the charge–discharge cycling, or to form a more effective protective layer, such as a solid electrolyte interface (SEI), on the negative electrode material.

We selected high power active electrode materials, and combined them with a *LILP* system and prepared two type vapor-free lithium ion polymer cells for demonstration purposes. The negative $Li_4Ti_5O_{12}$ electrodes (AKO-6), had a charge capacity of 0.42 mAh cm^{-2}, an area density of 3.00 mg cm^{-2}, and had an active electrode layer 25 micron-m thick on a copper foil. The other negative electrode used was a hard carbon electrode (AKT-2) with a charge capacity of 1.29 mAh cm^{-2}, an area density of 2.96 mg cm^{-2}, and a 33 micron-m in thick electrode active layer on a copper foil. The first positive electrode, a $LiMn_2O_4$ electrode (CKT-22) that was paired with the AKO-6 negative electrode, had a charge capacity of 0.49 mAh cm^{-2}, an area density of 6.60 mg cm^{-2} and an electrode active layer 36 -37 micron-m in thickness on an aluminum foil. The second $LiMn_2O_4$ electrode (CKT-9), paired with the AKT-2 had a charge capacity of 1.075 mAh cm^{-2}, an area density of 12.51 mg cm^{-2}, and a 65 micron-m thick electrode active layer on an aluminum foil. The specification of the electrodes are summarized in Table 1. The details of the preparation method have been described in our previous paper [9].

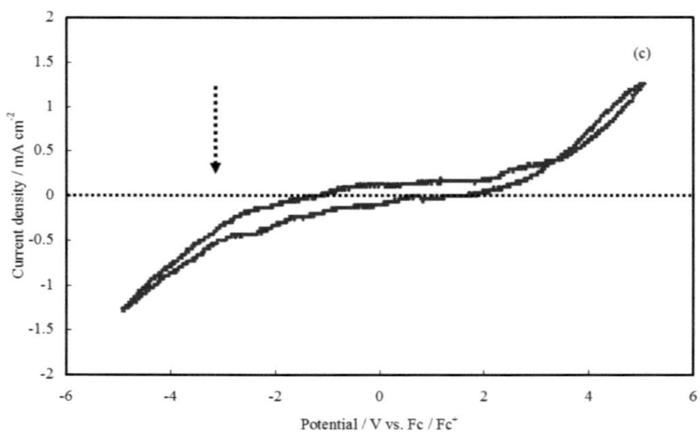

Figure 12. Cyclic voltammogram of poly(DEMM-TFSI) in PC at 25 °C. Scan rate: 10mVs^{-1}; platinum working and counter electrodes; Ag/Ag^{+} reference electrode. The potential value (V) was referenced to the ferrocene (Fc)/ferrocenium (Fc^{+}) redox couple in each salt. Polymer concentration, 0.002 mol kg^{-1} poly(DEMMTFSI) in PC solution. The concentration is defined in terms of molality = (mol solute/kg solvent). The dashed line arrow indicates the potential of the Li/Li^{+} couple.

Code	Active material	Polarity	Charge capacity (mAh cm$^{-2)}$	Area density (mg cm$^{-2)}$	Active layer thickness (μm)	Current collector/(μm)
AKO-6	Li^4Ti^5O12	Negative	0.42	3.00	25	Copper/13
CKT-22	$LiMn_2O4$	Positive	0.49	6.60	36-37	Aluminium/20
AKT-2	Hard carbon	Negative	1.29	2.96	33	Copper/13
CKT-9	$LiMn_2O^4$	Positive	1.075	12.51	65	Aluminium/20

Table 1. Electrode specification.

In Fig. 13, we show the first and second charge–discharge potential curves at 40 °C of the demonstration cells, consisting of hard carbon/*LILP* (1.3 mol of LiTFSI dissolved in poly(DEMM-TFSI)/ DEME-TFSI composite, giving a polymer concentration of 5.4 wt%)/

$LiMn_2O_4$. About 26% of the charge capacity was lost in the first cycle; however, from that point on, the cell exhibited an efficiency of 96% or more.

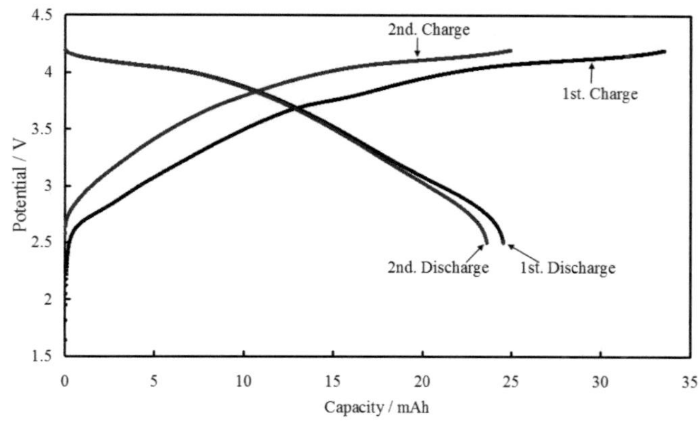

Figure 13. First and second charge/discharge curves of lithium polymer cell at 0.05C current at 40 °C. Positive electrode = $LiMn_2O_4$; negative electrode = hard carbon; electrolyte = 1.30 mol kg^{-1} of LiTFSI in a DEME-TFSI and poly(DEMM-TFSI) composite. The concentration is defined in terms of molality = (mol solute/kg polymer-ionic liquid composite). The polymer concentration for the composite electrolyte was 5.4 wt%.

The rate capability of this cell at 40 °C appears in Fig. 14. As the discharge current increased, the discharge capacity of this cell decreased significantly faster than that of a cell using conventional materials. The capacity at 1C discharge was approximately 63% of that at 0.1C discharge. Most likely, the greater decrease at large discharge currents in the capacity of the cell using the *LILP* electrolyte resulted from a large internal resistance in the cell. The cycling behavior of this cell, plotted in Fig. 15, was not suitable for practical use. When we set the upper limit voltage at 4.2V, the cell deteriorated sooner than with a voltage of 4.0V. The CV measurements suggested that degradation of the polymer or ionic liquid was occurring during the charge–discharge cycles. It is necessary to establish a method of forming an effective SEI if a practicable cycle performance is to be achieved.

We then prepared another type of lithium polymer cell, comprising $Li_4Ti_5O_{12}$ / *LILP* (1.3 mol of LiTFSI dissolved in poly(DEMM-TFSI) / DEME-TFSI composite, 5.4 wt% polymer concentration) / $LiMn_2O_4$. Because the intercalation-deintercalation potential of $Li_4Ti_5O_{12}$ is around 1.5V vs. Li / Li^+ potential, our prepared cell had about 3.0 V of charging potential with $LiMn_2O_4$. The overall cell reactions can be described as follows:

negative electrode,

$$Li[Li_{1/3}Ti_{5/3}]O_4 + \ xLi^+ + xe^- \ \Leftrightarrow Li_{1+x}[Li_{1/3}Ti_{5/3}]O_4 \qquad (2)$$

positive electrode,

$$LiMn_2O_4 \Leftrightarrow Li_{1-x}Mn_2O_4 + xLi + xe^- \quad (3)$$

overall reaction,

$$Li[Li_{1/3}Ti_{5/3}]O_4 + LiMn_2O_4 \Leftrightarrow Li_{1+x}[Li_{1/3}Ti_{5/3}]O_4 + Li_{1-x}Mn_2O_4 \quad (4)$$

The theoretical capacity of $Li_4Ti_5O_{12}$ was expected to be approximately 175 mAh g^{-1}. In our electrode, the discharge capacity was 171 mAh g^{-1}. The performances of the prepared cell are shown in Figs. 16-18.

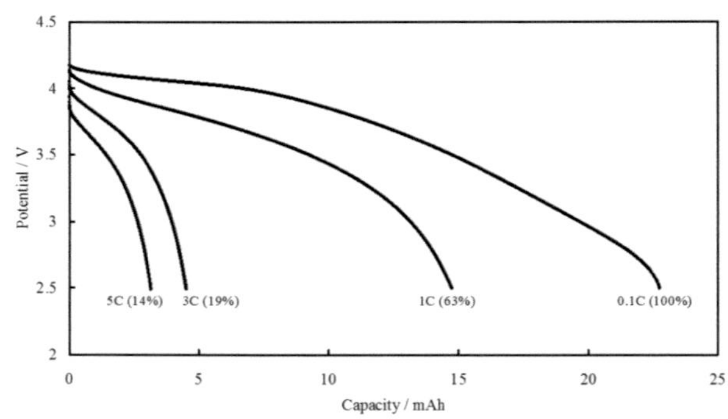

Figure 14. Discharge curves of the lithium polymer cell at various current densities at 40 °C. The positive electrode = $LiMn_2O_4$; negative electrode = hard carbon; electrolyte = 1.30 mol kg^{-1} of LiTFSI in DEME-TFSI and poly(DEMM-TFSI) composite. The concentration is defined in terms of molality = (mol solute/kg polymer-ionic liquid composite). The polymer concentration for the composite electrolyte was 5.4 wt%.

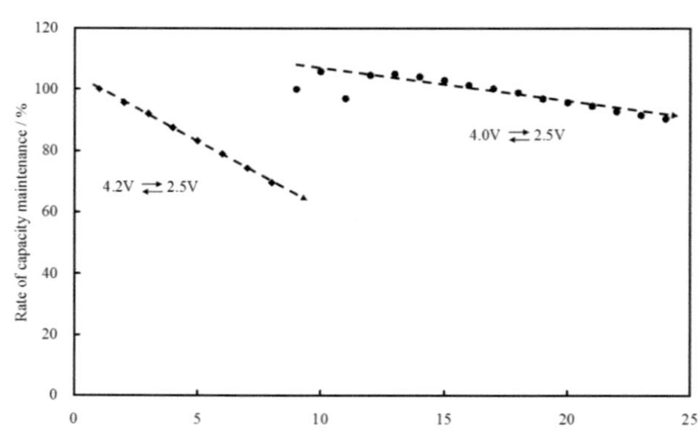

Figure 15. Cycle life of lithium polymer cell including hard carbon as negative active material. The charge–discharge process was performed at 0.1C at 40 °C. The cut-off voltages were 4.2V and 2.5V for nine cycles, and 4.0V and 2.5V from 10th to 24th cycle.

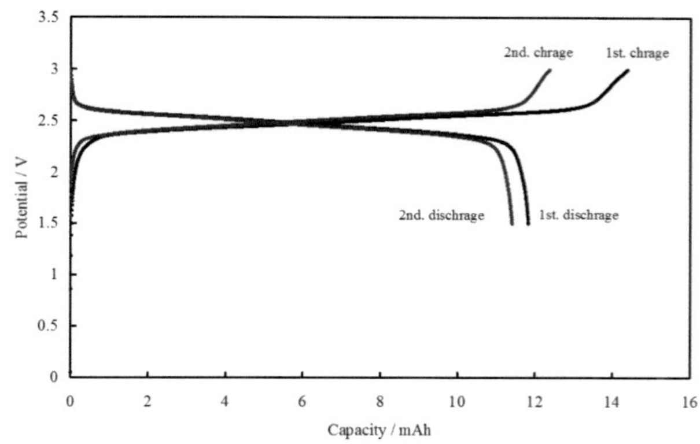

Figure 16. First and second charge/discharge curves of lithium polymer cell at 0.05 C current at 40 °C. The positive electrode = $LiMn_2O_4$; negative electrode = $Li_4Ti_5O_{12}$; electrolyte = 1.30 mol kg^{-1} of LiTFSI in DEME-TFSI and poly(DEMM-TFSI) composite. The concentration is defined in terms of molality = (mol solute/kg polymer-ionic liquid composite). The polymer concentration for the composite electrolyte was 5.4 wt%.

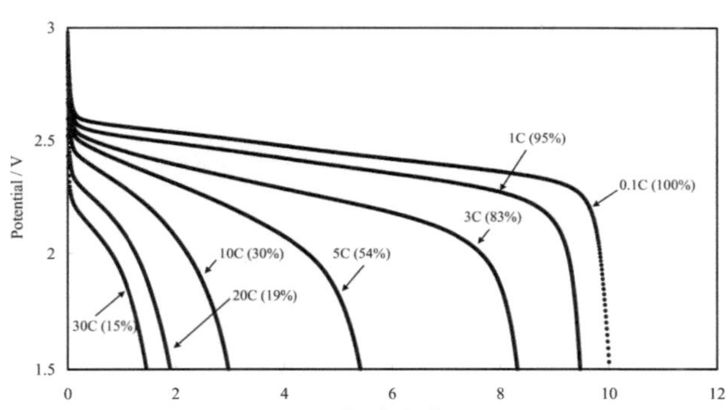

Figure 17. Discharge curves of lithium polymer cell at various current densities at 40 °C. The positive electrode = $LiMn_2O_4$; negative electrode = $Li_4Ti_5O_{12}$; electrolyte = 1.30 mol kg^{-1} of LiTFSI in DEME-TFSI and poly(DEMM-TFSI) composite. The concentration is defined in terms of molality = (mol solute/kg polymer-ionic liquid composite). The polymer concentration for the composite electrolyte was 5.4 wt%.

The discharge specific capacities and Columbic efficiency of the cell at the $1^{st.}$ was 11.8 mAh, and 82 %; at the second cycle, they were 11.4 mAh, and 92 %. After the $5^{th.}$ cycle, the Columbic efficiency of this cell remained at approximately 97-98 %.

The battery in Fig. 17 that combined the $Li_4Ti_5O_{12}$ anode with the $LiMn_2O_4$ cathode showed an excellent rate discharge character as for a lithium polymer battery. Evidently, this battery retains 83 % or more of the capacity maintenance rate at a 3C high power discharge. Thus, it was possible to create a new, leak-free battery with a vapor-free, practical discharge performance, and a prismatic cell design by selecting an electrode material with a high speed charge/discharge reaction when combined with an *LILP* system. The rate performances of the $Li_4Ti_5O_{12}$ negative cell and the hard carbon negative cells differ greatly, even though the electrolyte in each is the same *LILP* system. The interfacial compatibility of the active material and the electrolyte seem to have a bigger influence on the rate performance than the bulk ionic conductivity of the electrolyte itself. The cycling behavior of the cell, plotted in Fig. 18, indicated a good cycle durability, almost equivalent to that of a conventional lithium ion cell.

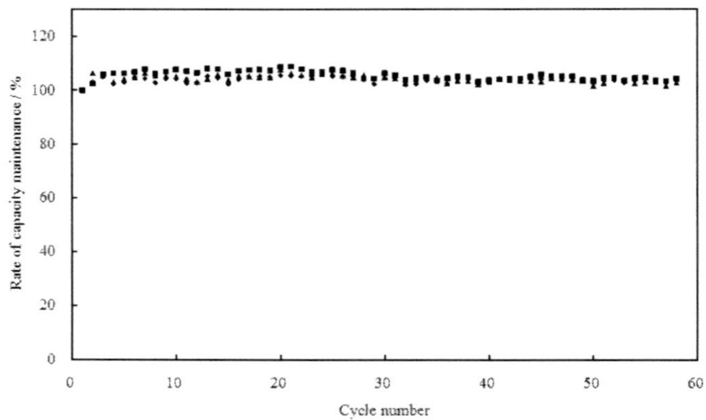

Figure 18. Cycle life of lithium polymer cell including $Li_4Ti_5O_{12}$ as negative active material. The charge- discharge process was performed at 0.1C at 40 °C. The cut off voltages were 3.0V and 1.5V. Three cells were tested.

As in the case of the hard carbon electrode, we expected that electrochemical degradation of the DEME-TFSI or poly(DEMM-TFSI) probably occurred. In our previous study, we suggested that some kinds of organic solvents, such as VC and ethylene carbonate, are effective as *SEI*-forming additives on the graphite used as an active material in the anode of a lithium ion cell with a binary Li-*IL* electrolyte. In this study, we did use a small amount of VC as a dilution solvent; however, because almost all the VC evaporates from the battery in the process of establishing the composite, it appears the small amounts of remained VC did not have large contribution to the SEI formation.

On the other hand, however, we should point out that the use of $Li_4Ti_5O_{12}$ as the negative electrode for the lithium-ion *LILP* cell avoids cathodic decomposition of the *LILP* composite during the charge/discharge cycling.

6. Conclusion

This chapter has reported the synthesis and physicochemcal properties of a quaternary ammonium polymerizable ionic liquid and of the corresponding polymer. In conclusion we find;

http://dx.doi.org/10.5772/51162

1. The polymerizable ionic liquid, DEMM-TFSI having a methacryl functional group as part of the ionic liquid's cation species has a remarkably low glass transition temperature (T_g = -68 °C), a relatively high ionic conductivity (0.6 mS/cm at 25 °C) and relatively wide potential windows.

2. The DEMM-TFSI monomer has a high reactivity for radical polymerization. We could prepare poly(DEMM-TFSI) with an ultra high molecular weight, with an M_w of over one million. However, in many cases, control of the molecular weight has been difficult because of the highly reactive nature of these monomers.

3. We succeeded in achieving molecular weight control by the radical polymerization of the ionic liquid monomer via three methods; atom transfer radical polymerization (ATRP), chain transfer radical polymerization (CTRP) and reversible addition-fragmentation chain transfer polymerization (RAFT). The control of the molecular weight occurred extremely easily in CTRP because of the addition of 3-mercapto-1-hexanol as a chain transfer reagent. In RAFT polymerization, molecular weight controlled poly(IL) with a low polydispersity index can be prepared.

4. A poly-cation type material such as poly(DEMM-TFSI) has bulky tetra-alkyl ammonium functional groups in each unit and has a relatively low T_g value due to it having a large free volume due to the effect of electrostatic repulsions of cationic groups. A polymer with a molecular weight that exceeds 100,000 has roughly a T_g of 50 °C while, on the other hand, a polymer with M_n about 10,000 had a 20 degrees lower T_g of 30 °C, presumably due to the chain ends effect.

5. The bulk poly(DEMM-TFSI) has a relatively low ionic conductivity that is not practical for use in electrochemical devices at room temperature. However, the temperature dependence of the ionic conductivity showed an interesting property, unlike that of a non-ionic polymer electrolyte, such as polyethylene oxide. Ionic conduction also appeared below the glass transition temperature in the polymer matrix. The anion seemed to show ionic conduction, independent of the segmental motion of the polymer chain. A new ionic conduction mechanism in the poly-cation matrix might be discovered by a further research.

6. We measured the self-diffusion coefficient value D of the poly-cation, anion and solvent molecule in a high concentration solution of poly(DEMM-TFSI) by PGSE-NMR. It was clarified that the anion had a D value that was comparatively close to that of the solvent molecule. We think that the anion has a high D value by escaping from a strong interaction with the poly-cation.

7. We obtained a polymer electrolyte by combining a small amount of the poly(ILs) and a binary Li-IL. The system of a lithium salt dissolved in an ionic liquid polymer and ionic liquid composite (*LILP*) could potentially comprise an electrolyte having zero vapor pressure. The discharge performance of a cell with an *LILP* system might be expected to show a poor discharge performance, because the solidified ionic liquid decreases the mobility of the ionic species. However, we have conceived a method of making a polymer battery with a practicable performance by combining electrodes that can offer a

high-speed charge/discharge reaction. The experimental battery that combined a $Li_4Ti_5O_{12}$ anode with a $LiMn_2O_4$ cathode and included an LILP electrolyte showed an excellent rate discharge character for a lithium polymer battery; at a 3C current rate, it retained 83% of its discharge capacity, and relatively good cycle performance. A lithium ion cell with a LILP system performed, in terms of cell performance and cycle durability, at a level of practical utility. This novel lithium polymer cell, non-flammable and leak-free, is a promising candidate as a safe, large size lithium secondary battery.

Acknowledgements

The authors thank Ms. Shoko Marukane and Mrs. Saika Honma at Tsuruoka National College of Technology for various assistance in the lithium polymer cell preparation and measurements.

Author details

Takaya Sato, Takashi Morinaga and Takeo Ishizuka

*Address all correspondence to: takayasa@tsuruoka-nct.ac.jp

Tsuruoka National College of Technology, Japan

References

[1] Rogers, R. D., & Seddon, K. R. (2001). Ionic Liquids: Industrial Applications for Green Chemistry (Acs Symposium Series). *American Chemical Society*, 0-84123-789-1.

[2] Fung, Y. S. Room temperature molten salt as medium for lithium battery and alloy electrodeposition- fundamental and application. Trends Inorg Chem. (1998).

[3] Fung, Y. S., & Zhou, R. Q. (1999). Room temperature molten salt as medium for lithium battery. *J. Power Sources*, 81.

[4] Caja, T. D. J., Dunstan, D. M., & Ryan, V. (2000). P.C. Trulove (Ed.). Molten Salts XII. *Pennington, NJ: Electrochem. Soc.*, 150.

[5] Fung, Y. S., & Zhu, D. R. (2002). Electrodeposited tin coating as negative electrode material for lithium-ion battery in room temperature molten salt. *J. Electrochem. Soc.*, 149, A319.

[6] Nakagawa, H., Izuchi, S., Kuwana, K., & Aihara, Y. (2003). Liquid and polymer gel electrolytes for lithium batteries composed of room-temperature molten salt doped by lithium salt. *J. Electrochem. Soc.*, 150, A695.

[7] Sakaebe, H., & Matsumoto, H. (2003). N-Methyl-N-propylpiperidinium bis(trifluoromethanesulfonyl)imide (PP13-TFSI)-novel electrolyte base for Li battery. *Electrochem. Commun.*, 5, 594.

[8] Shin, J.H., Henderson, W.A., Henderson, W. A., & Passerini, S. (2003). Ionic liquids to the rescue? *Overcoming the ionic conductivity limitations of polymer electrolytes. Electrochem. Commun.*, 5.

[9] Sato, T., Maruo, T., Marukane, S., & Takagi, K. (2004). Ionic liquids containing carbonate solvent as electrolytes for lithium ion cells. *J. Power Sources*, 138.

[10] Mc Ewen, A. B., Mc Devitt, S. F., & Koch, V. R. (1997). Nonaqueous electrolytes for electrochemical capacitors: imidazolium cations and inorganic fluorides with organic carbonates. *J. Electrochem. Soc.*, 144, L84.

[11] Mc Ewen, A. B., Ngo, H. L., Le Compte, K., & Goldman, J. L. (1999). Electrochemical properties of imidazolium salt electrolytes for electrochemical capacitor applications. *J. Electrochem. Soc.*, 146.

[12] Ue, M., & Takeda, M. (2002). Application of ionic liquids based on 1-ethyl-3-methylimidazolium cation and fluoroanions to double-layer capacitors. *J. Korean Electrochem. Soc.*, 5.

[13] Ue, M., Takeda, M., Toriumi, A., Kominato, A., Hagiwara, R., & Ito, Y. (2003). Application of low-viscosity ionic liquid to the electrolyte of double-layer capacitors. *J. Electrochem. Soc.*, 150, A499.

[14] Sato, T., Masuda, G., & Takagi, K. (2004). Electrochemical properties of novel ionic liquids for electric double layer capacitor applications. *Electrochim. Acta*, 49(2004), 3603-3611.

[15] Ohno, H., Yoshizawa, M., & Ogiwara, W. (2004). Development of new class of ion conductive polymers based on ionic liquid. *Electrochim. Acta*, 50(2004), 255-261.

[16] Nakajima, H., & Ohno, H. (2005). Preparation of thermally stable polymer electrolytes from imidazolium-type ionic liquid derivatives. *Polymer*, 46(2005), 11499-11504.

[17] Ogiwara, W., Washiro, S., Nakajima, H., & Ohno, H. (2006). Effect of cation structure on the electrochemical and thermal properties of ion conductive polymers obtained from polymerizable ionic liquids. *Electrochim. Acta*, 51(2006), 2614-2619.

[18] Lu, J., Yan, F., & Texter, J. (2009). Advanced applications of ionic liquids in polymer science. *J. Prog. Polym. Sci.*, 34(2009), 431-448.

[19] Mecerreyes, D. (2011). Polymeric ionic liquids: Broadening the properties and applications of polyelectrolytes. *J. Prog. Polym. Sci.*, 36-1629.

[20] Yuan, J., & Antonietti, M. (2011). Poly(ionic liquid)s: Polymers expanding classical property profiles. *polymer*, 52-1469.

[21] Benton, M. G., & Brazel, C. S. (2004). An Investigation of the Degree and Rate of Polymerization of Poly (methyl methacrylate) in the Ionic Liquid 1-Butyl-3- Methylimidazolium Hexafluorophosphate. *Polym. Int.*, 53.

[22] Sato, T., Marukane, S., Narutomi, T., & Akao, T. High rate performance of a lithium polymer battery using a novel ionic liquid polymer composite. *J. Power Sources*, 164(20042007), 390-396.

[23] Yu-H, Chang., Pei-Y, Lin., Ming-S, Wu., & King-F, Lin. (2012). Extraordinary aspects of bromo-functionalized multi-walled carbon nanotubes as initiator for polymerization of ionic liquid monomers. *Polymer*, 53(2012), 2008-2014.

[24] Sato, T., Morinaga, T., Marukane, S., Narutomi, T., Igarashi, T., Kawano, Y., Ohno, K., Fukuda, T., & Tsujii, Y. (2011). Novel solid-state polymer electrolyte of colloidal crystal decorated with ionic-liquid polymer brush. *Adv. Mater.*, 23(2011), 4868-4872.

[25] Appetecchi, G. B., Kim, G.T., montanino, M., Carewska, M., Marcilla, R., Mecerreyes, D., & De Meatza, I. (2010). Ternary polymer electrolytes containing pyrrolidinium-based polymeric ionic liquids for lithium batteries. *J. Power Sources*, 195(2010), 3668-3675.

[26] Kubisa, P. (2004). Application of ionic liquids as solvents for polymerization processes. *Prog. Polym. Sci.*, 29(2004), 3-12.

[27] Mori, H., Yahagi, M., & Endo, T. (2009). RAFT Polymerization of N-vinylimidazolium salt and synthesis of thermoresponsive ionic liquid block copolymers. *Macromolecules*, 42-8082.

[28] Santangelo, P. G., & Roland, C. M. (1998). Molecular weight dependence of fragility in polystyrene. *Macromolecules*, 31(1998), 4581-4585.

[29] Claudy, P., Letoffe, J. M., Camberlain, Y., & Pascault, J. P. (1983). Glass transition of polystyrene versus molecular weight. *Polym. Bull.*, 9(1983), 208-215.

[30] Teran, A. A., Tang, M. H., Mullin, S. A., & Balsara, N. P. (2011). Effect of molecular weight on conductivity of polymer electrolytes. *Solid State Ionics*, 203(2011), 18-21.

[31] Hayamizu, K. (2001). *Nihon-Denshi News*, 33(1), 6-10, http://www.jeol.co.jp/publication/ nihondenshi/j_backnumber/33/j33_all.pdf, (accessed 9 Jun 2012).

[32] Stejskal, E. O., & Tanner, J. E. (1965). Spin diffusion measurements: spin echoes in the presence of a time-dependent field gradient. *J. Chem. Phys.*, 42-288.

[33] Hayamizu, K. On Accurate Measurements of Diffusion Coefficients by PGSE NMR Methods -room- themperature ionic liquids- http://www.jeolusa.com/DesktopModules/Bring2mind/DMX/Download.aspx?EntryId=713&Command=Core_Download&PortalId=2&TabId=337 (accessed 9 Jun) (2012).

[34] Fuller, J., Breda, A. C., & Richard, R. T. (1998). Ionic liquid-polymer gel electrolytes from hydrophilic and hydrophobic ionic liquids. *J. Electroanal. Chem.*, 459.

[35] Abu-Bin, M., Hasan, S. T., Kaneko, T., Noda, A., & Watanabe, M. (2005). Ion Gels Prepared by in Situ Radical Polymerization of Vinyl Monomers in an Ionic Liquid and Their Characterization as Polymer Electrolytes. *J. Am. Chem. Soc.*, 127, 4876.

[36] Amine, K., Liu, J., Belharouak, I., & Park, S. H. (2006). Proceedings of advanced technology development review meeting at Sandia National Laboratory. *Carlsbad, NM, USA.*

[37] Holzapfel, M., Jost, C., & Nov´ak, P. (2004). Stable cycling of graphite in an ionic liquid based electrolyte. *Chem. Commun.*